Mass Transfer Processes

MASS TRANSFER PROCESSES

Modeling, Computations, and Design

P. A. Ramachandran

PRENTICE
HALL

Boston • Columbus • Indianapolis • New York • San Francisco • Amsterdam • Cape Town
Dubai • London • Madrid • Milan • Munich • Paris • Montreal • Toronto • Delhi • Mexico City
São Paulo • Sydney • Hong Kong • Seoul • Singapore • Taipei • Tokyo

For information about buying this title in bulk quantities, or for special sales opportunities (which may include electronic versions; custom cover designs; and content particular to your business, training goals, marketing focus, or branding interests), please contact our corporate sales department at corpsales@pearsoned.com or (800) 382-3419.

For government sales inquiries, please contact governmentsales@pearsoned.com.

For questions about sales outside the U.S., please contact intlcs@pearson.com.

Visit us on the Web: informit.com

Library of Congress Control Number: 2017956421

ISBN-13: 978-0-13-467562-6
ISBN-10: 0-13-467562-2

1   18

CONTENTS

PREFACE   xxix
ABOUT THE AUTHOR   xxxvii
NOTATION   xxxix

PART I   **FUNDAMENTALS OF MASS TRANSFER MODELING**   **1**

CHAPTER 1   **INTRODUCTION TO MODELING OF MASS TRANSFER PROCESSES**   **3**

    1.1   What Is Mass Transfer?   5

         1.1.1   What Is Interfacial Mass Transfer?   6

         1.1.2   What Causes Mass Transfer?   7

    1.2   Preliminaries: Continuum and Concentration   7

         1.2.1   The Continuum Assumption   7

         1.2.2   Concentration: Mole Units   8

         1.2.3   Concentration: Mass Units   9

         1.2.4   Concentration: Partial Pressure Units   10

    1.3   Flux Vector   10

         1.3.1   Molar and Mass Flux: Definition   10

         1.3.2   Convection Flux   12

         1.3.3   Diffusion Flux   12

1.4	Concentration Jump at Interface	15
	1.4.1 Gas–Liquid Interface: Henry's Law	15
	1.4.2 Vapor–Liquid Interface: Raoult's Law	17
	1.4.3 Liquid–Liquid Interface: Partition Constant	18
	1.4.4 Fluid–Solid Interface: Adsorption Isotherm	19
	1.4.5 Nonlinear Equilibrium Models	19
1.5	Application Examples	20
	1.5.1 Reacting Systems	20
	1.5.2 Unit Operations	21
	1.5.3 Bioseparations	24
	1.5.4 Semiconductor and Solar Devices	24
	1.5.5 Biomedical Applications	25
	1.5.6 Application to Metallurgy and Metal Winning	25
	1.5.7 Product Development and Product Engineering	26
	1.5.8 Electrochemical Processes	26
	1.5.9 Environmental Applications	28
1.6	Basic Methodology of Model Development	28
1.7	Conservation Principle	29
1.8	Differential Models	30
1.9	Macroscopic Scale	32
	1.9.1 Stirred Tank Reactor: Mixing Model	33
	1.9.2 Sublimation of a Solid Sphere: Mass Transfer Coefficient	34
	1.9.3 Model for Mixer-Settler	35
	1.9.4 Equilibrium Stage Model	36
1.10	Mesoscopic or Cross-Section Averaged Models	37
	1.10.1 Solid Dissolution from a Wall	38
	1.10.2 Tubular Flow Reactor	41
1.11	Compartmental Models	43

CHAPTER 2   **EXAMPLES OF DIFFERENTIAL (1-D) BALANCES**   **51**

   2.1   Cartesian Coordinates   52

          2.1.1   Steady State Diffusion across a Slab   52

          2.1.2   Steady State Diffusion with Reaction in a Slab   56

          2.1.3   Transient Diffusion in a Slab   62

          2.1.4   Diffusion with Convection   64

   2.2   Cylindrical Coordinates   67

          2.2.1   Steady State Radial Diffusion   68

          2.2.2   Steady State Mass Transfer with Reaction   70

          2.2.3   Transient Diffusion in a Cylinder   73

   2.3   Spherical Coordinates   73

          2.3.1   Steady State Diffusion across a Spherical Shell   73

          2.3.2   Diffusion and Reaction   75

          2.3.3   Transient Diffusion in Spherical Coordinates   76

CHAPTER 3   **EXAMPLES OF MACROSCOPIC MODELS**   **85**

   3.1   Macroscopic Balance   87

          3.1.1   In and Out Terms from Flow   88

          3.1.2   Wall or Interface Transfer Term   89

          3.1.3   Rate Term   90

          3.1.4   Accumulation Term   90

   3.2   The Batch Reactor   90

          3.2.1   Differential Equations for the Reactor   91

          3.2.2   ODE45 with CHEBFUN   93

   3.3   Reactor–Separator Combination   96

   3.4   Sublimation of a Spherical Particle   101

          3.4.1   Correlation for Mass Transfer Coefficient   102

   3.5   Dissolved Oxygen Concentration in a Stirred Tank   104

3.6		Continuous Stirred Tank Reactor	106
	3.6.1	First-Order Reaction	107
	3.6.2	Second-Order Reaction	108
3.7		Tracer Experiments: Test for Backmixed Assumption	110
	3.7.1	Interconnected Cells Model	111
	3.7.2	Model Composed of Active and Dead Zone	111
3.8		Liquid–Liquid Extraction	112
	3.8.1	Mass Transfer Rate	114
	3.8.2	Backmixed–Backmixed Model	115
	3.8.3	Equilibrium Stage Model	115
	3.8.4	Stage Efficiency	116

CHAPTER 4   **EXAMPLES OF MESOSCOPIC MODELS**   **123**

4.1		Solid Dissolution from a Wall	124
	4.1.1	Model Details	125
	4.1.2	Mass Transfer Correlations in Pipe Flow	128
4.2		Tubular Flow Reactor	129
	4.2.1	Plug Flow Closure	131
	4.2.2	Dispersion Closure	132
4.3		Mass Exchangers	134
	4.3.1	Single Stream	134
	4.3.2	Two Streams	136
	4.3.3	NTU and HTU Representation	143

CHAPTER 5   **EQUATIONS OF MASS TRANSFER**   **151**

5.1		Flux Form	153
	5.1.1	Mole Basis	153
	5.1.2	Mass Basis	155

5.2	Frame of Reference	156
	5.2.1 Mass Fraction Averaged Velocity	156
	5.2.2 Mole Fraction Averaged Velocity	158
5.3	Properties of Diffusion Flux	163
5.4	Pseudo-Binary Diffusivity	165
5.5	Concentration Form	166
	5.5.1 Mass Basis	166
	5.5.2 Constant-Density Systems	167
	5.5.3 Overall Continuity: Mass Basis	168
	5.5.4 Mole Basis	168
	5.5.5 Overall Continuity: Mole Basis	169
	5.5.6 Common Simplifications	170
5.6	Common Boundary Conditions	171
5.7	Macroscopic Models: Single-Phase Systems	172
5.8	Multiphase Systems: Local Volume Averaging	175

CHAPTER 6	**DIFFUSION-DOMINATED PROCESSES AND THE FILM MODEL**	**185**
6.1	Steady State Diffusion: No Reaction	186
	6.1.1 Combined Flux Equation	187
	6.1.2 Diffusion-Induced Convection	187
	6.1.3 Determinacy Condition	188
	6.1.4 Low Flux Model: The Laplace Equation	189
6.2	Diffusion-Induced Convection	193
	6.2.1 Conditions for the Validity of the Low Flux Model	193
	6.2.2 Analysis for UMD	193
	6.2.3 Drift Flux Correction Factor	195
	6.2.4 Mole Fraction Profiles in UMD	196

6.3   Film Concept in Mass Transfer Analysis   198

        6.3.1   Boundary Layer Concept for Fluid–Solid Mass
                  Transfer   198

        6.3.2   Film Model Approximation   201

        6.3.3   Film Model: Determinacy Correction Factor   204

6.4   Surface Reactions: Role of Mass Transfer   206

        6.4.1   Low Flux Model: First-Order Reaction   206

        6.4.2   Low Flux Model: Nonlinear Reactions   209

        6.4.3   High Flux Model: Effect of Product
                  Counter-Diffusion   210

6.5   Gas–Liquid Interface: Two-Film Model   212

        6.5.1   Mass Transfer Coefficients   212

        6.5.2   Overall Mass Transfer Coefficient   214

CHAPTER 7   PHENOMENA OF DIFFUSION   223

7.1   Diffusion Coefficients in Gases   224

        7.1.1   Model Based on Kinetic Theory   225

        7.1.2   Frictional Interpretation   232

        7.1.3   Multicomponent Diffusion   235

7.2   Diffusion Coefficients in Liquids   237

        7.2.1   Stokes-Einstein Model   239

        7.2.2   Wilke-Chang Equation   241

7.3   Non-Ideal Liquids   243

        7.3.1   Activity Correction Factor   243

        7.3.2   Activity Coefficient Models   244

7.4   Solid–Solid Diffusion   246

        7.4.1   Vacancy Diffusion   246

        7.4.2   Interstitial Diffusion   246

	7.5	Diffusion of Fluids in Porous Solids	248
		7.5.1 Single-Pore Gas Diffusion: Effect of Pore Size	248
		7.5.2 Liquid-Filled Pores: Hindered Diffusion	250
		7.5.3 Porous Catalysts: Effective Diffusivity	251
	7.6	Heterogeneous Media	254
	7.7	Polymeric Membranes	256
	7.8	Other Complex Effects	257
CHAPTER 8		**TRANSIENT DIFFUSION PROCESSES**	**265**
	8.1	Transient Diffusion Problems in 1-D	266
	8.2	Solution for Slab: Dirichlet Case	267
		8.2.1 Dimensionless Representation	268
		8.2.2 Series Solution	269
		8.2.3 Evaluation of the Series Coefficient	271
		8.2.4 Illustrative Results	272
		8.2.5 Average Concentration	274
	8.3	Solutions for Slab: Robin Condition	276
	8.4	Solution for Cylinders and Spheres	278
		8.4.1 Long Cylinder	278
		8.4.2 Sphere	279
		8.4.3 One-Term Approximation	280
	8.5	Transient Non-Homogeneous Problems	283
		8.5.1 D-D Problem in Slab Geometry	283
		8.5.2 Transient Diffusion with Reaction	284
	8.6	2-D Problems: Product Solution Method	285
	8.7	Semi-Infinite Slab Analysis	287
		8.7.1 Constant Surface Concentration	288

	8.7.2	Integral Method	290
	8.7.3	Pulse Response	292
8.8	Penetration Theory of Mass Transfer		294
8.9	Transient Diffusion with Variable Diffusivity		295
8.10	Eigenvalue Computations with CHEBFUN		297
8.11	Computations with PDEPE Solver		299
	8.11.1	Sample Code for 1-D Transient Diffusion with Reaction	300

CHAPTER 9	**BASICS OF CONVECTIVE MASS TRANSPORT**		**309**
9.1	Definitions for External and Internal Flows		310
9.2	Relation to Differential Model		311
9.3	Key Dimensionless Groups		313
	9.3.1	Other Derived Dimensionless Groups	314
9.4	Mass Transfer in Flows in Pipes and Channels		315
	9.4.1	Laminar Flow	315
	9.4.2	Turbulent Flow	315
	9.4.3	Channel Flow	316
9.5	Mass Transfer in Flow over a Flat Plate		316
	9.5.1	Laminar Flow	317
	9.5.2	Turbulent Flow	317
	9.5.3	The j-Factor	318
9.6	Mass Transfer for Film Flow		318
	9.6.1	Solid to Liquid	318
	9.6.2	Gas to Liquid	319
9.7	Mass Transfer from a Solid Sphere		320
9.8	Mass Transfer from a Gas Bubble		321
	9.8.1	Bubble Swarms and Bubble Columns	322
9.9	Mass Transfer in Mechanically Agitated Tanks		325

9.10 Gas–Liquid Mass Transfer in a Packed Bed Absorber 327

    9.10.1 Liquid Side Coefficient 328

    9.10.2 Gas Side Coefficient 328

    9.10.3 Transfer Area 328

CHAPTER 10 CONVECTIVE MASS TRANSFER: THEORY FOR INTERNAL LAMINAR FLOW 335

10.1 Mass Transfer in Laminar Flow in a Pipe 336

    10.1.1 Dimensionless Form 337

    10.1.2 Constant Wall Concentration: The Dirichlet Problem 339

    10.1.3 Concentration, Wall Mass Flux, and Sherwood Number 341

10.2 Wall Reaction: The Robin Problem 344

    10.2.1 Solution Using CHEBFUN 345

    10.2.2 Illustrative Results 346

10.3 Entry Region Analysis 348

10.4 Channel Flows with Mass Transfer 350

10.5 Mass Transfer in Film Flow 353

    10.5.1 Solid Dissolution at a Wall in Film Flow 354

    10.5.2 Gas Absorption from Interface in Film Flow 357

10.6 Numerical Solution with PDEPE 358

CHAPTER 11 MASS TRANSFER IN LAMINAR BOUNDARY LAYERS 365

11.1 Flat Plate with Low Flux Mass Transfer 366

    11.1.1 Concentration Equation 367

    11.1.2 Velocity Equations 368

    11.1.3 Scaling Results and the Analogies 369

    11.1.4 Exact or Blasius Analysis 371

11.2	Integral Balance Approach		376
	11.2.1	Integral Momentum Balance	376
	11.2.2	Integral Species Mass Balance	379
	11.2.3	Solution for No Reaction Case	379
	11.2.4	Solution for Homogeneous Reaction	381
11.3	High Flux Analysis		383
	11.3.1	Film Model	383
	11.3.2	Integral Balance Method	384
	11.3.3	Blasius Approach	386
11.4	Mass Transfer for Flow over Inclined and Curved Surfaces		388
	11.4.1	Pressure Variation Term	388
	11.4.2	Integral Balance Method for Inclined and Curved Surfaces	390
	11.4.3	Inclined Plates: Use of Similarity Variable	392
	11.4.4	Wedge Flow: Falkner-Skan Equation	393
	11.4.5	Stagnation Point (Hiemenz) Flow	394
	11.4.6	Flow over a Rotating Disk	395
	11.4.7	Flow past a Sphere	395
11.5	Bubbles and Drops		396
	11.5.1	Rigid Bubbles	396
	11.5.2	Spherical Cap Bubbles	397
CHAPTER 12	**CONVECTIVE MASS TRANSFER IN TURBULENT FLOW**		**403**
12.1	Properties of Turbulent Flow		404
	12.1.1	Transition Criteria	404
	12.1.2	Characteristics of Fully Turbulent Flow	405
	12.1.3	Stochastic Nature	405
12.2	Properties of Time Averaging		406

12.3	Time-Averaged Equation of Mass Transfer	408
	12.3.1 Turbulent Mass Flux	409
	12.3.2 Reynolds Stresses	409
	12.3.3 Reaction Contribution	410
12.4	Closure Models	411
	12.4.1 Turbulent Schmidt Number	411
	12.4.2 Prandtl's Model for Eddy Viscosity	412
12.5	Velocity and Turbulent Diffusivity Profiles	413
	12.5.1 Universal Velocity Profiles	414
	12.5.2 Eddy Diffusivity Profiles	415
	12.5.3 Wall Shear Stress Relations	416
12.6	Turbulent Mass Transfer in Channels and Pipes	417
	12.6.1 Simplified Analysis: Constant Wall Flux	418
	12.6.2 Stanton Number Calculation for Boundary Layers	421
	12.6.3 Analogy with Momentum Transfer	422
	12.6.4 Stanton Number for Pipe Flows	423
12.7	Van Driest Model for Large Sc	425
12.8	Turbulent Mass Transfer at Gas–Liquid Interface	427
	12.8.1 Damping of Turbulence	428
	12.8.2 Marangoni Effect	429
	12.8.3 Interfacial Turbulence	430
CHAPTER 13	**MACROSCOPIC AND COMPARTMENTAL MODELS**	**435**
13.1	Stirred Reactor: The Backmixing Assumption	436
13.2	Transient Balance: Tracer Studies	438
	13.2.1 Step Input	438
	13.2.2 Pulse or Bolus Input	439

	13.2.3	Age Distribution Functions	441
	13.2.4	Tracer Response for Tanks in Series Model	442
13.3		Moment Analysis of Tracer Data	444
	13.3.1	Moments from Laplace Transform of Response	446
13.4		Tanks in Series Models: Reactor Performance	449
13.5		Macrofluid Models	450
	13.5.1	Second-Order Reaction	451
	13.5.2	Zero-Order Reaction	451
13.6		Variance-Based Models for Partial Micromixing	453
13.7		Compartmental Models	454
	13.7.1	Matrix Representation	456
13.8		Compartmental Models for Environmental Transport	459
	13.8.1	Fugacity of Pollutants in Each Compartment	460
	13.8.2	Level I or Equilibrium Model	460
	13.8.3	Level II Model: Advection Effects	461
	13.8.4	Level III Model: Intermedia Transport Effects	461
	13.8.5	Level IV Model: Transient Effects	462
13.9		Fluid–Fluid Systems	462
	13.9.1	Backmixed–Backmixed Model	462
	13.9.2	Equilibrium Model	464
	13.9.3	Mixing Cell Model	465
13.10		Models for Multistage Cascades	465
	13.10.1	Equilibrium Model	466

CHAPTER 14   MESOSCOPIC MODELS AND THE CONCEPT OF DISPERSION   475

14.1		Plug Flow Idealization	476
14.2		Dispersion Model	478
	14.2.1	Boundary Conditions	480
	14.2.2	Solution for a First-Order Reaction	481

	14.2.3	Nonlinear Reactions	482
	14.2.4	Dispersion Model: Numerical Code Using CHEBFUN	482
	14.2.5	Criteria for Negligible Dispersion	483
14.3		Dispersion Coefficient: Tracer Response Method	484
	14.3.1	Laplace Domain Solution	485
	14.3.2	Moments of the Response Curve	485
	14.3.3	Time Domain Solution	487
14.4		Taylor Model for Dispersion in Laminar Flow	488
14.5		Segregated Flow Model	491
14.6		Dispersion Coefficient Values for Some Common Cases	493
14.7		Two-Phase Flow: Models Based on Ideal Flow Patterns	495
	14.7.1	Plug-Backmixed Model	496
	14.7.2	Non-Idealities in Two-Phase Flow	499
14.8		Tracer Response in Two-Phase Systems	503
	14.8.1	Single Flowing Phase	504
	14.8.2	Two Flowing Phases	506

CHAPTER 15   MASS TRANSFER: MULTICOMPONENT SYSTEMS   517

15.1		Constitutive Model for Multicomponent Transport	518
	15.1.1	Binary Revisited	518
	15.1.2	Generalization: The Stefan-Maxwell Model	519
15.2		Computations for a Reacting System	520
15.3		Heterogeneous Reactions	525
15.4		Non-Reacting Systems	528
	15.4.1	Evaporation of a Liquid in a Ternary Mixture	528
	15.4.2	Evaporation of a Binary Liquid Mixture	531
	15.4.3	Equimolar Counter-Diffusion	533
15.5		Multicomponent Diffusivity Matrix	535
	15.5.1	$\tilde{D}$ Matrix Relation to Binary Pair Diffusivity	536

CHAPTER 16   MASS TRANSPORT IN ELECTROLYTIC SYSTEMS   **543**

   16.1   Transport of Charged Species: Preliminaries   544

          16.1.1   Mobility and Diffusivity   544

          16.1.2   Nernst-Planck Equation   545

   16.2   Charge Neutrality   547

   16.3   General Expression for the Electric Field   548

          16.3.1   Laplace Equation for the Potential   549

          16.3.2   Transference Number   550

          16.3.3   Mass Balance for Reacting Systems   550

   16.4   Electrolyte Transport across Uncharged Membrane   551

   16.5   Transport across a Charged Membrane   553

          16.5.1   Interfacial Jump: Donnan Equation   553

          16.5.2   Transport Rate   554

   16.6   Transfer Rate in Diffusion Film near an Electrode   556

PART II   REACTING SYSTEMS   **565**

CHAPTER 17   LAMINAR FLOW REACTOR   **567**

   17.1   Model Equations and Key Dimensionless Groups   568

          17.1.1   Dimensionless Model Equations   568

          17.1.2   Boundary Conditions   571

   17.2   Two Limiting Cases   572

          17.2.1   Small B: Pure Convection Model   572

          17.2.2   Large B: Plug Flow Model   574

   17.3   Mesoscopic Dispersion Model   575

   17.4   Other Examples of Flow Reactors   577

          17.4.1   Channel Flow   577

          17.4.2   Non-Newtonian Fluids   577

Contents

17.4.3   Heat Transfer Effects   578

17.4.4   Turbulent Flow Reactor: 2-D Model   579

17.4.5   Axial Dispersion Model for the Turbulent Case   580

CHAPTER 18   MASS TRANSFER WITH REACTION: POROUS CATALYSTS   585

18.1   Catalyst Properties and Applications   586

  18.1.1   Catalyst Properties   587

18.2   Diffusion-Reaction Model   588

  18.2.1   First-Order Reaction   589

  18.2.2   Zero-Order Reaction   599

  18.2.3 nth-Order Reaction   603

18.3   Multiple Species   605

18.4   Three-Phase Catalytic Reactions   607

  18.4.1   Application Examples   608

  18.4.2   Mass Transfer Effects   609

18.5   Temperature Effects in a Porous Catalyst   610

  18.5.1   Equations for Heat and Mass Transport   610

  18.5.2   Dimensionless Representation   611

  18.5.3   Dimensionless Boundary Conditions   612

  18.5.4   Estimate of the Temperature Gradients   613

18.6   Orthogonal Collocation Method   615

  18.6.1   Basis of the Method   615

  18.6.2   Two-Point Collocation   616

18.7   Finite Difference Methods   617

  18.7.1   Central Difference Equations   618

  18.7.2   Zero-Order Reaction   619

  18.7.3   Nonlinear Kinetics   620

  18.7.4   Neumann and Robin Conditions   621

18.8   Linking with Reactor Models   622

        18.8.1   First-Order Reaction   623

        18.8.2   Second-Order Reaction   624

        18.8.3   Zero-Order Reaction   624

CHAPTER 19   REACTING SOLIDS   635

19.1   Shrinking Core Model   636

        19.1.1   No Solid Product   637

        19.1.2   Solid Product: Ash Layer Effects   641

19.2   Volume Reaction Model   644

        19.2.1   Kinetic Model   645

        19.2.2   Concentration Profile for Gas and Solid   646

        19.2.3   First-Order Reaction in B 646

        19.2.4   Zero-Order Reaction   649

19.3   Other Models for Gas–Solid Reactions   651

        19.3.1   Effect of Structural Changes   651

19.4   Solid–Solid Reactions   654

        19.4.1   Classical Models   654

        19.4.2   Dalvi-Suresh Contact Point Model   655

CHAPTER 20   GAS–LIQUID REACTIONS: FILM THEORY MODELS   661

20.1   First-Order Reaction of Dissolved Gas   662

        20.1.1   Boundary Conditions   663

        20.1.2   Dimensionless Version   664

        20.1.3   Flux Values at the Interface and into the Bulk 665

        20.1.4   Enhancement Factor   666

20.2   Bulk Concentration and Bulk Reactions   668

        20.2.1   Bulk Concentration   669

        20.2.2   Absorption Rate Calculation for $Ha < 0.2$ 670

20.3   Bimolecular Reactions   672

        20.3.1   Dimensionless Representation   673

        20.3.2   Invariance Property of the System   676

        20.3.3   Analysis for Pseudo-First-Order Case   677

        20.3.4   Analysis for Instantaneous Asymptote   678

        20.3.5   Second-Order Case: An Approximate Solution   678

        20.3.6   Instantaneous Case: Effect of Gas Film
                 Resistance   681

        20.3.7   Choice of Contactor Based on the Regimes of
                 Absorption   684

20.4   Simultaneous Absorption of Two Gases   684

        20.4.1   Model Equations   685

        20.4.2   Dimensionless Representation   685

        20.4.3   CHEBFUN Solution   686

20.5   Coupling with Reactor Models   688

        20.5.1   Semibatch Reactor   688

        20.5.2   Packed Column Absorber   691

20.6   Absorption in Slurries   692

        20.6.1   Particle Size Effect   693

        20.6.2   Instantaneous Reaction Case   694

20.7   Liquid–Liquid Reactions   697

CHAPTER 21   GAS–LIQUID REACTIONS: PENETRATION THEORY APPROACH   705

21.1   Concepts of Penetration Theory   706

        21.1.1   First-Order or Pseudo-First-Order Reaction   706

        21.1.2   Laplace Transform Method   707

        21.1.3   Flux and the Average Rate of Mass Transfer   710

        21.1.4   Relation between Film Theory and Penetration
                 Theory   711

21.2   Bimolecular Reaction   712

        21.2.1   Dimensionless Form of the Model   712

        21.2.2   Illustrative Results   713

21.3   Instantaneous Reaction Case   714

21.4   Ideal Contactors   717

        21.4.1   Laminar Jet Apparatus   718

        21.4.2   Wetted Wall Column   718

        21.4.3   Wetted Sphere   720

        21.4.4   Stirred Cells   720

CHAPTER 22   REACTIVE MEMBRANES AND FACILITATED TRANSPORT   727

22.1   Single Solute Diffusion   729

        22.1.1   Model Equations   729

        22.1.2   Dimensionless Representation   730

        22.1.3   Invariant of the System   732

        22.1.4   Instantaneous Reaction Asymptote   732

        22.1.5   Pseudo-First-Order Reaction Asymptote   735

22.2   Co- and Counter-Transport   736

        22.2.1   Model for Counter-Transport   737

        22.2.2   Model for Co-Transport   739

22.3   Equilibrium Model: A Computational Scheme   739

        22.3.1   Illustrative Results   741

22.4   Reactive Membranes in Practice   742

        22.4.1   Emulsion Liquid Membranes (ELM)   743

        22.4.2   Immobilized Liquid Membranes (ILM)   744

        22.4.3   Fixed-Site Carrier Membranes   744

CHAPTER 23   BIOMEDICAL APPLICATIONS   749

23.1   Oxygen Uptake in Lungs   751

        23.1.1   Oxygen-Hemoglobin Equilibrium   751

	23.1.2	Transport Steps for Oxygen Uptake	753
	23.1.3	Meso-Model for the Capillary	756
23.2		Transport in Tissues: Krogh Model	757
	23.2.1	Oxygen Variation in the Capillary	759
23.3		Compartmental Models for Pharmacokinetics	760
	23.3.1	Basic Framework	761
	23.3.2	Physiologically Based Compartments	762
23.4		Model for a Hemodialyzer	763
	23.4.1	Model Formulation	764
	23.4.2	Model for Patient-Dialyzer System	765

CHAPTER 24	**ELECTROCHEMICAL REACTION ENGINEERING**		**775**
24.1		Basic Definitions	776
	24.1.1	Anodic and Cathodic Reactions	776
	24.1.2	Half Reactions and Overall Reaction	777
	24.1.3	Classification of Electrode Reactions	779
	24.1.4	Primary Variables	781
24.2		Thermodynamic Considerations: Nernst Equation	781
	24.2.1	Equilibrium Cell Potential	785
24.3		Kinetic Model for Electrochemical Reactions	786
	24.3.1	Butler-Volmer Equation	788
	24.3.2	Tafel Equation	791
24.4		Mass Transfer Effects	791
	24.4.1	Concentration Overpotential	793
24.5		Voltage Balance	793
24.6		Copper Electrowinning	795
	24.6.1	Operating Current Density	795
	24.6.2	Voltage Balance	796
	24.6.3	Meso-Model for the Electrolyzer	798

24.7   Hydrogen Fuel Cell   798

24.8   Li-Ion Battery Modeling   800

    24.8.1   Charging   801

    24.8.2   Discharging   802

PART III   **MASS TRANSFER–BASED SEPARATIONS**   **809**

CHAPTER 25   **HUMIDIFICATION AND DRYING**   **811**

25.1   Wet and Dry Bulb Temperature   812

    25.1.1   The Lewis Relation   813

25.2   Humidification: Cooling Towers   815

    25.2.1   Classification   816

    25.2.2   General Design Considerations   817

25.3   Model for Counterflow   817

    25.3.1   Mass Balance Equations   818

    25.3.2   Enthalpy Balance Equations   819

    25.3.3   Merkel Equation   821

25.4   Cross-Flow Cooling Towers   825

25.5   Drying   827

    25.5.1   Types of Dryers   827

    25.5.2   Types of Solids   828

    25.5.3   Constant and Falling Rates   829

25.6   Constant Rate Period   830

25.7   Falling Rate Period   833

    25.7.1   Empirical Models   834

    25.7.2   Diffusion Type of Models   835

    25.7.3   Capillary Flow Models   838

    25.7.4   Choosing a Model   838

CHAPTER 26   CONDENSATION   **845**

26.1   Condensation of Pure Vapor   846

26.1.1   Laminar Regime: Nusselt Model   847

26.1.2   Wavy and Turbulent Regime   849

26.2   Condensation of a Vapor with a Non-Condensible Gas   850

26.2.1   Mass Transfer Rate   851

26.2.2   Heat Transfer Rate and Ackermann Correction
Factor   851

26.2.3   Interface Temperature Calculations   852

26.2.4   Condenser Model   854

26.3   Fog Formation   855

26.4   Condensation of Binary Gas Mixture   857

26.4.1   Condensation Rates: Unmixed Model   858

26.4.2   Calculation of the Interface Temperature   860

26.5   Condenser Model   861

26.5.1   Liquid and Vapor Phase Balances   862

26.6   Ternary Systems   864

26.6.1   Stefan-Maxwell Model   865

26.6.2   Condensation with Reaction   865

CHAPTER 27   GAS TRANSPORT IN MEMBRANES   **871**

27.1   Gas Separation Membranes   872

27.1.1   Membrane Classification   873

27.1.2   Transport Rate: Permeability   874

27.1.3   Transport Rate: Permeance   876

27.1.4   Selectivity   877

27.1.5   Sievert's Law: Dissociative Diffusion   877

27.1.6   Nonlinear Effects in Membrane Transport   878

27.2   Gas Translation Model   879

27.3   Gas Permeator Models   881

        27.3.1   Flux Relations   882

        27.3.2   Local Concentration   883

        27.3.3   Backmixed-Backmixed Model   884

        27.3.4   Countercurrent Flow   886

        27.3.5   Cross-Flow Pattern   889

27.4   Reactor Coupled with a Membrane Separator   890

CHAPTER 28   LIQUID SEPARATION MEMBRANES   **897**

28.1   Classification Based on Pore Size   898

28.2   Transport in Semi-Permeable Membranes   900

        28.2.1   Osmotic Pressure   900

        28.2.2   Reverse Osmosis   901

        28.2.3   Concentration Polarization Effects   903

        28.2.4   Kedem-Katchalski Model   905

        28.2.5   Equipment-Level Model   907

28.3   Forward Osmosis   907

28.4   Pervaporation   908

        28.4.1   Illustrative Applications   909

        28.4.2   Model for Permeate Flux   910

        28.4.3   Local Permeate Composition   911

CHAPTER 29   ADSORPTION AND CHROMATOGRAPHY   **919**

29.1   Applications and Adsorbent Properties   920

29.2   Isotherms   921

        29.2.1   Langmuir Model   921

        29.2.2   Competitive Adsorption Isotherm   923

        29.2.3   Freundlich Isotherms   923

        29.2.4   BET Isotherm   923

	29.3	Model for Batch Slurry Adsorber	924
		29.3.1 Model Equations	925
		29.3.2 Particle-Level Model	925
		29.3.3 Linear Driving Force Model	927
		29.3.4 Calculation of the Slurry Transients	928
		29.3.5 Simulation Using the Collocation Method	929
		29.3.6 Additional Complexities	931
	29.4	Fixed Bed Adsorption	931
		29.4.1 Equilibrium Model	932
		29.4.2 Axial Dispersion Effects	934
		29.4.3 Heterogeneous Model	936
		29.4.4 Klinkenberg Equation	936
		29.4.5 Scale-Up Aspects	937
	29.5	Chromatography	938
CHAPTER 30	**ELECTRODIALYSIS AND ELECTROPHORESIS**		**945**
	30.1	Technological Aspects	946
		30.1.1 When to Use Electrodialysis	948
		30.1.2 Membranes	948
		30.1.3 Electrodialysis Reversal Process	949
		30.1.4 Electrodialysis with Bipolar Membranes	950
	30.2	Preliminary Design of an Electrodialyzer	951
		30.2.1 Current and Voltage	952
		30.2.2 Limiting Current	954
		30.2.3 Detailed Models	955
	30.3	Principle of Electrophoresis	955
		30.3.1 Solutes with Fixed Type of Charge	956
		30.3.2 Solutes with Charge Dependent on pH	956

30.4   Electrophoretic Separation Devices   957

         30.4.1   Philpot Design   957

         30.4.2   Hannig Design   959

         30.4.3   Rotating Annular Bed   960

REFERENCES   965
INDEX   979

PREFACE

Mass transfer refers to processes involving transport of one or more components in a multicomponent mixture across two locations in a single-phase system or from one phase to another in a two-phase system. Examples are encountered in practically all fields of engineering and in biological systems. Production of chemicals starts with reactions often carried out in systems with two or more phases. The transport of reactants from one phase to a second reacting phase and the counter-transport of products play an important role in determining the rate of production, the reactor type to be used, and design of such equipment. The products formed in the reactor in turn have to be separated and purified. This is accomplished by various unit operations, most of which essentially involve mass transfer from one phase to another. Hence the study of fundamental principles of mass transfer and its application to reactors and separators is an essential skill set of the chemical engineering profession.

The study of principles of modeling of mass transfer processes is also essential to understand many other processes in biomedical engineering, environmental transport, dopant profiling in semiconductors, metallurgical processes, and so on. The underlying phenomena behind all these seemingly diverse applications is mass transfer and hence the analysis and modeling of this phenomena followed by computations of the developed model is an essential element in the analysis of all these systems. The central theme of this book is, therefore, to provide a basic introduction to modeling and computation of mass transfer processes, show their application to the design of reactors and separation systems, and also present some applications in environmental and biomedical areas. There is a need for a single volume textbook that embraces a wide range of topics in mass transfer and this book fulfills that need.

Key Distinguishing Features

The analysis of mass transfer processes is a well-studied problem and there are many excellent books on the subject. Each of these books emphasize some unique aspect of the overall subject but are not comprehensive. For example, texts on transport phenomena treat the theory and analytical solutions usually by analogy with heat transfer and provide useful and important analytical results for many cases. However, the multicomponent nature of the system, the role of chemical potential gradients and electrochemical transports, and so on, are not discussed and application of the theory to practice is often not clear.

Mass transfer effects are quite important in heterogeneous reactor analysis and design. But books on reaction engineering place emphasis mainly on the chemical kinetics and study of homogeneous reactions, which are obviously and important part of the subject. However, heterogeneous reactions are more common in industrial practice and the coupling of mass transfer with reaction is the most significant issue in design of these systems. The book shows how fundamental mass transport theory can be applied to simulation and design of these reactors.

Chemical reactors are followed by separation units in industrial production; mass transfer is underlying phenomena in most of the industrially important separation processes. The traditional books on unit operations deal mainly with equilibrium-based separation processes and do not provide adequate coverage of rate-based processes and some recent advances such as reactive separations, membranes, pervaporation, electrodialysis, electrophoresis, and so on. Hence it is reasonable to conclude that there is no text that is comprehensive enough to cover the fundamentals, and at the same time show the applications of mass transfer in reacting systems and in separations. This book meets that need and includes the fundamentals; the methodology of modeling mass transport processes; setting up the governing model equations; finding solutions through analytical methods for classic problems and through numerical methods using some simple MATLAB snippets; application examples for both heterogeneous reactions and separation; and some illustrative design examples.

Some novel topics and features of the current book are:

- An early introduction to models at three hierarchy levels and the connection between them.
- Solutions to illustrative problems using both numerical and analytical methods.

- Sample code in MATLAB for help in development of numerical problem-solving skills.
- Sample code using CHEBFUN methods, which is a powerful tool for solution of common ODEs and PDEs of importance in mass transfer. This is optional material but students may find it useful to study these tools.
- Detailed discussion on analysis of transport with chemical reactions. Coupling of local differential models with reactor models for macroscopic and mesocopic models is clearly illustrated.
- Detailed analysis of multicomponent diffusion using the Stefan-Maxwell model with many worked examples.
- Three chapters on electrochemical systems and ionic transport with illustrative application examples.
- A chapter on the role of mass transfer in illustrative biomedical problems.
- Introductory treatment of reactive separation processes and process intensification concepts.
- Sample MATLAB codes for many common mass transfer and separation processes so that students learn basic simulation and sensitivity analysis.

Intended Audience

The level and the sequence of presentation are such that the book is suitable for a junior- or senior-level course in chemical engineering. The book will also be useful for a first-level graduate course as well as an introductory learning source for students who intend to do research in this area. Industrial practitioners will find this to be a useful desktop reference tool due to the coverage of a broad range of topics.

Style of Presentation

The style of presentation is informal and represents more of a "classroom" conversational tone. Each chapter starts off with clearly defined learning objectives and ends with a summary of "must-know" information that should have been mastered from that chapter. Key equations are shown in boxes for easy reference. Short review questions are provided at the end of the chapter followed by a set of problems that reinforces the text material. Computer simulations are also illustrated together with analytical solutions.

Sample packages are also included to accelerate the applications of the computer-aided problem solving in the classroom. This sample code is presented in a separate subsection or boxed environment for easier reading within the main text. Additional supplementary material covering more worked examples, hints and answers to some exercise problems, illustration of the MATLAB code, and sample design case studies are available on the companion website: https://sites.wustl.edu/masstransfer/.

Topical Outline

The book is divided into three parts. Part I deals with the basic theory and fundamentals of mass transfer and modeling at three hierarchical levels. Part II deals with applications to reacting systems. Part III deals with applications to some selected separation processes.

A brief description of the contents and the scope of each part is presented here.

Part I

Part I deals with the basic theory of mass transfer and the fundamentals to be mastered so that you can apply these principles to simulation and design problems in a large range of application areas.

Chapter 1 introduces the basic methodology for modeling of mass transport processes and indicates three hierarchical levels of models, namely differential, macroscopic, and mesoscopic models. Chapters 2 to 4 take up introductory examples of modeling at these three levels, respectively, and you will learn the basic skills to model mass transport processes. Differential-level models are treated in further detail starting from Chapter 5, which develops the general differential equations for mass transfer. Application of the differential equations is then treated progressively in Chapters 6 to 12. Chapter 6 introduces the film model of mass transfer and shows how it is used to calculate the local rate of mass transfer. Chapter 7 provides the physical chemistry aspects of diffusion phenomena. Chapter 8 shows the application of differential equations for transient diffusion. Convective mass transfer is treated in detail in the next four chapters; empirical correlation in Chapter 9, internal laminar flows in Chapter 10, external laminar flows in Chapter 11, and turbulent flows in Chapter 12. Chapters 13 and 14 provide more details on macroscopic and mesoscopic models, respectively. Chapters 15 and 16 provide treatments of mass transfer in multicomponent systems and electrochemical systems, respectively.

Overall these chapters provide the foundations over which mass transfer models are built and provide the basics and the fundamentals needed in modeling of such processes. Reaction followed by separation forms the backbone of chemical industries. Mass transfer applications for reacting systems and separation systems are taken up in Parts II and III.

Part II

Part II starts with the problem of simulation and modeling of a laminar flow reactor in Chapter 17, which is a prototypical application of convective mass transfer with chemical reaction. It builds on the background provided in Chapter 10 and shows how these can be applied to reactor analysis. Chapter 18 presents the application of the diffusion-reaction model studied in Chapter 2 to the reaction in a porous catalyst and also shows how the local differential model for the particle can be built into a macro- or meso-model for the catalytic reactor. Chapter 19 deals with solids that undergo a chemical reaction with a component present in the gas phase. Chapters 20 and 21 deal with diffusional interaction in gas–liquid reaction systems and application to reactor selection and design. Chapter 22 deals with a similar topic and examines the role of diffusion of one or more solutes in membranes containing a reacting carrier. Chapter 23 examines the role of mass transfer in the some biomedical systems. Chapter 24 shows the application of transport in the ionic systems studied earlier in Chapter 16 to eletrochemical reactors. Overall Part II covers most of the essential topics in heterogeneous systems that the practitioner is likely to encounter.

Part III

Part III deals with application of the mass transfer model for some selected separation processes. In particular, systems where rate-based models are needed and systems with simultaneous heat and mass transfer are discussed in detail. Systems where equilibrium-based models are reasonably accurate, such as distillation, liquid extraction, and gas absorption are not covered as there are many excellent textbooks in this area.

Chapters 25 and 26 deal with processes where the simultaneous analysis of heat and mass transfer is needed and provides design calculations for humidification, drying, and condensation systems. Chapters 27 and 28 focus on membrane-based separations for gas and liquid systems, respectively. Chapter 29 deals with adsorption processes and chromatographic separations. Chapter 30 deals with separation processes where electrochemical mass transport effects are important. Overall Part III presents applications of the basic modeling tools introduced in Part I to these processes and also supplements the traditional books on separation processes.

For Instructors

Instructors will find the presentations novel and interesting and will be able to motivate students to appreciate the integrated structure of the field. They will also find the worked examples, short review questions, and exercise problems useful to amplify class lectures and illustrate theory. Solutions to the exercise problems and a PowerPoint deck of the figures in the book are available on Pearson's Instructor Resource Center.

The book has more material than can be covered in one semester and can be used in the following manner in teaching:

- For an integrated course for mass transfer fundamentals, Chapters 1 to 12 can be covered in one semester. Additional topics from Chapters 13 to 16 and Parts II and III may be selected and assigned as reading material or used in supplementary lectures. The chapters in these parts are written in a modular fashion and are more or less written in a stand-alone style. Hence these need not be followed in a sequential manner.
- For a course focused mainly on heterogeneous reactions, Chapters 1 to 4, 12, and 13 provide the needed foundation and can be followed by Part II of the book, which deals with applications of mass transfer to reacting systems.
- For a course focused mainly on unit operations, Chapters 1 to 4 provide the foundation and can be followed with Part III of the book. Chapters 25 to 30 supplement traditional books on equilibrium-stage models and provide a nice one-semester textbook for this.

Acknowledgments

First and foremost, I would like to acknowledge my teacher and mentor Professor M. M. Sharma, the guru of mass transfer applications. He initiated me to this field with his exciting undergraduate lectures and motivated me to do higher studies. It was a great opportunity to work under his advice for doctoral research and I am very grateful for the education and advice I received. This book is dedicated to Professor M. M. Sharma.

Washington University in St. Louis provided me an academic home and I express my gratitude for all the support and encouragement they provided throughout my long career of research and teaching. I would like to thank many colleagues with whom I worked on several projects and learned from, in particular Professors M. P. Dudukovic, A. Muthanna, and P. L. Mills. Working with graduate students in the Chemical Reaction Engineering Laboratory of Washington University and training them was a unique experience for me. I feel delighted that most of them are doing well in industrial practice.

Many summers were spent at Kasetsart University at Bangkok where I got an opportunity to teach some of these materials and interact with students on many projects, and I thank the university and Professors Limtrakul and Terdthai for their hospitality and interaction on many topics.

This is also an opportunity to thank many colleagues with whom I had a chance to interact and learn from. Close interactions with Professors L. K. Doraiswamy, R. A. Mashelkar, R. V. Chaudhari, R. Hughes, J. M. Smith, R. Krishna, S. Limtrakul, M. P. Dudukovic, P. L. Mills, A. Muthanna, and Bala Subramanian are acknowledged.

On the publishing side, I would like to thank Laura Lewin, executive editor at Pearson/IT Professional Group, who gave me the opportunity to publish this book, and Michael Thurston, development editor, who helped to review all chapters and gave me valuable suggestions on the content presentation.

I would also like to thank all the technical reviewers for identifying errors and for providing valuable feedback to this book. Special thanks to Professor A. K. Suresh, who encouraged me throughout the writing and offered valuable suggestions. Many thanks also to Professor R. V. Chaudhari for reading through several chapters in the book.

On the editing side many thanks are due to Vaibhav Kedar and Angela Weatherspoon for helping me with the preparation of many figures in the text.

Book writing is a difficult task and is not possible without the support of family and friends. I would like to express my appreciation to my immediate family in the USA, Nima, Josh, Maya, and Gabe, and my extended family in India and other parts of the world, particularly my brothers, sisters, and sisters-in-law for all their support and encouragement. Much appreciation is due to my friends in University City, Missouri, who encouraged me when the going got tough.

Register your copy of *Mass Transfer Processes* on the InformIT site for convenient access to updates and/or corrections as they become available. To start the registration process, go to informit.com/register and log in or create an account. Enter the product ISBN (9780134675626) and click Submit. Look on the Registered Products tab for an Access Bonus Content link next to this product, and follow that link to access any available bonus materials. If you would like to be notified of exclusive offers on new editions and updates, please check the box to receive email from us.

ABOUT THE AUTHOR

P. A. Ramachandran is a professor at Washington University in St. Louis in the Energy, Environmental, and Chemical Engineering Department. He holds bachelors and doctoral degrees in chemical engineering from the Bombay University Department of Chemical Technology. He has extensive teaching experience in transport phenomena, reaction engineering, and applied mathematics. His research interest is mainly in the application of transport phenomena principles to chemically reacting systems and development of continuum based models for multiphase reactor design. He is the author of *Boundary Element Methods in Transport Phenomena* and *Advanced Transport Phenomena*, and coauthor of *Three-Phase Catalytic Reactors*.

NOTATION

Commonly used notations are indexed here. Notations specific to certain sections are defined at appropriate places in those sections.

A	area for flow or mass transfer
A_1, A_2	usually integration constants
a_{gl}	interfacial area for gas-liquid mass transfer, m^{-1}
a_{ls}	interfacial area for liquid-solid mass transfer, m^{-1}
B_1, B_2	usually integration constants
Bi_h	Biot number for heat transfer, hL_{ref}/k_{solid}
Bi_m	Biot number for mass tranfser, $k_m L_{ref}/D$
C	total molar concentration of a multicomponent mixture, mol/m^3
$\bar{c}$	average molecular speed of gas molecules in kinetic theory
c_A	dimensionless concentration of species indicated in the subscript (A here), C_A/C_{ref}
C_A	local concentration of species indicated in the subscript (A here), mol/m^3
$\langle C_A \rangle$	cross-sectionally averaged concentration, mol/m^3
$C_{A,e}$	concentration of species indicated in the subscript at the exit
$C_{A,i}$	inlet concentration of species A for flow reactor, mol/m^3
c_{Ab}	dimensionless cup-mixed average concentration
C_{Ab}	concentration of species A indicated in the bulk phase, mol/m^3
C_{Ab}	cup-mixed (flow) average concentration of species for flow systems

C_{AG} concentration of species indicated in the subscript in the bulk gas, mol/m^3

C_{Ai} concentration of species A at the interface in Chapter 20, mol/m^3

C_{Ai} initial concentration of species A for transient problems, mol/m^3

C_{AL} concentration of species indicated in the subscript in the bulk liquid, mol/m^3

C_{AL}^* hypothetical concentration of A if in equilibrium with the bulk gas, mol/m^3

C_{As} concentration of species A at a solid surface, mol/m^3

C_D drag coefficient

C_G total molar concentration in the gas phase, mol/m^3

C_L total molar concentration in the liquid phase, mol/m^3

c_p specific heat of a species, mass basis, at constant pressure conditions, J/kg K

C_p specific heat of a species, mole basis, J/mol K

c_v specific heat of a species, mass basis, at constant volume conditions J/kg K

d diameter of the molecules treated as rigid sphere in Chapter 7

$\tilde{D}$ diffusivity matrix for multicomponent systems in Section 15.5, m^2/s

Da Damkohler number, ratio of mean residence time to reaction time

d_B bubble diameter in a gas-liquid dispersion

D_E axial dispersion coefficient, m^2/s

D_E^* dimensionless axial dispersion coefficient, $D_E / \langle v \rangle L$,

D_{eA} effective diffusitiy of species A in a heterogeneous media

D_i molecular diffusivity of species i, m^2/s

D_K Knudsen diffusion coefficient for small pores

d_p particle or solid diameter

d_t diameter of the tube or pipe

D_t turbulent mass diffusivity, m^2/s

E enhancement factor for gas-liquid mass transfer with reaction

$\boldsymbol{E}$ Electric field

$E(t)$ exit age distribution function in Chapters 13 and 14

e_x unit vector in the (x)-direction

f Fanning friction factor

F Faraday constant = 96,485 C/mol

$\mathcal{F}$ Drift flux correction factor

f_A fugacity of species A in Chapter 13

g acceleration due to gravity, m/s^2

G molar flow rate in mass exchanger; mol/s

$\dot{G}$ superficial gas velocity, kg/m^2s

G^E excess free energy for non-ideal liquids in Section 7.3.2, J/mol

Gr Grashoff number

$\hat{h}$ enthalpy per unit mass, J/kg

h heat transfer coefficient, W/m^2 K

Ha Hatta number for gas–liquid reactions

H_A Henry law solubility coefficient, usually atm m^3/mol

h_G heat transfer coefficient in the gas film

$\hat{h}_{gl}$ heat released on condensation of a species, J/kg

h_L heat transfer coefficient in the liquid film

$\hat{h}_{lg}$ heat of vaporization, J/kg

HTU   height of a transfer unit

$\boldsymbol{j}_A$ mass diffusion flux vector of A (mass reference), kg/m^2s

$\boldsymbol{J}_A$ molar diffusion flux vector of A (mole reference), mol/m^2s

J_{Ax} component of molar diffusion vector in direction x

j_D j-factor for mass transfer

j_h j-factor heat mass transfer

k thermal conductivity of a species, subscript l = liquid, g = gas, s = solid W/m K

k rate constant for reaction, general

k_0 rate constant for a zero-order reaction, mol/m^3 s

k_1 rate constant for a first-order reaction, 1/s

k_2 rate constant for a second-order reaction, m^3/mol s

k_ω mass transfer coefficient from an interface to bulk liquid (mass fraction driving force), mol/s m^2(massfrac)

K, K_{eq} equilibrium constant for a chemical reaction

K_A adsorption equilibrium constant

k_B Boltzmann constant = 1.38×10^{-23}J/K

k_G mass transfer coefficient from a gas to the interface (partial pressure driving force), mol/Pa m^2 s

K_G overall mass transfer coefficient from an bulk gas to bulk liquid (gas phase partial pressure driving force); mol/Pa m^2 s

K_L overall mass transfer coefficient from a bulk gas to bulk liquid (liquid concentration driving force), m/s

k_m mass transfer coefficient from a solid to fluid (concentration driving force), m/s

k_m° mass transfer coefficient under low mass flux conditions, m/s

k_s rate constant for a heterogeneous first order reaction, m/s

k_{sl} solid-liquid mass transfer coefficient, m/s

k_x mass transfer coefficient from an interface to bulk liquid (mole fraction driving force), mol/s m^2(molefrac)

K_x, K_y overall mass transfer coefficient from gas to liquid based on mole fraction driving force

k_y mass transfer coefficient from gas to interface (mole fraction driving force), mol/s m^2(molefrac)

L length of the plate or tube or catalyst slab, m

L liquid molar flow rate in mass exchanger, mol/s

$\dot{L}$ superficial liquid mass velocity, kg/m^2 s

L_C equivalent length parameter for a solid catalyst, V_p/S_e

Le Lewis number, ratio of thermal to mass diffusivity

m partition or solubility coefficient of a solute between two phases, y/x

$\dot{m}$ mass flow rate, kg/s

M ratio of diffusion time to reaction time (Chapters 21 and 22)

$\bar{M}$ average molecular weight of a mixture, kg/mol

$\bar{M}$ average molecular weight of a mixture

$\mathcal{M}$ total moles present in a control volume, g-mol

$\dot{\mathcal{M}}$ moles per sec entering the unit, i = inlet, e = exit

M_A molecular weight of species indicated in the subscript, kg/mol

$\mathcal{M}_A$ moles of A in the system or control volume

$m_{A,tot}$ total mass of A in an unit or control volume, kg

$\dot{\mathcal{M}}_{A,e}$ moles leaving control volume per unit time

$\dot{\mathcal{M}}_{A,i}$ moles entering a control volume per unit time

$\dot{m}_{Ae}$ mass flow rate of A exiting a unit, kg/s

$\dot{m}_{Ai}$ mass flow rate of A entering a unit, kg/s

$\dot{m}_{AW,tot}$ total mass of A transferred to walls from a unit or process, kg/s

$\boldsymbol{n}$ normal vector outward from a control surface

N number of mixed tanks into which the reactor is divided into in Chapter 13

$\boldsymbol{n}_A$ mass flux vector of species A, stationary frame, kg-A/m^{2}s

$\boldsymbol{N}_A$ combined mole flux vector of species A, stationary frame, mole of A/m^{2}s

N_{av} Avogadro number $= 6.23 \times 10^{23}$ molecules/mol

N_{Aw} molar flux at the tube wall, mol/m^2 s

n_{Ax} component of mass flux vector of A in the x-direction, kg-A/m^{2}s

N_{Ax} component of mole flux vector of A in the x-direction

ns total number of species in a multicomponent mixture

N_t total molar flux, $N_A + N_B$ in a binary

NTU number of transfers of unit parameter

p the concentration gradient in the p-substitution method

P power input for agitated vessels, W

P total pressure of a gas mixture, Pa

p, P fluid pressure, Pa

P_c critical pressure of a species, Pa

Pe Peclet number, $L_{ref}v_{ref}/D_A = ReSc$

Pe^* dispersion Peclet number, $1/D_E^*$

p_{vap} vapor pressure of a species, Pa

q concentration ratio parameter in Chapters 20 and 21

Q volumetric flow rate in a pipe, m^3/s

r radial coordinate in cylindrical and spherical system

$\mathbf{R}$ radius of cylinder or catalyst particle

R^* gas constant defined as R_G/M_w

r_A local rate mass of production of A by reaction per unit volume, mass units, kg/m^{3}s

R_A local rate of mole production of A by reaction per unit volume, mole units, mol/m^{3}s

R_A rate of production of a species A by reaction

$\langle R_A \rangle$ cross-sectional average of the rate of production of a species

R_A^v volumetric rate of absorption of A, mol/m^{3}s

Re Reynolds number, $L_{ref}v_{ref}\rho/\mu$

R_g gas constant, 8.314 Pa m^3/mol K

s shape parameter for Laplacian for transient diffusion, 0 = slab, 1 = long cylinder, 2 = sphere

S surface area of the control volume in Chapter 5

Sc Schmidt number, ν/D_A

S_e external surface area of a catalyst particle

Sh Sherwood number, $k_m L_{ref}/D_A$

S_i internal surface area of a catalyst particle per unit mass

St Stanton number, $k_m/v_{ref} = Sh/Re\,Sc$

t time variable

$\bar{t}$ mean residence time V/Q in Chapters 13 and 14

T_∞ temperature of the approaching fluid

T local temperature in the medium subscript c = coolant, G = gas, L = liquid

T_a temperature of the surroundings

T_c critical temperature of a species

t_E exposure time for a gas–liquid interface

T_i temperature of a gas–liquid interface

T_w temperature of a wall or tube

U overall heat transfer coefficient from hot fluid to cold fluid, W/m^2K

$\hat{u}$ internal energy per unit mass, J/kg

$\hat{U}$ internal energy per unit mole, J/mol

$\hat{v}$ specific volume, m^3/kg $= 1/\rho$

$\boldsymbol{v}$ velocity vector; also mass fraction–averaged velocity in a multicomponent mixture, m/s

$\boldsymbol{v}'$ fluctuating velocity vector in turbulent flow

$\bar{\boldsymbol{v}}$ time-averaged velocity vector in turbulent flow

$\boldsymbol{v}^*$ mole fraction–averaged velocity in a multicomponent mixture, m/s

$\langle v \rangle$ average velocity in the flow direction

v_θ velocity component in the tangential (θ) direction

V volume of a reactor or a macroscopic control volume

$\hat{V}$ molar volume, m^3/mol

$\boldsymbol{v}_A$ velocity of species A in a multicomponent mixture, stationary frame, m/s

V_b molecular volume at boiling point of solvent

v_e velocity component in the fluid outside the boundary layer, m/s

V_f friction velocity defined as $\sqrt{\tau_f/\rho}$ used in turbulent flow, m/s

v_p volume of a catalyst particle

v_x x-component of the velocity; v_y, v_z defined similarly

v_z axial (z-) component of velocity in cylindrical coordinates

x distance variable in the x-direction, y and z defined similarly

x_i mole fraction of species indicated by the subscript (usually in the liquid phase)

y distance variable in the y-direction

y^+ dimensionless length used in turbulent analysis near a wall

$y_B(l.m)$ log-mean mole fraction of the non-diffusing component

y_i mole fraction of species indicated by the subscript (usually in the gas phase)

z axial distance variable in cylindrical coordinates

Z frequency of molecular collisions in Section 7.4

z^* dimensionless axial distance variable in cylindrical coordinates, z/R

z_i number of charges on an ionic species

Greek Letters

α thermal diffusivity of a fluid or solid, m^2/s

γ_A activity coefficient of species A in Chapter 7

∇ gradient operator

∇^2 Laplacian operator

$\triangle$ ratio of boundary layer thickness, heat/mass to momentum

Δ difference operator, out-in

Δ ratio of mass and momentum transfer boundary layer thickness in Chapter 11

δ thickness of momentum boundary layer in general

$\Delta\Pi$ osmotic pressure difference in Section 28.2.1, Pa

δ_f film thickness for mass transfer; abbreviated as δ in Chapter 6

ΔH heat of reaction, J/mol

ΔH_c heat of condensation, J/mol

ΔH_v heat of vaporization, J/mol

δ_m thickness of mass transfer boundary layer

δ_t thickness of thermal boundary layer

ϵ gas or liquid holdup in a reactor

ϵ a parameter in the Lennard-Jones model in Chapter 7

ϵ power input to a vessel per unit mass

ϵ_p porosity of a solid or a catalyst particle

ζ dimensionless axial distance, z^*/Pe

η effectiveness factor of a porous catalyst in Chapter 18

η similarity variable defined by Equation 11.13 in Chapter 11

η similarity variable for convective heat transfer in Chapter 10

η dimensionless axial position in tubular reactor, z/L

θ angular direction in polar coordinates

θ latitude direction in spherical coordinates

θ dimensionless temperature in heat transfer examples

κ ratio of radius values, R_c/R_o, in the Krogh model in Section 23.2

κ Boltzmann constant, also denoted as k_B

λ mean free path defined by Equation 7.4

λ location of reaction plane for an instantaneous reaction, Chapters 20 and 21

μ coefficient of viscosity, Pa s

μ_A chemical potential of species A in Section 7.3, J/mol

μ_w chemical potential of water in Chapter 28

ν coefficient of kinematic viscosity, μ/ρ, m^2/s

ν stoichiometric coefficient of products formed from A

ν_e excess stoichiometric coefficient of products. $\nu - 1$

ν_t turbulent kinematic viscosity, μ/ρ, m^2/s

ν_T^+ dimensionless total (molecular + turbulent) kinematic viscosity

ξ dimensionless radial position, r/R or x/L

ρ density of the medium or the fluid, kg/m^3

ρ_A density of A in a multicomponent mixture, kg/m^3

ρ_A^0 pure component density of A, kg/m^3

σ surface tension, N/m

σ^2 variance of the tracer response in Chapter 13

$(\sigma*)^2$ dimensionless variance of the tracer response, $\sigma^2/\bar{t}^2$

τ dimensionless time in Chapters 2 and 8, t/t_{ref}

τ_w wall shear stress
ϕ longitude in the spherical coordinate system
ϕ Thiele parameter for a first-order reaction
ϕ_0 Thiele parameter for a zero-order reaction defined by Equation 18.14
ψ stream function defined by Equations 11.10 or 11.11
ω_A mass fraction of species indicated by the subscript, kg-A/kg-total

Common Subscripts

b bulk conditions
e exit values
g, G gas phase properties
i inlet values
i interface conditions
l, L liquid phase properties
s conditions at a surface of a solid or catalyst

Acronyms

AEM   anion exchange membrane in Chapter 30
BPM   bipolar membrane in Chapter 30
CEM   cation exchange membrane in Chapter 30
CFD   computational fluid dynamics
DEq   differential equation
ED   electrodialysis in Chapter 30
LHS   left-hand side of an equation
NRTL   non-random two-liquid model
PDE   partial differential equation
RHS   right-hand side of an equation
UNIFAC   universal functional-group activity coefficients

PART I

Fundamentals of Mass Transfer Modeling

CHAPTER 1

Introduction to Modeling of Mass Transfer Processes

Learning Objectives

After completing this chapter, you will be able to:

- Explain the role of modeling of mass transfer processes in a variety of diverse fields.
- Understand the steps needed in modeling of a given system or equipment involving mass transfer.
- Define control volume and mark its bounding control surface.
- Write a species conservation law (mass balance) first in words and then as a mathematical equation for the chosen control volume.
- Distinguish three hierarchical levels of models—that is, differential, macroscopic, and mesoscopic models—depending on the size of the control volume chosen.
- Identify the closure relations needed in addition to species mass balance to complete the mathematical description of the process.
- State Fick's law, which is a commonly used constitutive (closure) model, for the diffusion flux in the one-dimensional case and its extension to three-dimensional diffusion.
- State how the convective rate of mass transfer can be calculated.
- Define the mass transfer coefficient and explain its usefulness.
- Define flow weighted average (cup mixed average) and cross-sectional averaged concentrations and explain in which contexts these variables are needed.
- Explain what a compartmental model is and why it is needed for modeling large-scale systems.

Mass transfer refers to a net movement of a species A from one location to another, usually in a multicomponent mixture containing species A, B, C, and so on, usually caused by a concentration difference. Consider, for example, a pollutant introduced in air. It gets transported from the source point to another location by the action of two common mechanisms of mass transport—namely, diffusion (a result of concentration difference between two points) and convection (caused by, for example, by wind velocity). Another common example of considerable industrial applications is the situation in which species A moves from one phase to the interface of a second phase and then crosses to the second phase; this is commonly referred to as interfacial transport.

Mass transport phenomena are ubiquitous in nature and in industrial applications. In process industries, most of the common separation processes involve the process of mass transfer from one phase to another. Penicillin production is not possible without rapid extraction of the product of fermentation (penicillin) by liquid extraction; otherwise, the penicillin will decompose quickly to unwanted products. The catalytic converter in a car removes NO_x, CO, and other hydrocarbons from combustion gases and involves transport of NO_x and other products to the walls of the converter, followed by a heterogeneous reaction at the catalytic walls. Transport and absorption of oxygen into blood is vital to life itself. Mass transfer with simultaneous reaction with hemoglobin plays an important role here. Plants draw water from soil by osmosis, which is an example of mass transport driven by a chemical potential gradient. The formation of acid rain is a consequence of mass transfer and reaction of sulfur dioxide in the rain droplets. More examples are provided in Section 1.5.

Development of a mathematical model to describe the process is central to the analysis and design of these processes. The model is expected to provide information on the concentration profile of the species being transported in the system and the rate at which the species is transported across the interface, system, or equipment. The goal of this book is to teach the methodology of modeling mass transport processes. You will learn how to set up the model equations and develop the skill set needed to use appropriate analytic and numerical tools to solve the resulting mathematical problem.

Analysis of mass transport phenomena is based on continuum approximation, which assumes that the matter is continuously distributed in space. This assumption permits us to assign a value to the concentration at each and every point in space. Differential equations for the concentration variation can then be developed based on this continuum model using a differential control volume. Such differential models are formulated using the species conservation laws coupled with some constitutive models for transport of mass by diffusion. The resulting model is the most detailed model in the context of the continuum assumption. The basic formalism for constructing such a model is introduced in this chapter.

Differential models are not always used and simpler but less descriptive models based on a larger sized control volume are often used. As you will learn from this chapter that models can also be developed at two levels (meso and maco) using a larger sized control volume and average concentrations are needed in these models rather than pointwise values. Three common averages used are the volume average, the cross-sectional average and the flow-weighted average and this chapter defines these and indicates in what context they are used and the additional information needed to close the models based on larger sized control volume.

This chapter provides a roadmap for subsequent chapters in the book. A careful study and understanding of the basic concepts and definitions described here will provide the needed vocabulary for further chapters.

1.1   What Is Mass Transfer?

As indicated in the preamble, mass transfer refers to a net movement of a species *A* from one location to another in a multicomponent mixture containing species *A, B, C,* and so on. Note that mass can also be transported in a single-component system—for example, in water being pumped from one location to other—but these problems are more pertinent to hydraulics or fluid dynamics and not considered in the realm of mass transfer.

The analysis of the phenomena of mass transfer is important in engineering design applications and in various other fields. Examples of mass transport are readily seen. A lump of sugar put in a coffee mug dissolves and spreads completely into the liquid. A perfume sprayed in the air freshens the whole room. A crystal of $KMnO_4$ dropped in a jar of water leads to the color spreading over a column of water placed above due to crystal dissolution and subsequent mass transport by diffusion.

Mass transfer is often studied as part of the subject of transport phenomena, which deals with momentum, heat, and mass transfer. What distinguishes mass transport from momentum and heat is that mass transfer by diffusion can occur only in a multicomponent mixture; it cannot occur in a single-component system, unlike transport of momentum and heat. The multicomponent nature of the system needs some additional considerations in the analysis.

Mass transfer is the underlying phenomenon for a large number of industrially important separation processes in unit operations. Consequently, some introductory treatment of mass transfer is provided in courses and in textbooks on unit operations. Most often, however, each unit operations (e.g., distillation, absorption) are studied individually, rather than as a comprehensive topic. Hence the common underpinning in the modeling of various separation processes is often not clear without a proper study of fundamentals of mass transfer.

Mass transfer is often accompanied by a chemical reaction. A detailed study of this coupling is important in heterogeneous reaction (systems with two or more phases) analysis. It turns out that most of the industrially important chemicals are made by a heterogeneous reaction in which two or more phases are contacted. Hence the central theme in reactor design and scale-up is, most often, the evaluation of the coupling between mass transfer rate and chemical reaction rate.

1.1.1   What Is Interfacial Mass Transfer?

Interfacial mass transfer refers to a situation in which a species A crosses from one phase to another across the interface separating the two phases. Examples are widespread in the field of unit operations. In liquid–liquid extraction, for instance, species A in a mixture of A and B in one phase can be separated by contact with an immiscible solvent in which A is preferentially more soluble, resulting in an interfacial transport of A from the first phase to the solvent phase.

An important consideration here is that the concentration is a discontinuous function at the interface between the two phases (the concentration jump). Thus, if $x = 0$ is the location of the plane separating the two interfaces (for example, a gas phase and a liquid phase), the concentration of a species A at the gas side (say, $x = 0^-$) is different from that at $x = 0^+$, the liquid side of the interface. As an example, consider the air–water interface. The oxygen concentration on the air side of the interface is much larger then the oxygen concentration on the water side of the interface, and the two are related by the thermodynamics of phase equilibrium or the solubility relationship. Hence thermodynamic relations are needed in interfacial mass transfer analysis. This is another distinguishing feature of mass transport. In contrast, in momentum and heat transport, the corresponding quantities—namely, velocity and temperature—are continuous functions at an interface for most common situations.

1.1.2   What Causes Mass Transfer?

Mass transfer is caused by a combined process of diffusion and convection. Diffusion is an effect of molecular- or atomic-level interactions. For example, in a gas, the molecules are in a state of random motion, so there is tendency for concentration to equalize. Thus, if a concentration difference for species A exists between two points, then the molecular motion causes a net transport of A from a region of higher concentration to a region of lower concentration. (**Note:** There are a few cases in which mass transfer can occur from lower to higher concentration; these are discussed in Chapter 7.)

The phenomenon of diffusion is similar to heat conduction, in which a temperature difference causes a flow of heat from a higher-temperature location to a lower-temperature location. Diffusion may also be interpreted on the basis of the thermodynamics involved. If a chemical potential difference exists between two points, the natural tendency toward equilibrium is that the chemical potential should become the same at these points. Thus diffusion is caused by a gradient in chemical potential.

Convection refers to transport by bulk motion. If we add sugar to a mug of coffee and stir it, the dissolution rate is increased due to the fluid motion caused by stirring. Transfer of a pollutant from one location to another by wind is an another example of convection.

1.2   Preliminaries: Continuum and Concentration

The modeling of mass transfer process primarily involves the computation of the concentration profile and the rate at which species A, B, C, and so on are transported across any surface in the media. The intrinsic rate of transport (the transport rate per unit area) is quantified in terms of the flux vector, and this section introduces the preliminaries. The notion of the flux vector is then discussed in Section 1.3.

1.2.1   The Continuum Assumption

The key concept in transport modeling is the continuum assumption, which states that matter is distributed continuously in space. Thus we address the bulk properties rather than the molecular-level interactions. The continuum concept permits us to assign local values to variables at each and every point in an Euclidean space. We can then speak of a temperature field, concentration field, velocity field, and so on.

1.2.2   Concentration: Mole Units

In the context of the continuum approximation, the local concentration, C_A may be defined as

$$C_A = \lim_{V \to 0} \frac{\mathcal{M}_A}{V} \tag{1.1}$$

where V is the volume element and $\mathcal{M}_A$ is the moles of species A present in that volume. In this book, the S.I. system of units is used, with kg, m, and s being the basic units. However, the basic unit of concentration, is mol/m^3, where mol refers to g mol, rather than kg mol. The molarity, denoted as M, is also often used for concentration; it is equivalent to mol/L. Hence 1 M = 1000 mole/m^3—a useful conversion factor. Note that units of kmol/m^3 can also be used to represent concentration if convenient. In the final calculations, this will need to be converted to gmol/m^3 to maintain consistency of units.

The total molar concentration of the mixture C (at any position and at any time) is defined as the summation of all the species concentrations:

$$C = \sum_{A,B,..}^{ns} C_A \tag{1.2}$$

where ns is the total number of species present in the mixture.

The total concentration for a gas mixture C_G and the total pressure for an ideal gas mixture are related as

$$C_G = \frac{P}{R_g T}$$

Here R_g is the gas constant, which has a value of 8.314(Pa m^3/mol K in the S.I. system of units. Note that the units of R_g can also be expressed as J/mol K.

For example, for a gas at 1 bar pressure and 300 K, the total concentration is calculated as 1×10^5 Pa/[8.314(Pa m^3/mol K) $\times$ 300 K] and has a value of 37 mol/m^3.

Total concentration in liquid can be calculated if the partial molar volume of the components in the mixture solution is known. For dilute solutions, it can be based on the density of the solvent.

For example, for a dilute aqueous solution we can use the density of water as the representative density of the mixture. Hence $\rho = 1000$ kg/m^3. The molecular weight of water is expressed as 18×10^{-3} kg/mol rather than as 18 g/mol. Hence the total concentration in dilute solution can be approximated as 1000 kg/m^3/18 $\times 10^{-3}$ kg/mol = 55556 mol/m^3.

Very often it is necessary to refer to two or more phases—for example, a gas and a liquid. In such cases we use C_{AG} to represent the concentration of A in the gas phase and C_{AL} to represent the concentration in the liquid. Thus we use the first subscript to represent the species by either by numbers $i = 1, 2, \ldots$ or simply by names as $A, B, \ldots$, especially for binary mixtures. The second subscript is used to represent the phase (or the location) where it is present. Similarly, C_G rather than C is used for the total concentration in the gas phase and C_L for the total concentration in the liquid phase. The ratio of C_A to C is called the mole fraction, denoted by y_A for gas mixtures or x_A for liquid mixtures:

$$y_A \text{ or } x_A = C_A/C \tag{1.3}$$

The average molecular weight of the mixture $\bar{M}$ is calculated as

$$\bar{M} = \sum_{A,B,..}^{ns} y_A M_A \tag{1.4}$$

where M_A is the molecular weight of species A. Also note that

$$\bar{M} = \frac{\rho}{C} \tag{1.5}$$

where ρ is the density of the mixture.

1.2.3 Concentration: Mass Units

Another way of expressing the species concentration is to use the mass units. The mass concentration (also called the partial density) is designated by ρ_A and is defined as the mass of species A per unit volume of the mixture locally at a point in the fluid continuum. This quantity is related to C_A simply by the species molecular weight:

$$\rho_A = M_A C_A \tag{1.6}$$

Note that ρ_A is the density of A in the mixture (same as mass of A per unit volume of the mixture) and should not be confused with the pure component density of A. The latter will be denoted in this book with a superscript ρ_A° if we have to refer to both quantities in the same context.

The summation of this for all the species is the mixture density, ρ:

$$\rho = \sum_{A,B,..}^{ns} \rho_A \tag{1.7}$$

The ratio of ρ_A to ρ is the mass fraction of species A, denoted often by ω_A:

$$\omega_A = \rho_A/\rho \tag{1.8}$$

The average molecular weight of the mixture is calculated from the mass fraction values by the reciprocal weighting rule:

$$\frac{1}{\overline{M}} = \sum_{A,B,..}^{ns} \frac{\omega_A}{M_A} \tag{1.9}$$

1.2.4 Concentration: Partial Pressure Units

In gaseous systems, a measure of concentration is the partial pressure of a species. This follows from the ideal gas law:

$$C_A = \frac{p_A}{R_g T} \tag{1.10}$$

or

$$p_A = C_A R_g T \tag{1.11}$$

Partial pressure is therefore a direct measure of concentration in an ideal gas mixture and used commonly for such cases. The partial pressure difference can be used as a driving force for mass transfer instead of concentration differences for (ideal) gas-phase systems. The two sets of units can be interconverted easily, if needed, by the previous equations.

1.3 Flux Vector

Mass transfer can be caused by diffusion and bulk motion of the system as a whole. To describe the rate of mass transfer, we need the definition of flux.

1.3.1 Molar and Mass Flux: Definition

Consider a plane of a differential area ΔA with its normal pointing in the x-direction. The quantity (in moles or mass) of a species A crossing this plane per unit area per unit time is defined as the flux component in this direction. This is commonly denoted as N_{Ax} with units of moles A per unit area per unit time (see Figure 1.1). The first subscript indicates the species being transported, while the second subscript indicates the orientation of the plane or the surface under consideration. Similar definitions apply for the flux component

N_{Ay} is crossing from bottom to top

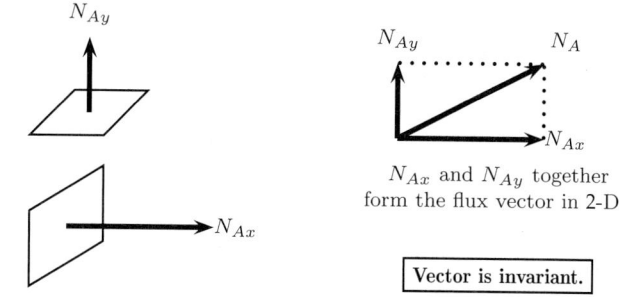

N_{Ax} and N_{Ay} together
form the flux vector in 2-D.

Vector is invariant.

N_{Ax} is crossing from left to right

Figure 1.1   Notion of flux vector and representation of its components. The flux vector is uniquely defined as any given point, while the components depend on the constructed coordinate system. Thus, if the coordinates are rotated, the components will be different but the flux vector will be the same.

in the y- and z-directions. Thus N_{Ay} represents the moles crossing a plane oriented in the $+y$ direction per unit time per unit area, and N_{Az} is defined in a similar manner. The three components together constitute a vector, with the molar flux vector usually denoted as N_A, and the units being mol of A/m^2 s.

If we use the mass units, we can define a mass flux vector, which is usually denoted as n_A, with the units being kg A/m^2 s.

The flux vector is unique at a given point, whereas the components N_{Ax} and so forth depend on the orientation of the x-axis and other dimensions. Thus the components can change in magnitude if, for example, the coordinates are rotated. This will not alter the results in any way since the flux vector is frame indifferent. Similarly, the flux components can be defined in cylindrical, spherical, or other body-fitted coordinate systems and used for the specified geometry. The flux results both from diffusion and due to the flow rate across the control surface and hence the flux are referred to as combined flux. For modeling purposes, it is necessary to split this entity into diffusion flux and convection flux. Thus

$$\text{Combined flux} = \text{convection flux} + \text{diffusion flux}$$

This partitioning, which is illustrated in Figure 1.2, is not unique. Rather, it will depend on which value is assigned to the convection flux, as will be shown in the next section.

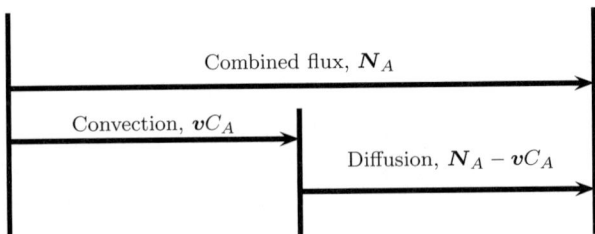

Figure 1.2   Partitioning of the combined flux into diffusion and convection flux. The total flux is fixed from the conditions prevailing at that point. In contrast, the partitioning is not unique and depends on the definition of the mixture velocity, v.

1.3.2   Convection Flux

Convection is transport due to flow, and convection flux can be easily calculated if the velocity of flow is known. Thus, if there is a velocity v_x in the x-direction, then a flux of $v_x C_A$ results from this velocity in the x-direction and is referred to as a convection flux. In general, for the 3-D case the convection flux depends on (a suitably defined) mixture velocity v and is defined as

$$\text{Convection flux vector} = vC_A$$

Note that the mixture velocity in a multicomponent system is not uniquely defined, unlike in a single-component system. This topic is discussed further in Chapter 5. The point to note here is that the convection flux depends on which definition of the mixture velocity is used. The diffusion flux is correspondingly adjusted since the total flux at any point is fixed.

1.3.3   Diffusion Flux

Diffusion is a tendency for a system to approach equilibrium conditions. This can be considered as a result of molecular-level motion when looked from a continuum point of view. Diffusion is therefore also referred to as molecular transport. It must be modeled based on empirical observation, with the resulting model being called a constitutive model. This result is system specific, as there is no universal law to describe the process of diffusive mass transfer. However, it is common to model diffusion by Fick's law, which states that the (molar diffusive) flux of A is proportional to the concentration gradient of A, dC_A/dx:

$$J_{Ax} = -D_A \frac{dC_A}{dx} \qquad (1.12)$$

Here J_{Ax} is the diffusion flux of A in the x-direction. Note that the law implies that the diffusion occurs in the direction of negative concentration gradient—that is, from a region of higher concentration to a region of lower concentration. This is similar to heat flowing from a hot body to a cold body. The constant of proportionality D_A is called the diffusion coefficient of A. It has the units of m^2/s.

More generally speaking, the gradients in the chemical potential can be viewed as causing mass transfer, such that use of these concepts leads to more complex expressions for the diffusive flux. These complexities are addressed in Chapter 7. One consequence of such models is that D_A may not have a constant value over the entire concentration range, but rather may be a function of concentration, especially for a non-ideal liquid mixture. Another consequence is that in some cases, the diffusion can occur counter to the direction of negative concentration gradient. These complex effects and models of them will be deferred to Chapter 7.

Note: A more precise definition of diffusion flux is provided in Chapter 5 and requires a more precise definition of the velocity of a multicomponent mixture. The point to note at this stage is that in a multicomponent system, the mixture velocity is not uniquely defined, unlike in a single-component system. The diffusion flux is defined correspondingly, with the mixture velocity as the frame of reference. Also, Fick's law (shown by Equation 1.12) applies to a system in which the total concentration C remains constant at various points in the system. A more general representation and the topic of diffusion in systems with total constant density are deferred to Chapter 5. Nevertheless, for many applications the simple Fick's model for diffusion shown here is widely used.

Generalization of Fick's law to the 3-D case is accomplished by using the notion of the gradient of a scalar:

$$\boldsymbol{J_A} = -D_A \nabla C_A \tag{1.13}$$

where ∇C_A is the concentration gradient vector.

Typical Values of Diffusivity

The range of values of the diffusion coefficient depends on the phase in which transport is taking place. The range of values are shown in Table 1.1.

Note that the convection flux is proportional to the concentration, while the diffusion flux is proportional to the concentration gradient.

Although convection and diffusion are the two fundamental modes of mass transport, additional phenomena such as turbulent diffusion and dispersion are also used. These are not independent mechanisms, but rather reflect the combined effects of diffusion and convection. A brief description

Table 1.1   Order of Magnitude Values of the Diffusion Coefficient

System	D_A Values in m^2/s
Gas in gas	10^{-5} to 10^{-4}
Gas in porous solid	10^{-6} to 10^{-5}
Gas in narrow pores	10^{-8} to 10^{-6}
Liquid in liquid	10^{-10} to 10^{-9}
Liquid in porous solid	10^{-11} to 10^{-10}
Solid in solid	10^{-12} to 10^{-17}

is provided here, but detailed discussion of these topics is deferred to later chapters (4 and 12).

Turbulent Diffusion

Convective transport is due to fluid motion. Fluid flow can be either laminar, turbulent, or transitional, depending on the flow rate, viscosity, and other parameters. For steady-state laminar flow, the velocity is time independent and the convection flux based on the time-independent velocity applies. However, in turbulent flow, random fluctuating components of velocity are superimposed on the main flow. The convective mass transport due to the fluctuating part of velocity, when averaged over time, is usually modeled as an additional mode of transport, referred to as eddy diffusion or turbulent diffusion. Thus the turbulent diffusion is a consequence of the concept of time averaging widely used in study of turbulent transport and is not an independent mass transport mechanism.

Diffusion can be viewed as manifestation of random molecular motion, while turbulent diffusion can be viewed as the result of random bulk motion of the eddies. Diffusion brings in species from a region of higher concentration to a region of lower concentration. Similarly, turbulent eddies bring in species from a region of higher concentration to a region of lower concentration and the process can be modeled by an equation similar to Fick's law:

$$\text{Flux due to turbulent eddies} = -D_t \nabla \overline{C}_A$$

where D_t is the turbulent diffusivity, also known as eddy diffusivity. Turbulent diffusivity is a flow-dependent property rather than a molecular property. The $\overline{C}_A$ is a time-averaged value of the fluctuating concentration, which is a characteristic of turbulent transport. Models to calculate D_t are provided in

Chapter 12. (This quantity is commonly defined as the time average of the product of velocity fluctuation and the concentration fluctuation.)

Dispersion

Consider the flow of a fluid in a pipe. There is a velocity profile across the tube radius. The velocity at the wall is zero, while that at the center is at its maximum. The contribution of the convection resulting from this velocity profile and the associated diffusion, when averaged across the pipe cross-section, can be modeled as another mode of mass transfer, referred to as dispersion. Thus the dispersion is a consequence of the concept of cross-sectional averaging, which is widely used in mesoscopic model analysis (described in more detail in Chapters 4 and 14) of mass transport, rather than an independent mass transport mechanism.

1.4   Concentration Jump at Interface

Another important point to understand is that the concentration is not continuous at a phase boundary (e.g., gas–liquid interface), unlike temperature. Consider air–water system as an example. Is the oxygen concentration on the air side of the interface the same as the oxygen concentration on the water side of the gas–liquid interface? The answer is no. Because oxygen has a poor solubility in water, the oxygen concentration in the water phase in much lower than that in the air phase. This difference in concentration is called the concentration jump at the interface. The thermodynamic relations needed to calculate the jump are for various cases are discussed next.

1.4.1   Gas–Liquid Interface: Henry's Law

The concentrations on the gas side and on the liquid side of the interface are often assumed to be linearly proportional—a concept summarized by Henry's law. The constant of proportionality is known as the Henry's law constant. Various definitions are used for this constant, depending on the units used to measure the concentrations. The common form is as follows:

$$p_A = H_A x_A \text{ at equilibrium conditions} \tag{1.14}$$

where p_A is the partial pressure of the species in the gas phase and x_A is the mole fraction in the liquid. The constant H_A has the units of atm or Pa or bars here. The values of the Henry's law constant for some common gases are shown in Table 1.2.

Table 1.2   Henry's Law Constant for Some Common Gaseous Species in Water at 298.15 K

Gas	H, atm
Hydrogen	7.099×10^4
Oxygen	4.259×10^4
Nitrogen	8.65×10^4
Ozone	4570
Carbon dioxide	1630
Sulfur dioxide	440
Ammonia	30

The larger the value of the Henry's law constant, H_A, the less soluble the gas is in the liquid phase. The effect of total pressure in the system is to increase the solubility in accordance with Henry's law. In general, increasing the temperature decreases the solubility. Hence the Henry's law constant is usually an increasing function of temperature. An exception occurs with hydrogen, which shows a retrograde behavior. Here the solubility increases at first with an increase in temperature, reaches a maximum, and then decreases thereafter.

Other Definitions of Henry's Law Constant

Other definitions of the Henry's law constant are also used, depending on which unit is used for the concentrations in the two phases. Two common definitions are as follows:

$$p_A = H_{A,pc} C_A \tag{1.15}$$

where $H_{A,pc}$ is the Henry's law constant (unit of Pa m^3/mol) with partial pressure, Pa, as the unit for the gas phase and mol/m^3 as the concentration unit for the liquid phase, respectively. Another form is

$$p_A H_{A,cp} = C_A \tag{1.16}$$

where $H_{A,cp}$ is the Henry's law constant (unit of mol/Pa m^3) and is the reciprocal of $H_{A,pc}$.

Since the concentration in the gas phase can also be used instead of the partial pressure unit, we have yet more ways of writing the Henry's law relationship! Example 1.1 shows the application of the Henry's law and the concentration jump at a gas–liquid interface.

Example 1.1   Oxygen Concentration in Water

Find the dissolved oxygen concentration in water, assuming equilibrium conditions.

Solution

Let A represent the oxygen species. The mole fraction in the liquid is then calculated using Henry's law:

$$x_A = p_A/H_A = 0.21 \text{ atm}/4.259 \times 10^4 \text{ atm} = 4.963 \times 10^{-6}$$

Hence the concentration in the liquid is $x_A C$, where C is the total concentration of the liquid mixture. For dilute systems, C is simply the solvent concentration. For water, the density is 1000 kg/m^3 and the molecular weight is 18×10^{-3} kg-mass/g-mole. Hence

$$C = 1000/(18 \times 10^{-3}) = 55,560 \text{ mol/m}^3$$

The oxygen concentration in the liquid is therefore 0.2739 mol/m^3. If air is bubbled for a long time in a batch of liquid, the oxygen concentration in the entire liquid will reach the previously given value, which is the final equilibrium value.

To illustrate the concentration jump at the interface, we can also calculate the concentration of oxygen in air using the ideal gas law:

$$C_{AG} = y_A P/R_g T = 0.21 \text{ atm} * 1.01 \times 10^5 (\text{Pa/atm})/8.314/298 = 8.467 \text{ mol/m}^3$$

The ratio of C_{AG} to C_{AL} is 30.9, which is the magnitude of the concentration jump at the interface.

Note: Mass transport is usually performed under non-equilibrium conditions. Hence Henry's law should be applied only at the interface, which is assumed to still be at equilibrium. Concentrations in phase 1 and phase 2, away from the interface, will not be the equilbrium values. Otherwise, no mass transfer will occur.

The schematic of the concentration variation for interfacial mass transfer is shown in Figure 1.3. Although the concentration varies in both phases, the interfacial concentration values are related by thermodynamic considerations. The subscript i in Figure 1.3 refers to the interface and y_{Ai} and x_{Ai} are related by Henry's law.

1.4.2   Vapor–Liquid Interface: Raoult's Law

For volatile liquid-phase species, Raoult's law is often used to relate the interfacial concentrations. It states that at the vapor–liquid interface, a pure liquid

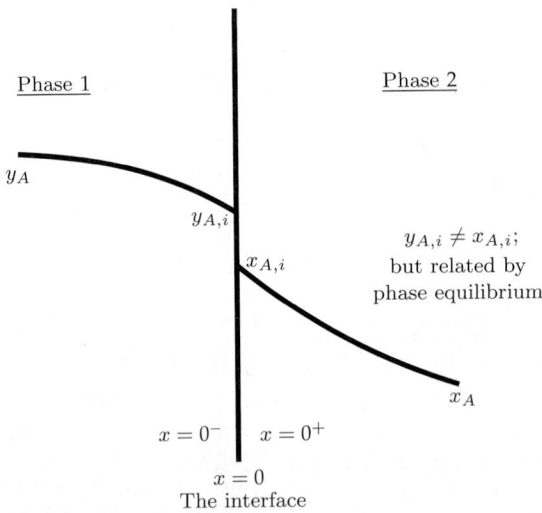

Figure 1.3   Schematic of the concentration jump at an interface.

exerts a partial pressure equal to the vapor pressure of the liquid at that temperature. For an ideal liquid mixture, the partial pressure at the interface is equal to the vapor pressure multiplied by the mole fraction in the liquid. Hence the interfacial relation is

$$p_{A,i} = x_{A,i}\, p_{vap,A}$$

The vapor pressure p_{vap} varies as a function of temperature and is often correlated by the Antoine equation:

$$\log_{10} p_{vap} = A - \frac{B}{C+T} \tag{1.17}$$

Values of the constants A, B, and C are tabulated in many books (for example, Reid et al., 1987) and websites. The units are often in mm Hg for vapor pressure and temperature (T) is in degrees Celsius rather than in standard S.I. units. Hence caution must be exercised when extracting these values from the literature.

1.4.3   Liquid–Liquid Interface: Partition Constant

A simple partition constant (denoted as m_A for species A) is often used to describe the interfacial equilibrium between two liquids:

$$y_{A,i} = m_A x_{A,i}$$

where $y_{A,i}$ is the interfacial mole fraction in one of the liquid phases and $x_{A,i}$ is the interfacial mole fraction at the interface of the second liquid. The partition coefficient, in turn, is related to ratio of activity coefficients of A in the two phases:

$$m_A = \frac{\gamma_A^I}{\gamma_A^{II}}$$

The value of this coefficient is often predicted using thermodynamic models for activity coefficients.

1.4.4   Fluid–Solid Interface: Adsorption Isotherm

Thermodynamic jump at gas–solid or liquid–solid interfaces is defined in a similar manner, using an adsorption equilibrium constant that has a same status as the solubility or partition coefficient. Often a linear relation is used: $q_A = K_A C_A$ where q_A is the equilibrium concentration of A in the solid phase and C_A is its concentration in the gas phase. K_A is referred to as the (linear) adsorption equilibrium constant.

1.4.5   Nonlinear Equilibrium Models

A note is in order regarding the linear models described in the previous sections. These models are widely used, even when they do not hold exactly for simplification of the models. However, more complex models are needed to describe equilibrium in non-ideal liquid mixtures—and in systems where there is a strong adsorbed layer due to differences between the interfacial tensions of various species dissolved in the liquids. This layer often has completely different property from the two liquids in contact and is sometimes referred to as a microphase.

On a similar note for fluid–solid systems, linear adsorption equilibria are often used, although these are actually the limiting case of nonlinear relations (the classical Langmuir equation and other isotherms are described in Section 29.3). The linear relation is a good approximation for dilute systems. More generally, the Langmuir isotherm is used to represent equilibrium for gas–solid and liquid–solid systems for concentrated solutions. For gases that undergo a reaction in the liquid, the solubility has to be viewed as the first step for the dissolution equilibrium constant—for example, A(gas) to A(aq) equilibrium. Dissolved gas may further react; hence the quantity of gas absorbed will also depend on the equilibrium constant for these reactions, which are accounted for separately. This value can exceed that calculated using the solubility parameter alone. An example of such system is SO_2 in water, where the dissolved SO_2 undergoes reaction to form HSO_3^- or SO_3^{--} ions depending on the pH of the solution.

1.5   Application Examples

Mass transport processes are ubiquitous in nature and engineering practice. One can cite numerous examples, which by themselves will fill an entire book. This section presents a few examples and includes the questions that an engineer may ask in an attempt to understand or design these processes. As the content of this book unfolds, we will be in a position to answer these questions and perhaps raise yet more questions. As you will appreciate from a perusal of these examples, the analysis and modeling of mass transfer processes (the main goal of this book) is widely useful in many fields. You will be well trained to tackle problems in your chosen engineering profession and be in a good position to do research and advance this area upon completing your study of this book.

1.5.1   Reacting Systems

Heterogeneous reactions are commonly encountered in chemical processing. These systems include two or more phases, with reaction taking place mainly in one of the phases while the reactants are usually present in the other phase. Hence a prerequisite for reaction to occur is mass transfer from one phase to the other. This coupling of mass transfer and reaction has many interesting consequences, which are studied in detail in Part II of the book. A simple classification of heterogeneous reacting systems is based on the type and number of phases contacted, as shown in Table 1.3 (along with one example for each case). Most reaction engineering books include sections on heterogeneous reactions and offer varying degrees of coverage of this area (e.g., Levenspiel, 1974). The extensive (almost encyclopaedic) monograph by Doraiswamy and Sharma (1984) is an useful reference book and deals exclusively with heterogeneous reactions.

Two examples of reacting systems where mass transfer plays an important role are shown here. Other examples are taken up in Part II.

Catalytic Converter

A common example of mass transfer accompanied by reaction is found in an automobile. A catalytic converter consists of a set of flow channels coated

Table 1.3   Examples of Heterogeneous Reactions Based on the Phases Being Contacted

Phases Contacted	Example
Gas + solid catalyst	oxidation of ethylene
Gas + solid reactant	combustion of coal
Gas–liquid	removal of CO_2 by reactive solvent
Liquid–liquid	biodiesel production
Gas–liquid + solid catalyst	removal of sulfur compounds from diesel
Gas–liquid + solid reactant	carbonation of lime
Gas–liquid–liquid	production of hydroxylamine from nitrobenzene

with an active layer of catalyst such as platinum (Pt). This is an example of a flow system accompanied by mass transfer and chemical reaction to reduce the release of pollutants such as CO and NO_x. The extent of pollutant removal depends on the flow rate of the gas, the temperature, the rate of mass transfer to the catalytic surface, and the rate of the reaction at the surface itself. The overall rate of reaction can be calculated by a mass transfer analysis and used to design the converter. The dynamic response of the system and the extent of pollutant removal during the initial cold start period (when the converter is cold) can also be found using these models.

Trickle Bed Reactor

A trickle bed reactor is an example of a three-phase reactor (gas–liquid–solid catalyst). It is similar to the packed column used in the unit operation of absorption of a gas. Here gas and liquid flow over a packed catalytic bed and a reaction occurs on the surface of the catalyst (Figure 1.4). Gas, liquid, and solid catalyst are the three phases present in the reactor. Such reactors are widely used in the chemical and petroleum industry. For example, they are used to remove sulfur compounds from diesel, an application that relies on a reactor with three phases: hydrogen, diesel, and a cobalt–molybdenum (Co-Mo) based catalyst. Today's urban air is often cleaner thanks to the trickle bed reactor. The book by Ramachandran and Chaudhari (1983) is a good starting source on this subject and covers the various types of three-phase catalytic reactors in detail.

1.5.2   Unit Operations

Unit operation is defined as a unified study of a particular separation technique used for a chemical engineering application. Distillation is an example

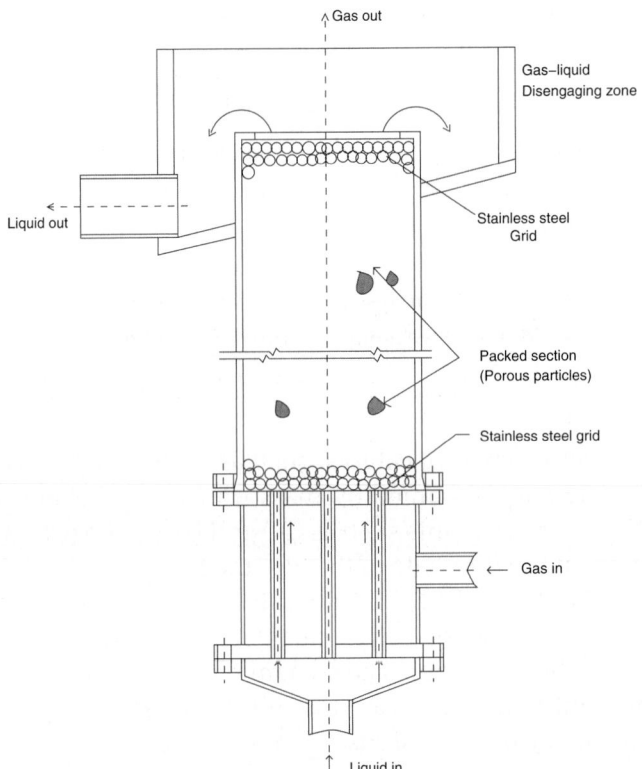

Figure 1.4   Schematic of a trickle bed reactor; mass transfer of species in the gas phase (e.g., hydrogen) to the liquid and then to the surface of the catalyst is followed by a surface reaction on the catalyst with a second species diffusing from the liquid. (An operation with upflow of both phases is shown here.)

in which the vapor pressure differences between two components are used to separate and purify these components. The unit operation approach takes the view that the analysis is same whether you are doing a distillation of a crude oil mixture or making brandy. Common mass transfer–based separation processes are listed in Table 1.4. All of these separations rely on interfacial mass transfer; hence the analysis and modeling of mass transport effects is a prelude to the design of such systems.

   An example of a unit operation is liquid–liquid extraction. The schematic of a simple single-stage extraction unit is shown in Figure 1.5, and we will use this unit as a prototype example for modeling stage contactors. In this unit operation, two phases are intimately mixed in the mixer section of the

Table 1.4   Common Mass Transfer–Based Separation Processes and the Phases Being Contacted

Phase 1	Phase 2	Operation
Vapor	Liquid	Distillation
Gas	Liquid	Absorption
Liquid	Gas	Stripping
Liquid	Liquid	Extraction
Gas	Solid	Adsorption
Liquid	Solid	Adsorption
Wet solid	Gas	Drying

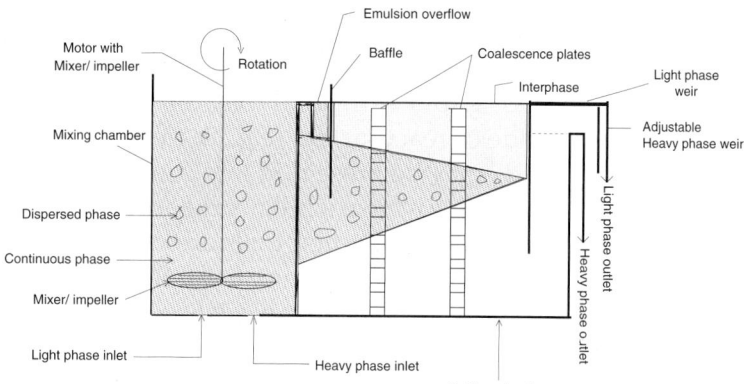

Figure 1.5   Schematic of a single-stage mixer-settler used in liquid–liquid extraction.

contactor to create an emulsion or dispersion, which promotes a high rate of mass transfer. The two-phase mixture is then allowed to settle in the settler section, and the enriched solvent and the lean solutions are separated. Determining the degree of mass transfer that can be achieved for a specified rate of agitation is one of the objectives of the mass transfer calculation.

The degree of separation that can be achieved in a single-stage contactor is usually limited; hence a multistage cascade is used, with the lean solution being treated further with fresh or recycled solvent. A designer may want to know how many stages are needed to achieve a certain level of purity. Modeling of such a multistage cascade is used to provide the answer.

Part III of the book covers the application of mass transfer principles to design some of the various unit operations listed in Table 1.3. The key idea is that the modeling methodology is common to all the unit operations and can be approached in an unified format. Thus the various aspects of individual unit operation, although very important on their own, can be brought together under one umbrella of modeling of mass transfer processes. Seader, Henley, and Roper (2011) offer valuable insights into separation process principles and provide a detailed analysis of the various operations.

1.5.3   Bioseparations

Bioseparation refers to separation of products produced by biochemical reactions; the separation of products from a fermentation broth is an often-cited example. These processes have a number of distinguishing features compared to the traditional separations practiced in the bulk chemical and petroleum industry. The distillation is the main workhorse in the chemical industry, but due to the heat-sensitive nature of bioproducts many alternative separation methods are needed. Rapid extraction may also be a requirement since the product may degrade or react further (e.g., in penicillin separation). In many cases the compounds may be present in low concentrations. The target compound to be separated may have similar properties to the other compounds in the broth, so that novel separation tools are needed. Common techniques used are extraction, adsorption, chromatography, and electrophoresis, and the use of mass transfer analysis for modeling these cases is illustrated in later chapters.

The book by Harrison, Todd, Rudge, and Petrides (2003) is a good introduction to this field. The book by Seader et al. (2011) has also considerable information on this topic. A review article published by Harrison (2014) provides an introductory reading, with the author stressing the importance of understanding the basic principles and theory as a prelude to design and control of purity of products obtained in bioprocessing.

1.5.4   Semiconductor and Solar Devices

The heart of the computer you use is made of a silicon chip, but the electronic activity arises due to the fact that the chip has undergone a diffusion process to incorporate phosphorus, boron, or other dopants. Semiconductor doped with group V metals are called n-type, while those with group III metals are called p-type. A junction is formed by contacting these two types of semiconductors and acts as a diode or transistor. The electronic behavior in such systems depends on the transport of the electrons from the n to p side and transport of holes from the p to n side, together with recombination. The diffusion-reaction analysis for porous catalysts presented in Chapter 18 can be readily adapted to this system.

A number of processing steps in this field involve chemical vapor deposition, in which a species (precursor) is transported from a vapor and reacts and forms a deposit or a film on a (substrate) surface. This process again involves mass transport of reactants, with control of the deposited material's properties being affected by the rate of transport. Oxidation of silicon to form an insulating layer is another example of mass transfer in fabrication of metal oxide semiconductor (MOS) devices. The book by Middleman and Hochberg (1993) is a classic in this area and a must-read for students who wish to get involved in this field.

1.5.5   Biomedical Applications

The focus of mass transfer analysis in the field of biomedical engineering is to bring together fundamentals of transport models and life sciences principles. Key areas where mass transport phenomena can be utilized include the following:

- Pharmacokinetics analysis, distribution, and metabolism of drugs in the body
- Understanding of transport of oxygen in the lungs and tissues
- Tissue engineering, including development of artificial organs
- Design of assistive devices such as dialysis units

Some application examples are briefly described in this book in Chapter 23. An early book by Lightfoot (1974) and the more recent books by Sharma (2010); Truskey, Yuan, and Katz (2004); and Fournier (2011) are illustrative of the mass transport applications in this field.

1.5.6   Application to Metallurgy and Metal Winning

Transport phenomena analysis and models are widely used in metallurgy and metal winning. Books by Szekely and Themelis (1991) and by Geiger and Poirier (1998) provide a number of applications in this field. Ore smelting in a blast furnace, gas–solid reactions in steel and copper making, and alloy formation by melt drop solidification are examples of applications in which mass transport principles are needed. Electrochemical processes are also used for metal winning (e.g., copper), in which transport of copper ions to the cathode is an important step in the overall process.

1.5.7   Product Development and Product Engineering

Transport phenomena are increasingly being exploited in product development. Example applications include the design of drug capsules that should provide a constant release rate and the design of polymer wrapping in food packaging to reduce oxygen diffusion. The book by Cussler and Moggridge (2001) is an useful resource for further reading in this field.

An example of drug release from a capsule is shown in Figure 1.6. In this case, if the drug has a uniform concentration inside the capsule, the release rate reaches its maximum in the beginning and decreases with time. Ideally we want a (nearly) constant rate, like that shown in the figure. The design of the pore structure of the capsule to achieve this type of steady release rate is an important application of transient mass diffusion principles.

1.5.8   Electrochemical Processes

Electrochemical processes have a wide range of applications, including batteries, solar cells, electro-deposition, thin films, and microfluidic devices.

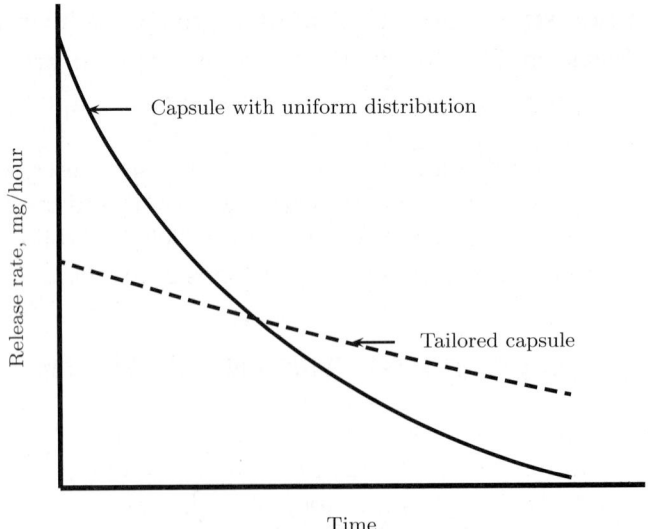

Figure 1.6   A product design application example. The drug release rate is shown for a drug with both a uniform distribution and a tailored capsule.

Transport phenomena principles are increasingly being used to design and improve these kinds of devices. The book by Newman and Thomas-Alyea (2004) is a good treatise on this subject.

In the energy sector, there is a need to store solar energy generated during nonpeak hours and to develop improved batteries for electric cars. A commonly used type of battery is the lithium-ion battery shown in Figure 1.7. Here Li ions are transported across the electrolyte separating cathode and anode. During the charging cycle, Li ions are transported and stored in the carbon matrix. During the discharging cycle, the transport takes place in the opposite direction, such that Li is stored in the metal oxide matrix. Mass transport considerations are an important component in the simulation of the performance of this device; a mass transfer–based model for this system is discussed in Chapter 24.

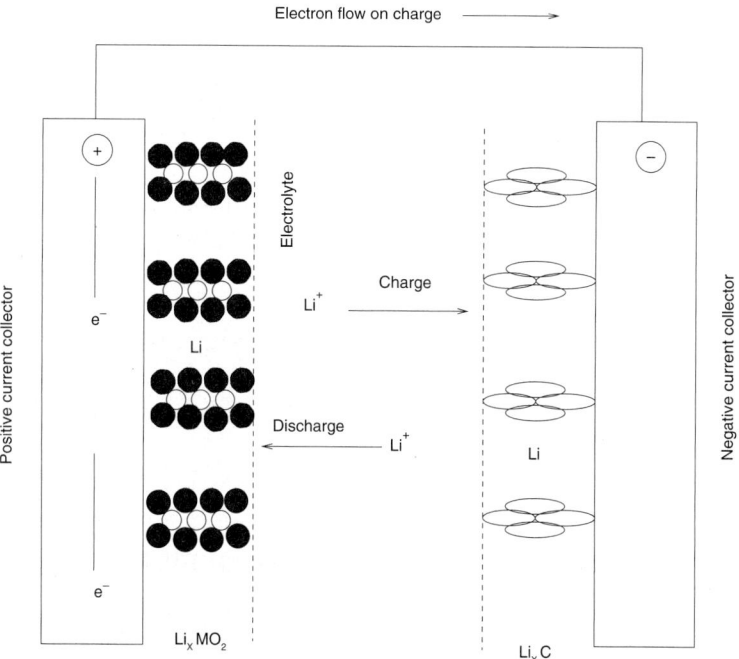

Figure 1.7   Schematic of a lithium-ion battery showing the various mass transfer and reaction steps occurring in the equipment. Lithium ions stored in the carbon "hotels" are released during the discharge cycle and diffuse through the electrolyte to the cathode, where they react with metal oxide matrix. This produces a current in the external circuit.

1.5.9   Environmental Applications

Transport phenomena principles and modeling have found extensive applications in environmental engineering, where they provide a modern perspective and new approach. Typical environmental problems that may addressed using the transport modeling methodology are as follows:

- Fate and contaminant transport in the atmosphere is usually simulated by dividing the system into four (air, water, soil, and biota) or more compartments and considering transport and reaction in each of the compartment. See, for example, Figure 1.14.
- Groundwater transport is another example. Leakage of contaminants from nuclear waste tanks into rivers could be a major problem, and some of these scenarios can be analyzed by transport models to provide information on the rate of leakage and measures needed to alleviate the problem.
- Transport of excess nutrients to water bodies leads to algae growth and destruction of other organisms, a process known as eutrophication. The rate of transport in such systems is needed to determine further remediation actions.
- Carbon dioxide sequestration in underground mines is contemplated as a solution to reduce the impact of global warming. Mass transfer analysis is needed to predict the leakage and long-term feasibility of this solution.

The book by Clark (1996) provides a nice introduction to some of the problems mentioned here.

Next, we discuss the general methodology involved in setting up models for mass transfer processes.

1.6   Basic Methodology of Model Development

In this section, we discuss the basic procedure for developing mass transfer models. The starting point for model formulation is to choose a control volume and mark its boundary surface (i.e., the control surface). A system or control volume can be defined as any part of the equipment, as the entire equipment, or even as the large-scale system as a whole. The basic conservation laws are then applied to the system. Conservation law is discussed in Section 1.7.

Control volume can be of any size and shape. Typically, models at three levels are developed based on the size of the control volume: differential models, macroscopic (or macroscale) models, and mesoscopic (or mesoscale) models. These three types of models are explained sequentially in the following sections. On the topmost level is the differential control volume and the differential models. This type of model contains the local information at each point in the continuum. However, for design of process equipment, macro or meso levels may also be used. Although models at these levels are not as detailed as the differential models, they are easier to solve and to apply in practical designs. Hence modeling at all three levels is covered in this book.

For modeling very large or complex systems, such as transport of a pollutant in the environment or drug metabolism in the human body, the system is divided into a number of interconnected compartments and a macroscale model is applied to each compartment. Applications of these so-called compartment models are provided in Chapter 13.

Conservation laws alone are not sufficient to complete the model, since the model will contain terms for the mass crossing into a control surface. Hence additional closure relations are needed. For example, a differential model using the conservation principle alone will contain the flux vector as a parameter. This system has to be closed and a flux versus concentration relation has to be applied using a constitutive model. Similarly, the macro- and meso-level models may contain terms such as the mass crossing a larger-size control surface or from one phase to another. Suitable transport laws must be applied to close these terms, thereby completing the model. Hence coupling the conservation laws together with suitable transport law is the basic methodolgoy in developing mass transfer models.

1.7   Conservation Principle

The conservation principle is the consequence of the law of mass conservation, which states that a species A cannot appear from nowhere and must be conserved. The species mass balance is therefore the starting point:

$$\boxed{In+Generation=Out+Depletion+Accumulation}$$

Species A can cross into the control volume from a part of the control surface (the *in* term) and leave out of the rest of control surface.

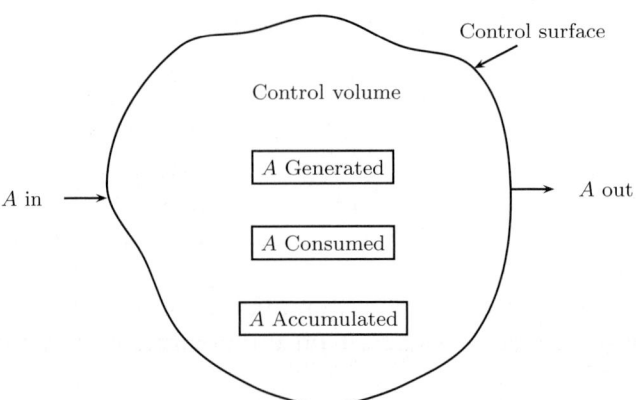

Figure 1.8   Schematic of the conservation principle applied to a control volume.

Species may also be consumed (depletion) or produced (generation) within the control volume by chemical reaction.

Finally, the species concentration in the control volume may change with time due to the accumulation of the species within the control volume. The various processes involved are shown schematically in Figure 1.8.

The depletion is normally taken as a negative generation term and is often not included separately. In such cases, the generation is to be understood as net generation, which is equal to generation − depletion.

Similarly the *out − in* term is often written as a *net efflux* term. Hence another form of the conservation statement is

$$\text{(Net) Generation} = \text{Net efflux} + \text{Accumulation}$$

We now illustrate the use of this principle for the three levels of models discussed earlier and indicate which additional relations are needed to complete the model.

1.8   Differential Models

In differential models, a differential control volume (e.g., a box with sides Δx, Δy, and Δz for Cartesian coordinates) is used to develop the transport models. The control volume is then made to tend to zero, thereby producing a set of differential equations for the primary field variable—namely, the concentration. These equations represent models at the highest level of hierarchy and contain the detailed information.

The simplest differential control volume is the one constructed in Cartesian coordinate volumes, which is shown in Figure 1.9. The net generation and accumulation depend on the control volume and, therefore, are known as volumetric terms. The net efflux term, known as the surface term, depends on the combined flux at that control surface. Application of the conservation principle, then, leads to the species mass balance equation involving the unknown fluxes. Nevertheless, this equation alone is not sufficient: We need additional relations because we ultimately want to solve for the concentration field. The flux term in the equation is the combined flux, which is the sum of the convection flux and the diffusion flux. The convection flux depends on the local velocity and local concentration, so it can be calculated if the velocity profile in the system is known. The diffusion flux results from the molecular interactions, which are not modeled at the continuum level. Hence a closure equation—that is, a constitutive model for diffusion—is needed to close the model. The most commonly used constitutive model is some form of Fick's law. Note that for multicomponent diffusion in mixtures, the Stefan-Maxwell model for diffusion is also used; it is discussed in detail in Chapter 15.

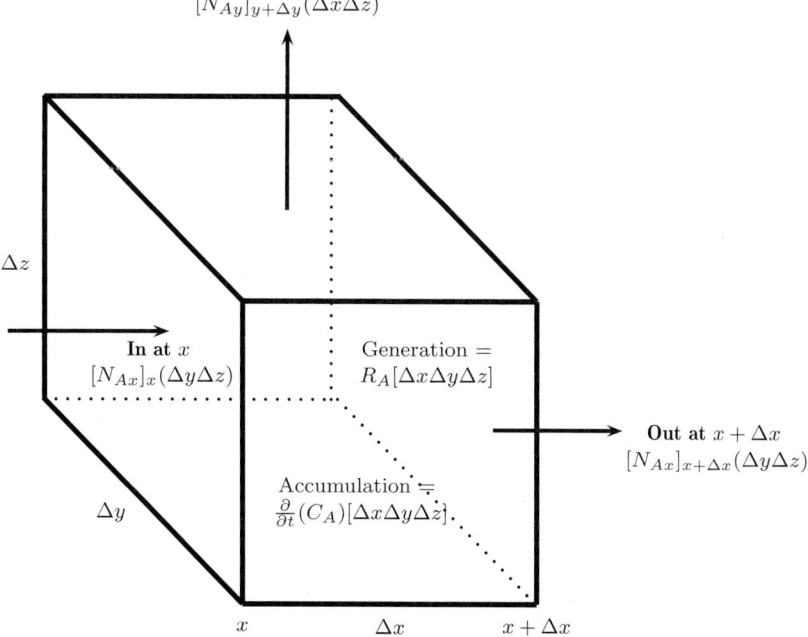

Figure 1.9 Differential control volume in Cartesian coordinates.
Note: Flux components from the hidden faces are not shown to avoid clutter.

In summary, mass transfer models at the differential levels are built through the combination of the conservation law and the constitutive law. The differential equations of mass transfer based on this approach are derived using this methodology in Chapter 5 and have general applicability. Many simpler problems can be deduced by simplification of these equations. In essence, this method is similar to that used in fluid mechanics, where the Navier-Stokes equations have general applicability for modeling flow of Newtonian fluids and where solutions to specific problems such as pipe flow can be obtained by simplification of these general equations. Likewise, we can go from the general to the particular in modeling of mass transfer. However, for simpler cases where the concentration varies only along one coordinate direction, it is instructive to derive and solve the equation by starting directly from the conservation laws and then using Fick's law, rather than by directly using the general differential equations. This approach is illustrated in Chapter 2 for some common problems in mass transfer. A careful study of Chapter 2 will provide you with a basic understanding of the process of setting up and solving differential models.

1.9   Macroscopic Scale

At the macro-level of modeling, a large-size control volume or the whole equipment is used. The conservation laws are, in turn, applied to this region. The volume element is not taken to tend toward zero, and the balances based on the selected larger volume are used directly. Such models are useful to obtain the relation between engineering quantities in an approximate way. Point-to-point information (details) is lost at this level, however, and the effects of these variations have to be incorporated through some sort of calculation (model averaging followed by some approximations) or by other means (e.g., empirical fitting of experimental data). This leads to the information loss principle:

> Information, on length (or time) scales lower than that at which analysis is done, is lost and has to be supplemented in some suitable manner.

Appreciation and awareness of this principle are important in modeling. In particular, for the macroscale analysis discussed in this section, we lose information on the point-to-point concentration values and the pointwise variation of the flux over a given control surface. This loss must be countered by making some assumptions and using some closure models.

In the following sections we discuss qualitatively the closures commonly used by examining some specific examples. More fully worked-out examples of macroscopic balances are provided in Chapter 3 and continued in Chapter 13.

1.9.1   Stirred Tank Reactor: Mixing Model

Consider a reactor that is stirred and operated continuously with in and out flows of a liquid. The control volume is the whole reactor shown in Figure 1.10; the various contributions to the mass conservation are also shown in a general manner in this figure. Now assume the reactor consists of a single phase and is operating at steady state. The accumulation term is zero. The loss term for the second phase shown in the figure can also be neglected here if we have only one phase. The conservation law simplifies to

$$\text{in} - \text{out} + \text{generation} = 0$$

where all the terms are in moles A per unit time.

Now let's use mathematical terms to describe this system. In and out can be formulated in terms of the volumetric flow rate times the corresponding concentration. The main assumption needed is for the generation term. This term is the total generation over the entire control volume and is the volume integral of the local rate of reaction (mole/m^3 local volume sec) R_A. The local rate, in turn, is a function of the local concentration. We assume that

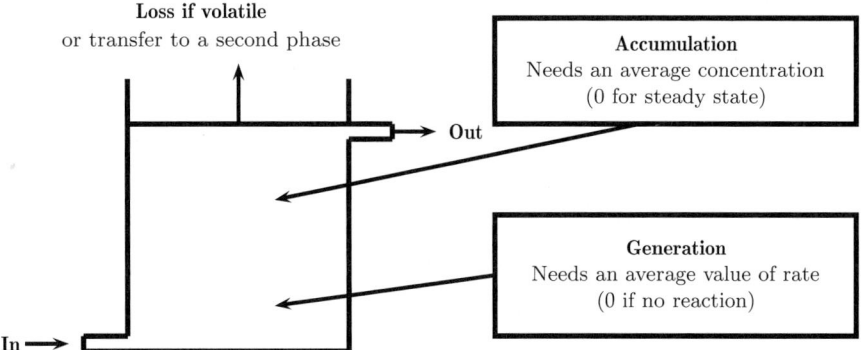

Figure 1.10   Illustration of a macroscopic control volume showing the various components that go into the species balance. The average values are the integrals of the (unresolved) pointwise values. The information on the pointwise values is lost at the macro-scale modeling level. Hence model assumptions or closures are needed.

the required relation is known from a kinetic model. Now the generation over the entire volume is the volume integral of the local rate:

$$\text{generation} = \int_V R_A(\text{local})\, dV \tag{1.18}$$

Here R_A (local) is some specified function of the local or pointwise concentration.

This local concentration is not known (not being modeled) in the macroscopic model, so it is not possible to evaluate precisely the generation term. Hence some assumption about the mixing level in the reactor (the concentration variation in the reactor) must be made. These details are studied in Chapters 3 and 13.

A common simplifying assumption is that the reactor is well mixed and the reactants present at a uniform concentration. This permits us to close the model, which is then referred to as a completely backmixed model. Many other models for closure may be used depending on the anticipated mixing level in the system (e.g., dividing the tank into a mixed zone and a dead zone or dividing the tank into two well-mixed subcompartments), and some of these models are explained in more detail in Chapter 13. The main point to note here is that some assumption about the mixing pattern must be made and the level of mixing needs to be tested and quantified by suitable experiments (e.g., by tracer methods).

1.9.2   Sublimation of a Solid Sphere: Mass Transfer Coefficient

Consider a solid that exerts sufficient vapor pressure and is subliming into a gas phase. The change in the radius of the solid is to be computed. A macroscopic balance for the solid is in order. *In* is zero and *out* is the mass transferred from solid to gas. In this scenario, the conservation law for the solid simplifies to

in (zero) − out (transferred from solid to gas) = accumulation

The rate of transfer from solid to gas is precisely the surface integral of the radially directed outward flux at the surface of the solid:

$$\text{transfer to gas} = \int_A (\mathbf{N}_A \cdot e_r)\, dA = \int_A (N_{As})\, dA$$

where N_{As} is the radial component of the flux vector at the surface.

This flux vector can be computed only if a differential model for the gas phase is solved, including the convection and diffusion of the solute in the gas phase. It is not always possible to do this (or such details may not be important) since this endeavor is computationally intensive and the velocity profile in the gas phase may not be precisely known for complex geometries and for turbulent flow. Obviously, we need a closure for the unknown term, which done by introducing the mass transfer coefficient, k_m. Using this parameter, we model the transfer rate as a product of the mass transfer coefficient times a concentration difference:

$$N_{As} = k_m(C_{As} - C_{Ab}) \qquad (1.19)$$

$C_{As} - C_{Ab}$ is taken as a driving force for mass transfer from the solid to the bulk gas. Note that C_{As} is the concentration of A in the gas phase at the gas side of the solid–gas interface, just near the solid. C_{Ab} is the concentration in the gas phase at a point far away from the solid, known as the bulk gas. This concentration is usually taken as zero.

Equation 1.19 is actually a definition, rather than a fundamental law. The mass transfer rate is given per unit area of transfer. Balancing the unit, we find that the mass transfer coefficient has the units of m/s. Values of the mass transfer coefficient are provided by numerous sources. For example, they may be empirically fit to experimental data or fit to detailed computational models. For many practical cases the values of the coefficient have previously been published; as a consequence, the concept of the mass transfer coefficient is widely used in practical design calculations. The main point here is that the use of the mass transfer coefficient completes the missing flux information in the context of macro-level (as well as meso-level) models.

1.9.3   Model for Mixer-Settler

Consider the mixer-settler shown in Figure 1.5. There are two phases present in this system: the dispersed phase (drops) and the continuous phase. Obviously, the differential model for both the dispersed and continuous phases is needed for a detailed analysis and can be quite challenging to construct. Here a two-phase flow model together with a local model for interfacial flux needs to be developed. In industrial design, however, macroscopic models are usually the models of choice. Here we assume that there are two control volumes: one for the solvent phase and one for the aqueous phase. These two control volumes have a common interface across which the solute is transferred. In other words, the system is modeled as two interconncected control volumes, one for the aqueous phase and one for the solvent phase. A schematic of such control volumes used in macroscopic analysis of two-phase systems is shown

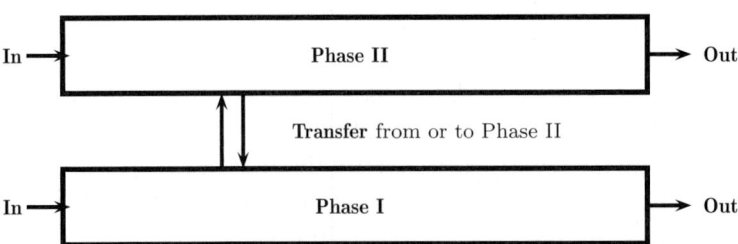

Figure 1.11   Schematic of a mixer-settler for liquid–liquid extraction modeled as two separate but interconnected control volumes.

in Figure 1.11. The transfer rate is modeled by using a suitably defined mass transfer coefficient. In addition, we must make an assumption about the mixing pattern in both phases as well as the value for the mass transfer coefficient. For example, are these phases completely mixed? In that case the backmixed assumption is sufficient. Also, how does the mass transfer coefficient change if some design parameter (e.g., agitation speed) is changed? These details needed to design the equipment are explored in more depth in Chapters 3 and 13. At this stage, the various components needed for a macroscale analysis and design of separation equipment simply need to be appreciated.

1.9.4   Equilibrium Stage Model

A simplifying assumption in the mixer-settler model is that the outlet streams are at equilibrium. This will be the case for an ideal contactor in which the mass transfer is complete. Such a simplified model, which is called equilibrium stage model, provides a benchmark for quick design calculations. The actual contactor may not have achieved equilibrium, so the results are often simply corrected by introducing a stage efficiency parameter. This can be an empirically fitted parameter, or it can be computed by including mass transfer effects into the model. Thus we note that a hierarchy of models with increasing levels of complexity can be built for this process—a practice that is common when modeling other separation processes as well. For example, a distillation column can be modeled using (1) a number of ideal or theoretical stages, (2) a number of stages with correction for stage efficiency, or (3) a model based on detailed mass transfer rates.

We now look at mesoscopic models.

1.10   Mesoscopic or Cross-Section Averaged Models

Mesoscopic models are useful when there is a principal direction over which the flow takes place, as in pipe flow, for instance. Control volume is taken as differential in this direction but assumed to span the entire cross-section in the cross-flow directions (Figure 1.12). The rationale for this assumption is as follows: The concentration variation in the flow direction is more significant, meaning it changes significantly from inlet to outlet, rather than in the radial direction. Changes in the radial direction at any fixed axial position may not be very large. Hence the mesoscopic control volume shown in Figure 1.12 is adequate in lieu of a complete differential model. The key point to note is that the information on the radial variation of concentration and its effect of the various terms in the conservation model is lost with this kind of model. Hence an appropriate closure model is needed the supplement the information lost due to area or radial averaging. More details on this approach are provided in Chapter 4. Here, we provide some simple examples of closure models used for this level of modeling.

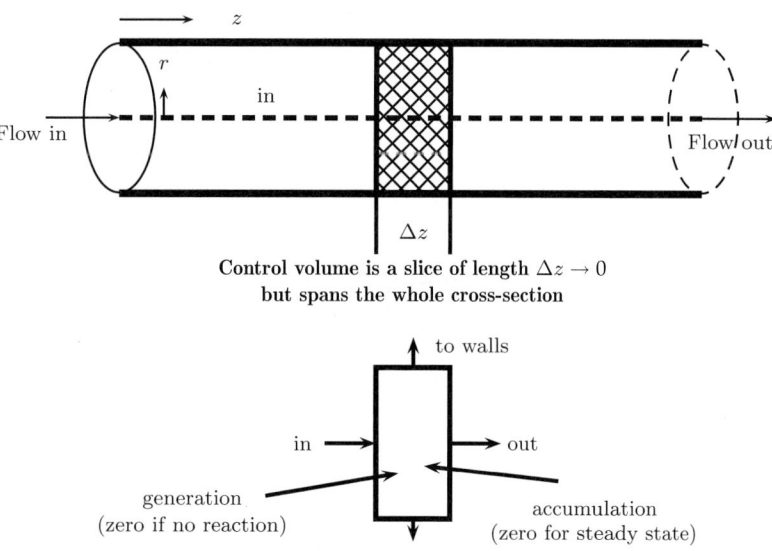

Figure 1.12   Illustration of the mesoscopic control volume for analysis of mass transfer processes in pipe flow. All of the terms in the mass balance are integrals of (unresolved) pointwise values. Hence model or closure for these terms is needed.

1.10.1 Solid Dissolution from a Wall

Consider the mesocopic model applied to a simple system of solid dissolution from a wall in a tube shown in Figure 1.12. Here we have a pipe coated with a dissolving or subliming solute material; a fluid is flowing in the pipe. The concentration of the solute is to be calculated as a function of the length. Two specific examples are a naphthalene-coated pipe with air flow; a pipe coated with benzoic acid (which is a solid at room temperature) with water flow. We assume steady state so that the accumulation is zero.

The conservation law ($in - out = 0$) is now represented as

in from flow at z + *in* by transported from the walls $-out$ from flow at $z + \Delta z = 0$

To calculate the *in* and *out* terms due to flow, we must recognize that the velocity can vary across the cross-section. Concentration also varies along the pipe, with the maximum concentration found at the dissolving pipe wall. The flow can be laminar or turbulent. Laminar flow of a Newtonian fluid has a parabolic profile, whereas the profile is very steep in turbulent flow as shown in Figure 1.13. The velocity for the turbulent flow is the time-averaged value, a concept that is discussed more fully in Chapter 12.

A cross-sectionally averaged velocity is an useful quantity and is defined as

$$\langle v \rangle = \frac{1}{A} \int_A \boldsymbol{v} \, dA \qquad (1.20)$$

which is the integral average of the local velocity. The volumetric flow rate is then equal to $\langle v \rangle A$; v is the axial velocity.

The plug flow shown in the Figure 1.13 is an idealization that is often used to simplify the models. Here the velocity is assumed to be the same at

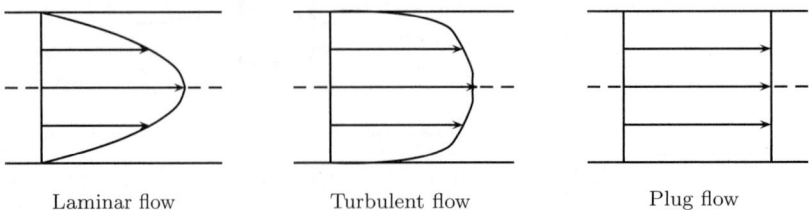

Laminar flow   Turbulent flow   Plug flow

Figure 1.13 Schematic of axial velocity profiles for laminar, turbulent, and plug flow in a pipe as a function of radial position.

each local radial position. Consequently, the cross-sectional average velocity is also equal to the local velocity.

The *in* and *out* terms are the integral average of the axial flux values. The flux component in the axial direction is primarily due to convection and is locally equal to vC_A. For a small differential area, ΔA, the moles of A crossing is therefore equal to $vC_A \Delta A$. Hence the moles crossing over the entire tube area is the integral of the product of local velocity and the local concentration:

$$\text{in, convection} = \int_A vC_A\, dA = \langle vC_A \rangle\, A$$

where the notation $\langle \dots \rangle$ is used as shorthand to indicate the averaged value of any quantity within the brackets. Thus we define

$$\langle vC_A \rangle = \frac{1}{A} \int_A vC_A\, dA \tag{1.21}$$

We need $\langle vC_A \rangle$ to evaluate the *in* and *out* terms, which cannot be calculated because local concentration values are not available at the meso-level. Hence we define and use an average concentration. This average concentration is called the cup mixing average concentration, of the flow weighted average. It is denoted as C_{Ab} and defined as follows:

$$C_{Ab} = \frac{\int_A vC_A dA}{\int_A v dA} = \frac{\langle vC_A \rangle}{\langle v \rangle} \tag{1.22}$$

Hence

$$\langle vC_A \rangle = \langle v \rangle\, C_{Ab} \tag{1.23}$$

The *in* and *out* terms are then represented simply as $A \langle v \rangle C_{Ab}$. Thus the cup mixing concentration is used as the representative concentration variable in the context of our mesoscopic model. The local concentration has no relevance because the local information is not available. **Note:** In the context of the meso-level model, the notation C_A is used in many books as a simpler notation; by implication, this is the cup mixed average value and should be interpreted accordingly. The results in such models are the variation of the cup mixing concentration with the axial distance.

Example 1.2 shows a calculation of the cup mixing concentration to clarify the definition. Here we assume that the radial variation is known from other methods, such as from a differential model or by experimental measurements.

Example 1.2   Cup Mixing Concentration

A pipe wall is coated with a solute that dissolves into the flowing liquid. The radial variation of the concentration of the dissolved solute in the liquid at a particular axial position was found to be

$$c_A = \frac{C_A(r)}{C_{As}} = (r/R)^2 - (r/R)^4/4 + 1/4 \qquad (1.24)$$

where C_A is the local concentration at radial position r, C_{As} is the saturation solubility of the solid, and R is the pipe radius. The term c_A is the scaled or dimensionless concentration.

Find the center concentration, the concentration near the pipe wall, and the cup mixing concentration. Flow is laminar and the liquid is Newtonian.

Solution

The (dimensionless) center concentration is $1/4$ by substitution of $r = 0$.

The concentration near the wall is equal to 1 by substitution of $r = R$. Liquid near the wall is saturated, so the dimensionless concentration is one as expected.

The cup mixing is done by using Equation 1.22. The weighting factor of $2\pi r v_z \Delta r$ is used in the integral because $\Delta A = 2\pi r \Delta r$ for a differential annular cross-section. Here v_z is the local axial velocity—that is, the velocity at location r. For laminar flow of a Newtonian fluid, this is a parabolic function of r. From fluid dynamics, the following relation is obtained:

$$v_z(r) = v_{max}[1 - (r/R)^2]$$

Using this information in Equation 1.22, the scaled cup mixing concentration is the ratio of the following integrals:

$$c_{Ab} = \frac{\int 2\pi r v_{max}[1 - (r/R)^2)]((r/R)^2 - (r/R)^4/4 + 1/4)dr}{\int 2\pi r v_{max}[1 - (r/R)^2]dr} = 13/24$$

Having defined the *in* and *out* terms using the cup mixing concentration, we need an expression for the transport rate to the walls so as to close the model. This is done by defining and using a mass transfer coefficient. It is common practice to use the cup mixing concentration as the representative concentration away from the solid; the driving force for mass transfer is then defined using $(C_{Ab} - C_{Aw})$ as the driving force. Hence the mass transfer rate is defined as

$$N_{Aw} = k_m(C_{Ab} - C_{Aw})$$

The mass transfer coefficients are available for a large number of flows (such as pipe flows) and, in turn, this approach to modeling is widely used in practice. The mesoscopic formulation can be cast into mathematical framework to account for variation of the concentration (cup mixing) as a function of axial position. Further details and the mathematical details follow in Chapter 4. At this stage, however, you should be able to appreciate the essential concepts that go into the construction of mesoscopic models.

1.10.2 Tubular Flow Reactor

Another common application of the meso-modeling concept is the tubular flow reactor. This is a pipe or channel in which a homogeneous chemical reaction is taking place. We will show that not only the cup mixing concentration is needed, but also another average, the cross-sectional average.

Let us start with the conservation principle in words:

$$\text{in} - \text{out} + \text{generation} = 0$$

The generation is the average of the local rate:

$$\text{generation} = (A\Delta x)\langle R_A \rangle$$

where

$$\langle R_A \rangle = \frac{1}{A} \int_A R_A dA$$

For a first-order reaction, for example, $R_A = -k_1 C_A$, where C_A is the local concentration. Hence

$$-\langle R_A \rangle = \frac{1}{A} \int_A k_1 C_A dA$$

To incorporate this into the conservation law, it is customary to define and use the the cross-sectional average concentration:

$$\langle C_A \rangle = \frac{1}{A} \int_A C_A dA$$

Hence the generation term (first order) to be used in the reactor model is $-(A\Delta x)k_1 \langle C_A \rangle$.

In and *out* terms are $A \langle v \rangle C_{Ab}$ as for the solid dissolution. Thus the conservation law in mathematical terms is

$$A \langle v \rangle C_{Ab}(x) - A \langle v \rangle C_{Ab}(x + \Delta x) - (A\Delta x)k_1 \langle C_A \rangle = 0 \qquad (1.25)$$

In the limit $\Delta x \to 0$, this produces the following differential equation:

$$\langle v \rangle \frac{d}{dx} C_{Ab} = -k_1 \langle C_A \rangle \tag{1.26}$$

We find that both the cup mixing and the cross-sectional average appear in the reactor model. Some closure relationship is needed, however, and the commonly used closure models are the plug flow model and the axial dispersion model.

Plug Flow Model

The plug flow model is an idealized reactor model in which the velocity is assumed to be constant, as shown in the Figure 1.13. The cup mixing and the cross-sectional averages are the same since the velocity profile is uniform and, therefore, $C_{Ab} = \langle C_A \rangle$. Only the cross-sectional concentration is needed; that is, no additional closure is needed. Equation 1.26 reduces to

$$\langle v \rangle \frac{d}{dx} \langle C_A \rangle = -k_1 \langle C_A \rangle$$

Axial Dispersion Model

The actual reactor may not be a plug flow model, in which case the deviation from plug flow can be modeled by including an additional correction term. The resulting model is known as the axial dispersion model. The representation for the *in* or *out* term is

$$A \langle v \rangle C_{Ab} = A \langle v \rangle \langle C_A \rangle + \text{correction term} \tag{1.27}$$

The first term on the right side is the mole flow crossing a cross-section if the plug flow prevails. The second term is a correction term. This system is modeled as though some diffusion type of mechanism is superimposed on the plug flow value. Hence

$$\text{correction term} \ = \ -AD_E \frac{d}{dx} \langle C_A \rangle$$

where D_E is called the axial dispersion coefficient. This is only a model, and not a fundamental principle, but nevertheless it provides an important crutch to proceed further in modeling these systems. The axial dispersion coefficient is a commonly used parameter when modeling both reactors and separation processes. Details of the origin of this term and the calculation and application of this concept are discussed in Section 14.2. For now, recognize that the dispersion model closure is

$$C_{Ab} = \langle C_A \rangle - \frac{D_E}{\langle v \rangle} \frac{d}{dx} \langle C_A \rangle \qquad (1.28)$$

The key points to appreciate from this section are the various definitions and closures that are needed in the context of meso-scale models.

Now we discuss models at an even higher level of simplicity—namely, the compartmental models.

1.11   Compartmental Models

In compartmental models, a set of macroscopic models are interconnected to form a system model for a complicated process. For example, the transport of pollutants in the environment is modeled in this manner. Here it is common to use four compartments as shown in Figure 1.14 to simplify an otherwise complex situation. An application of this model in environmental engineering is described in Chapter 13.

Compartmental models are also widely used in biomedical systems modeling. A simple two-compartment model to represent the body is shown in Figure 1.15. A simulation example based on this model is provided in Chapter 13.

Examples and solutions to compartmental models are taken up in Sections 13.7 and 13.8. It is common practice to simplify each compartment to be a well-mixed system with a constant composition. In addition, an exchange parameter needs to be assigned to model the transfer between the various parameters. In turn, the compartmental models typically use many

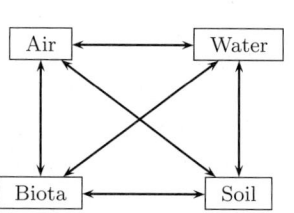

Figure 1.14   Illustration of a compartmental model for pollutant distribution in the environment. Species can be transferred across compartments, as indicated by the arrows. The transfer rate model uses an inter-compartmental exchange parameter.

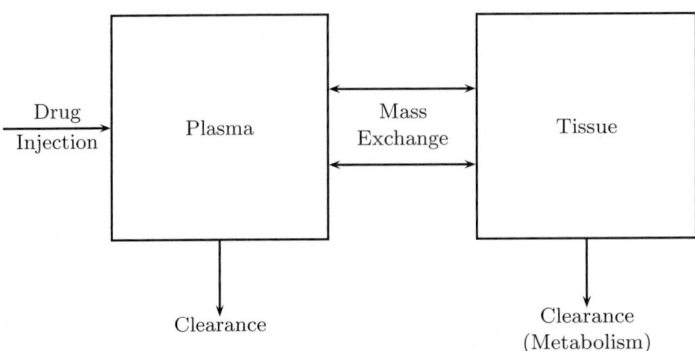

Figure 1.15   Schematic sketch of a simple two-compartment model to study drug uptake and metabolism.

simplifying (perhaps crude) assumptions. Despite this background, they have proven to have value in environmental and biomedical engineering.

Summary

- Mass transport phenomena are ubiquitous in nature and in industrial applications. In essence, the study of mass transfer is the main feature that distinguishes chemical engineering from other engineering disciplines. Mass transfer is caused by diffusion and convection.

- Development of a mathematical model to describe a process is central to the analysis and design of a mass transport process. The model is expected to provide information on the concentration in the system and the rate at which a species is transported across the system or equipment.

- Analysis of mass transport phenomena is based on the continuum approximation. This assumption permits us to assign a value for concentration at each and every Euclidean point in space.

- Concentration can be defined in various ways (mole basis, mass basis, partial pressure, mole fraction, mass fraction), and the relation and conversion from one unit to other should be noted.

- An important application of mass transfer is interfacial mass transfer, which refers to the transfer of a solute from one phase to another. For such problems, at a phase boundary (e.g., gas–liquid interface) the values of concentration are different on each side of the interface. This discontinuity is known as the concentration jump at the interface, and thermodynamic phase equilibrium relations are used to close the gaps between the two sides. Hence thermodynamic relations are always needed in the context of interfacial mass transfer.

- Models for mass transfer can be developed at many levels. At the top-most level are differential models, which contain the maximum information in the context of the continuum assumption. A differential equation for the field variable, the species concentration, is then developed using the species conservation laws coupled with some constitutive models for transport of mass by diffusion, usually Fick's law.

- Differential models are rather complex to solve for many problems, so models with less details are also used. In general, in transport phenomena analysis, one usually develops model at three levels:
  - Differential models provide the most detailed or pointwise information.
  - Macroscopic models provide overall information of the system. They typically require some closure parameters on the level of mixing in the system and use of mass transfer coefficients.
  - Mesoscopic models provide information on the variation of an averaged property as a function of the main flow direction. The main closure parameters are the transport coefficients and the dispersion coefficient to account for velocity profiles.

- The macro- and meso-levels are the volume averaged and cross-sectional averaged description of the differential models, respectively. The definitions of various averages are an important part of the chemical engineer's vocabulary. The concept of averaging is useful to inter-relate models at various levels of detail.

- Compartmental models offer a simplified platform through which to represent complex systems, such as calculation of a drug metabolic process in the human body. These models usually comprise a set of macroscopic control volumes connected in some specified manner.

Review Questions

1.1 What is meant by a concentration jump?

1.2 What is meant by the continuum assumption and what are its implications in model development?

1.3 Indicate some situations where the continuum models are unlikely to apply.

1.4 Give an example of a system in which the total concentration is (nearly) constant.

1.5 Give an example of a system in which the mixture density is nearly constant.

1.6 Can average molecular weight be a function of position?

1.7   Which information is missing in the differential models based on the continuum assumption?

1.8   Why are constitutive models needed in the context of differential models?

1.9   Does the Fick's law apply universally to all systems?

1.10   Write a form of Fick's law using partial pressure gradient as the driving force.

1.11   What is meant by the invariant property of the flux vector?

1.12   The combined flux is partitioned into the sum of the convection flux and the diffusion flux. Is this partitioning unique?

1.13   State the units of the gas constant if the pressure is expressed in bar instead of Pa, (ii) if expressed in atm.

1.14   Express the dissolved oxygen concentration in Example 1.1 in parts per million (ppm).

1.15   What is a macroscopic-level model? Which information needs to be added in such a model in addition to the conservation principle?

1.16   How is the mass transfer coefficient defined? Why is it needed?

1.17   What is meant by a cross-sectional average concentration? What is a cup mixing or the bulk concentration?

1.18   Which assumptions are involved in a plug flow model?

1.19   Which assumptions are involved in a completely backmixed model?

1.20   Which additional closure is needed for mass transport in turbulent flow? Why?

1.21   What is an ideal stage contactor? How do you correct the model if the stage is not ideal?

1.22   What is the dispersion coefficient and where is it needed?

Problems

1.1   **Mass fraction to mole fractions.** Show that mass fractions can be converted to mole fractions by the use of the following equation:

$$y_i = \frac{\omega_i}{M_i}\bar{M} \tag{1.29}$$

Derive an expression for dy_i as a function of $d\omega_i$ values. Do this for a binary mixture. Expression for multicomponent mixture becomes rather unwieldy!

1.2   **Mole fraction to mass fractions.** Show that mole fractions can be converted to mass fractions by the use of the following equation:

$$\omega_i = \frac{y_i M_i}{\bar{M}} \tag{1.30}$$

Derive an expression for $d\omega_i$ as a function of dy_i values for a binary mixture.

1.3   **Average molecular weight.** At a point in a methane reforming furnace, we have a gas of the composition $CH_4 = 10\%$, $H_2 = 15\%$, $CO = 15\%$, and $H_2O = 10\%$ by moles. Find the mass fractions and the average molecular weight of the mixture. Find the density of the gas.

1.4 **Average molecular weight variations.** Two bulbs are separated by a 20-cm-long capillary tube. One bulb contains hydrogen and the other bulb contains nitrogen. The mole fraction profile varies in a linear manner along the length of the capillary. Calculate the mass fraction profile and show that the variation is not linear. Also calculate the average molecular weight as a function of the length along the capillary.

1.5 **Mass fraction gradient.** For a diffusion process across a stagnant film, the mole fraction gradient of the diffusing species was found to be constant. What is the mass fraction gradient? Is it also linear? The mixture is benzene–air.

1.6 **Total concentration in a liquid mixture.** Find the total molar concentration and species concentrations of 10% ethyl alcohol by mass in water at room temperature.

1.7 **Effect of coordinate rotation on flux components.** In Figure 1.1, the flux vector is $2e_x + e_y$. Now consider a coordinate system that is rotated by an angle θ. Find this angle such that the flux component N_{Ay} is zero. What is the value of N_{Ax} in this coordinate system?

1.8 **Flux vector in cylindrical coordinates.** Define flux vector in terms of its components in cylindrical coordinates. Sketch the planes over which the components act. Show the relations between these components and the components in Cartesian coordinates.

1.9 **Flux vector in spherical coordinates.** Define the flux vector in terms of its components in spherical coordinates. Sketch the planes over which the components act. Show the relations between these components and the components in Cartesian coordinates.

1.10 **Different forms of the Henry's law constant.** Express the Henry's law constants reported in Table 1.2 as $H_{i,pc}$ and $H_{i,cp}$.

1.11 **Henry's law constants: Unit conversions.** Henry's law constants for O_2 and CO_2 are reported as 760.2 L · atm/mol and 29.41 L · atm/mol. Which form of Henry's law is being used? Convert these constants to values for the other forms of Henry's law shown in the text.

1.12 **Solubility of CO_2.** The Henry's law constant values for CO_2 are shown below as a function of temperature.

Temperature K	280	300	320
H. bar	960	1730	2650

Fit an equation of the type

$$\ln H = A + B/T$$

What is the physical significance of the parameter B? Find the solubility of pure CO_2 in water at these temperatures.

1.13 **Vapor pressure calculations: The Antoine equation.** The Antoine constants for water are $A = 8.07131, B = 1730.63, C = 233.426$ in units of mm Hg for pressure and degrees Celsius for temperature. Convert these constants to a form where pressure is in Pa and temperature is in kelvins. Also rearrange the Antoine equation to a form where temperature can be calculated explicitly. This represents the boiling point at that pressure. What is the boiling point of water at Denver, Colorado (the "Mile High City")?

1.14 **Vant Hoff relation.** Given the Antoine constants for a species, can you calculate the heat of vaporization of that species? Find this value for water from the data given in Problem 1.13.

1.15 **Concentration jump.** A solid rock of NaCl is in contact with water. Calculate the concentration of NaCl on the water side and salt side of the interface in mol/m^3.

1.16 **A well-mixed reactor: Mass balance calculation.** Consider a well-stirred reactor where a reaction $A \rightarrow B$ is taking place. The volumetric flow rate is 1 m^3/s and the reactor volume is 0.3 m^3. The inlet concentration of A is 1000 mol/m^3 and the exit concentration is 200 mol/m^3. What is the rate of reaction of A in the system? If the reaction is first order and the contents are well mixed, what is the value of the rate constant?

1.17 **Mass transfer coefficient calculation.** A naphthalene ball ($M_A = 128$ g/mol and $\rho_A = 1145$ kg/m^3) is suspended in a flowing stream of air at 347 K and 1 atm pressure. The vapor pressure of naphthalene is 666 Pa for the given temperature. The diameter of the ball was found to change from 2.1 cm to 1.9 cm over a time interval of 1 hour. Estimate the mass transfer coefficient from the solid to the flowing gas.

1.18 **A model for VOC loss from a holding tank.** Wastewater containing a VOC at a concentration of 10 mol/m^3 enters an open tank at a volumetric flow rate of 0.2 m^3/min and exits at the same rate. The tank has a diameter of 4 m and the depth of liquid in the tank is 1 m. The concentration of VOC in the exit stream and the rate of release of VOC is requested by the EPA. Use the conservation law to set a up a model. State further assumptions you may need to complete the model. List the parameters needed to solve the problem.

1.19 **Averaging.** Velocity profile in laminar flow is

$$v_z(r) = v_{max}[1 - (r/R)^2]$$

What does v_{max} represent? How is it related to the pressure drop? Find the average velocity.

The concentration distribution at a given axial position for a solid dissolving from the wall is given by Equation 1.24 Find the cross-sectional average concentration and compare the value with the cup mixing concentration calculated in Example 1.2.

1.20 **Cup mixing versus cross-sectional average.** The variation of scaled concentration in a laminar flow tubular reactor was measured and fitted to the following equation at a specified axial position:

$$c_A = 0.5[1 - (r/R)^2 + (r/R)^4/2]$$

Calculate the center, wall, cup mixing, and cross-sectional average concentrations.

1.21 **Turbulent flow velocity profile.** Velocity profile in turbulent flow is commonly modeled by the 1/7-th law:

$$v_z(r) = v_c[1 - (r/R)]^{1/7}$$

What does v_c represent? Find the average velocity. Compare the average velocity and the center line velocity.

CHAPTER 2

Examples of Differential (1-D) Balances

Learning Objectives

After completing this chapter, you will be able to:

- Develop differential models for problems in mass transfer where the concentration varies along only one spatial dimension.
- Derive differential equations for the species concentration by use of the conservation law followed by Fick's law in Cartesian coordinates.
- Derive the model equations in cylindrical and spherical coordinates and understand the differences arising due to geometry.
- State the equation in vector form and be able to use it in any coordinate system.
- Solve some simpler problems, especially those involving mass transfer with reaction, and examine the property of the solution.

This chapter illustrates the development of differential models by examining simple examples where the transport occurs in only one spatial direction. This simplifies both the model and the mathematics of deriving the solution. By focusing on such simple problems, we build our understanding of the model development. In Chapter 5, we will generalize this understanding to 3-D problems where we derive the complete differential equation for the concentration field and where the particular cases studied here can be obtained as a simplification of the general model. However, the model development is better understood by studying particular cases individually, as is done in this chapter.

To begin, we consider problems posed in Cartesian geometry. Here the concentration is assumed to be a function of only one coordinate, say x. We start off with a simple diffusion problem, followed by diffusion with reaction, transient diffusion, and diffusion with convection. Model development is illustrated in a step-by-step manner, followed by solution of some illustrative problems.

Problems posed in cylindrical and spherical coordinates where the concentration varies only in the radial direction are examined next. One distinguishing feature in contrast to the case involving Cartesian coordinates is that the area for diffusion varies as a function of the radial coordinates; hence the resulting equations appear to be slightly different from those in Cartesian coordinates. Illustrative problems are then solved for problems posed in these geometries. This chapter also indicates that the equations for all three geometries can be unified by using the divergence and the Laplacian operator from vector calculus. This approach provides the vector representation of the model, which is general and useful in any chosen coordinate system.

2.1   Cartesian Coordinates

The first set of problems we will cover are those concerned with mass transfer in a slab of finite thickness in the x-direction with rather large height and width in the y- and z-directions. This makes the transport one-dimensional. First we examine a problem with no generation, no accumulation, and no externally imposed flow across the slab. We can progressively include these terms in the analysis.

2.1.1   Steady State Diffusion across a Slab

Steady state diffusion across a slab is the simplest problem in mass transport analysis. It is so simple that a linear profile across the slab is often taken for granted in steady state analysis. Nevertheless, the steps and the assumptions involved in arriving the result should be examined and the problem should be analyzed step by step since the approach is similar to that used in other cases and in other coordinate systems.

An illustrative problem statement is as follows: A membrane or a medium of thickness L has concentrations C_{A0} on one side and C_{AL} on the other side. This can, for example, be accomplished by circulating a gas containing the solute A at the appropriate concentration levels on both sides of the membrane. The diffusion takes place along the x-direction only and the concentration profile in the slab and the quantity of species transported across the slab are to be computed. The schematic of the problem is shown in Figure 2.1.

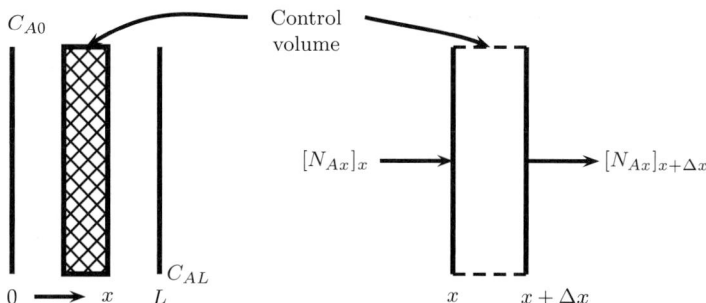

Figure 2.1 Schematic of steady state transport across a slab. The shaded section is the differential control volume used in the analysis.

We start with the conservation law, which reduces for this case as follows since there is no generation nor accumulation:

$$\text{in} - \text{out} = 0$$

Input at x is equal to the flux times the area evaluated at x and can be represented as $N_{Ax}A$. Here A is the area of cross-section in the direction perpendicular to diffusion (perpendicular to x here). Similarly, the output is $N_{Ax}A$ but evaluated at $x + \Delta x$.

The conservation principle therefore gives

$$[N_{Ax}A]_x - [N_{Ax}A]_{x+\Delta x} = 0 \tag{2.1}$$

If one divides the expression by Δx and takes the limit as Δx tends to zero, the following differential equation for the flux is obtained:

$$\frac{d}{dx}(N_{Ax}A) = 0 \tag{2.2}$$

Note: This expression is general and holds for all three geometries examined in this chapter when there is no generation due to chemical reaction and a steady state exists.

The areas for diffusion at x and $x + \Delta x$ are the same and hence this term cancels out.

$$\frac{d}{dx}(N_{Ax}) = 0 \tag{2.3}$$

So far we have dealt with the combined flux to keep the analysis general. In addition, the area term did not matter and canceled out. For some other cases (a cylinder and a sphere, for example), the area is a function of the position. These cases are examined in Sections 2.2 and 2.3.

Although the problem examined here is a diffusion-dominated process and there is no forced flow, convective flow can arise due to diffusion itself. The effect of such convection is examined in Chapters 5 and 6. For the time being, let us assume that the convection flux is negligible and the combined flux has only the contribution resulting from the diffusion flux. With this assumption we have

$$N_{Ax} \approx J_{Ax}$$

which is sometimes referred to as a low flux approximation. In other words, any convective flux (resulting from the diffusion process itself or due to any superimposed external flow) is neglected, and the conservation model reduces to

$$\frac{d}{dx}(J_{Ax}) = 0 \tag{2.4}$$

Note: The effect of diffusion-induced convection (examined in Chapter 6) is sometimes referred to the high flux model. The effect of a superimposed (external) flow in the direction of diffusion (the so-called blowing problem) is discussed in Section 2.1.4.

This is as far as we can proceed just by use of the conservation law alone. A model for diffusion flux is needed, and we use Fick's law as the model:

$$J_{Ax} = -D_A \frac{dC_A}{dx}$$

where D_A is the diffusion coefficient of A in the medium or the membrane. Substituting Fick's law into the flux conservation equation, Equation 2.4, the following equation is obtained for the concentration profile:

$$\frac{d}{dx}\left(D_A \frac{dC_A}{dx} \right) = 0 \tag{2.5}$$

We assume now that the diffusion coefficient is not a function of concentration, which turns out to be good assumption for many cases (e.g., diffusion of gases such as oxygen in polymeric membranes). This permits us to take D_A out of the derivative sign. It cancels out as well, leading to the following expression:

$$\frac{d^2 C_A}{dx^2} = 0$$

Integrating twice gives us a general expression for the concentration profile:

$$C_A = B_1 + B_2 x$$

Here B_1 and B_2 are two integration constants that are evaluated to satisfy the boundary conditions imposed at the two ends.

Using the values of C_{A0} at $x = 0$ and C_{AL} at $x = L$, we can find the integration constants and eliminate them from the previous equation. The final expression for the concentration profile is

$$C_A = C_{A0} - \left(\frac{C_{A0} - C_{AL}}{L} \right) x \qquad (2.6)$$

The concentration profile is simply a straight line connecting the concentration values at the two ends. The flux from the slope of this straight line is therefore

$$J_{Ax} = D_A \left(\frac{C_{A0} - C_{AL}}{L} \right) \qquad (2.7)$$

The rate at which the solute is transported across the media, $\dot{\mathcal{M}}_A$, can now be calculated as J_{Ax} times A and is equal to

$$\dot{\mathcal{M}}_A = A J_{Ax} = A D_A \left(\frac{C_{A0} - C_{AL}}{L} \right)$$

We now have both the concentration profile and the quantity transported across the slab, and the solution is complete. Let us apply the results to examine a simple problem dealing with membrane transport, which is shown in Example 2.1.

Example 2.1   Helium Leakage Rate across a Glass Wall

Light gases such as helium (He) has reasonable solubility in materials such as glass, so there may be leakage of this gas from a helium-containing glass shell that is used for a helium–cadmium laser in a copying machine. Calculate the helium leakage rate across a glass shell of thickness 1.5 mm with pure helium on one side and air containing negligible helium on the other side. Assume that the glass shell is at a constant temperature of 115 °C and the pressure inside the shell is 460 Pa at the given instant of time. (Problem adapted from Mills, 1995.)

The diffusion coefficient of He (m^2/s) in the solid glass is reported as (Mills, 1995)

$$D_A = 1.40 \times 10^{-8} \exp(-3280/T)$$

The solubility parameter has to be known as well and is given in a part of the solution.

Solution

Equation 2.7 is applicable for the flux. However, the concentrations of He are in the glass near the wall, not on the gas near the wall. Hence the concentration jump condition (introduced in Section 1.4) must to be used. We need to know the solubility relationship, which is the additional data needed to solve the problem.

The solubility parameter reported by Mills (1995) is

$$C_{As} = KC_{AG}$$

where K is reported as

$$K = 3.0 \times 10^{-5} \, T - 0.0012$$

For the given conditions, $T = 115 + 273 = 388$ kelvin, so K equals 0.0104.

On the gas side we have pure helium, whose concentration is calculated as

$$C_{AG}(x = 0) = \frac{P}{R_g T} = \frac{460}{8.314 \times 388} = 0.1426 \text{ mol/m}^3$$

Using the K value we find $C_{A0} = 0.0104 \times 0.1426 = 1.15 \times 10^3 \text{ mol/m}^3$, which is the concentration in the glass shell on the helium-exposed side.

On the air side, the helium concentration in air is zero. The glass-side concentration of helium C_{AL} is also equal to zero.

The diffusion coefficient has a value of $2.98 \times 10^{-12} \text{ m}^2/\text{s}$ at 388 K using the equation and the data given earlier.

The flux can be calculated as

$$J_{Ax} = \frac{D_A}{L}(C_{A0} - C_{AL}) = \frac{(2.98 \times 10^{-12})}{(1.5 \times 10^{-3})}(1.15 \times 10^3 - 0)$$

$$= 2.92 \times 10^{-12} \text{ mol/m}^2\text{s}$$

which is the leakage rate per unit surface area of the glass.

Note that thermodynamic relation was needed in addition to the diffusion flux expression to account for the concentration jump at the gas–glass interface. A similar approach is used to model the transport rate in gas separation membranes as shown in Section 27.1.

2.1.2   Steady State Diffusion with Reaction in a Slab

Now we extend the model by assuming that the diffusing solute undergoes a chemical reaction in the slab. The problem statement is as follows: Consider a material in the form of a long slab with a chemical reaction taking place at a rate of R_A per unit volume of the slab. An example would be a porous catalyst cast in the form of a long slab. A second example would be oxygen diffusing into a pool of liquid containing some biomass or a liquid-phase reactant that consumes the diffusing oxygen. Many other examples can be given.

Assume that the mass transfer of species A is only in the x-direction. In this section, we develop a differential equation for the concentration variation of A in the system as a function of x and use this in the flux expression to obtain the rate of mass transfer into the slab.

We proceed in a similar manner as in the earlier transfer problem, but include the chemical reaction term in the conservation law. This law should now read as follows:

$$\text{in} - \text{out} + \text{net generation} = 0$$

The input and output terms are the same as before and the additional term due to generation is equal to $R_A(A\Delta x)$. Hence

$$[N_{Ax}A]_x - [N_{Ax}A]_{x+\Delta x} + R_A A\Delta x = 0$$

Further simplification is done by taking the diffusion flux alone ($N_{Ax} \approx J_{Ax}$)—that is, by ignoring any effects due to diffusion-induced convection. Also A is constant and does not vary with x. We next divide through by Δx and take the limit as $\Delta x \to 0$ to obtain a differential equation for the variation of the diffusion flux in the system:

$$\frac{d}{dx}(J_{Ax}) = R_A \tag{2.8}$$

This is the basic model for the system based on the conservation principle. The model can be completed if a constitutive (transport) law for diffusion of a species A in a porous solid is available or developed. A Fick's law type of model is now used here to complete the problem:

$$J_{Ax} = -D_{eA}\frac{dC_A}{dx}$$

where D_{eA} is the (effective) diffusion coefficient of A in the porous medium.

Note: The rate of diffusion is modified due to the porous matrix. Specifically, a pore structure–dependent effective diffusivity is used in these cases in lieu of fluid-phase diffusivity. This topic is discussed in further detail in Section 7.6. We use the simplified notation D_A for the effective diffusivity D_{eA} in the remainder of this chapter.

Using a Fick's law type of model for pore diffusion as shown in the mass conservation equation and assuming constant diffusivity (Equation 2.8), we get

$$\boxed{D_A\frac{d^2C_A}{dx^2} = -R_A} \tag{2.9}$$

This equation is known as the diffusion-reaction equation and is widely applied in the analysis of transport in catalysts. Note that the second derivative term on the left side of the equation is the x-dependent part of the Laplacian equation. Hence the equation can be compacted as follows:

$$D_A \nabla^2 C_A + R_A = 0 \qquad (2.10)$$

Further information is needed on the kinetics of the reaction. In general, R_A is a function of the concentration of species A and temperature. It is also often a function of the concentration of other species in the system (e.g., bimolecular reactions, reactions with strong product inhibition). These dependencies are given by a kinetic model for the process. In Examples 2.2 and 2.3, we examine the (general) solution to the problem for simple forms for R_A—namely, for a zero-order reaction and a first-order reaction.

Example 2.2   Diffusion with Zero-Order Reaction

Derive a general expression for the concentration profile for diffusion accompanied by a zero-order reaction in a slab geometry.

Solution

For a zero-order reaction, the rate is given as

$$R_A = -k_0$$

where k_0 is the volumetric rate of consumption of reactant A (units of mol/m^3 s). Equation 2.9 now takes the following form:

$$D_A \frac{d^2 C_A}{dx^2} = k_0 \qquad (2.11)$$

The resulting differential equation can be integrated twice to obtain

$$C_A(x) = \frac{k_0}{2D_A} x^2 + B_1 x + B_2$$

which is the general solution for the concentration profile. The constants of integration B_1 and B_2 can be found if the boundary conditions are specified. The boundary conditions are problem specific and depend on what is happening at the two ends of the system.

A common application of this model is in trickling bed filters used in waste water treatment. It is also used for modeling oxygen transport in tissues, where the oxygen metabolism follows a zero-order rate law for most cases.

Example 2.3 shows the application to a first-order reaction.

Example 2.3   Diffusion with First-Order Reaction

Derive a general expression for the concentration profile for diffusion accompanied by a first-order reaction in a porous catalyst in a slab geometry.

Solution

In this case the rate is given as

$$R_A = -k_1 C_A$$

where k_1 is the rate constant for a first-order reaction in the units of s^{-1}. Equation 2.9 now takes the following form:

$$D_A \frac{d^2 C_A}{dx^2} - k_1 C_A = 0 \tag{2.12}$$

This equation can be solved by treating it as a linear second-order differential equation. The general solution can be expressed in two equivalent forms:

$$C_A(x) = A_1 \cosh\left(x\sqrt{k_1/D_A}\right) + A_2 \sinh(x\sqrt{k_1/D_A}) \tag{2.13}$$

with A_1 and A_2 being integration constants or

$$C_A(x) = B_1 \exp(x\sqrt{k_1/D_A}) + B_2 \exp(-x\sqrt{k_1/D_A})$$

where B_1 and B_2 now represent the integration constants.
    Note: The B_1 and B_2 will be a linear combination of A_1 and A_2, and vice versa.
    A common application is in constructing a diffusion model for a catalyst pore—an application that is discussed in Chapter 18. At this point, the model development and the nature of the resulting differential equation should be noted. The general solution with two integration constants is also noteworthy since it can be applied to many different boundary conditions. Various types of boundary conditions of common occurrence are summarized in Chapter 5.

In Example 2.4, we apply Equation 2.13 to a specific problem. We also illustrate the use of two common types of boundary conditions.

Example 2.4   Oxygen Profile in a Pool of Liquid

A pool of liquid is 10 cm deep and a gas A dissolves and reacts in the liquid. The solubility of the oxygen is such that the interfacial concentration is equal to 2 mol/m^3 in the liquid and the diffusivity of the dissolved gas is 2×10^{-9} m^2/s.

Assume a first-order reaction with a rate constant of 10^{-6} s^{-1} in the liquid. Find the concentration profile in the liquid. Express the solution in dimensionless form. Also find the flux of oxygen into the pool of liquid.

Solution

The schematic of the problem and the anticipated concentration profile of oxygen in the pool of liquid are shown in Figure 2.2. Equation 2.13 is applicable for the concentration profile. Also, Equation 2.13 (cosh and sinh form) is more convenient here (rather than the exponential form), and the constants of integration must be evaluated using the boundary conditions. You will find that taking $x = 0$ at the bottom will simplify the math in the cosh and sinh form of the solution. The location $x = L$ (10 cm here) is then the position of the gas–liquid interface. We need to evaluate the two constants by specifying appropriate boundary conditions.

Since no oxygen diffuses into the bottom of the tank, we take $dC_A/dx = 0$ at $x = 0$, the bottom of the tank. This type of boundary condition is generally known the boundary condition of the second kind, also known as the Neumann boundary condition.

The procedure to get one of the constants of integration is as follows. First, differentiate Equation 2.13 to get the derivative of the concentration:

$$\frac{dC_A}{dx} = A_1[\sqrt{k_1/D_A}]\sinh\left(x\sqrt{k_1/D_A}\right) + A_2[\sqrt{k_1/D_A}]\cosh(x\sqrt{k_1/D_A})$$

Then set this expression equal to zero at $x = 0$. We find $A_2 = 0$.

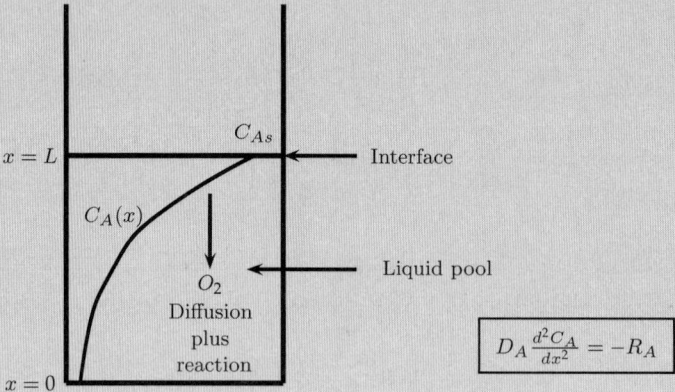

Figure 2.2 Schematic of oxygen diffusion with reaction in a pool of liquid and the expected nature of the concentration profile.

The solution now reduces to

$$C_A(x) = A_1 \cosh\left(x\sqrt{k_1/D_A}\right)$$

Now use the boundary condition at $x = L$ to get A_1. The concentration is set equal to C_{As}, the interfacial concentration of dissolved oxygen at $x = L$. This is known as the boundary condition of the first kind, also referred to as the Dirichlet condition. Using this information and some minor algebra yields

$$A_1 = \frac{C_{As}}{\cosh(L\sqrt{k_1/D_A})}$$

Both constants are evaluated and the solution for the concentration profile is completed. Backsubstituting of A_1 leads to the following result:

$$\frac{C_A}{C_{As}} = \frac{\cosh(x\sqrt{k_1/D_A})}{\cosh(L\sqrt{k_1/D_A})} \tag{2.14}$$

The dimensionless version of the solution requires the use of dimensionless concentration c_A defined as C_A/C_{As} and a dimensionless distance parameter, $\xi = x/L$. In addition, a dimensionless parameter called the Thiele modulus for diffusion-reaction systems is needed, which is defined here:

$$\phi = L\sqrt{k_1/D_A}$$

With these definitions, Equation 2.14 can be written in dimensionless form:

$$c_A = \frac{\cosh(\phi\xi)}{\cosh\phi} \tag{2.15}$$

The use of dimensionless quantities is a common practice in representation and solution to engineering models, since it makes the mathematical manipulations far easier and allows the final results to take a more compact form.

The rate of absorption of oxygen is calculated using Fick's law at the gas–liquid interface:

$$\dot{\mathcal{M}}_A = AD_A\left(\frac{dC_A}{dx}\right)_{x=L}$$

where A is the exposed surface area for oxygen transfer.

Note the plus sign used in Fick's law for the flux; it is appropriate because we are looking at the flux in a direction of decreasing x. Substituting for the derivative of the concentration leads to

$$\dot{\mathcal{M}}_A = A\sqrt{D_A k_1}\tanh(L\sqrt{k_1/D_A}) \cdot C_{As} \tag{2.16}$$

This result has wide applicability in reaction engineering.

Substituting the values, the following results are obtained:

$$\text{oxygen concentration at the bottom} = 1/\cosh(L\sqrt{k_1/D_{eA}})$$
$$= 0.4227 \text{ mol/m}^3$$

$$\text{oxygen transfer rate per unit area using Equation 2.16} = 8.74$$
$$\times 10^{-8} \text{ mol/m}^2 \text{ s}$$

2.1.3   Transient Diffusion in a Slab

Consider the problem examined in Section 2.1.2, but now assume the slab is initially at a uniform concentration of C_{Ai}. At $t > 0$, one end of the slab (say, $x = 0$) is exposed to a different concentration of C_{A0}. The other end is still at C_{Ai}. The problem is how to predict the temporal evolution of the concentration profile in the system and how use this information to get the instantaneous rate of mass transfer. Here we derive the governing differential equation for the transient diffusion process.

The conservation law should now include the accumulation term:

$$\text{input} - \text{output} = \text{accumulation}$$

The input and output terms are the same as before and the additional term due to accumulation is equal to the time derivative of moles of A in the control volume, $C_A A \Delta x$. Hence:

$$[N_{Ax}A]_x - [N_{Ax}A]_{x+\Delta x} \cdot = \frac{\partial}{\partial t}(C_A A \Delta x)$$

The final equation with the approximation of $N_A = J_A$ and the use of Fick's law for J_A leads to the following partial differential equation for the transient problem:

$$\boxed{D_A \frac{\partial^2 C_A}{\partial x^2} = \frac{\partial C_A}{\partial t}} \tag{2.17}$$

This equation is sometimes referred to as Fick's second law, but in reality is a combination of the conservation law and Fick's first law. Constant diffusivity is also implied. The equation needs two boundary conditions and one initial condition since it is now a partial differential equation (PDE). The solution method is described in detail in Section 8.2, but Example 2.5 shows an illustrative result. The example also shows how to represent the problem in dimensionless form.

Example 2.5   Temporal Evolution of a Concentration Profile in a Slab

A membrane, cast as a slab, has zero concentration at time zero. Suddenly, after time zero, one surface is exposed to a gas such that the concentration in the membrane at this point is C_{A0}. The other side of the membrane is kept at zero concentration.

State the governing equation and the boundary and initial conditions to be used. Also sketch the anticipated temporal evolution of the concentration profile in slab.

Solution

The governing equation is the transient model given by Equation 2.17, Fick's second law. Before stating the required conditions it is useful to write the equation in a dimensionless form. This can be done by using the following variables:

$$\text{dimensionless concentration } c_A = \frac{C_A}{C_{A0}}$$

$$\text{dimensionless distance } \xi = \frac{x}{L}$$

Notice that L^2/D_A has the same units as time, so we can use it to scale the actual time. Thus we can define a dimensionless time as

$$\text{dimensionless time } \tau = \frac{t}{(L^2/D_A)} = \frac{tD_A}{L^2}$$

The differential equation reduces to a simpler form with these variables:

$$\frac{\partial^2 c_A}{\partial \xi^2} = \frac{\partial c_A}{\partial \tau} \tag{2.18}$$

Note that there are no free parameters, which is a characteristic of a well-scaled problem. You should verify the dimensionless version starting from the original version using the chain rule.

The boundary conditions to be used are as follows: At $x = 0, C_A = C_{A0}$, which in dimensionless form is $c_A(\xi = 0) = 1$.

At $x = L, C_A = 0$, which in dimensionless form is $c_A(\xi = 1) = 0$

Both of these relations hold for all values of $\tau > 0$.

The initial condition is at $\tau = 0, c_A = 0$ for all ξ in the domain to 1, since the slab is maintained at zero concentration at the begining of the process.

The problem specification is now complete. This problem can be solved analytically by a modified method of separation of variables or by numerical methods (e.g., MATLAB PDEPE solver). These methods are elaborated in a later chapter. At this point you should focus on problem formulation and

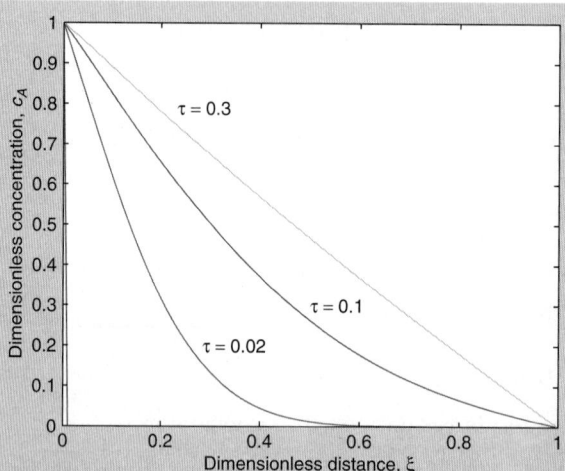

Figure 2.3 Evolution of concentration profile for transient diffusion across a slab.

dimensionless representation. The computed result using MATLAB PDEPE solver is shown in Figure 2.3. Note that for large time, a linear profile is obtained that is the steady state profile for diffusion across a slab.

L^2/D_A is the characteristic time scale for diffusion. In other words, it is the time scale for any disturbance to be smoothed out by diffusion, and one can expect that a steady state profile will be reached around this time. This parameter is called the diffusion time. In this particular case, the steady state is reached around $0.3 \times L^2/D_A$.

2.1.4   Diffusion with Convection

We now examine a simple problem where the convective flow is superimposed on diffusion. Specifically, the case involving steady state and no chemical reaction is analyzed. Consider again the problem discussed in Section 2.1.1, but with the addition of a superimposed flow across the system with a velocity of v_x. This information is used to derive the differential equation and to find the flux across the system. The problem is illustrated in Figure 2.4.

The application of the conservation law shows that the combined flux is the same at x (in) and at $x + \Delta x$ (out), since this is steady state and no reaction is taking place. Therefore Equation 2.3 applies here as well; it is repeated here for ease of reference:

$$\frac{d}{dx}(N_{Ax}) = 0$$

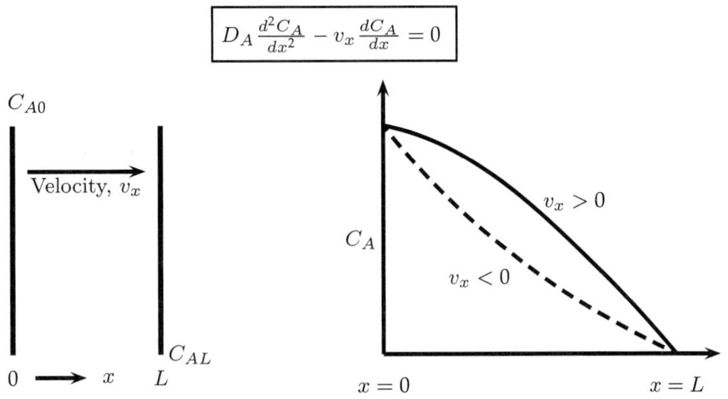

$$\boxed{D_A \frac{d^2 C_A}{dx^2} - v_x \frac{dC_A}{dx} = 0}$$

Figure 2.4 Schematic of steady state diffusion with superimposed convection in a Cartesian geometry. The profile will be a straight line for no convection. The dashed line is that for flow (convection) in the opposite direction to the concentration difference.

The combined flux has now contribution from both diffusion and convection:

$$N_{Ax} = -D_A \frac{dC_A}{dx} + v_x C_A$$

where v_x is the blowing or transpiration velocity. Substituting into the flux equation and assuming constant D_A and v_x, we obtain the following second-order differential equation:

$$D_A \frac{d^2 C_A}{dx^2} - v_x \frac{dC_A}{dx} = 0 \tag{2.19}$$

This is the simplified version of the general 3-D convection-diffusion equation that will be discussed in more detail in Chapter 5. This general equation is stated here to help you better appreciate the vector representation of mass transfer models:

$$D_A \nabla^2 C_A - \boldsymbol{v} \cdot \nabla C_A = 0$$

The general solution to Equation 2.19 can be written as follows:

$$C_A = A_1 + A_2 \exp(xv_x/D_A) \tag{2.20}$$

where A_1 and A_2 are two integration constants. (You may wish to verify this by direct substitution.)

The boundary conditions of the first kind are useful for this case since the concentrations are specified at the ends. At $x = 0$, the concentration is C_{A0}; at

$x = L$, the concentration is C_{AL}. Using these conditions, we can obtain the constants of integration and complete the solution. The details are left as an exercise. In this scenario, the profiles are no longer linear, in contrast to the result obtained in the pure diffusion case.

More useful than the concentration profile is the (combined) flux across the system. This can be calculated from the concentration profile as

$$N_{Ax} = -D_A \left(\frac{dC_A}{dx} \right)_{x=0} + v_x C_{A0}$$

or equivalently as

$$N_{Ax} = -D_A \left(\frac{dC_A}{dx} \right)_{x=L} + v_x C_{AL}$$

The result after some standard algebraic manipulations is as follows (which you should verify):

$$N_{Ax} = v_x(C_{A0} - C_{AL}) \frac{\exp(v_x L/D_A)}{\exp(v_x L/D_A) - 1} \tag{2.21}$$

The expression can also be written as

$$N_{Ax} = \frac{D_A}{L}(C_{A0} - C_{AL}) \left[\frac{(v_x L/D_A)\exp(v_x L/D_A)}{\exp(v_x L/D_A) - 1} \right] \tag{2.22}$$

where the bracketed term on the right side can be identified as the factor by which the blowing enhances the rate of mass transfer. This is illustrated in Example 2.6.

Example 2.6   Transport Rate in Presence of Convection

A porous plug 10 cm long has hydrogen at a partial pressure of 0.1 atm on one side and air at 1 atm on the other side at 293 K. Find the transport rate of hydrogen if the system is stagnant, if there is a superimposed gas flow at a rate of 0.1 cm/s in the same direction as that for diffusion, and if the flow is in the opposite direction to diffusion.

Solution

The parameter $v_x L/D_A$ is the key dimensionless group in Equation 2.22 showing the effect of the blowing velocity. This parameter is known as the Peclet number:

$$\text{Peclet number} = \frac{v_x L}{D_A} = \frac{\text{Convection rate}}{\text{Diffusion rate}}$$

It can be interpreted as the ratio of the convection rate to the diffusion rate. For no blowing, the pure diffusion model applies:

$$N_{Ax} = \frac{D_A}{L}(C_{A0} - C_{AL})$$

Equation 2.22 indicates that the pure diffusion flux should be corrected by the following factor:

$$\text{Correction for blowing} = \frac{(v_x L / D_A) \exp(v_x L / D_A)}{\exp(v_x L / D_A) - 1}$$

The correction factor is more compactly expressed using the Peclet number as

$$\text{Correction for blowing} = \frac{Pe \exp(Pe)}{\exp(Pe) - 1} \tag{2.23}$$

For numerical calculations, the value of the diffusion coefficient of hydrogen in air is needed. We will use a value of 0.7 cm^2/s here. Hence the Peclet number is $Pe = 0.1 * 10/0.7 = 1.4286$ and the correction factor for blowing calculated using Equation 2.23 is 1.8788.

The concentration of hydrogen at one end C_{A0} is calculated as $p_h/R_g T$, or 4.03 mol/m^3.

Flux in the absence of blowing is $D_A(C_{A0} - 0)/L$, which is found to be 0.0028 mol/m^2 s. The flux in the presence of blowing is $0.0028 \times 1.8788 = 0.0053$ mol/m^2 s.

If the flow is in the opposite direction, v_x is negative (suction condition) and hence $Pe = -1.4286$. The correction factor is now 0.45 (less than 1), so the flux is 0.0013 mol/m^2 s. Flux is reduced due to the blowing counter to the direction of diffusion.

We have now examined four prototypical problems in Cartesian coordinates: (1) diffusion alone, (2) diffusion balanced by reaction, (3) transient diffusion with accumulation, and (4) diffusion with superimposed flow. We now study similar problems posed in cylindrical coordinates but again with transport taking place only in the radial direction.

2.2   Cylindrical Coordinates

In this section, we repeat the analysis for the case of a cylindrical geometry to show some nuances that arise primarily due to the nature of the coordinate system. Specifically, we look at problems where the concentration is a function of only the radial position. The analysis is similar to that for a slab geometry, but the point to bear in mind is that the area over which diffusion takes place is a function of the radial position. We repeat some preliminary steps in the slab analysis and show the differences for clarity.

Steady state diffusion is considered first, followed by diffusion with reaction and then transient diffusion.

2.2.1   Steady State Radial Diffusion

The differential control volume, shown in Figure 2.5, is a ring element located at r and $r + \Delta r$. We start with the conservation law for this control volume, which reduces for the case of no generation and accumulation to

$$\text{in} - \text{out} = 0$$

Input at r is equal to the flux times the area evaluated at r and can be represented as $N_{Ar}A$. Similarly, the output is $N_{Ar}A$ but is evaluated at $r + \Delta r$. The conservation principle therefore gives

$$[N_{Ar}A]_r - [N_{Ar}A]_{r+\Delta r}. = 0 \tag{2.24}$$

Next, we note that the area across which diffusion is taking place is equal to $2\pi r L$. (This is different at r and $r + \Delta r$.) The conservation equation with the area A term substituted is

$$[2\pi r L N_{Ar}]_r - [2\pi r L N_{Ar}]_{r+\Delta r} = 0 \tag{2.25}$$

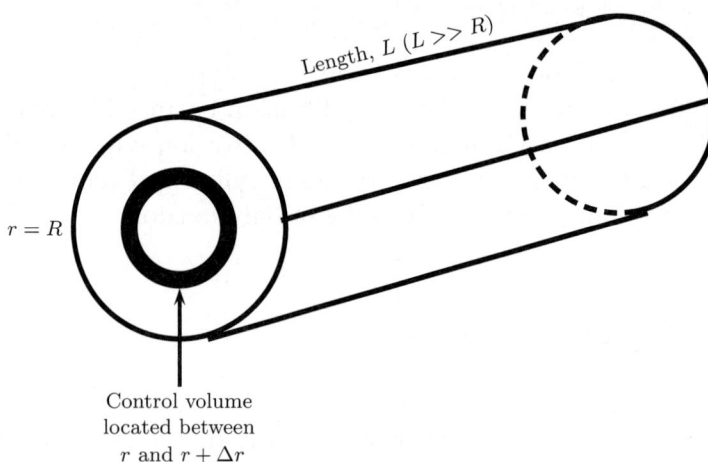

Length, L (L >> R)

$r = R$

Control volume
located between
r and $r + \Delta r$

Figure 2.5 Differential control volume used in the analysis of radial diffusion in a cylinder.

Dividing by Δr and taking the limit leads to

$$\frac{d}{dr}(r N_{Ar}) = 0 \tag{2.26}$$

The combined flux is taken as diffusion flux for further analysis. Using Fick's law to close the problem for J_{Ar}

$$J_{Ar} = -D_A \frac{dC_A}{dr}$$

leads to

$$\frac{d}{dr}\left(D_A r \frac{dC_A}{dr}\right) = 0 \tag{2.27}$$

For a constant diffusivity case, the following differential equation is obtained:

$$\frac{d}{dr}\left(r \frac{dC_A}{dr}\right) = 0 \tag{2.28}$$

This suggests that

$$r \frac{dC_A}{dr} = \text{constant, say } A_1$$

Now a second integration can be done and a general expression for the concentration profile can be obtained:

$$C_A = A_1 \ln r + A_2$$

The constants A_1 and A_2 can be found by fitting the boundary conditions.

Diffusion across a Cylindrical Shell

Consider the diffusion across a cylindrical shell with radius of R_i at the inner surface and R_o at the outer surface. If we set the concentrations as C_{Ai} and C_{Ao} at these points, we can evaluate the constants and obtain the following equation for the concentration profile in this shell:

$$C_A(r) - C_{Ao} = (C_{Ai} - C_{Ao}) \frac{\ln(r/R_o)}{\ln(R_i/R_o)} \tag{2.29}$$

The concentration profile is now logarithmic in r, rather than a linear profile across the shell.

The rate of transport across the system can be computed as

$$\dot{M}_A = -(2\pi R_i L) D_A \left(\frac{dC_A}{dr}\right)_{r=R_i}$$

or as

$$\dot{\mathcal{M}}_A = -(2\pi R_o L) D_A \left(\frac{dC_A}{dr} \right)_{r=R_o}$$

Both expressions will lead to the same result:

$$\dot{\mathcal{M}}_A = \frac{2\pi D_A L (C_{Ai} - C_{Ao})}{\ln(R_o/R_i)}$$

Mean Area for Diffusion

The expression for moles transported is often written in a format similar to that used in the Cartesian case:

$$\dot{\mathcal{M}}_A = A_m \frac{D_A}{t} (C_{Ai} - C_{Ao})$$

where t is the thickness for diffusion (equal to $R_o - R_i$ here) and A_m is the mean area for diffusion. Comparing the two expressions for $\dot{\mathcal{M}}_A$, we find the mean area for diffusion is the logarithmic mean area of the inner and outer cylinder:

$$A_m = 2\pi L \frac{(R_o - R_i)}{ln(R_o/R_i)}$$

2.2.2   Steady State Mass Transfer with Reaction

In our next scenario, we include the effect of reaction. This problem is representative of diffusion with reaction in a catalyst, which now takes the shape of a long cylinder or a concentric annular cylinder. We follow the method shown in Section 2.1.2, with the only addition being the reaction term, which is equal to control volume times the rate of reaction per unit volume, or $(2\pi L r \Delta r) R_A$. The final equation with the same assumption of constant D_A can be derived as

$$\boxed{\frac{D_A}{r} \frac{d}{dr} \left(r \frac{dC_A}{dr} \right) = -R_A} \tag{2.30}$$

Note that the differential term on the left side of the equation is the r-dependent part of the Laplacian operator. Also note the following two forms for this part:

$$\text{Laplacian, cylinder, } r\text{-only } = \frac{1}{r} \frac{d}{dr} \left(r \frac{dC_A}{dr} \right)$$

or equivalently

$$\text{Laplacian, cylinder, } r\text{-only} = \frac{d^2 C_A}{dr^2} + \frac{1}{r}\frac{dC_A}{dr}$$

We now examine the solution for simple cases for R_A. However, we also note that in general a numerical solution is needed for reactions with nonlinear dependency on concentration.

Zero-Order Reaction

For a zero-order reaction $R_A = -k_0$, the governing equation is

$$\frac{D_A}{r}\frac{d}{dr}\left(r\frac{dC_A}{dr}\right) = k_0 \tag{2.31}$$

This model holds for both a solid cylinder and an annular ring cylinder. The general solution is obtained easily by performing integration twice. The final solution is, however, different due to the change in the location where the boundary conditions are applied.

The geometry considered further here is a solid cylinder with a radius R. The domain of the solution is therefore $0 < r < R$.

The boundary condition of constant surface concentration C_{As} is imposed at $r = R$, while the condition $dC_A/dr = 0$ at $r = 0$ is imposed at the center. The final result for the concentration profile is as follows (which you should verify by doing some minor algebra):

$$C_A = C_{As} - \frac{k_0}{4D_A}(R^2 - r^2)$$

The final solution for a concentric cylinder is somewhat different since the domain of the solution is $R_i < r < R_o$; it is left as an exercise. An application of this geometry is in oxygen transport in tissues, a problem of importance in biomedical engineering that is discussed in Section 23.2.

First-Order Reaction

The solution for a first-order case where the rate is a linear function of concentration is somewhat complicated. The rate is given as

$$R_A = -k_1 C_A$$

Use of this rate form in Equation 2.30 leads to

$$\frac{D_A}{r}\frac{d}{dr}\left(r\frac{dC_A}{dr}\right) = k_1 C_A \tag{2.32}$$

The solution of this differential equation is stated in terms of the Bessel functions. The general solution is

$$C_A(r) = A_1 I_0 \left(r\sqrt{k_1/D_A} \right) + A_2 K_0(r\sqrt{k_1/D_A}) \tag{2.33}$$

The I_0 and K_0 components are the modified Bessel functions of the first and second kind, respectively.

This solution may be applied to diffusion with reaction in a catalyst having the shape of a long solid cylinder or long concentric cylinder (discussed further in Section 23.2). The constants of integration depend both on the geometry (solid or concentric cylinder) and on the imposed boundary conditions at the end points of the geometry. The solid cylinder case is described in Example 2.7.

Example 2.7 Diffusion with First-Order Reaction in a Long Cylinder

A porous catalyst is cast in the form of a long cylinder and is exposed to a gas concentration of C_{As} at the surface. State the boundary conditions and derive an expression for the concentration profile.

Solution

The general solution shown in Equation 2.33, applies but the concentration should be finite at $r = 0$. Note that this can be taken as one of the boundary conditions. Since the K_0 function goes to infinity at $r = 0$, we have to set $A_2 = 0$ so that the concentration does not becomes unbounded. The solution reduces to

$$C_A(r) = A_1 I_0 \left(r\sqrt{k_1/D_A} \right)$$

The second boundary condition at $r = R$ is applied. Here C_A is set as C_{As}, the prescribed value. A_1 can be estimated and substituted back to get the final result for the concentration profile. The result is compactly represented in dimensionless form as

$$c_A = \frac{I_0(\phi\xi)}{I_0(\phi)} \tag{2.34}$$

where c_A is a scaled or dimensionless concentration C_A/C_{As}, and ϕ is a dimensionless regrouping of the variables. You should verify that ϕ is defined as

$$\phi = R\sqrt{k_1/D_{eA}}$$

This dimensionless parameter is referred to as the Thiele modulus. Finally, $\xi = r/R$ is the dimensionless radial coordinate.

2.2.3   Transient Diffusion in a Cylinder

In the case of transient diffusion in a cylinder, we balance the $in - out$ term with the accumulation. Details are not shown but you should be able to derive the following PDE for transient diffusion in a long cylinder with diffusion in only the radial direction:

$$\frac{D_A}{r} \frac{\partial}{\partial r} \left(r \frac{\partial C_A}{\partial r} \right) = \frac{\partial C_A}{\partial t} \tag{2.35}$$

Solution methods for such PDEs are examined in detail in Chapter 8.

2.3   Spherical Coordinates

In this section we show examples of problems posed in spherical coordinates. Diffusion in a spherical shell, diffusion with reaction in a porous spherical catalyst, and transient diffusion from a sphere are the three examples explored here, as they are prototypical problems encountered in many applications.

2.3.1   Steady State Diffusion across a Spherical Shell

The control volume in spherical coordinates is a spherical shell contained between r and $r + \Delta r$, as shown in Figure 2.6. The key point to note is that the area for diffusion is $4\pi r^2$ at any location r. Similarly, the volume contained in the differential control volume is equal to $4\pi r^2 \Delta r$. With these modifications

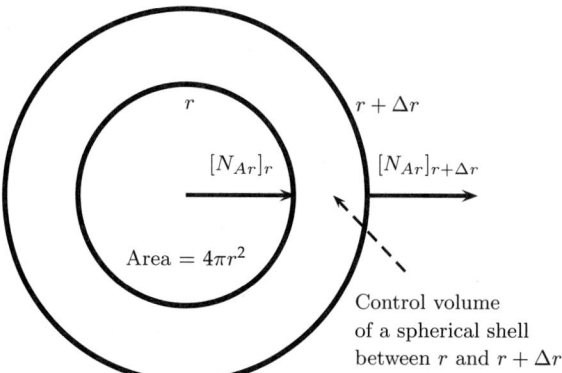

Figure 2.6   Control volume for 1-D differential analysis in a sphere.

the governing differential equations can be set up for various cases, following a similar procedure as the two earlier cases.

For the diffusion-only case, we obtain

$$\frac{d}{dr}\left(r^2\frac{dC_A}{dr}\right) = 0$$

This suggests that

$$r^2\frac{dC_A}{dr} = \text{constant, say } A_1$$

Now a second integration can be done and a general expression for the concentration profile can be obtained:

$$C_A = -\frac{A_1}{r} + A_2$$

This concentration profile is no longer linear. The constants A_1 and A_2 can be found by fitting the boundary conditions at $r = R_i$ and at $r = R_o$. Further details are straightforward and not shown.

The rate of transport across the shell is of more interest and can be computed as

$$\dot{M}_A = -(4\pi R_i^2)D_A\left(\frac{dC_A}{dr}\right)_{r=R_i}$$

Here Fick's law is used at $r = R_i$ and multiplied by the area for diffusion at this radius. Alternatively, we can use Fick's law at $r = R_o$:

$$\dot{M}_A = -(4\pi R_o^2)D_A\left(\frac{dC_A}{dr}\right)_{r=R_o}$$

Both expressions will lead to the same result. The final result can be put in a form similar to the Cartesian geometry solution, using a mean area parameter for the diffusion area. The result is

$$\dot{M}_A = A_m\frac{D_A}{t}C_{Ai} - C_{Ao})$$

where t is the diffusion length equal to $R_o - R_i$ and A_m is the area for diffusion. The latter can be shown to be equal to the "geometric" mean area of the inner and outer shell:

$$A_m = 4\pi R_i R_o$$

2.3.2   Diffusion and Reaction

In this section, a chemical reaction is assumed to take place together with diffusion. Again we consider concentration to vary as a function of r only. A shell balance approach leads to the following equation for diffusion with reaction in a spherical geometry:

$$\boxed{\frac{D_A}{r^2}\frac{d}{dr}\left(r^2\frac{dC_A}{dr}\right) = -R_A}$$ (2.36)

Again, the leading differential term on the left side of Equation 2.36 is found to be the Laplacian operator

$$\text{Laplacian, sphere, } r\text{-only} \; = \frac{1}{r^2}\frac{d}{dr}\left(r^2\frac{dC_A}{dr}\right)$$

or equivalently

$$\text{Laplacian, sphere, } r\text{-only} \; = \frac{d^2C_A}{dr^2} + \frac{2}{r}\frac{dC_A}{dr}$$

We find the diffusion-reaction problem for all the geometries can be generalized as

$$\boxed{D_A\nabla^2 C_A + R_A = 0}$$ (2.37)

A constant value for the diffusion coefficient is implied in this model.

Details and the implications of the solution for calculation of the rate of reaction in a porous catalyst are examined in Chapter 18.

For now, it is useful to present the final solution for a spherical catalyst that is exposed to a surface concentration C_{As}. Solutions are presented in dimensionless form, with their detailed derivation left as exercises.

Zero-Order Reaction

The solution for a zero-order reaction in dimensionless form is

$$c_A = 1 - \frac{\phi_0^2}{6}(1 - \xi^2)$$ (2.38)

where $\xi = r/R$ and $\phi_0 = R\sqrt{(k_0/(D_A C_{As}))}$.

The solution should be used only if $\phi_0 < \sqrt{6}$. If $\phi_0 > \sqrt{6}$, a concentration-depleted layer (a region of negative concentrations, which is unrealistic) develops near $\xi = 0$, the center of the sphere. A modified analysis is needed in such a case. Section 18.2.3 takes this issue up in more detail.

First-Order Reaction

The solution for a first-order reaction can be shown to be

$$c_A = \frac{\sinh(\phi\xi)}{\xi \sinh(\phi)} \tag{2.39}$$

where $\xi = r/R$ and $\phi = R\sqrt{(k_1/D_A)}$.

2.3.3   Transient Diffusion in Spherical Coordinates

Transient diffusion in a sphere in considered here briefly, as it has important applications in many areas. For example, this case may be used to find the rate of drug release from a capsule. A shell balance approach leads to the following partial differential equation for transient diffusion in a spherical geometry:

$$\frac{D_A}{r^2} \frac{\partial}{\partial r} \left(r^2 \frac{\partial C_A}{\partial r} \right) = \frac{\partial C_A}{\partial t} \tag{2.40}$$

Here we illustrate the case in which the sphere has a uniform concentration C_{Ai}, which provides the initial condition. The sphere is exposed to a constant surface concentration C_{A0} at $r = R$ after time zero. This provides one of the boundary conditions. The concentration is symmetric at $r = 0$ and hence $dC_A/dr = 0$ at this point, which provides the second boundary condition. The problem formulation is complete and analytical solutions or numerical solutions are used to solve for the temporal evolution of the concentration profile and the flux at the surface. These details are examined in Chapter 8.

For illustrative purposes, a sketch of dimensionless concentration versus position is shown in Figure 2.7. It is intended to help you get a qualitative feel for the result.

Figure 2.7 uses dimensionless variables—specifically, dimensionless distance, $\xi = r/R$; dimensionless time, $\tau = tD_A/R^2$; and dimensionless concentration (more of a ratio of concentration differences), defined as:

$$c_A = \frac{C_A - C_{A0}}{C_{Ai} - C_{A0}}$$

This concentration is actually a dimensionless concentration difference. Thus the initial value of c_A is 1, and the final value is equal to the surface concentration of 0. Key points to note from the solution shown in Figure 2.7 are the following: At the initial time, the concentration drop is confined to the region near the surface, so the initial profile is rather steep. A nearly steady

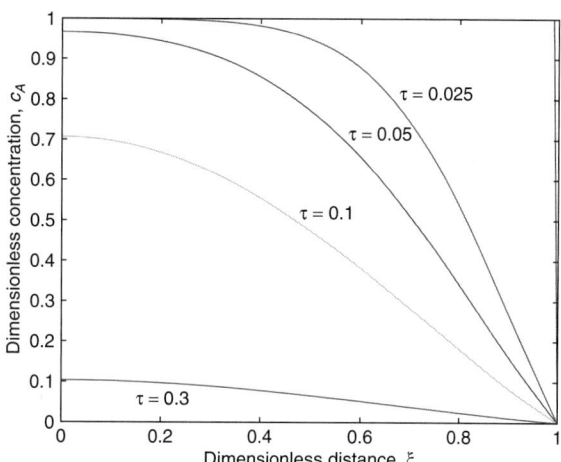

Figure 2.7 Evolution of the concentration profile for transient diffusion across a sphere. The final concentration should be 0. We reach almost 10% of the final value at time $\tau = 0.3$.

state profile is achieved at $\tau = 0.3$. In terms of actual time, the value is about $0.3R^2/D_A$. The final steady state (dimensionless) concentration should be 0.

Summary

- Diffusion in a slab of finite thickness with the ends held at different concentrations is the simplest problem in mass transfer. For constant diffusivity, the concentration profile is linear and the flux can be calculated from the slope of this linear profile.

- An important application of slab diffusion is to calculate the transport rate of a gaseous species across a membrane. Note that the solubility of the gas in the membrane is also needed to solve such a problem, because the concentrations given in Section 2.1 are in the membrane phase, rather than in the external fluid phase.

- The concentration profile for diffusion in long cylindrical shell is not a linear function of r, unlike the case in slab geometry. Instead, this profile is a logarithmic function. The difference arises due to the geometry or curvature effect—that is, due to the area of diffusion changing in the direction of diffusion. The curvature effect can be neglected for thin shells and the problem may be approximated as though a slab model applies.

- Likewise, the concentration profile for diffusion across a spherical shell is an inverse function of r due to the curvature effect.
- Diffusion with reaction is a well-studied and important problem in reaction engineering. The governing equation is the diffusion reaction equation given by Equation 2.9 for a slab, Equation 2.30 for a cylinder, and Equation 2.36 for a sphere. The equations can be compacted using vector notation and with the Laplacian operator of the concentration. The vector form applies to three-dimensional problems in any coordinate system.
- The solution of the equation for diffusion with reaction for a zero-order reaction is the quadratic function for a slab. It can be shown to be a combination of a quadratic function and a logarithmic function in cylindrical coordinates (with r-variation only), but is a combination of r and $1/r$ in spherical coordinates.
- Diffusion with a first-order reaction is amenable to analytical solution for the three common geometries considered in this chapter. Simple solutions in terms of exponential or hyperbolic functions are obtained for a slab. By comparison, the solutions in cylindrical coordinates are stated in terms of modified Bessel functions, while those for spherical coordinates are stated in terms of spherical Bessel functions.
- Transient diffusion is governed by a PDE (partial differential equation). An initial condition is also required. Results are usually shown using a dimensionless time. The definitions of time constant and dimensionless time are worth remembering.
- Convection imposed in the same direction as diffusion is one of the simplest problems in convective mass transfer. The effect of convection can be represented by a correction factor (blowing factor), which depends on a dimensionless number called the Peclet number.

Review Questions

2.1 Equation 2.6 shows that the concentration profile for diffusion in a slab is linear. State the assumptions that have gone into this result.

2.2 When can the concentration profile for diffusion in a slab geometry be nonlinear?

2.3 State the differential equation for concentration in a slab geometry for diffusion with a second-order reaction.

2.4 State the two commonly observed boundary conditions in diffusion-reaction problems. State which condition applies when.

2.5 What is the definition of the Thiele modulus for a first-order reaction and what is its significance?

2.6 How is the Thiele modulus defined for a zero-order reaction?

2.7 What is the definition of diffusion time and what is its significance?

2.8 What is the definition of the Peclet number and what is its significance?

2.9 Define blowing factor and state its use.

2.10 A catalyst is cast in the shape of a long concentric cylinder. A first-order reaction is taking place here. What is the general solution for the concentration profile?

2.11 State the solution to diffusion with a first-order reaction in a spherical catalyst particle.

Problems

2.1 **Membrane for oxygen separation.** Data for gas permeation in polymers have been reported by Seader et al. (2010). Table 2.1 shows these data for low-density polyethylene. Note that both diffusivity and solubility are needed to calculate the rate of diffusion.

   The solubility parameter H here is defined as the concentration in the solid (membrane) divided by the partial pressure of the gas at equilibrium. The concentration in the solid is defined as m^3 gas in STP perm^3 solid (q_A) rather than in mol/m^3 Thus the Henry's law constant is defined as $q_A = H_A p_A$.

   Find the rate of oxygen and nitrogen transport rate (in mol/m^2 s) for the previously described membrane when it is exposed to air with 150 psia and 78 °F on one side and atmospheric pressure on the other side. If a rate ratio of at least 5 is needed for good oxygen separation, is this a good membrane to enrich oxygen from air? The membrane thickness is 5 μm.

2.2 **Slab diffusion with diffusivity varying with concentration.** A membrane is set up such that the diffusion coefficient varies as a function of concentration according to the relation

$$D = D_0(1 + \alpha C_A)$$

The membrane is of thickness L with the concentrations of C_{A0} and C_{AL} as the end point values. Find an expression for the concentration profile and the flux across the system. Compare this expression with the flux value using the

Table 2.1 Diffusivity and Solubility Data for Gases in Low-Density Polyethylene

Gases	H_2	O_2	N_2	CH_4
$D \times 10^{10}$ m^2/s	0.474	0.46	0.32	0.193
$H \times 10^6$m^3 STP/m$^3 \cdot$ Pa	1.58	0.472	0.228	1.13

flux based on a constant value of the diffusion coefficient based on the average concentration—that is, the diffusivity based on $(C_{A0} + C_{AL})/2$.

2.3   **Hydrogen leakage from a steel tank.** A spherical steel tank of 1 L capacity and a 2-mm wall thickness is used to store hydrogen at 673 K. The initial pressure in the tank is 5 bar, and the hydrogen partial pressure outside the tank is zero. Hydrogen has finite solubility in steel and leaks out of the tank. Find the rate of leakage at the start of the process where the pressure in the tank is 5 bar.

The solubility and diffusivity of hydrogen are functions of temperature and given by the following relations:

$$C_A = 3.74 \times 10^3 \exp(-3950/T)\sqrt{p_A}$$

where C_A is the concentration of dissolved hydrogen, mol/m^3, in equilibrium with a pressure of p_A, with pressure unit in bars.

$$D = 1.65 \times 10^{-6} \exp(-4630/T)$$

with T in K and D in m^2/s.

Also calculate the leakage rate if the tank pressure is 10 bar.

Note that the solubility relation is nonlinear. This is often observed for gases such as hydrogen, which dissociates at the metal surface and is present as H-atoms in the solid matrix rather than as H_2 molecules in the gas phase. Hence doubling the pressure will not double the transport rate.

2.4   **Mean area for diffusion in a cylindrical shell.** Consider the diffusion across a shell of long cylinder of inner radius R_i and outer radius R_0. Often the results are expressed in the form similar to that for a slab:

$$\dot{M}_A = A_m D_A \frac{C_{Ai} - C_{Ao}}{t}$$

where t is the thickness, which is equal to $R_0 - R_i$

The area A_m is some representative average area for mass transfer. Verify that the area should be log mean of the inside area $2\pi R_i L$ and the outer surface area $2\pi R_o L$.

Also show that if the membrane thickness is small, either of the areas (inner or outer) can be used as an engineering approximation. The slab model is sufficient for such cases and the actual logarithmic profile is close to a linear profile.

2.5   **Mean area for diffusion in a spherical shell.** Repeat problem 4 for diffusion across a spherical shell. Show that the area should be the geometric mean of the inside area $4\pi R_i^2$ and the outer surface area $4\pi R_o^2$.

2.6   **General convection-diffusion equation.** Write out in detail the convection-diffusion equation for a 3-D problem in Cartesian coordinates. Show that Equation 2.19 will be obtained after making suitable simplifications.

Verify the general solution given by Equation 2.20 for the 1-D case shown in the text. Find the solution for a unit concentration at $x = 0$ and zero concentration at $x = L$.

2.7 **Convection augmentation of mass transfer in a slab.** Plot the augmentation factor due to convection (the blowing parameter) as a function of the Peclet number. Find the limiting values of this factor for large values of Pe. Plot the factor if the velocity is opposite to the direction of diffusion. What would be flux under large velocity values?

2.8 **Convection effects in a cylinder.** Consider a porous membrane cast as the shell of a long cylinder of inner radius R_i and outer radius R_0. Concentrations are maintained at different values C_{Ai} and C_{Ao} at each of these surfaces, and diffusion is taking place. What is the number of moles transported across the system? This provides the base value in the absence of convection.

Now to increase the transport rate, a gas at a flow rate of Q is forced across the system in the radial direction, and convection is also taking place in addition to diffusion. Set up the model for this system and derive the governing differential equation. Solve the equation and find the augmentation in the rate of mass transfer due to the flow.

2.9 **Effectiveness factor of a catalyst.** A catalyst is cast in the form a thin slab of thickness $2L$. Species A diffuses and undergoes a first-order reaction. Set up a 1-D differential model. Show that the solution can be expressed in dimensionless form using a Thiele parameter, ϕ, defined as

$$\phi = L\sqrt{k_1/D_{eA}}$$

Verify the following solution:

$$c_A = \frac{\cosh(\phi\,\xi)}{\cosh(\phi)}$$

where c_A is C_A/C_{As} and $\xi = x/L$.

The concentration variation along position causes the rate of reaction to vary along with the position. For design purposes, an average rate of reaction, $-\bar{R}_A$ is often interest. This is defined as

$$-\bar{R}_A = \frac{1}{L}\int_0^L k_1 C_A dx$$

Derive an expression for this concentration variation.

The average reaction rate is often scaled by the maximum reaction rate and the ratio is called the effectiveness factor of the catalyst. It is defined as

$$\eta = \frac{\bar{R}_A}{-k_1 C_{As}}$$

Show that the following expression can be derived for the effectiveness factor:

$$\eta = \frac{\tanh\phi}{\phi}$$

Show that this simplifies to $\eta = 1/\phi$ for $\phi > 3$.

Calculate the effectiveness factor for a catalyst slab with thickness $L = 3$ mm having an effective diffusion coefficient of 2×10^{-5} m^2/s. The rate constant has a value of 0.2 s^{-1}.

2.10   **Concentration profiles in a cylindrical catalyst.** A porous catalyst is used for CO oxidation and the process is modeled as a first-order reaction with a rate constant of 0.2 s^{-1}. The effective diffusion coefficient for pore diffusion was estimated as 4×10^{-6} m^2/s. The catalyst is exposed to a bulk gas at 600 K and 1 bar pressure with a CO mole fraction of 2%. External mass transfer coefficient is assumed to be not rate limiting, so a type I condition can be used. The catalyst is in the form of a long cylinder of radius 5 mm.

What is the solution for the concentration profile? Write the solution for concentration in terms of a Thiele parameter defined as $R\sqrt{k_1/D_{eA}}$. Plot the CO concentration within the catalyst as a function of radial position. Find the concentration value at the center of the catalyst.

2.11   **Zero-order reaction in catalyst of three shapes.** Derive an expression for the concentration profile for a zero-order reaction taking place in (a) a rectangular slab catalyst, (b) a long cylinder, and (b) a sphere.

2.12   **Transport of oxygen in a tissue.** Krogh (1919), in a Nobel Prize–winning paper, modeled oxygen transport in tissues by assuming the tissue to be an annular cylinder where a blood vessel of radius R_i is surrounded by a tissue region with an outer radius of R_o.

Oxygen diffuses into the tissues and undergoes a zero-order reaction. What is the governing differential equation? State the boundary conditions under the following assumptions:

$r = R_i$, the inner radius, is in contact with blood and the oxygen concentration at this point is fixed as the concentration in the blood.

Oxygen does not diffuse past $r = R_o$, the outer radius of the tissue. Solve the concentration profile in the tissue. If the oxygen concentration at the outer edge of tissue is nearly zero, what is the concentration in the blood?

2.13   **First-order reaction in a spherical catalyst.** A spherical catalyst is 6 mm in diameter and is maintained at a concentration of $20,000$ g $\cdot$ mol/m^3 at the surface. The effective diffusion coefficient is 0.02 cm^2/s and the rate constant is 0.2 s^{-1}. What is the center concentration?

An average rate of reaction can be defined as

$$\langle R_A \rangle = \frac{1}{V_p} \int_0^R 4\pi r^2 k_1 C_A dr$$

where V_p is the volume of the sphere equal to $4\pi R^3/3$. What is the average rate of reaction?

The ratio of the average rate of reaction to the value based on the surface concentration is called the effectiveness factor of the catalyst. Derive an expression for the effectiveness factor. Calculate the numerical value for the given data.

2.14 **Effect of particle size.** In problem 13, if the catalyst size is reduced by half what would be the effectiveness factor? If an effectiveness factor of 0.9 is needed, what should be the diameter of the catalyst?

2.15 **Drug release from a capsule.** A drug capsule is a 0.6-cm-diameter sphere and has an active component that diffuses out the capsule. If the diffusion coefficient is 3×10^{-6} cm^2/s, estimate the approximate time for which the active component will be effective.

CHAPTER 3

Examples of Macroscopic Models

Learning Objectives

After completing this chapter, you will be able to:

- Apply the conservation law to a larger control volume (macro-models).
- Interpret the resulting mathematical formulation of the conservation law as average values of local properties.
- Identify additional closures or assumptions needed to complete the model.
- Simulate batch reactors and batch reactors with simultaneous product removal using MATLAB tools.
- Understand what is meant by a backmixed system and how it is used to model continuous flow-stirred tank reactors.
- Model some transport problems in which the transport coefficient concept is used to close the conservation laws.
- Apply the macroscopic model to a simple separation process and explore the concept of an ideal stage contactor.
- Define the definition of stage efficiency and explain its use in separation process modeling.

Differential models provide pointwise or detailed information on the concentration distribution as well as the local values of the flux vector. Unfortunately, they are often difficult to solve, especially when there are two or more phases in the system. Hence macroscopic and mesoscopic models are commonly used in engineering practice and design applications.

This chapter introduces the methodology for modeling at the macroscopic level. In particular, these models need a number of assumptions or submodels to close the system. This is in line with the information loss principle, as these models reflect averages of the local values. Since local values are not available, however, we need to close these models. The commonly used closure assumptions are introduced in this chapter through a number of specific examples. By reviewing these examples, you will learn a number of basic concepts that go into building these models and be able to set up macro-scale models for any given process.

The first example is the modeling of a batch reactor. In this example, we demonstrate the conservation principle together with the rate law for chemical reactions to derive the differential equations needed to simulate the batch reactor. The computations are usually done by numerical tools for multiple reactions; we use a MATLAB-based tool for this purpose. In a variation of the batch reactor, the product is removed by transfer to a second phase. The model for this process requires the use of the mass transfer coefficient to evaluate the rate of transfer of the product to the second phase. Hence the batch reactor model must be coupled to a mass transport model; this coupling is illustrated in this chapter.

The next two examples (sublimation and oxygen absorption) illustrate conservation law coupled with the use of the mass transfer coefficient. We then look at continuous reactor modeling in homogeneous systems. First we introduce the concept of the backmixed reactor. In this analysis, the reactor is assumed to be well mixed, which means we can close the rate and thereby derive expression for the exit concentration. For nonlinear reactions, we show that even the well-mixed assumption needs additional closures for the reaction term. It is necessary to distinguish the level of mixing—that is, mixing at a large eddy scale (macromixing) versus mixing at a small eddy scale (micromixing). The difference in reactor performance is indicated (qualitatively) for these two levels of mixing for a second-order reaction.

We next analyze a two-phase system by examining a liquid–liquid extraction. Two macro-compartments, one for each phase, with an interfacial mass transfer are used as the control volumes to perform the analysis of such systems. The concept of ideal stage is often used here, where the assumption that the exit streams area at equilibrium is used to provide a quick estimate of the performance of the contactor. The mass transfer rate between the two phases is not needed in this model. Mass transfer considerations can then be added to develop a second-level model that includes the concept of stage efficiency. The ideal stage contactor and correction of it with application of stage efficiency are widely used in separation process modeling; the introductory treatment in this chapter provides the basic concepts underlying these models.

Through a careful study of these examples, you will be exposed to the main tools used in mass transfer modeling of these systems in a number of diverse fields. Further details and additional topics related to the macroscopic level of modeling are taken up in Chapter 13.

3.1   Macroscopic Balance

In the macroscopic model, the control volume spans any chosen large volume of the system under consideration—most often the whole equipment or the chemical reactor in which a transformation (physical or chemical) is taking place. The control volume and the control surfaces are shown schematically in Figure 3.1.

The control volume is enclosed by a control surface as shown in Figure 3.1. The control surface is usually split into four parts for convenience of model formulation: an inlet, an outlet, a permeable wall region or an interface or catalytic surface, and an impermeable or inert wall or interface from which no transfer is taking place. Mass can cross into or out of the control volume from these surfaces, except from the impermeable or inert part of the surface. In addition, species mass is generated or consumed by chemical reaction. Finally, species mass can accumulate within the control volume as well for non-steady state conditions.

These components are then organized into the conservation law:

In from incoming flow − out from outgoing flow − transferred to the wall/interface + produced by reaction = accumulation

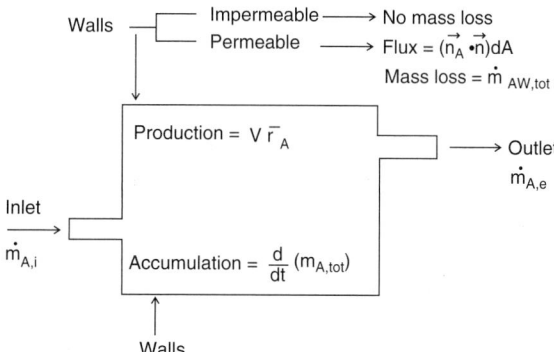

Figure 3.1 Simplified representation commonly used in macroscopic balances. Mass (shown in the figure) or mole units can be used.

Molar units are more convenient for chemically reacting systems since the various species react or form according to their stoichiometric proportions. We use $\mathcal{M}$ to denote moles here and $(\dot{\mathcal{M}})$ to denote the molar flow rates. Hence the macroscopic balance is represented as

$$\boxed{\dot{\mathcal{M}}_{A,i} - \dot{\mathcal{M}}_{A,e} - \dot{\mathcal{M}}_{A,W} + V \langle R_A \rangle = \frac{d}{dt}(\mathcal{M}_A)} \qquad (3.1)$$

which is a general form applicable as a starting point in macroscopic modeling. Equation 3.1 applies for all the species present in the system—*A*, *B*, and so on. Equation 3.1 can also be formulated on a mass basis.

The control volume does not tend to zero, so local information is therefore lost. Each term in Equation 3.1, then, is an integral of local values. This point must be kept in mind in completing the model. Since local information is not available, some approximations must be introduced when formulating each of these terms. Let us now define each term as an integral of the local values to see how the terms can be closed by defining and using additional transport parameters.

3.1.1   In and Out Terms from Flow

In and out terms are normally calculated based on the flow rates at the inlet and outlet. Thus

$$\text{in} = \dot{\mathcal{M}}_{A,i} = Q_i C_{A,i}$$

$$\text{out} = \dot{\mathcal{M}}_{A,e} = Q_e C_{A,e}$$

where Q represents the volumetric flow rate. A similar expression is used at the outlet. In these expressions, *i* represents the inlet and *e* the exit.

Note: This representation assumes a uniform velocity in the inlet and outlet pipes. Otherwise, the concentration values, $C_{A,i}$ and $C_{A,e}$, should be interpreted as the cup mixing (flow weighted) average values.

The volumetric flow rates at the inlet and exit can be calculated from the total molar flow rate $\dot{M}$ and the total molar concentration, C. The relation is

$$Q = \dot{\mathcal{M}}/C$$

where $\dot{\mathcal{M}}$ is the total molar flow rate (at inlet or outlet). The total molar concentration C is an intensive mixture property that can be calculated using some appropriate thermodynamic relation and depends on the mixture composition. Note that the volumetric flow rate at the inlet may be different from that at the exit, particularly for gas-phase systems with reactions that lead to appreciable mole change or for non-isothermal conditions. For liquid-phase systems, the volumetric flow rate at the inlet and outlet are usually assumed to be equal.

3.1.2   Wall or Interface Transfer Term

The local rate of mass transfer to the wall is related to the local flux vector. Thus, if n is a unit normal in the outward direction from the wall at a specified point, then the quantity transported is $[N_A \cdot n]dA$. The integral of this over the permeable wall area represents the quantity transported to the wall:

$$\dot{M}_{A,W} = \int_{A_w} (N_A \cdot n)dA$$

This information is based on the differential modeling concept, which needs the local or pointwise value of the flux vector and is not available in the macroscopic model. It is necessary to model this flux by using a mass transfer coefficient. Usually an overall mass transfer coefficient designated as $\bar{K}_m$ is used and the flux to the wall is modeled as

$$\dot{M}_{A,W} = \bar{K}_m A_w (H_A < C_A > - < C_A >_{ext}) \tag{3.2}$$

The driving force for mass transfer is based on some average concentration difference. Here $< C_A >$ is the average concentration in the control volume. It is corrected for the thermodynamic jump by multiplying by the Henry's law parameter H_A. Note that we define $H_A C_A(\text{phase 1}) = C_A(\text{phase 2})$ as the thermodynamic relation for A between the two phases. The quantity $< C_A >_{ext}$ is some average concentration in the external phase. Thus $H_A < C_A > - < C_A >_{ext}$ is taken as an effective driving force for mass transfer.

An estimate of the average concentration in the control volume and in the external fluid is also needed, which will require some assumptions about the mixing pattern in the control volume or the tank as well as in the external fluid. In addition, the mass transfer coefficient $\bar{K}_m$ is an average of the local values over the wall area. Local values may vary from point to point but an average value is used in the macroscopic model. $\bar{K}_m$ is also an overall mass transfer coefficient from phase 1 to phase 2 (a combination of individual values)—a concept that is explained in detail in Chapter 6. Thus this model contains quite a few unknowns that we must close by making further approximations. These are better explained on a case-by-case basis. Thus, at this stage, we simply identify the closure problems associated with the mass transfer rate in the context of macroscopic models.

3.1.3   Rate Term

Mathematically, the average rate can be defined by the following equation:

$$< R_A >= \frac{1}{V} \int_V R_A \,(\text{local}) \, dV$$

where R_A is the local rate (in molar units) at any point in the system corresponding to the local concentration, and V is the volume of the system. In the macroscopic analysis, there is no way to calculate this rate, since the local concentration values are not being computed. The information loss principle dictates that some assumption on the level of mixing is needed to approximate the average rate term. One assumption is that the reactor is well mixed so that the concentration in the reactor is the same at every point. This so-called back-mixed model assumption is discussed in Section 3.6. Other assumptions are discussed in Section 3.7.

3.1.4   Accumulation Term

The term on the RHS of Equation 3.1 is the accumulation term—that is, the time rate of change of total moles of A in the system. The total moles of A is given as the integral of the local concentration of A at any point:

$$\mathcal{M}_A = \int_V C_A \, dV = V \, \langle C_A \rangle$$

Again, some assumption about the mixing level is needed to assign the value for the average concentration.

    Application of macro-models is now illustrated with several examples in which we indicate how the models could be closed by making suitable assumptions and using some additional model parameters.

3.2   The Batch Reactor

The first and perhaps simplest example of a macroscopic model is a batch reactor. In a typical batch reactor, the reactants are charged initially and a specified amount of time, called the batch time, is set to allow the reaction to proceed. Batch reactors are widely used in laboratory evaluation of processes, pilot-scale testing, and even commercial production of specialty or fine chemicals produced at a small scale (usually 500 Mt/year). For larger scales of production, continuous reactors are more economical.

3.2.1   Differential Equations for the Reactor

The mathematical model for a batch reactor is simple because the *in* and *out* terms are zero. The reactor is assumed to be well mixed so that there is no concentration variation locally with position in the entire control volume. Hence the macroscopic balance for the control volume (usually the liquid in the tank or reactor) reads as follows:

$$\text{Accumulation} = \text{Net produced by reaction}$$

In this setting, the mathematical statement given by Equation 3.1 simplifies to

$$\frac{d}{dt}(\mathcal{M}_A) = V R_A \qquad (3.3)$$

where the <> notation for the rate term is suppressed because the reactor is assumed to be well mixed and is at a uniform concentration.

For a batch case, $\mathcal{M}_A = C_A V$ and hence

$$\boxed{\frac{d}{dt}(C_A V) = V R_A} \qquad (3.4)$$

Rate and Rate Function

The rate of production is a function of the concentrations and can be represented as

$$R_A = \nu_A \mathcal{R}(C_A, C_B, \cdots, T) \qquad (3.5)$$

for a single reaction of the type

$$\nu_A A + \nu_B B + \cdots = 0$$

where $\nu_A, \nu_B, \ldots$ are the stoichiometric coefficients of A in the reaction. These coefficients are assigned positive values for products and negative values for reactants.

The parameter $\mathcal{R}$ is called the rate function or the intrinsic kinetic model for the reaction or the species-independent rate of reaction. Note that the rate for any species, i, is simply given as

$$R_i = \nu_i \mathcal{R}$$

If the expression for the rate function is known, along with how the volume changes in the reactor as a function of time due to the change in the composition in the batch reactor, then Equation 3.4 can be integrated (usually numerically) to simulate the batch reactor.

For a first-order reaction, the rate function is represented as $k_1 C_A$. The k_1 value is a first-order rate constant that is a function of temperature:

$$\mathcal{R} \text{ (first order)} = k_1 C_A = A_f \exp(-E/R_g T) \, C_A$$

The Arrhenius law is used for the temperature dependency, where E is the activation energy and A_f is called the frequency factor. If the batch reactor is not isothermal, a heat balance equation for the temperature profile is needed in addition to the species mass balance equations. These equations are then solved simultaneously for C_A, C_B, ..., and T.

Rate for Multiple Reactions

If there are two reactions, we can define a rate function for each reaction—say, $\mathcal{R}_1$ and $\mathcal{R}_2$. These are sufficient to define the rate of production of any species i:

$$R_i = \nu_{1i} \mathcal{R}_1 + \nu_{2i} \mathcal{R}_2$$

or in general

$$R_i = \sum_{k}^{nr} \nu_{ki} \mathcal{R}_k$$

where ν_{ki} is the stoichiometric coefficient of species i in the kth reaction. Here nr is the total number of independent reactions.

Constant Volume Analysis

For liquid systems, the density does not change significantly during the batch. Hence a constant volume is assumed and the model is simplified even further:

$$\frac{dC_A}{dt} = R_A \text{ with } A, B, C, \ldots \tag{3.6}$$

For simple linear cases, the equations can be integrated analytically. Consider a single species A participating in a reaction represented as follows:

$$A \rightarrow B \text{ with } R_A = -k_1 C_A$$

The integrated form is

$$\frac{C_A}{C_{Ai}} = \exp(-k_1 t)$$

where C_{Ai} is the initial concentration of A in the reactor. The differential balance equation for B can be written and solved as well to obtain the product concentration as a function of time. However, an overall mass balance gives this concentration directly:

$$C_B = C_{Bi} + (C_{Ai} - C_A)$$

More generally, the batch reactor is solved using numerical methods. We now look at one such implementation in MATLAB.

3.2.2   ODE45 with CHEBFUN

The Runge-Kutta fourth- to fifth-order method is the workhorse for the numerical integration of a set of first-order differential equations. The MAT-LAB code ODE45 is an implementation of this method and is useful for this purpose. A sample driver is shown in Listing 3.1. The detailed mathematics underlying the Runge-Kutta method is not shown here, however; other books should be consulted for this explanation. The purpose here is to provide illustrative code that you can modify and use for many applications.

Listing 3.1   MATLAB Code for Solution of an Initial Value Problem

```
function odemain
global param1 param2 % parameters  to be given to function dydt.
param1 = 2.0; param2=0.0;;
 y0 =  [1]  % vector of initial values
tspan = linspace (0, 2, 11) ; % time intervals of solution
[ t y] = ode45 (@fun1, tspan, y0)  % calling routine
plot (t,y) % solutions; A plot is generated

% function subroutine follows:
% This could be output into a separate file.
function dydt = fun1 ( t, y)
global param1 param2
dydt = —param1*y; % simple first—order equation
% Add dydy(2), etc., for additional equations.
```

CHEBFUN Wrapping

This section may be omitted without loss of continuity.

An alternative implementation of ODE45 uses CHEBFUN, which provides a symbolic "feel" for the results with the efficiency of numerical computations. CHEBFUN helps to make the MATLAB codes easier and works with solutions as though they were analytic functions. The final results can be manipulated as though they were ordinary functions rather than numerical data. You will find it very useful once you start using these results.

CHEBFUN is a polynomial type of fitting of any given data or function using Chebecheff polynomials. Once fitted, these functions have the feel of a mathematical functions rather than a discrete set of data points. For more details, consult the papers by Battles and Trefethen (2004) and Trefethen (2007).

To get started, you need to type the following statement into the MATLAB command window:

```
unzip('http://www.chebfun.org/download/chebfun_v4.2.2889.zip')
cd chebfun_v4.2.2889, addpath(fullfile(cd,'chebfun')), savepath
```

This code installs the CHEBFUN directory and adds the required path to the file. At this point, you will have access to a number of new functions that make the MATLAB coding easier. The code will look much more compact with this functionality added to your MATLAB program. Some sample codes provided in some exercise problems in this book use these compact functions.

The use of ODE45 with CHEBFUN is straightforward. The initial conditions are specified by the *domain* statement, which overloads the ODE45 function and creates a pseudo-analytical representation of the results rather than a set of tabular values. Listing 3.2 provides example code for a batch reactor in which the following reaction is taking place:

$$A \to B \to C$$

Here A forms B but B reacts further to form C. If C is an unwanted or waste product, we need to optimize the production of B. Too large a batch time is not desirable, while a too small value produces very little B, which is likewise not desirable. Hence the optimal batch time must be calculated. Additional details are provided in the following example.

Listing 3.2   MATLAB Code ODE45 with CHEBFUN Wrapping

```
% Consecutive reaction using ODE45 and CHEBFUN
%    A ——> B ——> C; first—order reactions are used
% for illustration but the code can handle any nonlinear kinetics.
% Parameters used.
   k1= 1.0; k2=2.0; tmax = 3.0;
   C_init= [ 1 0 ] % vector of initial conditions
  % Define functions to be solved.
fun = @(t,C) [—k1*C(1); k1*C(1)—k2*C(2)];
C = ode45(fun,domain(0,tmax), C_init ); % solver
 plot(C) % generates the plot
```

Example of a Series Reaction Simulation

In the code presented in this example, we simulate an isothermal system with a first-order reaction. Hence $R_A = -k_1 C_A$ and $R_B = -k_2 C_B + k_1 C_A$ and the model equations to be solved are

$$\frac{dC_A}{dt} = -k_1 C_A$$

and

$$\frac{dC_B}{dt} = -k_2 C_B + k_1 C_A$$

with initial conditions of $C_A = C_{Ai}$ and $C_B = 0$. Analytical solutions can obtained and compared with the numerical solution. In the code the concentrations are scaled by C_{Ai}; hence the initial conditions are taken as 1 and 0 for A and B, respectively.

An illustrative result is shown in Figure 3.2. It should be compared to the analytical solution to the model equations.

Additional postprocessing can be readily accomplished with CHEB-FUN. For example, the concentration of B first increases with time and then decreases. The value of time at which the maximum concentration occurs is of importance, and here we show how to extract this value by postprocessing the concentration–time data. The lines of code shown in Listing 3.3 will do the postprocessing since CHEBFUB provides an analytic representation of the results.

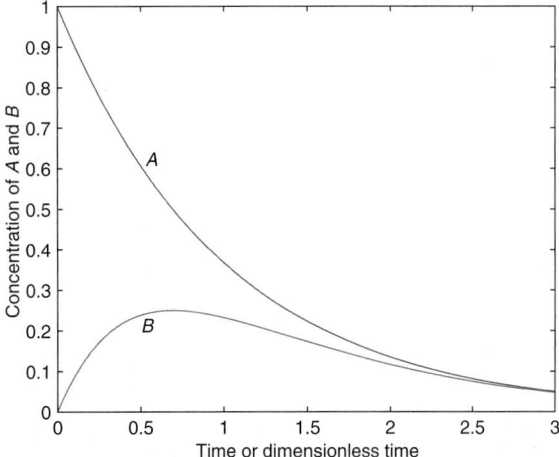

Figure 3.2   Concentration profiles for species A and B in a batch reactor simulated in MATLAB for a first-order reaction.

Listing 3.3 Example of Postprocessing of CHEBFUN Results

```
c_max = max ( C(:,2) ) % finds what the maximum value is
F = diff( C(:,2)) % takes the derivative of variable two (species B)
roots(F)  % finds where the maximum occurs
```

For the given example, the following analytical results can be calculated:

$$t_{opt} = \frac{\ln(k_2/k_1)}{k_2 - k_1} = 0.6931 \tag{3.7}$$

and

$$\frac{C_{B,max}}{C_{Ai}} = (k_1/k_2)^{[k_2/(k_2-k_1)]} = 0.25 \tag{3.8}$$

The simulations from the MATLAB-CHEBFUN program agree with the results of the analytical solution in an exact manner, thereby validating the code. Once the code has been validated, you can use it for more complex kinetics where the analytical solution is not generally available.

3.3   Reactor–Separator Combination

In this section, we consider a variation of the batch reactor, in which one of the products is volatile and is stripped to the gas phase by introduction of a purge gas. A schematic of this system is shown in Figure 3.3. Arrangements such as this are known as a reactor–separator combo (combination). In this example, we show that we can accomplish conversions larger than the equilibrium value because the product is stripped out of the reaction medium. This concept has many applications in reaction engineering. For example, a membrane reactor can be used to remove the product (e.g., hydrogen in steam reforming of methane), thereby improving the equilibrium conversion (see Section 27.4 for more details).

Consider a reversible reaction represented by the following expression:

$$A \rightleftharpoons B + C$$

Let the rate be given by elementary kinetics:

$$R_A = -(k_f C_A - k_b C_B C_C)$$

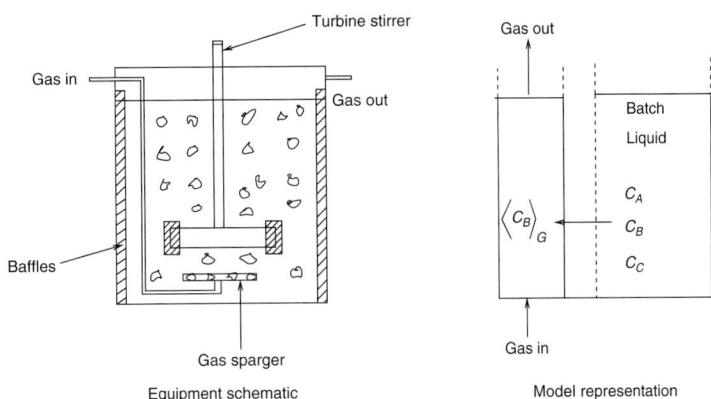

Equipment schematic   Model representation

Figure 3.3   Schematic of a batch reactor with gas purging for simultaneous product removal.

where k_f is the rate constant for the forward reaction and k_b is the rate constant for the backward reaction. The two are related by the law of mass action, and k_f/k_b is equal to K_{eq}, the equilibrium constant for the reaction.

We assume A and C are nonvolatile and stay in the liquid phase. The differential equations for these processes are therefore

$$\frac{dC_A}{dt} = -(k_f C_A - k_b C_B C_C)$$

and

$$\frac{dC_C}{dt} = +(k_f C_A - k_b C_B C_C)$$

Now assume that B is a volatile product and is removed from the liquid by introducing some gas phase in the system. In this scenario, the mass balance for B should include a term to account for the transfer to gas phase. Species balance can be expressed in words as

Accumulation = In − out + generation − transferred to gas phase

with the *in* and *out* terms being zero for a batch reactor.

Putting this relationship in mathematical terms we have

$$\frac{d}{dt}(V C_B) = +V(k_f C_A - k_b C_B C_C) - \dot{M}_{B,lg} \tag{3.9}$$

where the last term is the transfer rate of moles of B from the liquid phase to the gas phase. The volume of the system is assumed to remain nearly constant—a reasonable assumption for liquid-phase systems or for reactions carried out with a solvent. Hence

$$\frac{dC_B}{dt} = (k_f C_A - k_b C_B C_C) - \frac{1}{V}\dot{\mathcal{M}}_{B,lg} \tag{3.10}$$

A model for the for mass transfer from the liquid phase to the gas phase is now needed to close the transfer term. The concept of mass transfer coefficient is used here:

$$\dot{M}_{B,lg} = \bar{K}_m A_{lg}(H_B C_B - \langle C_B \rangle_g) \tag{3.11}$$

where H_B is the Henry's law constant for B. $\bar{K}_m$ is an average overall mass transfer coefficient from the liquid phase to the gas phase, and A_{lg} is the total gas–liquid interfacial area for mass transfer. $< C_B >_g$ is the average concentration in the gas-phase. Modeling this system will require a gas-phase mass balance and some assumption on the extent of mixing in the gas phase. The gas phase balance in words is

In − Out + transferred to gas phase from liquid = Accumulation in the gas phase

The *in* term is zero if an inert gas is being used as purge gas. A further assumption is to neglect the accumulation term, decision called a pseudo-steady state approximation. It is usually justified since the gas volume in the system is smaller than the liquid volume.

The simplest model is to assume that the gas phase is mixed and that the representative concentration in the gas phase is the same as the exit gas concentration. Hence the moles out is $Q_{G,e} < C_B >_g$. The gas-phase balance with this assumption of backmixing is

$$-Q_{G,e} < C_B >_g + \bar{K}_m A_{lg}(H_B C_B - < C_B >_g) = 0 \tag{3.12}$$

where a pseudo-steady state assumption is also used. This means that the gas phase concentration changes slowly over time and the accumulation term in the gas phase is neglected. The $< C_B >_g$ represents the value of the average concentration of B in the gas phase. The average value is also the exit concentration, if the gas is assumed to well mixed.

Rearranging Equation 3.12 leads to the following expression:

$$< C_B >_g = \frac{\bar{K}_m A_{lg} H_B}{\bar{K}_m A_{lg} + Q_{G,e}} \cdot C_B \tag{3.13}$$

When this definition is used in Equation 3.11, the mass transfer rate is modeled (after minor algebra) as

$$\dot{M}_{B,lg} = Q_{G,e} H_B C_B \left[\frac{\bar{K}_m A_{lg}}{\bar{K}_m A_{lg} + Q_{G,e}} \right] \tag{3.14}$$

The differential equation for the concentration of B (Equation 3.8) can be completed by substitution of this term. The three equations (for A, B, and C) can then be solved simultaneously by using the numerical solver to simulate the concentration profiles as a function of time.

Note that Equation 3.14 reduces to $Q_{G,e} H_B C_B$ when the mass transfer coefficient is relatively large. You may want to think about a physical interpretation of this observation.

Another point to note is that $< C_B >_g$ is a function of time, although Equation 3.13 does reflect this relationship. C_B is a function of time as per Equation 3.10 and hence $< C_B >_g$ is indirectly a function of time.

Example 3.1   Batch Reactor with Product Removal

Consider a constant volume batch reactor with a volume of reactor of 0.0982 m^3, charged with 2000 mol/m^3 of A, which reacts to form B and C. The reaction is reversible with the rate constants $k_f = 1$ hr^{-1} and $k_b = 5 \times 10^{-4}$ m^3/mol s at the operating temperature. Simulate this with ODE45. What is the final product composition?

Now consider that B is being removed by purging with an inert gas. The gas flow rate is $Q_G = 35$ m^3/hour.

The Henry's law coefficient for product B is $H_B = 2.2 \times 10^4$ expressed as a concentration ratio.

Also assume that the overall mass transfer coefficient as infinity. This provides an upper bound and gives an equilibrium model, in which the exit gas is assumed to be in equilibrium with the liquid.

Simulate the reactor for the conditions stated and show that all A can be converted under these conditions.

Solution

Let ζ moles of A be reacted at final condition. Then $C_A = 2000 - \zeta$, $C_B = \zeta$, and $C_C = \zeta$.

The law of mass action applies at equilibrium:

$$\frac{C_C C_B}{C_A} = \frac{\zeta \, \zeta}{2000 - \zeta} = K_{eq} = \frac{k_f}{k_b} = 1/5 \times 10^{-4} = 2000 \text{ mol/m}^3$$

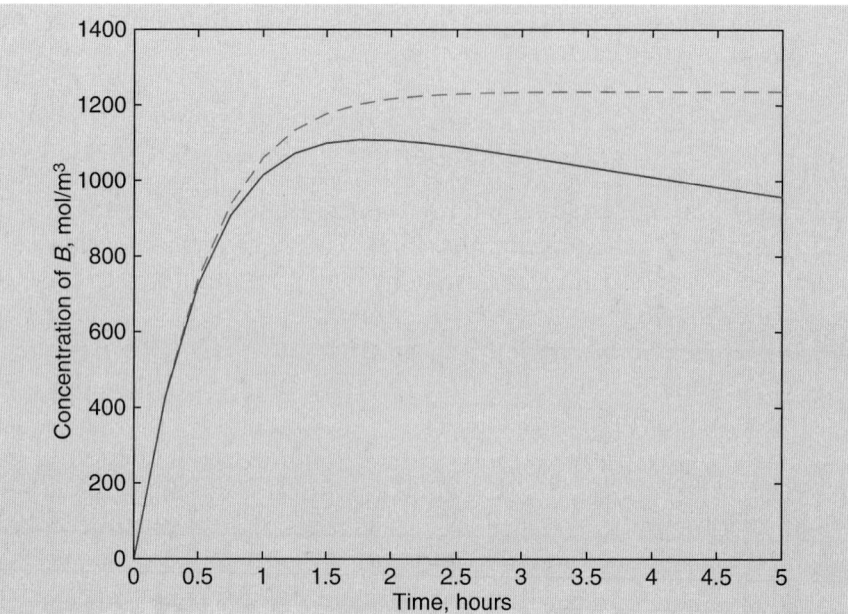

Figure 3.4   Concentration profile for product B in the batch liquid with (solid line) and without product (dotted line) removal.

Solving this expression as a quadratic equation, $\zeta = 1236 \text{ mol/m}^3$. The final composition should be then $C_A = 764$, $C_B = 1236$, and $C_C = 1236$, all in mol/m^3.

The result of the MATLAB simulation is shown as a dotted line in Figure 3.4. The concentration of B attains a plateau, which is the equilibrium value of 1236 mol/m^3.

Now consider the effect of a gas purge that strips B from the liquid. The differential equation for species B is now modified by subtracting the $\dot{M}_{B,lg}$ term. The results are shown in Figure 3.4. The concentration of B does not reach an asymptotic equilibrium value, but rather keeps decreasing with time as it is being transferred to the gas phase. Correspondingly, the conversion of A exceeds the equilibrium value.

The key points to notice when modeling this problem are that a macroscopic balance for the gas phase is used in conjunction with a macroscopic balance for the batch liquid. In addition, some average concentration in the gas phase needs to be assigned and this assignment will depend on the mixing pattern of the gas phase. The exit concentration was assigned in the example, which is appropriate if the gas phase is assumed to be backmixed. If the gas were in plug flow, however, the logarithmic average of the inlet and the outlet concentration would be used, as discussed in the context of the mesoscopic model in Chapter 4.

3.4   Sublimation of a Spherical Particle

Let us now consider a different problem, in which we once again demonstrate the use of the mass transfer coefficient. The problem deals with a spherical particle that has appreciable vapor pressure and sublimes into the gas phase. The initial radius is R_i and the change in radius due to sublimation is to be calculated as a function of time. Figure 3.5 provides a schematic description of this problem. The sphere at the current instant of time is taken as the control volume. Let us use Equation 3.1 to set up the basic model, keeping only the relevant terms. In this scenario, there is no *in* term. Mass is lost from the surface by evaporation, which serves as the *out* term. Hence

$$-\text{Out (from surface)} = \text{Accumulation}$$

The *out* term is represented as $\dot{\mathcal{M}}_A$, the evaporation rate.

Accumulation is the time rate of change of moles in the control volume. Moles equals mass divided by the molecular weight of the solid and is equal to $(4/3)\pi R^3\, \rho_s/M_A$, where R is the particle radius at the current value of time, ρ_s is the mass density of the solid, and M_A is the solid's molecular weight. In turn, accumulation is the time derivative of this term. This accumulation term (which will have a numerically negative value) is balanced by the loss from the solid:

$$\frac{d}{dt}\left(\frac{4}{3}\pi R^3 \cdot \frac{\rho_s}{M_A}\right) = -\dot{\mathcal{M}}_A \tag{3.15}$$

This provides the basic model based on the conservation principle.

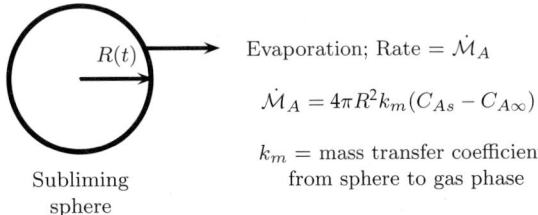

$$\dot{\mathcal{M}}_A = 4\pi R^2 k_m (C_{As} - C_{A\infty})$$

k_m = mass transfer coefficient from sphere to gas phase

Figure 3.5   Schematic of a sublimation of a solid.

The rate of evaporation appears as a term and has to be closed by using some transport law. Here, we define and use the following mass transfer coefficient:

Mass transfer rate from a surface per unit surface area
= Mass transfer coefficient times a driving force

Let k_m be defined as the mass transfer coefficient with units of m/s. The driving force is taken as the concentration (in the gas) near the solid C_{As} minus the concentration in the gas far away from the solid surface, in the bulk gas $C_{A\infty}$:

$$\dot{\mathcal{M}}_A = 4\pi R^2 k_m (C_{As} - C_{A\infty})$$

The value of C_{As} is the concentration on the gas side of the interface. It can be calculated from the vapor pressure data of the subliming solid. The value of $C_{A\infty}$ is usually taken as zero because the concentration of A gets very diluted in the regions away from the solid.

Using the expression for the transfer rate in Equation 3.15 and doing some minor simplification leads to

$$\frac{\rho_s}{M_A}\left(\frac{dR}{dt}\right) = -k_m(C_{As} - C_{A\infty}) \qquad (3.16)$$

This equation can be integrated if the dependency of the mass transfer coefficient on the radius is known. In general, k_m is not a constant, but rather a function of the particle radius R. Commonly empirical correlations are used to connect these quantities. Let us review a commonly used correlation.

3.4.1   Correlation for Mass Transfer Coefficient

In general, the mass transfer coefficient is expected to be a function of the parameters indicated in the following equation:

$$k_m = f(d_p, D_A, v_\infty, \rho_g, \mu_g)$$

Data are usually correlated in terms of dimensionless parameters:

Sherwood number, $Sh = \dfrac{k_m d_p}{D_A}$

Reynolds number, $Re = \dfrac{d_p \rho_g \, v_\infty}{\mu_g}$

Schmidt number, $Sc = \dfrac{\mu_g}{\rho_g D_A} = \dfrac{\nu_g}{D_A}$

A commonly used correlation is that developed by Froessling (1938) and Ranz and Marshall (1952):

$$Sh = 2 + 0.552 Re^{1/2} Sc^{1/3} \tag{3.17}$$

This correlation is valid for Re from 2 to 800 and for Sc from 0.6 to 2.7 (typical range for gases), and can be used to find k_m as a function of particle diameter or radius. For a stagnant gas, $Re = 0$ and therefore $Sh = 2$. Hence the mass transfer coefficient is given as

$$k_m(\text{ sphere to a stagnant gas }) = \frac{2D_A}{d_p} = \frac{D_A}{R} \tag{3.18}$$

Note that k_m is an inverse function of the particle radius. Equation 3.18 is a theoretical result and is derived in Example 6.2.

For a flowing gas, the second term in Equation 3.17 is more important. At high velocity, we can deduce from this equation that $k_m \propto d_p^{-1/2}$ at high flow rates. In general, therefore, we can expect k_m to vary as d_p^{-a}, where a is an exponent in the range of 1/2 to 1. The dependency of the mass transfer coefficient on the radius must be included when integrating Equation 3.16. The following example shows the integrated result for a no-flow condition.

Example 3.2   Rate of Shrinkage of a Subliming Sphere

A naphthalene ball ($M_A = 128$ g/mol and $\rho_s = 1145$ kg/m³) is suspended in **still** air at 347 K and 1 atm pressure. Estimate the time to reduce the diameter from 2 cm to 0.5 cm for the following parameter values: $D_A = 8.19 \times 10^{-6}$ m²/s and the vapor pressure of naphthalene is 666 Pa for the given temperature.

Solution

Equation 3.18 applies to a scenario involving still air, and we can replace the mass transfer coefficient by D_A/R in Equation 3.14. Also, far from the solid, the naphthalene concentration, $C_{A\infty}$, can be assumed to zero. Hence

$$\frac{\rho_s}{M_A} \left(\frac{dR}{dt} \right) = -\frac{D_A}{R} C_{As} \tag{3.19}$$

The variables can be separated:

$$\int_{R_i}^{R_f} R \, dR = \frac{D_A M_A}{\rho_s} C_{As} \int_0^t dt$$

This is followed by integration:

$$R_i^2 - R_f^2 = \left[\frac{D_A M_A}{\rho_s} C_{As}\right] \cdot t$$

For numerical calculations, it is easier to regroup the bracketed term on the LHS as K and write this expression as

$$R_f^2 = R_i^2 - Kt$$

The radius is seen to decrease as the square root of time.

Mathematical description of problems in which the size of a solid decreases due to chemical reaction with a gas has a similar structure and is discussed in Section 19.2. The dependency of the radius on time has a similar form with a square root dependency.

3.5   Dissolved Oxygen Concentration in a Stirred Tank

Now consider another application of the concept of mass transfer coefficient. The problem statement is as follows: Oxygen gas is continuously bubbled into a pool of liquid of volume V_L and the liquid becomes saturated with oxygen with the passage of time. The dissolved oxygen concentration can be measured with an oxygen probe as a function of time. We will develop a model to predict this transient oxygen concentration profile. The time needed for, say, 90% saturation can be calculated by this model. The problem can be approached by a macroscopic balance where we take the total liquid volume as the control volume and write the conservation law.

The input term in this case is the oxygen crossing into the liquid from the gas–liquid interface. Let us denote this term as $\dot{M}_A$ (mol/s).

There is no output term since there is no liquid leaving the system.

There is no chemical reaction, so the generation term is zero.

Finally, the accumulation term is the time derivative of dissolved oxygen in the tank. It is equal to d/dt of $(V_L C_A)$, where C_A is the oxygen concentration in the tank. We assume that the concentration is uniform everywhere in the liquid because the tank is well agitated. Hence there is no need to distinguish between $< C_A >$ and C_A.

For this scenario, the macroscopic conservation statement is

$$\dot{M}_A = \frac{d}{dt}(V_L C_A)$$

The liquid volume in the tank is assumed to remain constant and V_L can be pulled out, leaving us with

$$V_L \frac{dC_A}{dt} = \dot{\mathcal{M}}_A \qquad (3.20)$$

A model (a transport law) is needed to close $\dot{\mathcal{M}}_A$, the rate of gas–liquid mass transfer. Again, this is closed in the context of macroscopic models using a mass transfer coefficient. The driving force is the oxygen saturation solubility, C_A^*, minus the instantaneous oxygen concentration in the liquid, C_A. Hence the molar rate of transfer of oxygen is modeled as

$$\dot{\mathcal{M}}_A = k_L a_{gl}(C_A^* - C_A)V_R \qquad (3.21)$$

where k_L is the local mass transfer coefficient from a bubble surface to a bulk liquid, defined per unit area of the interface. We multiply this coefficient first by a_{gl}, the interfacial area per unit volume of the dispersion (gas + liquid), and then we multiply it by V_R, the total dispersion volume, to get $\dot{\mathcal{M}}_A$ as shown in the previous equation.

The parameter $k_L a_{gl}$ is called the overall or volumetric mass transfer coefficient since it gives the transfer rate per unit volume; in contrast, the parameter k_L is called the intrinsic (liquid side) mass transfer coefficient. The latter is based on the unit area for mass transfer.

Using the expression for $\dot{\mathcal{M}}_A$ from Equation 3.21 in Equation 3.20, we get

$$V_L \frac{dC_A}{dt} = k_L a_{gl} V_R (C_A^* - C_A) \qquad (3.22)$$

The ratio V_L/V_R is generally denoted as liquid holdup ϵ_l. Hence we have

$$\frac{dC_A}{dt} = \frac{k_L a_{gl}}{\epsilon_L}(C_A^* - C_A) \qquad (3.23)$$

The initial condition $C_A = 0$ at time $t = 0$ is used. The integrated solution is

$$\frac{C_A}{C_A^*} = 1 - \exp\left(-\frac{k_L a_{gl}}{\epsilon_L}t\right) \qquad (3.24)$$

The measurement of dissolved oxygen concentration thus provides a useful experimental tool to measure the overall mass transfer coefficient, $k_L a_{gl}$, in the equipment.

A question for discussion is how the model changes if air is bubbled into the liquid, rather than pure oxygen being the gas applied. The value of C_A^* will now depend on the representative oxygen partial pressure in the gas phase. This will, in turn, need a model for the gas phase and some assumption about the mixing pattern in the gas.

3.6   Continuous Stirred Tank Reactor

Section 3.2 considered the simulation of a batch reactor. In this section we model a continuous stirred tank reactor (CSTR). In this system, a reactant stream enters continuously in a tank that is well stirred. A chemical reaction takes place in the tank, and the product stream is withdrawn in a continuous manner. The schematic for this system is shown in Figure 3.6. We construct a macroscopic model and close it by using the backmixing assumption. The well-mixed model assumption is that the concentration in the tank is uniform and equal to the exit concentration; it is used to close the rate term as discussed further in this section.

The conservation law is:

in − out + generation = 0

The various terms can be calculated as follows and put into an algebraic equation:

Moles of A in = $QC_{A,i}$

Moles of A out = $QC_{A,e}$

Hence

$$QC_{A,i} - QC_{A,e} + V < R_A >= 0 \tag{3.25}$$

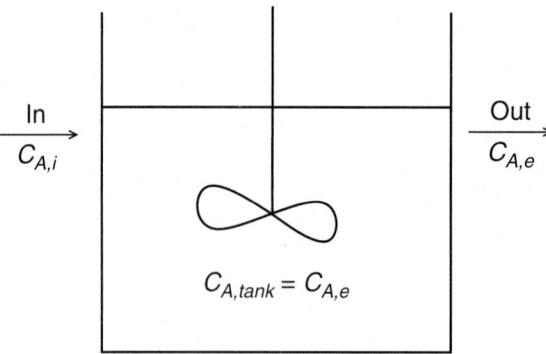

Figure 3.6   Schematic representation of a well-mixed reactor, the continuous stirred tank reactor (CSTR).

We need to close $< R_A >$ by making some assumptions about the level of mixing in the system. In this example, we assume complete backmixing. This assumption implies that the tank is well mixed, and that the concentration is the same everywhere and is equal to the exit concentration.

$$\text{Backmixed tank model: } C_A(\text{tank}) = C_{A,e} \qquad (3.26)$$

This simplification arises because the contents of a well-mixed tank have the same composition, which is also the same as the exit concentration. Model results for first-order and second-order reactions are presented next.

3.6.1   First-Order Reaction

The generation rate due to reaction is calculated as $-VkC_A(tank) = -VkC_{A,e}$. Here k is the rate constant of the first-order reaction.

The generation rate is assigned a negative sign because A is consumed by reaction. Also, the backmixed assumption permits us to write the rate in terms of the exit concentration, which we want to calculate.

Putting all the terms together, we have the following equation for the change in concentration between the inlet and outlet streams in the tank:

$$QC_{A,i} - QC_{A,e} - VkC_{A,e} = 0 \qquad (3.27)$$

This is the model for a backmixed reactor for a first-order reaction; it is widely used in chemical reaction engineering.

It is also convenient to work with dimensionless quantities. Let $c_A = C_A/C_{A,i}$, a measure of concentration. Let $Da = Vk/Q$, a dimensionless rate constant; this is usually referred to as the Damkohler number in reaction engineering parlance. The Damkohler number is the ratio of the holding time, V/Q, or the residence time to the reaction time, $1/k$.

The final solution to Equation 3.27 can be written compactly as follows:

$$c_{A,e} = \frac{C_{A,e}}{C_{A,i}} = \frac{1}{1 + Da} \qquad (3.28)$$

We now mention a complication as a consequence of averaging and due to the different levels of mixing prevalent in flow systems, especially under turbulent flow conditions.

3.6.2　Second-Order Reaction

In this subsection, we look at a second-order reaction to see the additional closures that may be needed. The model equation is similar to Equation 3.25:

$$QC_{A,i} - QC_{A,e} + V < R_A >= 0 \qquad (3.29)$$

The average rate $< R_A >$ is $-k_2 < C_{A,e}^2 >$, the integral of the square of the exit concentration. Although the concentration in the reactor is assumed to be the same as the exit concentration, locally there may be a variance in the concentration. Hence $< C_{A,e}^2 >$, the integral of the square, need not be the same as $(< C_{A,e} >)^2$, the square of the integral. The two are equal if the concentration at each and every point is the same as $C_{A,e}$ and, in addition, if there are no local fluctuations in the concentration. The backmixing assumption guarantees that the concentration, on average, at every point in the reactor is the same and also equal to the exit concentration. Nevertheless, that does not imply that there is no deviation from the average at every point. If there is no deviation (variance is zero) as well, the assumption is called a micromixed reactor. Hence the following additional assumption is made:

$$\langle C_{Ae}^2 \rangle = C_{Ae}^2$$

Once this assumption is made the rate term can be closed, leading to the following model for a second-order reaction:

$$QC_{A,i} - QC_{A,e} - Vk_2C_{A,e}^2 = 0 \qquad (3.30)$$

This can be solved as a quadratic equation for the exit concentration. It is often solved using a dimensionless parameter, the Damkohler number, which is defined as follows for a second-order reaction:

$$Da = \frac{Vk_2C_{A,i}}{Q}$$

The final result for the dimensionless exit concentration is

$$c_{A,e} = \frac{C_{A,e}}{C_{A,i}} = \frac{\sqrt{1 + 4Da} - 1}{2Da}$$

Similar models can be written for any other form of the rate equation and solved either by applying analytical methods or by using numerical algebraic solver routines.

The following discussion may be omitted on first reading.

A reactor analyzed as described in the preceding discussion is called a micromixed reactor with complete backmixing. In some cases, however, differences in the local concentration may exist, since the fluid elements entering at different times may not mix at a microscale and remain segregated. Models to account for this possibility can be developed at various levels of details. One such a model is called the segregated model or macrofluid model and is discussed here briefly.

Consider a point in the reactor and the time variation of the concentration as shown in Figure 3.7. The flow is usually turbulent and the concentrations fluctuate with time. If the fluctuations are small, their contributions to the rate can be neglected. The rate for a second-order reaction is then proportional to $(C_{Ae})^2$, as in the completely micromixed model. Fluid elements must be mixed on a microscale for this assumption to hold.

If the reactor contents are not completely micromixed, the fluctuating part of the concentration contributes to an extra rate in a time-averaged sense. Models incorporating the variance of the concentration distribution are needed to capture this effect (a topic briefly addressed in Chapter 13). A second limiting case, in which the fluid is assumed to be mixed only on a large scale, is known as macromixed or segregated model. In this model, the fluid elements are assumed to retain their identity from the time they enter the reactor to the time they exist from the reactor and to persist as separate segregated packets. These two limiting cases (micromixed and segregated) are used to bracket the performance for a non-first-order reaction.

The solution for the segregated model requires the concept of exit age distribution and is postponed to a later chapter in Section 13.5. To help you get a feel for the numbers, Table 3.1 gives the solution for the two models.

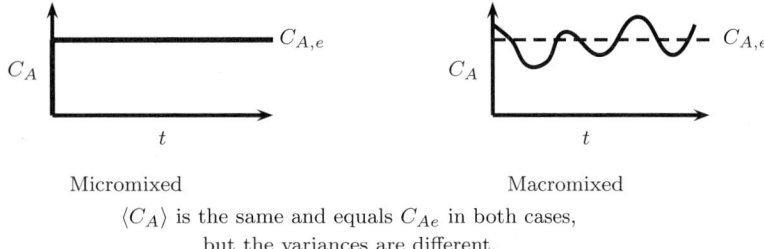

Micromixed   Macromixed

$\langle C_A \rangle$ is the same and equals C_{Ae} in both cases,
but the variances are different.

Figure 3.7   A qualitative illustration of macromixed and micromixed systems showing the presumed concentration variation at any point in the reactor as a function of time.

Table 3.1 Comparison of Micro Versus Segregation Model for a Backmixed Reactor for Second-Order Kinetics

Da	$c_{A,e}$, Microfluid	$c_{A,e}$, Macrofluid
0.1	0.9161	0.9156
1.0	0.6180	0.5963
10.0	0.2702	0.2015
100.0	0.0951	0.0408

The following differences between the two limiting approaches to model a well-mixed reaction are important to note:

- Whether the fluid is micromixed or macromixed makes no difference for a first-order reaction. Details are provided in Section 13.5.
- The actual reactor performance is somewhere in between the two limiting cases for reactions involving nonlinear kinetics. To model this performance in more detail, a segregation parameter must be introduced.
- The two models do not differ much at small values of Da, and the differences are significant only for fast reactions (larger values of Da).
- The segregated flow model gives a higher conversion rate $(1 - c_{A,e})$ for a second-order reaction, as shown in Table 3.1.
- The micromixed model gives a higher conversion rate for reaction orders less than 1 (results are not shown here but are discussed in Section 13.3).

3.7   Tracer Experiments: Test for Backmixed Assumption

The assumption of backmixing can be tested by performing tracer studies. In this kind of study, the feed solution is perturbed by adding a tracer and the exit concentration of the tracer is measured. A simple experiment, for instance, would be to replace water flowing in the tank with a NaCl solution. The concentration of NaCl in the exit flow is then measured. This type of experiment, which is called a step tracer experiment, is shown schematically in Figure 3.8.

The response will be an exponential function of time if the system is well mixed. The following equation (derived in Section 13.2) is applicable to the transient response:

$$\frac{C_{A,e}(t)}{C_{A,i}} = 1 - \exp\left(-t\frac{Q}{V}\right) \tag{3.31}$$

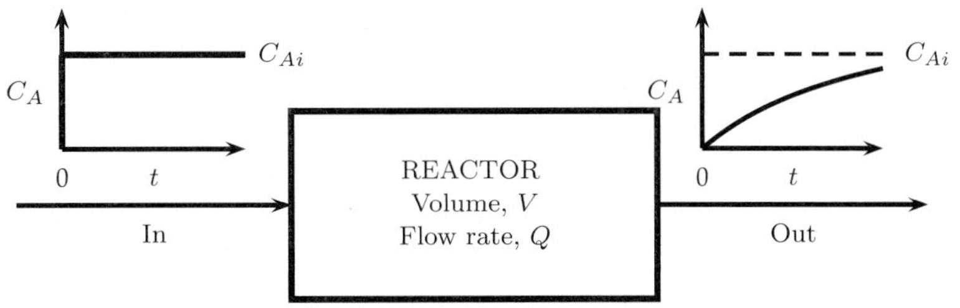

Figure 3.8   Illustrative representation of a step tracer experiment.

Any deviation of experimental data from this scenario indicates that the assumption of complete mixing may not be valid under the given conditions of flow rate. Other modes of injecting a tracer (e.g., pulse tracer) can also be used and the corresponding response examined. These details are taken up in Chapter 13.

3.7.1   Interconnected Cells Model

If a system is not backmixed, a model can be used in which the system is composed of a number of smaller macroscopic control volumes connected in series or parallel. Two examples of such models are described in this section. Modeling of each subunit is then done and the overall performance can be examined. More details are presented in Chapter 13.

In the first example, the reactor is modeled as two or more stirred tanks in series, shown schematically in Figure 3.9. The tracer data can be used to fit the equivalent number of well-mixed tanks needed to model the reactor. More details are presented in Section 13.3.

3.7.2   Model Composed of Active and Dead Zone

Another type of model used to account for deviations from a well-mixed tank incorporates a dead volume. Often the tank is nearly well mixed but mixing is poor in some regions in the tank—so-called dead zones. A model for such a system can be constructed as an active region and a dead region, as shown in Figure 3.10. A theoretical tracer response can be fitted to experimental data to ascertain the relative volume of the active zone and the dead zone as well as the rate of mass exchange between the two regions. The reactor model is then constructed by using two macroscopic control volumes connected with mass exchange.

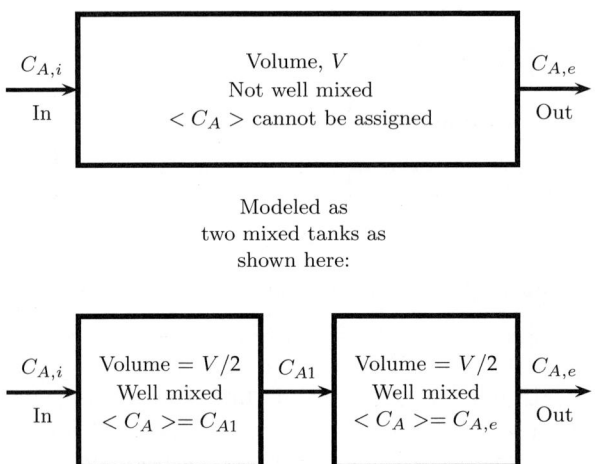

Figure 3.9   An equivalent two-tank representation of a reactor.

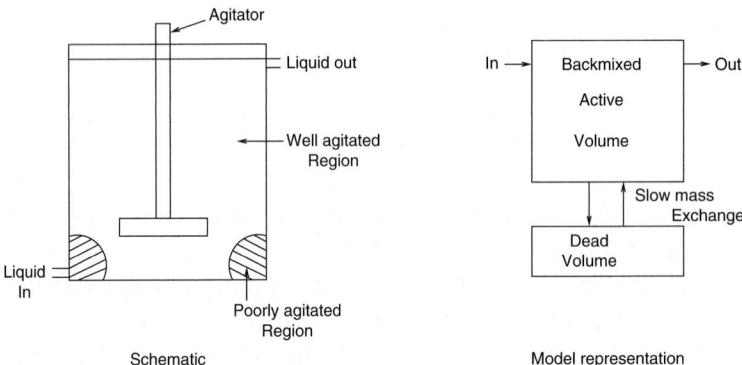

Figure 3.10   A stirred reactor with some dead zones modeled as two tanks connected in parallel. Flow occurs in and out of the active or mixed tank, and a slow mass exchange is assumed between the dead zone and the active zone.

3.8   Liquid–Liquid Extraction

In this section we model a two-phase system in which a liquid–liquid extraction takes place. The mixer-settler shown in Figure 1.4 is analyzed here. In this equipment, a feed mixture consisting of components A (the solute to be extracted) and C (a carrier) enters the contactor. The feed is contacted with a solvent S, and species A is transferred to the solvent phase. The two phases are intimately mixed and then separated. The solvent phase called the extract while the carrier phase is called the raffinate. The problem is to calculate the exit composition of the separated mixture.

This system is modeled by using two control volumes: one for the aqueous (feed) phase and one for the solvent phase, as shown in Figure 3.11. The phases share a common interface, and mass is exchanged across the interface. The macroscopic balances are used for each compartment. The model shown in this section is based on the assumption that C is not soluble in the S phase, and vice versa; that is, there is no mutual solubility of C and S.

The solute conservation is then applied to the two control volumes (feed phase and extract phase), as shown in Figure 3.11. It is preferable to use the mole ratio rather than the mole fractions since the total flow rate is changing from the inlet to the outlet due to the solute transfer. The mole ratio is defined as

$$X = \frac{x}{1-x} \text{ and } Y = \frac{y}{1-y}$$

Here x is the mole fraction of A in the feed phase and y is the mole fraction in the solvent phase. The subscript A for mole fraction is suppressed in this case, so, for example, $x = x_A$ in this section. Note that for a dilute mixture, the mole ratio is nearly the same as the mole fraction. $X = x$ and $Y = y$.

The balance of species A in the feed is

$$\text{in} - \text{out} - \text{transferred to solvent phase} = 0$$

Expressing this in mathematical terms, we have

$$F_C X_f - F_C X_e - \dot{M}_A = 0 \tag{3.32}$$

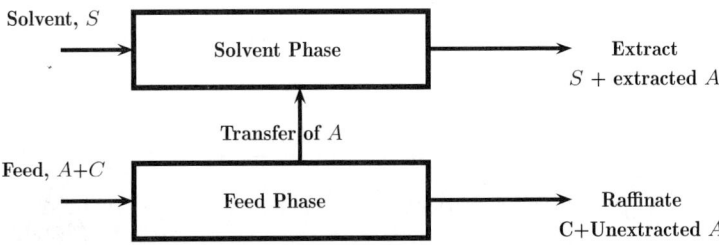

Figure 3.11   Schematic of a liquid–liquid extractor modeled as two macroscopic control volumes.

where F_C is the flow rate of the solute-free aqueous phase. We assume that the component C (the carrier) is not transferred to the solvent phase; hence its flow rate remains F_C invariant. (This is the reason for using the mole ratio instead of mole fractions.)

The balance of A in the solvent phase is formulated in a similar manner:

$$\text{in} - \text{out} + \text{received from the feed phase} = 0$$

This can be written as

$$SY_f - SY_e + \dot{\mathcal{M}}_A = 0 \tag{3.33}$$

which is the mass balance in mole ratio form. Here S is the solvent flow rate on a solute-free basis. Again, we assume no solvent goes to the feed phase; that is, S has no solubility in the aqueous phase.

3.8.1   Mass Transfer Rate

A mass transfer rate model is now needed to close the $\dot{\mathcal{M}}_A$ term. The mass transfer coefficient is defined in these systems using an effective mole fraction difference as the driving force, rather than the concentration difference. Let $<x>$ and $<y>$ be representative average mole fractions in the carrier phase and the solvent phase. The thermodynamic equilibrium relation is usually represented in terms of a partition coefficient m:

$$y^* = mx \tag{3.34}$$

where y^* is the mole fraction corresponding to x if equilibrium were attained. Thus $m<x> - <y>$ is the effective driving force.

The mass transfer rate is then the following term using the mole fraction difference as the driving force:

$$\dot{\mathcal{M}}_A = K_y a_{ll} V_d \left(m \langle x \rangle - \langle y \rangle \right)$$

where K_y is the intrinsic mass transfer coefficient, a_{ll} is the liquid–liquid transfer area per unit volume of the dispersion, and V_d is the volume of the dispersion.

The previous equation can also be expressed in terms of the mole ratio:

$$\dot{\mathcal{M}}_A = K_y a_{ll} V_d \left(m \frac{<X>}{1+<X>} - \frac{<Y>}{1+<Y>} \right)$$

3.8.2   Backmixed–Backmixed Model

In the previously described model, $\langle X \rangle$ and $\langle Y \rangle$ are representative concentrations of A in the system in the two phases. A further assumption about the mixing levels in the system is needed to assign values to these concentrations. One such assumption is that both phases are completely mixed, which leads to the backmixed–backmixed model. The $< X >$ and $< Y >$ concentrations are then assigned as the exit values and the mass transfer model is closed:

$$\dot{\mathcal{M}}_A = K_y a_{ll} V_d \left(m \frac{X_e}{1 + X_e} - \frac{Y_e}{1 + Y_e} \right) \qquad (3.35)$$

Model formulation is now complete, and Equations 3.32 and 3.33 are solved simultaneously using Equation 3.35 for the mass transfer rate. The MATLAB program FSOLVE or similar algebraic equation solvers can be used to solve the equations. The exit concentrations can thus be predicted for a given condition provided the values of the mass transfer coefficients are known. The overall mass transfer coefficient $K_y a_{ll}$ is sufficient, and the intrinsic coefficient K_y and the interfacial area a_{ll} need not be known separately.

3.8.3   Equilibrium Stage Model

A simpler model often used as a first approximation in practice is the ideal stage model, which assumes that the mass transfer rate is sufficiently large so that the exit streams reach an equilibrium value. Conservation laws together with thermodynamics of phase equilibrium are used in model building for such processes.

Adding Equations 3.32 and 3.33 results in elimination of the mass transfer rate term:

$$F_C X_f - F_C X_e + S Y_f - S Y_e = 0$$

This is the overall mass balance for A for the entire equipment.

If the solvent contains no dissolved A, we set Y_f equal to zero. Hence

$$F_C X_f - F_C X_e - S Y_e = 0 \qquad (3.36)$$

Now we assume that the exit streams are in equilibrium, thereby creating an equilibrium stage model. The equilibrium relation, which is usually expressed as the ratio of mole fractions in the two phases, is given by Equation 3.34. It can be recast in mole ratio form and expressed as

$$Y_e = K_A X_e$$

where K_A is the partition coefficient for A using the mole ratio units. This is related to the partition coefficient based on mole fraction units as follows:

$$K_A = m\frac{1-x}{1-y}$$

Inserting this result in Equation 3.36, we have

$$F_C X_f - F_C X_e - SK_A X_e = 0 \tag{3.37}$$

Solving for the solute mole ratio in the exit of the feed, we find

$$X_e = \frac{X_f}{(1 + SK_A/F_C)} \tag{3.38}$$

This model for liquid–liquid extraction assumes an ideal stage concept. Note that the conservation law and thermodynamics have been used, but no mass transfer effects are considered. The mass transfer is considered to be almost complete in the equilibrium stage model. Often separation processes are first modeled by making such an assumption to obtain a quick estimate of the performance of the system under ideal conditions. The dimensionless group SK_A/F_C is called the extraction factor, denoted as E. Equation 3.38 can then be written in terms of the extraction factor:

$$\frac{X_e}{X_f} = \frac{1}{1+E}$$

The fractional removal of A denoted as ϕ_A in a single state extraction device is then given as

$$\phi_A = \frac{E}{1+E}$$

3.8.4 Stage Efficiency

The ideal stage model can be corrected in an empirical manner by using the concept of stage efficiency, η:

$$\eta = \frac{X_f - X_e}{X_f - X_e^*} \tag{3.39}$$

where X_e^* is the solute concentration reached if equilibrium were achieved. This concentration is given by Equation 3.38. The stage efficiency depends on the mass transfer coefficient between the two phases, the interfacial area for mass transport, and the flow pattern of the two phases. Models for stage efficiency can be developed by solving the models including the rate mass transfer and comparing the results with the equilibrium model. More often, however, empirically based values for the stage efficiency (gained from prior design experience) are used.

For dilute systems with a constant value of m, the mass transfer model can be solved algebraically if both phases are assumed to be backmixed. Based on the result, the following relation can be derived for the stage efficiency.

$$\eta = \frac{\kappa(1 + E)}{1 + \kappa(1 + E)} \tag{3.40}$$

Here κ is a dimensionless mass transfer coefficient κ defined as follows:

$$\kappa = K_y a_{ll} S / F \tag{3.41}$$

In this expression, S is the solvent flow rate and F is the feed flow rate. Both are nearly constant for dilute solutions.

As κ tends to ∞ (i.e., with a large mass transfer rate), the stage efficiency tends to 1.

Summary

- Batch reactors are a well-studied problem in chemical kinetics, and the differential equations for the transient concentration profiles can be derived by balancing the accumulation to the rate of production. Spatial variations in the concentrations are usually small, so diffusional considerations are not important.

- Models for batch reactors involving multiple reactions are solved by numerical integration, with the ODE45 solver being a favorite tool. An improvement on this approach is the CHEBFUN embedded ODE45, which makes the postprocessing easier.

- Performance of a batch reactor for reversible reactions can be improved by simultaneous removal of the product. This strategy has important industrial applications. Mass transfer and mixing considerations play an important role in modeling such systems.

- The mass transfer coefficient concept is needed for many macro- (and meso-) modeling applications. It is often calculated using empirical correlations. A dimensionless parameter (Sherwood number) is often used as a dimensionless mass transfer coefficient; it is correlated in terms of the Reynolds and Schmidt numbers.

- For mass transfer from a solid sphere to a stagnant gas, the Sherwood number has a value of 2. An application to a subliming solid was described in Section 3.4, where an equation was developed for the change in radius as a function of time.

- Oxygen transfer into a stirred liquid is an important problem in many processes. The dissolved oxygen concentration can be calculated using an overall (volumetric) mass transfer coefficient, which is itself a product of the intrinsic coefficient and the gas–liquid transfer area. Only the overall coefficient is needed in this application.

- Continuous flow reactors are often modeled with the assumption that the tank or reactor is well mixed. This approach provides a closure to the (average) rate term needed in the conservation law.

- Backmixed reactors can be further classified as micromixed or segregated depending on the local scale of mixing. For a non-first-order reaction, there is a difference in the performance based on these two models, and the actual reactor can be bracketed between these limiting levels of mixing.

- A micromixed backmixed reactor can be modeled using the rate based on the exit concentration. Modeling of a segregated backmixed reactor requires the age distribution function (covered in Chapter 13).

- An important application of macroscopic balance is in the modeling of two-phase mass transfer in equipment such as a mixer-settler. The assumption that both flowing phases are backmixed is often used in such a system so that the driving force can be assigned for mass transfer calculations. The concentrations (or mole fractions) in the vessel for each phase can then be set equal to the exit values.

- For separation processes, a first-level model assumes that the exit streams are in thermodynamic equilibrium. Such a contactor is called an ideal stage and provides a benchmark for the maximum performance of the contactor, assuming complete mass transfer. The deviation from the ideal model can then be corrected by using the stage efficiency concept. Often stage efficiency is an empirically fitted parameter, but theoretical calculations are possible using mass transfer and mixing concepts.

Review Questions

3.1   What is the relation between the forward reaction rate constant and the backward reaction rate constant for a reversible reaction?

3.2   A series reaction has the following values for the rate constants: $k_1 = 100 \, \text{s}^{-1}$ and $k_2 = 100 \, \text{s}^{-1}$. What should be the batch time to get the maximum concentration of the intermediate?

3.3 Can you achieve a conversion rate greater than the equilibrium value in a reactor? How?

3.4 How does the mass transfer coefficient from a spherical solid to a gas vary with particle radius?

3.5 What is the difference between the intrinsic and volumetric mass transfer coefficients?

3.6 What does the assumption of a backmixed contactor or reactor mean? What is its usefulness?

3.7 What is Damkohler number? Define it for a first-order reaction as well as for a second-order reaction.

3.8 What is meant by a completely micromixed CSTR?

3.9 What is a segregated but backmixed reactor?

3.10 What is meant by the concept of an ideal stage?

3.11 Define extraction factor. Which parameter affects this factor?

3.12 If the extraction factor is 2, what is the percentage recovery of the solute?

3.13 What is stage efficiency? Which parameters affect the stage efficiency?

Problems

3.1 **Consecutive first-order reactions.** Solve the following case in a batch reactor analytically:

$$A \rightarrow B \rightarrow C$$

Verify the equations in the text (Equation 3.7) for the time at which the optimal concentration of B is reached. Compare your result with the numerical solution generated by ODE45.

3.2 **Multiple set of first-order reactions.** Consider the following series reaction scheme in a constant batch reactor:

$$A \xrightarrow{k_1} B \xrightarrow{k_2} C \xrightarrow{k_3} D$$

Assuming all reactions are first order and irreversible, set up the governing equations in matrix form for A, B, and C. Assume the rate constants $k_1 = 2$, $k_2 = 1$, and $k_3 = 2$ and an initial concentrations of $C_A = 1$, $C_B = 0$, and $C_C = 0$. **Note:** Time and concentrations are in arbitrary units for illustrative purposes.

Represent the solution in a compact form using matrix algebra.

Use the *expm* function (exponential of matrix) to obtain an analytical representation of the results.

Obtain numerical solutions using ODE45 and plot concentrations versus time profiles. Compare the analytical solution (using *expm*) with the numerical solution.

3.3 Non-isothermal batch reactor. A non-isothermal batch reactor can be modeled by including a heat balance term, thereby generating an additional differential equation for dT/dt. The heat balance is used here:

> generation of heat by reaction − heat transferred to the cooling medium = accumulation of enthalpy in the system

Complete the model by representing each term mathematically and show that the following differential equation holds for the batch reactor:

$$\left(\sum_i \mathcal{M}_i C_{pi}\right) \frac{dT}{dt} = (\Delta H)VR_A - UA(T - T_C)$$

Incorporate this additional equation into the ODE45 solver (Listing 3.1) and simulate a batch reactor for the following conditions: Reaction is first order with a frequency factor 7×10^{13} s^{-1} and an activation energy of 100 kJ/mol K and the heat of reaction is −100 kJ/mol (exothermic) reaction. The volume of the batch reactor is 10 L and has the same physical properties as water. The reactant is dissolved in water as a solvent; it has an initial concentration of 2000 mol/m^3 and an initial temperature of 300 K. The heat transfer coefficient is 1000 W/m^{2}K with a jacket area of 3 m^2. The coolant is kept at 320 K by circulating a large quantity of coolant in the jacket. Plot the temperature and the concentration in the reactor as a function of time.

3.4 Zero-order generation followed by a first-order reaction. Chemical A, a powdered solid, is slowly and continuously fed for half an hour into a well-stirred vat of water. The solid dissolves quickly and hydrolysis takes place for form B. B reacts by a first-order reaction to form a product C. The rate constant is estimated as 1.5/hour for B to C. The liquid volume in the tank remains constant at 3 m^3.

If no reaction of B occurred, the concentration of B in the vat would have been 100 mol/m^3 at the end of the half-hour addition of A.

Find the maximum concentration of B in the vat and the time it is reached.

What is the concentration of the product C at the end of one hour?

Hint: Model this system as a batch reactor for B with a zero-order generation from A and a first-order consumption to C. Use the given data (for no reaction of B) to find the zero-order rate constant first.

3.5 Reactor–separator combo: Inclusion of finite rate of mass transfer. Verify the results of Example 3.1 by writing MATLAB code and extend the results in Figure 3.4 for larger values of time. Note that the exit gas concentration was assumed to be the equilibrium value in the solution in the text. Now examine the effects of mass transfer by varying the parameter $\bar{K}_m A_{gl}$ in the range of 10^{-5} to 10^{-3} m^3/s.

3.6 Mass transfer coefficient from a sphere. Use the Froessling-Marshall equation for mass transfer from a solid sphere to calculate the mass transfer coefficient for transport from a solid of 3 mm diameter to air flowing at a velocity of 0.1 m/s at 300 K and 1 atm average pressure. The diffusion coefficient of the vaporizing species in air is 9.62×10^{-6} m^2/s. Also $\rho = 1.1769$ kg/m^3 and $\mu = 1.8464 \times 10^{-5}$ Pa s. Plot the mass transfer coefficient as a function of particle size in the range of 0.5 mm to 3 mm.

3.7 **Sublimation of a solid under high flow conditions.** Sublimation of a solid under a stagnant gas condition (no gas flow) was studied in Example 3.2. Under high flow conditions, the first term of 2 in Sherwood number correlation (Equation 3.17) can be neglected and the mass transfer coefficient can be regrouped as

$$k_m = \frac{A}{\sqrt{R}}$$

where A is a function of gas velocity and the properties of the gas phase. What is the expression for A when the equation for the Sherwood number (neglecting the factor 2) is regrouped in this manner?

Integrate Equation 3.16 with this dependency of the mass transfer coefficient on the radius (neglecting the factor 2) to find the radius as a function of time. What is the dependency on time now? Calculate and plot the change in radius as a function of time for a gas velocity of 20 cm/s with the other conditions remaining the same as in Example 3.2.

3.8 **Oxygen transfer to a pool of liquid.** Oxygen is bubbled through a pool of water and the dissolved oxygen concentration in the liquid is measured as a function of time. For a particular experiment, a 50% saturation is achieved after 3 minutes. Find the time needed to achieve a 90% saturation.

3.9 **Oxygen absorption with a first-order reaction.** Oxygen gas is continuously bubbled into a pool of liquid of volume V_L; the oxygen also reacts in the liquid with a first-order rate constant of 0.2 s^{-1}. Initially there is no dissolved oxygen in the system. Oxygen concentration increases with time and finally attains a plateau that is equal to the steady state concentration. Develop a model to predict the transient oxygen concentration as well as the steady state value.

3.10 **Zero-order reaction in a CSTR.** A zero-order reaction with a rate constant of k_0 is carried out in the backmixed reactor. Assume also that the vessel is micromixed. Show that the dimensionless exit concentration is given as

$$c_{A,e} = 1 - Da \text{ for } Da \leq 1$$

where Da is defined as

$$Da = \frac{V k_0}{Q C_{A,i}}$$

What is the exit concentration if Da is greater than 1?

3.11 **Rate constant from exit concentration data.** A liquid containing 1 M of species A enters a backmixed reactor of 1 L volume at a rate of 1 L/min. The exit liquid is analyzed for A and has a concentration of 0.5 M. Find the reaction rate constant if the reaction is first order in A, the reaction is second order in A, and the reaction is zero order in A.

3.12 **Tanks in series model.** A tracer responds to a CSTR by showing that it is not backmixed but can be modeled as two tanks in series. A liquid containing 1 M of species A enters a backmixed reactor of 1 L volume at a rate of 1 L/min. Find the conversion if (1) the reaction is first order with a rate constant 0.5 per hour, and (2) if it is second order with a rate constant of 0.01 L/mol h. For both cases assume that the mixing level in each tank corresponds to the micromixed condition.

3.13 **Bimolecular reaction in a CSTR.** An aqueous feed to a stirred tank reactor consists of a mixture of A and B with concentrations of 100 mol/m^3 and 200 mol/m^3 for A and B, respectively. The feed flow rate is 0.04 m^3/s and a reaction takes place as follows:

$$A + B \rightarrow \text{products}$$

The rate of reaction in mol/L m^3 follows the kinetics:

$$R_A = -0.4 C_A C_B$$

Find the exit concentrations of A and B if the reactor volume is 1 L.

3.14 **Step tracer response in a CSTR.** A step tracer is introduced into a reactor that is assumed to be completely backmixed. Write a macroscopic balance including the transient terms. Solve this model with an initial condition of zero tracer concentration in the reactor at time 0. Verify Equation 3.31 for the exit concentration of the tracer as a function of time.

3.15 **Step response of a CSTR for a reacting tracer.** A CSTR is charged with a liquid of volume V. At time 0, a feed containing A is fed at a volumetric flow of Q. A corresponding volume is then continuously removed from the exit of the reactor as well. The species A undergoes a first-order reaction in the system. Derive an expression for the exit concentration of A in the reactor as a function of time and the final steady state concentration in the system. Assume that the tank is well stirred and use the backmixing assumption.

3.16 **Use of extraction factor.** A feed containing 18,000 kg/h of 8% by weight of acetic acid is treated with methyl acetate, which is has a distribution coefficient of 1.28. Find the extraction factor for 5000 and 10,000 kg/h of solvent. Find the recovery of a solute in a single mixer-settler as a function of solvent flow rate.

3.17 **Liquid extraction: Equilibrium model.** Methyl-ethyl ketone is used as a solvent to extract methyl acetate from water. The incoming solution has 8% of acetate with a flow rate of 1500 kg/h and the concentration is to be reduced to 3%. Find the solvent flow rate needed, assuming equilibrium conditions at the exit. Use $m = 0.657$ in mass fraction units. If the raffinate is treated in a second mixer-settler with the same solvent flow rate, to what extent can the solute recovery improved?

CHAPTER 4
Examples of Mesoscopic Models

Learning Objectives

After completing this chapter, you will be able to:

- Apply the conservation law to a control volume that is differential only in the main flow direction.
- Formulate mesoscopic models used in the analysis of mass transfer processes in tubular or channel flows.
- Identify additional closures or assumptions to be made to complete the model.
- Formulate and solve the simplest problem in mesoscale analysis—namely, solid dissolution from a wall.
- Analyze tubular flow reactors and understand the model simplification using the concept of plug flow.
- Understand the concept of dispersion and show how the plug flow model can be corrected using this parameter.
- Set up models for systems with two flowing phases with interfacial mass transfer from one phase to another.
- Show the application of the meso-model with two flowing phases to the simulation of a countercurrent gas absorption column.

Chapter 3 discussed macroscopic models. These models are useful when the system is nearly well mixed and therefore at a nearly uniform concentration. However, in many systems, the concentration varies predominantly and strongly in one direction, the flow direction. In such cases it is more useful to set up mesoscopic models. This control volume is a differential only in the main flow direction but spans the entire cross-sectional area in the direction normal to the flow. The concentration variable is then a flow-weighted cross-sectional area concentration or the so-called cup mixing concentration, rather than the point-to-point values of concentration. A differential equation can, in turn, be set up and solved for the variation of the cup mixing concentration as a function of flow direction. An illustrative problem is solid dissolution from a wall. In this first example in this chapter, we will demonstrate the model formulation for this problem. Knowledge of the mass transfer coefficient from the wall is needed as a parameter to close the model.

We next look at modeling of a continuous-flow reactor carried out in a pipe or a square channel. By the averaging process, we show that in addition to cup mixing concentration, the cross-sectional average concentration is needed in the model. Hence some closure relations are needed to relate the two average concentrations and thereby complete the model. The simplest closure based on a plug flow assumption is discussed first and some solutions to reactor performance based on this model are shown. We then introduce the concept of dispersion, which relates the two averages, and thereby closes the model. The results for reactor performance based on this dispersion model are then presented for a first-order reaction.

Additional examples of mesoscopic models in mass exchanger analysis are then presented. The first example is a single-stream mass exchanger. Here we have a membrane from which a solute is removed to an external fluid that stays at a constant concentration. The next example is a two-stream mass exchanger in which two flowing streams, say a gas and a liquid, exchange mass from one phase to another. The study of this problem assumes both phases involve plug flow and the problem has important applications in unit operations—for example, in simulation of gas absorption columns.

In summary, this chapter introduces a range of problems in mesoscopic mass transfer analysis mainly by examining specific examples. Through a careful study of these examples, you will be exposed to all the main tools used in mass transfer modeling of these systems. Further aspects of mesoscale modeling and the phenomena of dispersion are taken up in Chapter 14.

4.1   Solid Dissolution from a Wall

The problem we study in this chapter is mass transfer from a wall to a flowing fluid. A prototype example is a pipe coated with a soluble material and a liquid flowing in the pipe. The concentration of the dissolved solute at the exit of the pipe is to be calculated. Problems of this type and its variations have many applications and were introduced in Section 1.10.1; we continue the discussion here. This convection–diffusion problem can be analyzed based on a differential model (Section 10.1). Such models give pointwise concentration

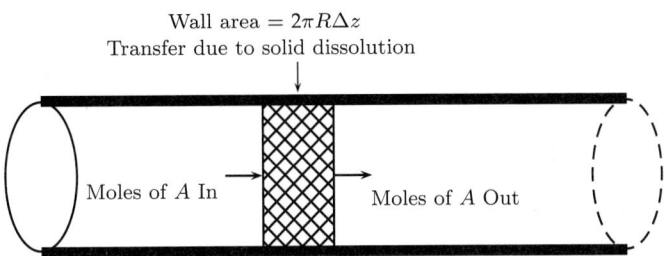

Figure 4.1   Schematic of a solute coated pipe: an example of mesoscopic modeling.

distributions—that is, C_A as a function of r and z. Often detailed information on concentration field may not be needed, and for such cases, we can simply use a mesoscopic control volume, as shown in Figure 4.1.

4.1.1   Model Details

The model formulation involves the use of the conservation law, followed by a closure for the rate of mass transfer from the wall using the mass transfer coefficient. The conservation law applied to this system can be written assuming no generation and reaction as

in − out + transferred from the walls by solid dissolution = 0

This relationship is represented in symbols as

$$\dot{\mathcal{M}}_{A,i} - \dot{\mathcal{M}}_{A,e} + \dot{M}_{A,W} = 0 \qquad (4.1)$$

In the mesoscopic model, the *in* $\dot{\mathcal{M}}_{A,i}$ and *out* $\dot{\mathcal{M}}_{A,e}$ terms will be based on some radial averages of the local concentration. The formal definition of the concentration to use is presented in the following paragraphs.

We assume there is a velocity profile at any cross-section and let v_z be the local velocity at any cross-section. For example, in pipe flow, v_z would be a function of r, the local radial position. Local flow rate is therefore $v_z \Delta A$ and local moles transferred across this area is $C_A v_z \Delta A$, where C_A is the local concentration. The rates at which the moles are crossing any given cross-section is therefore the integral of $C_A v_z \Delta A$ over the cross-section:

$$\text{Molar rate of transport of } A \text{ across a cross-section } = \int_A C_A v_z \, dA \qquad (4.2)$$

Note that the diffusive transport across any cross-section is usually small in presence of the superimposed flow and is neglected.

An average concentration is more useful for a simpler mesoscale representation since the local values will not be computed at this level of modeling. We define an average concentration by the following equation:

$$\text{Molar rate of transport of } A \text{ across a cross-section} = QC_{Ab} \qquad (4.3)$$

where Q is the volumetric flow rate, which is assumed to be constant. Equating the two expressions for moles crossing, we find that the average concentration is

$$C_{Ab} = \frac{1}{Q} \int_A C_A v_z dA \qquad (4.4)$$

Since Q is an area integral of the local volumetric flow rate,

$$Q = \int_A v_z dA$$

the average concentration can also be defined as a ratio of two integrals:

$$C_{Ab} = \frac{\int_A v_z C_A dA}{\int_A v_z dA} \qquad (4.5)$$

The average concentration defined in this manner is called the cup mixing concentration or flow weighted concentration. This definition was first introduced in Chapter 1, and the preceding discussion is an useful recap of the definition and its use.

With this definition, the *in* and *out* terms can be written as QC_{Ab} evaluated at z and $z + \Delta z$, respectively.

The moles transferred from the walls (per unit time per unit wall area) is expressed in terms of a mass transfer coefficient k_m for solid dissolution:

$$N_{As} = k_m(C_{As} - C_{Ab})$$

where C_{As} is the concentration in the liquid adjacent to the wall, rather than the concentration in the solid. The solubility of the wall in the liquid phase is needed to calculate this concentration.

The transfer rate defined in this way is based in the unit area for mass transfer. This transfer rate needs to be multiplied by the transfer area. The area for transfer is $P\Delta z$, where P is the perimeter of the flow cross-section. For example, for pipe flow, $P = 2\pi R$. Putting all terms together and allowing Δz to tend to zero, we obtain the following equation for the (cup mixing) concentration profile:

$$Q\frac{dC_{Ab}}{dz} = k_m P(C_{As} - C_{Ab}) \qquad (4.6)$$

The variables are now separated and the integral form of the equation is

$$\int_{C_{Ab,i}}^{C_{Ab,e}} \frac{C_{Ab}}{C_{As} - C_{Ab}} = \frac{P}{Q} \int_0^L k_m dz \qquad (4.7)$$

The mass transfer coefficient k_m is the local value and hence can be a function of z. (The functional dependency is examined further in Section 8.2.) The value of k_m is therefore retained within the integral over a pipe of length L. The use of an average mass transfer coefficient defined as follows is convenient:

$$\bar{k}_m = \frac{1}{L} \int_0^L k_m(z) dz$$

Using this expression in Equation 4.7, the following integrated form is obtained:

$$\ln \left(\frac{C_{As} - C_{Ab,i}}{C_{As} - C_{Ab,e}} \right) = \frac{PL\bar{k}_m}{Q} \qquad (4.8)$$

This equation relates the the cup mixing exit concentration $C_{Ab,e}$ to various parameters such as flow rate, length and diameter of the pipe, and mass transfer coefficient.

It is useful to write this relationship in terms of the total quantity of solute transferred across the entire pipe and recast it in terms of a overall driving force. The quantity of solute transferred over the whole pipe can be calculated by the overall mole balance from the exit and inlet:

$$\dot{M}_A = Q(C_{Ab,e} - C_{Ab,i})$$

Using this definition in Equation 4.8, the following equation for $\dot{M}_A$ can be derived:

$$\dot{M}_A = \bar{k}_m(PL) \left(\frac{C_{Ab,e} - C_{Ab,i}}{\ln(C_{As} - C_{Ab,i})/(C_{As} - C_{Ab,e})} \right) \qquad (4.9)$$

This can be expressed in terms of a log mean driving force. The driving force at the inlet is $C_{As} - C_{Ab,i}$, while that at the outlet is $C_{As} - C_{Ab,e}$. A log mean average of the two is defined as

$$\text{LMDF} = \frac{[C_{As} - C_{Ab,i}] - [C_{As} - C_{Ab,e}]}{\ln\left[(C_{As} - C_{Ab,i})/(C_{As} - C_{Ab,e})\right]}$$

Hence Equation 4.9 can be written as

$$\dot{\mathcal{M}}_A = \bar{k}_m (PL)[LMDF] \tag{4.10}$$

A similar equation is obtained for countercurrent separation processes in Section 4.3.2. The analysis there indicates that the log mean average is the appropriate overall driving force to be used in conjunction with the mesoscopic model. Equation 4.9 also implies that measurement of the exit solute concentration provides a method for estimating the mass transfer coefficient in the system.

4.1.2   Mass Transfer Correlations in Pipe Flow

The mesoscopic model described in the preceding subsection needs a value for the average mass transfer coefficient. Correlations, fitted either to results from a detailed theoretical model or to experimental data, are commonly used for this purpose. As an example, we look at mass transfer correlations for pipe flow. The correlations depend on whether the flow is laminar or turbulent. Correlations use the dimensionless groups of the Sherwood number, Reynolds number, and Schmidt number. For laminar flow, the following correlation is applicable:

$$\bar{Sh}_L = 3.66 + \frac{0.0668 Pe/L^*}{1 + 0.04[Pe/L^*]^{2/3}} \tag{4.11}$$

where $L^* = L/d_t$; $\bar{Sh}_L$ is defined as $\bar{k}_m d_t / D_A$, which is the average value of the Sherwood number; and Pe is defined as the product of the Reynolds number and the Schmidt number:

$$Pe = Re\, Sc = \frac{\langle v \rangle\, d_t}{D_A}$$

Note that as L^* becomes large (a long pipe), the average Sherwood number approaches an asymptotic value of 3.66.

A useful correlation for liquids for $Re > 4000$ (i.e., turbulent flow situation) is the Linton-Sherwood correlation (1950):

$$Sh = 0.023 Re^{0.83} Sc^{1/3}$$

More discussion of the correlations and the theoretically based models for convective mass transfer is deferred to Chapters 8, 9, and 10. The model is illustrated in Example 4.1 to calculate the exit concentration from a subliming pipe wall.

Example 4.1   Sublimation in a Pipe

A pipe of 2 cm inside diameter is coated with a thick layer of naphthalene, and air is flowing in the pipe at a flow rate of 50 cm^3/s. The pipe has a length of 1 m. Find the exit concentration of naphthalene in the pipe.

Solution

Physical properties of air are used to find the Reynolds, Schmidt, and Peclet numbers. The values used here are $\nu = 1.56 \times 10^{-5}$ m^2/s and $D = 9.62 \times 10^{-6}$ m^2/s. Hence $Sc = \nu/D = 0.63$.

The average velocity is computed as $Q/(\pi R^2) = 0.16$ m/s.

The Reynolds number is computed as $d_t <v>/\nu$ as 203.4, which indicates that the flow is laminar. Hence Equation 4.11 can be used to find the average Sherwood number.

This calculation requires the Peclet number, Pe, which is equal to $ReSc$ and has a value of 330.8 here. L^* is the dimensionless length, L/d_t, which equals 50.

Inserting these values in Equation 4.11, we find Sh is 4.04. We can then calculate the mass transfer coefficient: $\bar{k}_m = ShD/d_t = 0.0015$ m/s.

Equation 4.8 is now used. The inlet concentration is zero. The fractional saturation of naphthalene in the exit $C_{Ab,e}/C_{As}$ is found as 0.86 using Equation 4.8. The surface concentration C_{As} is on the gas side of the wall (concentration jump at an interface). Its calculation requires the value of the vapor pressure of naphthalene: $p_{vap} = 666$ Pa. Hence $p_{vap}/R_g T = 0.2309$ mol/m^3. The exit concentration is therefore 0.2 mol/m^3. This should be interpreted as the cup mixed exit concentration.

4.2   Tubular Flow Reactor

This section explores how mesoscopic models can be set up for a pipe (or a rectangular channel) where a chemical reaction is taking place. The model formulation involves two measures of average concentration, the cup mixing and area averaged values. These measures were introduced in Section 1.6.4 but the key aspects are reviewed again here.

The mesoscopic model is applied over a differential control volume of thickness Δz, as shown in Figure 4.2. As usual, we write the balance in words first:

$$\text{in} - \text{out} + \text{generation equals zero}$$

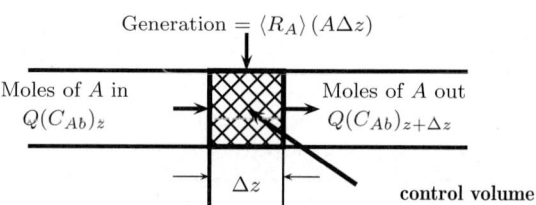

Figure 4.2   Control volume for analysis of a tubular reactor. *In* and *out* depend on the local cup mixing concentration, while the average generation depends on the cross-sectional average concentration.

In this case, the *in* and *out* components are based on the cup mixing concentration since the represent the moles crossing the *in* and *out* planes:

$$\text{in} = Q(C_{Ab})_z$$

$$\text{out} = Q(C_{Ab})_{z+\Delta z}$$

The local rate of reaction is $R_A \Delta V$, where R_A is the local rate of production. For the control volume, the total generation is

$$\text{generation} = \int_V R_A dV$$

Since $dV = \Delta z dA$, we have

$$\text{generation} = \Delta z \int_A R_A dA$$

Let us define a cross-sectional average as follows:

$$\langle R_A \rangle = \frac{1}{A} \int_A R_A dA$$

Hence the rate of generation is $\langle R_A \rangle A \Delta z$.

Putting all this information together in the conservation statement and taking the limit as Δz tends to zero yields

$$Q \frac{dC_{Ab}}{dz} = <R_A> A$$

or, using $<v> = Q/A$,

$$\langle v \rangle \frac{dC_{Ab}}{dz} = \langle R_A \rangle \tag{4.12}$$

Note that $< R_A >$ is the cross-sectional average rate, which is be based on the cross-sectional average concentration:

$$< C_A >= \frac{1}{A} \int_A C_A dA$$

This is different from the cup mixing concentration, and it is interesting to note that both the cup mixed and cross-sectional averages appear simultaneously in the model (as also shown in Chapter 1). This method of formulating mesoscopic models based on averaging does not appear to have been brought out in other texts in this subject. This approach identifies that a closure assumption connecting C_{Ab} and $< C_A >$ is needed. Two methods of closure are taken up in the following discussion.

4.2.1   Plug Flow Closure

With plug flow closure, we assume there is no variation in velocity or concentration as a function of radial position. In turn, there is no difference between the cup mixing and cross-sectional average concentrations and we have the following plug flow assumption:

$$C_{Ab} =< C_A >$$

The model is now closed. For a first-order reaction, we have

$$\langle v \rangle \frac{d < C_A >}{dz} = -k_1 < C_A >$$

The integrated form can be represented as

$$\langle C_A \rangle = C_{A,i} \exp\left(-k_1 z / \langle v \rangle\right) \tag{4.13}$$

where $C_{A,i}$ is the inlet concentration.

If the reaction is second order in A, the differential equation is:

$$\langle v \rangle \frac{d \langle C_A \rangle}{dz} = -k_2 \langle C_A \rangle^2 \tag{4.14}$$

which can be solved easily by separation of variables. Other kinetics can be handled in a similar manner. For complex rate forms, a numerical solution is needed, with ODE45 being the favorite tool for this purpose. The corresponding MATLAB code is very similar to Listing 3.1 for a batch reactor.

4.2.2   Dispersion Closure

The plug flow assumption is valid if the uniform velocity assumption is reasonable (e.g., turbulent flows; see Figure 1.13) and there are no other factors that can set up a radial concentration gradient.

What can set up a radial concentration gradient? The answer is different reaction rates at different radial positions. For example, if there is a velocity profile, fluid elements at different radial locations will have different residence times. Fluid at the center moves much faster than fluid near the wall, so it will spend less time in the reactor and will undergo a lesser extent of reaction. Such fluid elements will, therefore, have higher concentrations compared to a fluid at the wall. In this scenario, then, the plug flow assumption is not valid. The radial concentration variation prevails, which also causes a radial diffusion that tends to balance out some of the concentration variation.

The combined effect of velocity profile and radial diffusion will then influence the reactor performance. How should one proceed in modeling this system? As one way of performing the analysis, a full differential model can be set up and solved. (This approach is shown for laminar flow in Chapter 17.) If one wishes to use a simpler mesoscopic model, a closure relating the cup mixing and cross-sectional average is needed. The concept of dispersion (or axial dispersion in the present case) provides a commonly used approach to closure.

The concept of axial dispersion is widely used as a closure mechanism. In this approach, we apply a correction to the plug flow model:

moles crossing any cross-section =
moles crossing as if plug flow existed +
an additional diffusion type of transport superimposed on plug flow

The correction term (the last term) is modeled as though some diffusion type of mechanism is superimposed on the plug flow value:

$$\text{correction term per unit area} \ = -D_E \frac{d \langle C_A \rangle}{dz}$$

where D_E is called as the axial dispersion coefficient. Thus the closure applied is

$$\langle v \rangle \, C_{Ab} = \langle v \rangle \, \langle C_A \rangle - D_E \frac{d \langle C_A \rangle}{dz}$$

Note that the additional parameter introduced is not the diffusivity; instead, it is the dispersion parameter, although it has the same units as diffusivity.

Using this definition in Equation 4.12, the model equation for dispersed flow for a first-order reaction is

$$D_E \frac{d^2 \langle C_A \rangle}{dz^2} - \langle v \rangle \frac{d \langle C_A \rangle}{dz} - k_1 \langle C_A \rangle = 0 \qquad (4.15)$$

Because a second-order differential equation arises in the dispersion model, boundary conditions are needed at both the entrance and the exit of the reactor. These conditions and the solution of this model and applications are covered in Section 14.1. At this point you should note the additional term that appears as a correction to the plug flow model. It requires the axial dispersion coefficient parameter D_E. If $D_E = 0$, the plug flow model is recovered. The value for D_E for laminar flow can be calculated from theory, as described later in this section. Correlations are available for many types of equipment that are widely used in practice.

The concept of dispersion is useful not only in reaction engineering, but also in separation process equipment design. For example, countercurrent mass exchangers can be modeled as plug flows for simplicity. The plug flow model can be corrected if needed by adding the contribution due to axial dispersion. Values or estimates of dispersion coefficient for both phases will be needed in such a case.

Dispersion Coefficient in Laminar Pipe Flow

For laminar flow, the following relation is found from theoretical considerations:

$$\boxed{D_E = \frac{1}{48} \frac{<v>^2 R^2}{D}} \qquad (4.16)$$

This identity is derived in Section 14.3. We find here that D_E is inversely proportional to D!

Effect of Dispersion

The solution to the dispersion model for a first-order reaction can be obtained analytically and is presented in Chapter 14. For a qualitative understanding, it is useful to illustrate a sample result here; this is shown in Figure 4.3.

For a first-order reaction (and for dispersion models in general), two dimensionless parameters are needed: (1) the Damkohler number, Da, defined as $k_1 L / \langle v \rangle$; and (2) the dispersion number, defined as $D_E / (\langle v \rangle L)$. The effect of the dispersion number, when keeping Da fixed at 3, is shown in

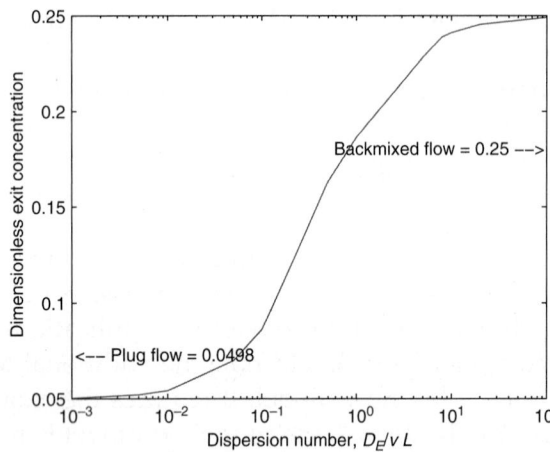

Figure 4.3 Effect of dispersion parameter on the exit concentration for a first-order reaction for $Da = 3$.

Figure 4.3. The plot shows two asymptotes for low and high values of the dispersion number. For low values of dispersion number, the plug flow limit is approached. The plug flow value for $Da = 3$ is $\exp(-3) =$ which is 0.0498 for the exit concentration. For at high values, the backmixed result is obtained. The backmixed value is $1/(1 + Da)$ and is equal to 0.25. Thus the dispersion model brackets the limits of conversion that can be realized in the reactor.

4.3 Mass Exchangers

In this section we examine models for mass exchangers where mass is exchanged from one phase to another. Two cases are considered. First, we describe a single flowing stream where mass is exchanged from this stream to a second phase (which is at a fixed concentration)—for example, a transport across membrane cast as a tube. Second, we discuss two flowing phases—for example, in a gas–liquid or liquid–liquid contactor.

4.3.1 Single Stream

This section presents a simple model for transport of a solute across a porous membrane. Such a setup is common in dialysis-type devices. A mesoscopic model is quite useful to correlate the data and to predict the extent of purification that is achievable in the device. The model analyzed is shown schematically in Figure 4.4.

Again, a bulk average concentration C_{Ab} is used such that the moles of A crossing any cross-sectional area (perpendicular to the flow direction) is

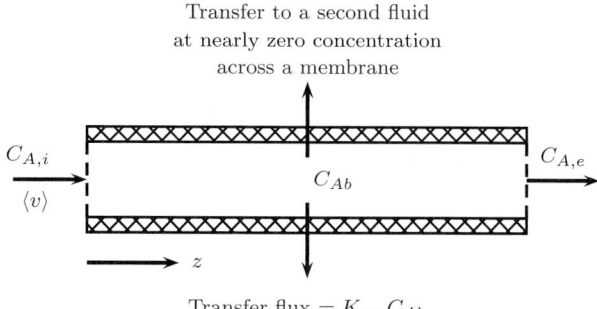

Transfer to a second fluid
at nearly zero concentration
across a membrane

$C_{A,i}$

$\langle v \rangle$

C_{Ab}

$C_{A,e}$

z

Transfer flux $= K_{om}C_{Ab}$

Figure 4.4   Schematic of a single-stream mass exchanger.

equal to QC_{Ab}, where Q is the volumetric flow rate. The application of the mass balance for the solute A then leads to

$$Q\frac{dC_{Ab}}{dz} = -(2\pi R)N_{Aw} \tag{4.17}$$

where N_{Aw} is the moles of species transferred across the membrane per unit area of the walls.

A model for mass transfer across the membrane is needed, and the use of a transport law will then complete the system model. A simple law may be

$$N_{Aw} = \mathcal{P}_{om}C_{Ab}$$

where $\mathcal{P}_{om}$ is an overall permeance parameter across the membrane. This is similar to an overall mass transfer coefficient.

In the previous equation, the driving force is the local cup mixed concentration minus the external fluid (permeate-side) concentration. We assume the external fluid concentration is zero here—an assumption that holds if the diffusing solute is collected in a large volume of liquid or is being washed away by a high flow rate of the external "sweep" fluid.

Substituting and integrating, we get the following performance equation for a single-stream mass exchanger:

$$\frac{C_{Ab}(z)}{C_{A,i}} = \exp\left(-\frac{2\mathcal{P}_{om}\,z}{\langle v \rangle\, R}\right)$$

This is an useful equation, for example, for determining the overall permeance value from the measured exit concentration.

Note that the permeation requires two transport steps: mass transfer from the bulk fluid to the wall, and permeation through the membrane walls. The overall permeance used in the preceding model is a combination of these two steps. It is equal to the true membrane permeance, $\mathcal{P}_m$, only if the transport to the wall from the bulk fluid is fast. This is usually the case, since the diffusion process in the membrane is much slower than that in the liquid phase. In general, a series resistance type of formula is often used:

$$\frac{1}{\mathcal{P}_{om}} = \frac{1}{\mathcal{P}_m} + \frac{1}{k_m}$$

where k_m is the mass transfer coefficient from the bulk liquid to the surface of the membrane.

4.3.2   Two Streams

Consider the modeling of the packed column absorber shown in Figure 4.5. This column is packed with some solids to promote mass transfer, with the gas and the liquid flowing (usually) countercurrent to each other.

   The gas and liquid phases have a common interface, and mass is transferred across the system. Each of these phases may be assumed to be in plug flow, which provides the simplest description of the mixing pattern. Alternatively, dispersion may be superimposed on the model. Also, we assume dilute systems, such that the gas flow rate does not vary significantly in the column. Constant temperature and pressure is also assumed. These assumptions allow for the simplest model for the system, which is useful for a quick design calculation. The corrections to account for these assumptions can then be applied progressively and as needed.

   Based on the presence of the two phases, the control volume shown in Figure 4.6 can be set up. It consists of a gas phase and a liquid phase, with mass transfer taking place from gas to liquid. The conservation law can be applied to the gas and liquid phases.

   The balance equations for the gas phase can be expressed in words:

$$\text{in} - \text{out} - \text{transferred to liquid} = 0$$

The *in* term is equal to $G\,y(z)$, where G is the molar flow rate of the gas and y is the species mole fraction at location z. Note that y is used for y_A for simplicity of notation here.

   Similarly, the *out* term is given by $G\,y(z + \Delta z)$. The molar flow rate of gas is assumed to be constant—an assumption that is valid for low concentrations of the solute in the gas phase. Hence G is not a function of z. This is not true for concentrated solutes, however, and in such a case the model needs to

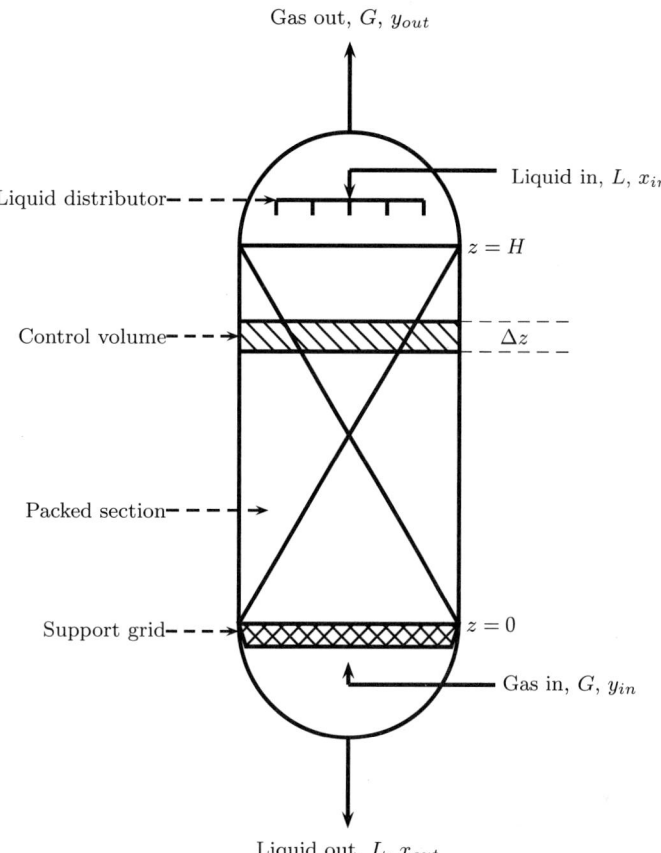

Figure 4.5 Schematic of a counterflow mass exchanger: a packed bed absorption column.

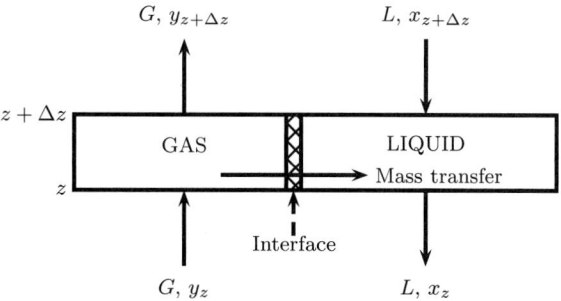

Figure 4.6 Control volume for gas and liquid for a counterflow gas absorption column.

be reformulated in terms of the mole ratio. Here we focus on dilute systems to illustrate the model building aspects.

Let the transfer rate term be denoted by $\dot{\mathcal{M}}_A$; this term will be modeled later in this section using a mass transfer coefficient. We will leave it as a term here to complete the first part of the modeling, the use of the conservation principle. The gas-phase balance equation in mathematical form is

$$G\,y(z) - G\,y(z + \Delta z) - \dot{\mathcal{M}}_A = 0$$

Similarly, the liquid-phase balance is

$$\text{in} - \text{out} + \text{transferred to liquid} = 0$$

where *in* term is $L\,x(z + \Delta z)$, The *in* location is now $z + \Delta z$ since the liquid is flowing in the opposite direction. The *out* term is $L\,x(z)$. The transfer term appears with a plus sign because mass is now received from the gas phase. The liquid balance is therefore

$$L\,x(z + \Delta z) - L\,x(z) + \dot{\mathcal{M}}_A = 0$$

Now we close the model by writing a transport model for the $\dot{\mathcal{M}}_A$ term. When doing so, we must consider the following points.

First, we will use an overall mass transfer coefficient K_y because there are two resistances to mass transfer: from gas to interface, and from interface to bulk liquid. Details of the relation between K_y and the individual mass transfer coefficients are provided later (Section 6.5.2). For now, we simply note that K_y represents the transfer coefficient from the bulk gas phase to the bulk liquid phase. This is based on the mole fraction driving force, so the subscript y is used.

Second, the mass transfer coefficient K_y is based on the unit transfer area (intrinsic coefficient), so we multiply it by the transfer area. The latter is equal to transfer area per unit column volume, a_{gl} times the column volume in the differential meso-volume $A_c\Delta z$. Here A_c is the column cross-sectional area. The combined coefficient $K_y a_{gl}$ is called the volumetric overall mass transfer coefficient.

The driving force must be corrected for equilibrium considerations. Note that $y - x$ is not the driving force. The driving force is the departure from equilibrium, so $y - mx$ is the proper driving force at any cross-section in the column. Here m is the partition coefficient, with y_{eq} being defined as mx. The departure from equilibrium is therefore $y - mx$ and is the correct driving force for mass transfer.

The mass transfer rate per unit contractor volume, then, is $K_y a_{gl}(y - mx)$ locally at position z. Hence the transfer rate from gas to liquid for the control volume is

$$\dot{\mathcal{M}}_A = K_y a_{gl}(y - mx)A_c \Delta z$$

where A_c is the cross-sectional area of the column.

Substituting this defintion in the mass balance equations for the gas and liquid models and taking the limit as Δz tends to zero, the following equations are obtained:

$$G\frac{dy}{dz} = -K_y a_{gl} A_c(y - mx) \tag{4.18}$$

and

$$L\frac{dx}{dz} = -K_y a_{gl} A_c(y - mx) \tag{4.19}$$

Two endpoint conditions are required since we have two first-order differential equations and the inlet values are used as the conditions:

At $z = 0$, gas inlet, $y = y_{in}$

At $z = L$, liquid inlet, $x = x_{in}$

Solution to the Model Equations

Equation 4.19 is multiplied by m and then subtracted from Equation 4.18, which gives:

$$\frac{d}{dz}(y - mx) = -K_y a_{gl} A_c \left(\frac{1}{G} - \frac{m}{L}\right)(y - mx) \tag{4.20}$$

This equation can then be integrated by separation of variables if we treat $(y - mx)$ as a variable:

$$\int_{in}^{out} \frac{d(y - mx)}{y - mx} = -K_y a_{gl}\left(\frac{1}{G} - \frac{m}{L}\right)A_c \int_0^H dz \tag{4.21}$$

The limits on the left are the inlet and exit values for $y - mx$.

The integration leads to the following equation, which relates the volume of the packed column, $V_c = A_c H$, to the exit mole fractions:

$$\boxed{\ln\left(\frac{y_{in} - mx_{out}}{y_{out} - mx_{in}}\right) = \frac{K_y a_{gl}}{G}\left(1 - \frac{mG}{L}\right)V_c} \tag{4.22}$$

This expression provides one relationship between the exit mole fractions. The second relationship is obtained by using an overall mass balance.

Overall Mass Balance

The second equation relies on the invariance property of the system—that is, the overall mass balance. Equations 4.19 and 4.18 can be combined to give

$$G\frac{dy}{dz} = L\frac{dx}{dz} \qquad (4.23)$$

The integrated form is the overall mass balance:

$$\boxed{G(y_{in} - y_{out}) = L(x_{out} - x_{in})} \qquad (4.24)$$

This equation is simply the statement that moles of A lost by the gas equals moles gained by the liquid.

Equations 4.24 and 4.22 permit calculation of two unknowns. Usually the inlet composition of the streams is known, which leaves three unknowns: the two exit compositions and the volume of the contacter. Two equations are available in our model, and one of the unknowns, is to be specified. Thus, if the volume of the absorber is specified, the exit mole fractions of both the gas and the liquid can be calculated (a simulation problem). Similarly, if one exit composition is specified, then the second exit concentration and the volume of the absorber needed to achieve the specified separation can be calculated (a design problem). This procedure is illustrated in Example 4.2.

Log Mean Driving Force

From Equation 4.24, we can show with a few steps of simple algebra that

$$\left(1 - \frac{mG}{L}\right) = \frac{[y_{in} - mx_{out}] - [y_{out} - mx_{in}]}{y_{in} - y_{out}} \qquad (4.25)$$

We can replace the bracketed term on the right side of Equation 4.22 with this identity. A log mean driving force (LMDF) is now defined to simplify the final result:

$$\text{LMDF} = \frac{[y_{in} - mx_{out}] - [y_{out} - mx_{in}]}{\ln\left([y_{in} - mx_{out}]/[y_{out} - mx_{in}]\right)}$$

This is the logarithmic average of the driving force at the inlet, $[y_{in} - mx_{out}]$, and that at the outlet, $[y_{out} - mx_{in}]$. In turn, Equation 4.22 can be rearranged as

$$\boxed{G(y_{in} - y_{out}) = K_y a_{gl}(A_C H)\ \text{LMDF}} \qquad (4.26)$$

The left side of Equation 4.26 represents the total moles/s of gas absorbed. The right side then says that this rate is equal to the volumetric mass transfer coefficient times the volume of the absorber times a driving force that is the log mean average of the driving force at the two ends.

Equation 4.26 has an exact analogue in heat transfer for a counterflow heat exchanger:

$$Q = U \; A \; \mathrm{LMTD}$$

where Q is the heat transferred from hot fluid to cold fluid, U is the overall heat transfer coefficient, A is the surface area of the heat exchanger, and LMTD is the log mean temperature difference.

Minimum Liquid Flow Rate

A quantity of interest is the minimum liquid flow rate. This rate can be calculated directly using the overall mass balance for A together with a thermodynamic condition. The overall mass balance is given by Equation 4.24, which is reproduced here for convenience:

$$G(y_{in} - y_{out}) = L(x_{out} - x_{in})$$

Assume the exit liquid is in equilibrium with the inlet gas, such that $x_{out} = y_{in}/m$. We can then solve for L, which gives the minimum liquid flow rate to be used.

If the entering liquid is fresh, then $x_{in} = 0$ and we get the following expression for the minimum liquid flow rate:

$$L_{min} = mG\frac{(y_{in} - y_{out})}{y_{in}} \tag{4.27}$$

Figure 4.7 shows the concept of the minimum liquid flow rate needed. The operating line touches the equilibrium line at the conditions of minimum L/G, known as the pinch point.

Typically the liquid flow rate is 1.4 times the minimum. This provides all the key information needed to design a two-phase mass exchanger based on the assumption that both phases are in plug flow and they flow countercurrent to each other. The calculations are illustrated in Example 4.2.

Example 4.2 Gas Absorption Column Design

A solute A is to be recovered from a gas stream, and the following conditions are specified: The gas flow rate is 0.062 kmol/s with 1.6% solute; the liquid flow rate is 1.4 times the minimum and the solubility coefficient, m, is 40.

The column should provide an outlet mole fraction of 0.004. Find the height of the absorber.

Use a value for the overall mass transfer coefficient of 0.05 kmol/m^3 s per mole fraction difference. Usually this value depends on the operating gas and liquid velocities and the type of packing used, and is estimated using empirical correlations.

Solution

We have $y_{in} = 0.016$ and $y_{out} = 0.004$. Also $x_{in} = 0$ since a pure liquid is being used.

The minimum liquid flow rate can be computed using Equation 4.27:

$$L_{min} = 40 * 0.062 * (0.0166 - 0.004)/0.016 = 1.86 \text{ kmol/s}$$

The value of L is usually 1.4 times the minimum flow. Hence $L = 2.6$ kmol/sec.

The mole fraction in the liquid exit is calculated using Equation 4.24. It gives x_{out} as 2.86×10^{-04}.

Now we can find the log mean driving force. The driving force at the gas inlet is $y_{in} - mx_{out} = 0.0046$. The driving force at the gas outlet is $y_{out} - mx_{in} = 0.004\ 0.043$. The logarithmic average of the two is calculated as $(0.0046 - 0.004)/\ln(0.0046/0.004)$ and is equal to 0.0043.

The volume of the column needed can then be calculated using Equation 4.26. The required volume is found to be 3.4772 m^3.

Column diameter is chosen based on flooding considerations, which will depend on the operating velocity. Too high a velocity will cause flooding in the column, so that the liquid will not find its way down. Too low a velocity will result in poor mass transfer. We will use an operating gas velocity of 1 m/s here for illustration. The gas volumetric flow rate is calculated as 1.5 m^3/s, so the column's cross-sectional area is 1.5 m^2. The column diameter is therefore equal to 1.2 m and the height of the column needed is 2.31 m.

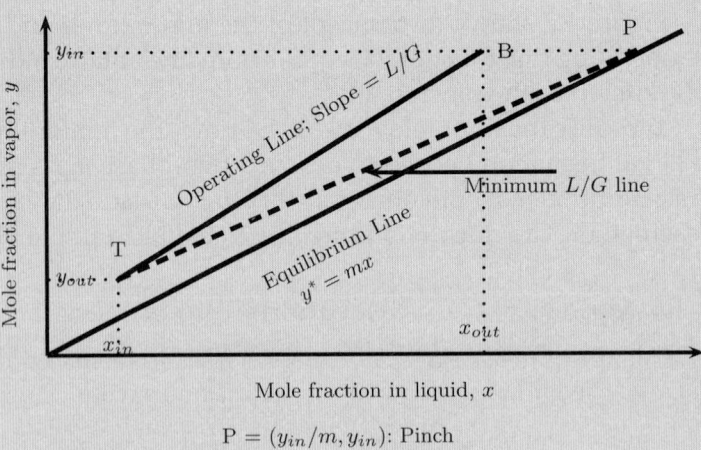

$P = (y_{in}/m, y_{in})$: Pinch
$B = (x_{out}, y_{in})$: Bottom
$T = (x_{in}, y_{out})$: Top

Figure 4.7 Gas absorption problem shown as a plot of operating line and equilibrium curve.

4.3.3   NTU and HTU Representation

The design equations can also be cast in an alternative form, known as the NTU, HTU representation. The gas-phase balance (Equation 4.18; repeated here) is the starting point for developing the HTU and NTU representation:

$$G\frac{dy}{dz} = -K_y a_{gl} A_c (y - mx)$$

Using y^* for mx, this can be generalized as

$$G\frac{dy}{dz} = -K_y a_{gl} A_c (y - y^*) \tag{4.28}$$

This form is applicable even to the nonlinear equilibrium relationship $y^* = y^*(x)$, where x is the local liquid mole fraction and y^* is the corresponding equilibrium value.

Equation 4.28 can be rearranged to

$$H = \left[\frac{G}{K_y a_{gl} A_c}\right]\left[\int_{in}^{out} \frac{dy}{y - y^*}\right] \tag{4.29}$$

The first bracketed term is called HTU (height of transfer unit), while the second term is called NTU (number of transfer unit).

$$\text{HTU} = \frac{G/A_c}{K_y a_{gl}}$$

where G/A_c is molar superficial gas velocity, in mol/m^{2}s. The HTU parameter has the units of height and represents the ease or difficulty of mass transfer. Larger values indicate poorer mass transfer. Data are available for a wide range of packing materials used in practice. The range of values is from 0.3 to 1.1 m.

The integral term in Equation 4.29 is referred to NTU:

$$\text{NTU} = \int_{in}^{out} \frac{dy}{y - y^*}$$

It is measure of the driving force available for mass transfer.

Hence the height of the column required is calculated as the product of HTU and NTU:

$$\text{Height of column} = \text{HTU} \times \text{NTU}$$

This form can be used for both linear and nonlinear equilibria. The only restriction is that it is valid just for dilute systems—that is, when the gas

velocity does not change significantly in the column. The NTU must be evaluated from numerical integration for the nonlinear case, whereas analytical
solutions can be found for the linear case as shown in Example 4.3. Since the
gas-phase mole fraction was used to define the driving force and mass transfer coefficients, Seader et al. designate HTU as H_{OG} and NTU as N_{OG}. Other
definitions, such as H_{OL} and N_{OL}, can also be used. These definitions apply
when the liquid mass balance is used as the starting point instead of Equation
4.28. The final results for the height of the column will be unchanged.

   For linear systems, the value of NTU can be obtained analytically as
shown in Example 4.3.

Example 4.3   NTU Equation for a Linear Equilibrium Case

Derive an expression for NTU for a linear equilibrium relation, and apply it
to Example 4.2.

Solution

For a linear relation, $y^* = mx$ and the integral for NTU is:

$$NTU = \int_{y_{in}}^{y_{out}} \frac{dy}{y - mx} \qquad (4.30)$$

At any point in the column, x is related to y by

$$L(x - x_{in}) = G(y - y_{out})$$

which is the mass balance from the top of the column to any arbitrary height.
Thus,

$$x = x_{in} + \frac{G}{L}(y - y_{out})$$

If this definition is substituted into Equation 4.30, the integration can be performed analytically. The final result for the common case of $x_{in} = 0$ is

$$NTU = \left(\frac{1}{1 - mG/L}\right) \ln\left[\frac{y_{in}(1 - mG/L) + y_{out}mG/L}{y_{out}}\right]$$

   For Example 4.2, the parameter mG/L is calculated as $(40 * 0.062/2.6) = 0.9524$. Now we can substitute the inlet and exit concentrations and this
parameter and estimate the NTU as 2.804.
   The HTU is calculated as $(0.062/1.5)/0.05 = 0.8267$ m. The column
height is the product of the NTU and the HTU and is equal to 2.318 m.

Summary

- Mesoscopic models are useful if there is a principal direction of flow along which the concentration changes significantly. Examples include a pipe flow reactor and a packed bed exchanger. The concentration variation in the cross-flow direction is not modeled; instead, its effect is captured by cross-sectional averaging. Thus the control volume spans a differential length in the flow direction and the entire pipe or reactor cross-section in the cross-flow direction.

- Two types of average concentrations can be defined: the flow weighted or the cup mixed average and the cross-sectional average. Their definitions are important to understand.

- The mass transfer coefficient concept is needed for meso-modeling applications, since the local wall flux value is not available. The coefficients are often calculated using empirical correlations. The cup mixed concentration is used as a measure of the fluid concentration in defining the driving force for mass transfer in pipe or channel flows.

- Solid dissolution from a pipe wall is the simplest problem in mesoscopic modeling. All the steps that go into the model formulation were shown in detail in this chapter, and form the basic pattern for mesoscopic modeling. An average mass transfer coefficient over the entire length of the pipe is used in these models since the local values can vary as a function of axial position.

- Mass transfer data can be correlated using dimensionless parameters, and such correlations are readily available for a large number of problems, including pipe flow. Commonly, a dimensionless parameter, the Sherwood number, is used as a dimensionless mass transfer coefficient. It is correlated in terms of the Reynolds and Schmidt numbers. The product of the Reynolds and Schmidt numbers is known as the Peclet number.

- Tubular flow reactors are often modeled using the concept of plug flow at a first level of approximation. The effect of non-uniform velocity is ignored in this model. The cup mixing and cross-sectional averages are the same here and no additional closure is needed.

- The dispersion model is an attempt to correct for the deviation from plug flow. The dispersion coefficient D_E is used as a parameter here. The value of D_E is zero for a plug flow case; in contrast, the value is close to infinity for a completely backmixed system.

- The dispersion coefficient should not be confused with the diffusion coefficient, although the units are the same for both (m^2/s.)

- Mass transfer units with one or two flowing phases can be modeled using mesoscopic models. The phases may be assumed to be in plug flow for a simple model. Often the dispersion concept is superposed on plug flow to account for the deviations from the ideal contacting pattern of plug flow.
- Mass transfer in a countercurrent mass exchanger is a well-studied problem and has many applications in the design of absorption and extraction processes and similar operations. This problem has an analogy with heat exchangers. The concept of log mean driving force provides a relation for the volume of the contactor required to achieve a given degree of separation.

Review Questions

4.1 Define cup mixing and cross-sectional average concentration.

4.2 How is the driving force defined for mass transfer in pipe flow?

4.3 Distinguish between local and average mass transfer coefficients.

4.4 What is the plug flow assumption and why is it useful?

4.5 Which parameter is commonly used to correct the deviation from plug flow?

4.6 When is the plug flow assumption inadequate?

4.7 Write the formula to calculate the dispersion coefficient for a pipe flow of Newtonian fluid.

4.8 Explain why the dispersion coefficient is inversely proportional to the diffusion coefficient.

4.9 What is the definition of the dispersion Peclet number?

4.10 Is plug flow approached for low or high values of the dispersion Peclet number?

4.11 Define logarithmic mean driving force for two-phase mass transfer.

4.12 What is meant by the pinch point?

4.13 How is the minimum liquid flow rate calculated in an absorption column?

4.14 Define HTU and NTU, and state the equation used to find the height of a countercurrent mass exchanger.

Problems

4.1 **Mass transfer coefficient calculations in pipe flow.** Calculate and plot the average mass transfer coefficient from bulk to pipe wall as a function of position from the entrance of the pipe for the following conditions. The pipe diameter is 1 cm.

 - Flow is laminar with a Reynolds number of 1000. Carrier fluid is air with $D_A = 2 \times 10^{-4}$ m^2/s.

- Flow is laminar with a Reynolds number of 1000. Carrier fluid is water with $D_A = 2 \times 10^{-9}$ m^2/s.
- Flow is turbulent with a Reynolds number of 5000. Carrier fluid is air with $D_A = 2 \times 10^{-4}$ m^2/s.
- Flow is turbulent with a Reynolds number of 5000. Carrier fluid is water with $D_A = 2 \times 10^{-9}$ m^2/s.

4.2 **Lead contamination in a corroding pipe.** A pipe has a diameter of 2.5 cm and a lead soldered section that is 15 cm long. The water flows in the pipe with an average velocity of 0.2 m/s. The saturation solubility of lead compound is 10 g/m^3. Find the dissolved lead concentration at the pipe exit if the mass transfer coefficient for solid dissolution has a value of 2×10^{-5} m/s.

4.3 **Microchannel reactor with a wall reaction.** A microchannel reactor consists of two parallel plates spaced 3 mm apart and 1 m wide. Gas at a flow rate of 0.05 m^3/s containing a pollutant flows in the gap between the two plates. The wall are coated with a catalyst, and a surface reaction is taking place on the pipe wall. Assume the surface reaction to be fast. Calculate the length of the channel required to remove 90% of the pollutant. Use 7.54 as the value of the average Sherwood number. Note that this is defined as $k_m d_h / D_A$, where d_h is the hydraulic diameter.

4.4 **Mass transfer value from measured absorption data.** Ammonia (5%) from air is absorbed into water falling as a thin film of liquid over a vertical wall of height 1 m. The flow rate of water is 5×10^{-3} kg/s per unit width of the plate. The thickness of the film is 1.4×10^{-4} m. The exit concentration of ammonia in the liquid is 200 mol/m^3 and the Henry's law coefficient is 30 atm. What is the value of the mass transfer coefficient in this system? Assume gas is in excess.

4.5 **Comparison of plug flow and backimixed flow reactors.** Show that the steady state exit concentration for a first-order reaction is $1/[1 + Da]$ for a backmixed flow. Determine the corresponding expression if the reactor is modeled as a plug flow reactor. Make a comparison plot of the exit concentration for these two cases as a function of Da. Show that if $Da <\approx 0.3$, then the mixing behavior is not important and the two models give nearly the same conversion.

4.6 **Second-order reaction in plug flow.** Derive an expression for the exit concentration for a second-order reaction taking place in a plug flow reactor.

4.7 **Zero-order reaction in plug flow.** A zero-order reaction with a rate constant of k_0 is carried out in the plug flow reactor. Show that the dimensionless exit concentration is given as

$$c_{A,e} = 1 - Da \text{ for } Da \leq 1$$

where Da is defined as

$$Da = \frac{V k_0}{Q C_{A,i}}$$

Find the exit concentration if Da is greater than 1.

4.8   Bimolecular reaction in a pipe. An aqueous feed to a pipe flow reactor consists of a mixture of A and B with concentrations of 100 mol/m^3 and 200 mol/m^3 for A and B, respectively. The feed flow rate is 0.04 m^3/s, and the following reaction takes places as:

$$A + B \rightarrow \text{products}$$

The rate of reaction (S.I. units) follows bimolecular kinetics:

$$R_A = 0.4 C_A C_B$$

Find the exit concentration of A and B if the reactor volume is 1 L. Assume plug flow.

4.9   Membrane aeration system. Membrane aeration systems are used in some applications to transfer oxygen to water or other liquids. They have applications in biomedical engineering. Such a system consists of an inner tube, with water containing no dissolved oxygen entering the system. The inner tube is surrounded by an annular tube, with a permeable wall separating the two tubes. Pure oxygen at a pressure of 2 atm is maintained in the annular tube. A 35% oxygen saturation was observed for a tube of 40 m for an average velocity of water of 50 cm/s. The inner tube diameter is 1 cm.

Determine the overall membrane permeance. Also calculate the mass transfer coefficient in the system and use the series resistance formula to find the true membrane permeance.

4.10   Height of an absorption column: LMDF method. A gas stream containing 8 mol % of ammonia is to be purified to an exit concentration of 0.5 mol % ammonia. The gas's superficial molar velocity is 100 kg mol/h m^2. Water is used as the absorbent, with the flow rate being 1.5 times the minimum flow rate. The mass transfer coefficient $K_y a_{gl}$ is estimated as 0.2 kg mol/h m^3. Use a constant value of 1 for the solubility parameter, m—although the equilibrium for ammonia–water is actually nonlinear. Find the height of the column needed using the log mean driving force method.

4.11   Height of an absorption column: NTU-HTU method. For problem 10, find NTU and HTU parameters and calculate the height of column again.

4.12   Simulation with MATLAB BVP4C solver. Numerical solutions of Equations 4.18 and 4.19 are examined in this problem. Write these equations in dimensionless form by introducing a dimensionless length $\zeta = z/H$. Show that the two dimensionless parameters defined here are needed. First is the dimensionless mass transfer parameter:

$$\kappa_{gl} = \frac{K_y a_{gl} H}{\mathcal{G}} \tag{4.31}$$

where $\mathcal{G}$ is the molar velocity, G/A_c. Second is the flow ratio parameter, G/L.

Show that the model equations now take the following simple forms:

$$\frac{dy}{d\zeta} = -\alpha_{gl}(y - mx)$$ (4.32)

and

$$L\frac{dx}{d\zeta} = -\alpha_{gl}\frac{G}{L}(y - mx)$$ (4.33)

Boundary conditions are needed at the two endpoints of ζ of 0 and 1. State the conditions. Write code to simulate this system using the MATLAB BVP4C solver. Use the code to model an absorber for the following conditions:

- Gas flow rate, $G = 0.062$ mol/s with a solute mole fraction of 1.6%
- Liquid flow rate, $L = 1.6$ mol/s with no dissolved solute
- Column area = 1.5 m^2
- Solubility coefficient, $m = 40$
- Overall mass transfer coefficient, $K_y a_{gl} = 0.05$ mol/m^3 s
- Height of the absorber, $H = 2.3$ m

Plot the liquid and gas mole fractions as a function of column height.

CHAPTER 5

Equations of Mass Transfer

Learning Objectives

After completing this chapter, you will be able to:

- Derive differential equations for mass transfer in Cartesian coordinates in flux form.
- Express these relationships in vector form so that they can be used for any coordinate system.
- Define the mixture velocity and the diffusion flux.
- Examine some properties of the diffusion flux.
- State some forms of Fick's law for binary systems depending on the frame of reference used.
- Derive the differential equation for the concentration field by combining the equation in the flux form with Fick's law.
- State the boundary conditions for many standard cases.
- Show how averaging of differential models leads to mesoscopic and macroscopic models, and describe the relationship between these models.
- Show how the differential models for two flowing phases arise by local volume averaging of the individual continuity equation.

Chapter 1 examined the basic philosophy behind mathematical modeling of mass transport processes, and Chapter 2 demonstrated the methodology with a few examples on a problem to problem basis. We examined how the mass conservation principle combined with a simple Fick's law as a constitutive equation can be used to set up differential models for some common problems posed in simple 1-D coordinates. Chapters 3 and 4 provided some examples of macroscopic and mesoscopic models, again on a problem to problem basis.

The goal of this chapter is to explore these topics in a more formal setting and derive general differential equations for concentration distribution. We first apply the conservation law to derive the equations in flux form. The diffusion flux must be closed at this point to get the equations in concentration form.

In dealing with mass transfer problems, an important concept is the system average velocity and the frame of reference used to define the diffusion flux of various species. The system average velocity can be defined in a number of ways, and the diffusion flux can be correspondingly defined in a number of ways. Thus, even the simple Fick's law for binary diffusion takes various forms depending on which frame of reference is used. These subtleties can be a source of confusion for the uninitiated. To clarify them, these concepts are defined in Section 5.2.

The next step in completing the differential models is to combine the constitutive model (usually a Fick's law model or some version of it) for the flux with the mass conservation equation. This then leads to a differential equation of mass transfer in concentration form. This equation represents the starting point for the analysis and modeling of a range of mass transport problems and has wide applications in many fields. Commonly used boundary conditions for various cases are then formulated, which completes the model formulation.

A volume average of the differential model over a larger control volume or the whole unit provides the macroscopic model for mass transfer. The necessary equations are formally derived starting from the differential models. Mesoscopic models can also be derived in a similar manner. Deriving these by averaging provides the link between the two levels of models and shows the interpretation of the various terms as integrals of pointwise values of the differential model. Closure assumptions needed in these models are clearly identified in the process.

Systems with two or more phases can be treated in the same manner, by starting with local volume averaging to obtain the differential equations for each of these phases. The interfacial transfer term arises naturally as a consequence of local volume averaging. Additional cross-sectional averaging leads to two-phase mesoscopic models. Similarly, averaging over a larger volume of local volume averaged models leads to two-phase macroscopic models. This systematic way of deriving the equations assigns the meaning to various terms and identifies the needed closure terms.

5.1   Flux Form

The equations for mass transfer can be derived in either mass units or mole units. In the combined flux form, these equations can be easily interconverted using the species molecular weight (mass = moles × the species molecular weight). We first show the equations on mole basis.

5.1.1   Mole Basis

Consider the control volume shown in Figure 1.9 in the shape of a box. This box has six control surfaces. We take them pairs and evaluate the net mole efflux from the control surfaces.

The faces at x and $x + \Delta x$ are taken as a first pair. The moles of A leaving the face $x + \Delta x$ per unit time is flux at the point times area, and is equal to N_{Ax} evaluated at $x + \Delta x$ times the area of the plane, $\Delta y \Delta z$.

$$\text{Moles of } A \text{ leaving in } x\text{-direction/time} = N_{Ax}(x + \Delta x)\Delta y \Delta z$$

Using Taylor series, we can approximate

$$N_{Ax}(x + \Delta x) = N_{Ax}(x) + \frac{\partial}{\partial x}(N_{Ax})\,\Delta x$$

Here, the derivative term on the right side is evaluated at x. Hence we have

$$\text{Moles leaving in } x\text{-direction} = N_{A,x}(x)\Delta y \Delta z + \frac{\partial}{\partial x}(N_{A,x})\Delta x \Delta y \Delta z$$

The first term on the right side is moles of A entering the plane at x.

The net mole efflux (leaving minus entering) from the two planes at x and $x + \Delta x$ is

$$\text{Net mole efflux' from } x\text{-faces} = \frac{\partial}{\partial x}(N_{Ax})\Delta x \Delta y \Delta z$$

The net mole efflux per unit control volume is therefore is obtained by dividing by the size of the control volume:

$$\text{Net mole efflux per unit volume; } x\text{-faces only} = \frac{\partial}{\partial x}(N_{Ax})$$

Similar expressions can be obtained for the y- and z-planes. Thus the net mole efflux per unit volume is equal to

Net mole efflux per unit volume; from all six faces $=$

$$\frac{\partial}{\partial x}(N_{Ax}) + \frac{\partial}{\partial y}(N_{Ay}) + \frac{\partial}{\partial z}(N_{Az})$$

Students proficient in vector calculus will quickly recognize this as the divergence of the flux vector N_A in Cartesian coordinates. The previous expression can be written compactly using vector notation as follows:

$$\text{Mole efflux per unit volume } = \nabla \cdot N_A \tag{5.1}$$

Although derived using Cartesian control volume, this expression in vector form is general and is applicable to any other coordinate system (e.g., cylindrical, spherical). The appropriate form of the divergence expression is used for a given coordinate system to get the detailed expression for the mole efflux in that coordinate system.

The accumulation of A in the control volume is $\partial C_A \partial t$ per unit control volume.

The mole rate of production of A is denoted as R_A per unit control volume.

The conservation statement is

$$\text{in} - \text{out} + \text{net generation} = \text{accumulation}$$

Alternatively, noting that

$$\text{out} - \text{in} = \text{net efflux}$$

we have

$$-\text{net efflux} + \text{net generation} = \text{accumulation}$$

Combining the derived forms for each of the terms, we get the the vector form of the species mass balance equation:

$$\boxed{-\nabla \cdot N_A + R_A = \frac{\partial C_A}{\partial t}} \tag{5.2}$$

This vector form is applicable to any coordinate system with the divergence defined appropriately for that coordinate system. The form of the divergence in three common coordinate systems is shown next for ease of reference.

Divergence operator in three common coordinate systems

Cartesian:

$$\nabla \cdot \boldsymbol{N}_A = \frac{\partial}{\partial x}(N_{Ax}) + \frac{\partial}{\partial y}(N_{Ay}) + \frac{\partial}{\partial z}(N_{Az})$$

Cylindrical:

$$\nabla \cdot \boldsymbol{N}_A = \frac{1}{r}\frac{\partial}{\partial r}(rN_{Ar}) + \frac{1}{r}\frac{\partial N_{A\theta}}{\partial \theta} + \frac{\partial N_{Az}}{\partial z}$$

Spherical:

$$\nabla \cdot \boldsymbol{N}_A = \frac{1}{r^2}\frac{\partial}{\partial r}(r^2 N_{Ar}) + \frac{1}{r \sin \theta}\frac{\partial}{\partial \theta}(N_{A\theta} \sin \theta) + \frac{1}{r \sin \theta}\frac{\partial N_{A\phi}}{\partial \phi}$$

5.1.2 Mass Basis

The equations can be also based on mass units using the following conversion: $N_A = n_A/M_A$. For example, Equation 5.2 can be expressed in mass units as:

$$\boxed{-\nabla \cdot \boldsymbol{n}_A + r_A = \frac{\partial \rho_A}{\partial t}} \tag{5.3}$$

where r_A is the mass rate of production (kilograms produced per unit time per unit volume), which is in turn equal to $R_A M_A$.

Equations 5.2 and 5.3 are both in flux form since the combined flux vector (either mass units or molar units) appears as a term the equations. To proceed further, we need to split the combined flux into a convection flux and a diffusion flux, and then incorporate a suitable expression for diffusion flux (the constitutive model). This will provide the equation in the concentration form rather than in the flux form, which is appropriate since we want to solve for the primary field variable, the species concentration.

One problem is that diffusion flux is not uniquely defined; that is, it can be defined in many ways. Our goal now is to define this component more precisely compared to the simple definition used in Chapters 1 and 2. Specifically, we need a definition of the frame of reference (mixture average velocity) because diffusion flux is defined assuming that the system has no net velocity—that is, the flux is assumed to be from diffusion only. Diffusion flux must therefore be defined with reference to a coordinate system with zero velocity.

5.2   Frame of Reference

One confusing aspect of a multicomponent mixture is that there are many ways to define the average velocity of a mixture, and correspondingly there are many definitions of diffusive flux. In general we can write the following expression in mole units:

$$J_A^{\text{ref}} = N_A(\text{fixed}) - C_A v^{\text{ref}}$$

Alternatively, we can write the expression in mass units by multiplying by M_A:

$$j_A^{\text{ref}} = n_A(\text{fixed}) - \rho_A v^{\text{ref}}$$

The diffusion flux (in either units) is therefore obtained by subtracting the convective part from the combined flux (which is fixed). The convective part, in turn, uses a reference velocity (the mixture velocity, v_{ref}), which can be defined in many ways. The combined flux N_A or n_A is fixed at a given point; in contrast, v and correspondingly J_A or j_A are not, but rather depend on how you define v.

Two common ways of defining the reference velocity are the mole fraction weighted average velocity and mass fraction weighted average velocity. These concepts are introduced now, and Fick's law for binary diffusion is revisited in this setting.

5.2.1   Mass Fraction Averaged Velocity

We start off by assigning a velocity to each of the diffusing species and then averaging these with a weighting factor to define a mixture velocity.

Species Velocity

Let n_A be the flux vector in mass units at a given point. Correspondingly, a velocity for species A, v_A, can be defined at this point, which is related to mass flux vector by the mass concentration of A:

$$\rho_A v_A = n_A$$

Similarly, the velocity of B can be defined (as well other components in the system):

$$\rho_B v_B = n_B$$

Mixture Average Velocity

The mixture velocity v is defined as a weighted average of the species velocities. Various definitions arise depending on which weighting factor is used to define this average. Here, we will use the mass fraction as the weighting factor. Thus the mixture velocity can be defined as follows for a binary mixture:

$$v^{(m)} = \omega_A v_A + \omega_B v_B \tag{5.4}$$

More generally, it can be defined as follows for a multicomponent mixture:

$$v^{(m)} = \sum w_i v_i = \left(\sum \rho_i v_i \right) / \rho \tag{5.5}$$

Here the summation is over all the species present in the system ($i = 1$ to ns). This mixture velocity is also referred to as mass-centric velocity.

A flux with respect to a moving frame can now be defined. Let the moving frame have velocity of $v^{(m)}$ (the mass averaged velocity). We will define relative velocity and then use this definition to define the fluxes. Let

$$j_A^{(m)} = \rho_A (v_A - v^{(m)})$$

represent the mass flux vector seen by an observer who is moving with the mixture velocity $v^{(m)}$. This mass flux can be attributed to only diffusion since the system (relative) velocity is zero in the case of a moving observer. Hence $j_A^{(m)}$ is one way of defining the diffusion flux. (The second way of defining the diffusion flux uses the mole centric velocity discussed in 5.2.2.) The flux defined in these two frames are shown schematically in Figure 5.1.

Note: In the following section and most places in the text, $j_A^{(m)}$ will be abbreviated as j_A and $v^{(m)}$ will be denoted as v.

The flux in a stationary frame is related to the flux in a moving frame as follows:

$$n_A = \rho_A (v_A - v) + \rho_A v = j_A + \rho_A v \tag{5.6}$$

or

$$\boxed{n_A = j_A + \rho_A v} \tag{5.7}$$

The two fluxes n_A and j_A and the corresponding observers are shown in Figure 5.1. Note that the left side of Equation 5.7 is the combined flux, the first term on the right side is the diffusion flux, and the second term on the right side is the convection flux.

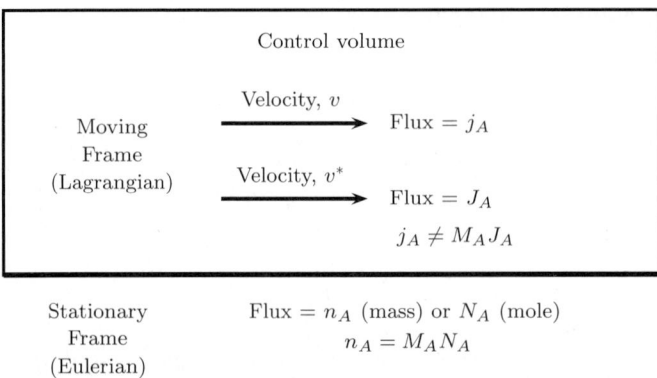

Figure 5.1 Schematic of the definitions of combined flux (frame of stationary observer) and diffusion flux (frame of moving observer).

It follows that $\rho_A v = \omega_A \rho v$ and $\rho v = n_t$, the total flux (mass units). Hence $\rho_A v = \omega_A n_t$ and the previous expression for the combined flux can be written as

$$n_A = j_A + \omega_A n_t \qquad (5.8)$$

The total flux n_t is equal to $n_A + n_B$ in a binary mixture.

Fick's Law: Version for j_A

For a binary system, the following constitutive model is applicable with the flux j_A seen to be proportional to the mass fraction gradient:

$$j_A = -\rho D_{AB} \nabla \omega_A \qquad (5.9)$$

In this version of Fick's law, D_{AB} is the diffusivity of A in the binary mixture of A and B. Note that we use D_{AB} for D_A. It turns out that for a binary system, $D_A = D_B$; this diffusivity is denoted as D_{AB}, meaning the binary pair diffusivity of A-B pair. This interpretation is proved in a later section.

5.2.2   Mole Fraction Averaged Velocity

The species velocity can also be defined using mole units. Thus, if N_A is the flux vector in mole units and C_A is the concentration, the following definition applies for the species velocity. This definition for v_A is the same as that obtained from using mass units:

$$N_A = C_A v_A \qquad (5.10)$$

Let $\boldsymbol{v}^*$ be the average (mole fraction weighted) velocity of the system defined as follows:

$$\boldsymbol{v}^* = \sum_{i=1}^{ns} y_i \boldsymbol{v}_i \qquad (5.11)$$

where y_i is the mole fraction of species i. This mole-centric velocity differs from the mass-centric velocity $\boldsymbol{v}^{(m)}$ (more simply abbreviated as $\boldsymbol{v}$, as indicated earlier). In general, mass fraction is not equal to mole fraction, so the two velocities are different.

For an observer who moves with a system with a velocity of $\boldsymbol{v}^*$, the flux observed will be

$$\boldsymbol{J}_A^* = C_A(\boldsymbol{v}_A - \boldsymbol{v}^*) \qquad (5.12)$$

Rearranging Equation 5.12, we have

$$C_A \boldsymbol{v}_A = \boldsymbol{J}_A^* + C_A \boldsymbol{v}^* \qquad (5.13)$$

We therefore find the following relation between the (molar) flux in a stationary frame and in a moving frame:

$$\boldsymbol{N}_A = \boldsymbol{J}_A^* + C_A v^* \qquad (5.14)$$

where the first term on the right side is referred to as the diffusive flux and the second term is the flux arising out of the net fluid motion (the convective flux). The two fluxes $\boldsymbol{N}_A$ and $\boldsymbol{J}_A$ and the corresponding observers are shown in Figure 5.1.

Equation 5.14 can also be expressed in terms of the mole fraction:

$$\boxed{\boldsymbol{N}_A = \boldsymbol{J}_A^* + y_A \boldsymbol{N}_t} \qquad (5.15)$$

where $\boldsymbol{N}_t$ is the total molar flux (stationary observer). This follows from Equation 5.14 since $\boldsymbol{v}^* = \boldsymbol{N}_t/C$.

Note: The star superscript on J will be dropped and we will denote $\boldsymbol{J}_A^*$ as $\boldsymbol{J}_A$ in further discussions.

Fick's Law: Mole Fraction Form

The diffusive flux in a binary system defined in the mole weighted frame is found to be proportional to the mole fraction gradient. This leads to another version of Fick's law:

$$\boxed{J_A = -C D_{AB} \nabla y_A} \qquad (5.16)$$

In this case, the mole fraction gradient is used as the driving force rather than the usual concentration gradient. Equation 5.16 has more general applicability because the concentration can vary simply due to a local difference in temperature (as per the ideal gas law). For instance, in a room filled with air, the total concentration may be different at two points due to temperature differences between these points (for example, the air may be colder near the window on a icy day in St. Louis). Hence the nitrogen (or oxygen) concentration (p_A/R_gT) is also different between two points but there is no diffusional flux of nitrogen (or oxygen) since there is no mole fraction difference between the two points.

For system with constant total concentration C = constant, (e.g., gaseous systems at a constant temperature and constant total pressure), we have

$$\boldsymbol{J}_A = -D_{AB}\nabla C_A \tag{5.17}$$

This is the traditional form of Fick's law, which was introduced in Chapter 1. This form is therefore valid for a system with constant total concentration, and the flux is stated with reference to a frame moving with v^*.

A few examples of calculation of the species and mixture velocity are now in order and shown in Examples 5.1 to 5.3.

Example 5.1   System Velocity in the Presence of Diffusion

Benzene is evaporating from a liquid in a tube with an exposed vapor space height, H, of 5 cm. The temperature is 300 K and the total pressure is 1 atm. The vapor pressure of benzene at this condition is estimated from the Antoine equation as 0.131 atm. The diffusion coefficient for benzene in air is 9.05×10^{-6} m^2/s.

Find the rate of evaporation assuming diffusion is the only mode of mass transport. Find v_A and v_B. Finally, find the mixture average velocities v and v^*.

Solution

The combined flux is assumed, as a first approximation, to be equal to diffusion flux; these values are then computed using a 1-D slab model. The 1-D slab model gives a linear profile for concentration, so Fick's law can be used in a finite difference form. Molar units are used here.

$$N_{Az} = D_A C \left(\frac{y_{As} - y_{Ab}}{H}\right) = 9.55 \times 10^{-4} \text{ mol } A/\text{m}^2 \text{ s}$$

where C, the total concentration, is calculated using ideal gas law, P/R_gT.

A velocity for species A can now be assigned based on this flux. Since $N_A = C_A v_A$ the corresponding species velocity at the interface is

$$v_A = N_A/C_A = 1.81 \times 10^{-4} \text{ m/s}$$

Since the air is not being transported into the liquid, $N_B = 0$ and, therefore, $v_B = 0$.

The velocity v^* at the interface is then equal to $y_{As} v_A$ (the contribution from A) and zero contribution from B. The value is 2.31×10^{-5} m/s.

Note that v^* is equal to N_t/C, which provides another way of calculating this value. Also v^* can be shown to be constant along the diffusion path since both N_t and C do not vary with the height.

Similarly, v can be calculated. To do so, we need n_A and the mass fraction at the interface. These can be calculated as follows:

$$n_A = N_A M_A = 7.45 \times 10^{-5} \text{ kg } A/\text{m}^2\text{s}$$

and

$$\omega_A = y_A M_A/\bar{M} = 0.2867$$

Hence the mass fraction weighted velocity is

$$v = \omega_A v_A + \omega_B v_B (zero) = 6.32 \times 10^{-5} \text{ m/s}$$

Note that v is equal to n_t/ρ, whcih provides another way of calculating this value. Also, since ρ is different at different points, v changes along the diffusion path. Both v^* and v are non-zero, but have different values.

In this system, a small but finite velocity arises due to diffusion. Since its magnitude is small, it is not intuitive that a velocity exists in this so called stagnant system.

Note: The combined flux calculated using diffusion turns out to be only an approximate value, and the effect of convection needs to be added as an adjustment. The detailed model is presented in Chapter 6. The actual flux accounting for the finite velocity is $10.4 \times 10^{-4} \text{ mol} A/\text{m}^2$ s, as shown in Chapter 6.

Example 5.2 shows a case where v^* is zero but v is not.

Example 5.2   Equimolar Counter-Diffusion

Consider an experiment in which two gases are separated by a porous membrane. The system is maintained at constant temperature and pressure, and diffusion takes place from one side to the other side (Figure 5.2). At the beginning, one side (the left bulb) contains pure hydrogen and the other side contains pure nitrogen. Show that v^* is zero but not v.

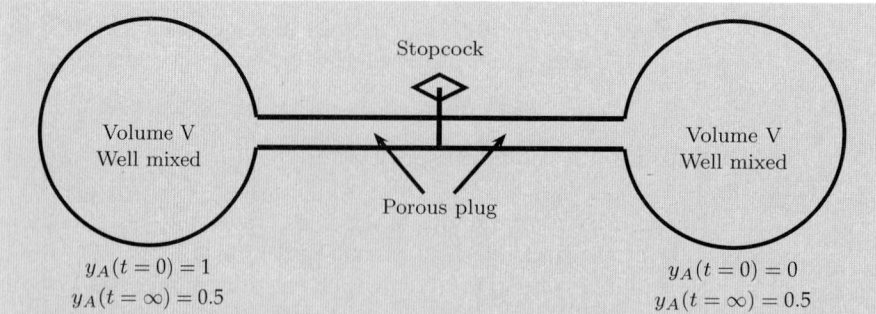

Figure 5.2 Two-bulb system for the study of diffusion coefficients. The entire system is at constant temperature and pressure.

Solution

To keep the total concentration in the bulk same, equal moles should counter-diffuse. Hence $N_A = -N_B$. $N_t = 0$, and $v^* = 0$.

The masses diffusing are not the same:

$$n_A = M_A N_A = 2N_A \text{ for hydrogen}$$
$$n_B = M_B N_B = -28N_A \text{ for nitrogen}$$

Hence

$$n_t = n_A + n_B = -26N_A$$

The net mass flow n_t is non-zero. The negative sign indicates that the flow is from the right side to the left side.

The corresponding velocity can be calculated as $v = n_t/\rho$, with its value depending on the total density of the mixture. Note that v is not zero.

The total density is different at the two ends. Hence the system velocity varies along the length and is directed toward the hydrogen side and increases as you approach the left bulb. Also note that the mass conservation concept requires ρv to be constant along the length.

Example 5.3 illustrates a case in which there is no mass averaged velocity but the molar average velocity is non-zero.

Example 5.3   Equimass Counter-Diffusion

Consider hydrogen and oxygen diffusing to a catalytic surface. The product is water vapor, which diffuses back to the bulk gas (Figure 5.3). Assume $2 \text{ mmol/m}^2 \text{ s}$ hydrogen is diffusing to the catalyst at 1 atm total pressure and a temperature of 300 K. Show that v is zero but not v^*.

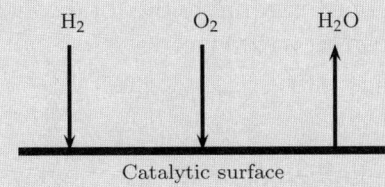

$$N_{O_2} = N_{H_2}/2 \text{ and } N_{H_2O} = -N_{H_2}$$

Figure 5.3 Diffusion followed by surface reaction to a catalytic surface.

Solution

By stoichiometry, we have the following relation between the fluxes since 2 moles of hydrogen reacts with 1 mole of oxygen to produce 2 moles of water:

Hydrogen: $N_A = 2$; oxygen: $N_B = 1$; water: $N_C = -2$. Units in $mmol/m^2s$.

Total moles diffusing: $N_t = 1$

Total concentration: $P/R_gT = 40 \, mol/m^3$

The velocity v^* can now be computed.

$v^* = N_t/C = 1/40 = 2.5 \times 10^{-4} \, mm/s$; a non-zero velocity.

Converting to mass units, we have $n_A = 4$; $n_B = 32$, $n_c = -36 \, mg/m^2 \, s$. Total n_t adds up to zero. Hence v is zero.

To get the equation of mass transfer in the concentration C_A or partial density ρ_A form, we start with the flux form. Then n_A or N_A can be split into convection and diffusion terms and substituted into the flux form of the conservation model. This will provide us with the needed equations. Before we do this, it is useful to show some properties of the diffusive flux.

5.3   Properties of Diffusion Flux

Diffusion flux is defined in the moving frame, and this definition leads to some properties that are discussed next.

Property 1: The sum of the diffusive fluxes of all species is equal to zero.

This is proved as follows. Summing the definitions of diffusive flux of any species i given by Equation 5.12, we have

$$\sum J_i = \sum (C_i v_i) - v^* \sum C_i \qquad (5.18)$$

The summation of C_i is equal to the total concentration C and hence the previous equation can be written as

$$\sum J_i = \sum (C_i v_i) - v^* C \tag{5.19}$$

From the definition of mixture average velocity v^*, we note

$$v^* C = \left(\sum_{i=1}^{ns} y_i v_i \right) C = \sum (C_i v_i) \tag{5.20}$$

The two terms on the right side of Equation 5.19 cancel out, proving that the sum of the diffusive flux taken over all the components is zero:

$$\boxed{\sum_{i=1}^{ns} J_i = 0} \tag{5.21}$$

A similar equation also holds for j_A, the diffusion flux in mass units, and for $\sum j_i = 0$.

For example, in a binary system we have $J_A = -J_B$ and there is only one independent diffusive flux. Likewise, for a multicomponent system there can be only $(ns - 1)$ independent diffusive fluxes. Thus, in a ternary system containing species A, B, and C, the following equation holds:

$$J_C = -(J_A + J_B)$$

There can be only two independent diffusion fluxes.

Property 2: In a binary system, a single parameter D_{AB} is sufficient to model the diffusion flux.

The proof is as follows. Let there be two parameters, D_A and D_B. Then the flux for each species is proportional to its mole fraction gradient in the system:

$$J_A = -C D_A \frac{dy_A}{dz} \tag{5.22}$$

and

$$J_B = -C D_B \frac{dy_B}{dz} \tag{5.23}$$

For a binary system, the diffusive fluxes of A and B are related as follows:

$$J_A = -J_B$$

For a binary system:

$$y_A + y_B = 1$$

or

$$\frac{dy_A}{dz} + \frac{dy_B}{dz} = 0$$

Hence it follows that

$$D_A = D_B = D_{AB}$$

Thus there is only one diffusion coefficient in a binary mixture, which is denoted as D_{AB}, the binary pair diffusion coefficient.

Property 3: In a multicomponent system, $ns(ns - 1)/2$ parameters are needed to model the diffusive fluxes.

This follows from Property 1. For example, in a three-component system there are only three independent diffusion fluxes. A model for flux can therefore be set up with the combination of three basic binary parameters: D_{AB}, D_{BC}, and D_{AC}. The Stefan-Maxwell equation, which we will study in Chapter 15, shows how the flux of each species can be related to these three basic binary parameters. A simpler model is to use the pseudo-binary diffusivity, which is discussed next.

5.4   Pseudo-Binary Diffusivity

Multicomponent diffusion does not follow Fick's law; it is discussed in detail in Chapter 15. A simplified concept is that of pseudo-binary diffusivity, which is defined as follows:

$$J_A = -C D_{A-m} \nabla y_A$$

where D_{A-m} is the diffusivity of A in the mixture defined as though the system were a binary setup—the source of the name pseudo-binary diffusivity.

The multicomponent diffusion is treated as though the binary Fick's law holds for each species taken individually. Note that this is only an approximation, but it turns out to be a good one if species A is present in dilute concentrations and diffusing in a mixture of $B, C, \ldots$, and is mainly required to track the diffusion of species A only. In such cases, the Wilke (1950) equation is often used:

$$\frac{1 - y_A}{D_{A-m}} = \frac{y_B}{D_{AB}} + \frac{y_C}{D_{AC}} + \cdots \tag{5.24}$$

The calculation of the pseudo-binary diffusivity is shown in Example 5.4.

Example 5.4   Pseudo-Binary Diffusivity in a Ternary Mixture

Find the diffusivity of 10% CO_2 in an equimolar mixture of hydrogen and water. Use the Wilke equation.

Solution

The binary pair values are $D_{12} = 0.164$ cm^2/s and $D_{13} = 0.55$ cm^2/s. For an equimolar mixture, $y_2 = y_3 = 0.45$. Also $y_1 = 0.1$ as per the given data.
    Hence from Equation 5.23, we obtain

$$\frac{1 - 0.1}{D_{A-m}} = \frac{0.45}{0.164} + \frac{0.45}{0.55}$$

Solving $D_{1-m} = 0.2527$ cm^2/s. For this particular case, the diffusivity of species 1 in the mixture is the harmonic mean of that in species 2 and 3.
    Note: The diffusivity calculated by the Wilke equation will be a function of the mixture composition; hence it is composition dependent. In practical applications, an average value based on an average mixture composition is used. The alternative is to use the more rigorous Stefan-Maxwell model discussed in Chapter 15. (This model was proposed independently by Stefan and Maxwell in 1871.)

5.5   Concentration Form

The basic differential equation for species mass transfer will now be derived. This section shows the derivations using the mass basis as well as the mole basis. The partial density ρ_A (same as mass concentration) is the field variable for the mass basis derivation, while the molar concentration C_A is the variable for the mole basis derivation.

5.5.1   Mass Basis

We repeat the mass conservation law derived earlier for convenience:

$$-\nabla \cdot \boldsymbol{n}_A + r_A = \frac{\partial \rho_A}{\partial t}$$

This is the basic differential species mass balance equation. Now since

$$n_A = \boldsymbol{v}\rho_A + j_A$$

as per Equation 5.7, the conservation equation can be written in terms of the diffusion flux (after moving some terms around):

$$\frac{\partial \rho_A}{\partial t} + \nabla \cdot (\boldsymbol{v}\rho_A) = -\nabla \cdot \boldsymbol{j}_A + r_A \tag{5.25}$$

The four terms represent the accumulation, convection, diffusion, and reaction.

If Fick's law is used for j_A (which is strictly true for an ideal binary system), we obtain

$$\frac{\partial \rho_A}{\partial t} + \nabla \cdot (\boldsymbol{v}\rho_A) = \nabla \cdot (\rho D_A \nabla w_A) + r_A \tag{5.26}$$

5.5.2   Constant-Density Systems

For constant-density systems, we can move ρ inside the differentiation in Equation 5.25:

$$\frac{\partial \rho_A}{\partial t} + \nabla \cdot (\boldsymbol{v}\rho_A) = \nabla \cdot (D_A \nabla \rho_A) + r_A \tag{5.27}$$

Liquid mixtures can often be approximated as constant-density systems. A gas mixture under isobaric and isothermal conditions with a small concentration of solute being transported can also be treated as constant-density system. Note that the density of a gas mixture is given as $\bar{M}P/R_gT$, where $\bar{M}$ is the average molecular weight. For mixtures with a small concentration of diffusing solute, the contribution to $\bar{M}$ from the changes in solute concentration is small; thus such mixtures can be treated as constant-density case.

If the diffusion coefficient is constant, the first term on the right side of Equation 5.27 can be expressed in Laplacian terms:

$$\frac{\partial \rho_A}{\partial t} + \nabla \cdot (\boldsymbol{v}\rho_A) = D_A \nabla^2 \rho_A + r_A \tag{5.28}$$

This is the equation on a mass basis. Note that the reference velocity is v here.

Dividing by M_A we get the molar form:

$$\frac{\partial C_A}{\partial t} + \nabla \cdot (\boldsymbol{v}C_A) = D_A \nabla^2 C_A + R_A \tag{5.29}$$

Note that r_A/M_A is replaced by R_A in arriving at this equation.

Equation 5.29 is commonly used to obtain species concentration for mass transfer analysis. Indeed, many applications studied in later chapters use this equation as a starting point. Note that the assumption of constant mixture density is implied in its derivation.

5.5.3   Overall Continuity: Mass Basis

If Equation 5.25 is summed over all the species, we obtain the overall mass balance, also known as the continuity equation. We note the following properties and use it in summation:

- $\sum \rho_A = \rho$, the mixture density
- $\sum j_A = 0$; the property of diffusion flux
- $\sum r_A = 0$; no total mass is formed by reaction

The overall continuity is then

$$\frac{\partial \rho}{\partial t} + \nabla \cdot (v\rho) = 0 \tag{5.30}$$

This is same as the continuity equation derived in fluid mechanics.

For constant-density systems, Equation 5.30 reduces to

$$\nabla \cdot v = 0$$

This is known as the incompressibility condition in fluid mechanics. Using this definition in Equation 5.26, we have

$$\frac{\partial \rho_A}{\partial t} + v \cdot \nabla \rho_A = \nabla \cdot (D_A \nabla \rho_A) + r_A \tag{5.31}$$

5.5.4   Mole Basis

Now we will repeat the same model derivations but on a mole basis. We will also use the molar average velocity v^* as the reference velocity.

We repeat the mass conservation law derived earlier for convenience:

$$-\nabla \cdot N_A + R_A = \frac{\partial C_A}{\partial t}$$

This is the basic differential species mass balance equation. Now since

$$N_A = v^* C_A + J_A$$

as per Equation 5.14, the equation can be written in terms of the diffusion flux as

$$\frac{\partial C_A}{\partial t} + \nabla \cdot (\boldsymbol{v}^* C_A) = -\nabla \cdot \boldsymbol{J}_A + R_A \tag{5.32}$$

If Fick's law is used for J_A (which is strictly true for an ideal binary system), we obtain

$$\frac{\partial C_A}{\partial t} + \nabla \cdot (\boldsymbol{v}^* C_A) = \nabla \cdot (C D_A \nabla y_A) + R_A \tag{5.33}$$

For constant concentration conditions and the constant diffusivity case, we obtain

$$\frac{\partial C_A}{\partial t} + \nabla \cdot (\boldsymbol{v}^* C_A) = D_A \nabla^2 C_A + R_A \tag{5.34}$$

5.5.5   Overall Continuity: Mole Basis

We now derive an overall continuity equation based on the mole basis and point out the difference between this equation and that obtained on a mass basis. The starting point is to sum Equation 5.32 over all the species. The sum of the diffusion fluxes remains zero. Although $\sum_i J_i = 0$, $\sum_i R_i$ may not be equal to zero.

Summing Equation 5.32 over all the species, we get the mole continuity equation:

$$\frac{\partial C}{\partial t} + \nabla \cdot (\boldsymbol{v}^* C) = \sum_i R_i$$

If there is no net change in the number of moles in the balanced chemical reaction, then $\sum_i R_i = 0$—for example, if $A + B$ gives $C + D$. More generally, we note that the following equation holds if a single reaction is taking place:

$$\sum R_i = \left(\sum \nu_i \right) \mathcal{R}$$

Here ν_i is the stoichiometric coefficient for the ith species. $\mathcal{R}$ is a rate function that measures the rate of production (defined as the rate of a product species with unit stoichiometry—that is, as R_A / ν_A). If $\sum \nu_i$ is equal to zero (there is no net change in the total moles), the $\sum R_i$ term will be equal to zero. Otherwise, this term should be retained in the overall mole continuity equation.

5.5.6   Common Simplifications

The commonly used governing equation for mass transfer is restated here for ease of reference:

$$\frac{\partial C_A}{\partial t} + \boldsymbol{v} \cdot \nabla C_A = D_A \nabla^2 C_A + R_A \qquad (5.35)$$

This equation applies to a constant-density system and is used as a starting point to solve many mass transfer problems. The four terms in this equation represent accumulation, convection, diffusion, and reaction. Depending on which of these terms balance, the equation can be simplified. The solutions of some prototypical problems with these simplifications are studied in later chapters.

For pure diffusion problems, Equation 5.35 reduces to

$$\nabla^2 C_A = 0$$

This model applies when there is no superimposed flow and when the diffusion-induced velocity is zero or can be neglected.

Diffusion-induced convection problems may be approached with the following form of Equation 5.35:

$$\nabla \cdot N_A = 0 \text{ for } A, B, C, \ldots$$

where N_A is given by Equation 5.14. Some additional problem-specific conditions needs to be superimposed to obtain the total flux $N_t = N_A + N_B + \cdots$ on the system.

Both pure diffusion and diffusion-induced convection problems are examined in Chapter 6.

For transient diffusion with no reaction, the differential equation reduces to Fick's second law:

$$\frac{\partial C_A}{\partial t} = D_A \nabla^2 C_A \qquad (5.36)$$

which states that the accumulation balances the net diffusion from the control surfaces. The solutions to illustrative problems are presented in Chapter 8.

In convection–diffusion problems, the diffusion term is balanced by the convection term, leading to the following simplified version of the general differential equation:

$$\boldsymbol{v} \cdot \nabla C_A = D_A \nabla^2 C_A \qquad (5.37)$$

which states that convection balances diffusion. A velocity profile has to be prescribed to proceed further. These types of problems are addressed in Chapters 10 and 11.

In diffusion–reaction problems, the diffusion term is balanced by the rate of production. This leads to the following governing equation:

$$D_A \nabla^2 C_A + R_A = 0 \tag{5.38}$$

which states that the net diffusion balances the rate of production. Diffusion with reaction is studied in more detail in Chapters 18 and 20.

5.6   Common Boundary Conditions

Boundary conditions for mass transfer are generally classified into three types:

1. Dirichlet: This applies if the concentration is specified at a point in the boundary. It is also known as the boundary condition of the first kind.
2. Neumann: This applies if the slope of concentration is specified at a point in the boundary. It is also known as the boundary condition of the second kind.
3. Robin: In this type, neither the concentration nor its slope is specified, but rather some relation connecting the two is specified. It is also known as the boundary condition of the third kind.

We now illustrate the various common cases and indicate what the appropriate boundary conditions are for these cases.

- At the interface (such as a dissolving solid surface), the concentration can be prescribed. Concentration in the liquid will be the saturation solubility of the solid. This leads to a Dirichlet type of condition.
- At a plane of symmetry or an impermeable wall, the flux is equal to zero, leading to a Neumann condition. Here, dC_A/dn is specified as zero.
- At a heterogeneous surface where a surface reaction occurs, a Robin condition results. The rate of diffusion N_A is equated to the rate of surface reaction $-R_{A,s}$ here:

$$N_A = -R_{A,s}$$

If Fick's law is used for N_A and a first-order reaction takes place at the surface, so that

$$R_{A,s} = -k_s C_A$$

then the boundary condition to be used is

$$-D_A \frac{dC_A}{dn} = k_s C_A \text{ at the surface} \qquad (5.39)$$

This can be seen as a boundary condition of the third kind (Robin condition). Here n is the direction toward the surface. Neither the concentration value nor the flux value is specified, but the relation in Equation 5.39 ties the two together. Hence the third kind of boundary condition is a mixed boundary condition.

- A limiting case for the heterogeneous reaction is a fast reaction at the surface. The concentration of the limiting reactant is set as zero for this case and the Dirichlet condition is used. Note that the concentration cannot be actually equal to zero, since then there would be no reaction. The zero value should be conceived as a limit when the rate constant k_s tends to infinity in Equation 5.39.

- At a gas–liquid interface, the concentration jump condition is used, but the fluxes are continuous at this surface. Thus N_{Ai} from the gas side is equal to N_{Ai} into the liquid (i means the interface here), while the concentration C_{Ai} at the interface on the gas side is related to C_{Ai} on the liquid side by an equilibrium constant or Henry's law constant.

- If the solid is in contact with a fluid, the flux by diffusion from the solid is equal to the convective transport from the solid surface to the bulk of the fluid. A Robin boundary condition is therefore used at the solid–fluid interface where a convective transport is taking place:

$$-D_A \frac{dC_A}{dx} = k_m (H_A C_A - C_{Ab})$$

with H_A being the Henry's law constant that relates the concentration in the solid phase to that in the fluid, k_m is the mass transfer coefficient, and C_{Ab} is the concentration in the fluid away from the solid (i.e., the bulk concentration).

These scenarios cover most of the commonly encountered boundary conditions in mass transfer problems.

5.7   Macroscopic Models: Single-Phase Systems

This section may be omitted with no loss of continuity.

This section provides the derivation of the macroscopic model used in mass transfer analysis by volume averaging of the differential equations of mass transport. Working models for design of mass transfer equipment are often based on such macroscopic models. Single-phase systems are shown here. Extension to multiphase systems requires some additional concepts of local volume averaging and is briefly discussed in the next section.

The starting point is integration of the differential equation given by Equation 5.2, which is reproduced here for ease of reference:

$$-\nabla \cdot N_A + R_A = \frac{\partial C_A}{\partial t} \tag{5.40}$$

This equation is integrated over a macroscopic control volume, which leads to the macroscopic species balance:

$$-\int_V (\nabla \cdot N_A)\, dV + \int_V R_A dV = \int_V \left(\frac{\partial C_A}{\partial t}\right) dV \tag{5.41}$$

We show the representation of each of these terms in the following paragraphs. The first term in Equation 5.41 is the combined flux term. To integrate this, we need the divergence theorem, also known as the Green-Gauss theorem in vector calculus. This theorem says that a surface integral can be converted to a volume integral. We use this idea in reverse to convert the volume integral to a surface integral:

$$-\int_V (\nabla \cdot \boldsymbol{N_A})\, dV = -\int_S (\boldsymbol{n} \cdot N_A) dS \tag{5.42}$$

The second term in Equation 5.41 is the reaction rate term $\int_V R_A dV$. It can also be written as $V\langle R_A \rangle$, where $\langle R_A \rangle$ is the average rate of reaction defined as

$$\langle R_A \rangle = \frac{1}{V}\int_V R_A dV$$

Finally, the accumulation term on the right side of Equation 5.41 can be written as

$$\text{accumulation} = \int_V \frac{\partial C_A}{\partial t} dV$$

This expression can be rewritten as follows, assuming that the control volume is not changing in time:

$$\text{accumulation} = \frac{d}{dt}\left(\int_V C_A dV\right)$$

In turn, this expression can be written in terms of an average concentration:

$$\text{accumulation} = \frac{d}{dt}\left(V\left\langle C_A\right\rangle\right)$$

where we use the following definition:

$$\langle C_A \rangle = \frac{1}{V} \int_V C_A dV$$

Combining all the terms, we obtain the macroscopic species balance equation:

$$\boxed{-\int_S (\boldsymbol{n} \cdot N_A)\, dS + V\left\langle R_A\right\rangle = \frac{d}{dt}\left(V\left\langle C_A\right\rangle\right)} \tag{5.43}$$

We can compare Equation 5.43 with the macroscopic balance shown earlier in Chapter 2, which is repeated here for ease of reference:

$$\boxed{\dot{\mathcal{M}}_{A,i} - \dot{\mathcal{M}}_{A,e} - \dot{\mathcal{M}}_{A,W} + V\left\langle R_A\right\rangle = \frac{d}{dt}(\mathcal{M}_A)} \tag{5.44}$$

The reaction terms and the accumulation terms are the same as in Equation 5.43 since $V\langle C_A \rangle = \mathcal{M}_A$.

The first three terms in Equation 5.44 can be formally established if the control surface is split into an inlet region S_i, an outlet region S_e, the permeable wall or interface S_w, and the impermeable walls S_{im}:

$$S = S_i + S_e + S_W + S_{im}$$

The surface integral term in Equation 5.43 is evaluated for each of these regions. The corresponding contributions of these integrals lead to the moles crossing the inlet region of the overall control surface, leaving the outlet part and the amount transferred to walls or to an interface or membrane.

For illustration, at the inlet the flux vector is mainly convective and can be approximated as vC_A. The inlet contribution of the integral in Equation 5.43 is

$$-\int_{S_i} (\boldsymbol{n} \cdot N_A)\, dS = \int_{S_i} (\boldsymbol{n} \cdot \boldsymbol{v} C_A)\, dS = S_i \langle vC_A \rangle$$

Here v (with no boldface) represents the velocity component in the direction of the inlet flow. (This is equal to $-\boldsymbol{n} \cdot \boldsymbol{v}$ since the normal direction $\boldsymbol{n}$ is usually chosen outward from the control volume.) Hence the inlet contribution can be represented as $S_i \langle vC_A \rangle$, and is seen to be the moles of A entering the inlet region, $\dot{\mathcal{M}}_{A,i}$ (the first term in Equation 5.44). The cup mixed average appears naturally as the representative concentration for the inlet (and outlet) streams.

Similar considerations apply for the integral over the exit region S_e and lead to the $-\dot{\mathcal{M}}_{A,e}$ term.

The permeable walls flux term likewise leads to the $-\dot{\mathcal{M}}_{A,W}$ term if one identifies this term as the surface integral of the local flux vector:

$$\dot{\mathcal{M}}_{A,W} = \int_{S_w} (\boldsymbol{n} \cdot \boldsymbol{N}_A)\, dS$$

Finally, the integral over the impermeable wall region is zero. Thus the first three terms in Equation 5.44 are the integrals of the corresponding flux components, and the macroscopic balance is derived by a formal integration of the differential balance.

5.8   Multiphase Systems: Local Volume Averaging

This section may be omitted with no loss of continuity.

The basis for the differential model for two-phase systems remains to be clarified. This section shows the basis for this model using the concept of local volume averaging.

The analysis presented here is somewhat advanced and not often covered in the undergraduate curriculum. The material will be useful to follow some current literature on this topic, and for understanding the basic ideas used in the CFD (computational fluid dynamics) model of mass transfer in multiphase systems. Hence the goal of this section is to provide an introductory background that supports further study in this field.

In a multiphase system, two phases flow past each other. One phase is usually dispersed in the other phase, and any selected control volume will contain both phases (Figure 5.4). We denote the phases as c and d. The species conservation law applies locally within each phase. Thus, for the dispersed phase:

$$\frac{\partial C_{Ad}}{\partial t} + \nabla \cdot \boldsymbol{N}_{Ad} = R_{Ad} \tag{5.45}$$

where the subscript d indicates that all quantities are applicable within the dispersed phase. A similar equation holds for the continuous phase (not shown for brevity).

If the precise locations of the phases are known, these equations can be solved with the boundary conditions applied at the interface between the two phases. The boundary conditions are the flux continuity and the concentration

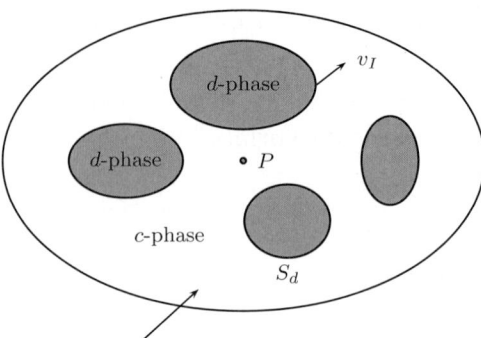

Volume, $V = V_c + V_d$

Figure 5.4 Local control volume showing two interacting phases and an averaging surface associated with any point P. S_d is the total surface area between the c-phase and d-phase within the volume. v_I is the relative velocity with which the surface of the dispersed phase may be moving.

jump at the interface. Examples of such cases include annular flow in a pipe, slug flow in a microchannel, and a falling film reactor. The two phases are segregated in these cases, and the species continuity equation can be solved for each phase, although this may be computationally intensive and challenging in some situations.

If the phases are dispersed as shown in Figure 5.4, the approach of solving each phase separately is not useful because this is now a dynamic system: The precise location of the interface is not known and may be changing with time. Solution of such a deterministic multiboundary problem is difficult. Some simplification is needed, which is provided by local volume averaging.

Averaging volume V is composed of V_c and V_d—that is, the volumes occupied by phases 1 and 2. Likewise, the averaging surface S consists of S_c and S_d.

The local volume average for the dispersed phase is defined as

$$\overline{C_{Ad}} = \frac{1}{V} \int_{V_d} C_{Ad}\, dV \tag{5.46}$$

A similar definition applies for the continuous phase as well. Here V_d is the volume occupied by the dispersed phase and V is a local total volume over which averaging is done.

We can also define an intrinsic volume average:

$$\tilde{C}_{Ad} = \frac{1}{V_d} \int_{V_d} C_{Ad} \, dV \tag{5.47}$$

The two local averages and the intrinsic average are related as follows:

$$\overline{C_{Ad}} = \epsilon_d \tilde{C}_{Ad} \tag{5.48}$$

where ϵ_d is the volume fraction of the dispersed phase.

$$\epsilon_d = \frac{V_d}{V}$$

Hence the volume averaged continuity is

$$\overline{\frac{\partial C_{Ad}}{\partial t}} + \overline{\nabla \cdot \mathbf{N}_{Ad}} = \overline{R_{Ad}} \tag{5.49}$$

To perform further manipulations, we need two theorems for volume average of the time derivative and divergence. These concepts are often referred to as Gray's (1975, 1983) theorem for volume averaging. (The theorem is on off-shoot of earlier papers by Slattery (1969) and Whitaker, 1969.)

Theorem for volume average of vector operations

- Volume average of the time derivative of a scalar:

$$\overline{\frac{\partial C_A}{\partial t}} = \frac{\partial \overline{C_A}}{\partial t} - \frac{1}{V} \int_{S_d} C_A \left(\mathbf{v}_I \cdot \mathbf{n} \right) dS \tag{5.50}$$

where $\mathbf{v}_I$ is the relative velocity with which the interface between c and d is moving.
- Divergence of a vector:

$$\overline{\nabla \cdot \mathbf{N}_A} = \nabla \cdot \overline{\mathbf{N}_A} + \frac{1}{V} \int_{S_d} \mathbf{N}_A \cdot \mathbf{n} \, dS \tag{5.51}$$

The first term in Equation 5.49 is the average of the time derivative of the average concentration. It is equal to the time derivative of the average concentration if the control surface of the dispersed phase is not changing with time. This will apply, for example, to dilute systems where the drop or the bubble size is not changing much during mass transfer. Otherwise, an extra term that includes the relative velocity of the interface (shown in Equation 5.50) must be taken into account.

The second term in Equation 5.49 is the volume average of the divergence. This is not equal to the divergence of the volume average, so the extra surface integral shown in Equation 5.51 must be added according to Gray's theorem.

The use of these theorems in Equation 5.49 then leads to the following expression:

$$
\frac{\partial \overline{C_{Ad}}}{\partial t} + \nabla \cdot \overline{N_{Ad}} = \overline{R_{Ad}} - \frac{1}{V} \int_{S_d} (\boldsymbol{N}_A \cdot \boldsymbol{n}) dS
$$
$$
+ \frac{1}{V} \int_{S_d} C_A(\boldsymbol{v}_I \cdot \boldsymbol{n}) dS = 0
$$

(5.52)

The last two terms are interpreted as the interfacial transfer term and the contribution due to change in volume of the dispersed phase due to shrinkage or growth. Note that the interface mass transfer term arises automatically as a consequence of local volume averaging combined with the use of Gray's theorem. The interfacial term is represented as the transfer rate $\dot{\mathcal{M}}_{dc}$ from the dispersed phase to the continuous phase per unit volume:

$$
\dot{\mathcal{M}}_{dc} = \int_{S_d} \boldsymbol{N}_A \cdot \boldsymbol{n} \, dS
$$

The use of these theorems in Equation 5.49 leads to the local volume averaged model for the dispersed phase:

$$
\frac{\partial \overline{C_{Ad}}}{\partial t} + \nabla \cdot \overline{N_{Ad}} = \overline{R_{Ad}} - \frac{\dot{\mathcal{M}}_{dc}}{V}
$$

(5.53)

where we have omitted the shrinkage/expansion term. This model assumes constant drop diameter, for instance.

The corresponding equation for the continuous phase completes the model:

$$
\frac{\partial \overline{C_{Ac}}}{\partial t} + \nabla \cdot \overline{N_{Ac}} = \overline{R_{Ac}} + \frac{\dot{\mathcal{M}}_{dc}}{V}
$$

(5.54)

Note that the mass transfer term now appears with a plus sign the direction of the unit normal is toward the dispersed phase.

It may be useful to summarize the various length scales involved and evaluate the local volume average on this basis. The following length scales are readily identified:

- Dispersed phase scale, d (e.g., bubble or drop diameter or particle diameter)
- Scale of averaging volume, r
- Total length scale (e.g., reactor or vessel diameter), L

The local volume averaging assumption assumes $d < r < L$. In other words, the averaging volume should include several particles of the dispersed phase but should be much smaller than the overall reactor length scale. Therefore, unless we are dealing with a highly dispersed system, the control volume cannot be reduced to an infinitesimal volume. This constraint is one difficulty in using this approach. Similarly, the rate of mass transfer depends on the the local velocity and local diffusion flux at the interface, which is not available from a volume averaged model. The volume averaging blurs the lines between the two phases and treats them as an interpenetrating continuum. Hence the equations must be supplemented with some transport model; commonly one defines and uses the local mass transfer coefficient for this purpose.

If each phase equation is now further averaged over a larger control volume, the two-phase macroscopic or mesoscopic models are obtained. This provides a formal way of developing the equation, identifies the information loss due to averaging, and suggests the required closures.

Summary

- The application of the mass conservation law to a species leads to the differential equation for mass transfer. Such equations can be written in either mass units or mole units. Mole units are more convenient for reactive systems.
- The differential equations based on the conservation law must be supplemented with a model for diffusion flux to obtain the corresponding equations in the concentration form. A definition of the mixture velocity is needed to model the diffusion flux.
- In a multicomponent mixture, a velocity can be associated with each species. The velocity is related to the fluxes in a stationary frame of reference. One can use mole or mass units freely here and convert from one set to the other.

- Averaging the species velocities gives us a value for the velocity of the mixture as a whole. However, there is no unique way of averaging, which leads to many definitions for this average velocity and the corresponding definitions for the diffusion flux.

- Two common ways of averaging are based on use of either the mass fraction or the mole fraction as the weighting factor. The results are the mass averaged and mole averaged velocities for the system. The notations v and v^*, respectively, are commonly used for these quantities.

- A flux—the diffusion flux—can be defined based on a coordinate system moving with either of these average velocities. Thus we can have a diffusion flux based on mass averaged velocity as a reference frame (j_A) or a diffusion flux based on mole averaged frame (J_A).

- In a binary mixture, there is only one independent diffusion flux although there are two components: $j_B = -j_A$ in mass units or $J_B = -J_A$ in mole units. In a multicomponent system, the sum of the diffusion fluxes taken for all the species is zero. Thus, in a ternary system, there are only two independent diffusion fluxes.

- Constitutive equations relate diffusion fluxes to the concentration gradient or, more generally, to the mole (or mass) fraction gradient. Fick's law is often suitable for binary systems or used as an approximation even for multicomponent systems. Depending on the frame of reference used to define diffusion flux, Fick's law can take different forms. The most common form relates J_{Ax} to dC_A/dx; this variant is used for isothermal and isobaric systems.

- The use of the diffusion flux permits us to develop differential equations of mass transfer in concentration form.

- The boundary conditions to be used in the differential equation can be classified as being of the Dirichlet, Neumann, or Robin type.

- At the phase interface, fluxes are matched on either side, while a concentration jump condition consistent with thermodynamic equilibrium is imposed for the concentrations.

- The integration of the differential models over an arbitrary control volume leads to the macroscopic mass balance models discussed in Chapter 3. Although the macroscopic models can be written starting from the basic conservation law, this averaging approach is more elegant and provides the link between differential and macroscopic models. It also provides a precise definition of all the terms appearing in the macroscopic model.

- Differential models for two-phase systems can be obtained by a procedure called local volume averaging. The mass exchange term between the two phases arises automatically by this procedure (which has to be closed by a transport law). The derived equations are applied locally for each phase at any given point; that is, the phases are assumed to be an interpenetrating continuum. If these "pointwise" models are averaged over a larger volume, then the macroscopic or mesoscopic two-phase models can be derived from first principles.

Review Questions

5.1 Is n_A equal to $M_A N_A$? Why?

5.2 Is $j_A^{(m)}$ equal to $M_A J_A^*$? Why?

5.3 State a (diffusion) problem where both v and v^* are non-zero.

5.4 State a (diffusion) problem where v is non-zero but v^* is zero.

5.5 State a (diffusion) problem where v is zero but v^* is non-zero.

5.6 State a (diffusion) problem where both v and v^* are zero.

5.7 In Example 5.1, does v^* change with distance along the vapor space? Does v change?

5.8 When is $\sum R_j$ equal to zero?

5.9 State equations for divergence of v and divergence of v^*.

5.10 How does the overall mole continuity equation simplify for the steady state and no change in moles in the reaction?

5.11 Are the fluxes the same on both sides of a gas–liquid interface? What about the concentrations?

5.12 How can mesocopic models be derived when starting from differential models?

5.13 What is meant by local volume averaging?

5.14 What is Gray's theorem, and when is it useful?

Problems

5.1 **Divergence of the flux vector interpretation.** If N_A is a flux vector at any point on the control surface and n is the unit normal outward from the surface, interpret the term $N_A \cdot n$. If dA is a differential area of surface at this point, write an expression for the mole efflux from this differential area. Then show that

$$\text{Net moles efflux from the control surface} = \int_A (N_A \cdot n) dA$$

Convert the area integration to a volume integral by using the Gauss divergence theorem. If the control volume now tends to zero, show that mole efflux per unit volume is equal to the divergence of the flux vector.

5.2   **Combined flux: r-diffusion only in cylinder and sphere.** How does the equation shown for the divergence operator in the text simplify to a case with cylindrical coordinates and diffusion only when there is only variation in the r-direction? Compare your answer with the result in Chapter 2.

How does the equation for the divergence simplify to a case with spherical coordinates and diffusion when there is only variation in the r-direction? Compare your answer with the result in Chapter 2.

5.3   **Comparison of v and v^*:** A gas mixture consists of hydrogen and nitrogen. The combined flux vector of hydrogen at a point is 10 mmol/m^2 s, where the gas composition is 20% by moles of hydrogen. The flux of nitrogen is zero. Calculate v and v^* at this point. Total pressure is 1 atm and temperature is 300 K.

5.4   **Relation between species velocities.** Show the following relations between the species velocities v_A and v_B referred to a stationary frame of reference:

$$v_A - v_B = j_A \left(\frac{1}{\rho_A} + \frac{1}{\rho_B} \right)$$

$$v_A - v_B = J_A \left(\frac{1}{C_A} + \frac{1}{C_B} \right)$$

5.5   **Velocity variation with position in the two-bulb apparatus.** In the two-bulb experiment shown in Figure 5.2, the porous plug is 1 m long and the diffusion coefficient is 7.0×10^{-5} m^2/s. The left bulb contains hydrogen and the right bulb contains nitrogen. Calculate and plot v as a function of length along the porous plug.

5.6   **Relation between j_A and J_A for a binary system.** Show that the following relationship holds for a binary mixture:

$$j_A = \frac{M_A M_B}{\bar{M}} J_A$$

5.7   **Alternative forms of Fick's law.** Derive the following form of Fick's law:

$$j_A = -\frac{C^2}{\rho} M_A M_B D_{AB} \nabla y_A$$

where y_A is the mole fraction. Show that this can also be expressed as

$$j_A = -C \frac{M_A M_B}{\bar{M}} D_{AB} \nabla y_A$$

5.8   **Inverted form of Fick's law.** For a binary mixture, derive the following equation based on Fick's law:

$$\nabla x_A = \frac{x_A N_B - x_B N_A}{C D_{AB}} \tag{5.55}$$

Show that the following equation holds as well:

$$\nabla x_A = \frac{x_A J_B - x_B J_A}{C D_{AB}} \qquad (5.56)$$

These are inverted forms; that is, they make flux implicit. Extension of these equations to multicomponent systems leads directly to the Stefan-Maxwell equation.

5.9 **Volume fraction weighting.** Mole fraction and mass fraction are commonly used weighting factors to find the mixture velocity. Volume fraction φ_v can also be used as the weighting factor and one can define the average velocity $\boldsymbol{v}^V$ as follows:

$$\boldsymbol{v}^V = \sum \phi_v \boldsymbol{v}_i$$

Verify that the volume fraction averaged velocity is the same as the mole fraction averaged velocity when the total molar concentration (molar density) is constant (e.g., in a gas mixture).

Verify that volume fraction averaged velocity is the same as the mass fraction averaged velocity when the total mass density is constant (e.g., in a liquid mixture).

Derive a form of Fick's law based on this definition of average velocity for a binary mixture.

Also show that the sum of the diffusion fluxes is not zero in this frame of reference but the following relation holds:

$$\sum_{i=1}^{ns} \hat{V}_i J_i^V = 0 \qquad (5.57)$$

5.10 **Boundary conditions.** State the boundary condition you would use for the following scenarios: a wall is coated with a dissolving solute; the wall is impermeable; a very fast catalytic reaction is taking place at the wall.

5.11 **Convection–diffusion equation for pipe flow.** A fluid is flowing in a pipe and some mass transfer process is also taking place. Expand the convection–diffusion equation and write a PDE for the concentration as a function of r and z.

5.12 **Convection–diffusion equation for boundary layer flow.** A fluid is flowing past a solid plate and some mass transfer process is also taking place. The flow is a boundary layer flow that has two velocity components, v_x and v_y. Expand the convection–diffusion equation and write a PDE for the concentration as a function of x and y. In this situation, x is the distance along the plate and y is perpendicular to it.

5.13 **Gray's theorem.** The theorems shown in Section 5.8 look complicated but are simply extensions of the Leibnitz rule to three dimensions. State the Leibnitz rule for differentiation under the integral sign. Now assume that the flux vector has only an x-component. Average the divergence of the flux vector, use the

Leibnitz rule, and confirm that the final results comply with the 1-D simplified version of Gray's theorem. Similarly, show the basis for the formula for average of a time derivative when the control surface is moving.

5.14   **Mesoscopic model for an absorber.** The mesoscopic model for an absorber commonly used in practice was shown in Section 4.3.2. List the assumptions needed to arrive at Equations 4.18 and 4.19. Discuss how these arise by averaging the local volume averaged differential equations.

5.15   **Macroscopic model for an liquid extractor.** The macroscopic models can be developed by averaging the local volume average model for the entire contactor volume. Show the development of a macroscopic model for liquid–liquid extraction column where one phase is dispersed as droplets within a second continuous phase. List the simplifying assumptions needed.

CHAPTER 6

Diffusion-Dominated Processes and the Film Model

Learning Objectives

After completing this chapter, you will be able to:

- Solve prototype problems involving steady state diffusion.
- Describe how diffusion can cause an associated flow (diffusion-induced convection).
- Account for the effect of convection in diffusion-dominated systems (i.e., stagnant systems).
- Describe the film theory concept of mass transfer and its usefulness in many applications.
- Evaluate how the rate of chemical reaction at a surface is affected by the rate of mass transfer.
- Explain the two-film theory for gas–liquid (or liquid–liquid) mass transfer and calculate the rate of interfacial mass transfer given the bulk composition of the two phases.

This chapter examines problems where the diffusion is the main mode of mass transport. Such systems are often referred to as diffusion in stagnant media for analogy with problems of purely heat conduction. However, this does not mean that the convection effects can be totally ignored. The term "stagnant" can therefore be misleading in some cases. Thus, the effect of convection-induced by diffusion is also addressed in this chapter.

A model widely used in mass transport analysis is the film model, which is based on the assumption of a "stagnant" film near a fluid–solid interface. Transport effects are analyzed using either diffusion or diffusion plus diffusion-induced convection as the mode of transport in this film. Detailed effects of the fluid motion near the fluid–solid interface are not considered in this model, but rather are assumed to be buried in the hypothetical parameter called the film thickness. Useful information can be gained from such a model, as illustrated in this chapter.

Film theory can be used to explore a variety of complex situations. An example is the transport of a reactant to a catalyst surface, followed by a heterogeneous reaction. (Another example is simultaneous heat and mass transport which is discussed in Chapter 25). This chapter shows how the heterogeneous reactions can be analyzed using the film model.

The film theory needs a modification for fluid–fluid systems since mass transfer is a sequence of two steps: fluid 1 to interface, and interface to fluid 2. Hence a two-film model (one film on each fluid side of the interface) is used. This model is discussed next and the concept of an overall mass transfer coefficient is introduced. This concept is widely used in modeling and design of separation processes.

Systems in which a chemical reaction takes place at a surface are referred to as heterogeneous reactions; they stand in contrast to systems in which the chemical reaction takes place in the bulk phase of a fluid. The latter are called homogeneous reactions. For heterogeneous reactions, the diffusion applies to the gas phase and the reaction appears only as a boundary condition; that is, the reaction occurs at the surface only and not in the bulk of the gas phase. Hence some aspects of heterogeneous reactions are also studied here, since they are amenable to analysis using film theory. In addition, the role of mass transfer in changing the overall rate is illustrated.

The problems analyzed in this chapter are quite important in practical applications and provide a framework for designing a wide variety of mass transfer and reactor equipment. After a detailed study of this chapter, you will be able to calculate the local rate of mass transport at any given point in mass transfer equipment. In later chapters we will combine this local rate with a macroscopic or mesoscopic model for a system such as a reactor or separator, and we will demonstrate such coupled models for design/simulation of such equipment.

6.1   Steady State Diffusion: No Reaction

This section summarizes the equations for steady state diffusion and discusses the role of diffusion-induced convection. If this convection effect is neglected and a constant diffusion coefficient is assumed, the concentration profile can be determined by the solution of the Laplace equation with appropriate boundary conditions.

6.1.1   Combined Flux Equation

The starting point is the equation of mass transfer. For the present case of steady state and no reaction, the equation simplifies to

$$\nabla \cdot \boldsymbol{N}_A = 0 \tag{6.1}$$

The flux term $\boldsymbol{N}_A$ is the combined flux, which can be expressed as a sum of the diffusion flux and a convection flux:

$$\boldsymbol{N}_A = \boldsymbol{J}_A + y_A \boldsymbol{N}_t \tag{6.2}$$

In this chapter we use Fick's law for diffusion flux $\boldsymbol{J}_A$, which is (strictly speaking) valid for ideal binary systems:

$$\boldsymbol{J}_A = -CD_A \nabla y_A$$

Here D_A is the binary pair diffusivity, which is equal to D_{AB} for a truly binary mixture. For multicomponent mixtures, D_A is interpreted as a pseudo-binary diffusivity D_{A-m}, which is defined in Equation 5.24.

Hence the equation for N_A can be written as follows:

$$\boxed{N_A = -CD_A \nabla y_A + y_A \boldsymbol{N}_t} \tag{6.3}$$

The first term is the diffusion flux, and the second term is the convection flux. This, together with the conservation requirement that the divergence of the flux vector should be zero, forms the basis for the solution of the problems examined in this chapter. We consider diffusion as the dominant mode of transport in this chapter. We assume that there is no superimposed flow caused by external means such as a pump or fan. The question then arises as to why the convection is included in this chapter, which specifically deals with diffusion-dominated processes. This question is answered in the following section.

6.1.2   Diffusion-Induced Convection

Consider a liquid evaporating from a container, as shown in Figure 6.1. There is no superimposed flow and the evaporation is mainly due to diffusion caused by the concentration difference in the vapor phase. The evaporation implies some movement of A and, therefore, there is a convective flow due to evaporation. In this case, the term $y_A \boldsymbol{N}_t$ in Equation 6.3 is retained in the model to account for the contribution due to convection. We examined

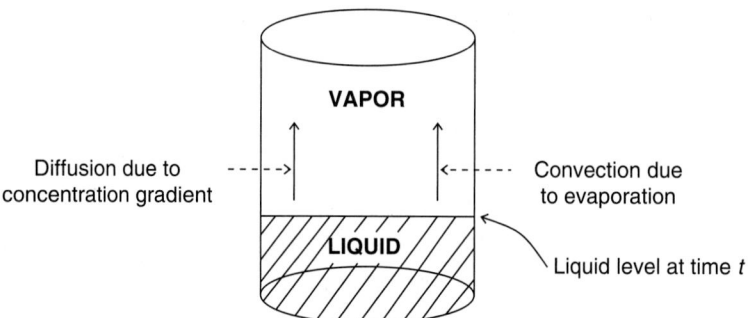

Figure 6.1 Liquid evaporating in a glass. Transport takes place both by diffusion and by convection, although the convective flow is not evident.

this case in Example 5.1 in Chapter 5, where we found a small but non-zero value for both the velocities v and v^*. The contribution due to this flow was not accounted for there, however, and will be considered later in this chapter.

6.1.3 Determinacy Condition

The term N_t is equal to sum of the combined flux of all the species. For example, in a binary system, it is equal to $N_A + N_B$. To close the problem, some additional relation for N_t or N_B has to be independently specified. This is known as the determinacy condition. The specification is situation dependent, and various cases are discussed in more detail in Section 6.3.3. Two common cases are indicated here for a binary mixture.

- Unimolecular diffusion (UMD): Second component has no transport. $N_B = 0$. An example would be a species (A) evaporating into an inert gas (B). Hence N_t is non-zero and diffusion-induced convection needs to be included to create a more exact model. The determinacy condition is therefore $N_t = N_A$.
- Equimolar counter-diffusion (EMD): Second component counter-diffusing with same magnitude of flux $N_B = -N_A$. An example would be the situation in which species A diffuses to a catalytic surface, reacts to produce an equal mole of B, and then B counter-diffuses to the bulk gas phase. Hence N_t is zero and diffusion-induced convection is not present (using a mole-centric frame of reference).

N_t *versus* n_t

An alternative way of writing the flux expression is in mass units:

$$\boldsymbol{n}_A = -\rho D_A \nabla \omega_A + \omega_A \boldsymbol{n}_t \tag{6.4}$$

This is a useful representation if ρ is constant across the system. Otherwise, the local variation of ρ has to be included in the calculations. Likewise, the mole unit version given by Equation 6.3 is suitable if the total concentration is a constant in the system.

A useful but sometimes confusing point to note is that if N_t is zero, it does not imply that n_t is also zero. This difference, which was also discussed in Chapter 5, is due to the difference in total mole balance versus total mass balance. It is also a consequence of the different ways of averaging the system velocity. A few examples are given here to illustrate this point.

An example where N_t is zero but n_t is non-zero is equimolar counter-diffusion. Consider a tube that has hydrogen at one end and nitrogen at the other end. The system is maintained at a constant temperature and pressure, so the total molar concentration in the tube is constant. The moles of hydrogen diffusing in one direction is equal to the moles of nitrogen diffusing in the opposite direction, leading to N_t equal to zero. However, there is a net mass flow rate from the nitrogen side to the hydrogen side since nitrogen has a higher molecular weight than hydrogen. Hence n_t is not zero.

Now consider a case where 1 mole of a species A diffuses to a catalytic surface and 2 moles is produced from diffusion in the opposite direction. Here N_t is not zero and n_t is zero, since on a mass basis the chemical reaction creates no change in mass. It is useful to reflect on and appreciate these subtle points. Examples 5.1 to 5.3 in Chapter 5 highlighted these points as well. Ultimately the choice of N_t versus n_t is dependent on which of these will lead to a simpler model.

6.1.4 Low Flux Model: The Laplace Equation

A commonly used model for diffusion-dominated systems is the low flux model, which is defined for our purposes as a model in which the effect of diffusion-induced convection is neglected. In this case, only diffusion flux is included. Thus

$$\boldsymbol{N}_A \approx \boldsymbol{J}_A = -C D_{AB} \nabla y_A \tag{6.5}$$

Combining Equation 6.5 and Equation 6.1, we have

$$\nabla \cdot (C D_{AB} \nabla y_A) = 0$$

This is the starting point for steady diffusion problems.

Consider a system at constant concentration—a condition that applies to isobaric and isothermal systems. Also assume that the diffusion coefficient does not vary much across the system, such that a constant value can be assigned to it. Then the equation simplifies to

$$\nabla^2 y_A = 0 \qquad (6.6)$$

Since the system is at constant concentration, we can also write this equation as

$$\boxed{\nabla^2 C_A = 0} \qquad (6.7)$$

Thus the concentration field satisfies the Laplace equation with the assumptions stated earlier. Solution of the Laplace equation is a well-studied problem in engineering mathematics. The form of the Laplacian in three common coordinate systems is shown next.

Expression for the Laplacian of Concentration in Common Coordinate Systems

1. **Cartesian coordinates:**

$$\nabla \cdot \nabla C_A = \nabla^2 C_A = \frac{\partial^2 C_A}{\partial x^2} + \frac{\partial^2 C_A}{\partial y^2} + \frac{\partial^2 C_A}{\partial z^2} \qquad (6.8)$$

2. **Cylindrical coordinates:**

$$\nabla^2 C_A = \frac{1}{r}\frac{\partial}{\partial r}\left(r\frac{\partial C_A}{\partial r}\right) + \frac{1}{r^2}\frac{\partial^2 C_A}{\partial \theta^2} + \frac{\partial^2 C_A}{\partial z^2} \qquad (6.9)$$

3. **Spherical coordinates:**

$$\nabla^2 C_A = \frac{1}{r^2}\frac{\partial}{\partial r}\left(r^2\frac{\partial C_A}{\partial r}\right) + \frac{1}{r^2 \sin\theta}\frac{\partial}{\partial \theta}\left(\sin\theta\frac{\partial C_A}{\partial \theta}\right)$$
$$+ \frac{1}{r^2 \sin^2\theta}\frac{\partial^2 C_A}{\partial \phi^2} \qquad (6.10)$$

Solutions to the Laplace equation for some simple situations are presented here to demonstrate the use of this model. Example 6.1 is applicable to a cylindrical geometry.

Example 6.1   Radial Diffusion in a Cylindrical Geometry

Consider the diffusion in cylindrical coordinates with the concentration varying only in the radial direction. Show that the concentration profile is a logarithmic function of the radial coordinate.

Solution

The Laplace equation in cylindrical coordinates is applicable here. Keeping only the variation in r, we get

$$\frac{1}{r}\frac{d}{d}\left(r\frac{dC_A}{dr}\right) = 0 \qquad (6.11)$$

The first integration gives

$$\left(r\frac{dC_A}{dr}\right) = \text{constant} = A_1$$

Dividing both sides by r and integrating again, we get

$$C_A = A_1 \ln r + A_2$$

Hence the profile is a logarithmic function of r. This was also shown in Chapter 2 starting from detailed balances. Here we show the same result starting directly from the general equation for steady state diffusion.

We now study another simple application to a spherical geometry and derive an expression for the transfer coefficient for mass transfer from a sphere to a stagnant gas in Example 6.2.

Example 6.2   Mass Transfer from a Sphere to a Stagnant Gas

A spherical solid (e.g., naphthalene) is subliming into a gas phase (e.g., air) that is stagnant. Derive an expression for the concentration profile in the gas phase and an equation for the rate of evaporation. Use them to obtain an expression for the mass transfer coefficient in the absence of any gas flow.

Solution

Here N_A, the flux of the sublimining solid, is non-zero and the flux of the gas phase (air) is zero. Hence the total flux N_t is non-zero here, as this is the UMD situation. Although diffusion-induced convection will have some effect in this system, if the vapor pressure of the subliming solid is small its contribution to the flux can be neglected and a low flux model used. Hence the Laplace equation in spherical coordinates applies for the gas-phase concentration of A. Keeping only the r-dependent term, we have

$$\frac{1}{r^2}\frac{d}{dr}\left(r^2\frac{dC_A}{dr}\right)=0 \tag{6.12}$$

This equation can be integrated twice to get a general expression for the concentration profile:

$$C_A=\frac{A_1}{r}+A_2$$

Now we impose two boundary conditions to get A_1 and A_2. We assume that far away from the subliming sphere, the concentration is zero. Hence as $r\to\infty$, we should have $C_A=0$. Hence $A_2=0$.

At the sphere surface $r=R$, the concentration in the gas phase is equal to C_{As} such that the thermodynamic equilibirum applies at this point:

$$C_{As}=\frac{p_{A,vap}}{R_g T}$$

where $p_{A,vap}$ is the vapor pressure of the subliming solid at the given temperature.

Using this condition the concentration profile in the gas phase is

$$C_A=C_{As}\frac{R}{r}$$

The flux at the solid surface is the rate of evaporation per unit area of the solid. It is calculated by applying Fick's law at the surface:

$$N_{As}=-D_A\left(\frac{dC_A}{dr}\right)_{r=R}$$

Using the concentration profile and its derivative at $r=R$, the following expression for the flux is obtained:

$$N_{As}=\frac{D_A}{R}\cdot C_{As}$$

If one writes the flux as a product of the mass transfer coefficient times the driving force, the following result is obtained for the mass transfer coefficient:

$$\boxed{k_m=\frac{N_{As}}{C_{As}}=\frac{D_A}{R}} \tag{6.13}$$

This equation sets a base value for the mass transfer coefficient for mass transfer from solid to gas in the absence of any superimposed external flow in the gas phase. The effect of the superimposed external flow is then fitted by adding an extra term, which depends on the Reynolds number. (See the discussion of the Ranz-Marshall correlation in Section 3.4.)

6.2   Diffusion-Induced Convection

In this section, we study the correction needed to account for diffusion-induced convection. First we discuss conditions in which the diffusion-induced convection can be neglected and the low flux model can be applied. This is followed by an analysis of diffusion-induced convection for the UMD case.

6.2.1   Conditions for the Validity of the Low Flux Model

The low flux model holds when the diffusive component of the combined flux is much larger than the convection flux—that is, when

$$\boldsymbol{J}_A >> y_A \boldsymbol{N}_t$$

This can happen, for example, when $\boldsymbol{N}_t$ is small (i.e., low mass transfer flux) or when y_A is small (dilute systems). In these cases it is fine to consider diffusion only.

Note: The diffusion component is the only term to be considered when $\boldsymbol{N}_t$ is zero due to the nature of the problem. In this case, which is known as EMD, two species diffuse at the same rate but in opposite directions. The net flux is then automatically zero even if y_A is large. The low flux model applies then, although the magnitude of the flux need not be low.

In general, it is necessary to define the total flux $\boldsymbol{N}_t$ independently, by specifying the determinacy condition. We now show this case for UMD. More examples of determinacy conditions for other cases are provided in Section 6.3.3.

6.2.2   Analysis for UMD

UMD (unimolecular diffusion) refers to the case in which only one species (say, A) has a non-zero flux; that is, the component B has no flux in a stationary frame of reference. Hence $N_B = 0$ and $N_A + N_B = N_t = N_A$, which is the determinacy condition for the UMD situation.

The expression for the combined flux can now be written as

$$\boldsymbol{N}_A = -CD_{AB}\nabla y_A + \boldsymbol{N}_A y_A \tag{6.14}$$

Rearranging, we have the flux expression for UMD:

$$\boldsymbol{N}_A = -\frac{CD_{AB}}{(1-y_A)}\nabla y_A \tag{6.15}$$

Note the presence of the $1 - y_A$ term in the denominator. This term is equal to 1 in the low flux model, if the diffusion-induced convection is neglected.

For the one-dimensional case, $\nabla y_A = dy_A/dz$, where z is the direction of diffusion. Hence the flux expression can be written after separating the variables as

$$N_A dz = -CD_{AB} \frac{dy_A}{1 - y_A} \tag{6.16}$$

N_A (no boldface) is used as an abbreviation for the z-component of the flux vector, N_{Az}.

The expression for the evaporation rate for the system shown in Figure 6.1 will now be derived including the diffusion induced correction, Integrating Equation 6.16 across the vapor space and keeping N_A constant (as required to satisfy Equation 6.1), we get the rate of evaporation:

$$\boxed{N_A = -\frac{CD_{AB}}{H} \ln\left(\frac{1 - y_{As}}{1 - y_{Ab}}\right)} \tag{6.17}$$

where y_{As} is the mole fraction at the evaporating surface $z = 0$ and y_{Ab} is the mole fraction in the bulk gas—that is, at $z = H$ and beyond. (H is the height of the vapor space in the evaporating tube.)

For the commonly used case of y_{Ab} equal to zero, Equation 6.17 reduces to

$$N_A = -\frac{CD_{AB}}{H} \ln(1 - y_{As}) \tag{6.18}$$

Note that the flux is not linear to the driving force.

Now if y_{As} is small we can make the following approximation for the log term:

$$\ln(1 - x) \approx -x \text{ as } x \to 0$$

The model then reduces to the low flux case:

$$N_A = \frac{CD_{AB} y_{As}}{H} \tag{6.19}$$

The flux is now linearly proportional to the mole fraction gradient.

The comparison of Equations 6.18 and 6.19 also helps us to ascertain the validity of the diffusion-only model. For example, if $y_{As} = 0.1$, then the convection model would give a driving force of $\ln(1 - y_{As}) = 0.1054$. Thus the convection effect can be neglected for dilute systems. If the value of y_{As} is 0.2, then the factors are 0.2231 (convection included) versus 0.2 (convection neglected) and the error is in the range of less than 10%.

Another way to compare the two approaches is by using the drift flux correction factor, as explained in the following subsection.

6.2.3   Drift Flux Correction Factor

The expression given by Equation 6.17 for a high flux model can be compared with the low mass flux model directly to ascertain the effect of "flow" in the system. The correction factor due to flow (this is ratio of the actual flux to the flux calculated neglecting the diffusion induced convection) can be expressed as a drift flux factor and the following expression can be derived:

$$\text{Correction factor } \mathcal{F} = \frac{N_A}{N_A^\circ} = \frac{\ln[(1 - y_{As})/(1 - y_{Ab}]}{y_{Ab} - y_{As}} \tag{6.20}$$

The correction factor can also be written in terms of the mole fraction of B using the relation $y_B = 1 - y_A$:

$$\mathcal{F} = \frac{\ln(y_{Bs} - y_{Bb})}{y_{Bs} - y_{Bb}} = \frac{1}{\text{log mean value of mole fraction of } B.}$$

where the log mean is defined as

$$\text{log mean } y_B - \frac{y_{Bb} - y_{Bs}}{\ln(y_{Bb}/y_{Bs})}$$

We can compare the effects of drift by examining a simple example, as shown in Example 6.3.

Example 6.3   Drift Correction to Low Flux Model

A volatile liquid is present in a tube with an exposed vapor height of 0.3 m. The vapor pressure is 0.15 atm at the given conditions of 35° C and 1 atm pressure. The diffusion coefficient is estimated as 8.8×10^{-6} m^2/s. Find the rate of evaporation.

Solution

The following quantities can be calculated:

* Total concentration, $C = P/R_gT$: 39.56 mol/m^3
* Flux from low flux model, given by Equation 6.19: 1.04×10^{-4} mol/m^{2}s

> - Flux from high flux model, given by Equation 6.18: $1.1318 \times 10^{-4} \text{ mol/m}^2\text{s}$
> - Correction factor: the ratio of the two and equal to 1.1230; thus convection augments diffusive transport by about 12%.
>
> The correction factor can also be ascertained by calculating the log mean mole fraction of the inert component. We have $y_{Bs} = 1 - 0.15 - 0.85$ and $y_{Bb} = 1.0$. Hence the log average of these is found as $(1 - 0.85) / \log(1/0.85) = 0.9230$.
> The correction factor is then equal to the reciprocal of this log mean: $1/0.9230 = 1.1230$. Both procedures give the same answer, as expected.

6.2.4 Mole Fraction Profiles in UMD

The flux expression was derived by integrating across the system—that is, from 0 to H. The mole fraction profile need not be computed in this case to get the flux. The calculation of the mole fraction profile requires a partial integration from 0 to z and is shown in this section.

To find the mole fraction profile, Equation 6.16 is integrated from 0 to any arbitrary position z, rather than across the complete system:

$$N_A \int_0^z dz = -CD_{AB} \int_{y_{As}}^{y_A} \frac{dy_A}{1 - y_A} \tag{6.21}$$

The integrated version is

$$z N_A = CD_{AB} \ln[(1 - y_A)/(1 - y_{As})]$$

This can be rearranged to the following form for the mole fraction profile:

$$y_A = 1 - (1 - y_{As}) \exp\left(\frac{z N_A}{CD_A}\right) \tag{6.22}$$

Note that the profile is no longer linear, in contrast to the result with the low flux model.

An illustrative comparison is shown in Figure 6.2 in which we use a y_{As} value of 0.5 (a rather large vapor pressure intended to show the convection effects). Note that the profile is not linear when the convection effect is included.

In Example 6.4, the flux model shown earlier is coupled with a macroscopic balance for the second phase (the liquid).

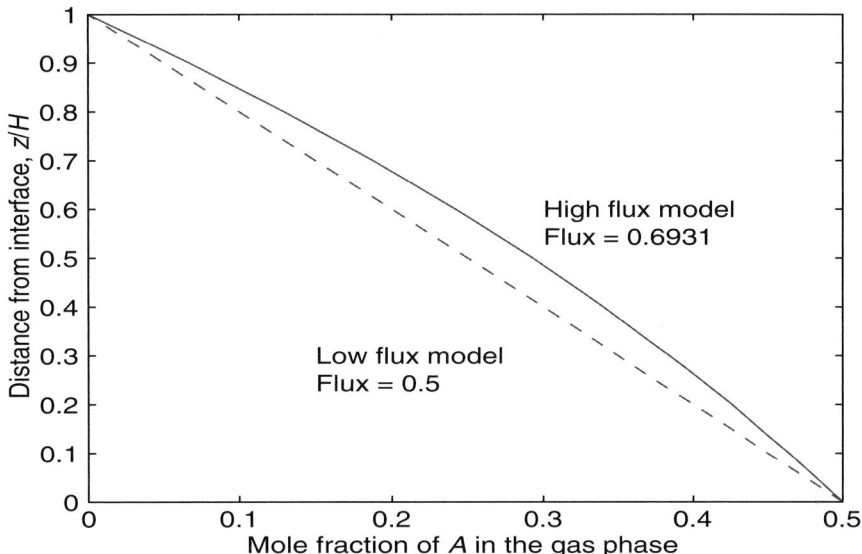

Figure 6.2   Evaporation in a tube: comparison of concentration (mole fraction) profiles with the low flux and high flux models. $y_{As} = 0.5$ and $y_{Ab} = 0$. Flux is scaled by $D_A C/H$.

Example 6.4   Liquid Evaporating in an Open Container

A schematic of the problem of evaporation in a tube was shown in Figure 6.1. Determine how the liquid level in the tube changes as a function of time.

Solution

An apparatus of this type is called an Arnold cell (or sometimes a Stefan cell). The level decreases as a function of time in the Arnold cell, and this change in level can be used to determine the diffusivity of the evaporating gas in the carrier gas. This is actually a transient diffusion problem, but for modeling purposes we focus on a particular instant in time when the vapor space has a height H and assume a steady state. This assumption, called the pseudo-steady state hypothesis, holds when the time scale for level changes is much larger than the time scale for diffusion. We will calculate the rate of mass transfer based on this assumption.

A transient mass balance for the liquid phase is now used:

$$\text{in} - \text{out} = \text{accumulation}$$

The *in* term is zero and the *out* term is the rate of evaporation in mol/s units. This rate is given as AN_A, where A is the cross-sectional area of the container. Accumulation is the time derivative of moles of A in the liquid at any instant of time. (This will be a negative quantity in this example.) The number of moles in the container at any instant of time is $\rho_L A(L - H)/M_A$, where L is the height of the container.

Putting all this information together, you should show that

$$\frac{dH}{dt} = \frac{M_A}{\rho_L} N_A$$

where N_A is the instantaneous rate of evaporation—that is, the rate of evaporation based on the current height of vapor space H:

$$N_A = \frac{D_A C y_{As}}{H} \mathcal{F}$$

where $\mathcal{F}$ is the drift correction factor. In this expression we assume that the bulk mole fraction of A (at the tube edge and outside the tube) is zero.

Substituting and integrating, the height change is given as

$$H^2 - H_0^2 = 2 \frac{M_A}{\rho_L} D_A C y_{As} \mathcal{F} t$$

6.3   Film Concept in Mass Transfer Analysis

The equation for flux—Equation 6.17 for instance—is derived for the evaporation of a liquid but finds wide application in the field of mass transfer. In this section, we discuss the classical film theory for mass transfer where such equations find use. The film model attempts to provide a simplified representation of mass transfer across a fluid–solid interface. The film model can be viewed as a simplified representation of the mass transport phenomena taking place near a fluid–solid boundary.

More detailed models use the convection–diffusion equation and the concept of a boundary layer near the solid–fluid interface (discussed in Chapter 11). The film model is simpler, however, and is also useful to incorporate the coupling between mass transfer and reaction. We review the boundary layer concepts here briefly to establish the background for understanding the film model.

6.3.1   Boundary Layer Concept for Fluid–Solid Mass Transfer

Near a solid surface there is a boundary layer where the velocity changes from zero to a bulk value over a small distance known as the hydrodynamic

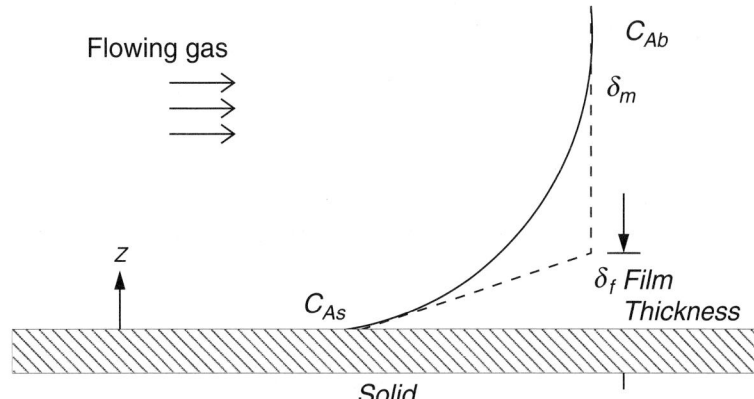

Figure 6.3   Concentration profile near a solid surface and its approximation by the film model. Solid lines are profiles as envisaged by boundary layer theory; the dashed line is an approximation according to the film model.

boundary layer. Similarly, there is a region where the concentration changes, called the mass transport boundary layer. Typical concentration profiles in the boundary layer are shown in Figure 6.3.

Results from Boundary Layer Theory for a Flat Plate

Key results for the mass transfer coefficient needed from the boundary layer theory are summarized here and more details are studied in Chapter 11. The results are useful to understand the film concepts in mass transfer. We will assume that we have a laminar boundary layer—that is, that the flow within the boundary layer is laminar and has no fluctuating velocity components.

The flux at the surface is approximated as a low flux value with Fick's law:

$$N_A = \text{ boundary layer model model} = -D_A \left(\frac{dC_A}{dy} \right)_{y=0} \tag{6.23}$$

The concentration gradient at the surface can be calculated by a detailed model, which is discussed in Chapter 11. This detailed approach is called Blasius analysis.

The concentration profile is often approximated as a cubic polynomial. The following approximation (known as the von Karman approximation and studied in more detail in Chapter 11) is a reasonable representation:

$$c_A = \frac{C_A - C_{AS}}{C_{Ab} - C_{As}} = \frac{3}{2}\frac{y}{\delta_m} - \frac{1}{2}\left(\frac{y}{\delta_m}\right)^3 \tag{6.24}$$

where δ_m is the thickness of the boundary layer for mass transfer. Note that this relation holds only if the flow in the boundary layer is laminar. An expression for the boundary layer thickness δ_m can be derived as shown later.

The flux at $y = 0$ can be calculated using this cubic profile:

$$N_{A0} = \frac{3}{2}\frac{D_A}{\delta_m}(C_{As} - C_{Ab})$$

Using the definition of the mass transfer coefficient, we can write this expression as

$$N_{A0} = k_m(C_{As} - C_{Ab})$$

Comparing the two expressions, the mass transfer coefficient is predicted from this equation as

$$k_m = \frac{3}{2}\frac{D}{\delta_m} \tag{6.25}$$

This is a local mass transfer coefficient at any position x along the flat plate. It varies along the plate since δ, the thickness of the hydrodynamic boundary layer, and δ_m, the mass transfer boundary layer, are both functions of x.

Relation to Momentum Boundary Layer

The mass transfer boundary layer thickness, δ_m, is often correlated as a function of the momentum boundary layer thickness, δ, and the Schimdt number defined as ν/D_A:

$$\delta_m = \frac{\delta}{Sc^{1/3}}$$

By using this definition in Equation 6.25, we can relate the mass transfer coefficient to the Schmidt number:

$$k_m = \frac{3}{2}\frac{D}{\delta}Sc^{-1/3} \tag{6.26}$$

An example of the use of this equation is provided in Example 6.5.

Example 6.5   Mass Transfer Coefficient for a Flat Plate with Laminar Flow

A gas stream is flowing past a plate at a velocity of 0.8 m/s. Estimate the value of the mass transfer coefficient at a distance 0.5 m from the leading edge of the plate. Assume the flow is laminar.

Solution

The physical properties needed are the kinematic viscosity and the diffusion coefficient. We use the following values: $D_A = 6 \times 10^{-6}$ m^2/s; $\nu = 1.57 \times 10^{-5}$ m^2/s. Hence the Schmidt number is $\nu/D_A = 2.6$.

Using momentum transfer theory, the boundary layer thickness for the momentum can be approximated as

$$\delta = 4.64\sqrt{\frac{x\nu}{v_\infty}}$$

Substituting the values of the parameters, the hydrodynamic boundary layer thickness, δ, can be calculated as 0.0145 m.

The mass transfer boundary layer thickness is $\delta/Sc^{1/3}$. In this example, its value is 0.0105 m. The mass transfer coefficient is now calculated using Equation 6.26; its value is 8.52×10^{-4} m/s. This is the local value at a position 0.5 m from the leading edge of the plate. The average value can be shown as twice this value.

Why Use the Film Model?

The convection model is a useful predictive tool for the mass transfer coefficient for ideal, well-defined hydrodynamic situations. However, this model becomes cumbersome for some complex situations, such as for turbulent flow, complex geometries, and similar situations. In the context of using the same model format for all these cases and to include additional complexities such as the effect of a chemical reaction, high flux correction, and so on, a simpler model for mass transfer is useful, in contrast to the detailed convection–diffusion model based on the boundary layer theory. The film model provides such a platform and is now discussed.

6.3.2   Film Model Approximation

The film model provides a simple description of mass transport from a solid to a fluid, or vice versa. If a straight line is used to represent the concentration profile in Figure 6.3, we can assume that there is a region of thickness δ_f over which the concentration profile changes. Outside this thickness, we assume a constant concentration equal to the bulk fluid concentration.

Using Fick's law, the flux can be represented as follows:

$$N_A^\circ = \frac{D_A}{\delta_f}(C_{As} - C_{Ab}) \qquad (6.27)$$

Representation of the concentration distribution by a linear function and the concept of the film thickness parameter δ_f forms the basis for the film model for mass transfer. This model was first proposed by Nernst (1904). Although the film is a hypothetical concept, it has enjoyed widespread success in modeling mass transport processes, as we will observe during the course of this chapter.

Equivalently, a mass transfer coefficient can be defined to characterize the mass transfer near a solid with a flow past the solid:

$$N_A^\circ = k_m^\circ(C_{As} - C_{Ab}) \qquad (6.28)$$

The mass transfer coefficient is therefore related to the film thickness by

$$k_m^\circ = \frac{D_A}{\delta_f} \qquad (6.29)$$

The superscript $\circ$ indicates that the coefficient is based on Fick's law and is therefore applicable to low flux conditions.

In Example 6.5, the average mass transfer coefficient is 1.7×10^{-3} m/s. We find a film thickness of 3.5 mm (D/k_m).

If the low flux mass transfer model is not applicable, the flux can be defined as the low flux rate times a correction factor:

$$N_A = k_m^\circ(C_{As} - C_{Ab})\mathcal{F}_m \qquad (6.30)$$

The correction factor depends on the prevailing situation—mainly on which other species are being transported. The correction factor is given by Equation 6.20 for UMD as before. It is equal to 1 for EMD since N_t is zero in this case.

An expression for the correction factor can be presented in a rather general way, as explained in the following discussion. Let us first see what the concentration profile looks like in presence of (diffusion-induced) convection, as shown in Example 6.6.

Example 6.6   Concentration Profiles in the Film

Derive an expression for the concentration profile for the film model that includes convection effects. Also relate the flux of A to the total flux in the system.

Solution

The problem and the associated boundary conditions are sketched in Figure 6.4. The mole fraction of A is y_{A0} in the bulk gas and $y_{A\delta}$ at the surface. The flux of A is to be computed. The expressions for the combined flux is used in presence of convection. Thus we have

$$N_A = -CD\frac{dy_A}{dz} + y_A N_t \tag{6.31}$$

Here N_A is the abbreviation for N_{Az} and not a vector.

Equation 6.31 can be put in a dimensionless form. If we use CD/δ as a scaling parameter for the fluxes, the resulting expression is

$$N_A^* = -\frac{dy_A}{d\zeta} + y_A N_t^* \tag{6.32}$$

Here ζ is the dimensionless distance in the film z/δ and the starred fluxes are dimensionless fluxes.

The component N_A^* is a constant and not a function of ζ (since we are assuming no reaction taking place simultaneously with diffusion). Hence we can integrate the previous differential equation. Integration and use of the boundary condition y_{A0} at $\zeta = 0$ lead to

$$y_A = \frac{N_A^*}{N_t^*} + \left(y_{A0} - \frac{N_A^*}{N_t^*}\right)\exp(N_t^*\zeta) \tag{6.33}$$

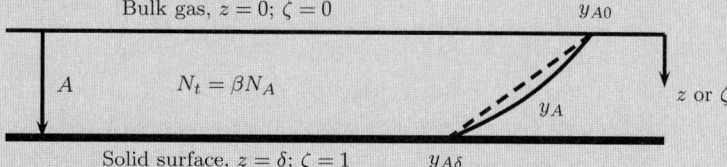

Figure 6.4   Schematic of the film model to derive a general flux expression. Species A is diffusing to a solid surface. Species B, C, and so on also diffuse or counter-diffuse, which sets up a net total flux N_t, which can be expressed as β times N_A. The dashed line is the profile for $\beta = 0$.

Now we use the boundary condition at $\zeta = 1$, which is $y = y_{A\delta}$, and then rearrange to obtain the following equation for N_A:

$$N_A^* = N_t^* \left(\frac{y_{A\delta} - y_{A0} \exp(N_t^*)}{1 - \exp(N_t^*)} \right) \tag{6.34}$$

This expression can be used for a wide range of problems with a suitable condition specified for N_t as shown below.

Further let

$$N_t^* = \beta N_A^*$$

Here β is a constant (determinacy correction factor) determined by the total flux condition. For example, $\beta = 1$ for UMD and $\beta = 0$ for EMD.

Rearranging Equation 6.34, we get a general relation for N_A^*:

$$\boxed{N_A^* = \frac{1}{\beta} \ln \left(\frac{1 - \beta y_{A\delta}}{1 - \beta y_{A0}} \right)} \tag{6.35}$$

This equation is useful for a number of applications, as shown in the following discussion. The actual (dimensional) flux can be calculated by multiplying N_A^* by $(D/\delta_f)C$ or by $k_m C$.

6.3.3   Film Model: Determinacy Correction Factor

The conditions for N_t and the corresponding value of β to use in Equation 6.35 can be stated on a case-by-case basis. Some common cases and the relations are shown here:

- Low flux mass transfer. Taking the limit of Equation 6.35 as $\beta \to 0$, we have

$$N_A^* = (y_{A0} - y_{A\delta}) \tag{6.36}$$

  which can also be obtained by direct application of Fick's law. The concentration (mole fraction) profile is linear here.

- UMD or evaporation of a pure liquid (also called Arnold diffusion). We use $N_t^* = N_A^*$ here, and $\beta = 1$ and the following equation hold from Equation 6.35:

$$N_A^* = \ln \left(\frac{1 - y_{A\delta}}{1 - y_{A0}} \right) \tag{6.37}$$

You should verify that this is a dimensionless version of Equation 6.17.

- EMD or equimolar counter-diffusion. Here $N_t = 0$ and correspondingly $\beta = 0$. The diffusion-induced convective flow is absent. The result for N_A^* is the same as that for low flux mass transfer.

- Distillation of a binary mixture. Here a species A evaporates from a liquid interface to the vapor and a species B condenses from the vapor. No heat is generally added in the distillation column itself. (All heat is added in the reboiler and removed in the condenser.) Hence the heat needed to vaporize A and B must balance, which provides the determinacy condition:

$$N_A \Delta H_{vA} + N_B \Delta H_{vB} = 0$$

which relates N_A and N_B. Now N_t can be found and β takes the following value:

$$\beta = 1 - \frac{\Delta H_{vA}}{\Delta H_{vB}}$$

If the heats of vaporization of A and B are equal, then the equimolar counter-diffusion model holds.

- Diffusion with a heterogeneous reaction at a surface. N_A and N_B are determined by the stoichiometry of the reaction. For example, consider a reaction scheme

$$A \rightarrow (\nu_E + 1)B$$

where the stoichiometric coefficient for B is written in a funny way for ease of later algebra. The stoichiometry requires

$$N_B = -(\nu_E + 1)N_A$$

Hence

$$N_t = N_A + N_B = N_A - (\nu_E + 1)N_A = -\nu_E N_A \qquad (6.38)$$

Equation 6.35 can now be used with $\beta = -\nu_E$ to find N_A. But we also need a surface reaction boundary condition to calculate the concentration of A at the reacting surface, $y_{A\delta}$. See Section 6.4.3 for details.

Thus, the various cases of diffusion-induced convection can be analyzed depending on how N_t and N_A are related. The film model provides a simple but useful platform to perform this analysis.

6.4   Surface Reactions: Role of Mass Transfer

In this section we study diffusion followed by a heterogeneous reaction, and we illustrate how the film model is useful to study the role of mass transfer. A schematic of the problem is shown in Figure 6.5, which illustrates the simple case in which species A diffuses to the surface and forms products that then counter-diffuse back to the bulk gas.

6.4.1   Low Flux Model: First-Order Reaction

If diffusion followed by a surface chemical reaction occurs under low flux conditions, then the mass transfer and reaction can be treated as two resistances in series. Let k_s be the rate constant for the reaction. The flux to the surface is equal to the rate of consumption by the reaction. Thus

$$N_A = k_m(C_{Ab} - C_{As}) = k_s C_{As} \tag{6.39}$$

The last two pairs can be used to eliminate C_{As}:

$$\frac{C_{As}}{C_{Ab}} = \frac{1}{1 + k_s/k_m} = \frac{k_m}{k_s + k_m}$$

The rate can then be expressed in terms of an overall rate constant k_o:

$$N_A = k_o C_{Ab} \tag{6.40}$$

where k_o is

$$k_0 = \frac{k_m k_s}{k_s + k_m} \tag{6.41}$$

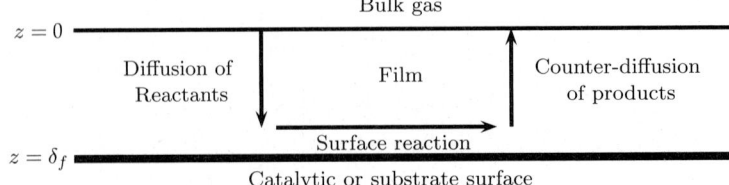

Figure 6.5   Schematic of the steps in mass transfer accompanied by a heterogeneous reaction or chemical vapor deposition on a substrate surface.

This can also be expressed as

$$\frac{1}{k_o} = \frac{1}{k_m} + \frac{1}{k_s}$$ (6.42)

In this expression, it appears as if two resistances are being added, which is indeed the case. Thus the mass transfer and the reaction resistances operate in series, and the overall resistance is the sum of these two resistances.

The following limiting cases are useful to understand:

- Reaction-limiting regime: If $k_m \gg k_s$, then $k_0 = k_s$; the process is limited by the rate of reaction.
- Mass transfer–limiting regime: If $k_s \gg k_m$, then $k_0 = k_m$; the process is limited by the rate of mass transfer.

The resistance concept does not apply for nonlinear kinetics, as shown in Section 6.4.2.

Measured *versus* True Kinetics

If the mass transfer resistance is much larger than the reaction resistance, it is not possible to determine k_s accurately from measured experimental data. What you measure is then an indication of the mass transfer coefficient k_m rather than the true rate constant.

In general, the measured rate of reaction can be expressed in terms of an apparent rate constant k_o that is equal N_A/C_{Ab} (assuming a first-order reaction). This is a combined parameter involving both the true rate constant k_s mass transfer coefficient k_m; it is not a true rate constant of the reaction. To isolate these values and report the true rate constant from the measured reaction rate data, one needs an estimate of the mass transfer coefficient and a correction for the effect of mass transfer. Increasing the gas velocity, and thereby increasing k_m, is one way of reducing the mass transfer resistance. An example of growth of an oxide layer over a silicon surface is shown in Example 6.7.

Example 6.7 Silicon Oxidation: Growth of an Oxide Layer

Growth of an oxide layer over a silicon surface is important in semiconductor processing. Here oxygen diffuses through a product layer of SiO_2 and reaches the SiO_2-Si interface reacts there. Assume a fast reaction. Find how the thickness of the oxide layer changes with time. The problem is shown in Figure 6.6.

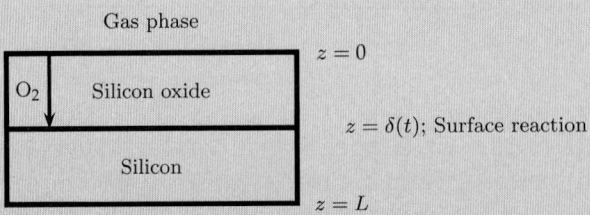

Figure 6.6   Schematic of the silicon oxidation problem and the study of the rate at which the oxide layer grows.

Solution

The oxygen concentration at the interface is nearly zero for a fast reaction. The process is therefore assumed to be limited by the mass transfer rate. In turn, the rate of oxidation is equal to the rate at which oxygen reaches the surface:

$$N_A = \frac{D}{\delta} C_{As}$$

The silicon oxide layer forms with time, and a mass balance for the silicon gives an expression for the thickness of the layer. A balance over the silicon remaining at any time, t, is now done.

$$\text{in} - \text{out} + \text{generation} = \text{accumulation}$$

No silicon is entering or leaving the control volume. Hence the *in* and *out* terms are zero. The balance simplifies to

$$\text{generation} = \text{accumulation}$$

The *generation* term will be negative here because silicon is actually being consumed by reaction. This, in turn, is equal to the negative of the amount of oxygen arriving at the surface since one mole of oxygen is needed to oxidize one mole of silicon. Hence

$$\text{Generation} = -N_A A$$

The accumulation is $\frac{d}{dt}[A(L - \delta)\rho_B/M_B]$, where L is the initial depth of silicon. Equating the two we get

$$-N_A A = -A \frac{d}{dt}[\delta\, \rho_B/M_B]$$

Using the diffusion model for N_A we get

$$(\rho_B/M_B)\frac{d\delta}{dt} = D_A\frac{C_{As}}{\delta}$$

Integration gives us the relation for the oxide growth layer:

$$\delta = \sqrt{\frac{2M_B D_A C_{As} t}{\rho_B}}$$

The film thickness increases with a square root dependency on time, which is also observed in many similar diffusion-limited growth or shrinkage processes.

6.4.2   Low Flux Model: Nonlinear Reactions

We now show how mass transfer models can be easily set up and solved for nonlinear reactions by taking two examples: one for a second-order reaction and another for a bimolecular reaction.

Second-Order Reaction

The surface reaction is now second order in the surface concentration. Equating the rate of mass transfer and the rate of reaction, we obtain

$$k_m(C_{Ab} - C_{As}) = k_s C_{As}^2$$

This can be solved as a quadratic equation to obtain C_{As}. The rate of reaction of A can then be calculated as $k_s C_{As}^2$. Note that the resistance concept does not apply.

Bimolecular Reaction

Consider two components diffusing to a catalyst surface and reacting with each other over that surface:

$$A + \nu B \rightarrow \nu_p P$$

One example is selective catalytic reduction of nitric oxide (NO) using ammonia. We will assume that A and B are present in low concentrations and, therefore, that diffusion-induced convection is not needed. The problem can be formulated by equating the rate of mass transfer to the rate of reaction for both A and B. For species A, we can write

$$N_A = k_{mA}(C_{Ab} - C_{As}) = k_s C_{As} C_{Bs}$$

The surface reaction is assumed to be second order (1,1) in this expression.
Similarly, for species B, we can write

$$N_B = \nu N_A = k_{mB}(C_{Bb} - C_{bs}) = \nu_B k_s C_{As} C_{Bs}$$

The unknowns are the two surface concentrations and the flux of A to the surface, which is also equal to the rate at which A reacts. The set of equations can be solved simultaneously to obtain these unknowns and the effect of mass transfer can be examined.

6.4.3   High Flux Model: Effect of Product Counter-Diffusion

Now consider the following reaction scheme:

$$A \to (\nu_E + 1)B$$

where ν_E is the excess stiochiometry parameter and is equal to zero for equal molar counter-diffusion. The stoichiometry requires

$$N_B = -(\nu_E + 1)N_A$$

Hence

$$N_t = N_A + N_B = N_A - (\nu_E + 1))N_A = -\nu_E N_A$$

Equation 6.35 can now be used with β set as $-\nu_E$. The flux of A to the surface is now given by

$$N_A = \frac{k_m C}{\nu_E} \ln\left(\frac{1 + \nu_E y_{A0}}{1 + \nu_E y_{A\delta}}\right) \tag{6.43}$$

An additional condition is required for fixing $y_{A\delta}$. This comes from the rate of reaction at the interface. We can equate the flux to the rate of surface reaction (per unit area). For a first-order surface reaction, the condition may be stated as

$$N_A = k_s C y_{A\delta} \tag{6.44}$$

or

$$y_{A\delta} = \frac{N_A}{k_s C}$$

This can be used in Equation 6.43, leading to the following nonlinear equation for N_A:

$$N_A = \frac{k_m C}{\nu_E} \ln\left(\frac{1 + \nu_E y_{A0}}{1 + \nu_E N_A/(k_s C)}\right) \tag{6.45}$$

Note that the addition of the resistance concept is no longer applicable even though this is a first-order reaction.

Fast Reaction

For the limiting case of a fast reaction, one can approximate $y_{A\delta}$ as zero. The following limiting case of Equation 6.43 then holds:

$$N_A = \frac{k_m C}{\nu_E} \ln(1 + \nu_E y_{A0}) \tag{6.46}$$

For fast reactions, the mass transfer coefficient is the key parameter and the rate constant is of no consequence.

Slow Reaction

For a slow reaction, $y_{A\delta} = y_{A0}$ and mass transfer resistance is no longer limiting the process. The rate is then calculated based solely on kinetic considerations and is equal to $k_s C y_{A0}$. The criterion for a slow reaction is the ratio k_s/k_m, which should be generally on the order of 0.1 or less. Likewise, if the ratio is greater than 10 or so, the fast reaction asymptote is expected to hold.

Example 6.8 shows an example in which a reaction is coupled with a macroscopic balance follows.

Example 6.8   Effect of Stoichiometry

Combustion of a spherical coal particle takes place over a carbon surface. The particle is suspended in air with no flow of air. The reaction conditions are such that 1.2 moles of product counter-diffuses for 1 mole of oxygen transported to the surface. The reaction may be assumed to be fast. What is the rate of reaction? At what rate is the particle shrinking?

Solution

The ν_E parameter is 0.2 and the dimensionless flux to the surface is

$$N_A^* = \frac{1}{0.2} \ln(1 + 0.2 \times 0.21) = 0.2055 \tag{6.47}$$

The actual flux is $N_A k_m C$. Further we use $k_m = D_A/R$ for a stagnant sphere. Hence the rate of oxygen uptake due to reaction at the surface of the coal particle is

$$\dot{M}_A = (4\pi RC D_A/0.2) \ln(1 + 0.2 \times 0.21) \tag{6.48}$$

This result can then be used for the carbon balance to derive an expression for carbon particle shrinkage:

$$\frac{d}{dt}(4\pi R^3 \rho_C / 3M_c) = \nu_C \dot{\mathcal{M}}_A \tag{6.49}$$

where ν_C is the moles of carbon reacted per mole of oxygen reacted.

If CO is formed as the sole product, we have:

$$2C + O_2 \rightarrow 2CO$$

In this case, ν_E is 1 and correspondingly $\nu_C = 2$ per mole of oxygen reacted. If CO_2 is the product, then

$$C + O_2 \rightarrow CO_2$$

which corresponds to an equimolar counter-diffusion with $\nu_E = 0$ and, from the stoichiometry above, $\nu_C = 1$.

The observed value is $\nu_E = 0.2$. Interpolating, we can use $\nu_C = 1.2$ for the current example.

Equation 6.49 can now be integrated to find the radius as a function of time.

6.5 Gas–Liquid Interface: Two-Film Model

Lewis and Whitman (1929) introduced the concept of two films at the interface—one on the gas side and one on the liquid side. A schematic of the concentration profiles based on this model is shown in Figure 6.7. Note that the profiles are linear in the figure which is, strictly speaking, applicable for the low flux model.

6.5.1 Mass Transfer Coefficients

The mass transfer coefficients on each side can be defined as follows.

Gas Side

$$k_m = D_{A,G} / \delta_G$$

where k_m is based on the concentration driving force. The flux on the gas side of the interface is therefore represented as

$$N_A = k_m(C_{AG} - C_{AG,i})$$

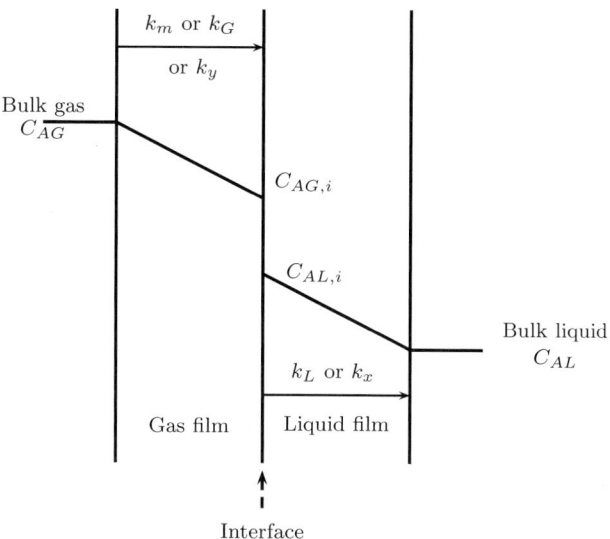

Figure 6.7 Two-film model. Two stagnant films are assumed on each side of the interface, and the resistance to mass transfer is assumed to be confined to these films.

Partial pressure units are more convenient to use for the gas phase. We can also use a mass transfer coefficient k_G, defined as follows:

$$N_A = k_G(p_{AG} - p_{AG,i})$$

The two mass transfer coefficients are related as

$$k_G = \frac{k_m}{R_g T}$$

The mole fraction difference can also be used as a driving force, which results in yet another definition:

$$N_A = k_y(y_{AG} - y_{Ai}) \tag{6.50}$$

and $k_G = k_y/P$.

Liquid Side

A similar definition can be used for the liquid film:

$$k_L = D_{A,L}/\delta_L$$

where k_L is based on the concentration driving force:

$$N_A = k_L(C_{AL,i} - C_{AL})$$

If the mole fraction driving force is used, we write

$$N_A = k_x(x_{Ai} - x_{AL}) \tag{6.51}$$

The relation $k_x = k_L C_L$ holds between the two definitions, where C_L is the total liquid concentration.

6.5.2   Overall Mass Transfer Coefficient

The interfacial concentrations are related by a Henry's law type of relation. This relation provides equations to eliminate the interfacial values and express the results in terms of an overall driving force and the definition of an overall mass transfer coefficient.

Let $p_{AG,i} = H_A C_{AL,i}$, where H_A is in pressure-concentration form. Then an overall driving force can be derived and used. This overall driving force can be expressed in two ways: (1) partial pressure (gas phase) driving force and (2) concentration (liquid phase) driving force.

Gas-Phase Driving Force

From the gas side

$$\frac{N_A}{k_G} = (p_{AG} - p_{AG,i})$$

Similarly, from the liquid side

$$\frac{N_A}{k_L} - (C_{AL,i} - C_{AL})$$

It is necessary to correct for the interfacial concentration jump. The preceding equation is multiplied by H_A throughout. On the right side, we can write the components in terms of partial pressures.

Let $p_{AG}^* = H_A C_{AL}$, a hypothetical partial pressure if the bulk liquid were to attain equilibrium. Then the driving force is $p_{AG} - p_{AG}^*$ and, correspondingly, the overall transfer coefficient can be defined as

$$N_A = K_G(p_{AG} - p_{AG}^*)$$

It can be shown that

$$\frac{1}{K_G} = \frac{1}{k_G} + \frac{H_A}{k_L} \tag{6.52}$$

The two terms on the right side of the equation may be interpreted as resistances—one on the gas side and the other on the liquid side.

Liquid-Phase Driving Force

The overall liquid concentration driving force is defined as $C^*_{AL} - C_{AL}$, where C^*_{AL} is a hypothetical concentration corresponding to the bulk gas conditions. The rate of transfer is then defined using an overall coefficient K_L:

$$N_A = K_L(C^*_{AL} - C_{AL}) \tag{6.53}$$

It can be shown that

$$\frac{1}{K_L} = \frac{1}{H_A k_G} + \frac{1}{k_L}$$

which relates the overall liquid-based mass transfer coefficient to the gas-side and liquid-side coefficients.

Mole Fraction Driving Force

The interfacial equilibrium is given as $y_i = m x_i$ to eliminate these values between Equations 6.50 and 6.51. The final result after a few steps of algebra, expressed in terms of the gas-phase mole fraction as the driving force, is

$$N_A = K_y(y - mx)$$

where K_y is the overall mass transfer coefficient:

$$\frac{1}{K_y} = \frac{1}{k_y} + \frac{m}{k_x}$$

Alternatively, we can use a mass transfer coefficient based on $y/m - x$ as the driving force. This is defined as

$$N_A = K_x(y/m - x)$$

where K_x is the overall mass transfer coefficient:

$$\frac{1}{K_x} = \frac{1}{m k_y} + \frac{1}{k_x}$$

The typical ranges for the values of the mass transfer coefficients are shown here to help you get a feel for the numbers:

- Liquid-side k_L is in the range of 10^{-5} to 10^{-3} m/s.
- Gas-side k_m in the range of 0.1 to 10 m/s.

The Henry's law parameter can vary widely, and hence the overall coefficient can vary over all these ranges.

Determining the Direction of Transfer

Absorption refers to transfer of mass from a gas to a liquid, while desorption refers to transfer of mass from a liquid to a gas. Whether the component is absorbing or desorbing depends on the overall driving force. Note that $y - mx$ is the driving force when K_y is the overall mass transfer coefficient. We can also use $y/m - x$ as the driving force when the mass transfer coefficient K_x is used. The value $y - x$ should not be inadvertently used.

If $y > mx$, then the gas is rich in solute and gas absorption takes place from gas to liquid.

If $y < mx$, then the gas is poor in solute and desorption takes place from liquid to gas.

Both absorption and desorption processes are of industrial importance. The two cases are shown schematically in Figure 6.8.

Controlling Resistance

It is also important to know which film offers the major resistance to mass transfer. This depends on the controlling resistance, which may be on the gas side or the liquid side depending on the value of H_A.

If H_A is a small number which corresponds to a highly soluble gas, then the controlling resistance is on the gas film and $K_G \approx k_G$.

If H_A is a larger large number which corresponds to poorly soluble gas, then the controlling resistance is on the liquid film and $K_L \approx k_L$.

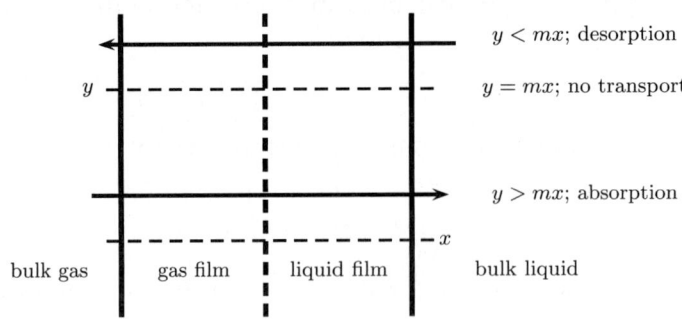

Figure 6.8   Driving force calculations for absorption and desorption. The difference $y - x$ is not the correct driving force.

Similarly, if m is large (low solubility case), then the liquid-side resistance is controlling and $K_x \approx k_x$. The reverse is true for small m. In that case, $K_y \approx k_y$. The calculation of the overall resistance is illustrated in Example 6.9.

Example 6.9   Calculation of the Overall Coefficient

SO_2 is absorbed into water in a packed column. At a point in the column the bulk conditions are $y_A = 0.1$ and $x_A = 0.01$. The Henry's law constant is 440 atm. Temperature is 300 K and the total pressure is 5 atm. The mass transfer coefficients are $k_G = 4 \times 10^{-5}$ mol/m^{2}s atm and $k_L = 2 \times 10^{-4}$ m/s. Find the rate of absorption at this point. Calculate and sketch the concentration profiles in the gas and liquid films.

Solution: The units of the Henry's law constant suggest that the equilibrium is represented as $p_A = H x_A$, with A being the solute SO_2. For the problem, $p_A = 0.1 \times 5 = 0.5$ atm. This sets the gas-phase value.

The liquid concentration is 0.01 and hence the equilibrium partial pressure is $p_A^* = 0.01 \times 440 = 4.4$ atm. The overall driving force is $0.5 - 0.44 = 0.06$ atm.

The overall mass transfer coefficient is calculated from Equation 6.52. A note of caution is that the Henry constant for this equation should be in pressure-concentration units. Hence the given value of 440 atm is converted by dividing it by the total liquid concentration. The corresponding flux value is obtained by multiplying this by the overall driving force.

The overall coefficient turns out to be nearly the same the as gas side coefficient indicating that all the resistance is on the gas side. The result for the flux ix then 2.4×10^{-6} mole/m^2.

Summary

- Stagnant systems are defined as systems with no superimposed external flow. They may not always be truly stagnant, as a flow can be generated in such cases due to diffusion itself. A common example is evaporation of a liquid in an inert gas. The evaporation causes a flow (often barely noticeable), which then enhances the rate of transport. The enhancement is often accounted by a drift flux correction factor.

- An Arnold cell is a simple device to measure the diffusion coefficient (in gas mixtures) by simple measurement of the change in the level of a liquid due to evaporation. The model for such systems is based on the concept of a pseudo-steady state. The evaporation is calculated as though the system were in (pseudo-) steady state at any given instant of time. The liquid level is then obtained as a function of time using the pseudo-steady state value for the evaporation.

- The assumption of a pseudo-steady state is used in many problems in mass transfer calculations. This assumption is reasonable whenever the time scale of diffusion is much smaller than the process time.

- Mass transfer near a fluid–solid interface is often modeled by using a film model. This model assumes there is a stagnant film near the interface within which all the resistance to mass transfer is contained. The diffusional rate of transport can then be calculated using the relation D/δ_f (or equivalently a mass transfer coefficient) multiplied by the overall concentration difference.

- The film model is very useful to study mass transfer accompanied by a surface reaction. For a simple first-order reaction with no mole change, an overall rate constant can be defined and used. The reciprocal of this rate constant is the sum of the reciprocal of the mass transfer coefficient and the reciprocal of the reaction rate constant. Thus it is not the true or intrinsic rate constant, but merely a convenient parameter for representing the data. The methodology is exactly equivalent to the law of addition of resistances.

- The simple concept of resistances fails even for a first-order reaction when there is a change in the number of moles present due to reaction. The counter-diffusion of the product can enhance or retard the mass transfer rate. For nonlinear kinetics, even with no change in moles, the resistance concept does not hold. In such a case, a nonlinear or transcendental implicit equation for the rate is generally obtained.

- The film model when applied to a fluid–fluid (gas–liquid or liquid–liquid) systems necessitates the use of two films—one on the gas side of the interface and the other on the liquid side. Each film offers its own resistance to mass transfer. These resistances can be added (for the low flux mass transfer case) to get an expression for the overall resistance. Correspondingly, an overall mass transfer coefficient can be defined and used to calculate the mass transfer rate.

- The mass transfer rate may be governed by the gas-side resistance, the liquid-side resistance, or both. Relative contributions can be evaluated, which is a very useful information for selection of appropriate equipment and design.

- For a highly soluble gas, the resistance is mostly in the gas film; the reverse is true for a poorly soluble gas.

Review Questions

6.1 What is meant by the determinacy condition? Why is it needed?

6.2 What is meant by UMD and EMD? Which factors are different in mass transfer calculations for each of these?

6.3 What is meant by the drift flux correction factor? Where is it used?

6.4 State the assumptions involved in the film theory of mass transfer.

6.5 What is the relationship between film thickness and the mass transfer coefficient?

6.6 State the difference between film thickness in the film model and the concentration boundary layer thickness in the convection–diffusion model.

6.7 If the mass transfer coefficient is measured at a low concentration of a solute, how would you correct it for high concentration? Assume UMD.

6.8 A surface reaction follows first-order kinetics. What is the unit of the rate constant?

6.9 A surface reaction follows second-order kinetics. What is the unit of the rate constant?

6.10 Distinguish between the apparent rate constant and the true rate constant.

6.11 A heterogeneous reaction is taking place under a rapid reaction condition. State the equation to calculate the flux to the surface and the rate of reaction.

6.12 What is the relation between K_G and K_L?

6.13 What is the Schmidt number? What is its role in boundary layer theory and mass transfer calculations?

Problems

6.1 **Variable diffusivity example.** A spherical capsule has an outer membrane thickness with inner and outer radii r_i and r_o, respectively. A solute is diffusing across this capsule. Consider the case where the diffusion coefficient is a function of concentration and can be represented in a general form as $D(C_A)$; it is not a constant. Derive an expression for the (total) rate of mass transport across this spherical shell. Use the low flux model.

6.2 **Evaporation from a beaker.** Benzene is contained in an open beaker that is 6 cm high and filled to within 0.5 cm of the top. Temperature is 298 K and the total pressure is 1 atm. The vapor pressure of benzene is 0.131 atm at these conditions and the diffusion coefficient is 9.05×10^{-6} m^2/s. Find the rate of evaporation based on the low flux model, the exact model, and the low flux model corrected by drift flux.

6.3 **Level change calculation.** For problem 2, find the time for the benzene level to fall by 2 cm. The specific gravity of benzene is 0.874. For this condition

($H = 0.5 + 2$ cm), find the mole fraction profile of benzene in the vapor phase and compare it with the linear approximation (which would be the prediction of the low flux model).

6.4   **Mass transfer across two bulbs.** Two bulbs are connected by a straight tube of 0.001 m in diameter and 0.15 m in length. Initially the first bulb contains nitrogen and the second bulb at the other end contains hydrogen. The system is maintained at a temperature of 298 K and a total pressure of 1 atm. The volume of each bulb is 8×10^{-6} m^3. Calculate and plot the mole fraction profile of nitrogen in bulb 1 as a function of time.

6.5   **Mass transfer with first-order surface reaction.** Species A is diffusing to a catalytic surface, where it undergoes a first-order surface reaction. Equal moles of product counter-diffuse from the catalytic surface to the bulk gas. A rate of 1.6×10^{-4} mol/m^{2}s was measured when the system was at 2 atm pressure and temperature of 300 K, with 10% A in the bulk gas. The mass transfer coefficient was estimated for the given flow conditions as 2×10^{-4} m/s. Estimate the true rate constant. If the gas velocity is doubled, find the rate of reaction. Assume that the mass transfer coefficient changes with gas velocity to the power of 0.8.

6.6   **Mass transfer with second-order surface reaction.** Species A is diffusing to a catalytic surface, where it undergoes a second-order surface reaction. Equal moles of product counter-diffuse from the catalytic surface to the bulk gas. A rate of 8×10^{-4} mol/m^2 s was measured when the system was at 2 atm pressure and a temperature of 300 K, with 10% A in the bulk gas. The mass transfer coefficient from the bulk gas to the catalyst was estimated for the given flow conditions 2×10^{-4} m/s. Estimate the true rate constant.

6.7   **Effect of product counter-diffusion.** An example of a problem in which a severe counter-diffusion of the products takes place is the deposition of SiO$_2$ from tetraethoxysilane (TEOS) on a solid substrate. The reaction is represented as

$$SiO(C_2H_5)_4(g) \rightarrow SiO_2(s) + 4C_2H_4(g) + 2H_2O(g)$$

Note that 6 moles have to counter-diffuse, which retards mass transfer unless the mole fraction of TEOS is small. Consider deposition from a gas at temperature of 400 K and 0.1 atm pressure, with a mole fraction of TEOS of 0.2. Diffusivity is 0.1 cm^2/s and the film thickness is 2 mm. Calculate the rate of reaction (equal to the rate of diffusion to the surface) and the film growth rate, assuming that the surface reaction is very rapid.

6.8   **Effect of stoichimetry on mass transfer rate.** In a combustion chamber, oxygen diffuses through air to a carbon surface, where it reacts. Depending on the reaction surface conditions and temperature, either CO or CO$_2$, or a mixture of both, is produced. The reaction at the surface can be assumed to be rapid. The conditions are mole fraction of oxygen in the bulk gas is 0.21; pressure = 2 bar, temperature = 600 K. The film model with a film thickness of 1 mm can be used for mass transfer, and the diffusion coefficient of oxygen in the system has a value of 0.2 cm^2/s. Calculate the rate of reaction assuming that only CO$_2$ is produced as the product. Calculate the rate of reaction assuming that only CO is produced as the product.

6.9 **Shrinking rate of a reactive particle.** Uranium solid is reacted with fluorine to produce a gas-phase precursor of uranium according to the reaction

$$U(s) + 3F_2(g) \rightarrow UF_6(g)$$

What is the expression for the combined flux of F_2 (denoted as species A here) in this system? If fluorine diffuses to an uranium surface through a film of thickness δ, derive the expression for the flux of fluorine to the surface. Assume the bulk mole fraction is y_{Ab} and the mole fraction at the slab surface is zero.

Now consider that the reaction is taking place on a solid **sphere** of radius R, and the transport of F_2 from the bulk gas to the surface through a thin boundary layer near the solid surface is the rate-controlling step. Again assume the bulk mole fraction is y_{Ab} and the mole fraction at the slab surface is zero. Also assume that the gas is stagnant. Derive an expression for the rate of fluorine transferred to the sphere ($\mathcal{M}_A$).

6.10 **Series reaction in a nonporous catalyst.** Consider a nonporous catalyst with a series reaction occurring at the surface:

$$A \rightarrow B \rightarrow C$$

Derive equations for the rates of reaction of A and B considering diffusional resistances. Use a low flux model.

6.11 **Various definitions of mass transfer coefficient.** At a certain point in an absorber, the value of k_y is 7×10^{-3} kmol/m^{2}s and that of k_x is 2×10^{-4} kmol/m^{2}s. The system has properties similar to air and water and is at 300 K and 1 atm. Find k_G, k_m, and k_L.

6.12 **Driving force and overall mass transfer coefficient.** At a certain point in a mass transfer system, the bulk mole fractions are $y_A = 0.04$ in the gas phase and $x_A = 0.004$ in the liquid phase. The mass density of the liquid is nearly the same as water. The Henry's law constant for A is reported as 7.7×10^{-4} atm m^3/g mol. Determine whether the species absorbing or desorbing. If the mass transfer coefficients are $k_G = 0.010$ g mol/m^2 s atm and $k_x = 1.0$ (mol/s m^2), find the overall transfer rate. Here k_x is the mass transfer coefficient on the liquid side based on the mole fraction difference as the driving force. Find the percentage of the resistance in each film.

6.13 **Overall mass transfer coefficient.** In an absorption column, the following parameters were observed: liquid-side mass transfer coefficient = 4×10^{-4} m/s; gas-side mass transfer coefficient = 2×10^{-4} gmol/m^{2}s atm. The Henry's law coefficient is 100 atm expressed as $p = Hx$. Find the overall mass transfer coefficient for the system K_L. Also state the relative percentage contribution of the resistances on the gas side and the liquid side.

6.14 **Contribution of gas-side resistance.** The value of k_G in an absorber is 3×10^{-3} kmol/m^{2}s atm and k_L is 5.44×10^{-3} m/s. Find the contribution of the gas side for Henry's law coefficients of 0.1, 1, and 10 m^3 atm/kmol.

CHAPTER 7

Phenomena of Diffusion

Learning Objectives

After completing this chapter, you will be able to:

- Explain the kinetic theory interpretation of diffusion in gases as due to random molecular motion and the basic equations needed for prediction of binary diffusivity.
- Calculate the value of diffusion coefficients in binary gas mixtures and describe the relationship between the diffusion coefficient and the temperature and pressure.
- Outline the concepts behind Einstein's model for diffusion of dilute solutes in liquids.
- Identify the complexities associated with diffusion and concentration effects on diffusion coefficients, especially in liquids.
- Associate the effects of pore structure and pore diameter with diffusivity of fluids in porous materials.
- Describe the basics of solid-phase diffusion and the mechanisms underlying interstitial and substitutional diffusion.
- Describe the basics of diffusion in polymeric solids, and explain the difference in diffusion in such solids compared to crystalline materials.

In this chapter, we look at the mechanisms of diffusion in gases, liquid, solids and porous solids. The molecular underpinnings of the diffusion mechanism are explained for all these cases, and some methods of *a priori* calculation of these coefficients are explored. We then address some complexities associated with diffusion, additional driving forces that can cause diffusion and the limitations of Fick's law. The Stefan-Maxwell model is introduced for diffusion in multicomponent mixtures.

Diffusion can be considered to be the result of random motion of molecules, and this interpretation is especially suitable for gaseous systems. The kinetic theory of gases is a model based on random motion of molecules, and two key parameters in this model are the mean speed and the mean free path of the molecules. A model for the diffusion coefficients in a binary mixture of gases with similar sizes can therefore be developed based on this approach, and we discuss this model first. Modification of the kinetic theory to account for short-distance molecular repulsion, however, is a more realistic model. Development of a diffusion model based on this concept leads to some predictive models for binary diffusivity. The relevant equations are directly presented in this chapter without delving into the detailed molecular theory.

Adoption of a chemical potential gradient as a driving force provides an alternative platform for models of diffusion phenomena. The force due to the chemical potential gradient is assumed to be balanced by a frictional force, leading to an frictional interpretation for diffusion. We show how this provides the interpretation of multicomponent diffusion in gases, leading to the Stefan-Maxwell model for diffusion. The friction concept also leads to a model for diffusion of solutes in liquids which in turn provides the Einstein and Wilke-Chang model for diffusion in a liquid. In addition, we consider the effect of thermodynamic nonidealities on binary diffusion in a liquid mixture and the concept of activity correction to diffusivity.

We then move on to study solid–solid diffusion in crystalline solids and show the mechanisms and correlating equations for such a system. Diffusion of a fluid in a porous solid is also an important problem. In this context, we discuss how the diffusivity is affected by the pore structure, and we contrast diffusion in large-diameter pores with diffusion in small-diameter pores. A model for calculation of the effective diffusion coefficient in porous solids then follows. Diffusion in composite media and in polymeric solids is discussed as well. The chapter concludes with examples of some additional complex effects that influence diffusion.

7.1   Diffusion Coefficients in Gases

The kinetic theory of gases provides us with a framework to understand the phenomena of diffusion and offers some simple correlations for the diffusion coefficients for a gas phase binary mixture. This is discussed in Section 7.1.1. A frictional interpretation of the phenomena of diffusion is then presented in

Section 7.1.2. Extension of the frictional concept to a multicomponent system leads to the Stefan-Maxwell model, which is introduced in Section 7.1.3.

7.1.1   Model Based on Kinetic Theory

The kinetic theory states that gas molecules are in a state of random motion. The physical properties of a gas that we observe on a continuum scale are associated with this random motion at the molecular level. For example, the pressure exerted by a gas may be viewed as the force exerted on the walls of a container by the gas molecules impinging on the wall and being reflected back. Similarly, the temperature of the gas, a continuum property, can be associated with the mean kinetic energy of molecular motion.

The kinetic theory interpretation of diffusion can be understood by looking at Figure 7.1. Molecules are in random motion and molecules from a plane at $y - a$ and at $y + a$ arrive at a plane a with equal probabilities. Here a is defined to be the distance at which molecules lose their identity from the point of their origin. This distance is related to the mean free path, as shown later in this section.

If the gas is pure, there is no net transport because the mass moving across from each side cancels out. In contrast, if there is a mass (or mole) fraction gradient, a net transport occurs from the region of higher concentration to the region of lower concentration. This relationship provides an interpretation of Fick's law based on the kinetic theory concepts and a theory to predict the diffusion coefficients in binary gas mixtures. Let us walk through some of the mathematics that arises when there is a mass (or mole) fraction gradient in the system.

Diffusivity of A in a Medium of Similar Proprieties

Consider two planes separated by a distance $2a$. The planes are located at $y + a$ and $y - a$. Let us consider the molecules arriving at y from the top and

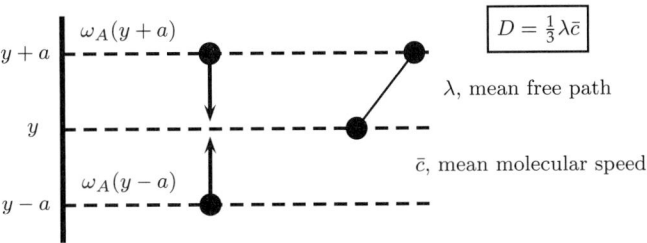

Figure 7.1   Model for diffusion in gases based on kinetic theory of gases.

bottom planes. The mean speed of the molecules is denoted as $\bar{c}$ in random directions. The speed of a molecule as it passes across an area $\Delta x \, \Delta z$ is taken as $\bar{c}/4$. Using this mean speed, the transport to the plane y from $y - a$ and $y + a$ can be calculated using the expressions for the convective flux:

$$\text{Mass flux of } A \text{ crossing from } y - a \text{ to } y = (\bar{c}/4)\rho \, w_A(\text{at } y - a)$$

Hence $w_A(y - a)$ is the mass fraction of A in the mixture at $y - a$.

Similarly, mass coming from $y + a$ and arriving at y is

$$\text{Mass from } y + a = (\bar{c}/4)\rho \, w_A(\text{at } y + a)$$

If there is a mass fraction gradient, then a net transport occurs from a plane at higher concentration to a plane at lower concentration. The corresponding mass flux is the difference between the fluxes given by the previous two equations. Hence the net mass flux due to diffusion is

$$\text{Flux at } y = \frac{\bar{c}}{4}\rho \left[w_A(\text{at } y - a) - w_A(\text{at } y + a) \right]$$

The difference in mass fractions can be related to the mass fraction gradient as follows:

$$w_A(\text{at } y - a) - w_A(\text{at } y + a) \approx -2a\frac{dw_A}{dy}$$

Hence flux at y is

$$\text{Flux at } y = -\frac{\bar{c}}{4}\left(2a\rho\frac{dw_A}{dy} \right)$$

This suggests that the net flux is proportional to the mass fraction gradient and can be modeled by a Fick's law type of equation:

$$\text{Flux} = -D_A \rho \frac{dw}{dy}$$

The parameter $2a\bar{c}/4$ may be defined as the diffusion coefficient of A in a medium with similar molecular properties:

$$D_{AA} = 2a\frac{\bar{c}}{4} \tag{7.1}$$

The parameter a represents the distance between two collisions in the normal direction. From kinetic theory, this is related to λ as the mean free path:

$$a = (2/3)\lambda$$

Hence the diffusion coefficient is related to the mean free path and mean molecular speed as follows:

$$D_{AA} = \frac{1}{3}\lambda\bar{c} \tag{7.2}$$

The following additional results from the kinetic theory can now be used. The average velocity is given as

$$\bar{c} = \sqrt{\frac{8\kappa T}{\pi m}} \tag{7.3}$$

while the mean free path is given as

$$\lambda = \frac{1}{\sqrt{2}\pi d^2 n} \tag{7.4}$$

Here κ is the Boltzmann constant, 1.38×10^{-23} J/K; m is the mass of the molecule; d is defined as the diameter of the molecule treated as a rigid sphere; and n is the number concentration, in units of molecules/m^3.

We can now use ideal gas law to express n (number of molecules per unit volume) in terms of pressure, P, and temperature, T:

$$P = n\kappa T$$

Hence the mean free path can also be expressed as

$$\lambda = \frac{\kappa T}{\sqrt{2}\pi d^2 P} \tag{7.5}$$

Using these relations for $\bar{c}$ and λ in Equation 7.2, we get the following expression for the diffusion coefficient for gases with similar properties, which is also called the self-diffusion coefficient:

$$D_{AA} = \frac{2}{3}\sqrt{\left(\frac{\kappa^3}{\pi^3 m}\right)}\frac{T^{3/2}}{P d^2} \tag{7.6}$$

Equation 7.6 is based on many approximations but nevertheless predicts the pressure dependency correctly. The diffusion coefficient is seen to be inversely proportional to the total pressure of the gas. The temperature dependency is predicted somewhat less approximately. The equation indicates that the diffusivity is proportional to $T^{1.5}$, but experimental data indicate that the exponent is actually closer to 1.75.

Diffusivity of A in a Medium of a Second Gas B

Development of the formula for D_{AB} based on rigid sphere theory for two gases A and B with unequal molecular diameters and unequal molecular weights is more complex. Here, we will simply state the parameters needed and the final formula from other sources.

The extension of Equation 7.6 is based on a model in which molecules attract each other in general, but at short distances repel each other. Thus the assumption of molecules hitting each other (classical kinetic theory) is not correct because at short distances they repel each other. The basic kinetic theory was modified and corrected to account for this phenomenon by Chapman and Enskog.

Chapman and Enskog Model

The key modification in this model is the inclusion of the potential energy of the interacting molecules' collisions. The intermolecular potential energy has the functionality shown in Figure 7.2 as a function of the distance of separation between the molecules.

The potential energy distribution is not exactly known but can be approximated by the 6-12 function, which is a good model for nonpolar molecules:

$$\phi(r) = 4\epsilon \left[\left(\frac{\sigma}{r}\right)^{12} - \left(\frac{\sigma}{r}\right)^{6} \right] \tag{7.7}$$

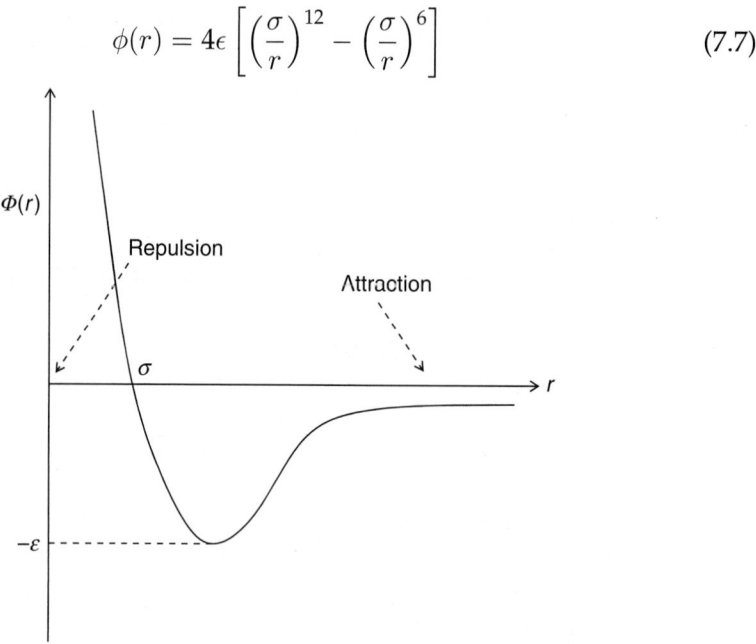

Figure 7.2   Potential energy as a function of the distance between two molecules.

This function was introduced by Lennard-Jones (1924). Here σ is the effective diameter of the molecules and has a value close to the molecular diameter d in the earlier version of the kinetic theory, but not exactly the same. The second parameter, ϵ, is the minimum potential energy value.

The values for σ are tabulated for many gases in many books, and values for some common gases are shown in Table 7.1. More extensive data are available in the book by Reid et al. (1977). In the absence of data, the value of σ can be calculated as an approximation from the critical properties data as follows:

$$\sigma = 2.44(T_c/P_C)^{1/3}$$

Similarly, the values of ϵ are tabulated for many substances but, as a quick approximation, can be determined based on the properties of the fluid at the critical point as follows:

$$\epsilon/\kappa = 0.77T_C$$

The units here are K for ϵ/κ and A° for σ. The critical pressure is in atm.

Thus, the two parameters σ and ϵ/κ for each of the gases in the binary pair were introduced into the molecular transport model, and the transport properties were predicted using this model in the work of Chapman and Enskog. Details are beyond the scope of this book, but the key result for the diffusion coefficient is

$$D_{AB} = 1.8583 \times 10^{-7} \sqrt{T^3 \left(\frac{1}{M_A} + \frac{1}{M_B} \right)} \frac{1}{P\sigma_{AB}^2 \Omega_{D,AB}} \tag{7.8}$$

Here P is the total pressure in atm and D_{AB} is in m^2/s. Molecular weights are in g/mol here, rather than in kg/mol.

Two molecular-level parameters are needed to calculate the diffusivity using this formula. The first parameter σ_{AB} is taken as the average of the individual species values:

$$\sigma_{AB} = (\sigma_A + \sigma_B)/2$$

Table 7.1   Lennard-Jones Constants for Some Gaseous Species

Gas	Air	CO$_2$	Hydrogen	Ethane	Methane
σ in $^\circ A$	3.617	3.996	2.968	4.388	3.780
ϵ/κ in K	97	190	33.3	232	154

The second parameter $\Omega_{D,AB}$ is the collision integral for diffusivity. For rigid spheres, the values are unity. The values for many gas pairs are tabulated in the literature (see, for instance, Bird, Stewart, and Lightfoot (1962)). This parameter is usually correlated with an energy parameter ϵ_{AB} or the equivalent temperature parameter ϵ_{AB}/κ. This is taken as the geometric mean of the individual energy parameters:

$$\epsilon_{AB}/\kappa = \sqrt{(\epsilon_A/\kappa)(\epsilon_B/\kappa)}$$

The collision integral Ω_D, in turn, is correlated with a function of a dimensionless temperature, Θ, defined as $T/(\epsilon_{AB}/\kappa)$:

$$\Omega_D = \frac{1.06036}{\Theta^{0.15610}} + \frac{0.193}{\exp(0.47635\Theta)} + \frac{1.03587}{\exp(1.52996\Theta)} + \frac{1.76474}{\exp(3.89411\Theta)} \quad (7.9)$$

The calculation of the diffusivity in binary gas mixture is shown in Example 7.1.

Example 7.1　Binary Diffusivity Calculation

Calculate the binary pair diffusivity of the methane–ethane pair at 293 K and 1 atm pressure.

Solution

The collision cross-sections for each species from Table 7.1 are as follows:
$\sigma_A = 3.780°A$ and $\sigma_B = 4.388°A$
Hence σ_{AB} on average $= (3.780 + 4.388)/2 = 4.084°A$
Energy parameters from Table 7.1 for each species:

$$\epsilon_A/\kappa = 145K \text{ and } \epsilon_B/\kappa T = 232K$$

The average energy parameter is the geometric mean average of the ϵ values and is calculated as $\epsilon_{AB}/\kappa = \sqrt{145 \times 232} = 183.41$ K.

The dimensionless temperature Θ is the actual temperature divided by the above value for the energy parameter $= 293/183.41 = 1.5975$.

The collision integral can then be calculated using Equation 7.9 using the above value for Θ. Its value is found to be $\Omega_D = 1.168$.

All of the quantities needed for the Chapman-Enskog equation (Equation 7.8) are now known. The diffusion coefficient is calculated by direct substitution as 0.1421 cm^2/s. This is within 5% of the experimental value quoted in Reid, Prausnitz, and Sherwood (1977).

The MATLAB snippet shown in Listing 7.1 may be useful to perform these calculations since they are lengthy. The needed parameters for the gas pairs have to be entered as input to the program.

Listing 7.1   Binary Diffusion Coefficient Calculation

```
% Binary diffusivity in a gas mixutre.
 sigmaAB = (sigmaA+ sigmaB)/2.
 ebyK = ( ebykA * ebykB)^0.5
 % dimensionless temperature for collision integral calculations
 tstar = T/ebyK
 % Equation 7.9
 Colli = 1.06036/tstar^0.15610 + 0.193/exp(0.47635 * tstar) ...
     + 1.03587/exp(1.52996 * tstar)+ 1.76474/exp(3.89411 * tstar)
 % Equation 7.8 D in m^2/s
    D_AB = 1.8583E—07 * (T^3 *(1/MA + 1/MB) )^(1/2) ...
    /P/sigmaAB^2/ Colli
```

Correlation of Fuller et al. for Diffusivity

An empirical model developed by Fuller et al. (1966) is also useful to predict diffusion coefficients:

$$D_{AB} = 10^{-3}\frac{T^{1.75}}{P}\frac{(1/M_A + 1/M_B)^{1/2}}{[(\sum_i V_{i1})^{1/3} + (\sum_i V_{i2})^{1/3})]^2} \qquad (7.10)$$

Note that the format of the Chapman-Enskog correlation is retained here except for two changes:

- Temperature dependency is changed to have an exponent of 1.75, which is closer to experimental values, and the collision integral is not used.
- The sigma parameter is replaced by the cube root of a volume parameter, the diffusion volume. This in turn is calculated as the contribution of the atoms constituting the molecules. Diffusion volume for each molecule is obtained as a summation of the groups comprising the molecules. Values for many atoms are tabulated and some illustrative values are shown in Table 7.2.

Table 7.2   Group Contribution of Atoms for Diffusion Volume Calculation

Group	C	H	O	N	Cl	S	Aromatic ring
Contribution	16.5	1.98	5.48	5.169	19.5	17.0	−20.2

Table 7.3   Diffusion Volume for Some Common Molecules

Molecule	Hydrogen	Helium	Oxygen	Air	CO_2	CO
Contribution	7.02	2.88	17.9	16.6	26/9	22.8

Values for some simple molecules are also available and are shown in Table 7.3. These values can be directly used for such cases; that is, summation over the atoms making up the molecules is not needed.

Example 7.2 illustrates the use of the Fuller et al. correlation.

Example 7.2   Diffusivity by Fuller et al. Method

Find the diffusion coefficient of the methane–ethane pair at 1 atm and 273 K using Fuller et al.'s correlation.

Solution

Let $1 = CH_4$. The molar volume for methane is calculated by summing the contributions of the atoms that make up the molecules: $1C + 4H = 16.5 + 4 \times 1.98 = 24.42$

Let $2 = C_2H_6$. The molar volume for ethane is calculated as $2C + 6H = 2 \times 16.5 + 6 \times 1.98 = 44.88$.

The pressure P needed in units; of $P = 1$ atm. The temperature is $T = 298$ K.

The molecular weights are $M_A = 16$ and $M_B = 30$, in units of g/mol as needed in the correlation.

Substituting all the quantities into Equation 7.10, we find the diffusivity for this binary pair is $D = 0.1542 \text{ cm}^2/\text{s}$.

Diffusion at high pressure is important in many applications. Such measurements are often difficult, with only atmospheric pressure values being easily measured. In such cases, the relation $D \propto 1/P$ is applied and the atmospheric values are scaled accordingly.

7.1.2   Frictional Interpretation

A second phenomenological model for diffusion is based on the understanding that the diffusion rate is related to the frictional interaction between the diffusing species. This interpretation is useful for modeling diffusion in multicomponent systems, as shown in the next subsection. It is also useful for a first-level model of diffusion of a dilute solute in a liquid, which is discussed in the next section.

A frictional force is assumed to exist at the molecular level due to relative motion of A with respect to B (or vice versa). This frictional force per mole of A (1) acting on A exerted by B (2) is represented as

$$F = y_2 f_{12}(v_1 - v_2)$$

where f_{12} is a friction coefficient parameter.

The velocities can be related to the diffusion fluxes:

$$v_1 - v^* = \frac{J_1}{y_1 C}$$

A similar expression holds for species 2:

$$v_2 - v^* = \frac{J_2}{y_2 C}$$

Therefore the velocity of species 1 with respect to species 2 is

$$v_1 - v_2 = \frac{J_1}{y_1 C} - \frac{J_2}{y_2 C} \tag{7.11}$$

Hence the frictional force is represented as

$$F = y_2 f_{12} \left(\frac{J_1}{y_1 C} - \frac{J_2}{y_2 C} \right)$$

This expression for force is rearranged as

$$F = y_2 f_{12} \frac{(y_2 J_1 - y_1 J_2)}{y_1 y_2 C} \tag{7.12}$$

The frictional force is assumed to be balanced by a driving force. For the diffusion process, the force can be identified as the negative gradient of the chemical potential per mole. Thus

$$F = -\nabla \mu_1 \tag{7.13}$$

Figure 7.3 is a schematic representation of the balance of the two forces acting on a diffusing molecule.

For ideal systems

$$\nabla \mu_1 = R_g T \nabla \ln y_1 = R_g T \frac{\nabla y_1}{y_1} = -F \tag{7.14}$$

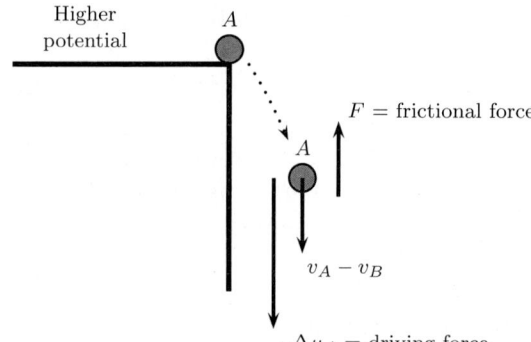

Figure 7.3   Frictional force interpretation of diffusion.

Equating the two expressions for F and with some slight rearranging, we obtain

$$-\nabla y_1 = \frac{f_{12}}{R_g T} \frac{(y_2 J_1 - y_1 J_2)}{C} \tag{7.15}$$

Fick's law for a binary system can be rearranged (after some algebra) to an "inverted" form:

$$-\nabla y_1 = \frac{(y_2 J_1 - y_1 J_2)}{C D_{12}}$$

This suggests that the friction coefficient can be replaced by diffusivity, with the latter defined as follows:

$$D_{12} = \frac{R_g T}{f_{12}} \tag{7.16}$$

This provides a frictional interpretation of the diffusion coefficient. Diffusivity is inversely proportional to the friction coefficient. The extension of this concept provides a model for diffusion in multicomponent systems—namely, the Stefan-Maxwell model discussed in Section 7.1.3.

The reciprocal of the frictional resistance is referred to as mobility μ_m. Hence the following relation holds between mobility and diffusivity:

$$\mu_m = \frac{1}{f_{12}} = \frac{D_{12}}{R_g T} \tag{7.17}$$

Equation 7.17 is known as the Einstein equation.

7.1.3   Multicomponent Diffusion

The frictional force on species 1 is calculated as an additive contribution to the friction due to all other species (2,3, ...) present in the mixture:

$$F_1 = \sum_{j \; j \neq 1}^{ns} y_j f_{1j}(v_1 - v_j)$$

The frictional resistance f_{1j} can be related to each binary pair diffusivity as per Equation 7.16. Using this relation, we obtain

$$\frac{F_1}{R_g T} = \sum_{j \; j \neq 1}^{ns} y_j \frac{(v_1 - v_j)}{CD_{ij}}$$

The left side of this equation is equal to $-\nabla y_1 / y_1$ as per Equation 7.14 for the ideal gas. On the right side of the equation, the relative velocity of species 1 can be related to the diffusion fluxes as per Equation 7.11. Use of these relations then leads to following equation (known as the Stefan-Maxwell model) which is useful for ideal gas mixtures:

$$-\nabla y_i = \sum_{j=1}^{ns} \frac{y_j \boldsymbol{J}_i - y_i \boldsymbol{J}_j}{C \, D_{ij}} \tag{7.18}$$

where ns is the number of components and D_{ij} is the binary pair diffusivity in the gas phase. (Note: Summation for $j = i$ cancels out and is not needed.) This equation can also be written in terms of the combined flux since

$$J_i = N_i - y_i N_t$$

The resulting equation is

$$-\nabla y_i = \sum_{j=1}^{ns} \frac{y_j \boldsymbol{N}_i - y_i \boldsymbol{N}_j}{C \, D_{ij}} \tag{7.19}$$

This is another version of the Stefan-Maxwell equation. Equation 7.19 relates the fluxes $\boldsymbol{N}_i$ to the concentration gradients as expected from a constitutive model. Note that the fluxes here are implicit; they are not provided directly as a function of concentration gradients and have to solved as a part of the solution procedure together with the species mass balance equations. Hence the solution procedure is more involved (and is taken up in more detail in Chapter 15).

A further complexity is that the fluxes are all coupled, which is evident from the examination of the Stefan-Maxwell equation.

The concentration gradient of any particular species is a function of the fluxes of all the components present in the system.

Some interesting and unexpected effects can arise in multicomponent systems as a result of this coupling. These effects, which are not seen in ordinary diffusion in binary systems, are briefly discussed here:

- **Reverse diffusion:** A species may diffuse in a direction opposite to its concentration gradient.
- **Osmotic diffusion:** Species may diffuse even though the concentration gradient is zero.
- **Barrier diffusion:** Species may not diffuse even though there is a favorable concentration gradient.

Figure 7.4 depicts these complex effects.

These pathological cases arise mainly when the binary pair values are vastly different—for example, in a ternary mixture of hydrogen, air, and a heavy hydrocarbon such as heptane.

Example 7.3 applies the Stefan-Maxwell equation to a binary mixture.

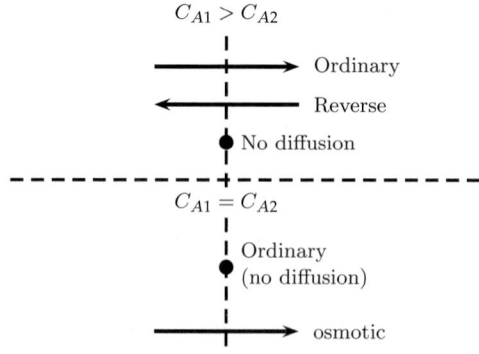

Figure 7.4   Some complexities observed in multicomponent diffusion.

> ### Example 7.3   Stefan-Maxwell Model for a Binary Mixture with UMD
>
> Gas A diffuses into an inert gas B. Apply the Stefan-Maxwell model to derive an expression for the flux of A. Assume no forced flow.
>
> ---
>
> **Solution**
>
> We use Equation 7.19 as the constitutive model for species A. In this case, $ns = 2$.
>
> $$-\nabla y_A = \frac{y_B \mathbf{N}_A - y_A \mathbf{N}_B}{C D_{AB}} \tag{7.20}$$
>
> Equation 7.20 cannot be applied to species B, because $\nabla y_A = -\nabla y_B$ and, therefore, no independent equation is generated. Hence we need a determinancy condition. In this case we have UMD and $N_B = 0$. Also $y_B = 1 - y_A$.
>
> Using this information and rearranging, we get
>
> $$N_A = -\frac{C D_{AB}}{(1 - y_A)} \nabla y_A \tag{7.21}$$
>
> which is the model for flux in this case. Equation 7.21 was derived by starting with the version of Fick's law given earlier and adding the convection contribution.
>
> Extension of these equations to the multicomponent case and their solution are taken up in Chapter 15.

7.2   Diffusion Coefficients in Liquids

Because molecules are tightly packed in liquids, the kinetic theory approach of random collisions is not very appropriate for this form of matter. The mechanistic model for diffusion in a liquid is the so-called jump model, which the solute jumps from a vacant site to another vacant site. This model, which is shown schematically in Figure 7.5, is referred to as the Eyring theory. This model is more suitable for diffusion in solids, a topic discussed in Section 7.4.

A second theory applicable to liquids is the hydrodynamic theory, which states that the diffusivity is proportional to the mobility of the solute. In this model, the movement of the solute is treated as a sphere moving in an infinite pool of liquid with viscosity μ_L. The frictional interpretation of diffusion discussed in Section 7.1.2 is used here. Some remarks from that section are repeated here for ease of reading.

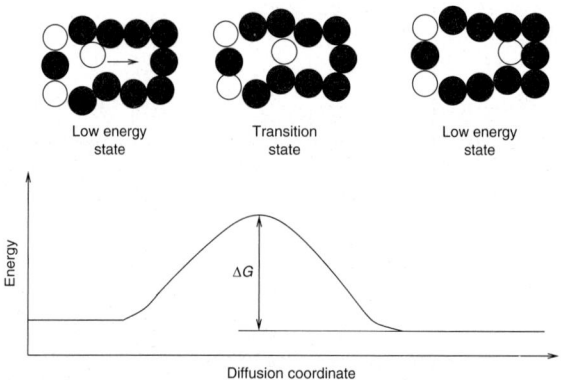

Figure 7.5 Eyring model for diffusion in liquids.

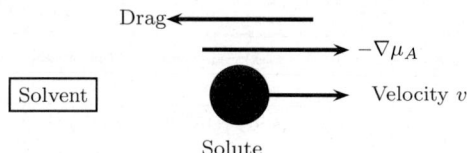

Figure 7.6 Frictional model for diffusion for liquids.

The force needed to move a particle with a velocity of v_A (relative to B) is given, in general, by

$$F = f v_A$$

where f is the coefficient of friction and v_A is the terminal velocity of A.

For the diffusion process, the force can be identified as the negative of the gradient of the chemical potential per mole. Thus

$$F = -\nabla \mu_A$$

The model and the balance of forces on a diffusing solute are illustrated in Figure 7.6.

The balance of the forces is represented in terms of the gradient of the chemical potential as

$$-\nabla \mu_A = f v_A$$

From the flux considerations, the relative velocity of motion of A is given as:

$$v_A = \frac{J_A}{C_A} = \frac{J_A}{C x_A}$$

Hence

$$J_A = -\frac{Cx_A\nabla\mu_A}{f}$$

The term $1/f$ is identified as D_A/R_gT. In turn, the preceding expression can be rearranged to

$$J_A = -CD_A\frac{x_A\nabla\mu_A}{R_gT} \tag{7.22}$$

Equation 7.22 shows that the gradient of the chemical potential is more appropriately considered the driving force for mass transfer, rather than the mole fraction gradient used in the traditional Fick's law.

For ideal systems, this expression can be shown to be the same as Fick's law. This follows from the definition of the chemical potential. The following relation from basic thermodynamics holds:

$$\mu_A = \mu_A^0 + R_gT\ln x_A$$

Taking the derivative to get the gradient of the chemical potential, we get

$$\nabla\mu_A = R_gT\frac{\nabla x_A}{x_A} \tag{7.23}$$

Substituting in Equation 7.22, we obtain

$$J_A = -CD_A\nabla x_A$$

which is Fick's law for an ideal mixture of liquids. The more general expression given by Equation 7.22 then provides an extension for diffusion in non-ideal systems. Moreover, the frictional interpretation provides a link to the molecular-level interpretation for diffusion and leads to a methodology to predict the diffusion coefficient. These two effects are explored in detail in the following sections.

7.2.1   Stokes-Einstein Model

The frictional resistance and the diffusivity are related as

$$D_A = \frac{R_gT}{f} \tag{7.24}$$

where f is the frictional coefficient based on the number of moles of A. For the molecular model, it is more useful to write this relation as follows:

$$D_A = \frac{\kappa T}{f_{12}}$$ (7.25)

where f_{12} is the force per molecule. Note that $\kappa = R_g/N_{avg}$ has been used here to get from one form to the other.

At this point we use concepts from the fluid dynamics of slow flow, also called Stokes flow. The velocity of motion on a spherical particle in a liquid under slow flow conditions is related to the force F acting on the particle, and this relation is known as Stokes law:

$$F = 6\pi \mu_L r_A v$$ (7.26)

where r_A is the solute radius. Hence $6\pi \mu_L r_A$ can be identified as the friction coefficient f_{12}.

Note: μ_L is the viscosity of the liquid medium (as indicated by the subscript L) and should not be confused with chemical potential.

Using the expression for f_{12} from Stokes law (Equation 7.25), we obtain

$$\boxed{D_A = \frac{\kappa T}{6\pi r_A \mu_L}}$$ (7.27)

Equation 7.27 is the Stokes-Einstein equation, which is widely used as a starting model for liquid-phase diffusivity predictions. Predictions made using Equation 7.27 are fairly accurate under the conditions used in the model derivation—namely, a small rigid sphere (solute) moving in a continuum (solvent). Hence the model is representative of experiments in which the solute radius to solvent radius ratio is smaller than 5. For solute molecules with complex structure, a correction factor is often applied and a modified equation used. Nevertheless, a key result that emerges from Equation 7.27 is that

$$\frac{D_A \mu_L}{T} = \text{constant}$$

which fits the data for a large class of molecules. Indeed, this relation may hold even with molecules for which the Stokes-Einstein description may not be a good fit, such as nonspherical molecules and long-chain proteins.

7.2.2   Wilke-Chang Equation

An extension to the Stokes-Einstein equation is the Wilke-Chang equation:

$$D_A = 7.4 \times 10^{-12} \frac{(\phi M_S)^{1/2} T}{\mu_L \hat{V}_A^{0.6}}$$

where $\hat{V}_A$ is the molar volume of the solute A (in cm^3/mol) and M_S is the molecular weight of the solvent (in g/mol). Also the viscosity, μ_L, is in centipoise which is equivalent to 0.001 Pa · s, the S.I. unit for viscosity. The calculated value of D is given in units of m^2/s here.

The empirical parameter, ϕ, in the preceding equation is known as the association parameter. It has a value of 1 for most organic solvents, 1.9 for methanol, 1.5 for ethanol, and 2.6 for water. Note that the equation has a form similar to that of the Stokes-Einstein equation ($D_A \mu_L / T$ is a constant), and the dependencies of the temperature and the liquid viscosity are identical. The equation is not very accurate for concentrated systems, as the diffusion coefficient varies strongly with concentration. An activity correction is usually applied in such cases, as discussed in the Section 7.3.

The data for molar volume have been tabulated for many compounds, and some are provided in Table 7.4.

In the absence of such data, the following correlation developed by Tyn and Calus (1975) can be used:

$$V_A = 0.258 V_c^{1.048}$$

where V_c is the critical volume of the solute in $cm^3/gmol$.

Group contribution methods are also available to calculate this parameter, as discussed by Le Bas (1915). In this method, atomic volume increments are added together as per the molecular formula to obtain the molecular volume. The use of Wilke-Chang equation is shown in Example 7.4.

Table 7.4   Molar Volume for Some Common Species in Liquid Phase

Solute	Hydrogen	Oxygen	CO_2	NH_3	Water
$\hat{V}_s$, cm^3/mol	14.3	5.6	34.0	25.8	18.9

Example 7.4   Diffusion Coefficient in Liquids

Calculate the diffusivity of ethanol in water at a temperature of 298 K. Ethanol is present in small concentration.

Solution

Since ethanol is present in small concentration, it is treated as the solute and water is considered to be the solvent. The density of ethanol is 0.8 g/cm^3 Hence the molar volume of ethanol is $M_A/\rho_A = 46$ g/mol$/0.8$ g/cm$^3 = 57.5$ cm^3/mol.

The viscosity of water is to be used in the Wilke-Chang equation because water is in excess and treated as the solvent. The value is $\mu_L = 9 \times 10^{-4}$ Pa s. We need to get this value in centipoise (c.p.). The conversion factor is 1000 c.p./Pa s. Hence μ (centipoise) = 0.9 c.p.

Molecular weight of solvent (water) is 18 gm/gmol.

Association factor, ϕ, is equal to 2.6.

Substituting in the Wilke-Chang equation, we get the diffusion coefficient: $D_A = 1.47 \times 10^{-9}$ m^2/s. Since ethanol was considered the solute, this is the value for a dilute solution of ethanol in water.

The Wilke-Chang correlation is suitable only for dilute concentrations of the solute. One consequence is that the binary diffusivity in an A-lean system (denoted here as D_A) is not the same as the binary pair diffusivity in an A-rich system (denoted as D_B here). This is because the environment (solvent) seen in an A-lean system (full of B) is different from that in an A-rich system (full of A). The values D_A and D_B are referred to as infinite dilution values.

For concentrated solutions, the following Vignes (1966) equation is often useful:

$$D_{AB} = D_A^{x_B} D_B^{x_A} \tag{7.28}$$

where D_A is the diffusion coefficient of A in a B-rich mixture and D_B is the diffusion coefficient of B in an A-rich mixture, with x being the mole fraction as usual. This relation can be viewed as logarithmic averaging of the two endpoint values.

An alternative way is to take the simple linear average:

$$D_{AB} = x_B D_A + x_A D_B \tag{7.29}$$

In both cases, we find the diffusion coefficient is concentration dependent.

Having discussed some predictive methods for liquid-phase diffusivity, we now look at the case of non-ideal liquid mixtures, for which where a thermodynamic correction is needed.

7.3   Non-Ideal Liquids

Note: This section may be omitted on first reading.

The chemical potential gradient can be considered to be the driving force for diffusion rather than the concentration gradients. This leads to a modified form of Fick's law that includes a correction factor known as the activity correction. The details are shown next.

7.3.1   Activity Correction Factor

The chemical potential can be represented as a function of the activity of species A:

$$\mu_A = \mu_A^\circ + R_g T \ln a_A \tag{7.30}$$

The gradient for the chemical potential can be written as

$$\nabla \mu_A = R_g T \frac{d \ln a_A}{d x_A} \nabla x_A \tag{7.31}$$

In these equations, a_A is the activity of species A in the mixture; x_A is the mole fraction as usual. Using the expression for $\nabla \mu_A$ in Equation 7.22, the Fick's law type of model takes the following form:

$$J_A = -C D_{AB} \left(\frac{d \ln a_A}{d \ln x_A} \right) \nabla x_A \tag{7.32}$$

For an ideal solution, the term $d \ln a_A / d \ln y_A$ is equal to 1; hence Fick's law holds in the form stated earlier. For non-ideal solutions, Equation 7.32 may be written in a form analogous to Fick's law:

$$J_A = -C D_{AB}^\circ \nabla x_A \tag{7.33}$$

where D_{AB}° is the representative observed diffusion coefficient, which becomes equal to

$$D_{AB}^\circ = D_{AB} \left(\frac{d \ln a_A}{d \ln x_A} \right) \tag{7.34}$$

Also $a_A = \gamma_A x_A$, where γ_A is the activity coefficient of A. Hence

$$\frac{d \ln a_A}{d \ln x_A} = 1 + \left(\frac{d \ln \gamma_A}{d \ln x_A} \right)$$

Using this definition in Equation 7.34, we find

$$D^\circ_{AB} = D_{AB} \left[1 + \left(\frac{d \ln \gamma_A}{d \ln x_A} \right) \right]$$

The term in the square brackets is called the (thermodynamic) activity correction factor.

The term D_{AB} on the right side of the equation is also a function of concentration, since it should be some average of the infinite solution values (see Equation 7.28 or 7.29). Using Equation 7.29 and combining it with the previous equation leads to the the Darken relation:

$$D^\circ_{AB} = (x_B D_A + x_A D_B) \left[1 + \frac{d \ln \gamma_A}{d \ln x_A} \right] \tag{7.35}$$

This relation is commonly used to interpret and model diffusion in liquids.

Note: In many design applications, these complexities are ignored and absorbed in other fitting parameters. In other words, a fitted value of diffusion coefficient is used. Nevertheless, one has to bear these considerations in mind for more accurate modeling of mass transport in liquid–liquid systems.

7.3.2   Activity Coefficient Models

To incorporate the thermodynamic correction, the activity has to be fitted by a thermodynamic relation. A brief discussion of the models for the activity coefficient is useful here, although the detailed study of this topic is deferred to texts in thermodynamics. A number of models are available for this purpose, such as the Margules equation (the simplest), the van Laar equation, the Wilson equation, the NRTL three-constant model, and the UNIQUAC model.

In general, the activity coefficient is related to the excess free energy G^E_A as follows:

$$\frac{G^E_A}{R_g T} = \ln \gamma_A \tag{7.36}$$

The excess free energy is given by various models for non-ideal behavior due to differences in molar sizes, intermolecular forces, degree of hydrogen bonding, and other factors. These details are given in thermodynamic books. For binary systems, the Redlich-Kister equation is often used:

$$\frac{G_A^E}{R_g T} = x_A x_B [A + B(x_A - x_B) + C(x_A - x_B)^2]$$

where A, B, and C are fitted parameters.

The simplest model for γ is the obtained by keeping only the first term, which leads to the one-parameter Margules equation. This produces the following expression for the activity coefficient:

$$\ln \gamma_A = A x_b^2 = A(1 - x_A)^2$$

The model presents a symmetric variation of the activity coefficient that is rarely observed. Thus this model is not suitable for thermodynamic correction for certain ranges of A. For example, the diffusion coefficient can become negative for certain range of mole fractions if the parameter A is greater than 2.

A two-parameter model is often used in which A is replaced by a linear function of the mole fraction:

$$\ln \gamma_A = (1 - x_A)^2 [A_{12} + 2x_A (A_{21} - A_{12})]$$

This change is often sufficient to incorporate the effect of non-idealities on the prediction of the diffusivity. A more detailed NRTL model using three parameters may be more accurate, though. We now show the use of activity correction in Example 7.5.

Example 7.5   Activity Correction

Calculate the diffusion coefficient of acetone in water for a 50 mole % mixture.

Solution

We use the Wilke-Chang equation to find the infinite dilution values first. If acetone is treated as the solute, we get $D_A = 1.26 \times 10^{-5}$ cm^2/s. Similarly, if water is treated as the solute, we get $D_B = 4.68 \times 10^{-5}$ cm^2/s.

If no thermodynamic correction is applied, a linear interpolation of these values for $x_A = 0.5$ leads to $D_{AB} = 2.97 \times 10^{-5}$ cm^2/s.

To apply the thermodynamic correction, $\ln \gamma$ vs x_A was computed from the NRTL model. Taking the slope at $x_A = 0.5$, we get $d\ln \gamma_A / d\ln x_A = -0.7$. The thermodynamic correction is $1 + d\ln \gamma_A / d\ln x_A = 0.3$.

Hence the diffusion coefficient for a 50 mole % mixture is $0.3 \times 2.97 \times 10^{-5} = 0.8910 \times 10^{-5}$ cm^2/s.

The value will be 0.7285×10^{-5} cm^2/s if the Vignes equation (Equation 7.28) is used, followed by the thermodynamic correction.

7.4   Solid–Solid Diffusion

In this section, we discuss diffusion in crystalline solids. In this scenario, atoms are placed in a regular array and closely packed. Diffusion is obviously a slow process in such a case, and the effects are observed only at high temperatures. Two mechanisms of solid–solid diffusion are discussed in the following sections.

7.4.1   Vacancy Diffusion

In a crystalline solid, some of the lattice positions are unoccupied and referred to as vacancies. This phenomenon is illustrated in Figure 7.7. Diffusion of a solute or impurity atom is then visualized as a jump of the diffusing molecule to a vacancy in the solid. A theoretical model based on this depiction leads to the following equation for crystalline materials:

$$D = R_o^2 f_v \omega$$

where R_o is the spacing between atoms, f_v is the fraction of vacant sites in the crystal, and ω is the jump frequency.

7.4.2   Interstitial Diffusion

A second mechanism is the interstitial diffusion. In this case, the solute occupies an intersite of the substrate atoms and moves to an adjoining interstitial site. This mechanism is illustrated in Figure 7.8.

Interstitial diffusion is modeled as

$$D = D_0 \exp(-E/\kappa T)$$

where E is the activation energy needed to jump to the adjacent site and D_0 is the pre-exponential factor. The activation energy is usually measured in

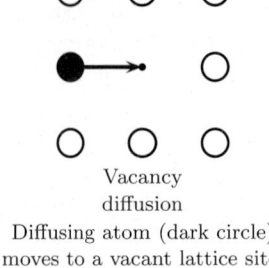

Vacancy
diffusion
Diffusing atom (dark circle)
moves to a vacant lattice site.

Figure 7.7   Vacancy or substitutional diffusion.

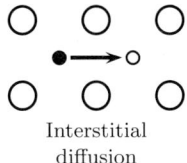

Interstitial
diffusion

Diffusing atom (dark circle) moves
from one interstitial site to another.

Figure 7.8 Interstitial diffusion.

electron-volts (eV). Hence the Boltzmann constant κ is used in the exponential term; it is equal to 8.617×10^{-5} eV/K.

Substitution diffusion is an activated process but requires a lattice vacancy to form or exist. This requires an extra activation energy, denoted as E_v. Thus substitution diffusion is modeled by including two activation parameters, E and E_v:

$$D = D_0 \exp(-(E + E_v)/\kappa T)$$

Many dopants can exist in both interstitial and substitutional sites. The diffusion can then occur by both mechanisms. Although the substitutional concentration is usually larger than the interstitial concentration, the interstitial jump is much faster than the substitutional component. Consequently, the overall diffusion is often governed by interstitial concentration.

Concentration-dependent diffusion is often observed with dopants in silicon. A power law type of model is commonly used for these cases. More details are given by Ghandhi (1983) and Middleman (1993). The mechanisms can be complex, and these sources should be consulted for more details. A few examples are given here to give a feel for the values of diffusion of some dopants in silicon.

Boron in Silicon: The data at concentrations of boron less than 10^{19} atoms/cm^3 are correlated by an Arrhenius type of model:

$$D = 3.17 \exp\left(\frac{-3.59}{\kappa T}\right)$$

where $\kappa = 8.62 \times 10^{-5}$ eV/K. Note that the activation energy is usually reported in electron-volts in these systems, whereas the diffusivity is in cm^2/s. For higher concentrations, diffusivity is a linear function of boron concentration.

Arsenic in Silicon: The arsenic diffusion follows a similar pattern with the following values:

$$D = 23 \exp\left(\frac{-4.1}{\kappa T}\right)$$

Phosphorus in Silicon: Phosphorus diffusion follows a similar pattern, with the diffusion coefficient remaining constant at low concentrations. It then decreases with concentration first and subsequently increases at larger concentration levels.

Note: Diffusion of phosphorus induces an electric field, and the combined mechanism should consider the effect of migration due to the electric field. Details are given in the paper by Fair and Tsai (1997) and in some other sources (Middleman and Hochberg, 1993).

7.5   Diffusion of Fluids in Porous Solids

Diffusion of a fluid in a porous medium is important in several applications. For example, reactants have to diffuse into the pores of a catalyst so that the reaction can take place; the products then have to diffuse out. The observed rate of reaction is affected by the rate of diffusion. In turn, the calculation of the effective diffusion coefficient in a porous solid is required to design these reactors. A model commonly used to describe this phenomenon is discussed in this section. Obviously, this process needs some model for the porous media. Typically, the average porosity is used to characterize the media. First we examine the diffusion in a single straight cylindrical pore.

7.5.1   Single-Pore Gas Diffusion: Effect of Pore Size

Let us consider the case of diffusion in a gas-filled pore. Assume that the pore size can be described by assigning a pore diameter d_p to it. The operative mechanisms are different in large pores and in small pores, as shown in Figure 7.9. For large-diameter pores, molecular gas–gas collision is the dominant mechanism and the diffusion can be modeled by the binary gas pair diffusivity D_{AB}. Pore size is not important here. In contrast, for small-diameter

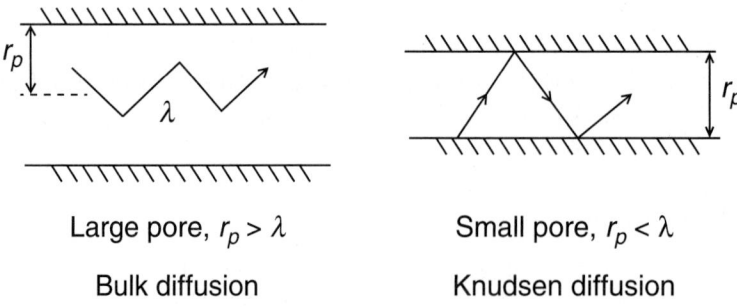

Large pore, $r_p > \lambda$

Bulk diffusion

Small pore, $r_p < \lambda$

Knudsen diffusion

Figure 7.9   Mechanisms for diffusion in pores of a catalyst.

pores, the molecules are likely to collide more with the walls of the pores rather than with each other. The diffusion is therefore a result of gas–wall collision and the phenomenon at work is known as Knudsen diffusion.

The criterion used to distinguish between the two mechanisms is comparison of the pore diameter, d_p, to the mean free path of the gas molecules, λ. If $d_p << \lambda$, the Knudsen diffusion is the main mode of diffusion. The diffusion coefficient for this case is modeled using concepts similar to those found in the kinetic theory of gases:

$$D_{KA} = (d_p/3)\bar{c} \tag{7.37}$$

Here $\bar{c}$ is the mean molecular velocity given earlier in Equation 7.3. The only difference between this equation and that from kinetic theory is that the mean free path λ is not used, but instead is replaced by d_p, the pore diameter.

Using the expression for the mean velocity from kinetic theory and substituting some universal constants, we can derive the following equation from Equation 7.37 for the Kundsen diffusion coefficient:

$$D_{KA} = 4850 d_p \sqrt{\frac{T}{M_A}} \tag{7.38}$$

The units are cm for d_p g/mol for M_A, and cm^2/s for D.

A combined model is used for the overall diffusion:

$$\frac{1}{D_{pore}} = \frac{1}{D_K} + \frac{1}{D_{AB}} \tag{7.39}$$

where D_K is the Knudsen diffusion coefficient (Equation 7.38) and D_{AB} is the gas-phase (binary pair) diffusivity. For small pores, the first term on the right side of Equation 7.39 dominates, so the overall diffusion coefficient depends on the pore diameter. For large pores, the second term dominates, so the diffusion coefficient is not a function of the pore diameter. Example 7.6 provides an example of the calculation of diffusivity in a single pore.

Example 7.6   Diffusion in a Single Pore

Hydrogen is diffusing into a cylindrical pore filled with methane. The pore diameter is 10 μm. Find the diffusion coefficient in this pore. Determine the dominant mechanism for diffusion.

Solution

The Kundsen diffusion coefficient is calculated from Equation 7.38. The pore diameter in centimeters is to be used: $d_p = 10 \times 10^{-4}$ cm. Also $T = 900$ K and $M = 2$ g/mol. The value calculated is 102 cm^2/s.

To find the dominant mechanism, we find the bulk diffusion coefficient at 273 K when the value of $D_{AB} = 0.625$ cm^2/s. The temperature correction with an exponent of 1.75 is applied. We find $D_{AB} = 5.94$ cm^2/s at 900 K, a much smaller value than the Knudsen diffusion coefficient. Hence the process is controlled by bulk diffusion and the Knudsen diffusion does not contribute to the transport rate. Molecules hit each other rather than hitting the wall.

We can also calculate the mean speed as given earlier and use the relation $(d_p/3)$ times the mean speed to get the Knudsen diffusion coefficient. S.I. units are to be used here. The mean speed $\bar{c}$ is 3096 m/s calculated using Equation 7.3. Correspondingly, the Knudsen diffusion coefficient is $D_K = d_p \bar{c}/3 = 0.0103$ m^2/s. The pore diameter in meters should be used here.

Pressure Buildup in Small Pores

Consider two gases A and B that are counter-diffusing in a narrow porous capillary under conditions where Knudsen diffusion is the dominant mechanism. The diffusion coefficients are not equal; they are inversely proportional to the square root of the molecular weight. As a result, the sum of the diffusion fluxes do not add up to zero. The mass balance constraint for EMD requires that the net flux should be zero, but this condition is not being satisfied due to the unequal diffusion coefficient values for species A and B. In such a case, there must be a compensating mechanism, and a pressure profile builds up in the capillary. The viscous flow due to the self-generated pressure profile then takes care of the mass balance requirement. Hence the effect of the pressure-driven viscous flow must be included for cases where the pore size is sufficiently small for Knudsen diffusion to be dominant and when the two diffusing species have significantly different molecular weights. The overall effect of the pressure profile is usually within 10%, as confirmed by Evans (1972), and is often neglected.

For liquid pores, there is no Kundsen diffusion since the mechanism of diffusion is not collision, but rather frictional flow. Nevertheless, liquid-filled pores of a small size exhibit the phenomenon known as hindered diffusion, which is discussed in the following section.

7.5.2   Liquid-Filled Pores: Hindered Diffusion

Consider diffusion of a molecule whose size is comparable to the size of the pores. The frictional resistance due to the pore wall should play a role, with the diffusivity being reduced compared to that for a larger-sized pore. This phenomenon is referred to as restricted diffusion or hindered diffusion. The extent of restriction depends on the size of the solute in comparison to the size of

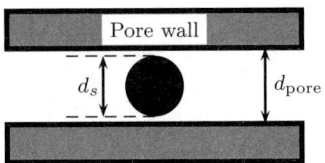

Figure 7.10   Mechanisms for diffusion in the pores of a liquid-filled catalyst.

the pores. The possibility of hindered diffusion also opens up an engineering opportunity to design materials with controlled pore size so as to achieve selective diffusion of only the chosen solute.

A schematic of a small molecule diffusing in a small pore is shown in Figure 7.10.

The ratio of the solute diameter to the pore diameter is denoted by φ. The hindered diffusion becomes important if φ is close to 1 and is often modeled by introducing two correction factors, F_1 and F_2, that depend on φ. The first correction factor is simply a geometric factor known as the steric partition constant and is defined as follows:

$$F_1(\varphi) = (1 - \varphi)^2$$

The second correction factor is known as the hydrodynamic hindrance factor:

$$F_2(\varphi) = 1 - 2.104\varphi + 2.09\varphi^3 - 0.95\varphi^5$$

The diffusion coefficient in small pores is obtained by multiplying the large pore value by $F_1 F_2$. This is known as the Renken (1954) correction.

Hindered transport has been reviewed by Deen (1992). This paper is a useful reading for students who wish to explore this field further.

Adsorbed species can also move along the surface of the pore walls. This phenomenon, known as surface diffusion, and may operate in parallel with the pore diffusion.

7.5.3   Porous Catalysts: Effective Diffusivity

Porous catalysts cannot be modeled solely as diffusion in a single cylindrical pore since they constitute a network of interconnected pores. Instead, the pore size distribution plays an important role in the calculation of the effective diffusion coefficient.

A simple model is to define an effective diffusivity with the following equation:

$$D_e = D_{pore}\epsilon_p/\tau$$

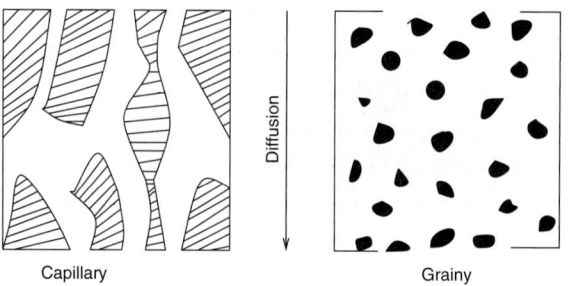

Figure 7.11 Two commonly assumed pore structures for a porous catalyst.

where ϵ_p is the catalyst (particle) porosity and τ is a tortuosity factor. The factor ϵ_p accounts for the reduced area for diffusion, while the factor τ is the correction for non-straight pores. These factors depend on the pore structure. Two common models for pore structure are shown in Figure 7.11. The first model assumes pores are composed of capillaries of various sizes connected in a random manner. The second model is applicable when the catalyst is formed by compaction of small, nonporous grains to form a porous pellet.

The pore diameter is often calculated by the following relation:

$$d_p = \frac{4\epsilon_p}{S_p}$$

where ϵ_p is the particle porosity and S_p is the surface area of the pores per unit volume. The basis for this equation is the hydraulic radius concept, which is widely used in fluid mechanics for studying flow in noncircular channels. This concept applies to materials in which the pores show a unimodal distribution.

Example 7.7 shows some calculations for these quantities.

Example 7.7   Effective Diffusivity in a Porous Catalyst

A porous catalyst has a porosity of 0.60 and a surface area of 100 m²/g and is used for CO oxidation. The pellet has a density of 1.2 g/cm³. Find the effective diffusivity. Temperature is 400 K and pressure is 1 atm.

Solution

Surface area per unit volume is (100 m²/g)(1.2 g/cm³)/(10^{-6} m³/cm³) = 1.2×10^8 m²/m³.

The average pore diameter is calculated as $4 \times 0.60/1.2 \times 10^8 = 3.3 \times 10^{-8}$ m. Using this, the Knudsen diffusion coefficient is calculated from Equation 7.38 as 3.66×10^{-6} m²/s.

> The bulk gas diffusivity is 3.2×10^{-5} m^2/s calculated from the Chapman-Enskog model given by Equation 7.8 for a CO–air mixture. Knudsen diffusion is the dominant mechanism and hence $D_{pore} \approx 3.66 \times 10^{-6}$ m^2/s.
>
> Now we correct for the porosity and the tortuosity factors. Assuming a value of 2 for the tortuosity factor, we find $D_e = D_{pore}\epsilon_p/\tau = 3.66 \times 10^{-6} \times 0.2/2$, which is equal to 1.06×10^{-6} m^2/s.

The tortuosity factor is in the range of 2 to 3 for many catalysts with unimodel pore size distribution. Wakao and Smith (1964) suggest a value of $1/\epsilon_p$ could be used as an approximation in the absence of experimental measurements. Carniglia (1986) suggested the following equation for tortuosity:

$$\tau = (2.23 - 1.13 v_p \rho_B) \left(0.92 \frac{4}{S_p} \sum \frac{\Delta v_p}{d_p} \right)^{1+\alpha}$$

where v_p is the total pore volume, Δv_p is the pore volume within an interval with an average pore diameter of d_p, S_p is the BET surface area, and α is an exponent that depends on the pore shape. To use this equation, information is needed regarding the pore size distribution, which can be obtained from mercury porosimetry.

Bimodal Distribution of Pores

Many solids exhibit a macro–micro pore structure (a bimodal pore size distribution). For for such solids, the model proposed by Wakao-Smith (1964) is suitable. This model was developed by using the conceptual scheme shown in Figure 7.12. Model uses two average porosity values, ϵ_m for macro and ϵ_μ for micropores.

The following equation can be derived based on this concept; it is shown directly without derivation here:

$$D_e = \epsilon_m^2 D_m + \epsilon_\mu^2 \frac{1 + 3\epsilon_m}{1 - \epsilon_m} D_\mu$$

The pore scale diffusivity D_m and D_μ are estimated by the single-pore consideration given by Equation 7.39. This requires an average diameter for the macropores and micropores. These diameters reflect the surface area of the macropores and micropores. Note that Kundsen diffusion may be the controlling mechanism in micropore transport, whereas bulk diffusion may dominate the transport in macropores. Hence the pore-scale diffusivity values must be computed separately for the micropores and the macropores.

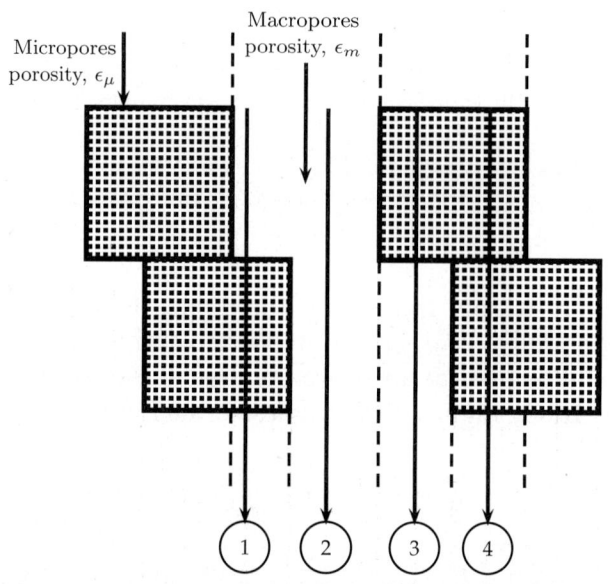

1. Macro–micro diffusion   2. Macro-only diffusion

3. Micro–macro diffusion   4. Micro-only diffusion

Figure 7.12   Illustration of the concepts leading to the Wakao-Smith model for solids with macro–micro pores.

7.6   Heterogeneous Media

Transport in a heterogeneous medium is an important consideration in many applications. For example, a membrane may have a second material embedded within it.

Consider a material made of two types of matrices, with permeability P_c and P_d, respectively. Let ϵ_d be the volume fraction made of material d; $1 - \epsilon_d$ is then the volume fraction of the second material. An effective permeability of the material P_e is to be computed. Here we discuss first two simple models: parallel and series arrangement of the laminate series. They provide two limiting cases to the general case (Figure 7.13).

For a series arrangement, the following equation—which is based on adding resistances in series—is used:

$$\frac{1}{P_e} = \frac{1 - \epsilon_d}{P_c} + \frac{\epsilon_d}{P_d}$$

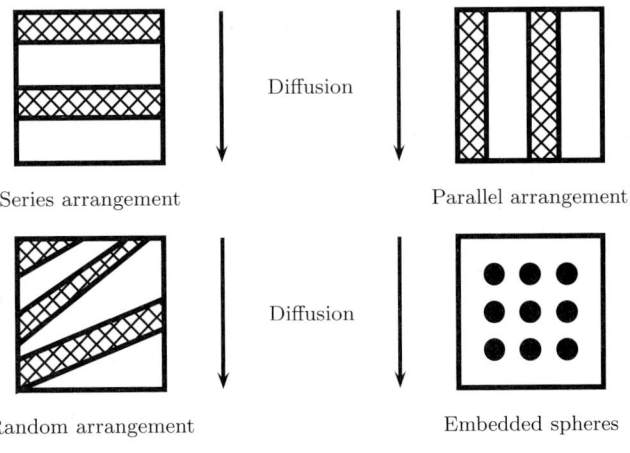

Series arrangement

Parallel arrangement

Random arrangement

Embedded spheres

Cross-hatch = dispersed phase
Open area = continuous phase

Figure 7.13   Models for diffusion or permeation in a composite material.

This can be rearranged to

$$P_e = \frac{P_c P_d}{P_d(1 - \epsilon_d) + P_c \epsilon_d}$$

The equation for the calculation of the permeability of parallel arrangement of laminates is

$$P_e = P_c(1 - \epsilon_d) + P_d \epsilon_d$$

The predictions of the parallel and series modes bound the expected behavior of the composite matrix. Students studying heat transfer will be able to relate this to conduction in series versus in parallel.

An equation for the general case (in which the two materials are distributed in a random manner) was derived by Petropoulos (1985):

$$P_e = P_c \left[\frac{nP_d + (1 - n)P_c - (1 - n)\epsilon_d(P_c - P_d)}{nP_d + (1 - n)P_c + n\epsilon_d(P_c - P_d)} \right]$$

Here n is the shape parameter of the dispersed phase. The parallel arrangement corresponds to $n = 0$, a series arrangement to $n = 1$, and an embedded sphere (shown next) to $n = 1/3$.

In another useful model, one of the phases, denoted as d or the dispersed phase, is embedded in the matrix as solid spheres in a second phase, the continuous phase. The permeability of this composite can be calculated using an equation originally derived by Maxwell in 1873 in the context of electromagnetism to describe a composite dielectric medium:

$$P_e = P_c \left[\frac{P_d + 2P_c - 2\epsilon_d(P_c - P_d)}{P_d + 2P_c + \epsilon_d(P_c - P_d)} \right]$$

A special case occurs when P_d is zero; it is applicable to the effective diffusivity in gas pores of a porous catalyst. In this special case:

$$D_e = D_c \frac{\epsilon_c}{3 - \epsilon_c}$$

Compare this relation with the expression for the effective diffusivity. The preceding equation predicts that the tortuosity factor is $3 - \epsilon_c$, which gives credence to the values of 2 to 3 for the tortuosity factor used in the reaction engineering literature.

7.7   Polymeric Membranes

Small permeates diffusing through a polymeric matrix are common in membrane separations. Examples are found in gas separation processes and discussed further in Chapter 27. Oxygen separation from air is one example.

In such gas separation processes, we require both a high rate of diffusion and selectivity for a particular gas. Solute (water) diffusion in polymeric food wrapping is an important application of this process, in which we need the polymer to act as a barrier. Transport of ions in a conducting polymeric matrix is important in electrochemistry. Oxygen permeability coefficients in polymers used to make hard and soft contact lenses are an important biomedical application, as is diffusion of drugs across a swollen polymeric capsule.

The glass transition temperature is an important parameter affecting the rate of diffusion. This is the temperature at which the polymer changes from a brittle solid structure to a glassy rubbery structure. If the glass transition temperature is exceeded, the diffusivity may change by an order of magnitude. Size of the permeate is an important factor as well. Swelling is another factor, as swelling typically increases the diffusion coefficient.

Recall that permeability is a product of diffusivity and solubility. Diffusivity increases with temperature, whereas solubility decreases under the same condition. The combined effect is that the permeability often shows a maximum with temperature.

Free volume theory is often used as a platform to predict diffusion. A free volume is a space that the solute molecule can occupy, similar to the holes observed in liquid diffusion. The free volume increases with an increase in temperature. The temperature at which the free volume increases rapidly is the glass transition temperature.

Neogi (1996) and Peppas and Meadows (1983) are good starting references that provide an introduction to the various theories that apply to polymeric membranes. These theories should be interpreted as a fitting tool rather than a complete predictive tool, since this research remains in its infancy. Many parameters can be adjusted to fit the data; however, the theoretical underpinnings in these models provide a solid platform for future development in this area. More coupling with molecular-level models can be expected in this field. The discussion provided in this section is rather brief and intended to give just an overview of this topic.

7.8   **Other Complex Effects**

In this section, we discuss a number of complex effects that can affect the rate of diffusion.

- **Dissociation Effects:** Dissociation effects are of importance in liquid–liquid systems where a solute associates with the solvent molecules and the associated pair diffuses together. The solute may associate with itself (a type of dimer) and the diffusion may be controlled by that of the dimer rather than the original solute. Also the solute may dissociate and the two ions formed may then diffuse as a pair together with the diffusion of any undissociated solute. Diffusion of acetic acid in water is an example:

$$CH_3COOH \rightleftharpoons CH_3COO^- + H^+$$

  The diffusion rate of acetic acid is therefore determined by the diffusion of the undissociated species CH_3COOH and the diffusion of the ions, CH_3COO^-, and H^+. Strong and complicated dependency on the concentration and the pH of the solution can be expected in many of these cases (since pH changes the extent of dissociation).

- **Facilitated Diffusion:** Facilitated diffusion is a complexity usually associated with transport in membranes where a simultaneous chemical reaction is also taking place. A chemical reaction binds the diffusing species and enhances its transport. Life itself would not exist but for oxygen–hemoglobin interactions that proceed by facilitated diffusion. Similarly, many membranes (especially biological) have carriers, and the diffusing species binds to the carrier. Both the species and the carrier diffuse across the membrane , such that the transport is considered "facilitated." This topic is studied in further detail in Chapter 22.

- **Active Diffusion:**   Active transport refers to diffusion against a concentration gradient at the expense of some work done on the system. A common example is the Na^+-K^+ pump, which is common across all living systems. This "pump" involves transport of sodium ions against a concentration gradient using the hydrolysis of ATP (adenosine triphosphate) as the energy source. The free energy released in the hydrolysis is used to overcome the difference in chemical potential of Na^+, and a concentration difference on the order of 110 mol/m^3 may exist for the steady state conditions across a cell membrane.

- **Thermal Diffusion:**   In thermal diffusion, a species diffuses due to the presence of a temperature effect. The additional mass flux resulting from thermal diffusion is usually modeled as

$$J_A, \text{ Thermal diffusion } = -C D_{AB} k_T \nabla \ln T \qquad (7.40)$$

where k_T is the thermal diffusion factor. In general for binary systems, species with larger weight move to the colder region, and vice versa.

- **Pressure Diffusion:**   In pressure diffusion, a species diffuses as a result of a pressure gradient. The contribution of pressure to diffusion is modeled as

$$J_A, \text{ Pressure diffusion } = -D_{AB} \frac{y_A}{R_g T} \left(1 - \frac{M_A}{\overline{M}} \right) \nabla P \qquad (7.41)$$

- **Diffusion of Charged Species:**   Diffusion of charged species (e.g., ions) is important in systems such as electrochemical processes and fuel cells. The transport rate is augmented by adding the electrochemical potential as an additional driving force. Simultaneous solution of the diffusion equation and the electric field is needed for such problems. Chapter 16 provides a more detailed discussion of this subject.

Summary

- The simple kinetic theory of gases provides a solid foundation to interpret diffusion in gases as well as a first-level model for predicting the diffusion coefficient. Diffusion is described as resulting from random molecular motion that brings species from a region of higher concentration to a region of lower concentration, consistent with Fick's law.

- Basic kinetic theory may be modified to account for short-range molecular repulsion, described by the Lennard-Jones model. Two molecular-level parameters for each species (σ and ϵ) are needed in this model, and the model is in reasonable agreement with experiments for a large class of molecules. The Lennard-Jones model may not be suitable for long-chain molecules, which can not be represented by an effective collision cross-section. For such molecules with complex structure, molecular dynamics tools (not discussed in this text) are useful to predict the diffusion coefficients.

- Another description of diffusion is based on the frictional interpretation. In this model, the driving force for diffusion is the gradient in chemical; this force is balanced by the frictional resistance created by molecule A squeezing past a solvent phase. The frictional resistance is, in turn, modeled as Stokes law for the liquid phase, which then leads to the Stokes-Einstein equation. This equation also forms the basis for the Wilke-Chang equation, which is widely used for calculating liquid diffusion coefficients.

- For non-ideal liquid mixtures, the chemical potential depends on the activity coefficients. In turn, an activity correction factor is used in calculating the diffusion of non-ideal liquid mixtures. The Darken equation is widely used for this case.

- Application of the frictional approach to a gaseous system leads to an inverted form of Fick's law. This approach, when extended to a multicomponent system, leads to the Stefan-Maxwell model for multicomponent systems.

- Additional effects due to solute–solvent molecular interactons are also important in diffusion in liquids. In such a setting, the nature of the species in solution (whether associated or dissociated, for example) can affect the rate of diffusion.

- The Stefan-Maxwell model, in some situations, can cause strong coupling between the fluxes of various species. Complex effects such as reverse diffusion, osmotic diffusion, and barrier diffusion can arise due to these couplings.

- Diffusivity of a fluid in a porous catalyst is an important parameter in modeling of fluid–solid reactions. For single straight pores, the diffusion can be retarded when the size of the solute is comparable to the pore diameter, a phenomenon known as Knudsen diffusion. For liquids, the frictional interpretation supports models for hindered diffusion, in which the diffusivity otherwise calculated for a larger pore is corrected for steric and hydrodynamic effects.

- For industrial catalysts, the straight pore diffusivity value is corrected by porosity and divided by a tortousity factor. This model is applicable to materials with a unimodal pore distribution. For catalysts with a bimodal distribution, both the micropore porosity and the macropore porosity must be considered; the Wakao-Smith model is commonly used for this purpose.

- Diffusivity of a solute in a second solid is important in microelectonic applications. The diffusing species can occupy the substitutional sites, the interstitial sites, or both. An activation energy type of model is used for these cases, especially to extrapolate the effects of temperature.

- Diffusion in a heterogeneous medium depends on how the second phase (disperse phase) is distributed within the matrix. Simple models are based on either a parallel and series arrangement or an embedded sphere distribution of one phase in the second continuous phase. The relevant equations to calculate the overall effective permeability were given in this chapter; the Petropoulis equation is commonly used for a random distribution of the two phases.

- For diffusion in a polymer, the glass transition temperature is an important property of the polymer. Th diffusion coefficient increases rapidly beyond this temperature. The free volume theory is widely used to model diffusion in polymeric systems.

- Many other complex effects are often exhibited by diffusing systems. For example, reactive species present in the system can couple with the species being transported, leading to facilitated diffusion. Solute–solvent interactions can affect the diffusion. Additional diffusion can be caused by thermal and pressure gradients, although the contributions of these factors are small compared to the diffusion caused by the concentration gradient.

Review Questions

7.1 State the numerical value of the Boltzmann constant with units and show that it is equal to R_g/N_{av}.

7.2 How does the mean free path change with temperature and pressure?

7.3 How does the average speed of random molecular motion change with temperature and pressure?

7.4 What are the effects of temperature and pressure on binary diffusivity in a gas mixture?

7.5 Verify the dimensional consistency of Equation 7.6.

7.6 What is the effect of temperature on the diffusivity of a solute in a liquid?

7.7 Why and when is a thermodynamic correction needed to calculate the diffusion coefficient?

7.8 What is the Vignes relation?

7.9 What is the Darken equation and when is it used?

7.10 Can the parameter A be negative in the one-parameter Margules equation? If so, for which case?

7.11 What is Knudsen diffusion? When is it important?

7.12 What is the effect of temperature on the Knudsen diffusion coefficient? How does it compare to bulk (gas–gas) diffusion?

7.13 When does a pressure gradient build up in a porous solid?

7.14 What is hindered diffusion? When is it important?

7.15 What are the two factors to correct for hindered diffusion of a solute in a liquid-filled pore?

7.16 What is meant by the tortousity factor? State a simple formula to calculate this factor that can be used in the absence of experimental data.

7.17 How is the effective diffusion in a catalyst with a bimodal pore distribution calculated?

7.18 What is the glass transition temperature, and how does it affect diffusion in a polymer?

7.19 What is the effect of temperature on the diffusion coefficient of a gas in a polymeric solid?

7.20 What is facilitated diffusion?

7.21 What are thermal diffusion and pressure diffusion?

Problems

7.1 **Maxwell-Boltzmann distribution and the average molecular speed.** The speed of molecules according to kinetic theory is given by the Maxwell-Boltzmann distribution function f. Thus $f(c)dc$ represents the probability that the speed c lies between c and $c + dc$, and the distribution function is modeled as

$$f(c) = \frac{4}{\sqrt{\pi}} \alpha^{3/2} c^2 \exp(-\alpha c^2)$$

where $\alpha = m/(2\kappa T)$.

Plot the function for CO_2 at 300 K. Show that the area under the distribution function is unity, as would be expected for any probability distribution function.

Find the mean speed, defined as

$$\bar{c} = \int_0^\infty cf(c)dc$$

7.2 **Kinetic theory parameters for an oxygen molecule.** Calculate the mass of an oxygen molecule and the average speed of the molecules $\bar{c}$ are based on kinetic theory at a temperature of 273 K and 1 atm pressure. Compare this speed to the speed of sound in the medium. Compare this speed to the speed of Superman. Hint: Superman can reach the top of the St. Louis Arch 2.5 seconds. Also calculate the mean free path, assuming a molecular diameter of 3 $^\circ A$ (0.3 nm). Determine the self-diffusion coefficient of oxygen. Compare it with the oxygen–air binary diffusivity value of 1.75×10^{-5} m^2/s.

7.3 **Mean free path and mean speed.** Calculate the mean free path of a nitrogen molecule at 1 atm, at a low vacuum (1000 Pa), and at a high vacuum (100 Pa). Assume a molecular diameter of 0.3 nm. The molecular Knudsden number is defined that a dimensionless number defined as the ratio of the molecular mean free path length to the molecular diameter. Find its value for nitrogen. Also calculate the mean molecular speed at 1 atm and a temperature of 300 K. How does it compare with the speed of sound?

7.4 **Lennard-Jones parameter from critical properties.** The critical temperature for CO_2 is 304.25 K and the critical pressure is 7.35 MPa. Estimate the σ and ϵ parameters and compare them with the values given in Table 7.1.

7.5 **Binary pair diffusivity in the gas phase.** Estimate the diffusion coefficient of CO_2 in air at 20° C and 1 atm pressure using the Chapman-Enskog equation. Compare the results using the Fuller correlation.

7.6 **Stefan-Maxwell equation for a ternary gas mixture.** Write out in detail the Stefan-Maxwell equation for a gas A diffusing in a mixture of B and C. Now consider the case in which only species A has a net non-zero combined flux— In other words, N_B and N_C are zero. Simplify the Stefan-Maxwll model for this case. Compare the result with the Wilke equation for a pseudo-binary diffusivity.

7.7 **Liquid-phase diffusivity of a dissolved gas.** Calculate the diffusion coefficient of dissolved CO_2 in water at 300 K.

7.8 **Liquid-phase diffusivity in infinite dilution conditions.** Calculate the diffusion coefficient of ethanol in water (dilute solution with a large mole fraction of water) at 298 K. using the molar volume based on the density and the molar volume based on the group contribution method. The atomic increments are carbon = 14.8, hydrogen = 3.7, and oxygen = 7.4 (all in cm^3/mol), as reported in Le Bas. Also calculate the diffusivity of water in ethanol (dilute solution with a large mole fraction of ethanol).

7.9 **Liquid-phase diffusivity as a function of composition.** Calculate and plot the diffusivity as a function of the mole fraction for an ethanol–water mixture using either the linear interpolation or the Vignes equation. Ignore the thermodynamic correction. Now find the activity coefficient of ethanol as a function of composition using thermodynamic data and correct the results for the diffusivity using the thermodynamic correction factor.

7.10 **Diffusivity in an acetone–water mixture.** Verify the calculations for the diffusion coefficient of acetone in water for a 50 mole % mixture, as shown in Example 7.5. Now find the values at various compositions and plot the diffusivity as a function of mole fraction of acetone.

7.11 **One-parameter model for activity coefficient.** Consider the simplest model for the activity coefficient:

$$\ln \gamma_A = A x_B^2 = A(1 - x_A)^2$$

Calculate γ_A and plot it as a function of the mole fraction of A for different values of A ($A = 0$, which is the ideal case; $A = 1$; $A = 2$ and 3). The correction becomes negative for certain value of A, which means the observed diffusion coefficient becomes negative! Provide an explanation for this outcome.

7.12 **Use of Gibbs-Duhem equation.** State the Gibbs-Duhem equation. Using this equation, show that the thermodynamic correction for diffusivity for species A is the same as that for B in a binary mixture. In other words, verify that the sum of the diffusion fluxes is zero at any specified mole fraction value. Also verify that $D_{AB} = D_{BA}$ at a fixed concentration, although the values may be concentration dependent.

7.13 **Diffusion in a straight pore.** Calculate the value for silane diffusing in a 10-μm cylindrical pore at 900 K, which is representative of the deposition of solid silicon in thin tubes. Determine the dominant mechanism for diffusion in this case.

7.14 **Hindered diffusion in a nanosized pore.** Glucose is diffusing across a microporous membrane with pores of 2-nm diameter. The mean diameter of the glucose molecule is about 0.86 nm. Temperature is 300 K. Find the diffusion coefficient using the Stokes-Einstein equation. To what extent is the hindered diffusion correction important in this case?

7.15 **Size exclusion–based separation.** A protein solution is separated from water by allowing it to diffuse across a porous nanosize membrane. The size of the protein is 25 nm. Find the diameter of the membrane to have a steric partition coefficient of 0.64. If a second protein has a a size of 50 nm, find the relative exclusion of this protein in the membrane.

7.16 **Effective diffusivity in a porous catalyst.** Calculate the effective diffusivity of ethylene diffusing in hydrogen gas at 1 atm pressure at a temperature of 298 K given the following properties of the catalyst: porosity = 0.4; surface area = 100 m^2/g; bulk density of the catalyst = 1.4 g/cm^3.

7.17 **Bimodal pore distribution.** A porous catalyst has a bimodal distribution with porosity of 0.233 in the micropores and 0.492 in the macropores. The average pore radii values are 40 nm in the micropores and 2 μm in the macropores. Find the effective diffusivity for a methane–air mixture.

7.18 **Diffusion in a composite matrix.** A polymeric membrane is made of two material with permeability of 1×10^{-5} and 4×10^{-5}, respectively. The second material is embedded in the first, with a distribution of 15%. Calculate the overall permeability assuming (a) a series arrangement, (b) a parallel arrangement, and (c) the second material embedded as small spheres in the first material.

CHAPTER 8

Transient Diffusion Processes

Learning Objectives

After completing this chapter, you will be able to:

- Understand and use the separation of variables method for the solution of homogeneous linear partial differential equations.
- Use the modified method of separation of variables for non-homogeneous problems.
- Understand the superposition method for the solution of multidimensional transient diffusion problems.
- Solve transient problems posed on a semi-infinite domain.
- Understand the concepts behind the penetration model for mass transfer from a gas–liquid interface.
- Solve eigenvalue problems with the CHEBFUN program.
- Solve partial differential equations numerically with the PDEPE program in MATLAB.

In Chapter 2 we briefly examined some problems on transient diffusion, albeit on a qualitative basis. In this chapter we examine unsteady state problems in more detail. This chapter introduces and solves such problems and illustrates a number of important mathematical and computational techniques to obtain the solutions.

For linear partial differential equations, analytic solutions based on the direct separation of variables method are commonly used. They can be applied to problems where the differential equations and the boundary conditions are homogeneous (explained in the text) and apply to problems posed in a finite spatial domain. This solution procedure is developed and demonstrated here, focusing more on a physical perspective rather than a rigorous mathematical viewpoint.

The method of separation of variables cannot be used directly for non-homogeneous problems and needs some modifications. We study such problems next and then present a number of examples. A more general solution method is based on the eigenfunction operator and known as the finite Fourier transform method. This can be applied to a wider class of (still linear) problems, but this method is not discussed in this text.

Many problems in transient diffusion are posed on a semi-infinite domain. For such problems a solution method based on similarity transformation is useful. This method is introduced and applications to a number of problems are described. An important application of this is the penetration model for mass transfer, which is often used in lieu of the classical film theory. Another application is to predict the effect of a reaction on mass transfer under transient conditions, which is taken up in detail later in the context of gas–liquid reactions.

For nonlinear problems, numerical solutions are required and we discuss one program based on MATLAB code, PDEPE, as an illustrative and useful method for nonlinear problems.

This chapter therefore provides an introduction to solution methods for transient mass diffusion problems together with some useful worked examples.

8.1   Transient Diffusion Problems in 1-D

The governing equation for transient diffusion follows from the general equation for mass transfer derived in Chapter 5:

$$\frac{\partial C_A}{\partial t} + \nabla \cdot \boldsymbol{N}_A = R_A \tag{8.1}$$

If a low flux mass transfer model is used for N_A, which is then equal to J_A, and if Fick's law is used for diffusion (with a constant value for D_A), we have:

$$\frac{\partial C_A}{\partial t} = D_A \nabla^2 C_A + R_A \tag{8.2}$$

The Laplacian can be simplified for common 1-D problems as presented in Figure 8.1. These problems are 1-D diffusion in a slab, and diffusion only in the radial direction in a long cylinder or sphere. These are important and useful problems and will be analyzed in detail in the following sections. We examine the solutions to these 1-D problems for the cases where the separation

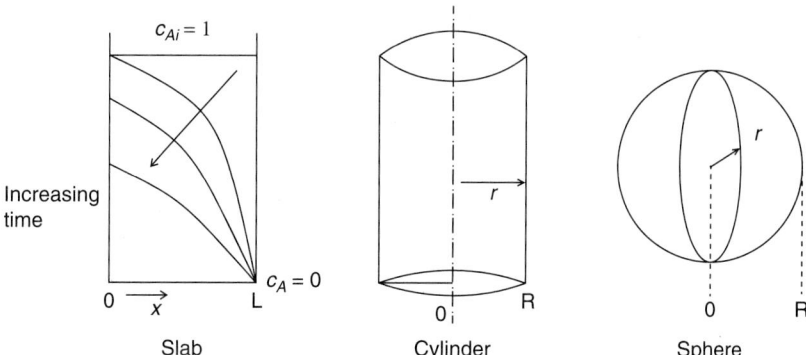

Figure 8.1   Three simple geometries where the transient problem can be treated in one spatial coordinate (and time).

of variables can be directly applied. Then we show some cases where a modified method is required.

8.2   Solution for Slab: Dirichlet Case

The problem analyzed in this section is transient diffusion in a porous slab of thickness $2L$; it has an initial concentration of C_{Ai} and is exposed to a concentration of C_{As} on the surface. No reaction occurs in the slab and the transient diffusion with reaction case is analyzed later in Section 8.5. We will predict the transient concentration profile in the slab. The problem is shown schematically in Figure 8.2 as problem (a). This can be cast as a homogeneous problem and direct separation of variables can be used. The coordinate system is defined such that $x = 0$ is the midpoint of the slab. This is a mathematical convenience as will become clear during the derivation.

A variation of this problem, shown as (b) in Figure 8.2, is analyzed later in Section 8.5. Here the surfaces of the slab are exposed to two different concentrations. The solution requires a modification in the separation of variables method by subtracting out the steady state solution.

For the slab case the governing equation is Equation 8.2 using the Laplacian in Cartesian with only x-dependency:

$$\frac{\partial C_A}{\partial t} = D_A \frac{\partial^2 C_A}{\partial x^2} \tag{8.3}$$

It is useful to put this in dimensionless form, which is discussed next.

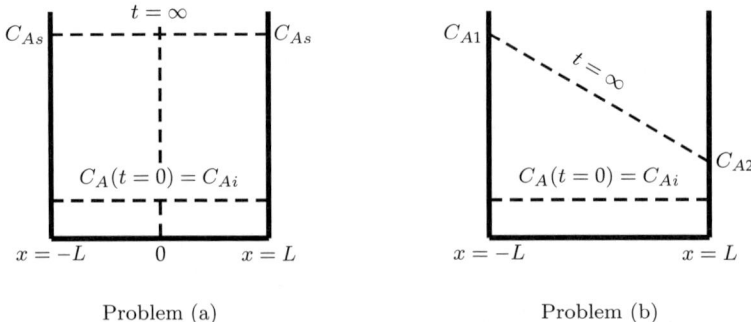

Problem (a)   Problem (b)

Figure 8.2 Schematic of the illustrative slab transient problems analyzed for transient diffusion. Two cases are shown. Problem (a) is solved by direct separation of variables where the same surface concentration exists at both edges of the slab while problem (b), where the two surface concentrations are different, requires some pre-manipulations prior to solution by the separation of variables method.

8.2.1   Dimensionless Representation

The dimensionless representation begins by defining a scaled concentration and distance:

$$c_A = \frac{C_A - C_{As}}{C_{Ai} - C_{As}}$$

where C_{As} is the surface concentration, which remains constant with time for the Dirichlet problem. C_{Ai} is the initial concentration, which is assumed to be uniform here and not a function of the spatial variable, x. The above dimensionless definition makes the initial concentration one. The surface concentration is now zero, which makes the boundary conditions homogeneous. This is an important consideration when using direct separation of variables.

Note: The initial concentration need not be uniform and can be a specified function of position, in which case the average value of the initial concentration has to be used to define the dimensionless concentration.

The dimensionless distance is defined by ξ as x/L. This also makes the distance go from zero to one. These choices for scaling reduce the equation to the following form using the chain rule:

$$\frac{L^2}{D_A} \frac{\partial c_A}{\partial t} = \frac{\partial^2 c_A}{\partial \xi^2} \tag{8.4}$$

The time was not made dimensionless as we do not know what the time scales are. However the equation suggests that L^2/D_A should be chosen as the reference time to scale all terms equally. Hence the dimensionless time τ is defined as

$$\tau = \frac{tD_A}{L^2} = \frac{t}{L^2/D_A}$$

L^2/D_A is the reference time, also called the diffusion time as noted earlier in Example 2.5, and is approximately the time for internal gradients to reach a steady profile for a surface change in the concentration. The τ is thus the ratio of the elapsed time to the diffusion time.

The dimensionless version for further analysis is therefore

$$\frac{\partial c_A}{\partial \tau} = \frac{\partial^2 c_A}{\partial \xi^2} \tag{8.5}$$

Note that there are no free parameters, which is a characteristic of a well-scaled problem in general.

The boundary and initial conditions are as follows:

- At $\xi = 0$, we have symmetry and hence $\partial c_A/\partial \xi = 0$.
- At $\xi = 1$, we use the specified concentration, which makes $c_A = 0$.
- At $\tau = 0$ we assume a specified initial concentration profile: $c_A(\tau = 0) = c_{Ai}(\xi)$ in general. For the case of a uniform starting concentration profile, $c_{Ai} = 1$.

The boundary conditions and the differential equation are homogeneous and we can use the separation of variables method directly.

8.2.2   Series Solution

The separation of variables method starts with proposing a solution as a product of separate functions in time τ and distance ξ. It seeks solutions of the following form:

$$\text{Assumed form of the solution} = G(\tau)\, F(\xi)$$

The time-dependent part can be shown to be $\exp(-\lambda_n^2 \tau)$, where λ_n is a constant known as the eigenvalue. (The subscript n is needed since there are many (infinite) values of λ, as shown later.) Hence we seek solutions of the following form:

$$\text{Form of the solution} = \exp(-\lambda_n^2 \tau) F_n(\xi)$$

Substituting the assumed solution in Equation 8.5 we find that the F function should satisfy the following ordinary differential equation (verify the algebra):

$$\frac{d^2 F_n}{d\xi^2} + \lambda_n^2 F_n = 0 \tag{8.6}$$

The boundary conditions for F follow from those for c_A. Hence we use the following homogeneous conditions:

- At $\xi = 0$, $dF/d\xi = 0$.
- At $\xi = 1$, $F = 0$.

Equation 8.6 and the associated homogeneous boundary conditions constitute an eigenvalue problem. The solution shown next result in the generation of the F functions, which are known as the eigenfunctions.

Eigenvalues and Eigenfunctions

The general solution to Equation 8.6 is

$$F_n(\xi) = A_n \cos(\lambda_n \xi) + B_n \sin(\lambda_n \xi) \tag{8.7}$$

where A_n and B_n are the integration constants. Now we will evaluate these using the boundary condition.

Use of the symmetry condition at $\xi = 0$ will lead to $B_n = 0$. Hence

$$F_n = A_n \cos(\lambda_n \xi)$$

Use of the condition at $\xi = 1$ will lead to

$$A_n \cos(\lambda_n) = 0 \tag{8.8}$$

A_n cannot be zero as it will lead to a trivial solution. Hence this equation provides us the means to find the values for λ_n. Since the cos function is zero at $\pi/2, 3\pi/2, 5\pi/2$, and so on, we have many discrete values for λ_n:

$$\lambda_n = \left(n + \frac{1}{2}\right)\pi \tag{8.9}$$

Here $n = 0, 1, 2, 3, \cdots$, generating an infinite set of eigenvalues.

The F function is therefore of the form $F_n = A_n \cos(\lambda_n \xi)$. These are known as the eigenfunctions, as indicated earlier.

Series Solution

As there are infinite number of eigenvalues for this problem and all these values are justified, a series solution that incorporates all values of n seems logical; we can write the solution as an infinite series:

$$c_A(\xi, \tau) = \sum_{n=0}^{\infty} A_n \exp(-\lambda_n^2 \tau) \cos(\lambda_n \xi) \qquad (8.10)$$

The only thing missing is that the initial conditions are not yet satisfied. The series constants A_n should be chosen to satisfy the initial conditions as shown in the following section.

8.2.3   Evaluation of the Series Coefficient

Using the initial conditions we have

$$c_{Ai}(\xi) = \sum_{n=0}^{\infty} A_n \cos(\lambda_n \xi) \qquad (8.11)$$

Now, in order to find explicit expressions for A_n, we have to use the orthogonal property of the F function.

First we state the orthogonality property of the eigenfunctions:

$$\int_0^1 F_n(\xi) F_m(\xi) d\xi = 0 \text{ if } n \neq m$$

In our particular case of this problem, we have

$$\int_0^1 \cos([n + 1/2]\pi\xi) \cos([m + 1/2]\pi\xi) d\xi = 0 \text{ if } n \neq m$$

which can be verified using the table of integrals, MAPLE, or MATHEMATICA. But it is useful to note that the orthogonality property is a general property for all the eigenvalue problems of the Strum-Liouville type. If the differential operator is linear and symmetric then the eigenfunctions are orthogonal (with a weighting function), which is a basic property of such operators. Interested students should refer to the book by Ramkrishna and Amundson (1985) on linear operator methods. We now proceed to evaluate the series coefficients.

If we use the orthogonality property of the eigenfunction, then the series unfolds itself and the coefficients can be found one at a time using the following ratio of two integrals:

$$A_n = \frac{\int_0^1 c_{Ai}(\xi) \cos(\lambda_n \xi) d\xi}{\int_0^1 \cos^2(\lambda_n \xi) d\xi} \tag{8.12}$$

This is applicable to any specified initial condition given by $c_{Ai}(\xi)$. For a particular case of a constant initial concentration $c_{Ai} = 1$, the series coefficients are

$$A_n = \frac{\int_0^1 \cos(\lambda_n \xi) d\xi}{\int_0^1 \cos^2(\lambda_n \xi) d\xi} = \frac{2(-1)^n}{\pi(n + 1/2)} \tag{8.13}$$

This completes the solution to the Dirichlet problem in a slab.

The method holds for all similar problems, for example, the other 1-D geometries mentioned in Figure 8.1. The only changes will be in the eigenfunction, eigencondition, eigenvalues, and series coefficient. If expressions for these are derived, the series solution is complete. In fact the solution to a number of homogeneous PDE problems can be represented in series form similar to that given by Equation 8.10 and similar solution steps can be used. Let us return to the slab problem and look at some features of the solution.

8.2.4   Illustrative Results

An illustrative plot of concentration versus distance is presented in Figure 8.3; it is obtained by summing the series up to 50 terms in MATLAB for some selected values of time.

Concentration profiles are rather sharp for small values of τ. The series will converge rather slowly here and a large number of terms is needed. An alternative is to model this as a semi-infinite slab as shown in Section 8.7. The following solution (derived in Section 8.7) will then be applicable:

$$c_A = \text{erf}\left(\frac{1 - \xi}{2\sqrt{\tau}}\right)$$

where erf represents the error function. This solution is marked as * in Figure 8.3. Note the excellent comparison with the above erf solution for $\tau < 0.05$.

The series converges rapidly for values of $\tau > 0.2$. Only one term may be sufficient to get a reasonable estimate of the concentration profile. Using the term approximation ($A_0 = 4/\pi$ here) the concentration can be described as a function of time as

$$c_A = (4/\pi) \exp(-\pi^2 \tau/4) \cos(\xi \pi/2)$$

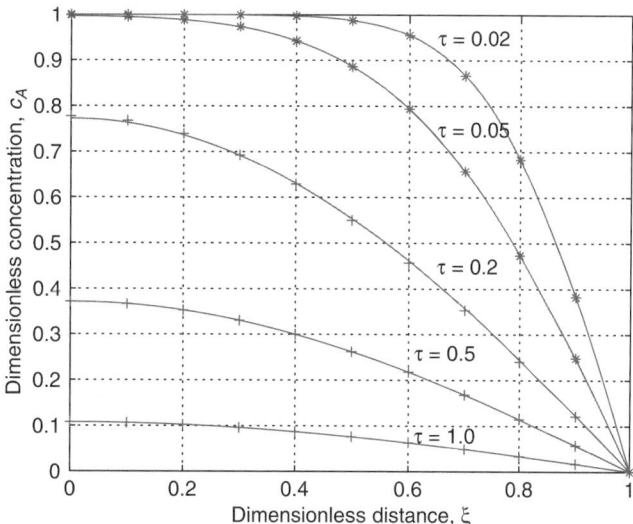

Figure 8.3 Time-concentration plot for slab case with Dirichlet boundary condition at the surface. The points marked * represent the semi-infinite solution for comparison; points marked + show the series solution computed with only one term.

which is valid for $\tau > 0.2$ or so. The one term approximation is marked as + in Figure 8.3 for comparison and we find that this is sufficient for $\tau > 0.2$. Steady state is almost reached around a τ of one. Hence the actual time needed to reach steady state is of the order of the diffusion time, L^2/D_A.

The expression for the mass flux at the surface is important in design calculations. This is obtained by applying Fick's law at the surface. For a constant initial concentration of C_{Ai}, the expression can be shown to be

$$N_{As}(\tau) = 2D_A \frac{(C_{Ai} - C_{As})}{L} \sum_{n=0}^{\infty} \exp\left[-(n+1/2)^2\pi^2\tau\right] \qquad (8.14)$$

Students may wish to verify the algebra leading to this expression.

A plot of mass flux as a function of dimensionless time is presented in Figure 8.4. A flux scaled by $D_A(C_{Ai} - C_{As})/L$ is used in this plot. The mass flux is very large at time near zero and decreases with time with square root proportionality in the early stages of the process. At early stages for $\tau < 0.1$, the dimensionless flux can be calculated more simply as $1/\sqrt{\pi\tau}$, which follows from the alternative semi-infinite slab solution shown in Section 8.7. For larger values of time ($\tau > 0.2$ or so), the one-term approximation is sufficient

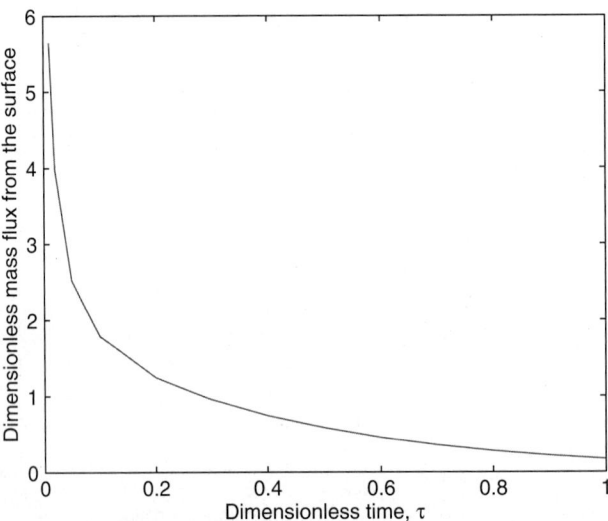

Figure 8.4   Dimensionless mass flux from the surface as a function of time for the slab problem.

to calculate the fluxes. The flux decreases as an exponential function of time during this period.

8.2.5   Average Concentration

The volume average concentration in the solid is defined in a general manner for all three geometries shown in Figure 8.1:

$$\langle c_A \rangle = (s+1) \int_0^1 \xi^s c_A(\xi) d\xi \tag{8.15}$$

where s is the shape factor. Note that $\xi^s d\xi$ is a measure of the local volume at any given ξ position.

For the slab case $s = 0$ and the average concentration is simply the integral under the curve in Figure 8.3 for each time value. The integrated expression is

$$\langle c_A \rangle(\tau) = \sum_{n=0}^{\infty} A_n B_n \exp\left[-(n+1/2)^2 \pi^2 \tau\right] \tag{8.16}$$

The term B_n arises from the spatial integration and is given as

$$B_n = \int_0^1 \cos(\lambda_n \xi)d\xi = \frac{\sin(\lambda_n)}{\lambda_n}$$

A related quantity is the fractional mass loss parameter, Φ, which is defined as

$$\Phi = \frac{\mathcal{M}_0 - \mathcal{M}(t)}{\mathcal{M}_0} = 1 - \langle c_A \rangle$$

This represents the ratio of moles of A that have been removed at the given time to the total amount of moles of A that have to be removed (which is the same as the initial moles in the system $\mathcal{M}_0$). An application of the solution is shown in Example 8.1.

Example 8.1   Transient Diffusion of Oxygen

A pool of liquid of 10 cm deep is exposed to air and oxygen diffuses into the pool. Take 1 m^2 as the surface area. Find the time interval up to which the oxygen concentration is nearly zero at the bottom of the pool. Find the time interval over which the liquid is nearly saturated (90%) with oxygen. For the time equal to half this value, find the oxygen concentration at the bottom of the pool.

The diffusion coefficient of oxygen in water is 2×10^{-9} m^2/s. The saturation solubility of oxygen is found from Henry's law as 0.27 mol/m^3.

Solution

The diffusion time is calculated as $(0.1 \text{ m})^2/[2 \times 10^{-9}\text{m}^2/\text{s}] = 5 \times 10^6$ seconds.

From Figure 7.3 we find the concentration at the bottom remains nearly one until about $\tau = 0.05$. Hence this is the time up to which oxygen concentration remains zero at the bottom. The corresponding actual time is 2.5×10^5 seconds.

The time at which oxygen concentration is 90% of the final value corresponds to $\tau = 1$ from Figure 7.3. This corresponds to an actual value of time of 5×10^6 seconds, which is also equal to the diffusion time. A long time is needed to saturate in a stagnant liquid with oxygen and some type of aeration is generally needed to quicken the process.

For half this time (i.e., $\tau = 0.5$) a one-term approximation is sufficient. The dimensionless concentration is

$$c_A(\text{bottom}) = (4/\pi)\exp(-0.5 * \pi^2/4) = 0.3708$$

For an initial concentration of zero, the definition of c_A is $1 - \frac{C_A}{C_{As}}$. Using this definition, the actual concentration C_A is calculated as 0.17 mol/m^3 for a time of 2.5×10^6 seconds.

8.3   Solutions for Slab: Robin Condition

If the surface is losing mass by convection to the surroundings, the surface concentration is unknown. What we know is the external fluid or ambient concentration, denoted as $C_{A\infty}$ here. For such problems the convective boundary condition should be used; this is obtained by a balance of mass flux at the surface:

$$-D_A \left(\frac{\partial C_A}{\partial x} \right)_{x=L} = k_m [C_A (\text{at } x = L) - C_{A\infty}]$$

The left-hand side is the mass arriving by diffusion while the right-hand side is the mass leaving the surface by convection to the surroundings with k_m defined as the (solid-to-fluid) mass transfer coefficient. The dimensionless temperature c_A is now defined as $(C_A - C_{A\infty})/(C_{Ai} - C_{A\infty})$. The boundary condition in dimensionless form at the surface is

$$\left(\frac{\partial c_A}{\partial \xi} \right)_{\xi=1} = -Bi \, c_A \text{ at } (\xi = 1) \tag{8.17}$$

The dimensionless parameter appearing in this boundary condition is Bi, defined as $k_m L / D_A$. This is known as the Biot number. Thus we have a mixed (Robin) boundary condition with the flux appearing on the left-hand side and the surface concentration appearing on the right-hand side. The surface concentration is no longer constant and varies as a function of time.

   Note: $C_{A\infty}$ assumes that the diffusing solute has a partition coefficient of one in the bulk liquid. Otherwise the term $C_{A\infty}$ should be replaced by $K_A C_{A\infty}$, where K_A is the ratio of the solid concentration of A to its liquid concentration at equilibrium.

   The differential equations and the center boundary condition remain the same. The initial condition is taken as unity for further analysis.

   The series given by Equation 8.10 is still used but with modified values for eigenvalues, which now are specific to each Biot number. The series coefficients are also different now.

   The following relations hold for this case and the student should verify the details:

- The eigenvalues are the solutions to the following equation:

$$Bi \cos(\lambda) - \lambda \sin(\lambda) = 0 \tag{8.18}$$

Note that the eigenvalues now depend on the Biot number and must be calculated by the solution of this transcendental equation. They are not simple multiples of quantities like $\pi/2$ in contrast with the Dirichlet case.

- The eigenfunctions F_n are $\cos(\lambda_n \xi)$. These functions have the orthogonality property.
- The series expansion coefficient for a uniform initial condition of one is

$$A_n = \frac{2 \sin(\lambda_n)}{\lambda_n + \sin(\lambda_n) \cos(\lambda_n)} \tag{8.19}$$

- The expression needed to find the average temperature is the same as before:

$$B_n = \frac{\sin(\lambda_n)}{\lambda_n} \tag{8.20}$$

However, the B_n values are now Biot number dependent since the eigenvalues depend on the value of the Biot number.

An illustrative example of the effect of the Biot number is shown in Figure 8.5.

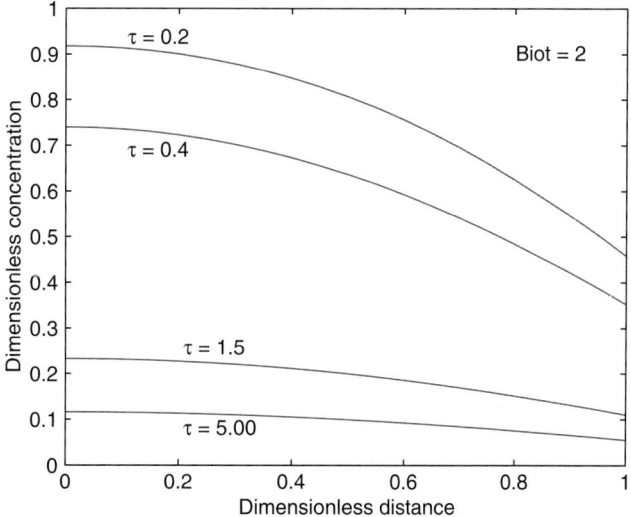

Figure 8.5 Time-concentration plot for the slab case with a Robin boundary condition at the surface. A Biot number of two is used in this illustration.

Two limiting cases are noted below:

- For large Bi (> 20), the eigenvalues reduce to $\lambda_n = (n + 1/2)\pi$ and the Dirichlet solution is recovered for large Bi. The physical interpretation is that the surface concentration becomes nearly the same as the bulk concentration.

- For small Biot numbers ($Bi < 0.3$), a lumped parameter approximation is sufficient. The physical interpretation is that there is no significant internal concentration gradient and the external mass transfer rate becomes the limiting factor. An overall macroscopic mass balance for the solid is sufficient here. The volume-averaged concentration predicted by such a model is

$$< c_A >= \exp(-Bi \ \tau) \tag{8.21}$$

Derivation of the lumped parameter model is left as en exercise. Drying a porous solid is an example where such a model is likely to apply because the external partial pressure difference in the gas phase is the limiting driving force.

8.4   Solution for Cylinders and Spheres

The solution for cylinders and spheres can be obtained using the same procedure. The series solution can be represented as

$$\boxed{c_A(\xi, \tau) = \sum_{n=0}^{\infty} A_n \exp(-\lambda_n^2 \tau) F_n(\lambda_n \xi)} \tag{8.22}$$

where λ_n are the eigenvalues and F_n is the corresponding eigenfunction. The expressions for these and the series coefficient for long cylinders and spheres are given in the following subsections without derivation.

Note: The coefficient $A_0 = 0$ for cylinders and spheres and the above series starts with $n = 1$.

8.4.1   Long Cylinder

The solution for a cylinder is in terms of the Bessel functions of the first kind and can be enumerated as follows:

- The eigenvalues are the solutions to

$$Bi J_0(\lambda) - \lambda J_1(\lambda) = 0$$

The values can be found by the solution of the above nonlinear equation for λ for a fixed Biot number.

- The problems with large values of the Biot number are often of more interest in mass transfer problems because the internal diffusion is often rate limiting compared to external mass transfer. For this case the eigenvalues can be directly found as the roots of

$$J_0(\lambda) = 0$$

The first three eigenvalues are 2.4048, 5.5201, and 8.6537 from tabulated or calculated values of the J_0 Bessel function.

- The eigenfunctions F_n are $J_0(\lambda_n \xi)$.
- The eigenfunctions are orthogonal to each other with a weighting factor of ξ.
- The series expansion coefficient for a uniform initial condition of one is

$$A_n = \frac{2 J_1(\lambda_n)}{\lambda_n [J_0^2(\lambda_n) + J_1^2(\lambda_n)]}$$

Note that the series coefficients simplify for the large Biot case to

$$A_n = \frac{2}{\lambda_n [J_1(\lambda_n)]}$$

- The coefficients B_n needed to calculate the average concentration are

$$B_n = 2 \frac{J_1(\lambda_n)}{\lambda_n}$$

8.4.2   Sphere

The eigenfunctions for a sphere belong to a class of Bessel functions of the spherical kind. The results are as follows:

- The eigenvalues are the solutions to the following transcendental equation:

$$\lambda \cos(\lambda) + (Bi - 1) \sin(\lambda) = 0$$

- For the large Biot number case, the eigenvalues can be directly found as the roots of

$$\sin(\lambda) = 0$$

Hence the first two eigenvalues are π and 2π.

- The eigenfunctions F are $\sin(\lambda\xi)/(\lambda\xi)$.
- The eigenfunctions are orthogonal with a weighting factor of ξ^2.
- The series expansion coefficient for a uniform initial condition of one is

$$A_n = 2\frac{\sin(\lambda_n) - \lambda_n \cos\lambda_n}{\lambda_n - \sin(\lambda_n)\cos(\lambda_n)}$$

The series coefficients simplify for the large Biot case to

$$A_n = \frac{2}{\cos(\lambda_n)}$$

- The coefficient needed to find the average concentration is

$$B_n = 3\frac{\sin(\lambda_n) - \lambda_n \cos(\lambda_n)}{\lambda_n^3}$$

Note that the above coefficient simplifies for the large Biot case to

$$B_n = -3\frac{\cos(n\pi)}{n^2\pi^2}$$

An illustrative plot of the concentration profile for a sphere with the Dirichlet condition is shown in Figure 8.6. Note that a nearly steady state value is attained for dimensionless time of $\tau = 0.3$. Also, for $\tau > 0.1$ a one-term approximation discussed in Section (8.4.3) is sufficient.

The surface flux calculated for the sphere problem is presented in Figure 8.7, which shows the effect of Biot number. This represents a case where there is a barrier to diffusion near the surface, for example, having an inert core near the surface. The flux is reduced at lower Biot numbers as expected, but the release rate is more uniform. In other words a large rate at times near zero, a characteristic of the Dirichlet problem, is no longer observed. This has implications on the design of drug capsules which can provide a nearly uniform release rate.

8.4.3   One-Term Approximation

Often the leading term of the series is accurate enough for large values of time, $\tau > 0.1$ for spheres and 0.3 for slabs. This is known as the one-term approximation. Often we are interested in the concentration at the center as the response is slowest at this point. The one-term approximation for the center concentration is

$$c_A(\xi = 0) \approx A_1 \exp(-\lambda_1^2\tau)$$

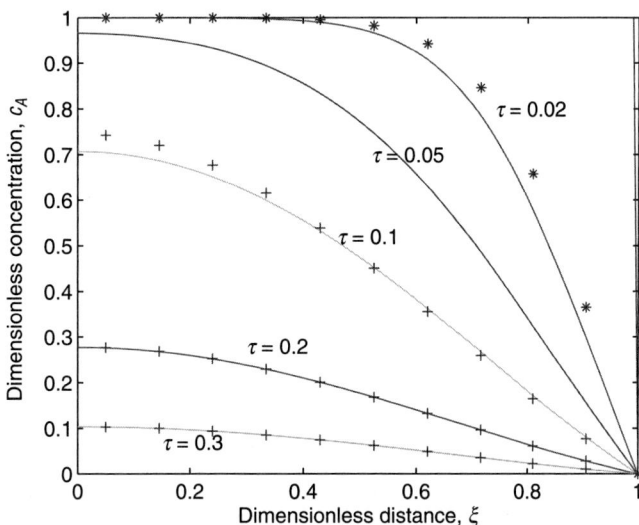

Figure 8.6 Time-concentration plot for transient diffusion in a sphere with Dirichlet boundary condition at the surface. The points marked * represent the semi-infinite solution for comparison; points marked + show the series solution computed with only one term.

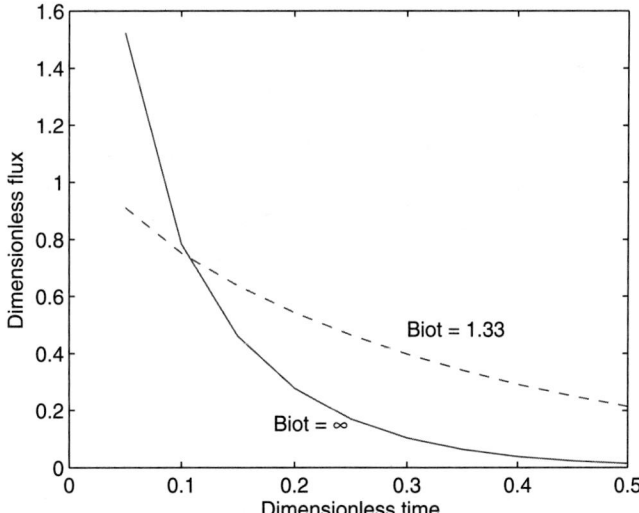

Figure 8.7 Effect of Biot number on the instantaneous flux from a sphere; in the simulation Biot = 1.33. Flux is dimensionless and a scale flux of $D_A C_{Ai}/R$ is used as a reference. Note that the release rate has become somewhat more uniform due to the diffusion barrier.

Table 8.1 Coefficients for use in the One-Term Approximation. Large Biot Number Case.

–	λ_1^2	A_1	B_1
Slab	2.467	1.273	0.6366
Cylinder	5.784	1.602	0.4317
Sphere	9.869	2.000	0.3040

Similarly the average temperature is another quantity of interest; the one-term approximation for this is

$$\langle c_A \rangle \approx A_1 B_1 \exp(-\lambda_1^2 \tau)$$

Note that the average concentration is equal to the center concentration multiplied by B_1. Only the first eigenvalues and the first series coefficients are needed for the one-term model. Further the large Biot case is of interest in mass transfer processes. The values of the parameters needed in the one-term approximation for this Dirichlet problem are shown in Table 8.1; they are for quick calculation for design purposes. The following problem shown in Example 8.2 illustrates the use of the one-term approximation.

Example 8.2   Drug Release Rate from a Capsule

A capsule has a radius of 0.3 cm and has an initial drug concentration of 68.9 mg/cm^3. The diffusion coefficient of the drug in the porous matrix is 3×10^{-6} cm^2/s. Calculate and plot the center concentration and the amount of drug released as a function of time. Use one-term approximation as a simplification, but note in general that this is not valid for short times.

Solution

First we estimate the time constant to calculate how long the process will last:

$$\text{Time constant} = \frac{R^2}{D_A} = \frac{(0.3^2)\ \text{cm}^2}{3 \times 10^{-6}\ \text{cm}^2/\text{s}} = 3 \times 10^4 \text{ sec} = 8.33 \text{ hours}$$

Hence we look for time values in this range. Consider a time $t = 1$ hour, which corresponds to $\tau = 1/8.33 = 0.12$. The center concentration is calculated from the one-term approximation to the series solution as $2.0 \exp(-9.899\tau) = 0.6119$ using the tabulated value of A_1 and λ_1^2.

The average concentration is calculated as 0.1860 using the tabulated value of B_1 and multiplying the center concentration by this number.

The fraction of drug released is $1 - \langle c_A \rangle$, which is 81.4%.

The quantities for other time values can be calculated in a similar manner and the data on the amount of drug released as a function of time can be generated.

8.5   Transient Non-Homogeneous Problems

In this section, we consider how the transient diffusion equation can be solved for non-homogeneous cases. Clearly use of separation of variables directly is not possible and we have to manipulate the equations somewhat before we can proceed further. One common method is to subtract the steady state solution. The solution is then represented as the sum of the steady state part and the deviation from the steady state:

$$C_A(x, t) = C_{A,s}(x) + Y(x, t)$$

The variable Y can be interpreted physically as the deviation of the concentration from the steady state. The governing PDE for Y and the associated boundary conditions can be derived from the original problem. The problem for Y is homogeneous and can be solved by direct separation of variables. We present two examples to demonstrate the method. The book by Crank (1975) is an useful source for a wide range of transient diffusion problems and many other examples may be found here.

8.5.1   D-D Problem in Slab Geometry

Consider a slab of thickness L that has a uniform initial concentration of C_{AL} at time zero. At time zero one end of the slab at $x = 0$ is raised to a concentration of C_{A0} while the other end is maintained at C_{AL}. The problem is presented in Figure 8.2(b).

The dimensionless variable c_A defined as

$$c_A = \frac{C_A - C_{AL}}{C_{A0} - C_{AL}}$$

is appropriate. The boundary conditions for c_A are as follows: at $\xi = 0, c_A = 1$ and at $\xi = 0, c_A = 0$. **Note:** $\xi = 0$ starts at one edge of the slab and not at the center. Since the Dirichlet condition applies at both ends, this is tagged as a D-D problem. The initial condition is $c_A = 0$ for all ξ at $\tau = 0$.

The boundary conditions are non-homogeneous and hence direct separation of variables is not possible. The problem has a steady state solution:

$$c_{A,s}(\xi) = 1 - \xi$$

This can be used to formulate the problem in terms of the deviation variable Y. The governing equation is now the same as the transient diffusion with no generation:

$$\frac{\partial Y}{\partial \tau} = \frac{\partial^2 Y}{\partial \xi^2} \tag{8.23}$$

The homogeneous boundary conditions result for the Y problem since Y is in zero at both ends $\xi = 0$ and 1.

The modified initial conditions are as follows:

$$Y(0, \xi) = -(1 - \xi)$$

A point of caution is to change the initial conditions for the Y problem. The original conditions specified for c_A should not be inadvertently used for Y.

The Y problem can be solved by separation of variables and the composite solution is the sum of the steady state solution and the transient part given by the Y solution. The details are left as an exercise and the final solution is presented here:

$$c_A = 1 - \xi - \sum_{n=1}^{\infty} \frac{2}{n\pi} \sin(n\pi\xi) \exp(-n^2\pi^2\tau) \tag{8.24}$$

8.5.2   Transient Diffusion with Reaction

A second example where a similar prior change of variables is needed before we can use separation of variables is presented in the following. We consider the transient diffusion with reaction in a slab. For the reacting system a source term is added to the transient diffusion equation, resulting in the following equation for a first-order reaction in a slab geometry:

$$\frac{\partial C_A}{\partial t} = D_A \frac{\partial^2 C_A}{\partial x^2} - kC_A \tag{8.25}$$

The boundary conditions are the usual: no flux at $x = 0$ and a concentration of $C_A = C_{As}$ at $x = L$. The initial condition examined here is $C_A(\tau = 0) = 0$, the so-called start-up problem.

It should be noted here that the differential equation is homogeneous while the boundary condition is not. Hence the method of subtraction of the steady state solution is needed. Again it may be more useful to work in dimensionless coordinates. The governing equation in dimensionless form is:

$$\frac{\partial c_A}{\partial \tau} = \frac{d^2 c_A}{d\xi^2} - \phi^2 c_A \tag{8.26}$$

where ϕ is the Theile modulus defined as $L\sqrt{k_1/D_A}$. The dimensionless concentration c_A is defined here as C_A/C_{AS}.

The steady state solution was part of Section 2.3 and is

$$c_{A,s} = \frac{\cosh(\phi\xi)}{\cosh(\phi)}$$

Hence the transient part of the problem $Y = c_A - c_{A,s}$ is governed by

$$\frac{\partial Y}{\partial \tau} = \frac{d^2 Y}{d\xi^2} - \phi^2 Y \tag{8.27}$$

This can be verified to be a homogeneous differential equation; the boundary conditions for the Y problem are the following: $dY/dx = 0$ at $\xi = 0$ and $Y = 0$ at $\xi = 1$. Note that the original boundary conditions for the c_A problem were not homogeneous but the Y problem is. Hence separation of variables can be used for the Y problem.

The initial condition for the Y problem is

$$Y(0, \xi) = -\frac{\cosh(\phi\xi)}{\cosh(\phi)}$$

This initial condition is obtained by subtracting the steady state solution from prescribed initial condition for c_A. The original initial condition should not be applied inadvertently.

The Y problem can be solved by separation of variables. The solution details are left as an exercise since our goal here is to show how to recast non-homogeneous problems.

8.6   2-D Problems: Product Solution Method

Consider a transient diffusion problem in a 2-D rectangular slab schematically shown in Figure 8.8. The boundary conditions of the homogeneous type are prescribed along the perimeter.

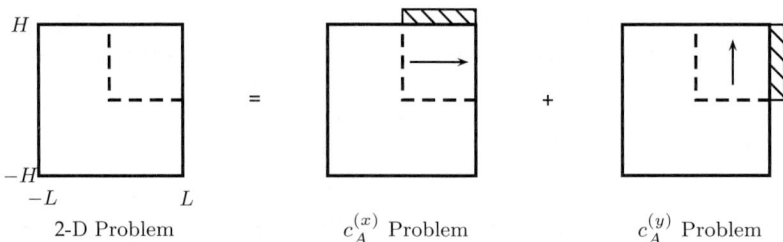

$c_A^{(x)}$ Problem $c_A^{(y)}$ Problem

2-D Problem

Figure 8.8   Product solution method: a problem in 2-D transient diffusion split into two simpler problems.

The differential equation is the 2-D transient diffusion equation:

$$\frac{\partial c_A}{\partial t} = D_A \left(\frac{\partial^2 c_A}{\partial x^2} + \frac{\partial c_A}{\partial y^2} \right) \tag{8.28}$$

This is a PDE in two spatial variables, x and y, and time and the solution by separation of variables is lengthy. A simpler method is available: the product solution method. The method applies if the differential equation and the boundary conditions are homogeneous.

We claim that a product solution of the following type holds:

$$c_A(x, y, t) = c_A^{(x)}(x, t) \times c_A^{(y)}(y, t) \tag{8.29}$$

where $c_A^{(x)}(x, t)$ is the subproblem with concentration varying in only the x-direction while $c_A^{(y)}(y)$ is the subproblem posed in the y-direction only. (Superscripts with brackets are used here for clarity since often the subscripts denote differentiation and superscripts can also be confused for the exponents.)

The preceding claim can be verified by direct substitution in the governing equations. Since the boundary conditions on c_A have been made homogeneous, we can also show that each subproblem satisfies the specified boundary conditions subproblems in the x- and y-directions.

Some caution has to be exercised in calculating the solutions to the subproblems as the length scales in the x- and y-directions may be different. Let $2L$ be the length along the x-direction and $2H$ be the length along the y-direction. We know in general that the concentration profile depends on the dimensionless distance, time, and the Biot number. These have to calculated separately for each subproblem.

Thus for the x-direction, we use the characteristic distance L and represent the subproblem in the x-direction parametrically as

$$c_A^{(x)}(x, t) = c_A^{(x)}(x/L, tD_A/L^2; k_m L/D_A) \tag{8.30}$$

that is, the dimensionless distance of x/L, the dimensionless time of tD_A/L^2 and the Biot number of $k_m L/D_A$ should be used for the calculation of $c_A(x,t)$.

Similarly for the y-direction the parametric representation is

$$c_A^{(y)}(y,t) = c_A^{(y)}(y/H, tD_A/H^2; k_m H/D_A) \tag{8.31}$$

The average concentration can be demonstrated to be the product of the solution to the subproblems as well:

$$\langle c_A \rangle (t) = \langle c_A \rangle^{(x)} (t) \langle c_A \rangle^{(y)} (t) \tag{8.32}$$

The fractional mass loss $(1 - \langle c_A \rangle)$ from the slab at any time t can be calculated as

$$\Phi = \Phi^{(x)} + \Phi^{(y)} - \Phi^{(x)}\Phi^{(y)} \tag{8.33}$$

where $\Phi^{(x)} = 1 - \langle c_A \rangle^{(x)}$ with $\Phi^{(y)}$ defined similarly.

The method is also applicable to a finite cylinder, that is, a case where the length of the cylinder is comparable to the radius. For a finite cylinder the concentration profile can be found by product solutions of the form

$$c_A(r,z,t) = c_A^{(r)}(r,t) \times c_A^{(z)}(z,t) \tag{8.34}$$

The r-subproblem is computed by using the transient diffusion solution for a long cylinder. The radius of the cylinder is used as the characteristic length for this case. The z-subproblem is computed using the 1-D slab solution. Half the height of the cylinder is used as the reference length here.

The method can also be used for 3-D transient diffusion problems in the geometry of a cube. The solution to the 3-D case is the product of the solutions to the subproblems in x, y, and z.

8.7   Semi-Infinite Slab Analysis

The series solution is valid for all time values. However, at sufficiently short times the concentration changes are localized near the surface of the slab or solid. The mass does not diffuse out from the interior of the solid during this time and the concentration stays nearly at its initial value. For these small values of time ($\tau < 0.05$ is taken as the criteria for the finite slab case analyzed in Seciton 8.2), the series solution converges rather slowly and many terms need to be included. In such situations, it is easier to treat the problem as a semi-infinite media and separate closed form analytical solutions can be derived. This is taken up in the following discussion.

8.7.1   Constant Surface Concentration

The problem considered here is stated as follows: a semi-infinite slab has an initial concentration C_{Ai} and at $t > 0$, the surface of the slab at $x = 0$ is exposed to a concentration C_{As} and maintained at that value subsequently. The transient concentration profile developed in the slab and these profiles have to be computed.

The dimensionless concentration for this problem is commonly defined as

$$Y = \frac{C_A - C_{Ai}}{C_{As} - C_{Ai}} \tag{8.35}$$

The definition is different from that in Section 8.2 where c_A was used. Note that $Y = 1 - c_A$. The initial concentration for Y is thus made zero here, which is convenient for mathematical manipulations.

Note: The variable Y in the previous equation should not be confused with Y the deviation concentration in Section 8.5. Use of the same symbol in different contexts is common in engineering mathematics as long as the contexts are separate and no confusion can result.

The governing equation for Y is

$$\frac{\partial Y}{\partial t} = D_A \frac{\partial^2 Y}{\partial x^2} \tag{8.36}$$

Note that the distance and time has not been normalized, only concentration. This is because there is no reference length scale or time scale obvious for the problem. But we note that the following group of variables is dimensionless:

$$\eta = \frac{x}{2\sqrt{D_A t}} \tag{8.37}$$

The factor 2 in the denominator is only for later numerical convenience. You should verify that η is dimensionless. It is called the similarity variable.

Now the solution should be expected to a function of η only. This is known as a similarity principle. Equation 8.36 can be shown to reduce to an ordinary differential equation of the following form:

$$\frac{d^2 Y}{d\eta^2} + 2\eta \frac{dY}{d\eta} = 0 \tag{8.38}$$

Detailed derivation of this equation is obtained from Equation 8.36 by applying the chain rule but the derivation is not quite important to the current theme of deriving an equation for the transient profiles.

The boundary and initial conditions for the Y problem are the following:

- At $\eta = 0$ (surface), $Y = 1$.
- At $\eta \to \infty$ (far away from surface), $Y = 0$.
- For $t = 0$, which also corresponds to $\eta = \infty$, $Y = 0$.

Note that the initial condition and the far away distance condition merge into a single condition, which is a characteristic feature of problems with similarity transformation. Hence there are only two conditions used.

The first integration of Equation 8.38 is straightforward and leads to

$$\frac{dY}{d\eta} = A \exp(-\eta^2)$$

with A the integration constant. The second integration has no analytical closed form solution and is written as

$$Y = A \int_0^\eta \exp(-u^2)du + B$$

where u is a dummy variable. The integration constant B is found to be 1 by applying the boundary condition at $\eta = 0$. Hence the solution is

$$Y = 1 - A \int_0^\eta \exp(-u^2)du$$

Applying the boundary conditions at $\eta = \infty$ we find

$$A = \int_0^\infty \exp(-u^2)du$$

This integral has a value of $2/\sqrt{\pi}$, which provides the value of A. Hence the solution is

$$Y = 1 - \frac{2}{\sqrt{\pi}} \int_0^\eta \exp(-u^2)du$$

The final result can be expressed in a compact form using the "error" function (erf):

$$\mathrm{erf}(\eta) = \frac{2}{\sqrt{\pi}} \int_0^\eta \exp(-u^2)du$$

The final solution is

$$Y = 1 - \mathrm{erf}(\eta) = 1 - \mathrm{erf}\left(\frac{x}{2\sqrt{D_A t}}\right) \qquad (8.39)$$

Since $1 - \text{erf}(\eta)$ is the same as $\text{erfc}(\eta)$, the complementary error function, the solution can also be expressed as

$$Y = \text{erfc}(\eta) \tag{8.40}$$

Note that the MATLAB function *erfc* can be used to calculate the complementary error function.

This completes the solution for the concentration profile in a semi-infinite slab with a constant surface concentration. The solution expressed in terms of dimensional variables is

$$\frac{C_A - C_{Ai}}{C_{As} - C_{Ai}} = \text{erfc}\left(\frac{x}{\sqrt{4D_A t}}\right) \tag{8.41}$$

The instantaneous flux into the surface is given as

$$N_{As} = (C_{As} - C_{Ai})\sqrt{\frac{D_A}{\pi t}} \tag{8.42}$$

The surface flux decreases with a square root of time dependency.

The average flux over a period of time t_E is given as

$$\bar{N}_{As} = \frac{1}{t_E}\int_0^{t_E} N_{As}dt = (C_{As} - C_{Ai})\left(2\sqrt{\frac{D_A}{\pi t_E}}\right) \tag{8.43}$$

This expression is important in the context of mass transfer at interfaces exposed to a short contact time. The resulting model is known as a penetration model and elaborated in Section 8.8.

An illustrative plot of the concentration profile is shown in Figure 8.9. The profile is of a boundary layer type with only a certain region near the surface coming under the influence of the surface change in concentration. This thickness of this region increases as time progresses.

We now discuss the integral method, another useful solution method. The Laplace transform method is another useful method discussed in Chapter 21.1.3 in the context of transient diffusion in a semi-infinite region with a first-order reaction. This can be studied at this stage as well.

8.7.2   Integral Method

The integral method assumes that there is a penetration depth of λ beyond which the solute concentration is zero. The λ is a function of time (see Figure 8.10). The boundary condition of $Y = 0$ is used at λ. Further, since

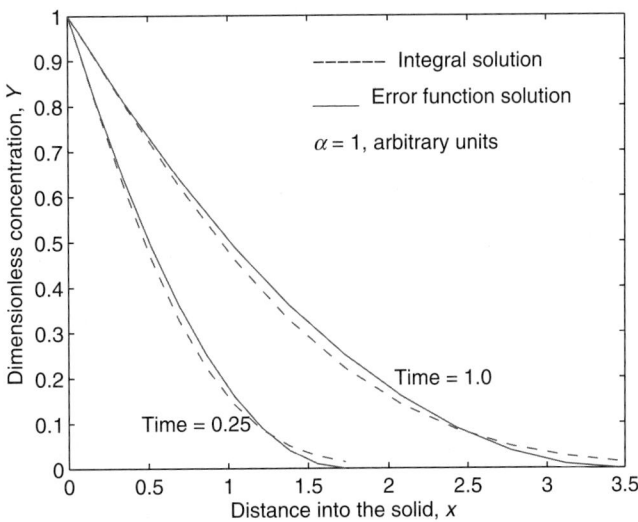

Figure 8.9 Concentration profiles in a semi-infinite slab; time in arbitrary units.

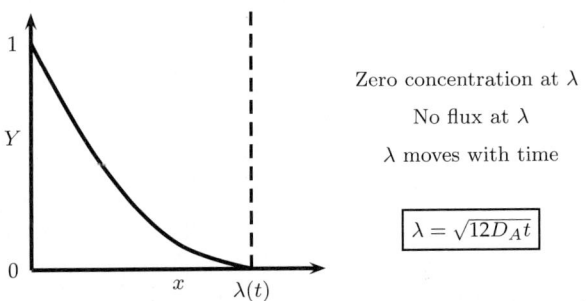

Figure 8.10 Basis for integral method: a concentration boundary layer of thickness λ is proposed; this moves inward with time as more of the slab becomes contaminated.

no solute moves beyond this, dY/dx is also set as zero at $Y = \lambda$. Of course $Y = 1$ at $x = 0$. These three conditions permit us to approximate the concentration profile using a quadratic function. A function which satisfies all three conditions is

$$Y = \left(1 - \frac{x}{\lambda}\right)^2 \tag{8.44}$$

Equation 8.36 is now applied in an integral sense:

$$\int_0^{\lambda(t)} \frac{\partial Y}{\partial t} \, dx = \int_0^{\lambda(t)} D_A \frac{\partial^2 Y}{\partial x^2} \, dx \tag{8.45}$$

Substitution of the Y approximation in the integral representation given by Equation 8.45 and further algebraic manipulations using the Leibnitz rule for the left-hand side leads to the following differential equation for λ:

$$\frac{1}{3} \frac{d\lambda}{dt} = D_A \frac{2}{\lambda} \tag{8.46}$$

Integrating, we obtain an expression for variation of the penetration depth with time:

$$\lambda = \sqrt{12 D_A t} \tag{8.47}$$

Thus the concentration profile creeps inward with a square root dependency on time. Note that such a square root dependency is common to many moving boundary problems. The corresponding concentration profile is now obtained from Equation 8.44.

The surface flux can also be calculated. dY/dx at $x = 0$ is given as $-2/\lambda$ from the quadratic profile given by Equation 8.44. Substituting $\sqrt{12 D_A t}$ for λ, we find dY/dx at the surface to be $-1/\sqrt{3 D_A t}$. This gives a surface flux value of $\sqrt{D_A/3t}$ times the concentration difference. The numerical factor is 3.14 (π) in the exact solution given by Equation 8.42 instead of 3 in the integral method. Hence the error in mass flux at the surface flux is on the order of 3% by comparing with the exact result based on the error function. Therefore, the results of the integral method, although approximate, provide a simple solution method for problems posed in the semi-infinite domain. Further the integral method can be more easily applied to other cases such as variable diffusivity, simultaneous chemical reaction, and so on, where simpler analytical solutions based on the similarity method may not be possible.

8.7.3   Pulse Response

The pulse response is also of interest in some applications. The boundary condition at the surface is now a pulse that can be mathematically represented as a Dirac delta function. The solution is stated as follows without derivation:

$$C_A - C_{Ai} = \frac{S}{\sqrt{\pi D_A t}} \exp\left(-\frac{x^2}{4 D_A t}\right) \tag{8.48}$$

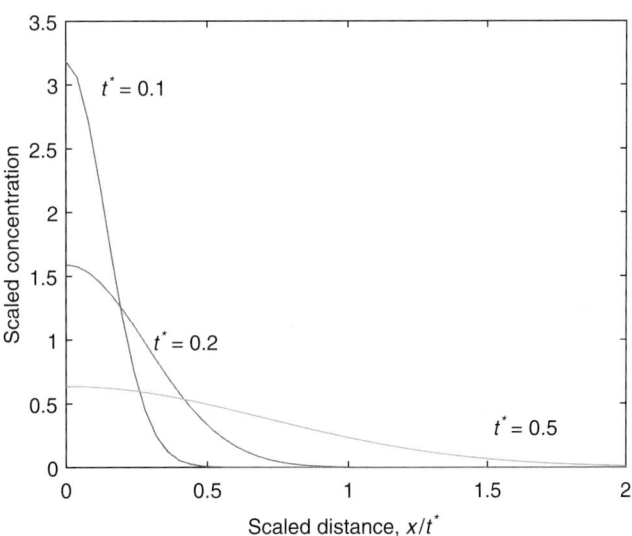

Figure 8.11 Concentration (scaled by pulse strength) response to a pulse applied at the surface of a semi-infinite region. $t^* = \sqrt{D_A t}$.

Here S is the pulse strength, the quantity of solute placed on the surface as pulse at time zero.

Surface concentration is maximum at time zero and decreases thereafter. The surface mass flux is zero at all times greater than zero. Figure 8.11 illustrates typical concentration profiles for a pulse response. As time progresses, the concentration front moves inward but the value of the concentration decreases along the distance as well. The surface concentration is maximum at the initial time and decreases with time as the concentration front starts to move inward. Example 8.3 shows an application of to electronic material processing.

Example 8.3   Doping of Silicon with Phosphorus

Silicon is doped with phosphorus under constant surface conditions and as a pulse introduced at time zero. Calculate the concentration profiles as a function of time. Assume the initial concentration C_{Ai} is zero and the surface concentration is one, that is, all concentrations are scaled.

Solution

The diffusion coefficient of phosphorus in silicon is reported as 6.5×10^{-13} cm^2/s at 1100°C by Middleman and Hochberg (1993).

The penetration depth is where the error function takes a value of 0.99. This corresponds to an argument of η of 1.8. Hence at time $t = 3600$ sec the phosphorus would have diffused up to $1.8 \times 2\sqrt{Dt}$, which is equal to $1.74\ \mu$m.

At a location half this distance the parameter η is 0.9. The corresponding value of erfc is 0.2031. Hence phosphorus concentration at this location is 20.31% of the surface value. Similar calculations can be done at other positions and the profiles can be computed for any given value of time.

The total quantity of phosphorus diffusing (per unit area) is equal to $\bar{N}_{As} \times t$, which is equal to $\sqrt{4Dt/\pi}$, which is 5.45×10^{-5} times the actual surface concentration.

If the doping were done with a pulse of strength $S = 5.45 \times 10^{-5}$, then the concentration at $0.87\ \mu$m would be calculated from Equation 8.48 as 0.2832. The concentration with the pulse source is larger than that with the step source at the same location for a given value of time; this provides a means of controlling the junction thickness.

8.8   Penetration Theory of Mass Transfer

In Section 6.3 we discussed a film model for mass transfer. This assumes that a steady state concentration profile is established near the interface where mass transfer is taking place. There are however many situations where the steady state profile does not quite get established. For example, consider the case of mass transfer from a bubble to a liquid. Mass transfer can take place only when the bubble is in contact with the liquid, which receives mass only for this time; hence the mass transfer is viewed as a transient process. The penetration model proposed by Higbie in 1932 is an attempt to describe mass transport using the transient diffusion model.

In this model the liquid element is assumed to be in contact with a bubble (or a gas phase) for a certain time t_E, the exposure time. Mass transfer takes place up to this time. At the end of this time, the eddies mix the liquid, a fresh liquid is exposed, and the process starts again with the arrival of the next bubble. The process envisioned in the modeling is illustrated in Figure 8.12.

The mass transfer rate over this "exposure" time can then be calculated using the transient diffusion model in a semi-infinite domain. The mass transfer coefficient is calculated by dividing the average flux given by

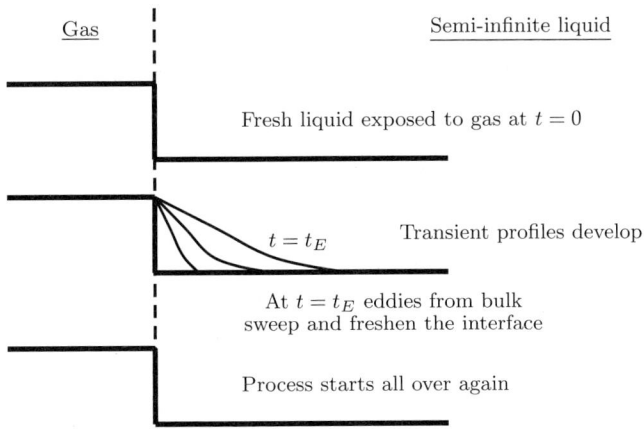

Figure 8.12 Snapshot of events visualized in the penetration model for mass transfer.

Equation 8.43 by the driving force; the following expression results for the mass transfer coefficient:

$$k_L = 2\sqrt{\frac{D_A}{\pi t_E}} \qquad (8.49)$$

An application of this theory is to calculate the mass transfer coefficient from a bubble to a liquid. An estimate of the exposure time is needed and in the context of bubbles it is taken as the diameter of the bubble divided by the bubble rise velocity, that is, the time it takes for the bubble to travel past a distance equal to its diameter. The liquid is assumed to be mixed again after this and a fresh bubble rises through the pool of liquid. This method provides values for the mass transfer coefficient that are in reasonable agreement with the experimental data. Another important application is to interpret the effect of a reaction based on the penetration theory; this is discussed in some detail in Chapter 21.

8.9   Transient Diffusion with Variable Diffusivity

Problems where the diffusivity is a function of concentration is of importance in many applications, including fabrication of micro-electronic devices. In this section we introduce a mathematical method useful for solving these problems.

The problem can be stated in a slab geometry as

$$\frac{\partial C_A}{\partial t} = \frac{\partial}{\partial x}\left(D(C_A)\frac{\partial C_A}{\partial x}\right) \tag{8.50}$$

A power law model for the diffusivity is often used in semi-conductor device analysis:

$$D(C_A) = D_0[1 + BC_A^n]$$

where D_0 is the limiting value of diffussivity at low concentration, B is a constant for concentration dependency, and n is an index; $n = 0$ corresponds to a constant diffusivity. For example, arsenic diffusion in Si is modeled with an index of one (Middelman and Hochberg, 1993). A solution in a semi-infinite region is often needed and analytical solutions are rare.

The dimensionless form for a power law with first-order diffusion variation ($n = 1$) is

$$\frac{\partial c_A}{\partial t} = \frac{\partial}{\partial x}\left((1 + \beta c_A)\frac{\partial c_A}{\partial x}\right) \tag{8.51}$$

where the diffusion coefficient is assumed to be a linear function of concentration. The coefficient of proportionality is β and the constant diffusivity model is represented with $\beta = 0$.

The boundary conditions are $x = 0, c_A = 1$, and as $x \to \infty, c_A = 0$, which is also the initial condition.

The problem posed in a semi-infinite region can be cast in terms of the similarity variable η defined as $x/\sqrt{4D_0 t}$. Equation 8.51 reduces to an ordinary differential equation. Using the chain rule for differentiation on both sides of Equaiton 8.51 we find

$$-2\eta\frac{dc_A}{d\eta} = \frac{d}{d\eta}\left((1 + \beta c_A)\frac{dc_A}{d\eta}\right)$$

which can be rearranged to the following second-order nonlinear but ordinary differential equation:

$$\frac{d^2 c_A}{d\eta^2} + \frac{\beta}{1 + \beta c_A}\left(\frac{dc_A}{d\eta}\right)^2 + 2\frac{\eta}{1 + \beta c_A}\frac{dc_A}{d\eta} = 0$$

Unlike the constant diffusivity case this equation can not be solved analytically and a numerical solution is needed. The equation was solved in MATLAB using BVP4C and the plot is shown in Figure 8.13.

One key idea here was to demonstrate use of the similarity transformation to simplify the computations even for nonlinear diffusivity as the

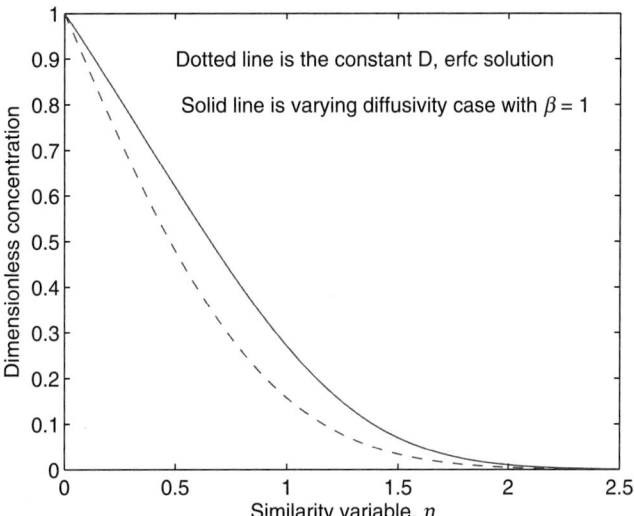

Figure 8.13 Concentration profile for nonlinear diffusion as a function of the similarity variable. A linear variation is used and compared with the constant diffusivity case.

procedure reduces the PDE to an ODE. The ODE is easier to solve numerically than a PDE. Also note that the initial condition and the condition at ∞ merge into a single condition, which is needed for the similarity method to apply.

8.10   Eigenvalue Computations with CHEBFUN

This section may be omitted without loss in continuity.

Eigenfunctions can be readily computed using CHEBFUN with MATLAB as it has an overloaded *eig* function. The following code solves for the eigenfunctions of the Robin problem shown in the text in Section 8.3. The code with minor modifications can be used for many other cases, for example, cylinder and sphere geometries. Code can also readily evaluate the series coefficients. Thus the whole procedure of the method of separation of variables can be automated with code similar to this. You will be able to solve or verify many of the problems in this chapter with this code and also be able to use it in your research.

Listing 8.1 Series Solution with CHEBFUN

```
% Solves A y = L B y $ where A and B are linear differential operators.
xi = chebfun('xi',[0,1]);
A = chebop(0,1);
A.op = @(xi,u) diff(u,2) ;

A.lbc = @(u) diff(u,1)   % Neumann
Biot = 1.0
A.rbc =  @(u) diff(u,1) + Biot* u % Robin
B = chebop(0,1);
B.op = @(xi,u) —u ;

% Then we find the eigenvalues with eigs.
[F,L] = eigs(A,B)
omega = sqrt(diag(L)) % eigenvalues

% Series coefficient can be evaluated
for i = 1:6
    center = F(0,i);
N1 =center* int ( F(:,i)  );
 N2 = int ( F(:,i).* F(:,i)  );
 B1(i)= int ( F(:,i)  )/center
 A1(i)= N1/N2 % Answer C1=  1.119
end
```

The problem coded here is the slab with the Robin condition examined in Section 8.3. On running the code you should get the following results for Biot = 1.

The first six eigenvalues will be computed as

$$0.8603 \quad 3.4256 \quad 6.4373 \quad 9.5293 \quad 12.6453 \quad 15.7713$$

The first six series coefficients for Biot = 1 will be calculated as

$$1.1191 \quad -0.1517 \quad 0.0466 \quad -0.0217 \quad 0.0124 \quad -0.0080$$

It is only a matter of summing the series for any chosen time to find the concentration profiles. You can write a small piece of code and generate time-concentration plots for any given Biot number.

Extension to the cylinder and sphere and other complicated PDEs is trivial and therefore you will find this to be a powerful method to solve linear PDEs.

8.11   Computations with PDEPE Solver

The PDEPE routine solves the initial-boundary value problems for parabolic-elliptic PDEs in 1-D. Small systems of parabolic and some elliptic PDEs in one space variable x and time t can be solved to modest accuracy and are a good practice tool to understand the behavior of such systems.

The general structure of the program is a solver for PDEs of the following type:

$$c\frac{\partial u}{\partial t} = \frac{1}{x^s}\frac{\partial}{\partial x}\left(x^s F\right) + S = 0 \qquad (8.52)$$

Here s is equal to 0, 1 and 2 for slab, infinitely long cylinder, and sphere geometries, respectively. The solution is sought for u as a function of distance x and time t. Note that these can be actual or dimensionless variables and the same symbol is used here for generality. In the present context u is the concentration, either actual or dimensionless. Note that working with dimensionless variables is often easier and the results are also more general.

Note that multiple PDEs can be solved and where u becomes the "solution vector" with components u_1, u_2, and so on. For example simultaneous solution of concentration and temperature profile within a porous catalyst can be accomplished using this code. The other terms in Equation 8.52 are explained in the following.

The coefficient in the time derivative, c, is a capacity term and is in general a function of x, t, u, and p where p is the gradient of u defined as du/dx. In general, c is a diagonal matrix when multiple equations are being solved. The diagonal elements of the matrix c are either identically zero or positive. An entry that is identically zero corresponds to an elliptic equation and otherwise to a parabolic equation. There must be at least one parabolic equation. For the transient diffusion problems examined here $c = 1$.

F is the mass flux term and is defined as $F = D(\partial u/\partial x)$ It is the negative of the Fick's law flux. Note the value of the diffusion coefficient appears inside Equation 8.52 and hence variable diffusivity cases can also be solved. Hence D can be defined as a function of position and the concentration for such problems.

Finally $S = S(x, t, u, p)$ is the source term, which can be any linear or nonlinear function of the variables. This corresponds to the R_A term in this chapter. Thus the general nonlinear reaction case can be handled.

The calling statement is

```
SOL = PDEPE (shape, PDEFUN, ICFUN, BCFUN, XMESH,TSPAN)
```

The variables in the calling statement are defined as follows:

1. *shape* is the parameter that defines the geometry and must be equal to either 0, 1, or 2, corresponding to slab, cylindrical, or spherical symmetry as earlier in Equation 8.52.
2. PDEFUN is the function that evaluates the quantities defining the differential equation.
   The calling statement is [C,F,S] = PDEFUN(X,T,U,DUDX). The input arguments are scalars X and T and vectors U and DUDX, which approximate the solution and its partial derivative with respect to x, respectively. PDEFUN returns column vectors C (containing the diagonal of the matrix C(x,t,u,dudx)), F, and S (representing the flux and source term, respectively). For constant diffusivty, the "flux" term is usually written as F = dudx. Note that here we assume a dimensionless formulation and hence the diffusivity does not appear as an explicit term.
3. ICFUN is the function that defines the initial conditions and the structure of this function is as follows: U = ICFUN(X) . When called with a scalar argument X, ICFUN evaluates and returns the initial values of the solution components at X in the column vector U.
4. BCFUN defines the boundary conditions. These are defined as the left and right boundaries in the general form A + B F = 0, where A and B are in general functions of the values of x, t, and u and B is a diagonal matrix.
5. XMESH defines the mesh over which the solution is calculated. XMESH can be generated using the LINSPACE function if equal spacing in x is used.
6. TSPAN is the time values at which the solution is to be calculated.

8.11.1 Sample Code for 1-D Transient Diffusion with Reaction

Sample code is presented in the following that is useful for getting started. You can modify this for a wide range of problems. The problem considered is transient diffusion with a first order chemical reaction in a sphere:

$$\frac{\partial u}{\partial t} = \frac{1}{x^2}\frac{\partial}{\partial x}\left(x^2 F\right) - \phi^2 u \tag{8.53}$$

Here $F = \partial u/\partial x$, assuming a constant diffusion coefficient.

Note all variables are dimensionless although the symbols t and x are used.

The boundary conditions are (1) symmetry at $x = 0$ and (2) constant concentration, Dirichlet condition at $x = 1$. The initial condition is taken as $u = 0$.

The components to set up the problem and the function routine needed are shown in the following. The main calling block is presented as Listing 8.2 and can be used as a common block for many problems.

Listing 8.2 Main Section for PDEPE SOLVER

```
% test problem given by Equation 8.53
shape = 2;  % sphere.
% meshes in the x-direction created with linspace function
xnodes = 21 ; xmesh= linspace (0,1,xnodes);
% time intervals at which solution is ought
t = [0   0.2  0.4  0.6   0.8 1.0   1.25 1.5 1.75 2.00];
% call the function to generate the solution
sol = pdepe(shape,@pdeproblem,@pdeic,@pdexbc,xmesh,t);
```

Note solution (*sol*) is a vector of three dimensions. It is of the form $u = sol(i,j,Neq)$. The first index is the time node, the second index j is the x node, and the third is the dependent variable number. If there is only one variable, then $Neq = 1$. If two PDEs are solved simultaneously then Neq can be called with either 1 or 2 to find the u_1 and u_2 solutions.

The derivative of the solution can be calculated using the PDEVAL function. An alternative is to use the three-point finite difference formula shown in the following calling statements:

$$\left(\frac{du}{dx}\right)_{\text{at node } i} = \frac{u(i-2) - 4u(i-1) + 3u(i)}{2\Delta x}$$

where Δx is the node spacing. The following lines of code can be used to implement this.

Listing 8.3 Three-Point Finite Difference to Find the Derivative at the Surface

```
% flux at the surface
  Delx = 1/(xnodes−1);
  u1 = sol(j,xnodes−2,1);
  u2 = sol(j,xnodes−1,1); %
  u3 = sol(j,xnodes,1); % surface node
  f1(j) = (u1−4*u2+3*u3 )/2/Delx;
```

This will calculate the derivative of the solution at time j using the solution values at three nodes near the surface. Derivatives are not very accurate if the x-mesh is small even though the concentration predictions are closer. (See the results section as well.)

The other m-files needed are listed below.

Listing 8.4　PDEPROBLEM File for the Test Problem

```
%————Problem definition ——————
function [c,f,s] = pdeproblem(x,t,u,DuDx)
% This defines the capacity term, f (flux term) and the source term.
c = 1 ;    % capacity term
f =   DuDx;        % flux term is defined (constant D case)
phisq = 4;
s = —phisq*u ;          % source term
```

Initial Conditions

The initial conditions are defined as a function of x as shown in the following m-file:

```
function u0 = pdeic(x);
u0 = 0.000 ;    % initial conditions
```

Finally, boundary conditions are defined in a file pdebc as shown in the following. For the left boundary pl and ql are used in the definition and the boundary condition is restated as

$$pl + ql f = 0$$

where f is the flux vector. For example, $pl = 0$ and $ql = 1$ will provide a no flux boundary condition.

Similarly the right-side $(x = 1)$ conditions are specified by providing values for pr and qr. For example, $pr = ur - 1$ and $qr = 0$ will make the boundary condition at $x = 1$ go to $u = 1$.

Listing 8.5　Boundary Condition m-file for the Test Example

```
function [pl,ql,pr,qr] = pdebc(xl,ul,xr,ur,t)
% Left boundary conditions (Neumann; symmetry)
pl = 0;
ql = 1;    % no flux condition
```

```
% Right boundary condition (Dirichlet type)
 pr = [  ur(1)-1.0 ];
 qr = 0;
% If Robin applied use the following:
% Bulk concentration is cb here.
% pr = [biot* ur(1)-cb ];
% qr = 1;
```

Results for Transient Diffusion with Reaction in a Sphere

The computed results are shown in Table 8.2 for a first-order reaction.

The center concentration should be 0.5514 but the code gives 0.5418 showing some discretization error. This is due to crude spatial discretization. We used only six points here; if you increase to 101 points the exact solution is recovered. One should always perform these mesh dependency studies and check the results. Testing on a simpler problem while having an analytical solution in hand is useful for such studies before undertaking a more complex case.

The steady state flux values calculated for different levels of x-discretization are shown in Table 8.3.

Table 8.2   PDEPE Results for Transient Diffusion with Reaction in a Sphere; $\phi = 2$

τ	$\xi = 0$	0.2	0.4	0.6	0.8	1.0
0	0	0	0	0	0	1.0000
0.1	0.2155	0.2445	0.3436	0.5035	0.7258	1.0000
0.5	0.5398	0.5546	0.6051	0.6887	0.8160	1.0000
1.0	0.5418	0.5564	0.6065	0.6897	0.8164	1.0000
2.0	0.5418	0.5564	0.6065	0.6897	0.8164	1.0000

Table 8.3   Steady State Surface Flux Values as a Function of Level of Spatial Nodes

N	11	21	101
flux	1.0701	1.0733	1.0746

The flux at the surface should have an analytical value of 1.0746 and the results are quite close. The built-in PDEVAL calculator did not give the proper value of the fluxes and the three-point finite difference formula appears to be more accurate.

Summary

- Transient problems lead to partial differential equations involving time and spatial variables. The simpler problems use only one spatial variable and are posed in ideal slab, long cylinder, and sphere geometries. Further constant physical properties and a constant or linear rate of production is assumed, which leads to a linear PDE. The governing equations can be represented in a compact form in terms of dimensionless variables.

- Linear PDEs with homogeneous boundary conditions can be solved by separation of variables. The general properties of such systems are the eigencondition (the solution of which gives the eigenvalues of the problem), eigenfunctions, and representation of the solution as a series. The series coefficient can be obtained by Fourier expansion of the initial condition in terms of the eigenfunctions. The procedure can be applied to all three geometries. The eigenfunctions and other properties are geometry and the boundary condition dependent but the general solution procedure form of the series solution are the same.

- The eigenvalues for the Dirichlet problem in a slab are simple but the eigenvalues for the Robin problem have to be obtained as the solution of a transcendental equation. This difference is worth noting. The eigenvalues for the Robin problem are functions of the Biot number. For small Biot numbers the concentration profile within the slab is nearly uniform and a lumped parameter model (a macroscopic model for the solid as a whole) can be used as a simple model.

- Important properties of the series solution should be noted. The series converges rather slowly for small values of time while for large values of dimensionless time only one term is sufficient. Since the series converges slowly, a semi-infinite model is more suitable for small values of time.

- Many 2-D problems can be split into two 1-D problems and the results for the separated 1-D problems can be combined and used for such cases. The 2-D result is a product of the separated 1-D problems; hence the method is also called the product solution method.

- Series solutions can be implemented with computer algebra tools such as MAPLE. This removes the tedium of doing lengthy algebraic manipulations. The computation of the eigenvalues can also be accomplished using CHEBFUN. The use of these tools however requires experience; hence the study of some standard problems should be attempted first.

- Many transient problems are posed in a semi-infinite domain. This approach is suitable when the depth of penetration of the concentration front is much smaller than the domain dimension. The mathematical problem can then be reduced to an ordinary differential equation in terms of the transformed variable. For constant surface concentration the solution is the error function of a similarity variable. For a pulse input at the surface, the solution is a half-Gaussian function. These solutions find applications, for example, in electronic material processing to find dopant profiles.

- Mass transport between a gas-liquid can be modeled by a transient diffusion equation if the contact time between the gas and liquid is short. This leads to a model for interfacial mass transfer known as the penetration model. This has wide application, especially for reacting systems, and is often used as an alternative to the film model.

- Transient diffusion with a variable diffusion coefficient in a semi-infinite medium can be modeled using a similarity transformation method. There is no true similarity and hence the final computations have to be done numerically. However, the transformation makes the numerics somewhat easier.

- The MATLAB tool PDEPE is a versatile method for solving transient diffusion problems. Nonlinear diffusion, nonlinear source terms, and nonlinear boundary conditions can be readily incorporated. The solution can be used for all three basic geometries by simply changing the shape parameter, s, in the code. The code can also handle multiple differential equations. A worked example presented in the chapter provides a learning tool for this case. Students will be in a position to analyze and solve a wide range of transient transport problems with the modification of the code shown in the text.

Review Questions

8.1 How is diffusion time defined? What is the significance?

8.2 What is meant by an eigenvalue problem?

8.3 What is meant by an eigencondition? State one example.

8.4 What is meant by orthogonal functions? How do they help us in solving PDEs?

8.5   How does the mass flux vary with time in the initial time period for the slab problem examined in Section 8.2?

8.6   How does the mass flux vary with time in the longer time period for the slab problem examined in Section 8.2?

8.7   State the physical meaning of the Biot number.

8.8   Is external mass transfer the main resistance for large Biot numbers?

8.9   The Sherwood number has a similar grouping of variables as the Biot number. What is the difference between these two dimensionless groups?

8.10 Can the separation of variables method be used for non-homogeneous problems?

8.11 What is the product solution for 2-D transient problems?

8.12 What is the definition of the error function?

8.13 State the definition of the similarity variable used in transient diffusion in a semi-infinite medium.

8.14 If a pulse is applied at a surface of a semi-infinite region, what would the transient concentration profile look like?

8.15 How does the mass transfer coefficient vary with the diffusion coefficient in the context of the penetration model?

Problems

8.1   **Separation of variables: exponential decay in time.** Assume that the product solution indicated in the following holds as shown in classical mathematical books:

$$c_A = F(\xi)H(\tau)$$

where F is a function of position only and H is a function of time only. Use this in Equation 8.5 and verify that the time dependency is indeed an exponentially decaying function as assumed in the text.

8.2   **Eigenvalues for the Robin problem.** Solve the eigenvalue problem for the slab with the Robin condition at the surface and verify that the eigenvalues are the solutions to the transcendental equation given by Equation 8.18.

8.3   **Eigenvalue differential equation for long cylinder and sphere.** Derive the differential equations for the F function for the long cylinder and sphere cases. Refer to your engineering mathematics textbooks and write out the general solution to the equation. State the boundary conditions for F and verify the expression for the eigenvalues and the eigenfunctions shown in the text.

8.4   **Lumped model for low Bi.** For low Biot numbers the concentration inside the system is nearly equal to $\langle c_A \rangle$. The convection loss at the boundary is the controlling resistance. Show that a macroscopic balance for the solid leads to the following equation:

$$V_p \frac{d \langle C_A \rangle}{dt} = -k_m S_e (\langle C_A \rangle - C_{A\infty})$$

V_p is the volume of the solid and S_e is the external surface area. Integrate this equation to get the average concentration as a function of time. Write your results in dimensionless form and verify Equation 8.21 for a slab geometry.

8.5 Use of diffusion time. Find the time needed to remove 95% of a solvent in polymeric film. The film is attached to an impermeable surface at the bottom and has a thickness of 2 mm. The diffusion coefficient of the solvent in the polymer matrix is estimated as 4×10^{-11} m^2/s.

8.6 Measurement of diffusivity by fitting the transient release data. The following data were obtained for drug released from a spherical capsule of 4 mm as a function of time. Initially the drug concentration in the capsule is uniform at 70 mg/cm^3. The data can be used to calculate the diffusion coefficient of the drug. How would you plot the data to estimate the value of the diffusion coefficient? Determine its value.

t, hrs	3.7	7.4	11.1	14.8	18.5	22.2	25.9
mg/hr	1.16	0.596	0.350	0.212	0.1298	0.0797	0.048

8.7 Center concentration in a cylinder. A porous cylinder, 2.5 cm in diameter and 80 cm long, is saturated with alcohol and maintained in a stirred tank. The alcohol concentration at the surface of the cylinder is maintained at 1%. The concentration at the center is measured by careful sampling and is found to drop from 30% to 8% in 10 hours. Find the center concentration after 15 and 20 hours.

8.8 Slab with Dirichlet-Dirichlet condition. Consider the problem studied in Subsection 8.5.1. Solve the resulting eigenvalue problem for Y. Verify that the eigenvalues are multiples of π and the eigenfunctions are $\sin(n\pi\xi)$. Evaluate the series coefficients as well and verify the solution given by Equation 8.24.

8.9 Effect of chemical reaction. Consider the problem of transient diffusion in a slab accompanied by a first-order chemical reaction. and focus on the Y problem given by Equation 8.27. Show by direct substitution that the time-dependent part of the solution should be $\exp(-[\lambda^2 + \phi^2]\tau)$. Thus $\exp(-\lambda^2\tau)$, which is used for the no-reaction case, should not be used. Use this to derive the eigenvalue problem and find the eigenvalues, eigenfunctions, and series coefficient, thereby completing the solution.

8.10 Use of product solution: release rate from a cube. Consider the drug release problem in the text (Example 8.2) but now the drug is in a form of a cube with 6 mm sides. Find the center concentration after 3 hours and the fraction of drug released.

8.11 **Use of product solution for a short cylinder.** A drug tablet has a diameter of 4 mm and a length of 4 mm and has an initial drug concentration of 68.9 mg/cm^3. The diffusion coefficient of the drug in the porous matrix is 3×10^{-6} cm^2/s. Calculate and plot the center concentration at a function of time using the product solution method. Also find the fraction of the drug released as a function of time.

8.12 **Semi-infinite slab: surface flux.** Verify Equation 8.42 by applying Fick's law at $x = 0$. Integrate the flux over a contact time t_E and verify Equation 8.43 for the average flux over this time period.

8.13 **Transient oxygen profiles in a deep tank.** A deep tank filled with oxygen has a initial concentration of oxygen of 2 mol/m^3. The surface concentration on the liquid side of the interface is changed to 9 mol/m^3 and maintained at this value. The diffusion coefficient of oxygen is 2×10^{-9} m^2/s in water. Calculate and plot the concentration profile for time = 3600 and 36,000 s. Find the rate of oxygen transport at these times. Find the total oxygen transferred starting from the beginning to the end of the above time period.

The penetration depth of oxygen is defined as $\sqrt{12 D_A t}$ and is the measure of the depth up to which the oxygen concentration has changed. Calculate the values at the two times indicated in the problem statement.

8.14 **Arsenic diffusion into silicon.** An arsenic thin film is laid down on the surface of silicon and arsenic diffuses into silicon, forming a p-type material with a diffusion coefficient of 5×10^{-13} cm^2/s. The saturation solubility of arsenic in silicon is 2×10^{21} atoms/cm^3. The initial concentration of arsenic is 10^{12} atoms/cm^3. Find the arsenic concentration at 2 μm from the surface and the quantity of arsenic that has diffused into silicon. Determince the location where the arsenic concentration is 2×10^{17} atoms/cm^3 for a process time of one hour.

8.15 **Mass transfer from a bubble.** A gas bubble of 3 mm diameter is rising in a pool of a liquid. Calculate the mass transfer coefficient if the diffusion coefficient is 2×10^{-5} cm^2/s. Note the contact time needed for finding the mass transfer coefficient. One way of approximating this is

$$\text{Contact time} = \frac{\text{Bubble diamter}}{\text{Rise velocity}}$$

The rise velocity can be calculated using Stokes' law. Small bubbles are spherical and Stokes' law is a reasonable approximation here.

8.16 **Transient diffusion in a composite slab with PDEPE.** Consider the problem of transient diffusion in a composite slab with two different diffusion values. Thus region 1 extending from 0 to a dimensionless length κ has a diffusivity D_1 while region 2 from κ to 1 has a diffusivity of D_2. The slab is initially at a dimensionless concentration of 1 and both ends are exposed to a zero concentration. Set up the problem and solve by separation of variables. Solve by PDEPE as well and compare the results.

8.17 **Eigenvalues with CHEBFUN.** Compute the eigenvalues and the eigenfunctions for all three geometries shown in Figure 8.1. Compare these with the analytical values shown in Section 8.4.

CHAPTER 9

Basics of Convective Mass Transport

Learning Objectives

After completing this chapter, you will be able to:

- Understand the definition of local versus average mass transfer coefficient.
- Define the key dimensionless groups in convective transport.
- Use common correlations for internal flows and external flows.
- Evaluate mass transfer rate into a liquid flowing as a film.
- Calculate the mass transfer rate from a solid sphere and from a gas bubble to a fluid.
- State and use common correlations for the mass transfer coefficient for some industrial process equipment.

The mass transfer coefficient is an important parameter to calculate the contribution from flow to interfacial mass transfer, and is a key parameter needed to close macroscopic and mesocopic models. We have presented several examples of the use of this parameter in Chapters 3 and 4. The next four chapters deal with evaluation of this parameter in detail. Two common approaches to calculating the mass transfer coefficients are theoretical results obtained from differential models and empirical correlations that are fitted to experimental measurements. Both approaches are obviously complimentary. The theoretical approach is good for simple geometries under laminar flow conditions (since the flow field is now well characterized) while the empirical approach is good for complex geometries and for turbulent flow conditions. This chapter takes more of an empirical approach and provides useful correlations for some common situations. More discussion and analysis of differential models for the convection+diffusion process will be taken up in the next two chapters (10 and 11) where we show how the basic convection-diffusion theory can be applied to predict mass transfer coefficients, at least for cases where the velocity profile is known exactly. Mass transfer under turbulent flow conditions requires additional closure for eddy diffusivity and some semi-theoretical models are then discussed in Chapter 12.

In this chapter we first revisit the definition of mass transfer coefficients introduced in earlier chapters and distinguish between local and averaged values. We then present the common dimensionless groups used to correlate mass transfer data. This is followed by correlations for a number of common standard cases and for some common industrial reactors.

9.1   Definitions for External and Internal Flows

Consider a species A being transported from a solid to a fluid phase. The flux can be represented as a combination of the diffusion and a convection flux:

$$N_{As} = J_{As} + v_s C_A(s)$$

where J_{As} is the diffusion flux from the surface. The second term is the contribution of convection induced by mass transfer. Here v_s is the normal component of the velocity at the surface and depends on how the other species present in the mixture are being transported from the surface. For example, it is zero for EMD. This term can also be neglected for low flux conditions and for dilute systems.

The diffusive part of flux from a surface is then modeled as a product of the mass transfer coefficient and a driving force:

$$J_{As} = k_m \Delta C_A$$

This leads to the following basic definition of the mass transfer coefficient:

$$k_m = \frac{J_{As}}{\Delta C_A}$$

Here ΔC_A is a suitably defined driving force. The choice of the driving force ΔC_A is often not clear. The definition commonly used for external and internal flows is presented in the following.

External flows are flow past solid bodies. Examples are flow over a flat plate, flow around a spherical solid, and so on. Flow past a flat plate with mass transfer from (or to) the plate to the gas shown in Figure 6.3 is an example. The problem is of a boundary layer type and there is a clearly defined external concentration here outside the boundary layer, denoted as $C_{A\infty}$, which is the concentration far away from the solid. This is also called the bulk gas concentration. The difference, $C_{As} - C_{A\infty}$, is used as the driving force. (The C_{As} is the concentration in the fluid phase adjacent to the solid.)

Internal flows are flows in confined geometries, such as flow in a pipe or flow in channels, thin liquid films, and so on. A pipe with a dissolving wall or a catalytic wall is an example of mass transfer in internal flows. The concentration at the wall itself is C_{As}, but C_A varies across the pipe as a function of r. The driving force is defined as C_{As} minus some average concentration in the pipe. This average concentration is taken as the cup mixing concentration, defined as

$$C_{Ab} = \frac{\int_A v_z C_A dA}{\int_A v_z dA}$$

Here v_z is the local velocity in the flow direction and C_A is the local concentration.

Hence $C_{Ab} - C_{As}$ is used as the driving force and the mass transfer coefficient in internal flows is defined as

$$J_{As} = k_m(C_{Ab} - C_{As})$$

Here N_{As} is the flux from the bulk of the fluid to the wall. Note that C_{Ab} is an average concentration and not the concentration at any particular radial location. This is in contrast to external flow, where $C_{A\infty}$ is the actual concentration at the edge of the boundary layer.

Near the entrance to a pipe, the flow is not fully developed and is of a developing internal flow. The mass transfer coefficient here is defined differently from that for a fully developed flow. The reason is that in developing flow there is a mass transfer boundary layer near the wall. Most of the pipe (in the center region) is still at the inlet concentration $C_{A,i}$ and the concentration varies only in a thin region near the wall and reaches C_{As} at the wall itself. The driving force used is now defined as $C_{A,i} - C_{As}$. The $C_{Ab} - C_{As}$ used for fully developed internal flows is not used in the developing region.

9.2   Relation to Differential Model

How is k_m related to differential description? The mass transfer coefficient is usually defined on the basis of the low flux case where the diffusion at the surface is the only mode of transport. The normal component of the velocity at the wall is taken as zero. Hence locally near the wall the mass transfer rate can be computed using Fick's law at the wall. The flux to the wall is given by

$$J_{Aw} = -D_A \left(\frac{dC_A}{dy} \right)_{y=w}$$

where w is the location of the wall. If one divides the flux by the local driving force one can get the value of the mass transfer coefficient:

$$k_m(Local) = \frac{-D_A \left[(dC_A/dy)_{y=w} \right]}{\Delta C_A}$$

This relates the mass transfer coefficient to the concentration gradient at the wall. The latter in turn is from a differential model (also called convection-diffusion models) and the above relation is useful to extract the values of mass transfer coefficients from detailed convection-diffusion models. In fact we will use the relation in later chapters to evaluate mass transfer coefficients based on theory.

What if there is high flux, that is, the normal component of the velocity at the wall is significantly different from zero? In this case, a correction factor is applied to the low flux value and the mass transfer rate is calculated as

$$N_A = k_m C \Delta y_A \, \mathcal{F}$$

For UMD, $\mathcal{F}$ is the drift flux correction, which is equal to the reciprocal of the log mean mole fraction of the non-diffusing component. For EMD the correction factor is not needed and set as one. For the reacting case with reactants diffusing to and products counter-diffusing from the surface, the correction factor will depend on the excess stoichiometric coefficient, ν_E.

Often the average value over a length of plate or pipe is sufficient for design calculations, especially in the context of mesoscopic models; this is defined as

$$\bar{k}_m = \frac{1}{L} \int_0^L k_m(\text{local}) dx \tag{9.1}$$

For flow over a flat plate, it can be shown that k_m local is proportional to $x^{-1/2}$ if the flow is laminar. This is an useful result to remember and we will derive it later based on laminar flow theory in Chapter 11. If k_m local is represented as $A_1/\sqrt{x}$, where A_1 is the coefficient of proportionality, we find that the average value is $2A_1/\sqrt{L}$.

For laminar flow in a pipe, there is an entry region near the pipe entrance for mass transfer and the local value is a function of axial position, z, in this region. A dependency of $z^{-1/3}$ is predicted from theory here. After the entry length the mass transfer coefficient does not change significantly with z and reaches an asymptotic value. For a long pipe the asymptotic value contributes most to the average value. Hence the average value may be taken as a constant equal to the asymptotic value. The following result applies for this parameter as will be proved by detailed analysis in Chapter 10:

$$\bar{k}_m = 3.66 \frac{D_A}{d_t} \tag{9.2}$$

This value is applicable to a case where the wall is maintained as a constant concentration (e.g., solid dissolving from a wall or a rapid catalytic reaction at the wall). It is a rather interesting result; the mass transfer coefficient does not depend on the flow velocity for long pipes in laminar flow!

For turbulent flow in a pipe, the mass transfer coefficient is nearly constant along the pipe length. Thus the local value and the average values are nearly the same. Also it is a strong function of the flow velocity, usually to the power of 0.8.

9.3   Key Dimensionless Groups

Mass transfer data can be correlated using three key dimensionless groups: the Sherwood number, the Reynolds number, and the Schmidt number. These groups were introduced earlier in Section 3.4 and we recap them here for ease of reference.

The Sherwood number defined in the following is the dimensionless representation of the mass transfer coefficient:

$$Sh = \frac{k_m L_{ref}}{D_A}$$

Here L_{ref} is a reference length scale. The pipe diameter is used as the length scale for flow in a pipe while the distance from the leading edge x is used for external flow past a flat plate. For internal flow in non-circular channels, the hydraulic diameter is used as the length scale. This is defined as

$$\text{hydraulic diameter} = 4 \frac{\text{flow area}}{\text{wetted perimeter}}$$

The Reynolds number is a measure of the effect of flow velocity and is defined in general as

$$Re = \frac{L_{ref} \langle v \rangle}{\nu}$$

The reference length scale is the pipe diameter for pipe flows. Hence

$$Re = \frac{d_p \rho <v>}{\mu} \text{ for pipe flows}$$

The reference length is x, the distance from the leading edge for flow over a flat plate; hence

$$Re_x = \frac{x \rho v_\infty}{\mu} \text{ for flow over a plate}$$

The third group is the Schmidt number:

$$Sc = \frac{\mu}{(\rho D_{A-m})}$$

It represents the ratio of momentum diffusivity to mass diffusivity. This group is a physical property and does not depend on the operating parameters. For gases the values are close to one while for liquids the values are high, in the range of 500 to 1500.

9.3.1   Other Derived Dimensionless Groups

The basic groups shown above can be combined to get other groups that can be used in lieu of the Sherwood number. Thus the group $Sh/(ReSc)$ is often used instead of Sh. This group is called the Stanton number:

$$St = \frac{Sh}{ReSc} = \frac{k_m}{v_{ref}}$$

where v_{ref} is the reference velocity, either $<v>$ or v_∞.

The j-D factor is also used in correlation. This is defined as

$$j_D = St(Sc)^{2/3} = \frac{k_m}{v_{ref}}(Sc)^{2/3}$$

where St is the Stanton number for mass transfer. One use of this is that j-factor is closely related to the friction factor in momentum transport, as we will see later.

Another widely used group is the Peclet number, Pe, defined as

$$Pe = ReSc$$

In some cases (e.g., laminar flow in a pipe) Sh does not depend on Re and Sc separately and is the function of the product of these quantities. In such cases use of the Peclet number as the correlating group is more convenient.

Now we show several examples of the use of the dimensionless groups for correlation of mass transport data for common flow situations and in process equipment.

9.4   Mass Transfer in Flows in Pipes and Channels

In this section the correlation for mass transfer coefficients for flows in pipe and channels is summarized.

9.4.1   Laminar Flow

The laminar flow is amenable to a theoretical analysis discussed in Section 10.2. The following equation for the average Sherwood number over a length L of pipe can be derived from theory:

$$\bar{Sh}_L = 3.66 + \frac{0.0668 Pe/L^*}{1 + 0.04[Pe/L^*]^{2/3}} \tag{9.3}$$

Here $L^* = L/d_t$. As L^* becomes large, the average Sherwood number approaches an asymptotic value of 3.66. This equation was also used earlier in Section 4.1 in conjunction with mesoscopic analysis and was used in a model for mass transfer from a dissolving wall. Flow is laminar up to a Reynolds number of 2100.

9.4.2   Turbulent Flow

Gilliland and Sherwood (1934) correlation is commonly used for mass transfer to or from a dissolving wall under turbulent flow conditions:

$$Sh = 0.023 Re^{0.83} Sc^{0.44}$$

For liquid the correlation of Linton and Sherwood (1950) is found to fit the data:

$$Sh = 0.023 Re^{0.83} Sc^{1/3}$$

The equation is very similar to the Dittus-Boelter (1930) equation used for heat transfer, indicating the commonality in the transport of heat and mass. Flow is turbulent when the Reynolds number is greater than 4000.

9.4.3 Channel Flow

For flow between two parallel plates, the Sherwood number is defined using the hydraulic radius as the reference length. This is defined in general as four times the area for flow divided by the wetted perimeter. Using this definition d_h for flow in a channel is equal to twice the spacing between the two plates. Hence the Sherwood number is defined as $2k_m h/D_A$.

For laminar flow, the average Sherwood number for a channel of length L is given as

$$\bar{Sh}_L = 7.54 + \frac{0.03 Pe/L^*}{1 + 0.016[Pe/L^*]^{2/3}} \tag{9.4}$$

Here L^* is equal to L/d_h. As L or correspondingly L^* becomes large, the average Sherwood number approaches an asymptotic value of 7.54. This equation is based on the theoretical result of the convection-diffusion model and is not empirical.

For turbulent flow, the correlations for pipe flow (e.g., the Dittus-Boelter equation) is generally used with the pipe diameter now replaced by the hydraulic diameter. The rationale for using the same correlation is that the viscous sublayer near the wall is very thin and this layer offers most of the resistance to mass transfer. The concentration drop in the core fluid is rather small. Hence the precise shape of the wall region is not very critical.

9.5   Mass Transfer in Flow over a Flat Plate

The dimensionless mass transfer coefficient is the local Sherwood number $Sh_x = k_m x/D$, and this should be correlated in terms of the local Reynolds number and the Schmidt number Sc. Note that the local distance from the leading edge point is used in the definition of the Reynolds number, Re_x.

A power law correlation type is used. Thus the equation

$$Sh_x = A Re_x^\alpha Sc^\beta$$

appears to be a good form for correlation based on dimensionless analysis. Dimensional analysis does not resolve what exponents α and β in this equation should be. It is merely a platform for the correlation of the experimental data.

9.5.1   Laminar Flow

The following equation derived from theory shown in Chapter 11 is applicable for laminar flows:

$$Sh_x = 0.332 Re_x^{1/2} Sc^{1/3} \tag{9.5}$$

Note that the exponent on the Sc is $1/3$. This is characteristic of many other problems as well where mass transfer takes place from a no-slip (fluid-solid) boundary.

The previous expression gives the local mass transfer coefficient. An integrated average value is often used in engineering design calculations. Integrating over a length L of plate, an average Sherwood number can be defined and is given by the following expression:

$$Sh_L = \frac{1}{L} \int_0^L Sh_x dx = 0.664 Re_L^{1/2} Sc^{1/3} \tag{9.6}$$

Note that the above equation should not be used if $Re_L > 2 \times 10^5$. The flow ceases to be laminar under these conditions. The boundary layer becomes fully turbulent if $Re_L > 2 \times 10^6$.

9.5.2   Turbulent Flow

For turbulent flow over a plate, the following correlation for local value is commonly used:

$$Sh_x = 0.0292 Re_x^{0.8} Sc^{1/3} \tag{9.7}$$

Note the change in the dependency to 0.8 on Re, which is now much stronger in turbulent flow. Thus the flow velocity has a much stronger effect on the mass transfer coefficient. The exponent on Sc is still the same as in laminar flow.

The average value is given by integration as

$$Sh_L = 0.0365 Re_L^{0.8} Sc^{1/3} \tag{9.8}$$

Turbulent mass transfer in both pipes and in flow over flat plates is often analyzed using an analogy with momentum transfer. The analysis is semi-theoretical and discussed in detail in Chapter 12.

9.5.3   The j-Factor

The j-factor form of the correlations for external flows is now shown and the relation to the drag coefficient is indicated.

For laminar flow, Equation 9.6 for Sh_L can be rearranged to

$$j_D = \frac{0.664}{Re_L^{1/2}} \tag{9.9}$$

The drag coefficient for friction averaged over a length of L has a value of

$$c_{DL} = \frac{1.228}{Re_L^{1/2}}$$

Comparing the previous two equations, we find that the j-factor is to the $c_D/2$. This is one form of the analogy between momentum and mass transfer that is discussed in more detail in Chapter 11.

A similar relation applies for turbulent flow:

$$j_D = \frac{0.0365}{Re_L^{0.2}} \tag{9.10}$$

9.6   Mass Transfer for Film Flow

Film flow over a flat plate under laminar flow conditions is amenable to a theoretical analysis, which is presented in Section 10.4. Although film flow is an idealized case, the results form a basis for correlation of data over more complex equipment such as a packed bed where the liquid flows as thin rivulets over the packing surface. Here we summarize the key correlation and the effect of parameters on the mass transfer coefficient. Two cases are presented: mass transfer from the solid (flat plate surface) to the liquid film and mass transfer from a gas to the film.

9.6.1   Solid to Liquid

An example of this is mass transfer from a dissolving wall over which a thin liquid film is flowing. A second example is a catalytic wall to which mass transfer is taking place from the liquid film. The mass transfer from or to a solid surface is given by the following equation:

$$k_{sl} = 0.5384 D^{2/3} \beta^{1/3} z^{-1/3} \tag{9.11}$$

Here β is the slope of the velocity profile at the solid–liquid interface. The expression that can be derived for this quantity based on momentum transfer considerations is

$$\beta = \frac{\rho g}{\mu} \delta \qquad (9.12)$$

Here δ is the film thickness, which depends on the flow rate and is given as

$$\delta = \left(\frac{3 Q \mu}{\rho g} \right)^{1/3} \qquad (9.13)$$

with Q the flow rate per unit width of the plate. The equation is valid only if the Reynolds number is less than 20. For larger values of Re, the film becomes wavy and eventually turbulent.

The average value of the mass transfer coefficient over a length L is obtained by averaging and the final result is

$$\bar{k}_{sl} = 0.8075 D^{2/3} \beta^{1/3} L^{-1/3} \qquad (9.14)$$

The following parametric effects can be deduced by substituting for β from the above equation:

- The mass transfer coefficient varies as diffusivity to the power of 2/3. This is characteristic of mass transfer from a solid–fluid boundary where a no-slip boundary condition for velocity is applicable.
- It increases with film thickness to the power of 1/3, since β is proportional to δ.
- It decreases with length with an exponent of 1/3.
- It is proportional to 1/6 exponent of the liquid flow rate per unit width.

9.6.2   Gas to Liquid

The mass transfer from a gas phase to a liquid flowing as a thin film over a solid wall under laminar flow conditions is also amenable to theoretical analysis (Section 10.5). The penetration theory discussed in Section 8.6 is used here and the result is

$$k_L = 2 \sqrt{\frac{D_A v_{max}}{\pi L}}$$

Here v_{max} is the maximum velocity at the surface, which is given from hydro-dynamic considerations as

$$v_{max} = \frac{\rho_L g \delta^2}{2\mu}$$

with δ being the film thickness given by Equation 9.13.

Mass transfer in wavy and turbulent film requires somewhat different considerations and is discussed in later chapters. Turbulent flow needs additional correlations for eddy diffusivity.

9.7   Mass Transfer from a Solid Sphere

Mass transfer from a solid sphere to gas is a well studied problem and usually the Froessling-Marshall correlation introduced in Section 3.4 is used. This is reproduced below for completeness:

$$Sh = 2 + 0.552 Re^{0.5} Sc^{1/3} \qquad (9.15)$$

Note the limiting value of 2 for $Re \rightarrow 0$, that is, for a stagnant gas. This is a theoretical result that was shown in Section 6.3.

For mass transfer into a liquid stream the correlation of Brian, Hales, and Sherwood (1969) is often used:

$$Sh = \left(4 + 1.21 Pe^{2/3}\right)^{1/2}$$

The Peclet number, which is the product of Re and Sc is used in this correlation. The above correlation is suitable for $Pe < 10^4$. The first term represents the contribution under stagnant conditions: $Sh \rightarrow 2$ as $Pe \rightarrow 0$. The second term captures the contribution due to flow.

For larger Peclet numbers ($> 10^4$) the simpler correlation of Levich is considered suitable since the first term representing the stagnant flow contribution can be neglected:

$$Sh = 1.01 Pe^{1/3}$$

Again a 1/3 exponent dependency on the Schimidt number is observed; this is similar to that for flow over a flat plate.

In general, two points are worth noting. First, there is a considerable variation of the local mass transfer coefficient over the angular direction. The mass transfer correlations shown above represent azimuthal averaged values. Second, the natural convection effects often accompany forced convection transfer. The dimensionless group to account for natural convection effects is the Grashof number:

$$Gr = \frac{d^3 \rho_L g \Delta \rho}{\mu_L^2}$$

Natural convection is usually neglected if $Re > 0.4 Gr^{1/2} Sc^{1/3}$, as indicated by Steinbeger and Treybal (1960).

Some theoretical analysis is also possible based on the convection-diffusion model. Again hydrodynamics play an important role. For small Re the flow is analyzed using what is known as Stokes' law. The average mass transfer coefficient can be fitted within 2% accuracy by the following correlation based on theory:

$$Sh = 1 + (1 + Pe)^{1/3} \tag{9.16}$$

This correlation is useful if the Reynolds number range is less than one. For a stagnant fluid ($Pe = 0$), Sh has a value of two, which is also the limiting value obtained from the above equation.

As the Reynolds number increases beyond one, the improvement over the Stokes flow known as the Oseen (1910) model is useful. The convection-diffusion equation can be solved under these conditions and theoretical results for the local mass transfer coefficient can be obtained. The local values are averaged over θ and the results are usually reported in terms of an average Sherwood number. A number of studies have been conducted using this approach and the results are summarized by Leal (2007) and Clift, Grace, and Weber (1978). Interested readers may wish to study these books and the papers cited thereof for further discussion.

9.8   Mass Transfer from a Gas Bubble

It is useful at the outset to point out some differences in mass transfer from a gas bubble and from a solid sphere to a liquid. A gas bubble is a deformable surface and the spherical shape may not hold for all conditions. Even assuming a spherical shape, there may be internal circulation within the bubble itself. In contrast, there is no internal circulation in a solid sphere. Whether the internal circulation is present or not depends on the properties of the liquid, mainly whether there is a surface active agent present in the liquid or not. The internal circulation is reduced in the presence of surface active agents. They adsorb in the rear of the bubble and create a surface tension–driven flow that counters the internal circulation within the bubble. (The surface tension–driven flow is also known as Marangoni flow). The drop or bubble then behaves more like a rigid sphere. For these conditions the rigid sphere correlation given in the earlier section is applicable. The mass transfer coefficient is usually proportional to diffusivity to the 2/3 power in this case.

The regime with significant internal circulation is called the Hadamard-Rybczynski (1911) regime, named after the scientists who proposed this effect. This, applies for clean liquids, that is, in the absence of surface active agents. For these conditions, the following correlation is found to be suitable:

$$Sh = 1 + (1 + 0.564 Pe^{2/3})^{3/4} \tag{9.17}$$

where $Pe = Re\,Sc$ with the Reynolds number based on bubble diameter. Hence we need additional correlation to calculate the bubble size. Note that the mass transfer coefficient now becomes proportional to diffusivity to the power of one half.

We now present correlations for industrial contactors where a gas is sparged into a liquid.

9.8.1   Bubble Swarms and Bubble Columns

Mass transfer from a swarm of bubbles to a liquid finds many important applications in industrial processes. For example, bubble column reactors, where a gas is sparged into a tall pool of liquid, find use in a wide variety of processes including oxidation, hydroformylation, Fisher-Tropsh synthesis, and so on (see Ramachandran and Chaudhari, 1983). In this contactor we often have a swarm of bubbles that create the circulation in the liquid; hence the prediction of the mass transfer coefficient can be a difficult task since the velocity field near a bubble cannot be accurately calculated. (Note that CFD simulations can be useful in this context but they involve many closure parameters and provide more of a smeared value based on the local volume averaging.) Further one needs to classify the flow regime since the hydrodynamics is different for small spherical bubbles, large ellipsoidal bubbles, and large spherical cap bubbles. The results also depend on whether there is an internal circulation within the bubbles or not. We can only provide a brief (empirical) discussion on these topics. Clift et al. (1978) provide additional details. The correlations suggested here are for a quick estimation and to get a feel for the numbers. They are also useful to understand the interrelations between various parameters. There is a vast body of literature as well as measurements on these parameters and for more accurate estimation, consult these sources. The paper by Akita and Yoshida (1974) is an useful reference on this topic.

Mass Transfer Coefficient

The correlation of Calderbank and Moo-Young (1961) is widely used in the literature for bubbles of diameter greater than 2.5 mm:

$$Sh = \frac{k_L d_b}{D_A} = 0.42 Gr^{1/3} Sc^{1/2}$$

where Gr is defined as

$$Gr = \frac{d_B^3 \rho_L g \Delta \rho}{\mu_L^2}$$

A penetration model is used as an alternative with contact time taken as d_B/v_B, where v_B is the effective rise velocity of the swarm of bubbles in the liquid:

$$k_L = \sqrt{\frac{4 D_A v_B}{\pi d_B}}$$

For bubbles with diameter less than 2.5 mm the following correlation is commonly used:

$$Sh = 0.31 Gr^{1/3} Sc^{1/3}$$

Small bubbles of this size do not have internal circulation and hence the mass transfer process is closer to that of a solid. Hence the exponent of Sc is 1/3 and correspondingly the dependence on the diffusivity is 2/3. In contrast, for large bubbles the dependency is 1/2.

A knowledge of bubble diameter is needed in order to use these correlations. Also, in order to use the penetration theory, the relative velocity between the gas and liquid is needed. Some correlations for these quantities are summarized in the following.

Bubble Diameter

The maximum bubble size is an important parameter and the average bubble size can be related to this. A force balance model for estimation of this parameter was proposed by Kolmogorov (1949) and Hinze (1955). In a turbulent field, the local fluctuations are responsible for bubble breakup. The surface tension forces oppose this deformation. A balance of these forces can be used to estimate the maximum bubble size. The key dimensionless parameter resulting from these balances is the Weber number, defined below:

$$\text{Weber number} = \frac{\tau}{\sigma/d_{max}} \left(\frac{\rho_d}{\rho_c} \right)^{1/3}$$

where τ is the stress, which can be correlated to the power input per unit mass (defined as ϵ here) by the following equation:

$$\tau = \rho_c[2(\epsilon d_{max})^{2/3}]$$

Rearranging we get the maximum bubble diameter in terms of a critical Weber number:

$$d_{max} = \left(\frac{We_{crit}}{2}\right)^{3/5} \left(\frac{\sigma}{\rho_c}\right)^{0.6} \left(\frac{\rho_d}{\rho_c}\right)^{0.2} \epsilon^{-0.4}$$

The bubble diameter is proportional to the power input to the system by the power of -0.4. This forms the basis for many correlations for the bubble diameter, two of which are summarized below.

Bhavaraju et al. (1978) suggested the following correlation:

$$d_B = 0.53 \left(\frac{\mu_c}{\mu_d}\right)^{0.1} \left(\frac{\sigma}{\rho_c}\right)^{0.6} \epsilon^{-0.4}$$

The following correlation of Calderbank (1958) is also often used:

$$d_B = 4.15\epsilon_G^{0.5} \left(\frac{\sigma}{\rho_c}\right)^{0.6} \epsilon^{-0.4}$$

This is similar to the Bhavaraju correlation but needs an estimate of the gas holdup ϵ_G as an additional input.

Power Input

The power input to the system is needed to find the average bubble diameter. This is taken as the work done in expansion of the gas as it travels from the column bottom to the top; the following simplified formula is often used:

$$\text{Power input} = Q_G \rho_L g h_L$$

where Q_G is the gas flow rate and h_L is the clear liquid height. Hence $\epsilon = u_G g$ is used as a simple model for the power input.

Relative Velocity

The terminal velocity of bubble rise is often used for the relative velocity (needed for instance in the penetration model) and the following correlation suggested by Clift et al. (1978) is useful:

$$v_B = [2.14\sigma/\rho_L d_B] + 0.505(g d_B)]^{1/2}$$

This correlation applies if the bubble holdup is small (< 0.1), in which case the bubble motion is not affected by the presence of neighboring bubbles. Note that a correction is needed for a large holdup, which is known as the Richardson-Zaki correction.

Interfacial Area

The interfacial area for mass transfer depends on the bubble diameter and the gas holdup. Assuming an average bubble diameter d_B, the following expression is commonly used:

$$a_{gl} = \frac{6\epsilon_g}{d_B}$$

The gas holdup is needed in order to use this equation.

Gas Holdup

A correlation proposed by Yamashita and Inoue (1975) is a commonly used and simple equation suitable for estimation of the gas holdup:

$$\epsilon_G = \frac{u_G}{2.2u_G + 0.3(gd_t)^{0.5}}$$

Here u_G is the superficial gas velocity.

9.9   Mass Transfer in Mechanically Agitated Tanks

Another common piece of equipment for gas–liquid contacting is mechanically agitated tanks. These are tanks equipped with an impeller or agitator with gas sparged at the bottom. Mechanical stirring causes the gas to be dispersed into bubbles and also keeps the contents of the tanks well-mixed. Usually the volumetric mass transfer coefficient is measured and correlated. van't Riet (1979) proposed the following correlation:

$$k_L a_{gl} = 2.6 \times 10^{-2}(P/V)^{0.4}u_G^{0.5}$$

Here P/V is the power input per unit volume of liquid and u_G is the superficial gas velocity. Again a 0.4 dependency on power input is seen, similar to the bubble columns.

The power consumption in an agitated tank in turn is correlated using the power number, which is defined as

$$Po = \text{Power number} = \frac{P}{\rho \Omega^3 d_i^5}$$

where Ω is the speed of agitation in revolutions per second (r.p.s). The power number is usually correlated as a function of the agitator Reynolds number, defined as

$$Re = \frac{d_I^2 \Omega \rho}{\mu}$$

The suggested form of the correlation for power consumption is

$$Po = f(Re)$$

Charts of Po versus Re are available in many common reference books (for example, Nagata, 1975). For large Reynolds numbers ($Re > 10^5$), the power number becomes independent of the Reynolds number. The explanation is that the viscous effects well captured by the Reynolds number are not so important at these conditions. An asymptotic value of 6.3 is often observed for the power number. Hence the power consumption can be calculated as

$$P = 6.3 \rho \Omega^3 d_I^5$$

This provides a simple formula for power consumption and also provides a useful scale-up rule. We find the power consumption is proportional to d_I^5 while the dependence on agitation speed is third power.

For gas–liquid dispersions in agitated vessels, the gas flow rate Q_G is also important. An additional group, the flow number, defined in the following, is also needed to correlate the data:

$$\text{Flow number} = \frac{Q_G}{\Omega d_I^3}$$

Hence the power number is then correlated as a function of the Reynolds number and the flow number.

The power consumption in a gas aerated system is smaller than that for a gas free tank. Calderbank (1958) suggested the following correlation to include the effect of gas sparging:

$$P = \psi P_0$$

where ψ is a correction factor to include the effect of gas flow. This factor in turn was correlated as

$$\psi = 1.0 - 12.6\frac{Q_G}{\Omega d_I^3} \text{ for } \frac{Q_G}{\Omega d_I^3} < 3.5 \times 10^{-2}$$

and

$$\psi = 0.62 - 1.85\frac{Q_G}{\Omega d_I^3} \text{ for } \frac{Q_G}{\Omega d_I^3} > 3.5 \times 10^{-2}$$

The following correlations for the mass transfer coefficient are commonly used at the first level of analysis:

$$k_L = 0.42\left[\frac{(\rho_L - \rho_G)\mu_L g}{\rho_L^2}\right] Sc^{-.5}$$

$$a_{gl} = 1.44\sigma^{-0.6}\rho_L^{0.2}(u_G/v_B)^{0.5}(P/V)^{0.4}$$

Note that the intrinsic mass transfer coefficient does not depend on the power input while the interfacial area depends on the power per unit volume (P/V) to an exponent of 0.4. An equal value of P/V is often used as a criteria for scale-up from small vessel to large vessel since it then provides the same value for the volumetric mass transfer coefficient.

9.10 Gas–Liquid Mass Transfer in a Packed Bed Absorber

Gas absorption columns are usually packed with irregularly shaped solid particles whose role is to break up the liquid into flow as rivulets and thereby promote gas–liquid mass transfer. Common packings include Raschig rings, Intalox saddles, Pall rings, and Berl saddles. The mass transfer phenomena is complex and depends on many parameters, for example packing geometry and the resulting flow pattern, wettability of the packing, and so on. Hence the approach is more empirical rather than theoretical and a number of correlations have been proposed. For illustration, we show the equation developed by Onda et al. (1968), which is usually within 20% of the experimental data. The coefficients for the liquid side and the gas side are correlated separately and an overall coefficient is calculated as

$$\frac{1}{K_y} = \frac{1}{k_y} + \frac{m}{k_x}$$

This is a formula based on two-film theory, discussed in Chapter 6. In addition a correlation for the interfacial transfer area is also needed. These are summarized next.

9.10.1 Liquid Side Coefficient

The parameter k_x is correlated as

$$k_x = 0.051 C_L (\nu_L g)^{1/3} (a_p d_p)^{0.4} Re_L^{2/3} Sc_L^{-1/2} \tag{9.18}$$

The variables used in the above equations are defined in the following. C_L is the total molar concentration of the liquid phase. a_p is the specific surface area of the packing per unit column volume m^{-1} and d_p is the nominal size of the packing. Data for common packings are available from the packing manufacturers.

The Reynolds number for the liquid Re_L is based on a_p and is defined as

$$Re_L = \frac{u_L \rho_L}{a_p \mu_L}$$

where u_L is the superficial liquid velocity.

9.10.2 Gas Side Coefficient

The parameter k_y is correlated as

$$k_y = 5.23 C_G (\nu_G a_p)(a_p d_p)^{-2} Re_G^{0.7} Sc_G^{-2/3} \tag{9.19}$$

C_G is the total molar concentration of the gas phase. The Reynolds number for the gas phase Re_G is also based on a_p and is defined as

$$Re_G = \frac{u_G \rho_G}{a_p \mu_G}$$

where u_G is the superficial liquid velocity.

9.10.3 Transfer Area

Onda et al. (1967) proposed the following correlation for the gas–liquid interfacial area:

$$\frac{a_{gl}}{a_p} = 1 - \exp\left[-1.45 Re_L^{0.1} Fr_L^{-0.05} We_L^{0.2} (\sigma_c/\sigma)^{0.75}\right]$$

Note that the left-hand side represents the fraction of the surface area of the packing that is wetted by the liquid. The additional parameters needed in the preceding equation are defined as follows:

$$\text{Froude number} = Fr_L = \frac{a_p u_L^2}{g}$$

$$\text{Weber number} = We_L = \frac{u_L^2 \rho_L}{\sigma a_p}$$

where σ is the surface tension of the liquid and σ_c is a critical surface tension of the packing material for which values are available. Typical values for illustration are 0.061 N/m for ceramic, 0.033 for polyethylene and 0.075 for steel.

Summary

- The mass transfer coefficient is defined as flux (under low flux conditions) divided by a suitably defined concentration difference. The difference between the surface concentration and the bulk concentration is used for external flow while the difference in surface concentration and cup mixing concentration is used for internal flows.
- The mass transfer coefficient may vary along the surface and a local value can be ascribed as a function of the length along the surface. In practice however it is convenient to use an average value for a given length or geometry in conjunction with mesoscopic or macroscopic balances. The variation along the length is strong in laminar flow but very small in turbulent flows.
- In laminar pipe flows an asymptotic value of the mass transfer coefficient is reached for long pipes and the value does not depend on the flow velocity.
- Key dimensionless parameters to correlate mass transfer data are the Sherwood number, the Reynolds number, and the Schmidt number. Data can also be correlated using the Stanton number or the j-factor.
- The convective mass transfer in boundary layers in laminar flow over a flat plate is a well studied problem and is amenable to a complete theoretical analysis, which is presented in Chapter 11. In this chapter the correlations are shown for laminar and turbulent flows for both local and average values.

- Mass transfer into a falling film is an important problem in convective mass transfer. The solid dissolution at the walls and the gas absorption at the interface are the two important prototype problems. The length of the wall (contactor) is usually small and hence an asymptotic entry region model type is generally used for these cases. The local and average value of the mass transfer coefficient can be predicted from such a model as shown in the next chapter.

- It is important to note the dependency on the diffusion coefficient in film flow. In the solid dissolution (no slip boundary) case the dependency is to the power of 2/3 while in the gas absorption (no shear boundary) case it is 1/2.

- Mass transfer from a solid sphere is a complex problem to compute theoretically but in practice many empirically based correlations (e.g., Brian, Hales, and Sherwood 1969; Levich, 1962) are available; these can be used to estimate this parameter.

- Gas–liquid mass transfer from a single bubble is complex as the bubble surface is a deformable surface, in contrast to a solid surface. Further, there may be internal circulation flow within the bubble. The correlation of Calderbank and Moo-Young (1961) is commonly used which is presented in this chapter.

- For gas–liquid mass transfer in process equipment, a volumetric mass transfer value is often used. This is a product of the intrinsic mass transfer coefficient and the gas–liquid interfacial area per unit volume of the reactor. Illustrative correlations for bubble column reactors and stirred tank reactors are presented in this chapter.

- For the gas–liquid mass transfer in a packed column, separate correlations are proposed for the intrinsic liquid side and gas side coefficient and for the transfer area. The correlation of Onda et al. (1967, 1968) shown in this chapter is useful to get an estimate of these quantities and the overall volumetric coefficient can then be calculated by combining these.

Review Questions

9.1 What representative concentration difference is used in pipe flow mass transfer?

9.2 Define the cup mixing concentration and indicate its use.

9.3 How does the local coefficient vary with the distance along a plate for external flow over a flat plate? Assume laminar conditions.

9.4   How does the local coefficient vary with the axial distance in a long pipe? State the answer for both laminar flow and turbulent flow.

9.5   Given an expression for the local mass transfer coefficient, how can the average mass transfer coefficient be calculated?

9.6   Name and define the various groups used in convective mass transfer.

9.7   What is the relation of the hydraulic diameter to the length or width in a square channel?

9.8   What is the j-factor for a mass transfer coefficient? Why is it commonly used?

9.9   What is the asymptotic value for the Sherwood number for solid dissolution from a long pipe?

9.10   At what point does the flow become unstable in an external flow over a plate?

9.11   How does the mass transfer coefficient vary with local Reynolds number for laminar flow and for turbulent flow over a plate?

9.12   How does the mass transfer coefficient vary with the diffusion coefficient for solid–fluid mass transfer and for fluid–fluid mass transfer?

9.13   When does a gas bubble show strong internal circulation?

9.14   What is Marangoni flow?

9.15   What is the Hadamard-Rybcznski regime?

9.16   For mass transfer from a gas bubble to a liquid, what is the dependency of the mass transfer coefficient on the diffusion coefficient?

9.17   What is the dependency in general of the mass transfer coefficient on power input per unit volume?

9.18   Which parameter changes strongly on power input, k_L or a_{gl}?

9.19   An agitated vessel is scaled up to ten times the lab scale size. By what factor does power consumption increase? By what factor does power consumption per unit volume increase? Assume that geometric similarity is maintained.

9.20   What is the physical interpretation of the Froude and Weber numbers?

Problems

9.1   **Mass transfer cofficients for a pipe.** Calculate the mass transfer coefficient form a bulk liquid to pipe wall for the following conditions: pipe diameter = 1 cm; length of the pipe = 100 cm; $D_A = 2 \times 10^{-9}$ m^2/s. The physical properties are similar to water and the range of velocity from 1 to 20 cm/sec is to be examined. Repeat if the fluid is a gas with properties similar to air with $D_A = 2 \times 10^{-5}$ m^2/s.

9.2   **Mass transfer over the surface of a pond.** An open tank is rectangular with a length of 100 m and long in the other direction. Air at 300 K and 1 atm flows parallel to the surface of the pond with a bulk velocity of 6 m/sec. Determine the position in the tank at which the flow is no longer laminar. For further calculations. assume that the flow is fully turbulent beyond this point although there is a transition flow region in between. Calculate and plot the local and average value of the gas side mass transfer coefficient from the leading edge to the edge of the plate. Use $\nu = 0.15$ cm^2/s and $D = 0.085$ cm^2/s.

9.3 **Mass transfer for flow over a plate.** A thin 1.0 mm coat of fresh paint has just been sprayed over a 1.5 m by 1.5 m square steel body part that can be modeled as a flat surface. The paint contains benzene as the solvent and exerts a partial pressure of 0.137 atm at the process temperature of 300 K. Air at 1 atm pressure at a velocity of 1 m/s is blown over the surface. Determine the average mass transfer convective coefficient and the solvent evaporation rate from the surface in g/min.

9.4 **Mass transfer from a solid and from a gas in film flow.** Consider a liquid flowing down a vertical wall at a rate of 1×10^{-5} m^2/s per meter unit width. The wall is coated with benzoic acid. Find the local mass transfer coefficient as a function of distance from the wall. Find the average mass transfer coefficient for a wall of 50 cm. Use $D = 2 \times 10^{-9}$ m^2/s, $C_{As} = 20$ mol/m^3, and the physical properties of water.

Repeat the analysis of the this problem if instead a gas such as oxygen is being absorbed into the liquid film.

9.5 **Film thickness and mass transfer.** Water trickling down a thin vertical wall is used in pollution treatment (an example of a trickle bed filter). Consider water flowing at a rate of $1. \times 10^{-6}$ m^2/s per unit meter width of vertical wall. Water is exposed to air and there is mass transfer of oxygen into the liquid from the interface. Find the thickness of the film. Find the value of the local mass transfer coefficient at a point 20 cm from the entrance. Based on this value find the hypothetical thickness of the mass transfer film thickness. Compare this to the hydrodynamic film thickness and explain the difference. Use the following values for the physical properties: $\mu = 0.001$ $Pa.s$; $D = 2 \times 10^{-9}$ m^2/s.

9.6 **Intrinsic mass transfer coefficient in a bubble column.** Consider a bubble column with gas and liquid properties similar to that for air and water. The diffusion coefficient of the gas in the liquid is 2×10^{-9} m^2/s. The gas superficial velocity is 5 cm/sec. Estimate the following parameters: the bubble size d_B; the bubble rise velocity v_B; exposure or contact time d_B/v_B; mass transfer coefficient based on the penetration model; and mass transfer coefficient using the equation of Calderbank and Moo-Young.

9.7 **Volumetric mass transfer coefficient in a bubble column.** For a system with properties similar to air and water and $D = 2 \times 10^{-9}$ m^2/s calculate and plot the volumetric mass transfer coefficient $k_L a_{gl}$ as a function of superficial gas velocity in the range of 5 to 15 cm/s.

9.8 **Power consumption and mass transfer in an agitated tank.** Calculate the power consumption and the gas–liquid mass transfer coefficient as a function of speed of agitation for the following conditions: tank diameter = 10 cm, impeller diameter = 5 cm, and superficial gas velocity = 0.2 cm/s. Physical properties may be assumed to be the same as that for an air-water system.

9.9 **Scale-up of an agitated tank.** The mass transfer coefficient in an agitated vessel of 10 cm diameter was measured as $0.2 \, \mathrm{s}^{-1}$ at a speed of revolution of 10 rps. The impeller diameter is usually half the tank diameter and the height of clear liquid is 10 cm. What should be the speed of agitation to have the same mass transfer coefficient in a larger vessel of 100 cm diameter? Assume geometric similarity.

9.10 **Mass transfer in a packed bed reactor.** The correlation of Gupta and Thodos (1963) is suitable for mass transfer to a solid in a packed bed:

$$\epsilon j_D = \frac{2.06}{Re^{0.505}}$$

Here ϵ is the void fraction or the bed porosity. The Reynolds number Re is evaluated using the packing diameter and superficial velocity. Consider a tube of 2.5 cm internal diameter packed with alumina pellets of 3 mm diameter where the void fraction of the bed is 0.38. Air flows through the tube with a superficial mass velocity of $1.3 \, \mathrm{kg/m^2 s}$ at 600 K. Find the j_D factor and the mass transfer coefficient if the diffusivity of the reacting gas is $D_A = 2 \times 10^{-5} \, \mathrm{m^2/s}$.

9.11 **Mass transfer in a fluid bed reactor.** Mass transfer data from a solid to a gas in a fluidized bed is given by the following correlation:

$$\epsilon j_D = 0.010 + \frac{0.863}{Re^{0.585} - 0.483}$$

Based on this equation, calculate the gas to solid mass transfer coefficient in a fluidized bed reactor operating with 1 mm catalyst particles at a gas velocity of 15 cm/sec. The bed porosity ϵ is 0.42.

9.12 **Mass transfer coefficient in a packed absorber.** Evaluate k_x, k_y, and the transfer area a_{gl} in a packed column of $1.2 \, \mathrm{m^2}$ cross-sectional area. The column is packed with ceramic Berl saddles of 0.038 m nominal diameter and provides a specific surface area of $125 \, \mathrm{m^{-1}}$. The flow rate of the gas and the liquid are 0.35 kg/s and 1.90 kg/s with properties similar to air and water. The tower operates at 1 atm pressure and 300 K temperature. The critical surface tension of ceramic is 0.061 N/m.

CHAPTER 10

Convective Mass Transfer: Theory for Internal Laminar Flow

Learning Objectives

After completing this chapter, you will be able to:

- Formulate models for mass transfer under laminar flow in confined geometries.
- Express the model in dimensionless form and identify dimensionless groups.
- Solve for mass transfer in fully developed flow from a pipe using the hypergeometric function.
- Model and solve for mass transfer in the entry region of a pipe using the similarity solution method.
- Solve for convective mass transfer with wall reaction using the CHEBFUN code.
- Model mass transfer rate in solid dissolution and gas absorption from a falling film.
- Use the PDEPE solver for numerical solution of convective mass transfer problems in a more general setting.

Convective mass transfer problems are best studied in two categories of flow: internal flows in confined geometries where flow becomes fully developed and external flows, which are essentially of the boundary layer type. This chapter deals with problems with internal flows, for example, flow in a pipe, flow in a square channel, and similar situations, while the next chapter deals with problems where a boundary layer type of analysis is needed.

The first problem studied in this chapter is that of mass transfer in pipe flows. The velocity profile is assumed to be fully developed and a parabolic function for Newtonian fluids. Using this profile, the convection diffusion equation is solved. The concentration is now a function of two coordinate directions, namely the axial distance and the radial (or cross-flow) direction and therefore the governing equations are now partial differential equations (PDEs). Mathematically speaking the PDEs are similar to that for transient mass transfer studied in an earlier chapter; hence the method of separation of variables can be used. But in the context of convective transport these PDEs, although still linear, have variable (position dependent) coefficients. This adds additional mathematical complexity to the solution, mainly in the form of the eigenfunctions and the calculations of the eigenvalues. These details will be examined. Nature of the series, convergence properties, mass flux, and the predicted mass transfer coefficients will be examined and illustrative examples are presented. The methodology is the same for channel flows and duct flows and we show some illustrative results for this case as well.

An alternative method using the similarity variable is more useful in the entry region to a pipe. The boundary layer nature of the concentration distribution near the entrance region permits this approach. We will define the similarity variable and derive a closed form analytical solution to the entry region problem for pipe flow.

The next class of problems studied is those in convective mass transfer in film flows. The mass transfer into a liquid flowing as film over a vertical (or inclined) surface is an important prototype problem in convective mass transfer since it simulates the flow in more complex industrial equipment such as packed beds. We examine two important cases here: solid dissolving from the wall and gas being absorbed from the interface of the liquid film. We explore the similarity solution approach to predict the mass transfer coefficients for the two cases. We will also compare and contrast the two problems.

The final section of the chapter deals with numerical solution of convective transport problems. Mainly the use of the PDEPE solver in MATLAB and CHEBFUN is demonstrated. Some analytical results will be used to benchmark the numerical results. Next numerical results for some additional problems will be presented.

10.1   Mass Transfer in Laminar Flow in a Pipe

The starting point is the convection-diffusion equation introduced in Chapter 5:

$$\boldsymbol{v} \cdot \nabla C_A = D_A \nabla^2 C_A \tag{10.1}$$

In this section we apply this to solve mass transfer in a fully developed flow in a circular pipe. In such cases the velocity has only the z-component and the velocity profile (for a Newtonian fluid) is given as

$$v_z = 2 \langle v \rangle \left[1 - \left(\frac{r}{R}\right)^2\right]$$

Here $\langle v \rangle$ is the average velocity. The Laplacian term is used with the r- and z-directions as there is no variation of the concentration in the θ direction. The governing equation is therefore

$$2 \langle v \rangle \left[1 - \left(\frac{r}{R}\right)^2\right] \frac{\partial C_A}{\partial z} = D_A \left[\frac{\partial^2 C_A}{\partial z^2} + \frac{1}{r}\frac{\partial}{\partial r}\left(r\frac{\partial C_A}{\partial r}\right)\right] \qquad (10.2)$$

The solution to this problem is now studied with suitable boundary conditions at the walls. Before proceeding further, a note on the entry region for flow is in order. In this formulation (10.2), we assume that $z = 0$ is at a location in the pipe where the velocity profile is already fully established. This requires a hydrodynamic entry length, L_e, given by the following equation:

$$\frac{L_e}{R} = 0.07 \, Re$$

Hence $z = 0$ is not the actual entrance of the flow and represents a starting point after the hydrodynamic entry region. We assume mass transfer conditions start beyond this point. If the mass transfer starts at a position earlier than this length l_e^*, the Equation 10.2 has to be modified since the velocity profile is not unidirectional in the entry region for flow. Flow is of the boundary layer type here and both v_z and v_r contribute to convective transport for that case.

10.1.1 Dimensionless Form

In dimensionless form the following equation can be derived:

$$(1 - \xi^2)\frac{\partial c_A}{\partial z^*} = \frac{1}{Pe}\left[\frac{\partial^2 c_A}{\partial z^{*2}} + \frac{1}{\xi}\frac{\partial}{\partial \xi}\left(\xi\frac{\partial c_A}{\partial \xi}\right)\right] \qquad (10.3)$$

Here z^* is the dimensionless axial distance defined as z/R and ξ is the dimensionless radial distance defined as r/R. Also, c_A is a suitably defined dimensionless concentration.

Hence the parametric representation of the problem is

$$c_A = c_A(z^*, \xi; Pe)$$

The parameter Pe appearing in dimensionless form (Equation 10.3) is defined as follows:

$$Pe = \frac{\langle v \rangle \, d_t}{D_A} = \frac{\langle v \rangle \, d_t}{\nu} \frac{\nu}{D_A} = ReSc$$

where Re is the Reynolds number based on the pipe diameter d_t and Sc is the Schmidt number defined as ν/D_A.

This dimensionless group has the following significance:

$$Pe = \frac{\text{mass transport by convection}}{\text{mass transport by diffusion}}$$

This can also be realized by writing Pe as

$$Pe = \frac{\langle v \rangle \, d_t}{D_A} = \frac{<v> \Delta C_A}{D_A \Delta C_A / d_t}$$

where ΔC_A is some representative concentration difference in the system. The numerator is a measure of convective transport while the denominator is a measure of diffusive transport; therefore the Peclet number is the ratio of the two modes of transport.

Simplification for Large Peclet Number

If $Pe > 10$, the convection term will dominate over diffusion in the axial direction. Hence the following approximation can be made:

$$\frac{\partial^2 c_A}{\partial z^{*2}} << \frac{1}{\xi} \frac{\partial}{\partial \xi} \left(\xi \frac{\partial c_A}{\partial \xi} \right)$$

and the model can be reduced considerably. This can be illustrated by a detailed scaling analysis, which is not shown here. Hence for large Peclet numbers the axial diffusion can be dropped, resulting in

$$(1 - \xi^2) \frac{\partial c_A}{\partial z^*} = \frac{1}{Pe} \left[\frac{1}{\xi} \frac{\partial}{\partial \xi} \left(\xi \frac{\partial c_A}{\partial \xi} \right) \right] \tag{10.4}$$

This suggests that a rescaling of the dimensionless axial distance may be useful. Let us define ζ (zeta) as

$$\zeta = \frac{z^*}{Pe} \tag{10.5}$$

This is also called the contracted axial distance parameter.

Equation 10.4 reduces to

$$(1 - \xi^2)\frac{\partial c_A}{\partial \zeta} = \frac{1}{\xi}\frac{\partial}{\partial \xi}\left(\xi\frac{\partial c_A}{\partial \xi}\right) \tag{10.6}$$

The parametric representation now simplifies to

$$c_A = c_A(\zeta, \xi)$$

No free parameter appears, which is a characteristic of a well scaled problem.

The solution to the model for the case of constant wall concentration will now be examined.

This problem is also known as the Dirichlet problem for laminar convection or the Graetz problem for mass transfer. Note that the problem has an exact analogy to the problem of heat transfer with constant wall temperature. This was first studied by Graetz in 1883.

10.1.2 Constant Wall Concentration: The Dirichlet Problem

For the case of constant wall concentration, the dimensionless concentration is defined as

$$c_A = \frac{C_A - C_{As}}{C_{A,i} - C_{As}}$$

The dimensionless inlet concentration is now 1 while the wall concentration is zero with this definition. This is convenient to make the problem homogeneous. The symmetry condition is applied at $\xi = 0$ as usual. Hence the boundary conditions for a constant wall concentration case for Equation 10.6 are as follows:

- Entrance, $\zeta = 0$: $c_A = 1$
- Wall, $\xi = 1$: $c_A = 0$
- Center, $\xi = 0$: $\partial c_A/\partial \xi = 0$

Note that only one condition in the axial direction ζ (entrance condition) is needed. This would not be the case if the axial diffusion term were to be retained in the model. Two boundary conditions are needed in the radial direction ξ.

The problem, being linear, can be solved by separation of variables. Mathematical details are skipped and the final series solution is shown here.

The solution can be represented as a series of the form

$$c_A(\zeta, \xi) = \sum_{n=1}^{\infty} C_n \exp(-\lambda_n^2 \zeta) F_n(\xi) \tag{10.7}$$

where λ_n are the eigenvalues to the associated eigenproblem and F_n are the eigenfunctions. The associated eigenvalue problem is the differential equation for F. This equation can be derived by substituting the proposed solution Equation 10.7 in Equation 10.6; you should verify that the following eigenvalue problem is obtained:

$$\frac{\partial}{\partial \xi}\left(\xi \frac{\partial F_n}{\partial \xi}\right) = -\lambda_n^2 \xi(1 - \xi^2) F_n \tag{10.8}$$

This has homogeneous boundary conditions and the solutions to this equation are the eigenfunctions to the Graetz problem. It can be shown by a series expansion that the eigenfunctions are

$$F_n(\xi) = \exp(-\lambda_n \xi^2/2) M\left(\frac{1}{2} - \frac{\lambda}{4}, 1, \lambda_n \xi^2\right) \tag{10.9}$$

Here M is the confluent hypergeometric function or the Kummer functions. These are power series in ξ^2 similar to an exponential function. See Abramowitz and Stegun (1964) for more details on these types of functions. Computations of these functions can easily be done with the *hypergeom* in MATLAB.

The function represented above has the symmetry property at $\xi = 0$ since only terms of ξ^2, ξ^4, and so on, appear. Hence the boundary condition at $\xi = 0$ is satisfied. Now using the boundary condition at $\xi = 1$ we obtain the eigencondition

$$\exp(-\lambda/2) M\left(\frac{1}{2} - \frac{\lambda}{4}, 1, \lambda\right) = 0 \tag{10.10}$$

The solution of this non-linear algebraic equation provides a set of (infinite) eigenvalues. The series coefficients can be calculated from the orthogonality property of the F function:

$$C_n = \frac{\int_0^1 \xi(1 - \xi^2) F_n(\xi) d\xi}{\int_0^1 \xi(1 - \xi^2)[F_n(\xi)]^2 d\xi}$$

The formula above is applicable for an inlet condition of one. Note that the eigenfunctions F_i and F_j (with $i \neq j$) are not orthogonal with themselves but

Table 10.1   Constants Needed for the Solution to the Graetz Problem with Constant Wall Concentration

	n = 1	n = 2	n = 3	n = 4	n = 5
Eigenvalue λ_n	2.704	6.6790	10.6733	14.6710	18.6698
Coefficient C_n	0.9774	0.3858	−0.2351	0.1674	−0.1292
$F_n(\xi = 0)$	1.5106	−2.0895	−2.5045	−2.8426	−3.1338
$F_n'(\xi = 1)$	−1.5322	−2.8192	3.9379	−4.9631	5.9256

need a weighting function $\xi(1 - \xi^2)$ to make them orthogonal. Hence this term is also needed in the integral for the calculation of the series coefficient. A few values of the series coefficients are also given in Table 10.1 together with the corresponding eigenvalues.

This completes the analytical solution to the classical Graetz problem. Properties of the solution and calculation of the local Sherwood number are presented next.

10.1.3 Concentration, Wall Mass Flux, and Sherwood Number

The concentration profiles are shown in Figure 10.1 as a function of radial position at four axial positions. Note that the mass transfer is nearly complete for an axial distance, ζ, around 0.5.

The center concentration is often of interest since this is where the concentration has a maximum deviation from the value at the wall. The leading term in the solution for the center concentration is obtained from Equation 10.7 by setting $\xi = 0$ and taking $n = 1$:

$$c_A(\xi = 0) \approx 0.9774 \exp(-2.704^2 \zeta) F_1(0)$$

This is called the one-term approximation. The value of $F_1(0)$ is also tabulated in Table 10.1 for quick calculations of the center concentration and is equal to 1.5106. We find the mass transfer is almost complete at $\zeta = 0.5$. For this length the value of the center concentration is 0.0381. The one-term solution should not be used near the entrance ($\zeta < 0.02$ or so) as the concentration in the entry region is rather steep. More terms in the series should be used or an alternative similarity solution discussed in Section 10.2 is more suitable.

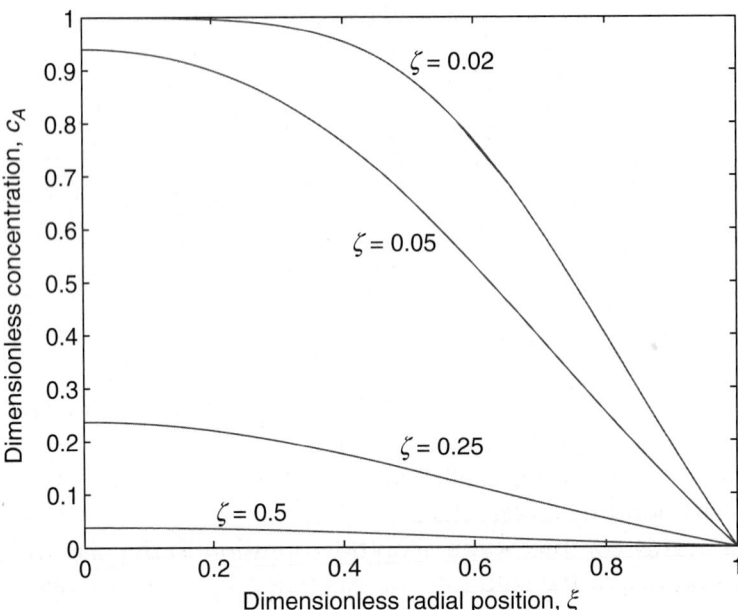

Figure 10.1 Concentration profile for Graetz problem as a function of radial position; the solution is shown at four axial locations, ζ.

Local Sherwood Number

The quantity of primary engineering interest is usually the mass flux N_{Ar} or mass transfer rate to the wall. This is obtained by using Fick's law at the wall:

$$N_{Ar} = -D_A \left(\frac{\partial C_A}{\partial r} \right)_{r=R} \tag{10.11}$$

The results are expressed in terms of a local mass transfer coefficient k_m defined as

$$N_{Ar} = k_m (C_{Ab} - C_{As})$$

In this expression, C_{Ab} is the cup-mixing or flow-averaged concentraion, defined as

$$C_{Ab} = \frac{\int_0^R 2\pi r v_z(r) C_A(r) dr}{\int_0^R 2\pi r v_z dr} \tag{10.12}$$

The mass flux to the wall or equivalently the mass transfer coefficient is expressed in dimensionless form in terms of the Sherwood number, defined as

$$Sh = \frac{k_m d_t}{D_A} = \frac{d_t N_{Ar}}{D_A(C_{Ab} - C_{As})} \tag{10.13}$$

From the relation for N_{Ar} given by Equation 10.11 the following expression for the Sherwood number is obtained:

$$Sh = -\frac{2}{c_{Ab}} \left(\frac{\partial c_A}{\partial \xi} \right)_{\xi=1} \tag{10.14}$$

Here c_{Ab} is the dimensionless cup-mixing average temperature, defined as

$$c_{Ab}(\zeta) = 4 \int_0^1 \xi(1 - \xi^2) c_A d\xi \tag{10.15}$$

The mass flux to the wall and the corresponding local Sherwood number can be calculated by taking the derivative of the F function at the pipe wall and also by evaluating c_{Ab} (see Equation 10.15). The result for the local Sherwood number, after some clever algebra that avoids integration of the Kummer functions, is

$$Sh(\zeta) = \frac{\sum_{n-1}^{\infty} C_n F_n'(1) \exp(-\lambda_n^2 \zeta) F_n'(1)}{\sum_{n=1}^{\infty} 2C_n [F_n'(1)/\lambda_n^2] \exp(-\lambda_n^2 \zeta)}$$

The first five numerical values of F_n' are also shown in Table 10.1 as a quick aid to calculation of the Sherwood number. For one-term approximation, the value of Sh reduces to $\lambda_1^2/2$. Hence the Sherwood number takes an asymptotic value of 3.657. The reader may wish to note that the entire calculation can be easily done in MALTAB with CHEBFUN as shown later in this chapter.

Average Sherwood Number

Note that the local Sherwood number will be a function of the axial position ζ. The quantity of interest, especially for the 1-D or mesoscopic model discussed in Section 4.2, is the average Sherwood number over a length L^* of the pipe. This is defined as

$$\bar{Sh}_L = \frac{1}{L^*} \int_0^{L^*} Sh d\zeta$$

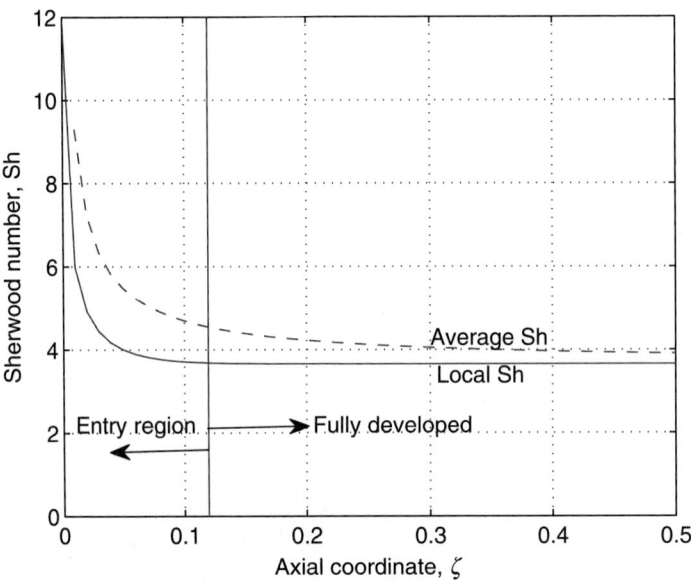

Figure 10.2   Plot of local Sherwood number as a function of axial distance.

The average Sherwood number over a length L^* of pipe can be calculated by integration and is given as

$$\bar{Sh}_L = 3.66 + \frac{0.0668 Pe/L^*}{1 + 0.04[Pe/L^*]^{2/3}}$$

where $L^* = L/d_t$. As L^* becomes large the average Sherwood number approaches an asymptotic value of 3.66.

A plot of the local and average Sherwood number as a function of the axial coordinate ζ is shown in Figure 10.2.

10.2   Wall Reaction: The Robin Problem

A variation to the constant concentration problem is the Robin problem, which is applicable when there is a heterogeneous reaction at the wall. The reaction is assumed to follow first-order kinetics in this discussion so that the separation of variables method is applicable. Nonlinear kinetics require a numerical solution and the PDEPE code in Section 8.11 can be used for such cases. For linear reactions, the eigenproblem can be solved using CHEBFUN, which is illustrated in Section 10.2.1.

   The wall boundary condition for a first-order surface reaction at the wall is obtained by balancing the flux to the wall (Fick's law) to the rate at which A is consumed by the wall reaction:

$$-D_A \left(\frac{dC_A}{dr} \right)_{r=R} = k_s C_A(r = R)$$

The inlet condition is taken as C_{A0}. The dimensionless concentration is now scaled by the inlet concentration and defined as $c_A = \frac{C_A}{C_{Ai}}$. Note that the wall concentration is not used in this definition unlike the earlier problem, as this is unknown.

   The wall boundary condition in dimensionless form is

$$\frac{\partial c_A}{\partial \xi} + Da_w c_A = 0 \text{ at } \xi = 1$$

where Da_w is the wall reaction Biot number defined as $k_s R/D_A$. The boundary condition is homogeneous, which is a requirement for application of direct separation of variables. The problem is linear and amenable to an analytical solution using separation of variables but the computation of eigenvalues and eigenfunctions are even more formidable than the classical Graetz problem. The computations can be completed more easily using CHEBFUN to extract the eigenvalues and simultaneously calculate the eigenfunctions; the code is shown next.

10.2.1 Solution Using CHEBFUN

The use of CHEBFUN for the transient diffusion problem was discussed in Section 8.10 and the process is similar here. The code listing for the Graetz and the wall reaction problems is shown in Listing 10.1. The statement *A.rbc* in the code needs to be changed accordingly for these two cases.

Listing 10.1 Solution of the Linear Convection-Diffusion Problem Using CHEBFUN

```
% CHEBFUN code for Graetz problem.
xi = chebfun('xi',[0,1]);
A = chebop(0,1);
A.op = @(xi,u) xi.*diff(u,2)+diff(u,1);
A.lbc = @(u) diff(u,1)    % Neumann
% % Dirichelt problem
A.rbc =  @(u) u           % Dirichlet
% % Robin problem
```

```
Da = 1.0;
A.rbc =  @(u) diff(u,1)+Da*u   % Robin
B = chebop(0,1);
B.op = @(xi,u) —xi.*(1—xi.^2).*u ;
% then we find the eigenvalues with eigs
[F,L] = eigs(A,B)
omega = sqrt(diag(L)) % eigenvalues
% series coefficient can be evaluated
for i = 1:6
   N1 = int (xi.*(1—xi.^2).* F(:,i) );
 N2 = int (xi.*(1—xi.^2).* F(:,i).* F(:,i) );
 A1(i)= N1/N2 ;
end
%  solution for a given axial distance
zeta1= 0.5;  % axial distance parameter
% one term solution for comparison
i = 1
theta_c = A1(i)*exp(—omega(i)^2*zeta1)* F(xi,i)% answer 0.0381
  %  six term solution
sum = 0.0 ;
for i = 1:6
     sum = sum +  A1(i)*exp(—omega(i)^2*zeta1)* F(xi,i) ;
end
c_zeta = sum %
plot (c_zeta) %  plot of the solution
%% cup—mixing concentration
c_cup = 4* int(xi.*(1—xi.^2).* c_zeta)
%% local Sherwood number
 Fdash = diff(c_zeta) % derivative of the concentration
Sh = —2.* Fdash(1)/c_cup  % answer =  3.6568.for Graetz
```

10.2.2 Illustrative Results

Some illustrative results for the Sherwood number and wall concentration at four axial locations are shown in Table 10.2 for three values of the wall Biot number. A point worth noting is that for low values of Da_w the asymptotic value of the Sherwood number approaches 4.36. This represents a case of constant mass input to the wall (constant flux condition at the wall) and the result can also be proved from a theoretical analysis. At high values of Da_w, the Sherwood number approaches an asymptotic value of 3.66, the constant concentration asymptote.

The concentration profile is shown in Figure 10.3 at an axial location of $\zeta = 0.25$ as a function of the radial position.

Table 10.2   Results for First-Order Wall Reaction

ζ	0.0500	0.2500	0.5000	1.0000
		$Da_w = 0.1$		
Sh, local	4.9402	4.3339	4.3309	4.3309
$c_A(\xi = 1)$	0.9427	0.8685	0.7893	0.6519
		$Da_w = 1.0$		
Sh, local	4.7178	4.1264	4.1242	4.1242
$c_A(\xi = 1)$	0.6044	0.3368	0.1717	0.0447
		$Da_w = 10$		
Sh, local	4.1947	3.7633	3.7629	3.7629
$c_A(\xi = 1)$	0.1117	0.0283	0.0058	0.0002

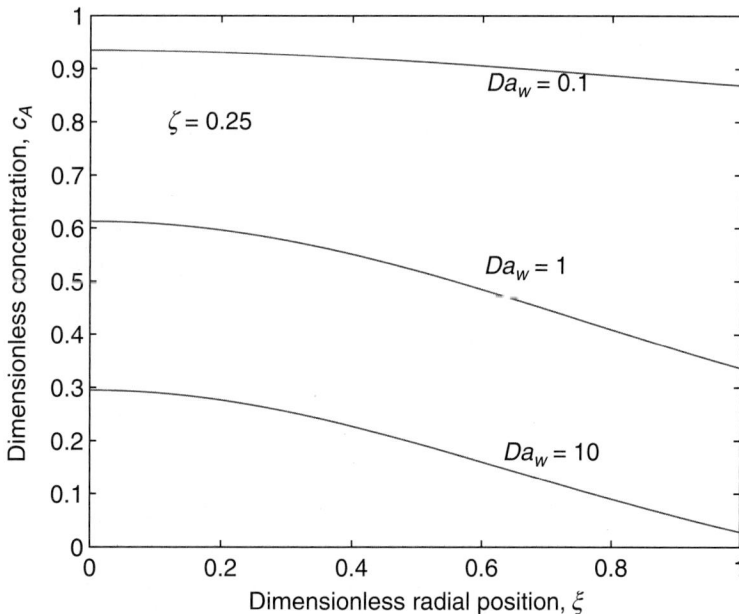

Figure 10.3 Concentration profile for the wall reaction problem as a function of radial position; the solution is shown at axial location ζ of 0.25 for three values of the wall Biot number.

The key points to note are the following. At low values of Da_w, there is no significant radial variation and mass transfer effects are not important. The reaction resistance is dominant. At higher values of Da_w the profiles are steep and the wall concentration tends to zero. This represents a case where the mass transfer to the wall is the dominant resistance.

10.3   Entry Region Analysis

Application of the series solution described in Section 10.1.2 requires a large number of terms for small values of ζ, that is, near the entrance region. The reason is that the concentration profile is rather steep and confined to a region near the wall. Much of the core fluid remains at the inlet concentration. Hence an alternative method is to seek a similarity solution similar to the error function solution discussed for transient mass transfer.

The starting differential equation 10.6 is rewritten, expanding the Laplacian as

$$(1 - \xi^2)\frac{\partial c_A}{\partial \zeta} = \frac{\partial^2 c_A}{\partial \xi^2} + \frac{1}{\xi}\frac{\partial \theta}{\partial \xi} \tag{10.16}$$

The distance from the wall, defined as $s^* = 1 - \xi$, is now more convenient. The velocity profile term $1 - \xi^2$ is now approximated as $2s^*$, that is, we keep only the linear term. Also we keep only the first term for the Laplacian. Thus the second term on the right-hand side is dropped. This amounts to ignoring the curvature effects and representing the geometry using simpler rectangular coordinates.

The differential equation now is

$$2s^*\frac{\partial c_A}{\partial \zeta} = \frac{\partial^2 c_A}{\partial s^{*2}} \tag{10.17}$$

We find that $2s^{*3}/\zeta$ should be used as a similarity variable (or its cube root) since there is no well-defined length scale in the entry region. Note that the radius of the pipe is not a relevant length scale here as the concentration changes are now confined to a thin region near the wall. It is now convenient to use a similarity variable defined as

$$\eta = s^*\left(\frac{2}{9\zeta}\right)^{1/3} \text{ or } = s^*\left(\frac{2Pe}{9z^*}\right)^{1/3}$$

The factor 2/9 is included for the convenience of a simplified form of the results at the end.

The differential equation 10.17 now reduces to an ordinary differential equation in terms of the similarity variable η:

$$\frac{d^2 c_A}{d\eta^2} + 3\eta^2 \frac{dc_A}{d\eta} = 0 \tag{10.18}$$

The boundary conditions to be used now are as follows. At $\eta = 0$, the wall, $c_A = 0$. At $\eta \rightarrow \infty$, the center or the inlet, $c_A = 1$. We also note that the three conditions merge into two, which is a characteristic of similarity solution methods.

The solution is found by integrating this equation twice. The first integration gives an expression for the concentration gradient. A second integration gives the concentration profile. The details are left as an exercise. The second integration has no closed analytical form and the final answer is expressed in terms of the incomplete gamma function:

$$\boxed{\text{Entry region: } c_A = \frac{1}{\Gamma(4/3)} \int_0^\eta \exp(-t^3) dt} \tag{10.19}$$

where η is the similarity variable defined earlier and t is a dummy variable for integration. Equation 10.19 is also known as the Leveque solution.

The results can be expressed compactly as an incomplete gamma function as $c_A = \text{gammainc}(\eta^3, 1/3)$, which is readily calculated in MATLAB. A plot of the incomplete gamma function is presented in Figure 10.4 as an aid to quick calculations.

The corresponding local mass flux is calculated as

$$N_{Aw} \frac{R}{D_A} = \left[\frac{(2Pe/z^*)^{1/3}}{9^{1/3}\Gamma(4/3)} \right] (C_{As} - C_{A,i}) \tag{10.20}$$

Rearranging and basing the Sherwood number on diameter we have

$$Sh = 1.077(2Pe/z^*)^{1/3} \tag{10.21}$$

Note that the driving force for mass transfer is taken as $(C_{As} - C_{A,i})$ rather than $(C_{Ab} - C_{A,i})$ in the definition of the Sherwood number. This was also mentioned earlier in Chapter 9.

The Graetz number parameter, defined as $Q/(D_A z)$, is used in laminar mass transfer literature, where Q is the volumetric flow rate. Thus turns out to be equal to $Gz = (\pi/4)(2Pe/z^*)$. Using this in Equation 10.21, the Sherwood

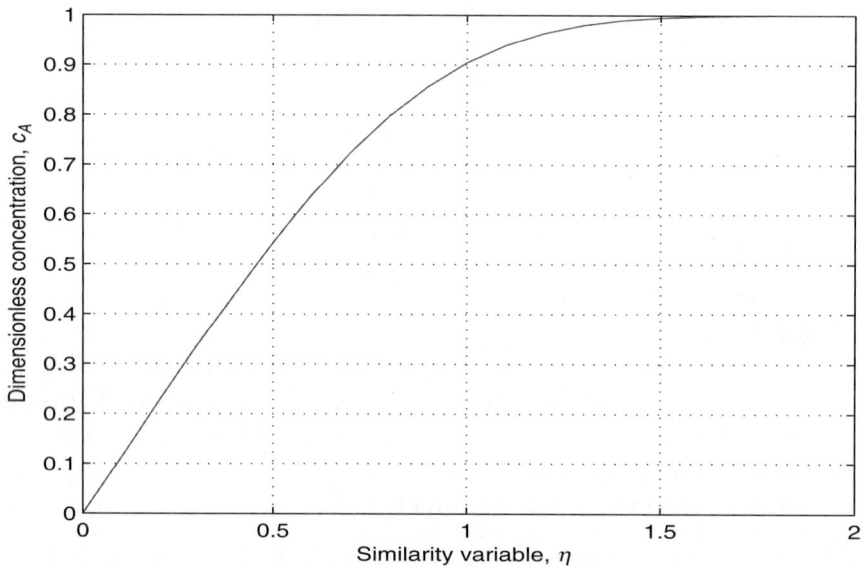

Figure 10.4 Plot of concentration profile for the entry region as a function of the similarity variable. This is a plot of gammainc($\eta^3, 1/3$).

number in the entry region can also be expressed in terms of the Graetz number as

$$Sh = 1.877 Gz^{1/3} \tag{10.22}$$

10.4 Channel Flows with Mass Transfer

In this section, we examine briefly the problem of convective mass transfer with flow between two parallel plates. Flow is in the x-direction. The plate surfaces are at $y = \pm h$, that is, h is half the gap width. $y = 0$ represents the center and is a plane of symmetry. The plates are assumed to be long in the z-direction (perpendicular to the x-y plane) and hence the problem is solved in Cartesian x-y coordinates. The governing equation is

$$v_x \frac{\partial C_A}{\partial x} = D_A \frac{\partial^2 C_A}{\partial y^2} \tag{10.23}$$

The diffusion in the axial (x) direction is ignored in the above model. Thus the convection in the x-direction balances the diffusion in the y-direction.

The velocity profile in channel flow is given from fluid mechanics as

$$v_x = v_{max}\left[1 - \left(\frac{y}{h}\right)^2\right] = \frac{3}{2}\langle v\rangle\left[1 - \left(\frac{y}{y}\right)^2\right]$$

The flow is assumed to be fully developed here and the fluid to be Newtonian. Note that the maximum velocity is $3/2$ times the average velocity in channel flow.

Equation 10.23 reduces to the following when dimensionless variables are used:

$$\frac{3}{2}(1 - \xi^2)\frac{\partial c_A}{\partial \zeta} = \frac{\partial^2 c_A}{\partial \xi^2} \tag{10.24}$$

The parametric representation is now

$$c_A = c_A(\zeta, \xi)$$

The ζ is the dimensionless coordinate in the flow direction, defined as

$$\zeta = \frac{4(x/h)}{ReSc}$$

The Reynolds number is defined based on the hydraulic diameter, which is equal to $4h$:

$$Re = \frac{4h\langle v\rangle}{\nu}$$

Equation 10.24 has the series solution similar to that for pipe flow (Equation 10.7), which is reproduced below for ease of reference:

$$c_A(\zeta, \xi) = \sum_{n=1}^{\infty} C_n \exp(-\lambda_n^2\zeta)F_n(\xi) \tag{10.25}$$

However, the eigenvalues and so on are now different since the diffusion operator is now in Cartesian coordinates rather than in cylindrical coordinates. These can be generated easily using the CHEBFUN code shown in Listing 10.2.

Listing 10.2   CHEBFUN for Eigenvalues and Eigenfunctions for Channel Flow Mass Transfer

```
xi = chebfun('xi',[0,1]);
A = chebop(0,1);
A.op = @(xi,u) diff(u,2);
```

```
A.lbc = @(u) diff(u,1)   % Neumann
A.rbc =   @(u) u         % Dirichlet

B = chebop(0,1);
B.op = @(xi,u) -3/2*(1-xi.^2).*u ;
%%
neig = 6 % first six values
[F,L ] = eigs(A,B, neig)
```

The square of the eigenvalues is found in the diagonal L while the eigenfunctions are found in F as Chebechev polynominals. The series coefficients are then calculated using the code segment shown in Listing 10.3. Note that the eigenfunctions are now orthogonal with a weighting factor of $1 - \xi^2$.

Listing 10.3   Code to Find Series Coefficients for Channel Flow

```
i = 1:neig
  N1 = int ((1-xi.^2).* F(:,i) );
 N2 = int ((1-xi.^2).* F(:,i).* F(:,i) );
A1(i)= N1/N2 ;
 B1(i)  = N1
 C1(i)   = A1(i)*B1(i)*3/2
end
```

The solution can now be calculated using Equation 10.25. Once the concentration distribution is known, the Sherwood number is calculated as

$$Sh = \frac{4hk_m}{D_A} = -\frac{4(\partial c_A/\partial \xi)_{\xi=0}}{c_{Ab}}$$

where c_{Ab} is the dimensionless cup-mixing concentration defined as

$$c_{Ab} = \frac{3}{2} \int_0^h (1 - \xi^2)c_A(\xi)\, d\xi$$

Note that the hydraulic diameter ($4h$) is again used as the length scale.

The plot of the Sherwood number as a function of axial distance is shown in Figure 10.5.

The Sherwood number reaches an asymptotic value of 7.541. Laminar flow holds until a Reynolds number of 2800 for channel flows and the model is not applicable in turbulent flow conditions.

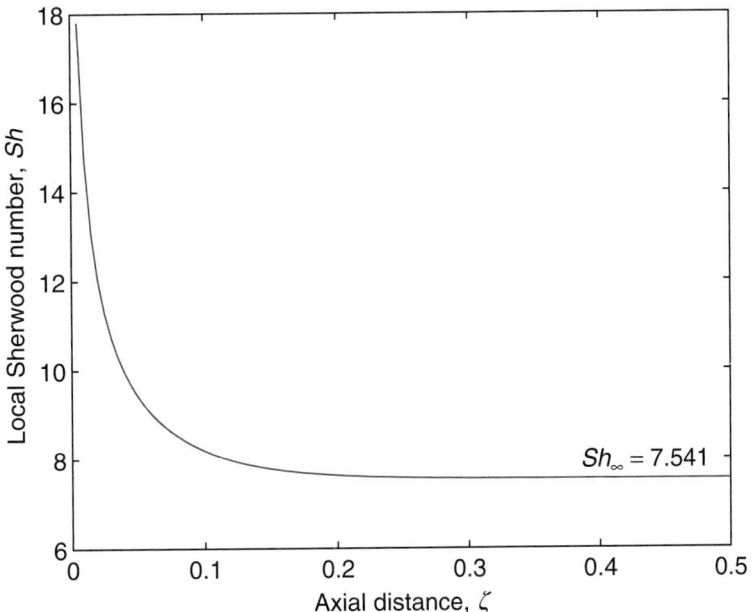

Figure 10.5   Mass transfer coefficient (Sherwood number) for flow in a channel with constant wall concentration.

The case of wall reaction can also be easily generated using the CHEBFUN-MATLAB code. Details are not presented here since it is an extension of the corresponding scheme for pipe flows.

10.5   Mass Transfer in Film Flow

An important class of problems involves mass transfer to or from a liquid flowing as a thin film over a vertical plate. The velocity profile for film flow is obtained by balancing the gravity forces with viscous forces. The result from fluid dynamic literature for laminar flow of the film is

$$v_z(y) = \frac{\rho g}{\mu} \left[\delta y - \frac{y^2}{2} \right] \tag{10.26}$$

Here z is the flow direction, (vertically downward for film flow over a vertical wall). The y is the cross-flow direction starting from the wall. The laminar flow assumption is valid if the Reynolds number based on film thickness and average velocity is less than 20.

The film thickness δ can be computed from an overall integral mass balance (integral from 0 to y) and is related to the liquid flow rate Q/W (per unit width of the wall) as follows:

$$\delta = \sqrt[3]{\frac{3\mu Q/W}{\rho g}}$$

Now we analyze the mass transport of a species dissolving from the wall or absorbed from the interface into the film.

10.5.1 Solid Dissolution at a Wall in Film Flow

The governing equation for mass transfer in film flow is

$$v_z(y)\frac{\partial C_A}{\partial z} = D\frac{\partial^2 C_A}{\partial y^2} \tag{10.27}$$

Here we neglect diffusion in the flow direction. Thus the left-hand side represents convection in the direction of flow (z) and the right-hand side represents diffusion in the cross-flow direction (y). The Sc, defined as ν/D, is the key parameter affecting the mass transfer penetration distance. For liquids, the Sc is large and we expect a boundary layer type of concentration distribution; the solute concentration changes will be confined to only a small part of the hydrodynamic film thickness.

Anticipating a concentration boundary layer near the wall, we use a linearized form as an approximation for the velocity. The approximation is shown in Figure 10.6:

$$v_z = \beta y$$

where β is the slope of the velocity profile near the wall, given as

$$\beta = \frac{\rho g}{\mu}\delta$$

This will be the velocity profile "seen" by the diffusing species and hence the convection-diffusion model (Equation 10.27) reduces to

$$\beta y\frac{\partial C_A}{\partial z} = D\frac{\partial^2 C_A}{\partial y^2} \tag{10.28}$$

The domain of the solution is set as semi-infinite since the concentration profile is assumed to be confined to a thin region near the wall. A similarity

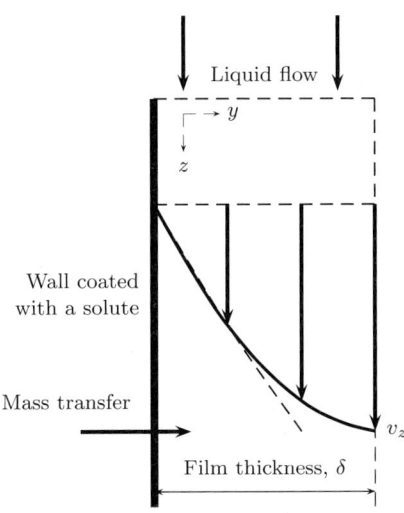

Wall coated
with a solute

Mass transfer

Film thickness, δ

Liquid flow

v_z

– – – Assumed linear velocity profile near the solid

——— Actual parabolic velocity profile

Figure 10.6 Solid dissolution in film flow. In the solid dissolution case, the region near the wall is important.

variable η is defined as follows:

$$\eta = y \left(\frac{\beta}{9Dz} \right)^{1/3}$$

With this transformation, Equation 10.28 reduces to an ordinary differential equation for the dimensionless concentration as a function of η:

$$\frac{d^2 c_A}{d\eta^2} + 3\eta^2 \frac{dc_A}{d\eta} = 0 \tag{10.29}$$

The inlet condition and the condition as $y \rightarrow \infty$ merge into a single condition that states that at $\eta = \infty$, $c_A = 0$.

The wall boundary condition is the second boundary condition and at $\eta = 0$, $c_A = 1$.

The solution to Equation 10.29 with these boundary conditions can be derived as

$$\frac{C_A}{C_{As}} = 1 - \frac{\int_0^\eta \exp(-\eta^3 d\eta}{\Gamma(4/3)} \tag{10.30}$$

The solution can be readily computed using the built-in function in MATLAB as

$$\frac{C_A}{C_{As}} = 1 - \text{gammainc}(\eta^3, 1/3) \tag{10.31}$$

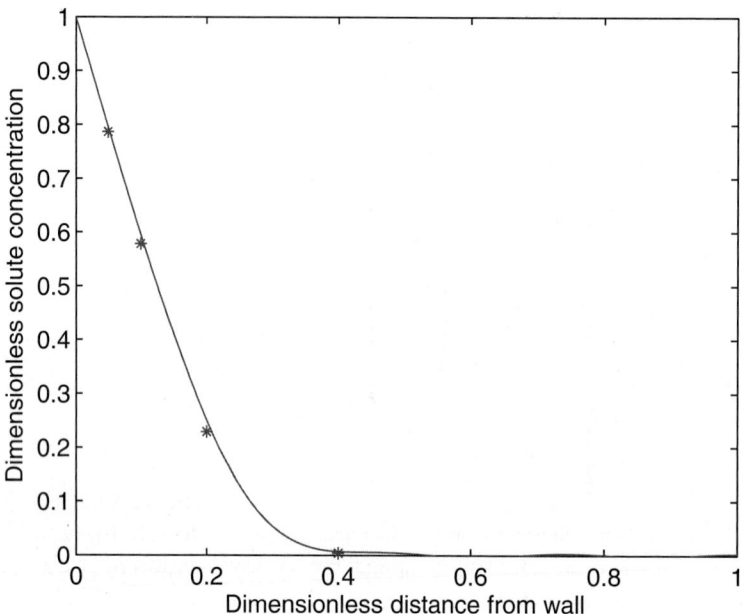

Figure 10.7 Concentration profile for solid dissolution in a falling film and comparison of the similarity solution with the numerical solution, marked with *.

which gives the dimensionless concentration as a function of η. The concentration profile is shown in Figure 10.7 where a comparison with the numerical solution is also shown.

The local rate of mass transfer is obtained using Fick's law at the wall:

$$N_A = \frac{DC_{As}}{\Gamma(4/3)} \left(\frac{\beta}{9Dz} \right)^{1/3}$$

The average rate of mass transfer over a length L is given by

$$\bar{N}_A = \frac{2DC_{As}}{\Gamma(7/3)} \left(\frac{\beta}{9DL} \right)^{1/3}$$

If one divides this by C_{As}, the driving force, one can calculate the average value of the mass transfer coefficient over a length L:

$$k_{sl} = \frac{2}{\Gamma(7/3)} D^{2/3} \beta^{1/3} L^{-1/3}$$

The model can be extended to film flow of non-Newtonian fluids. The corresponding velocity profile is to be included. Mashelkar and Chavan (1973) provide the results for solid dissolution mass transfer for pseudoplastic fluids.

10.5.2 Gas Absorption from Interface in Film Flow

Here we consider the mass transfer of a species present in the gas into the liquid film. Again a concentration boundary layer is proposed near the interface. The velocity profile seen is now flat and equal to v_{max} rather than a linear profile as in the previous subsection. The maximum velocity in turn is given from Equation 10.26 by setting $y = \delta$ and is

$$v_{max} = \frac{\rho g\, \delta^2}{2\mu}$$

The diffusing species now sees this flat velocity profile rather than the full parabolic profile. This is also illustrated in Figure 10.8.

The convection-diffusion equation can now simplified by using v_z as a constant equal to v_{max}:

$$v_{max}\frac{\partial C_A}{\partial z} = D\frac{\partial^2 C_A}{\partial y^2} \tag{10.32}$$

Note that the left-hand side of Equation 10.32 has no functional dependence on y and therefore the problem is closer to the transient diffusion

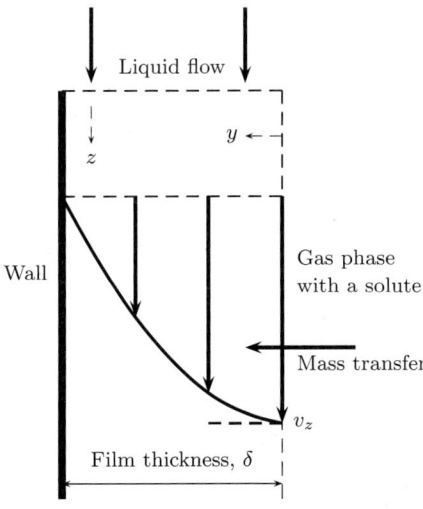

Figure 10.8 Gas absorption in film flow: the region near the interface is important.

– – – Assumed constant velocity profile near the interface

—— Actual parabolic velocity profile

problem rather than a convection-diffusion problem. In fact z/v_{max} is a time-like variable and represents the exposure time of an element of fluid moving down the interface. The concentration profile is now represented in terms of an error function:

$$\frac{C_A}{C_{As}} = 1 - \text{erf}\left(\frac{y}{\sqrt{4Dz/v_{max}}}\right) \qquad (10.33)$$

Here y is the now defined as the distance measured from the interface.

The local rate of mass transfer is obtained by using Fick's law at the wall. The result is the same as the penetration theory model discussed in Section 8.8 with t replaced by the exposure time, which is equal to z/v_{max}:

$$N_A = C_{As}\sqrt{\frac{Dv_{max}}{\pi z}}$$

The average mass transfer coefficient over a length L is given as

$$\boxed{k_L = 2\sqrt{\frac{Dv_{max}}{\pi L}}} \qquad (10.34)$$

This is again the same as the penetration theory model with the contact time set as L/v_{max}.

10.6   Numerical Solution with PDEPE

The PDEPE program shown in Section 8.11 is a useful tool to simulate these classes of problems as well. A brief discussion is provided together with sample results for the Graetz problem.

This problem is readily handled by PDEPE by simply defining the capacity term as a function of radial position. In the notation used in the sample code, shown in Section 8.11, x represents dimensionless radial position, ξ, while t represents the dimensionless axial position η. The capacity term is now defined as

$$C = 1 - x^2$$

in order to match Equation 10.6. The geometry parameter s should be set as one for pipe flow to match the Laplacian in cylindrical geometry. Setting it to zero will solve the channel flow case.

A sample code file graetz.m can be written with these modifications of the code for transient diffusion shown in Section 8.11.1.

Table 10.3 Results for the Graetz Problem Simulated with PDEPE

ζ	$1 - c_{A(0)}$	N_{Ar}	$1 - c_{Ab}$	Sh
.05	0.0605	1.158	0.4213	4.00
0.1	0.2988	0.733	0.6047	3.71
0.15	0.5081	0.501	0.7265	3.665
0.20	0.6582	0.347	0.8103	3.658
0.25	0.7628	0.241	0.8684	3.657
0.30	0.8355	0.167	0.9087	3.657
0.35	0.8858	0.116	0.93679	3.657
0.40	0.9208	0.0803	0.9561	3.657

Some results are given in Table 10.3 so that students can benchmark the programs and verify the results. The results are also shown for the case of constant wall concentration examined in Section 10.1.1.

The value from the one-term solution for $\zeta = 0.4$ is 0.9207 for the center concentration, which matches well with the PDEPE-generated solution shown in this table.

The Robin problems, nonlinear rate, and other cases can be easily handled by this procedure. Only the boundary conditions need to be modified. For channel flow we set the s parameter to zero.

The advantage of PDEPE over the CHBEBFUN method is that it can handle nonlinear boundary conditions. For linear problems the CHEBFUN method is most accurate since there is no spatial or time discretization.

Summary

- Mass transfer from the walls to a fluid in a pipe under laminar flow conditions is the most widely studied problem in convective mass transfer. A number of assumptions are usually made to simplify the problem, such as constant viscosity and fully developed flow. Even then the problem is mathematically complicated and results in a partial differential equation in r and z with a variable coefficient term. The variable coefficient arises due to the velocity variation in the r direction. It is useful to understand the mathematical structure of the model equation.

- The boundary condition at the wall can be of constant concentration (first kind or Dirichlet) or given by a wall reaction case (third kind or Robin). A series solution can be obtained for both cases. It is straightforward for the constant wall temperature case but the eigenfunctions are not simple well-known common functions; they to a class of functions called hypergeometric functions. These functions can however be readily evaluated in MATLAB or MAPLE or CHEBFUN.

- From a mesoscale modeling point of view the local and average value of the transfer coefficient are needed. These can be predicted from the detailed concentration profiles based on the differential model. The results are expressed in terms of a dimensionless group, the Sherwood number. The Sherwood number is an function of axial distance and decreases with distance, reaching an asymptotic value for large distances from the entrance.

- The convergence of the series is rather poor near the entrance. An alternative method can be used here where we visualize a concentration boundary layer near the wall and hence it is more convenient to obtain the similarity solution for the entry region.

- The effect of wall reaction can be analyzed by changing the wall boundary condition from Dirichlet to Robin. The eigenvalues can be obtained by using the CHEBFUN routine and the Sherwood number in the presence of a reaction can be computed.

- Mass transfer in channel flow from dissolving walls is a similar problem and can be solved by the separation of variables method. Again CHEBFUN can be used to obtain the full solution without any mathematical pain.

- Mass transfer into a falling film is an important problem in convective mass transfer. Solid dissolution at the walls and gas absorption at the interface are the two important prototype problems. The length of the wall (absorber) is usually small and hence an asymptotic entry region type of model is generally used for these cases. The local and average value of the mass transfer coefficient can be predicted from such a model. It is important to note the dependency on the diffusion coefficient. In the solid dissolution case the dependency is to the power of 2/3 while in the gas absorption case it is 1/2.

- The PDEPE software studied in Chapter 8 is a useful tool for convective mass transport problems as well. Only minor code adjustments are needed to match the PDE being solved and the appropriate boundary condition. The use of this was demonstrated for the classical Graetz problem and the results for other cases can be similarly computed.

Review Questions

10.1 What is hydrodynamic entry length? Is the velocity unidirectional in this region?

10.2 What is the definition of a contracted axial distance parameter?

10.3 What is the asymptotic value of the Sherwood number for constant wall conditions?

10.4 What is the contracted length for which the mass transfer is nearly complete?

10.5 What is the asymptotic value of the Sherwood number for constant mass removal conditions, for example, a zero-order reaction at the wall?

10.6 How is the wall Biot number defined?

10.7 Does the Sherwood number increase or decrease with the wall reaction Biot number?

10.8 What is the dependency of the mass transfer coefficient on the mass flow rate in the entry region?

10.9 How does the mass transfer coefficient vary with axial length in the entry region?

10.10 What is the definition of the Graetz number?

10.11 What is the dependency of the mass transfer coefficient with diffusivity at a no-slip boundary?

10.12 What is the dependency of the mass transfer coefficient with diffusivity at a no-shear boundary?

Problems

10.1 **Eigenvalues with CHEBFUN.** Run the code given in Listing 10.1 and generate the results shown in Table 10.1 for the Dirichlet problem. Modify and run the code with a wall reaction and generate the results shown in Table 10.2. Run the code shown in Listing 10.2 for channel flow and verify the results in Figure 10.5. Extend the channel flow code for a linear wall reaction and plot Sh as a function of Damkohler number. Choose any fixed axial position for this plot.

10.2 **Subliming wall.** Water is flowing in a laminar flow reaction in a 1 cm diameter tube coated with benzoic acid. The value of $D = 1 \times 10^{-9}$ m^2/2. The velocity is 10 cm/sec. Calculate and plot the mass transfer coefficient as a function of axial position in the pipe. If the pipe is 3 m long, plot the solute concentration at the exit as a function of radius. Find the cup-mixing concentration and compare the results with the mesoscopic model given in Section 4.2.

10.3 **Laminar flow with wall reaction.** Find the exit concentration and the conversion for a laminar flow with wall reactor under the following conditions: fluid properties are similar to that for water; radius = 1 cm; length = 500 cm; mass flow rate = 0.1 kg/s; surface reaction constant $k_s = 2 \times 10^{-4}$ m/s.

10.4 **Mass transfer under constant wall flux conditions.** A constant wall flux boudary condition is applied if there is a constant mass removal rate at the walls. For example if a zero-order reaction is taking place, the wall boundary condition can be stated as follows: show that the wall boundary condition in dimensionless form is

$$\frac{\partial c_A}{\partial \xi} = -Da_{w0} \text{ at } \xi = 1$$

Here Da_{w0} is the wall reaction Biot number for a zero-order wall reaction. How is this defined?

The asymptotic value of the mass transfer can be derived by the following procedure. Assume that the dimensionless concentration is linear invariant in the axial direction and propose a solution of the form

$$c_A = A\zeta + F(\xi)$$

where A is a constant. Substitute this in Equation 10.6 and derive an equation for F. Integrate this equation to find F with the boundary condition at the wall and the center. Find the cup-mixing concentration and use this in the definition for Sherwood number given by Equation 10.13. Show that the resulting value of the asymptotic Sherwood number is 48/11, which is is a classic result.

10.5 **Solid dissolution in a falling film.** Consider a liquid flowing down a vertical wall at a rate of 1×10^{-5} m^2/s per meter unit width. Find the concentration at a height 25 cm below the entrance for a dissolving wall such as a wall coated with benzoic acid as a function of distance from the wall. Also find the local mass transfer coefficient.

Find the average mass transfer coefficient for a 50 cm wall. Assume $D = 2 \times 10^{-9}$ m^2/s and other properties similar to that of water. Use $C_{As} = 20$ mol/m^3.

10.6 **Gas absorption in a falling film.** Repeat the analysis from problem 5 if instead, a gas such as oxygen is being absorbed into the liquid film.

10.7 **Laminar flow and mass transfer for a power law fluid.** Extend the analysis of the laminar flow mass transfer in Section 10.1 for a power law liquid. The velocity profile is now given as

$$v_z = \frac{3n+1}{n+1} \langle v \rangle \left[1 - \left(\frac{r}{R} \right)^{(n+1)/n} \right] \tag{10.35}$$

where n is the power law index. Perform some computations using the PDEPE solver and show how the power law index affects the mass transfer coefficient.

10.8 **Mass transfer in a counterflow exchanger.** A feed solution containing a solute enters the inner tube of a mass exchanger. A membrane separates the inner tube with an annular outer cylinder. The solute diffuses across the membrane into the annular region. A sweep solution enters from the other side in the annular section and carries away the diffusing solute. The schematic is presented in Figure 10.9. The problem is representative of transport across dialysis equipment.

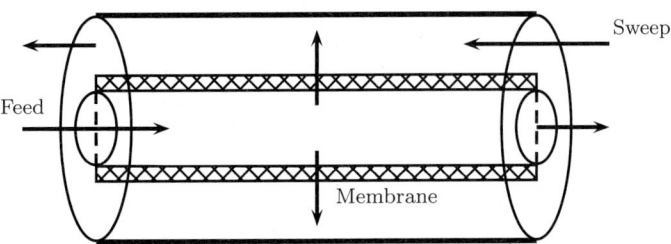

Figure 10.9 Mass transfer in a counterflow mass exchanger.

Assume laminar flow in both the inner and outer pipe. Set up the convection-diffusion model for both the inner pipe and the annular pipe. Set up the boundary conditions showing all the details clearly.

10.9 **Mass transfer in oscillating flow.** Mass transfer can be enhanced by flow oscillation. Assume that the flow oscillation is caused by a sinusoidal variation of the inlet pressure. Perform a scaling analysis based on the key time constants to find conditions under which the enhancement is likely and also investigate the problem using numerical tools such as PDEPE. Determine how the time-averaged Sherwood number varies with the dimensionless amplitude of oscillation.

CHAPTER 11

Mass Transfer in Laminar Boundary Layers

Learning Objectives

After completing this chapter, you will be able to:

- Set up the model for mass transfer for flow of a fluid over a flat plate.
- Understand the differences in model formulation for low flux and high flux mass transfer.
- Solve the flat plate problem by the Blasius method.
- Understand the integral solution method for boundary layer problems and use it for a number of prototypical problems in mass transfer.
- Correct the results of the low flux case for the high flux case using the film model as well as the Blasius and integral methods.
- Model the mass transfer for flow over an inclined plate using the Falkner-Skan method.
- Understand the basics of model development for mass transfer for flow past a sphere and from bubbles and drops.

In this chapter we study from a theoretical viewpoint mass transfer in external (unconfined) flows. A characteristic feature of such flows is the existence of the boundary layers near the solid surface. These are regions where the velocity and the concentration vary steeply. Outside the boundary layer the velocity and the concentration are uniform and equal to the free stream values. Some features were discussed in Section 6.3.1 for flow over a flat plate and it is useful to review that section again. In that section, we provided some formulae for calculation of the boundary layer thickness and the mass transfer coefficient. Here we analyze the problem in more detail and provide the mathematical underpinnings behind these key formulae.

A key distinguishing feature of external flow compared to internal flow is that the flow is never fully developed in external flow. Thus the velocity has both x- and y-components and hence the mathematical analysis is somewhat more involved. The problem is first analyzed using a low flux mass transfer. The equations for the velocity are reviewed from momentum balance. These are then used in the convective mass transport model and two methods of solution are shown together with the final result for the mass transfer coefficient. Then we indicate the corrections needed for the high flux case.

This chapter deals with laminar flow. For a flat plate, the flow starts off laminar and a region near the leading edge would be under laminar conditions. The models shown here are applicable to this region. For large distances from the leading edge the flow becomes turbulent and analysis of turbulent boundary layers is treated in the next chapter.

Laminar boundary layers are also encountered in complex geometries such as flow past a sphere. One distinguishing feature, compared to flat plates, is the possibility of boundary layer separation. Analysis of such a problem is more complex. We briefly discuss the methodology and supplement it with some illustrative results. Although CFD-based tools are now common for solution of such problems, it is important to understand the theoretical underpinnings in the model formulation so that the CFD tools can be more effectively used and the results interpreted in terms of a physical concepts. Hence the discussion in this chapter is very useful as prelude to using these tools.

The final section briefly discuss mass transfer from drops and bubbles. This is an even more complex problem compared to a solid boundary since the interface is now mobile and there can be flow both inside the drop/bubble phase as well as in the external phase. In the literature, the mathematical models have been proposed and solved for a number of standard cases, but a detailed discussion of these studies is beyond the scope of the current text. Hence the discussion provided here is mostly qualitative and some references are provided for those interested in pursuing this topic further.

11.1   Flat Plate with Low Flux Mass Transfer

Consider a fluid flowing past a solid plate with some mass transfer taking place from or to the plate as well. The flow direction is marked x while the cross-flow direction is marked as y. The analysis of such mass transfer problems may be classified into two types depending on whether the rate of mass transfer from/to the surface to/from the flowing fluid is large enough to affect the flow. Note that the mass transfer can create a net mass flux n_t (in the

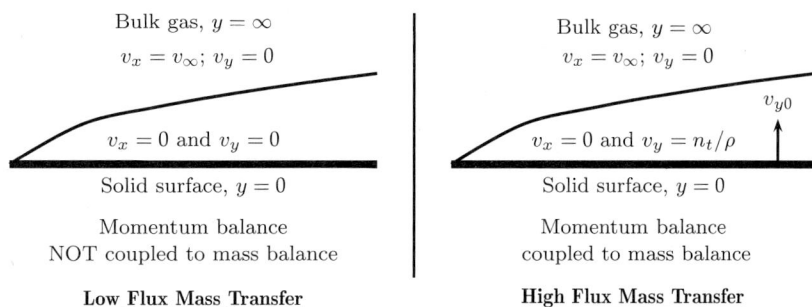

Figure 11.1   Difference in momentum boundary condition for the low flux versus high flux mass transfer; v_y at $y = 0$ is non-zero for the high flux case and the value has to come from mass transfer considerations.

y-direction) in some cases as discussed in Chapter 6. A flow in the y-direction is introduced as the result of this flow:

$$v_y(y = 0) = n_t/\rho \tag{11.1}$$

where n_t is the component of total mass flux in the y-direction at the surface. Hence the no-slip boundary condition for momentum transfer at $y = 0$ has to be modified accordingly. The x-component of velocity at the surface v_x is still zero at $y = 0$ but the y-component is equal to n_t/ρ. Hence the momentum balance boundary conditions can be altered by the rate of mass transfer. This is referred to as the high flux case or detailed model. The difference in these two problems is illustrated in Figure 11.1.

The point to note is that the high flux case is a two-way coupled problem between momentum transfer and mass transfer. In contrast if n_t is small (or even zero as a consequence of equal counter diffusion), then the momentum transfer can be solved independently of mass transfer. This is referred to as a low flux case and is the focus of this section. The problem is now one-way coupled and the flow analysis for boundary layer theory can then be directly used in mass transfer analysis. The momentum transfer is not coupled to mass transfer. The high flux case in contrast is a two-way coupled problem and is somewhat complicated in terms of the mathematics. Some aspects of it are discussed in Section 11.3.

11.1.1 Concentration Equation

The concentration field is now given by the convection-diffusion equation, which is repeated here for convenience:

$$\boldsymbol{v} \cdot \nabla C_A = D_A \nabla^2 C_A$$

The velocity has both x- and y-components here. Hence the expanded version of this equation is

$$v_x \frac{\partial C_A}{\partial x} + v_y \frac{\partial C_A}{\partial y} = D_A \left(\frac{\partial^2 C_A}{\partial x^2} + \frac{\partial^2 C_A}{\partial y^2} \right) \tag{11.2}$$

The two terms on the left-hand side represent the convection due to flow in the x-direction and that due to flow in the y-direction. The two terms on the right-hand side represent the diffusive transport in the flow direction (x) and the cross-flow direction (y).

It turns out that the diffusion term in the flow direction is smaller than that in the cross direction (y), that is,

$$\frac{\partial^2 C_A}{\partial x^2} << \frac{\partial^2 C_A}{\partial y^2}$$

Hence this term may be neglected in Equation 11.2 reducing the model to

$$v_x \frac{\partial C_A}{\partial x} + v_y \frac{\partial C_A}{\partial y} = D_A \frac{\partial^2 C_A}{\partial y^2} \tag{11.3}$$

This is the concentration equation for the flat plate mass transfer problem.

To proceed further we need the momentum balance considerations so that relations for v_x and v_y can be used in this equation. The relevant equations are shown next.

11.1.2 Velocity Equations

The x-momentum balance can be simplified for flat plate flow:

$$v_x \frac{\partial v_x}{\partial x} + v_y \frac{\partial v_x}{\partial y} = \nu \frac{\partial^2 v_x}{\partial y^2} \tag{11.4}$$

The key simplification is the pressure gradient term is set to zero. Note that for inclined plates this term will not be zero and an extra term has to be added in the x-momentum balance as shown in Section 11.4.

The x-momentum equation needs to be solved together with the continuity equation:

$$\frac{\partial v_x}{\partial x} + \frac{\partial v_y}{\partial y} = 0 \tag{11.5}$$

The x-momentum can also be written using the continuity equation as

$$\frac{\partial(v_x\, v_x)}{\partial x} + \frac{\partial(v_y v_x)}{\partial y} = \nu\frac{\partial^2 v_x}{\partial y^2} \tag{11.6}$$

This is an easier form to obtain the integral momentum balance shown in the next section, and is known as the conservation form of the momentum balance.

Likewise the concentration equation can also be written for conservatives as

$$\frac{\partial(v_x C_A)}{\partial x} + \frac{\partial(v_y C_A)}{\partial y} = D_A\frac{\partial^2 C_A}{\partial y^2} \tag{11.7}$$

11.1.3 Scaling Results and the Analogies

One point to note is that Equations 11.6 and 11.7 for v_x and C_A have very similar structure. The kinematic viscosity appears on the right-hand side of the velocity equation while the diffusion coefficient appears in the concentration equation. Thus the profiles (in dimensionless form) are expected to be identical if $\nu = D_A$ or equivalently if $Sc(=\nu/D_A) = 1$ (assuming similar wall boundary conditions). The momentum boundary layer thickness is now the same as the mass transfer boundary layer thickness. Correspondingly we can show that

$$St - \frac{C_f}{2}$$

where St is the Stanton number, a measure of mass transfer, and C_f is the drag coefficient, a measure of momentum transfer. This is one example of analogy between momentum and mass transfer, known as the Reynolds analogy.

Scaling analysis refers to a technique where the order of magnitude of various terms are compared. In a properly scaled equation the order of each term should be the same. Terms that are much smaller relative to others can be neglected. Results of engineering interest can often be obtained by scaling analysis without the need to solve the full equations. A useful result arises from scaling analysis of boundary layer flow where we compare the order of magnitude of various terms. This result provides an estimate of the momentum boundary layer thickness δ_M:

$$\delta_M = O\left(\sqrt{\frac{\nu x}{v_\infty}}\right)$$

where O denotes that the two terms have nearly the same order of magnitude.

Scaling analysis also provides a relation for the ratio of the mass transfer boundary layer thickness, δ_m, to the momentum boundary layer thickness:

$$\frac{\delta_m}{\delta_M} = \frac{1}{Sc^{1/3}} \tag{11.8}$$

which is often referred to as the Prandtl analogy. This again shows that the two boundary layer thicknesses are close if $Sc \approx 1$ (gases) while the mass transfer boundary layer becomes very thin at large Schmidt numbers (e.g., in liquids).

The mass transfer coefficient or equivalently St is proportional to D/δ_m. Similarly, the drag coefficient $C_f/2$ is proportional to μ/δ_M. Hence the two dimensionless groups can be shown to be related through the Schmidt number:

$$St = \frac{C_f}{2} Sc^{-2/3} \tag{11.9}$$

Expressing in terms of the j-factor we find

$$j_D = St Sc^{2/3} = \frac{C_f}{2}$$

which provides an analogy between mass and momentum transfer.

A heat transfer factor, j_H can be defined as

$$j_H = \frac{h}{\rho c_p v_\infty} Pr^{2/3}$$

We also find

$$j_H = j_D$$

This is referred to as the Chilton-Colburn analogy and provides an analogy between heat and mass transfer.

These analogies are useful even for complex geometries and are not restricted to flat plates. The key postulate is that even for these complex geometries, the concentration variation is confined to a thin boundary layer near the solid surface and hence the precise curvature of the solid surface is not an important factor.

If there is boundary layer separation, the analogy between mass transfer and momentum transfer will not hold in the region of separation. But the analogy between heat and momentum transfer still holds. Hence heat transfer data can often be used to predict mass transfer coefficients!

11.1.4 Exact or Blasius Analysis

Now we look at a semi-analytical solution to the problem of flat plate mass transfer under low flux conditions. Note that the momentum is uncoupled here and we therefore look at the solution of the momentum problem first. This is then followed by the mass transport equation.

Momentum Equation

Blasius (1908) combined the momentum and continuity equations into a single third-order ordinary differential equation. Let us follow his footsteps and see how he did it.

First, he introduced the stream function so that the continuity equation would automatically be satisfied:

$$v_x = \frac{\partial \psi}{\partial y} \tag{11.10}$$

$$v_y = -\frac{\partial \psi}{\partial x} \tag{11.11}$$

Using this in Equation 11.4, the x-momentum balance equation becomes:

$$\frac{\partial \psi}{\partial y}\frac{\partial^2 \psi}{\partial x\,\partial y} - \frac{\partial \psi}{\partial x}\frac{\partial^2 \psi}{\partial y^2} = \nu\frac{\partial^3 \psi}{\partial y^3} \tag{11.12}$$

He then introduced a new stretched coordinate (dimensionless) in the y-direction. This coordinate η is defined as

$$\eta = \frac{y}{2}\left(\frac{v_e}{\nu x}\right)^{1/2} \tag{11.13}$$

Note: Here v_e is the x-velocity at the edge of the boundary layer, which is the same as the approach velocity v_∞ for a flat plate. This change of notation is not needed here but is useful for the more general case of inclined plates considered in Section 11.4.

He also introduced a dimensionless stream function f defined by

$$f = \frac{\psi}{(\nu v_e x)^{1/2}} \tag{11.14}$$

He postulated that f is a function of η.

The velocity profiles in turn can be related to f as follows:

$$v_x = \frac{v_e}{2} \frac{df}{d\eta}$$

$$v_y = \frac{1}{2} \left(\frac{\nu v_e}{x}\right)^{1/2} \left(\eta \frac{df}{d\eta} - f\right)$$

The derivatives of the velocity can also be written in terms of the f-function. With all of these chain rule substitutions, Equation 11.12 reduces to the following form:

$$\boxed{\frac{d^3 f}{d\eta^3} + f \frac{d^2 f}{d\eta^2} = 0} \tag{11.15}$$

This is called the Blasius equation for laminar boundary layers. We find neither x nor y appears explicitly and hence f is a function of η only as postulated by Blasius. η is called a similarity variable.

Let us now look at the boundary conditions to be used for this equation. Since the differential equation for f is third order, three boundary conditions are needed. We start from the boundary conditions for the primitive variables and map these into the $f - \eta$ domain, which is shown next.

- At the surface $y = 0$, the x-velocity v_x is zero and since v_x is proportional to f', this leads to $f' = 0$ at $\eta = 0$.
- At the surface $y = 0$, the y-velocity v_y is also zero in the low flux model. Since v_y is proportional to $\eta f' - f$, this leads to $f = 0$ at $\eta = 0$. Note that this boundary condition will need to be modified for the high flux model analyzed in Section 11.3.3.
- Far away from the plate the free stream velocity v_e is approached. Hence as $y \to \infty$, v_x should approach v_e. Thus at $\eta = \infty$, we should set f' to two.

Blasius solved the f differential equation by series expansion. It is easier to solve this nowadays numerically and MATLAB code using ODE45 is shown later (Listing 11.1). Note that an ODE in f is being solved rather than a PDE, which is one of the simplifications in the Blasius approach. Also note that the infinite domain has to be truncated to a finite domain for numerical solution. Usually $\eta = 5$ is sufficient.

This completes the velocity problem and the profiles can be computed independently of the mass transfer case as we have a one-way coupling here. The above boundary layer solution fits very well the experimental data of Nikuradse (1932).

Thickness of Boundary Layer

The solution for the velocity indicates that the x-velocity becomes 0.99 of the bulk value when $\eta = 2.5$. This can be checked by looking at the numerical results for f' values generated by the code in Listing 11.1. The f' is nearly equal to 2 at $\eta = 2.5$ and hence v_x/v_e is nearly one at this point. Hence the boundary layer thickness is 2.5 when viewed from η as the coordinate.

Using the definition of η, the actual distance at which the boundary layer nearly ends can be shown to be

$$\delta_M = 5\sqrt{\frac{\nu x}{v_e}} \tag{11.16}$$

and the results are also in agreement with the scaling estimate indicated earlier.

Stress on the Wall

The stress exerted by the flowing fluid on the wall, τ_0, can be calculated using Newton's law of viscosity:

$$\tau_0 = \mu \left(\frac{\partial v_x}{\partial y}\right)_{y=0} \tag{11.17}$$

Using dimensionless variables this can be rearranged as

$$\tau_0 = \frac{\mu v_e}{4} \left(\frac{v_e}{\nu x}\right)^{1/2} f''(0) \tag{11.18}$$

The value of f'' at $\eta = 0$ (the solid surface), is 1.3282 from the numerical solution. Using this in Equation 11.18 we obtain

$$\tau_0 = 0.332 \mu v_e \left(\frac{v_e}{\nu x}\right)^{1/2}$$

This can be related to the drag coefficient, which is defined as

$$C_f = \frac{\tau_0}{\rho v_e^2/2}$$

Combining the previous two equations, the following relation is obtained for the local drag coefficient:

$$C_{fx} = \frac{0.664}{\sqrt{Re_x}}$$

Now we use the velocity results to simulate the concentration profiles within the boundary layer.

Mass Transfer Equations

For the mass transfer problem defined by Equation 11.7 we introduce a dimensionless concentration c_A, defined as

$$c_A = \frac{C_A - C_{A\infty}}{C_{As} - C_{A\infty}} \tag{11.19}$$

With this definition the surface concentration is unity while the far stream concentration is zero. The convection-diffusion equation is now changed in terms of the f function and the dimensionless concentrations. After some algebra (use of the chain rule) the following equation is obtained:

$$\boxed{\frac{d^2 c_A}{d\eta^2} + Scf\frac{dc_A}{d\eta} = 0} \tag{11.20}$$

which can be solved numerically together with the equation for f given by Equation 11.15.

The boundary conditions for c_A are as follows:

- At $\eta = 0$, $c_A = 1$.
- At $\eta = \infty$, $c_A = 0$.

The MATLAB code BVP4C is useful for these calculations and is presented in Listing 11.1. This solves for both f and c_A simultaneously.

An illustrative result for the concentration profile is shown in Figure 11.2 for three values of the Schmidt number. As Sc increases, the concentration boundary layer becomes thin.

Having computed the concentration profiles, the local Sherwood number can be calculated; this is discussed next.

Local Sherwood Number

The concentration gradient at the surface evaluated at $y = 0$ can be related to Sherwood number as

$$k_m = \frac{-D_A \left(\frac{\partial C_A}{\partial y}\right)_{y=0}}{(C_{As} - C_{A\infty})}$$

If this is converted into dimensionless form, the following expression is obtained for the local Sherwood number:

$$Sh_x = -\frac{\sqrt{Re_x}}{2}\left(\frac{dc_A}{d\eta}\right)_{\eta=0} \tag{11.21}$$

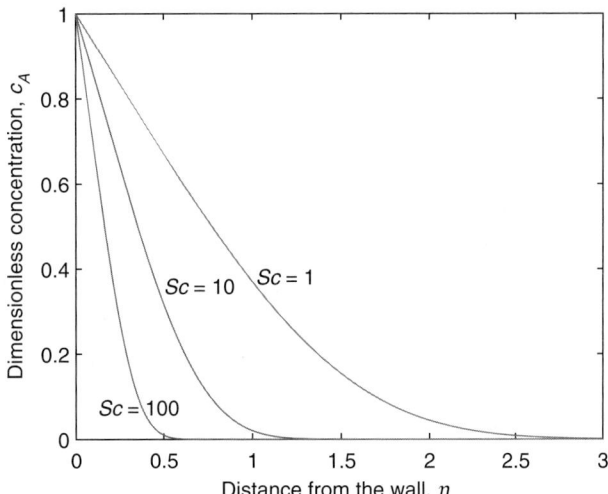

Figure 11.2 Concentration distribution in laminar boundary layer over a flat plate with solute coated wall; note how the boundary layer thickness gets smaller as Sc increases.

Thus the local Sherwood number depends on the slope of the (dimensionless) concentration gradient at the surface.

For $Sc = 1$ the velocity and concentration profiles are identical. The slope at the surface is -0.6641. Hence the value of Sh is $0.332\sqrt{Re_x}$. The local Stanton number is obtained by dividing this by $Re_x Sc$ or simply by Re_x here as $Sc = 1$. The results confirm that the Stanton number is equal to $C_f/2$, which is the Reynolds analogy.

For $Sc = 100$, the slope at the surface is now 3.1438. Similarly the slope can be found at various values of Sc and the data can be fitted closely as

$$St = \frac{C_f}{2} Sc^{-2/3}$$

This is the basis for the Prandtl analogy.

Reacting System

The Blasius approach can be extended for the case where the reacting species undergoes a chemical reaction in the boundary layer. Using the same variable

η, the following equation can be derived for the concentration profile within the boundary layer:

$$\frac{d^2 c_A}{d\eta^2} + Scf\frac{dc_A}{d\eta} - 4Da_x Sc\, c_A = 0 \qquad (11.22)$$

where Da_x is defined as $k_1 x / v_e$. Note the appearance of x in this definition; this is a local Damkohler number at any fixed position x. The appearance of x indicates that a true similarity is not achieved and the concentration is now a function of both η and x. The profiles can be simulated using the code shown in Listing 11.1 and the effect of the reaction on the boundary layer thickness can be examined. The details are left as an exercise problem.

Blasius' work is a classic in the field of fluid dynamics and the corresponding extension to mass transfer is another follow-up classic. The Blasius method can be extended to a few other cases (for example a reacting case as above), high flux mass transfer with some assumptions (11.3.3), inclined plates (11.4.4), and so on. But for more general cases and complex geometries, other methods are useful. One such method is the integral method, which is taken up next.

11.2   Integral Balance Approach

The idea behind the integral balance approach is that a mesoscopic version of the boundary layer equation is solved rather than the detailed differential model. The equations can be developed by integrating the boundary layer model with respect to y from 0 to δ_M for momentum and from 0 to δ_m for mass transfer. First we show the derivation for momentum balance and then proceed to the integral mass balance.

11.2.1 Integral Momentum Balance

The differential x-momentum balance is given by Equation 11.6 in the conservative form and is repeated here for ease of reference:

$$\frac{\partial(v_x v_x)}{\partial x} + \frac{\partial(v_y v_x)}{\partial y} = \nu\frac{\partial^2 v_x}{\partial y^2} \qquad (11.23)$$

The equation is integrated with respect to y from 0 to δ_M (the momentum boundary layer thickness) and the following equation results, which is known as the integral momentum balance equation:

$$\int_0^\delta \left[\frac{\partial(v_x v_x)}{\partial x}\right] dy + (v_y v_x)]_0^\delta = \nu \left[\frac{\partial v_x}{\partial y}\right]_0^\delta \tag{11.24}$$

A notational abbreviation of δ is used here for δ_M.

The Liebnitz rule for integration under the differential sign is used in the first term. For the second term we can set v_y equal to zero at $y = 0$ for the low flux case. The value of v_y at $y = \delta$ is obtained from the continuity equation:

$$v_y(\delta) = -\int_0^\delta \left(\frac{\partial v_x}{\partial x}\right) dy \tag{11.25}$$

The term on the right-hand side of 11.24 is equal to $-\tau_0/\rho$, where τ_0 is the stress exerted on the wall. Using these considerations, the following integral momentum balance (for a flat plate case) is obtained:

$$\boxed{\frac{d}{dx}\left(\int_0^\delta v_x(v_x - v_e)dy\right) = -\frac{\tau_0}{\rho}} \tag{11.26}$$

This is equivalent to a mesoscopic momentum balance for the boundary layer. But as δ is unknown, some approximation for v_x is needed. Commonly a cubic form is used:

$$v_x = a + by + cy^2 + dy^3$$

where the constants a, b, c, and d have to be evaluated by imposing some "boundary" conditions. Four conditions are needed for the cubic form:

- $v_x = 0$ at $y = 0$ as required from the no-slip condition.
- $v_x = v_e$ at $y = \delta$ as required from imposing the velocity v_e at the edge of the boundary layer.
- dv_x/dy is set equal to zero at $y = \delta$. This is required so the velocity blends in smoothly to the flow outside the boundary layer.
- d^2v_x/dy^2 is set equal to zero at $y = 0$. This is a derived condition obtained from the x-momentum balance since v_x and v_y are both zero at $y = 0$. Hence the second derivative must be therefore zero at this point.

Note: A simpler quadratic approximation for the velocity profile can also be used to simplify the algebra. In this case the last condition is not used.

Fitting the four constants, the cubic profile can be represented as

$$\frac{v_x}{v_e} = \frac{3}{2}\frac{y}{\delta} - \frac{1}{2}\left(\frac{y}{\delta}\right)^3 \tag{11.27}$$

Please verify that this profile satisfies all four conditions stated above.

For further manipulations it is useful to write this in dimensionless form as

$$v_x^* = \frac{v_x}{v_e} = \frac{3}{2}y^* - \frac{1}{2}y^{*3}$$

Here $y^* = y/\delta$, the dimensionless distance within the boundary layer.

If one substitutes this velocity profile in the integral balance equation (Equation 11.26), the following equation for the variation of the momentum thickness with x is obtained:

$$I_1 \rho v_e^2 \frac{d\delta}{dx} = -D_1 \frac{\mu v_e}{\delta} \tag{11.28}$$

where I_1 is

$$I_1 = \int_0^1 v_x^*(v_x^* - 1)dy^* = -\frac{39}{280}$$

and

$$D_1 = \left(\frac{dv_x^*}{dy^*}\right)_{y^*=0} = \frac{3}{2}$$

Integration of Equation 11.28 gives an expression for the momentum thickness as a function of x:

$$\delta(x) = \sqrt{\frac{280}{13}}\sqrt{\frac{\nu x}{v_\infty}} = 4.64\sqrt{\frac{\nu x}{v_\infty}} \tag{11.29}$$

Here the integration constant has been set to zero since $\delta = 0$ at $x = 0$ is zero.

Equation 11.29 is very similar to the Blasius solution with a factor of 4.64 instead of 5. The expression for the wall shear stress can likewise be obtained from the slope of the velocity profile at the surface:

$$\tau_0 = D_1 \frac{\mu v_e}{\delta}$$

Using the expression for δ given by 11.29, the following expression for the wall stress is obtained:

$$\tau_0 = 0.323 \mu v_e \left(\frac{v_e}{\nu x}\right)^{1/2}$$

This differs from the Blasius solution by only 3%. Other approximations for the velocity have also been used, for example, a quadratic approximation, sine function approximation, and so on. These are suggested as an exercise problem and you will find that the results are close to the that obtained from a cubic approximation. A quadratic profile is especially useful for the reacting systems studied in the next section as the condition for the second derivative at the wall gets somewhat complex to implement with a cubic profile. (The value of the second derivative is non-zero for a reacting case.)

This completes the momentum analysis and now we show how we can derive a similar equation for mass transfer boundary layer thickness as a function of x.

11.2.2 Integral Species Mass Balance

Here we can directly proceed with the species mass balance, which is written for a binary system in terms of the concentration of A, denoted as usual by C_A. A first-order reaction and low flux model is assumed. The governing differential equation in conservative form is used as starting point; it is repeated here for convenience:

$$\frac{\partial(v_x C_A)}{\partial x} + \frac{\partial(v_y C_A)}{\partial y} = D_A \frac{\partial^2 C_A}{\partial y^2} - k_1 C_A \tag{11.30}$$

The integral version of this equation is obtained by integration with respect to y from 0 to δ_m:

$$\frac{d}{dx}\left[\int_0^{\delta_m} v_x(C_A - C_{A\infty})dy\right] = N_{A0} - \int_0^{\delta_m} k_1 C_A dy \tag{11.31}$$

where N_{A0} is given as $-D_A(dC_A/dy)_{y=0}$.

11.2.3 Solution for No Reaction Case

The solution can be determined if some form for the velocity and the concentration profile is assumed to be a function of y. To avoid clutter we examine a no reaction case, that is, $k_1 = 0$. A reaction case is shown briefly in the next subsection.

The concentration profile is now approximated by a cubic equation as

$$c_A = \frac{C_A - C_{As}}{C_{A\infty} - C_{As}} = \frac{3}{2}y^* - \frac{3}{2}y^{*3} \tag{11.32}$$

Here y^* is now defined as y/δ_m, that is, y is now scaled by the mass transfer boundary layer thickness.

The velocity approximation is rewritten in terms of this new y^* as

$$v_x^* = \frac{3}{2}y^*\triangle - \frac{1}{2}y^{*3}\triangle^3$$

Here $\triangle = \delta_m/\delta$, the ratio of the thickness of the two boundary layers, mass by momentum.

Further integration using these velocity and concentration profiles in Equation 11.31 leads to the following equation for the variation of $\triangle$ with x:

$$v_e\frac{d(\delta I_1)}{dx} = D_1\frac{D_A}{\delta\triangle} \tag{11.33}$$

Here I_1 is

$$I_1 = \int_0^1 v_x^*(c_A - 1)dy^* = -\frac{3\triangle^2}{20} + \frac{3\triangle^3}{280}$$

and

$$D_1 = \left(\frac{dc_A}{dy^*}\right)_{y^*=0} = \frac{3}{2}$$

The term with $\triangle^3$ in the I_1 equation is usually neglected, which leads to

$$\frac{1}{10}\frac{d}{dx}[\delta\triangle^2] = \frac{D}{v_\infty\delta\triangle} \tag{11.34}$$

The momentum thickness δ, represented as

$$\delta(x) = \sqrt{\frac{280}{13}}\sqrt{\frac{\nu x}{v_\infty}}$$

is now substituted as a function of x in Equation 11.34 and a linear differential equation for $\triangle^3$ is obtained:

$$\frac{4}{3}x\frac{d}{dx}\triangle^3 + \triangle^3 = \frac{13}{14}\frac{1}{Sc} \tag{11.35}$$

This is a (linear) differential equation of the Cauchy-Euler type with a non-homogeneous constant term; the solution is

$$\triangle^3 = c_1 x^{-3/4} + \frac{13}{14}\frac{1}{Sc} \tag{11.36}$$

Here the integration constant c_1 is obtained by applying the proper initial conditions for $\triangle$. If the mass transfer starts right at $x = 0$ then $\triangle = 0$ at $x = 0$ is used as the boundary condition. This leads to

$$\triangle = \sqrt[3]{\frac{13}{14}} \frac{1}{Sc^{1/3}} = 0.9076 \frac{1}{Sc^{1/3}}$$

and the resulting expression for the Sherwood number is

$$Sh_x = 0.36 Re_x^{1/2} Sc^{1/3}$$

which is close to the exact solution shown in the earlier section. The coefficient was 0.332 and the error in the integral solution is 4%.

Note that the mass transfer boundary layer may not start at the same point where the momentum boundary layer starts. If the mass transfer starts at some point $x = X$, then $\triangle = 0$ at this point. Equation 11.36 for this case is

$$\triangle = 0.9076 Sc^{-1/3}[1 - (x/X)^{3/4}]^{1/3} \tag{11.37}$$

Note that starting the mass transfer somewhere past the leading edge gives leeway to adjust the mass transfer coefficient to the surface and has important implications in chemical vapor deposition processes. The variation of the mass transfer boundary layer thickness is not so sharp in that case compared to the case where the mass transfer starts right at the leading edge. This leads to a more uniform deposit thickness along the plate.

11.2.4 Solution for Homogeneous Reaction

In this section the integral analysis is extended to a case where a species dissolves at the wall, diffuses toward the bulk liquid, and simultaneously undergoes a first-order reaction in the boundary layer. An example will be a benzoic acid–coated wall in contact with a flowing NaOH solution.

The analysis is an extension of the method shown in the previous section. The reaction term now contributes an extra integral term as shown in Equation 11.31. However the cubic approximation for c_A given by Equation 11.32 is more complicated to implement. For the no reaction case, $d^2 C_A/dy^2 = 0$ at $y = 0$ was used as a condition in obtaining this relation. For a reacting case, this term is equal to kC_A/D and a minor correction is needed. Hence a quadratic equation is often used for the concentration to simplify the analysis (e.g., Plawsky, 2010):

$$c_A = 2y^* - y^{*2}$$

The velocity can still be approximated as a cubic. The following equation is obtained using these profiles in Equation 11.31 followed by some algebra:

$$x\frac{d}{dx}[\Delta^3] + \frac{3}{4}\Delta^3 = 4\frac{kx}{v_\infty}\Delta^2 + \frac{24}{25}\frac{1}{Sc} \tag{11.38}$$

The no reaction case leads to

$$\Delta = \left(\frac{32}{25}\right)^{1/3}\frac{1}{Sc^{1/3}} = 1.085\frac{1}{Sc^{1/3}} \tag{11.39}$$

which is a slightly different result from the 13/14 in the cubic approximation; this is due to the quadratic approximation. The difference is not very large and also shows that the quadratic approximation is close to the cubic approximation.

Now focusing on the reaction case, Equation 11.38 has no analytical solution since the differential equation is now nonlinear due to the presence of the kx/v_∞ term in Equation 11.38. A numerical solution is needed (not shown here for brevity) but it is useful to show the two limiting cases of slow reaction and fast reaction.

The slow reaction case applies for small Da_x and Equation 11.38 can be solved by assuming a power series expansion for Δ:

$$\Delta = \frac{1}{Sc^{1/3}}\left[1 + a_1 Da_x + a_2 Da_x^2 + \cdots\right]$$

The coefficients can be determined by substitution in the differential equation. The method is a variation of what is known as the regular perturbation method in the differential equation literature. The leading term $a_1 = -5/7$. Hence we conclude that the chemical reaction thins the boundary layer.

The fast reaction case applies when Da_x is large. The leading term of the solution can be obtained by setting the entire left-hand side of Equation 11.38 as zero. The diffusion balances the reaction in this case:

$$\Delta = \sqrt{5}(ScDa_x)^{1/2}$$

Note that the exponent representing the dependency on Sc is now $1/2$. Thus the exponent depends on the value of the reaction rate constant k. For low rate constants it is proportional to Sc to the one-third power, while it is the one-half power for high rate constants.

11.3   High Flux Analysis

In this section we study the high flux problem, where there is an induced velocity at the surface due to mass transfer. The problem is somewhat complicated in the sense that both momentum and mass balance have to be solved together. However, there are a number of simplified approaches that are useful and the primary focus is to point out these methods. The classical film model is widely used and we revisit this to start.

11.3.1   Film Model

The main feature of the film model is that there is a region near the surface (called the film thickness) where v_x is approximated as zero. The other component v_y need not be zero. (This is zero as well only under the low mass flux assumption or in the absence of blowing or suction across the plate.) If one uses continuity, one can conclude that v_y should be constant in the film. In contrast, for the boundary layer theory, the v_y vary as a function of y. Thus the film model can be viewed as a simplified version of the boundary layer model. Another difference is that the boundary layer thickness varies along the length of the plate in the detailed boundary layer model. Such refinement is not considered in the film model and the film thickness is assumed to be constant and not a function of x.

The convection-diffusion equation now simplifies to

$$D_A \frac{d^2 C_A}{dy^2} - v_0 \frac{dC_A}{dy} = 0 \tag{11.40}$$

where v_0 is used as the abbreviation for v_{y0}, the y-component of the velocity. This should be treated as a constant in the film model. Note that v_0 can be positive for blowing conditions, that is, if mass transfer is from the surface, that is, $n_t > 0$, and negative for suction, that is, when the mass flux is toward the surface.

The model in Equation 11.40 is the same as that discussed for 1-D convection-diffusion problem studied in Section 2.1.4 and it may be appropriate to revisit that section. The solution satisfying the required boundary conditions of prescribed concentrations at $y = 0$ and $y = \delta_f$ is readily found. The effect of the blowing parameter on the concentration profile is presented in Figure 11.3.

The diffusion flux at the surface can be calculated and compared with the base case of v_0 equal to zero, which represents the low flux case. The ratio

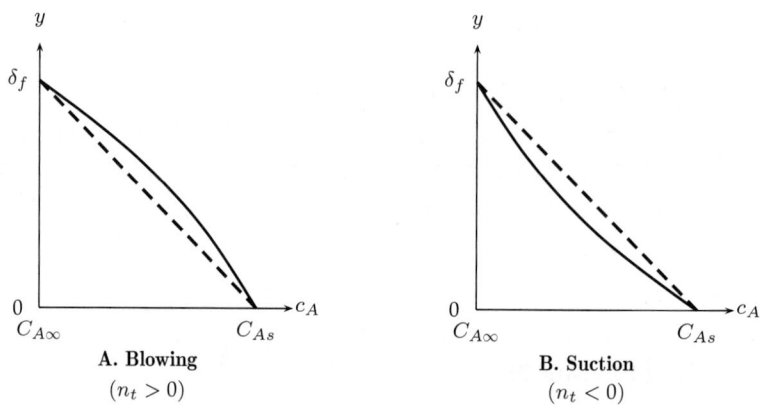

Figure 11.3 Comparison of concentration profiles for low flux ($n_t = 0$) and high flux ($n_t \neq 0$); dashed line is for $n_t = 0$.

of the two can be represented as an augmentation factor and is given as

$$\mathcal{E} = \frac{Pe_f}{\exp(Pe_f) - 1}$$

where Pe_f has the same grouping as the Peclet number and is defined as $v_0 \delta_f / D_A$. This parameter is often referred to as the blowing parameter.

Note that $\mathcal{E}$ represents the enhancement in the diffusive part of the flux, that is, diffusive flux is equal to $\mathcal{E}$ times $-D_A \Delta C_A / \delta_f$. Additional contribution to convective flux $v_0 C_{As}$ should be added to get the total flux.

11.3.2 Integral Balance Method

Integral analysis can also be used for the high flux model and the equations are derived in the same manner as the low flux model, that is, by integrating the transport equations with respect to y from zero to the corresponding boundary layer thickness. The only difference is that v_y at the plate surface is not set as zero and is retained in the integral balance. Here we present the set of integral equations for the high flux case.

Momentum Integral

We now consider a simplified boundary layer model and solve it to show the effect of the blowing parameter on the hydrodynamic boundary layer thickness. This parameter is assumed to be independent of x to simplify the model. (The effect of variation of this with distance can be included with minor modifications.)

The integral momentum balance is shown here for the high flux case. The only difference from the integral balance for low flux given by Equation 11.26 is that the expression for v_y at $y = \delta$ is now different from that given by Equation 11.25. It should now read

$$v_y(\delta) = v_{y0} - \int_0^\delta \left(\frac{\partial v_x}{\partial x}\right) dy$$

The additional term v_{y0} accounts for the fact that v_y is not zero at $y = 0$. This adds an extra term $v_{y0}v_e$ to the momentum balance. Hence the integral balance in the presence of the induced velocity is

$$\frac{d}{dx}\left(\int_0^\delta v_x(v_x - v_e)dy\right) = -\frac{\tau_0}{\rho} - v_{y0}v_e \qquad (11.41)$$

v_{y0} is treated as a constant to focus on the effect of suction or blowing. This represents a case where the plate is porous and mass transfer is taking place from the plate at a uniform rate all along the plate.

The quadratic profile is used for v_x for simplicity rather than a cubic and the following differential equation for δ_M can be derived. The details are left as an exercise.

$$\frac{d\delta^*}{dx^*} = \frac{15}{\delta^* Re_X} - \frac{15}{2}\frac{v_e}{v_{y0}}$$

where $\delta^* = \delta/X$ and $x^* = x/X$ where X is an arbitrary fixed distance from the leading edge, the observation point.

The integration provides an implicit equation for the boundary layer thickness. The integrated expression x is

$$x^* = \frac{2v_e}{15v_{y0}}\delta^* + \frac{4}{15Re_X}\left(\frac{v_{y0}}{v_e}\right)^2 \ln\left(1 + \delta^* Re_X \frac{v_e}{2v_{y0}}\right)$$

A variation of the problem occurs when v_0 varies as the inverse square root of x; this is left as an exercise problem.

Integral Mass Balance

The corresponding integral balance for mass transfer is

$$\frac{d}{dx}\left[\int_0^{\delta_m} v_x(C_A - C_{A,\infty})dy\right] - v_{y0}(C_A - C_{A,\infty})$$
$$= -D\left(\frac{dC_A}{dy}\right)_{y=0} - \int_0^{\delta_m} k_1 C_A dy \tag{11.42}$$

The two equations are now coupled. The solution proceeds by assuming suitable profiles for velocity and concentration. Often to simplify the problem a quadratic profile is assumed rather than a cubic profile, since the second derivative condition at the surface leads to additional mathematical complexities. Using the assumed profiles, two differential equations, $\frac{d\delta_M}{dx}$ and $\frac{d\delta_m}{dx}$, for the two boundary layer thicknesses can be derived. In general these are coupled ordinary first-order equations and can be numerically integrated and solved. The procedure is general and a useful tool to understand and use in practical applications. Further details are left as an exercise problem.

11.3.3 Blasius Approach

The case of high mass flux can also be analyzed using the similarity variable. This is similar to the Blasius approach used for momentum transfer, which is Equation 11.15 for f. The only difference is in the boundary conditions.

The true similarity exists only if $v_y(y = 0)$, denoted as v_0, varies inversely as the square root of x. This can be ascertained by looking at the definition of v_y in terms of the f function. Recall from Section 11.1.4 that the y-velocity is given in terms of f as

$$v_y = \frac{1}{2}\left(\frac{\nu v_\infty}{x}\right)^{1/2}[\eta f' - f]$$

At $y = 0$ we have $\eta = 0$ and hence

$$v_0 = -\frac{1}{2}\left(\frac{\nu v_\infty}{x}\right)^{1/2} f$$

which is one of the boundary conditions at $\eta = 0$. The variable x remains in the boundary condition unless v_0 equals zero hence complete similarity is not achieved for the high flux case. Let us assume that v_0 varies inversely as the square root of x. Let

$$v_0 = K\frac{1}{2}\left(\frac{\nu v_\infty}{x}\right)^{1/2}$$

The assumed inverse variation of v_0, although needed for the similarity to hold, has some practical application. It corresponds to a case where the surface velocity v_0 is proportional to the flux at the surface and is therefore applicable to the problem where the surface is maintained at a constant concentration.

The boundary condition (at $\eta = 0$) is now $f = -K$ where K is a constant called the blowing or sucking factor. (x no longer appears in the boundary condition and true similarity is approached.) The case of K positive refers to positive v_0, that is, mass transfer from the surface (blowing) while K negative would be the case of mass transfer away from the surface (sucking). This effect of mass transfer on the momentum transfer can be examined using the similarity solution method. The velocity and concentration equations can be simultaneously solved using BVP4C with the program in Listing 11.1.

An illustrative result is shown for $Sc = 1$ in Figure 11.4. The boundary layer gets thicker if K is positive (blowing) and thinner if K is negative (sucking). The concentration profile is equal to $1 - v_x/v_e$ for $Sc = 1$. (These superimpose on each other for $Sc = 1$ and are therefore not marked separately in this figure.)

Profiles for other values of the Schmidt number can be easily simulated with the program. The proportionality of $Sc^{-1/3}$ is still maintained for the

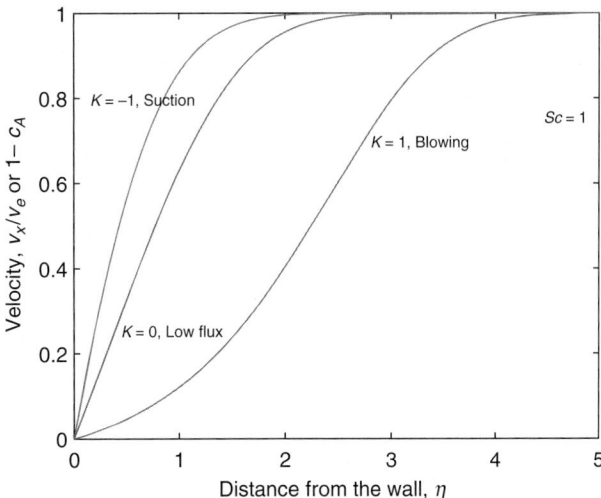

Figure 11.4   Velocity and concentration profile for high flux mass transfer. Results are for $Sc = 1$; note $c_A = 1 - v_x/v_e$ for this case.

concentration profiles. The Sherwood number can be represented as

$$Sh = -\frac{c_A'(0)}{2} Re_x^{1/2} Sc^{1/3}$$

where $c_A'(0)$ is the slope of the concentration profile at $\eta = 0$.

11.4   Mass Transfer for Flow over Inclined and Curved Surfaces

So far we have analyzed the flat plate case. In this section we discuss flow over inclined plates and curved surfaces and provide more details for flow over an inclined flat surface. The coordinate direction x is marked along the surface and y is taken perpendicular to the surface. The effect of curvature is neglected and locally we use the x-momentum equation without consideration of centripetal forces. The external flow velocity at the edge of the boundary layer v_e now varies as a function of x and is not constant. This imposes an additional term in the x-momentum balance, which is the pressure gradient term. The modifications needed and the x-momentum to be solved within the boundary layer are discussed now. It may be noted that for a flat plate v_e is not a function of x and the pressure gradient term was not included in the analysis.

11.4.1   Pressure Variation Term

The external flow can be modeled using the potential flow theory and variation of pressure in the external flow is related to velocity by the Bernoulli equation, which states that the pressure energy must balance the kinetic energy:

$$\frac{P_e}{\rho} + \frac{1}{2}v_e^2 = \text{constant}$$

where v_e is the velocity in the external flow at a solid surface (i.e., just outside the boundary layer). Note that this can be calculated from the potential flow model, which applies for external flow, that is, outside the boundary layer. Differentiating the previous equation, we find the following relation holds for the pressure gradient.

$$-\frac{dP_e}{dx} = \rho v_e \frac{dv_e}{dx} \tag{11.43}$$

This extra pressure term should be included in the equation for the x-component of velocity within the boundary layer

$$\frac{\partial(v_x v_x)}{\partial x} + \frac{\partial(v_x v_y)}{\partial y} = v_e \frac{dv_e}{dx} + \nu \frac{\partial^2 v_x}{\partial y^2} \qquad (11.44)$$

which needs to be solved together with the continuity equation.

The pressure variation in the external flow (or equivalently the v_e variation term) is needed to solve the problem. This can in turn be obtained from the potential flow theory or experiments conducted outside the boundary layer. The sign of the pressure gradient has a significant effect on the boundary layer flow as discussed in the following.

Inclined Flat Plates

For a flat plate that is inclined upward to the flow direction, the velocity increases with distance along the plate. The pressure gradient is therefore negative here in accordance to Equation 11.43. The second derivative of velocity is proportional to the pressure gradient as can be seen from the x-momentum balance equation (Equation 11.44). The second derivative is therefore positive and does not change sign for the entire boundary layer. The boundary layer remains attached to the plate here. The situation is shown as Figure 11.5a. The negative pressure gradient is also called as favorable gradient. For a flat plate the pressure gradient is zero (Figure 11.5c).

For a plate which is inclined downward (Figure 11.5d) the fluid decelerates and the pressure gradient is positive. The second derivative is positive at the wall now, and it approaches zero at large y from the negative side. Hence the second derivative becomes zero at some point in the boundary layer. This is associated with an inflexion point in the velocity profiles. Thus the boundary layer can separate from the plate if the pressure gradient is positive (Figure 11.5d). The solution for an inclined plate can be determined using the similarity solution method and is discussed in Section 11.6.

Curved Surfaces

The prototype example of a curved surface is the flow past a cylinder (see Figure 11.6). The pressure gradient term is negative for 0 to $\pi/2$ and positive thereafter for $\pi/2$ to π. Hence the boundary layer can separate from the solid at some point where $\theta > \pi/2$. Regions of wake can exit behind the trailing edge of the cylinder. The prediction of local mass transfer boundary layer thickness and the mass transfer coefficient is a more challenging problem. Numerical simulation is used here. One such study is by Conner and Elghobashi (1987);

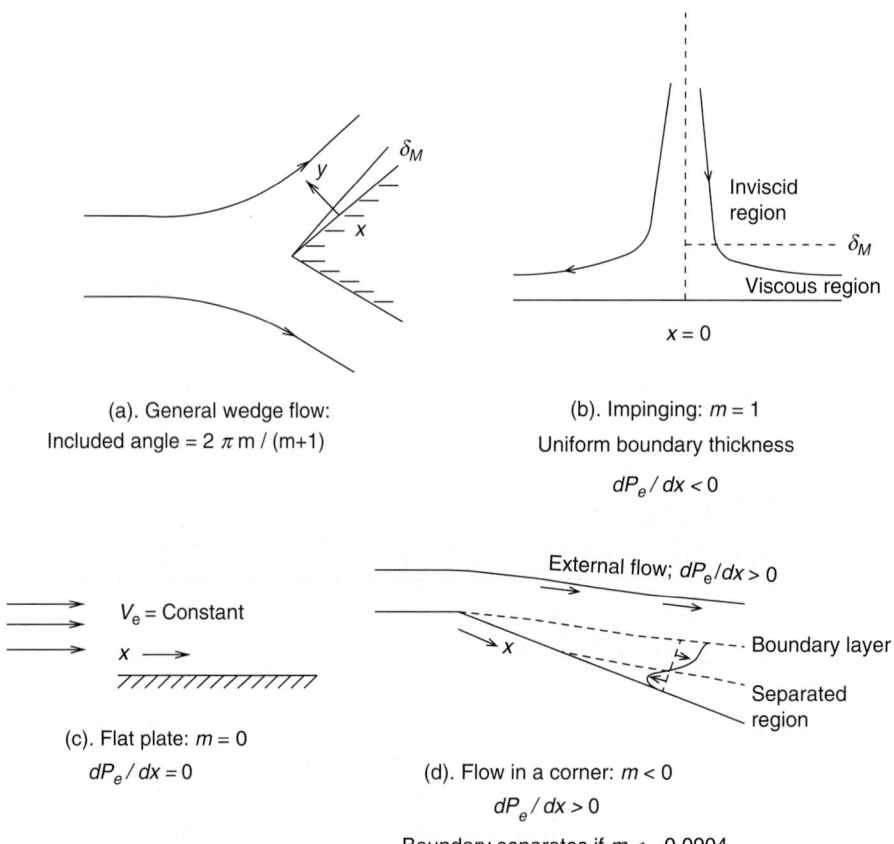

(a). General wedge flow:
Included angle = $2\,\pi\, m\,/\,(m+1)$

(b). Impinging: $m = 1$

Uniform boundary thickness

$dP_e\,/\,dx < 0$

(c). Flat plate: $m = 0$

$dP_e\,/\,dx = 0$

(d). Flow in a corner: $m < 0$

$dP_e\,/\,dx > 0$

Boundary separates if $m < -0.0904$

Figure 11.5   Examples of boundary layer flow over an inclined plate. These flows can be solved by the Falkner-Skan method presented in Section 11.4.4. (a) General wedge flow with the wedge angle β, (b) impinging flow studied in Section 11.4.5 with wedge angle π, (c) flat plate with wedge angle zero, and (d) flow where boundary layer separation can occur.

they used techniques developed by Patankar (1980) for general heat and mass transfer simulation. COMSOL-based solvers are also useful. It is important to note that the computed flow field used in the mass transfer model should accurately portray the wake region.

11.4.2 Integral Balance Method for Inclined and Curved Surfaces

Integration of Equation 11.44 with respect to y from 0 to δ_M leads to the integral balance for flow past a curved surface, which is similar to that for the flat

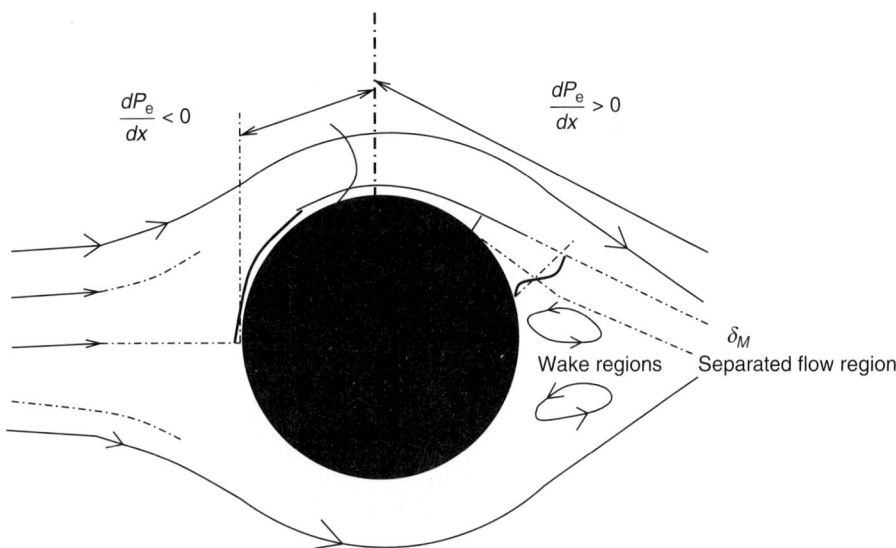

$\dfrac{dP_e}{dx} < 0$ $\dfrac{dP_e}{dx} > 0$

δ_M

Wake regions   Separated flow region

Figure 11.6 Examples of a boundary layer over a curved surface (e.g. flow past a cylinder) showing the attached and detached boundary layer and wake region behind the object.

plate. The only addition is the extra term arising from the pressure variation along the flow direction. The resulting equation is

$$\frac{d}{dx}\left(\int_0^\delta (v_x(v_x - v_e))dy\right) = -\frac{\tau_w}{\rho} + \frac{dv_e}{dx}\left(\int_0^\delta (v_e - v_x)dy\right) \qquad (11.45)$$

The last term is the correction term for the change in external flow with x. This term is assumed to be known from the potential flow analysis for the specified geometry. The solution of the integral equation proceeds in the same manner once v_e is calculated as a function of x from potential flow theory. The detailed analysis is presented by Deen (2012), who used a quartic approximation for the velocity. Note that one extra condition needs to be imposed for a quartic profile; usually the second derivative of velocity at δ_M is taken as zero, which provides the extra condition. Once the momentum thickness is known, the scaling laws can be used to find the mass transfer coefficient by using the $Sc^{-1/3}$ dependency of Δ. This is a simpler approach and gives reasonable results. We simply use a Prandtl type of analogy here. Alternatively an integral mass balance is solved in conjunction with the integral momentum balance. The integral mass balance has the same form as in Section 11.2

The boundary layer flow over inclined but flat plates is amenable to a more detailed analysis and is now discussed.

11.4.3 Inclined Plates: Use of Similarity Variable

In this section we show how the use of a similarity variable is useful in simplification of the boundary layer equation. Full similarity is not achieved in general but the method is useful for inclined flow where a full similarity is achieved. In this case, the Blasius approach can be modified for an inclined plate and a brief discussion is provided in the following.

First we eliminate the continuity equation through use of the streamfunction ψ defined earlier. Using the stream function, the x-momentum balance equation (Equation 11.44) becomes

$$\frac{\partial \psi}{\partial y}\frac{\partial^2 \psi}{\partial x\,\partial y} - \frac{\partial \psi}{\partial x}\frac{\partial^2 \psi}{\partial y^2} = v_e\frac{dv_e}{dx} + \frac{\mu}{\rho}\frac{\partial^3 \psi}{\partial y^3} \qquad (11.46)$$

Two variables, η and f, are introduced as for the flat plate case shown in Section 11.1.4. The definitions are identical but v_e is no longer a constant and becomes a function of x. In addition a dimensionless pressure gradient parameter m defined as

$$m = \frac{x}{v_e}\frac{dv_e}{dx} \qquad (11.47)$$

is required as well. With these substitutions and much algebra, Equation 11.46 reduces to the following form:

$$f''' + (m+1)ff'' + m[1 - (f')^2] = x\left[f'\frac{\partial f'}{\partial x} - f''\frac{\partial f}{\partial x}\right] \qquad (11.48)$$

where the prime $'$ refers to differentiation with respect to η.

The equations have been transformed in the form $f = f(\eta, x)$. The presence of x in Equation 11.48 shows that complete similarity has not been achieved. It appears that nothing has been gained by going from x-y coordinates to x-η coordinates. However, the advantage is that the solution is easier using η than y and a distance marching along x can be used for any arbitrary shaped geometry. The range of η is nearly constant, unlike y. m as a function of x is needed and this is obtained separately from potential flow theory outside the boundary layer. Marching in the x-direction is the suggested computational scheme.

A case where Equation 11.48 simplifies and leads to a complete similarity representation is the wedge flow, discussed in the following section.

11.4.4 Wedge Flow: Falkner-Skan Equation

The schematic of a wedge flow was shown in Figure 11.5. Here the fluid is blowing past a plate inclined at an angle β to the approach direction.

If m (in Equation 11.48) is a constant, then the flow does not depend explicitly on x. (In this case, the right hand side of the Equation 11.48 can be shown to become equal to zero.) Such flows are called similar flows since the velocity profiles look the same when plotted in terms of η. Similar flows are defined as flows where f is a function of η only (and not a function of x). If m has to be a constant then the external flow v_e has to vary in a power law manner with respect to x:

$$v_e = Cx^m \tag{11.49}$$

Here C is a constant. It can be shown that the above equation applies to wedge flow with $m = \beta/(\pi - \beta)$, where β is **half** the wedge angle. For similar flows Equation 11.48 becomes a third-order ordinary differential equation for f as a function of η only:

$$f''' + (m+1)ff'' + m[1 - (f')^2] = 0 \tag{11.50}$$

with the boundary conditions given Section 11.1.4. These are repeated here for ease of continuity. The $f(0) = 1; f'(0) = 0$ and $f(\infty) = 2$. This can then be supplemented with the mass transfer equation shown earlier:

$$\frac{d^2 c_A}{d\eta^2} + Scf\frac{c_A}{d\eta} = 0 \tag{11.51}$$

Both velocity and concentration can be solved simultaneously. A MATLAB program to solve for wedge flow together with mass transfer is presented in Listing 11.1. This program is also useful for the high flux case where the boundary condition for f' should be changed. Note that the inverse square root dependency on flux is to be used for the high flux mass tranfer; otherwise similarity does not exist. (In case this case the boundary condition is $f(0) = -K$ where K is the blowing factor as discussed in Section 11.3.3.) In general, the program shown in Listing 11.1 is therefore useful in many applications. It uses the BVP4C routine with CHEBFUN wrapping.

Listing 11.1   Program to Simulate Velocity and Concentration for Flow over an Inclined Plate

```
function FalknerSkan
global Sc Da m
```

```
Sc=  1; Da=0.0; m=0;
endpt=10
do  = [ 0 endpt ] % stands for do(main)
x = chebfun ('x', do)
% inital guess function; a quasi—matrix
y0 = [ x.^0   x.^0 exp(—x)   x.^0 x.^0 ]
y = bvp4c (@odes, @bcs, y0)
% values at surface and at infinity
    y(0.0,:)
    y(endpt,:)
%— function block————————————————————————
function dydx = odes ( x, y )
global Sc Da m
dydx = [ y(2)
         y(3)
         —y(1)*y(3)*(m+1)—m*(1—y(2)^2)
         y(5)
         —Sc*y(1)*y(5)+4*Da*Sc*y(4) ];
%% boundary condition block
    function res = bcs ( ya , yb)
res = [ (ya(1))
        ya(2)
        yb(2)—2
        ya(4)—1
        (yb(4)) ];
```

11.4.5 Stagnation Point (Hiemenz) Flow

Consider the problem sketched schematically in Figure 11.5b. A fluid is approaching a plate and the flow diverts in two directions around the origin. The problem is to investigate the nature of the flow and boundary layer that develops around the plate. The Falkner-Skan model is applicable directly with the wedge angle equal to π and hence the m parameter is set to one to simulate the flow. The velocity profile in the external flow just at the edge of the boundary layer is then

$$v_e = Cx$$

Since v_e is now proportional to x, we find that m is a constant and not a function of x. This suggests that the thickness of the boundary layer is independent of x. The surface provides the same mass transfer and is therefore useful in practice, for example to grow films by chemical vapor deposition with a uniform thickness all across the surface. This property is also shared by flow over a rotating disk, discussed in the following section.

The thickness of the boundary layer can be determined from numerical solution to have the following value:

$$\delta_M = 2.4\sqrt{\nu/C}$$

11.4.6 Flow over a Rotating Disk

The geometry here is similar to the Hiemenz flow where a jet is impinging on a disk now rather than on a plate. Also the disk is rotated. The boundary layer thickness on the plate can be shown to be a constant:

$$\delta_M = 5.42\sqrt{\nu/\Omega}$$

where Ω is the angular velocity of rotation of the disk. Correspondingly the rotating disk provides an equi-accessible surface for mass transfer. The mass transfer coefficient is independent of the location and can be obtained using the Prandtl analogy; it is given as

$$k_L = 0.2\Omega^{1/2}\nu^{-1/6}D^{2/3}$$

Strictly speaking, the model applies to an infinite disk but can be applied to a finite disk of radius R providing the condition $R > \delta_m$ or equivalently $R > \sqrt{\nu/\Omega}$ is satisfied. A rotating disk finds many applications in chemical vapor deposition, measurement of kinetics of heterogeneous reactions, electrochemical processes, and so on, as the mass transfer coefficient is known independently and can be controlled by the rate of rotation of the disk.

11.4.7 Flow past a Sphere

An illustrative diagram of the variation of mass transfer coefficient as a function of the angle from the stagnation point from the work of Connor and Elgobashi (1987) is shown in Figure 11.7. Their results represent flow past a sphere at a Reynolds number of 40 and match the experimental data of Froessling closely. Large variations in local coefficient are observed in this figure with the mass transfer dropping significantly in the trailing edge.

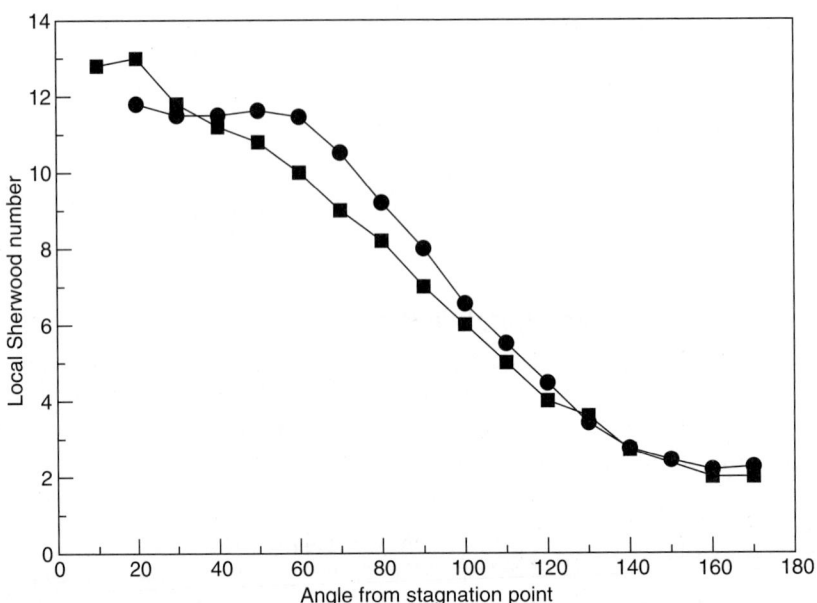

Figure 11.7 Local Sherwood number on a sphere with a flowing fluid at $Re = 48$ and $Sc = 2.5$; Filled circles = Numerical solution of Conner and Elgobashi (1987); Filled squares = Data of Froessling.

11.5   Bubbles and Drops

Theoretical prediction of mass transfer from a bubble to a fluid is more complex as the interface is now mobile rather than a solid. If the interface is assumed to be rigid, then the modeling is similar to that for fluid–solid systems. If surface active agents are present the rigid interface assumption is usually justified.

11.5.1 Rigid Bubbles

Mass transfer from a rigid bubble to a fluid can then be approached from a theoretical angle. The velocity profile is computed and substituted into the convection-diffusion equation. For axisymmetric flows around a sphere the governing equation is

$$v_r \frac{\partial C_A}{\partial r} + \frac{v_\theta}{r} \frac{\partial C_A}{\partial \theta} + \frac{D}{r^2} \left[\frac{\partial}{\partial r} \left(r^2 \frac{\partial C_A}{\partial r} \right) + \frac{1}{\sin \theta} \frac{\partial}{\partial \theta} \left(\sin \theta \frac{\partial C_A}{\partial \theta} \right) \right]$$

This equation has to be solved with the velocity profile computed from the momentum transfer considerations. Note that the mass transfer boundary layer is usually thin and hence an accurate description of the velocity around the bubble is needed. Also the flow field for the wake region has to be accurately captured. A further complication is the presence of surface contaminants, which affect the interfacial boundary conditions due to surface tension gradients. A computational study by Dani et al. (2006) is useful in this context.

The computational study results indicate that considerable variance in the local mass transfer coefficient is observed as a function of θ. Illustrative results are shown by Clift et al. (1978) and other sources. For example, for $Pe = 1000$ and $Re = 1$, the local Sherwood number is about 15 at the forward stagnation point ($\theta = \pi$) and decreases to a value close to 2 near the rear stagnation point ($\theta = 0$). Note that the value of the Sherwood number is 2 for no flow conditions and the results show that the solute is swept away near the rear stagnation point. The average value is about 11 for this case. The results of Conner and Elgobashi (1987) are similar.

For practical application, an average mass transfer coefficient is used. The results for the average mass transfer coefficient can be represented in terms of a Sherwood number and are expressed as a function of Pe. The resulting correlation was presented in Sections 9.7 and 9.8.

11.5.2 Spherical Cap Bubbles

Bubbles can be of different sizes and shapes while in motion; small bubbles are usually spherical, intermediate-sized bubbles are elliptical, and large bubbles are a spherical cap shape, as shown in Figure 11.8. The parameters of importance in determining the shape in general are as follows:

- Reynolds number, $Re_B = d_e v_B \rho_L / \mu_L$, where v_B is the rise velocity of the bubble, and d_e is the equivalent diameter, which is defined as the diameter of a sphere with the same volume as the bubble.
- Eotvos number, Eo, defined as $g d_e^2 \Delta\rho / \sigma$. Here $\Delta\rho$ is the density difference between gas and liquid; this is nearly equal to ρ_L under conditions of moderate pressure.
- Morton number, Mo, defined as $g \mu_L^4 \Delta\rho / \rho_L^2 \sigma^3$.

A spherical cap is observed if $Re > 20$ and $Eo > 40$. A detailed chart delineating the various regimes are given by Clift et al. (1978). The equivalent diameter of such bubbles is in the range of 1 to 2 cm. An interesting feature of

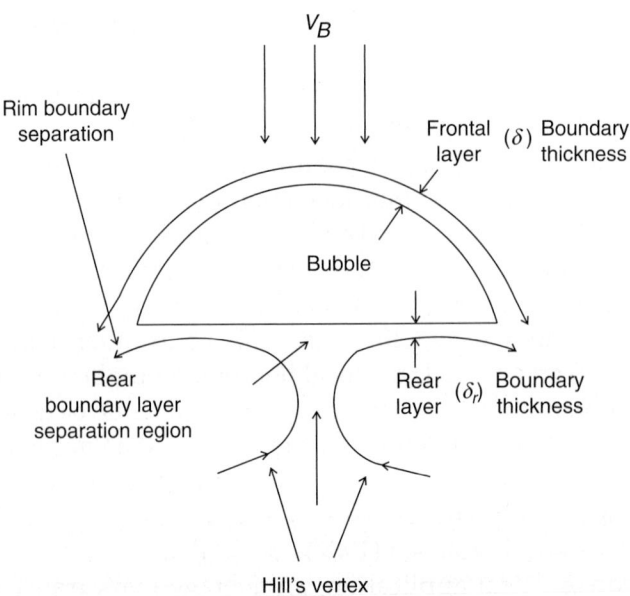

Figure 11.8 Spherical cap bubble and sketch of boundary layer around the bubble.

the bubble dynamics is that the rise velocity is independent of the properties of the fluid for a spherical cap bubble and can be approximated as $0.67\sqrt{(gd_e)}$. The theoretical analysis of convective mass transfer to spherical cap bubbles was studied by Kendoush (1994). The potential flow was assumed to be outside the bubble and the boundary layer approximation for mass transfer was used at the bubble surface. The curvature effects were neglected and the diffusive transport in the θ direction (transvere) to the boundary layer was neglected. These are standard assumptions as used in any curved boundary layer analysis and permit analytical solutions.

Both local and average rate of mass transfer from the front and rear surface of the cap were evaluated. Local distribution was shown to be

$$Sh(\theta) = 1.683 \frac{\sin^2\theta}{\sqrt{\cos^3\theta - 3\cos\theta + 2}}(ScRe)^{0.5}$$

where Re is defined in terms of bubble rise velocity v_B and d_B, the equivalent diameter of the bubble.

The average value, assuming the half cap angle of 50°, was

$$\bar{Sh} = 2.113(ScRe)^{0.5}$$

This agreed reasonably closely with the data of Calderbank et al. (1970) for glycerol, but overestimated the value for water.

Summary

- External flows are characterized by a boundary layer and the flow is never fully developed in contrast to internal flows. Thus the velocity has both x- and y-components; hence the mathematical analysis is somewhat more involved. Convective mass transfer in external flows is thus mathematically more challenging than the internal flow case studied in the last chapter.

- Mass transfer can lead to a normal component of velocity at the plate surface (unless you have an equi-mass counterflow situation). The momentum and mass transfer analysis becomes coupled for such cases and leads to yet another complexity in mass transfer modeling. Under low flux conditions this is neglected; the velocity profiles are obtained first and then used in the mass transfer analysis.

- Flow analysis in boundary layers in laminar flow over a flat plate is a well studied problem and is amenable to a complete theoretical analysis. A boundary layer develops and grows near the plate and at some point the flow becomes unstable and finally turbulent. A Reynolds number value of 2×10^5 based on the distance from the plate is used as a criterion for transition.

- The Schmidt number is the key dimensionless group that determines the relative thickness of the momentum and mass boundary layers and is similar to the Prandtl number for heat transfer problems. The mass transfer boundary layer thickness decreases as the Schmidt number is increased.

- For flat plates with low flux assumptions, the x-momentum and the continuity equation can be combined and reduced to a third-order ordinary differential equation using a similarity variable η defined in the chapter. The resulting equation is called the Blasius equation. The solution of the third-order equation can be done numerically, using BVP4C for instance, and provides the velocity profile. This can then be used in the convection-diffusion equation to predict the concentration profile as well as the local and the average mass transfer coefficient.

- An additional consideration unique to mass transfer is that the velocity v_y at the surface can be non-zero, for example, unimolecular diffusion (UMD). The effect of this becomes important for high mass flux cases.

In many cases this effect may dominate. In such cases it is reasonable to assume a constant boundary layer thickness and use the simpler film model. An alternative is to use the Blasius type of method, which is useful if v_{y0} varies inversely as the square root of x.

- The integral method is another solution to the boundary layer problem. The governing equations are integrated across the boundary layer and provide the integral momentum and mass balance equations. These equations can be solved if an approximate velocity and concentration profile is assumed in the boundary layer. A quadratic or cubic equation is commonly used. The integral method is applicable to a wide range of solutions to boundary layer problems including reacting systems, high flux mass transfer, and so on.

- For flow and mass transfer over an inclined or curved surface, the velocity at the edge of the boundary layer is not a constant, unlike the flat plate case. The flow outside the boundary layer can be solved separately using the potential flow theory. This then provides the pressure gradient term to be added to the boundary layer x-momentum equation. The solution to the flow problem can be determined using either the integral method or a modified version of the Blasius method. This modified method is useful for flat inclined surfaces and the resulting model is known as the Falkner-Skan equation, which is a third-order ordinary differential equation. The calculated velocity profiles can then be used with the mass transfer equations to estimate the rate of mass transfer.

- Two flows where the mass transfer coefficient is independent of position are the stagnation point flow and flow over a rotating disk. These find extensive application in laboratory measurements of mass transfer with heterogeneous reaction and chemical vapor deposition since the data interpretation becomes easy due to the known and constant value of the mass transfer coefficient.

- Theoretical prediction of mass transfer from a bubble to a fluid is more complex as the interface is now mobile rather than solid. If the interface is assumed to be rigid, then the modeling is similar to that for fluid–solid systems. If surface active agents are present, the rigid interface assumption is usually justified. Bubbles free of surface active agents undergo an internal toroidal circulation and hence the velocity profiles both inside and outside the bubble are needed for mass transfer computations.

- Large bubbles take a spherical cap shape and move with velocity independent of the fluid properties. Mass transfer models permit prediction of the local and average mass transfer coefficient along the angular coordinate on the bubble surface.

Review Questions

11.1 Why are convective high flux mass transfer problems more difficult to solve than the corresponding low flux case?

11.2 When can a momentum equation be solved independently of a mass transfer equation for convective mass transfer problems?

11.3 What is the relative ratio of the momentum and mass transfer boundary layer thickness if the fluid mixture has properties close to air with $Sc = 1$?

11.4 What is the relative ratio of the momentum and mass transfer boundary layer thickness if the fluid mixture has properties close to water with $Sc = 1000$?

11.5 State the Reynolds and Prandtl analogies.

11.6 State the Chilton-Colburn analogy.

11.7 Define the j-factor for heat and mass transfer and state a relation between the two.

11.8 How can heat transfer data be used for mass transfer problems?

11.9 Verify that the continuity equation is automatically satisfied when the velocities are expressed in terms of stream function.

11.10 What are the key steps in integral analysis of transport in boundary layers?

11.11 Point out the key differences in the hydrodynamics envisioned in the film model versus the boundary layer model.

11.12 Define the f and η used in the Blasius solution for flow over a flat plate.

11.13 What requirement is needed for the similarity method (Blasius method) for high flux mass transfer?

11.14 What additional considerations are needed to model flow over an inclined or curved surface?

11.15 State the Falkner-Skan equation and indicate where it can be used.

11.16 State geometries that provide an equal rate of mass transfer at all locations along the surface.

11.17 When can a bubble surface be assumed to be rigid?

11.18 Does the surface tension have to be high or low for bubbles to have a spherical cap shape?

Problems

11.1 **Derivation of the Blasius equation.** Show all the steps leading to the Blasius equation (Equation 11.5), which is a third-order ordinary differential equation. The algebra leading to this is lengthy but no new concepts are involved, and the mathematical tedium can be alleviated using computational algebra tools such as MAPLE or MATHEMATICA.

11.2 **Drag coefficient over a flat plate.** Show the missing steps leading to Section 11.1.4 for the drag coefficient over a flat plate. Also show that the average drag over a length L of a plate is equal to twice the local value at L.

11.3 **Boundary layer for a reacting solute.** Show the steps leading to Equation 11.22. Simulate for Da_x equal to 1 and 10 and compare the boundary layer thickness with and without a chemical reaction.

11.4 **Naphthalene-coated plate.** Air at 300 K and 1 atm flows at 10 m/s along a flat plate. The plate is bare for $0 < x < 10$ cm and is coated with a thin layer of naphthalene at $x > 10$ cm. At a point 20 cm from the leading edge, calculate the values of Sh_x and the local evaporation of naphthalene. Find the value of Sh if the plate was coated right from the leading edge. If the transition occurs at an Re^* of 2×10^5, find the transition point. Calculate the total naphthalene evaporated up to this point.

11.5 **Boundary layer with suction.** Show the details leading to Equation 11.41 and verify the equation is obtained for the boundary layer thickness. Plot the boundary layer thickness for the following parameters: $v_0 = 0.2$ m/s and $v_e = 1$ m/s. If the mass transfer boundary layer is still assumed to have a Sc dependency of 1/3, plot the mass transfer coefficient in the presence of suction for $Sc = 1000$.

11.6 **Boundary layer with suction variable v_{y0}.** Assume v_{y0} varies as the inverse square root in x. Repeat the analysis of the problem shown in Section 11.3.2. Compare your results with the Blasius solution.

11.7 **Porous plate mass transfer.** Air at 300 K and 1 atm flows at 10 m/s along a porous plate through which air containing 1% CO_2 is injected. At 20 cm from the leading edge calculate the mass transfer coefficient and the concentration of CO_2 at the plate surface for the following two cases: constant injection at a rate of 4×10^2 m^2/sec and position-dependent injection at a rate of $4 \times 10^2 x^{-1/2}$ m/sec.

11.8 **Mass transfer from spherical cap bubbles.** Calculate and plot the local and average mass transfer coefficient over spherical cap bubbles with equivalent diameters in the range of 1 to 5 cm for fluid properties close to water and fluid properties close to glycerol.

CHAPTER 12

Convective Mass Transfer in Turbulent Flow

Learning Objectives

After completing this chapter, you will be able to:

- Understand basic properties of turbulent flow and the concept of time averaging.
- Appreciate that there is a closure problem in modeling of turbulent flows.
- Define and use the eddy diffusivity model for closure of the problem.
- Apply the eddy diffusivity model to find the flux and mass transfer coefficients in typical turbulent solid–liquid mass transfer problems.
- Understand the basis of the analogy of mass transfer with momentum transfer.
- Understand common models of turbulent mass transfer from a gas–liquid interface.

This chapter introduces and solves problems where mass transfer occurs under turbulent flow conditions. We first review some properties of turbulent flow. Turbulent flow is characterized by random velocity fluctuations superimposed on the main flow as well as concentration fluctuations. Hence the flow is always transient in nature. It is difficult to resolve these random fluctuations. To resolve this, the concept of time averaging is introduced, where the time-averaged values are used as representative parameters. The time-averaged form of the convection-diffusion equation for mass transfer and the Navier-Stokes equation for flow are then derived, which forms the starting point of most turbulent transport models.

The averaging introduces additional unknowns known as cross-correlation terms, which have to be closed. This is referred to as the closure problem in turbulent flow analysis. These additional terms can be interpreted as additional transport contributions resulting from the fluctuations. One method of closure is to introduce an eddy viscosity for flow problems and eddy diffusivity for mass transfer problems. These are defined and then used for the solution of illustrative problems in solid–fluid mass transfer in channel flows, pipe flows, and flat plate turbulent boundary layers. A close analogy exists between momentum and mass transfer in such flows and some of the common analogies along with the basis for these are shown.

Turbulent mass transfer from a gas–liquid interface follows a similar approach. Some closure relations based on the eddy diffusivity concept suggested in the literature are reviewed and applied to a problem of mass transfer in turbulent film flow. Overall the chapter provides the necessary background information to follow the more advanced literature in the field and also helps you in understanding the basis of the models that are incorporated into CFD codes for simulation of turbulent transport.

12.1 Properties of Turbulent Flow

Turbulence is the result of flow instability. Any small disturbance could lead to a chaotic type of flow with velocity fluctuating on a small time scale around a mean value. This happens generally when the viscous forces are much smaller than the inertial forces. The viscosity effect stabilizes the flow and dampens any disturbance. In the absence of significant viscous forces, any disturbance persists and leads to a continuous fluctuation of velocity, leading to turbulent flow. Thus turbulent flows are characterized by small random fluctuations around a mean value and essentially are chaotic unsteady state phenomena. The velocity fluctuations cause corresponding fluctuations in concentration for mass transfer problems and in temperature for heat transport situations. Thus the concentration (or temperature in heat transfer problems) fluctuates in a random manner around a mean value.

12.1.1 Transition Criteria

The viscous forces have to be larger than inertial forces for flow to be laminar and these provide a damping effect to any flow disturbance. The relative magnitude of these forces is characterized by a dimensionless group, the Reynolds number:

$$\text{Reynolds number} = \frac{\text{Inertia force}}{\text{Viscous force}}$$

Thus a low Reynolds number means viscous forces are larger than inertial forces and the flow will be laminar. At higher Reynolds numbers the flow becomes turbulent. For pipe flow, the transition Reynolds number is 2100–2300 and the flow is fully turbulent above a Reynolds number of 4000.

For flow in a boundary layer over a flat plate, the transition to turbulence occurs at a Reynolds number of around 2×10^5 and the flow is fully turbulent above 2×10^6. Note that the distance from the leading edge is used in the Reynolds number here and defined as $x v_\infty \rho / \mu$.

12.1.2 Characteristics of Fully Turbulent Flow

The flow becomes oscillatory at some critical Reynolds number, which is the onset of instability. Beyond this stage the flow eventually becomes fully turbulent. A fully turbulent flow field has the following characteristics (Warsi, 1999):

- Random: The flow field is a random field. For example, two identical experiments may not leave an identical streamline track of a marker particle released at the same position.
- Diffusive: The flow disturbance spreads over a larger distance as if it were a diffusion process. A streak of dye released in a pipe under turbulent conditions spreads over the whole pipe.
- Dissipative: There is a large energy loss in turbulent flow. For example, the pressure drop has to be increased by 3.5 times in order to double the volumetric flow rate.
- Three dimensional: Even a fully developed unidirectional flow (in a time-averaged sense) has some fluctuating velocity components (superimposed on the main flow) in all other three directions.

12.1.3 Stochastic Nature

An important property of turbulent flow is that no two identical experiments give the same velocity profile track, but the statistical properties remain the same. The flow itself may appear steady if the velocity is measured with devices that are not sensitive to small changes (e.g., a pitot tube). The applied pressure drop and the resulting flow rate will be constant in the system in a time-averaged sense (if the system is maintained at constant inlet and outlet pressures). Such flows are called "steady-on-average" in order to distinguish them from turbulent flow, which is inherently transient in nature due to random fluctuations in the velocity. Similarly the concentration at any point will be a constant if measured by sampling the liquid or by probes that are not sensitive to detecting small fluctuations. In engineering analysis these "steady" values are of more interest. Hence it seems logical to look at time-averaged properties of flow rather than the instantaneous velocity versus time profiles. This can be done by a model-reduction procedure called time averaging, discussed next.

12.2   Properties of Time Averaging

In turbulent flow conditions, the instantaneous velocity at any point can be represented as a sum of a mean value and a fluctuating component:

$$v_x(t) = \overline{v}_x + v_x'(t) \tag{12.1}$$

Similar equations hold for the y- and z-directions. The bar indicates a mean or time average of a quantity, which is defined, for example, as

$$\overline{v}_x = \frac{1}{T} \int_0^T v_x(t)\, dt$$

where T is a sufficiently large value of time for the results to be statistically meaningful. The symbol $'$ is used for fluctuating values.

Similarly the concentration (or the temperature) varies in turbulent flow and can be represented as

$$C_A(t) = \overline{C}_A + C_A'(t) \tag{12.2}$$

Here C_A' is the fluctuating part of the concentration and $\overline{C}_A$ is the time-averaged mean value at the given point. The time averaging defined here is a way of modeling the average properties rather than worrying about every minor fluctuation.

Flow and mass transport can be classified into two types depending on whether the time-averaged values change with time or not:

- Steady on average case: $\overline{v}_x$ and so on and $\overline{C}_A$ do not change with time. For example, consider solid dissolution in a pipe. If the inlet flow velocity does not change with time, the time-averaged concentration measured at any point will be found to be constant. Here we assume that the concentration is measured with a device that does not detect small fluctuations, which are characteristics of turbulent flow.

- Unsteady on average: If the inlet condition is perturbed, for example, by changing the feed condition, then $\bar{v}_x$, $\overline{C}_A$, and so on change with time due to these changes in the external conditions. The time scale of these changes is in general relatively large compared to the time scale of fluctuations. Hence the time averaging indicated by, for example, Equation 12.2, can still be used. In particular we will use the property that the time average of any fluctuating quantity is zero. Thus, for example, the time average of C'_A is zero.

With this background you should note the following properties of time averaging:

- The time average of any fluctuating quantity is zero.
- The time average of the product of a time-averaged quantity and fluctuating quantity is zero. Thus, for example, the time average of the product of $\bar{v}_x$ and v'_x is zero. This rule can be applied for both the steady on average and unsteady on average cases noted previously.
- The time average of the product of two fluctuating quantities is non-zero, for example, the time average of the product of v'_x and v'_y is non-zero. We say that there is a cross-correlation between v'_x and v'_y .

Cross-correlation is an important concept in turbulent flow analysis. If two random variables, say v'_x and v'_y, are completely independent then the cross-correlation is zero while if there is some dependency on each other (as is the case is in turbulent flow) then the cross-correlation is non-zero. For instance, the fluctuating components have to satisfy the continuity equation

$$\frac{\partial v'_x}{\partial x} + \frac{\partial v'_y}{\partial y} = 0$$

and hence are therefore interdependent, giving rise to a cross-correlation.

Similarly the cross-correlation average of the product of v'_x, for instance, and C'_A is non-zero. This has important implications in turbulent modeling of mass transfer processes. This is elaborated next and leads to what is known as the closure problem in turbulent flow.

Using these rules we can show that

$$\overline{v_x C_A} = \bar{v}_x \, \overline{C}_A + \overline{v'_x C'_A}$$

with similar expressions for the other velocity terms. The first term on the right-hand side is the product of two time-averaged quantities. The second term is the extra term that arises due to cross-correlation. This is the time average of the product of two fluctuating quantities.

Similarly the product of two velocities when time-averaged produces an extra term as indicated in the following:

$$\overline{v_x v_y} = \overline{v}_x \, \overline{v}_y + \overline{v'_x v'_y}$$

The extra term on the right-hand side of the above equations can be interpreted as additional transport contributions to the fluctuations. These terms have to be modeled and there is no fundamental way of calculating them. This is referred to as the closure problem in turbulent transport analysis.

12.3   Time-Averaged Equation of Mass Transfer

We show here how the equations of mass transfer are time averaged and how extra terms arise due to time averaging. The equation for mass transfer for the convection-diffusion case is written as

$$\frac{\partial C_A}{\partial t} + v_x \frac{\partial C_A}{\partial x} + v_y \frac{\partial C_A}{\partial y} = -\left(\frac{\partial J_{Ax}}{\partial x} + \frac{\partial J_{Ay}}{\partial y}\right) \qquad (12.3)$$

The transient terms are retained in turbulent flow. Also the z-terms are not shown to avoid clutter and can be readily added. The diffusion terms are written in flux form for convenience of later interpretation of the turbulent diffusion flux term. These are molecular diffusion terms.

In view of the following continuity condition we have

$$\frac{\partial v_x}{\partial x} + \frac{\partial v_y}{\partial y} = 0$$

The velocity terms in the concentration equation (Equation 12.3) can be moved into the differential, leading to

$$\frac{\partial C_A}{\partial t} + \frac{\partial (v_x C_A)}{\partial x} + \frac{\partial (v_y C_A)}{\partial y} = -\left(\frac{\partial J_{Ax}}{\partial x} + \frac{\partial J_{Ay}}{\partial y}\right) \qquad (12.4)$$

Each term is integrated over a chosen period of time (to cover sufficient fluctuations), leading to a time-averaged equation. The bar will be used to denote the time-averaged quantities. All terms with C_A appearing alone can be replaced by the time averaged value since the time average of the fluctuations are zero. This is the consequence of property of time averaging.

However the time average of terms such as $v_x C_A$ leads to extra terms:

$$\text{Time average of } \frac{\partial (v_x C_A)}{\partial x} = \frac{\partial (\overline{v_x} \times \overline{C}_A)}{\partial x} + \frac{\partial (\overline{v'_x C'_A})}{\partial x}$$

The second term on the right-hand side is the time average of the product of the velocity fluctuation and the concentration fluctuation. Similar terms appear for the v_y and v_z terms. These terms are usually moved to the right-hand side of the equation; the resulting equation is

$$\frac{\partial \overline{C}_A}{\partial t} + \bar{v}_x \frac{\partial \overline{C}_A}{\partial x} + \bar{v}_y \frac{\partial \overline{C}_A}{\partial y} = - \left(\frac{\partial J_{Ax}}{\partial x} + \frac{\partial J_{Ay}}{\partial y} \right) - \frac{\partial}{\partial x}(\overline{v'_x C'_A}) - \frac{\partial}{\partial y}(\overline{v'_y C'_A}) \tag{12.5}$$

12.3.1 Turbulent Mass Flux

We find extra terms on the right-hand side of Equation 12.5 (last two terms) that can be viewed as contribution of fluctuations to mass transfer. These terms have the appearance of a divergence of a flux vector and hence a turbulent mass flux vector can be defined whose component j is

$$J^t_{A,j} = \overline{C'_A v'_j}$$

Equation 12.5 can then be written as

$$\frac{\partial \overline{C}_A}{\partial t} + \bar{v}_x \frac{\partial \overline{C}_A}{\partial x} + \bar{v}_y \frac{\partial \overline{C}_A}{\partial y} = - \left(\frac{\partial J_{Ax}}{\partial x} + \frac{\partial J_{Ay}}{\partial y} \right) - \frac{\partial}{\partial x} J^t_{A,x} - \frac{\partial}{\partial y} J^t_{A,y} \tag{12.6}$$

The turbulent flux terms are then added to the molecular flux term (first two terms on the right-hand side) for modeling turbulent mass transfer. These terms have to be modeled and therefore we need some constitutive model for turbulent transport.

12.3.2 Reynolds Stresses

The velocity appearing in the mass transfer model (Equation 12.6) is the time-averaged velocity. In principle it can be computed from the time average of the Navier-Stokes equation. When the N-S equation is averaged extra terms appear; they are represented as turbulent stress tensor. These stresses are also known as Reynolds stresses and are related to the time average of the product of the fluctuating component of velocity. They are defined as

$$\tau_{ij}^{(t)} = -\rho \overline{v_i' v_j'}$$

The resulting x-momentum equation upon time averaging of the Navier-Stokes equation is

$$\rho \left(\frac{\partial(\overline{v}_x \overline{v}_x)}{\partial x} + \frac{\partial(\overline{v}_x \overline{v}_y)}{\partial y} \right) = -\frac{\partial \overline{P}}{\partial x} + \frac{\partial \overline{\tau}_{xx}^{(v)}}{\partial x} + \frac{\partial \overline{\tau}_{yx}^{(v)}}{\partial y} + \frac{\partial \tau_{xx}^{(t)}}{\partial x} + \frac{\partial \tau_{yx}^{(t)}}{\partial y} \quad (12.7)$$

The terms in the z-direction are not shown to avoid clutter and can be readily added. Only the x-component of the equation is shown above. Similar expressions hold for the y- and z-components. Also note that in the above equation we have also assumed that the flow is steady on average; hence the time derivative of mean flow is not shown. These equations are known as the RANS (Reynolds-averaged Navier-Stokes) equation.

The extra terms in the RANS equation arise naturally as a consequence of time averaging, but there is no way to predict them. Some extra closure laws are needed; these are discussed in the next section. Before we proceed we show additional coupling that can arise in mass transfer problems with reaction.

12.3.3 Reaction Contribution

For reacting systems, additional contribution to the reaction rate arises due to concentration fluctuations if the reaction is a nonlinear function of concentration. This can be demonstrated by time averaging the reaction terms as shown in the following paragraph.

Consider a second-order reaction with rate R_A defined as $-k_2 C_A^2$. Here C_A is the instantaneous concentration. You should time average this and show that the rate of reaction can be represented as

$$\text{Rate, time averaged} = -\left(k_2 \overline{C}_A^2 + k_2 \overline{C_A' C_A'} \right)$$

The first term on the right-hand side is simply the rate based on the time-averaged concentration. The second term is the average of the product of fluctuating concentrations and leads to additional contribution to the rate. This term is non-zero and cannot be calculated. Again we have the closure problem of turbulence for reacting systems (whenever the kinetics are nonlinear). Some phenomenological models are needed at this stage. This is closely related to the macromixing concepts introduced in Chapter 3. See Figure 3.7 as well.

Similar effects can be seen when the reaction rate is expressed as an exponential function of the temperature. Terms of the form $\overline{C'_A T'}$ will appear as additional terms in the rate equation.

12.4   Closure Models

The simplest way to look at the closure is to use some equation similar to Newton's law of viscosity. Here the turbulent stresses are correlated as a linear function of the mean velocity gradient for 1-D flows (or rate of strain based on mean velocity for 3-D flows). We simply use the analogy with Newton's law of viscosity and define an eddy diffusivity parameter, ν_t, by the following relation:

$$\tau_{yx}^{(t)} = \rho \nu_t \frac{d\overline{v}_x}{dy} \tag{12.8}$$

for one-dimensional shear type of flows. The parameter introduced, ν_t, is called the kinematic eddy viscosity. Also, the term $\rho \nu_t$ may be viewed as a "turbulent viscosity," denoted often by μ_t. However, these quantities are property of the flow and not of the fluid. Thus these will be a function of the local Reynolds number, distance from the wall, and so on, and NOT a simple property like viscosity. The closure problem still remains but is now transferred to ν_t or μ_t.

Similarly the mass diffusivity due to turbulent flow is closed by defining a turbulent mass diffusivity parameter, D_t:

$$J_{A,y}^t = \overline{C'_A v'_y} = -D_t \cdot \frac{d\overline{C}_A}{dy} \tag{12.9}$$

This is for the y-direction with similar definitions for the x- and z-directions. Note that the closure model is similar to Fick's law.

12.4.1 Turbulent Schmidt Number

Since more data are available on turbulent momentum diffusivity, it is common to scale the turbulent mass diffusivity by scaling by this value. This leads to a parameter, the turbulent Schmidt number:

$$Sc_t = \frac{\nu_t}{D_t}$$

This is a flow-dependent parameter and not a physical property. A value of 0.9 is commonly used and the problem is closed as an approximation. Thus turbulent diffusivity is assumed to be proportional to eddy momentum diffusivity and is not modeled separately. The idea is that both transports are similar and closely related to each other. Hence data on turbulent flow can be used to calculate the eddy diffusivity.

12.4.2 Prandtl's Model for Eddy Viscosity

The definition of eddy viscosity does not provide the closure for turbulence since we now need a model for this. It simply transfers the cross-correlation problem to eddy viscosity. The simplest model was provided by Prandtl in 1925 and is even today still widely used. We discuss the model and the resulting equations for eddy viscosity now. Other models are briefly reviewed later.

Prandtl (1925) used an analogy with the kinetic theory of gases to obtain a closure for the eddy viscosity. Fundamental to his theory was the concept of a "mixing length," denoted usually by l. This is an average length to which an eddy can travel without losing its identity. An eddy can be assumed to retain its momentum up to a distance of l. The (time-averaged) velocity difference between two points separated by a distance l determined using Taylor's series is

$$\overline{v}_x(y + l) = \overline{v}_x(l) + l\frac{d\overline{v}_x}{dy}$$

Hence an estimate of the magnitude of the velocity fluctuation is

$$v_x' \approx \pm l\frac{d\overline{v}_x}{dy}$$

The $\pm$ sign is used since the eddy can move in both the $+y$ and $-y$ directions.

If one assumes that v_y' is of the same order of magnitude as v_x' (i.e., turbulence is isotropic or uniform in all directions) then an estimate for the product $v_x'v_y'$ can be obtained:

$$\overline{v_x'v_y'} \approx l\frac{d\overline{v}_x}{dy} \cdot l\frac{d\overline{v}_x}{dy}$$

Noting that the turbulent shear stress is equal to $\rho\overline{v_x'v_y'}$ in magnitude, the following equation can be used to represent the turbulent shear stress:

$$\tau_{yx}^{(t)} = \rho l^2 \left|\frac{d\overline{v}_x}{dy}\right|\frac{d\overline{v}_x}{dy} \tag{12.10}$$

The absolute value is used to fit the sign convention that the stress is positive if the velocity gradient is positive and vice versa.

The mixing length is found to depend linearly on the distance from the wall and the relation is expressed as

$$l = \kappa y$$

where κ is a constant (usually taken as 0.4). Hence the turbulent shear stress can also be represented as

$$\tau_{yx}^{(t)} = \rho \kappa^2 y^2 \left| \frac{d\bar{v}_x}{dy} \right| \frac{d\bar{v}_x}{dy} \tag{12.11}$$

The previous equation is known as the Prandtl closure. Comparing Equations 12.8 and 12.11, we have the following closure equation for ν_t:

$$\nu_t = l^2 \left| \frac{d\bar{v}_x}{dy} \right| = \kappa^2 y^2 \left| \frac{d\bar{v}_x}{dy} \right| \tag{12.12}$$

The velocity profile for channel flow, pipe flow, boundary layers, and so on can be calculated using this model. The essential details needed for mass transfer applications are presented in the next section.

12.5   Velocity and Turbulent Diffusivity Profiles

The velocity changes rapidly near the wall in turbulent flow in contrast to laminar flow, where it is a smooth parabolic function. This suggests that different length and velocity scales are needed to make the problem dimensionless. The traditional average velocity and pipe radius used for normalization in laminar flow are not useful in turbulent flow. Detailed analysis (discussed, for example, by Ramachandran, 2013) leads to the following reference velocity:

$$v_{ref} = v_f = \sqrt{\tau_w/\rho}$$

v_f is called the friction velocity. In this definition, τ_w is the wall shear stress.

Similarly the following length scale is suggested:

$$L_{ref} = \frac{\mu}{\rho v_f}$$

The dimensionless velocity in the flow direction is now defined as $\bar{v}_x/v_{ref}$ and is denoted as v^+. Note that this is a time-averaged velocity. The dimensionless length is defined as y/L_{ref} and is denoted as y^+ where y is the distance measured from the wall. The velocity profile of v^+ versus y^+ derived on the basis of the Prandtl eddy viscosity closure are now presented without derivation. These are called universal velocity profiles since they are observed for a wide class of shear driven flows, for examples, pipes, channels, turbulent boundary layers and so on.

12.5.1 Universal Velocity Profiles

The flow domain in turbulent flow can be divided into three regions: a laminar sublayer also known as viscous layer, a buffer zone, and a turbulent core; velocity profiles are given separately for each of these regions and are as follows:

Near Wall Region: Viscous Sublayer

The turbulence is suppressed near the wall and the flow is laminar in this region. This is known as the laminar sublayer and normally extends to a length equal to 5 when expressed in dimensionless form as y^+.

The velocity profile in this region is given as

$$v^+ = y^+ \text{ for } y^+ < 5$$

This equation is found to be valid up to a y^+ of five, which is called the laminar (viscous) sublayer. This equation can be shown to arise from Newton's law of viscosity, assuming that the shear stress is nearly constant.

Outside the Viscous Layer: Turbulent Core

Outside the viscous region, usually starting at $y^+ = 30$, the flow is mainly turbulent and the laminar contribution to the stress can be neglected. The velocity profile in this region is a logarithmic function of distance and is given as

$$v^+ = 2.5 \ln y^+ + 5.5 \text{ for } y^+ > 30 \tag{12.13}$$

This is called the turbulent core region. This equation can be derived using the Prandtl mixing length model together with some simplifying assumptions.

Buffer Zone

The region in between the laminar layer and the turbulent core, $5 < y^+ < 30$, is called the buffer layer. The velocity here is correlated as

$$v^+ = 5 \ln y^+ - 3.05 \text{ for } 5 < y^+ < 30$$

This is more of a curve fit rather than a model-based equation.

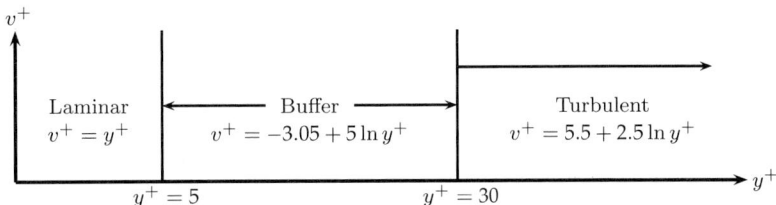

Figure 12.1 Illustration of the three regimes for velocity profile equations in turbulent flow.

The various regions and the expressions for the corresponding profiles are summarized in Figure 12.1.

Velocity Defect near the Center

Equation 12.13 for the fully developed core fits the data well almost up to the center of the channel. At the center the symmetry condition has to be satisfied. that is, dv^+/dy^+ must be zero at $y^+ = H^+$ (the center of the channel or pipe), but it is not as per Equation 12.13. Corrections to account for this error has been suggested and these are in general referred to as the velocity defect law. The error caused by this defect is small for use of these profiles in mass transfer analysis. This is because the concentration change is usually confined to a smaller region near the wall and does not extend all the way to the center of the channel or pipe. However, for study of homogeneous reactions in turbulent flow, this defect can cause errors in overall mass balance and more detailed models for the center region may be warranted in such cases.

12.5.2 Eddy Diffusivity Profiles

For mass transfer analysis, the primary parameter is the mass eddy diffusivity or equivalently the turbulent Schmidt number. We now demonstrate how this parameter can be calculated based on the velocity profiles. For this we need ν_t, the turbulent eddy viscosity, first. It is useful to define a dimensionless total momentum diffusivity as

$$\nu_T^+ = \frac{\nu + \nu_t}{\nu} = 1 + \frac{\nu_t}{\nu} \tag{12.14}$$

Note that this is not a physical property and is a turbulence parameter. Thus it varies as a function of y. This parameter is related to the local velocity gradient by the following relationship:

$$\nu_T^+ = \frac{1}{(dv^+/dy^+)} \tag{12.15}$$

Using this expression the following relations can be obtained by differentiating the velocity profiles for the three regions:

- Laminar layer: $\nu_T^+ = 1$
- Buffer layer: $\nu_T^+ = 0.2y^+$
- Turbulent core: $\nu_T^+ = 0.4y^+$

It is also common to define a dimensionless total diffusivity as

$$D_T^+ = \frac{D + D_t}{D}$$

This is related to the turbulent Schmidt number Sc_t by

$$D_T^+ = 1 + \frac{Sc}{Sc_t}(\nu_T^+ - 1) \tag{12.16}$$

Thus if the turbulent Schmidt number is assigned some value, then D_T^+ and correspondingly the turbulent diffusivity D_t can be estimated. The value of 0.9 is usually assigned for the turbulent Schmidt number.

12.5.3 Wall Shear Stress Relations

It should be noted here that the wall shear stress is needed as a primary parameter to complete these calculations (since it was used to scale the velocity and the length). A commonly used equation for this is summarized here from the fluid mechanics literature. The wall shear stress is correlated to the friction factor for internal flows:

$$\tau_w = f \frac{\rho \langle v \rangle^2}{2}$$

where $\langle v \rangle$ is the average velocity in the pipe. The friction factor in turn is correlated empirically as a function of the Reynolds number and the following correlation is commonly used for pipe flows (a number of other correlations are available):

$$f = \frac{0.0791}{Re^{1/4}}$$

This is commonly used for smooth pipes. Similar correlations are available for rough pipes (see for example in the book by Welty et al., 2008).

For external flow the wall shear stress is known if the drag coefficient is known, and they are related by the following equation:

$$\tau_w = C_{fx}\frac{\rho v_e^2}{2}$$

Here v_e is the velocity in the external flow, that is, at the edge of the boundary layer. The local drag coefficient in turbulent external flow is correlated as

$$C_{fx} = \frac{0.0576}{Re_x^{1/5}}$$

With this background we can examine turbulent mass transfer in channel flow and pipe flows. Again some further approximations are needed so that simple analytical solutions can be obtained. More complete analysis requires a numerical solution. The analysis is first done for gases where the molecular Schmidt numbers are on the order of one. For liquids. a modified analysis is useful since most of the concentration drop occurs mostly near the wall due to the large value of the Schmidt number.

12.6   Turbulent Mass Transfer in Channels and Pipes

The convection-diffusion equation of mass transfer in a channel under turbulent flow conditions can be represented for the time averaged value as

$$\bar{v}_x(y)\frac{\partial \overline{C_A}}{\partial x} = -\frac{\partial}{\partial y}(J_y^{(m)} + J_y^{(t)}) \tag{12.17}$$

The main difference between this and the laminar flow is the inclusion of the extra term (last term) arising from turbulent fluctuations, the so-called eddy mass flux term. Also, the velocity is the time-averaged value. Appropriate boundary conditions are applied at the channel walls. A similar equation can be formulated for pipe flows and the resulting problem is often referred to as the turbulent Graetz problem. The concentration profile in the turbulent boundary layer for mass transfer from a plate, for example, follows a similar model.

In general this problem requires a numerical solution and a number of studies (e.g., Larson and Verazunis, 1973) have addressed this approach. Here a simpler approach is presented where the concentration variation only in the cross-flow direction is considered. This approach involves some simplifications to the preceding differential equation. But this is adequate to predict the wall mass transfer coefficient and is therefore widely used in engineering practice. The simplified model is now presented.

12.6.1 Simplified Analysis: Constant Wall Flux

We analyze here the problem under conditions of constant wall mass flux J_w. Since the variation of concentration in the y-direction is more significant, we make a simplifying assumption that the mass flux varies linearly in the y-direction. This means that the convective term (left-hand side) of Equation 12.17 is simply assumed to be a constant. This assumption of a linear variation of the total mass flux, J_y, with y can be represented as

$$J_y = J_y^{(m)} + J_y^{(t)} = J_w(1 - y/H)$$

where J_w is the mass flux at the wall.

The molecular term is written using the Schmidt number, Sc, as

$$J_y^{(m)} = -D\frac{d\overline{C}_A}{dy} = -\frac{\nu}{Sc}\frac{d\overline{C}_A}{dy} \tag{12.18}$$

The turbulent term is written using the turbulent Schmidt number, Sc_t, as

$$J_y^{(t)} = -D_t\frac{d\overline{C}_A}{dy} = -\frac{\nu_t}{Sc_t}\frac{d\overline{C}_A}{dy} \tag{12.19}$$

ν_t is equal to $\nu(\nu_T^+ - 1)$ where ν_T^+ is the dimensionless total momentum diffusivity. Hence the turbulent mass is represented as

$$J_y^{(t)} = -\nu\frac{(\nu_T^+ - 1)}{Sc_t}\frac{d\overline{C}_A}{dy} \tag{12.20}$$

The total flux can be written as

$$-\nu\left(\frac{1}{Sc} + \frac{\nu_T^+ - 1}{Sc_t}\right)\frac{d\overline{C}_A}{dy} = J_w(1 - y/H) \tag{12.21}$$

Let us introduce the dimensionless distance y^+ here. Recall that this was defined as $y^+ = y\rho v_f/\mu = yv_f/\nu$.

Hence Equation 12.21 can be written as

$$-v_f\left(\frac{1}{Sc} + \frac{\nu_T^+ - 1}{Sc_t}\right)\frac{d\overline{C}_A}{dy^+} = J_w\left(1 - \frac{y^+}{H^+}\right) \tag{12.22}$$

This calls for a dimensionless (time-averaged) concentration c_A^+, which should be defined as

$$c_A^+ = (C_{As} - \overline{C}_A)v_f/J_w \qquad (12.23)$$

to simplify the model representation. Also, c_A^+ is zero when y^+ is zero. H^+ is the dimensionless width of the channel, defined as H/L_{ref}, which is equal to $Hv_f\rho/\mu$; this has the same grouping as the Reynolds number.

Using the above definition of the dimensionless concentration, Equation 12.22 is written as

$$\left(\frac{1}{Sc} + \frac{\nu_T^+ - 1}{Sc_t}\right)\frac{dc_A^+}{dy^+} = \left(1 - \frac{y^+}{H^+}\right) \qquad (12.24)$$

The integrated form of the above equation is

$$\boxed{c_A^+ = \int_0^{y^+} \frac{(1 - y^+/H^+)}{1/Sc + (1/Sc_t)(\nu_T^+ - 1)}dy^+} \qquad (12.25)$$

This is the formal representation to calculate the concentration distribution. If a model for ν_T^+ and for Sc_t as a function of y is proposed, the above equation can be integrated to find the concentration profiles.

Solution for Different Regions

It is convenient to integrate Equation 12.25 for each region separately. Also, the term $(1 - y^+/H^+)$ is often taken as 1 for simplification. This assumption certainly holds near the walls where most of the concentration changes occur. Hence Equation 12.25 is simplified as

$$c_A^+ = \int_0^{y^+} \frac{dy^+}{1/Sc + (1/Sc_t)(\nu_T^+ - 1)} \qquad (12.26)$$

This may not be good near the center but the effect of this "center defect" on the calculation of wall mass flux and the Sherwood number is generally small. The solution for c_A^+ for three regions (laminar, buffer, and turbulent) will now be presented.

Near Wall Region: Viscous Sublayer

Here ν_t is small and therefore $\nu_T^+ = 1$. Hence Equation 12.26 simplified and it is easy to show that the concentration profile in the viscous region is

$$c_A^+ = Scy^+ \qquad (12.27)$$

This is valid for $y^+ < 5$. Note that the (dimensionless) concentration profile is now scaled by Sc.

Since $v^+ = y^+$ we find

$$c_A^+ = Sc\, v^+ \tag{12.28}$$

and if $Sc = 1$, c_A^+ is the same as v^+, which justifies the Reynolds analogy.

Buffer Zone

The region between the laminar layer and the turbulent core, $5 < y^+ < 30$, is called the buffer layer. The velocity here was correlated as

$$v^+ = 5\ln y^+ - 3.05$$

Differentiation of this gives an expression for ν_T^+; it was shown earlier that

$$\nu_T^+ = 0.2 y^+$$

for the above velocity profile. Hence the formal expression for the concentration profile is

$$c_A^+ - 5Sc = \int_5^{y^+} \frac{dy^+}{1/Sc + (1/Sc_t)(0.2y^+ - 1)} \tag{12.29}$$

The value of Sc_t in the buffer zone is taken as one. Using this value and integrating, we obtain

$$c_A^+ = 5Sc + 5\ln\left[Sc[(y^+/5) - 1] + 1\right] \tag{12.30}$$

The value of c_A^+ at the edge of the buffer zone (at a y^+ of 30) is

$$c_A^+(30) = 5Sc + 5\ln(5Sc + 1) \tag{12.31}$$

which is needed as the integration constant for the next step.

Turbulent Core

Here the velocity profile in the turbulent core is given as

$$v^+ = 2.5\ln y^+ + 5.5 \tag{12.32}$$

which is found to be valid for $y^+ > 30$. This is called the turbulent core region. Differentiating this, we represent ν_T^+ as $0.4 y^+$ as shown earlier. Also, Sc_t is taken as 0.9. Further, $1/Sc - 1/Sc_t$ is much smaller than ν_T^+/Sc_t, and this term can be neglected. Hence the integral for the concentration profile in the turbulent core in Equation 12.6 simplifies to

$$c_A^+ - c_A^+(\text{at}\, y^+ = 30) = \int_{30}^{y^+} \frac{dy^+}{(1/0.9)(0.4y^+)} \tag{12.33}$$

Integrating and using Equation 12.31 for c_A at $y^+ = 30$, the concentration profile in the turbulent core is obtained as

$$c_A^+ = 5Sc + 5\ln(5Sc + 1) + 2.25\ln(y^+/30) \tag{12.34}$$

where the factor 2.25 arises from $Sc_t/0.4$, which is equal to $0.9/0.4 = 2.25$.

The concentration profile has a discontinuity in the slope at the edge of the turbulent core. Also, dc_A^+/dy^+ does not reach zero at $y^+ = H^+$, that is, at the center. But these are not considered to be of great concern in the predictions of mass transfer coefficients.

The results for the concentration profile shown for the channel flow are also applicable to pipe flow with y defined as the distance from the wall as well as for turbulent boundary layers. We now demonstrate the use of this to derive an expression for the calculation of the Stanton number.

12.6.2 Stanton Number Calculation for Boundary Layers

Using the expression for the concentration profile and its gradient, an expression for the mass transfer coefficient can be developed. Note that $C_{As} - C_{A\infty}$ is used as the driving force for external flows and hence

$$k_m = J_w/(C_{As} - C_{A\infty})$$

This is represented usually in terms of the Stanton number, defined as k_m/v_∞. Now, using the definition of dimensionless concentration defined by Equation 12.23, the local Stanton number can be shown to be related to the drag coefficient as

$$St_x = \left(\frac{C_{fx}}{2}\right)^{1/2} \frac{1}{c_{Ae}^+} \tag{12.35}$$

where c_{Ae}^+ is the dimensionless concentration at the edge of the boundary layer. This is defined as

$$c_{Ae}^+ = (C_{As} - C_{A\infty})v_f/J_w \tag{12.36}$$

The expression for c_{Ae}^+ can be obtained by applying Equation 12.34 at $y = \delta$ or correspondingly at $y^+ = \delta^+$. The latter is in turn related to v_e^+, which is in turn related to the drag coefficient. The resulting expression for c_{Ae}^{+} after some algebra (which is left as an exercise problem) is

$$c_{Ae}^{+} = 5Sc + 5\ln(5Sc + 1) + Sc_t \left[\sqrt{2/C_{fx}} - 14 \right] \qquad (12.37)$$

Note that the factor 2.25 in 12.34 is equal to $Sc_t/0.4$, which is used in getting to this form; $Sc_t = 0.9$ is used here.

Substituting in the local Stanton number expression from Equation 12.35, we have

$$St_x = \frac{\sqrt{C_{fx}/2}}{5Sc + 5\ln(5Sc + 1) + Sc_t[\sqrt{2/C_{fx}} - 14]} \qquad (12.38)$$

The denominator can be viewed as the sum of three resistances in series, the viscous sublayer, the buffer region, and the fully developed turbulent core. The relative magnitude of these will depend on the molecular Prandtl number and the Reynolds number. The equation is a good representation of data for $0.5 < Sc < 30$.

Equation 12.38 can be rearranged to

$$St_x = \frac{C_{fx}/2}{Sc_t + 5\sqrt{C_{fx}/2}\,[Sc + \ln(5Sc + 1) - 14Sc_t/5]} \qquad (12.39)$$

This form is easier to compare with the various analogies proposed between momentum and heat transfer. We now illustrate the various analogies proposed in the literature and show that these arise as special cases of the above equation.

12.6.3 Analogy with Momentum Transfer

The simplest is the Reynolds analogy, which states that

$$St_x = \frac{C_{fx}}{2}$$

Comparing with Equation 12.39, we find that this holds if $Sc = 1$, and $Sc_t = 1$, since the denominator can be shown be equal to one for these values.

Then there is the Prandtl analogy. This follows from Equation 12.39 by setting Sc_t equal to one:

$$St_x = \frac{C_{fx}/2}{1 + 5\sqrt{C_{fx}/2}\,[Sc - 1]} \qquad (12.40)$$

The von Karman analogy is obtained by setting $Sc_t = 0.9$, and is shown next:

$$St_x = \frac{C_{fx}/2}{1 + 5\sqrt{C_{fx}/2}\,[Sc - 1 + \ln(1 + 5(Sc - 1)/6]} \tag{12.41}$$

All the analogies differ only in the way the denominator term in Equation 12.39 is handled and what value is assigned to Sc_t.

Role of Molecular Sc

For the larger Sc case the mass transfer boundary layer is very thin and can be buried within the laminar sublayer and the buffer region. The thickness of the order of $15/Sc^{1/3}$ is considered to be an indicative magnitude of this region. The velocity profile in these regions affect the mass transport rate significantly for this case. Hence, a more accurate value of velocity profile in the buffer region and also in the viscous region may be needed. A modified equation is suggested in Section 12.7 where the von Karman profile for velocity (rather than the Prandtl profile with $v^+ = y^+$) is used.

Role of Turbulent Sc

A second point of caution is the value assigned to Sc_t. A constant value of 0.9 is usually assigned but it appears from the literature that a value that depends on the Re is more realistic. One explanation is that as Re is increased the turbulent energy spectrum moves to larger eddies, which may not be contributing in an equally proportional manner to mass transfer. A survey of various methods for assigning a proper value to Sc_t has been reviewed by Combest et al. (2011). For a further study of this topic this paper may be useful.

12.6.4 Stanton Number for Pipe Flows

The calculation of the Stanton number for pipe flow is similar but the difference is that the cup-mixing concentration rather than the external concentration should be used in the driving force. Thus the definition of mass transfer coefficient is

$$J_w = k_m(C_{As} - C_{Ab})$$

where C_{Ab} is the average concentration in the pipe (at any given axial location) and hence the Stanton number ($k_m/<v>$) can be shown to be

$$St = \sqrt{\frac{f}{2}} \cdot \frac{1}{c_{Ab}^+} \tag{12.42}$$

which is an expression similar to that for the boundary layer (Equation 12.35). Here f is the friction factor. However c_{Ab}^+ is an integral of the concentration profile and is formally given in terms of dimensionless variables as

$$c_{Ab}^+ = \frac{2}{v_b^+(R^+)^2} \int_0^1 v^+ c_A^+ (R^+ - y^+) dy^+ \tag{12.43}$$

In practice, it is difficult to estimate c_{Ab}^+. Note numerical integration can be done using Equation 12.43 if detailed v^+ and C_A^+ profiles are computed. But it is easy to find c_{Ac}^+, the dimensionless center concentration, by simple substitution in the concentration profile expression for the turbulent core region. Hence St is expressed as

$$St = \sqrt{\frac{f}{2}} \left(\frac{c_{Ac}^+}{c_{Ab}^+} \right) \left(\frac{1}{c_{Ac}^+} \right)$$

From the expressions derived earlier (Equation 12.34), the dimensionless center concentration is easily obtained:

$$c_{Ac}^+ = 5Sc + 5\ln(5Sc + 1) + 2.25\ln(R^+/30)$$

Since $R^+ = Re/(2v_b^+) = Re\sqrt{f/2}/2$, this can be expressed as

$$c_{Ac}^+ = 5Sc + 5\ln(5sc + 1) + 2.25\ln(Re\sqrt{f/2}/60)$$

Hence the Stanton number is given as

$$St = \frac{c_{Ac}^+}{c_{Ab}^+} \cdot \frac{\sqrt{f/2}}{5Sc + 5\ln(5Sc + 1) + 2.25\ln(Re\sqrt{f/2}/60)} \tag{12.44}$$

The first term c_{Ac}^+/c_{Ab}^+ is usually close to unity in turbulent flow due to the fact that the concentration profiles are flatter in the turbulent case than the laminar case and confined more closely to the wall region. This term is therefore approximated as one as a further approximation. (**Note:** A correction factor for the center to bulk concentration ratio can be developed, see, for instance, Mills, 1993).

Hence Equation 12.44 for the Stanton number is simplified as

$$St = \frac{\sqrt{f/2}}{5Sc + 5\ln(5Sc + 1) + 2.25\ln(Re\sqrt{f/2})/60)} \tag{12.45}$$

This expression relates the Stanton number (a measure of mass transfer) to the friction factor (a measure of momentum transfer) and is another example of the analogies used in mass transfer.

These models are particularly good compared to experimental data for gases and useful for $Sc < 30$ or so. For large Sc fluids the mass transfer boundary layer is rather thin and a slightly modified treatment described in the following section is used.

12.7   Van Driest Model for Large Sc

In the treatment of mass transfer in turbulent liquid flow an additional complication arises for mass transfer in liquids where the Sc can be as high as 1000! Here the concentration gradients are confined mostly to the viscous layer and to some extent to the buffer zone and any error in the velocity profile in this region will affect the predictions considerably. We illustrate the difference by looking at alternative models for mixing length near the wall.

It can be shown that the eddy viscosity varies as y^3 rather than as y^2 as per the Prandtl mixing length model (exercise problem 7). This can be accommodated by a more precise form of l, the mixing length. One such model was suggested by van Driest (1956), who proposed a simple exponential damping factor for l:

$$l = \kappa y \left[1 - \exp \left(-\frac{y^+}{26} \right) \right] \qquad (12.46)$$

This permits a smooth transition from the viscous layer to the turbulent core without the need to introduce an artificial buffer zone. This is one of many equations suggested for the near wall region and many improvements have been suggested. For example, Cebeci and Smith (1974) proposed that the constant 26.0 should be replaced by a function of dimensionless quantities involving factors such as pressure gradients, mass transfer rate, fluid compressibility, and other factors. Note that the van Driest model predicts the cube variation of the eddy viscosity near the wall.

A modified van Driest equation has also been proposed by Hanna et al. (1981). According to their model a denominator term is added to the original van Driest model:

$$l = \kappa y \frac{[1 - \exp(-y^+/26)]}{\sqrt{1 - \exp(-0.26 y^+)}} \qquad (12.47)$$

This expression also predicts the correct behavior that the turbulent viscosity will be proportional to y^3 near the wall. Taking the limit as $y \to 0$, the eddy viscosity near the wall can be represented as

$$\nu_t = \nu \left(\frac{y^+}{C}\right)^3 \tag{12.48}$$

where C includes all the constant terms in the modified model. An adjusted value of 14.5 is assigned to this based on experimental results. The relation for mass transfer coefficient based on the modified van Driest model is now examined.

The mass flux is given by

$$J_A = -(D + \nu_t/Sc_t)\frac{dC_A}{dy} \tag{12.49}$$

where Sc_t is the turbulent Schmidt number. Note that ν_t/Sc_t is equal to D_t, the turbulent diffusivity.

The flux J_A is assumed to be a constant equal to wall mass flux J_{Aw} in the boundary layer.

We also introduce a dimensionless total momentum diffusivity as done earlier:

$$\nu_T^+ = 1 + \frac{\nu_t}{\nu} \tag{12.50}$$

Equation 12.49 can now be expressed as

$$\frac{J_{Aw}}{\nu} = -\left[\frac{1}{Sc} + \frac{(1}{Sc_t}(\nu_T^+ - 1)\right]\frac{dC_A}{dy} \tag{12.51}$$

Dimensionless length $y^+ = y(\tau_w/\rho)^{1/2}/\nu$ is also introduced and Equation 12.51 can be formally integrated across the boundary layer to obtain an expression relating the wall mass flux and the wall shear stress:

$$(\sqrt{\tau_w/\rho})\frac{C_{A\infty} - C_{As}}{J_{AW}} = \int_0^{\delta^+} \frac{1}{1/Sc + (1/Sc_t)(\nu_T^+ - 1)}dy^+ \tag{12.52}$$

The relation $\tau_w = C_{fx}\rho v_e^2/2$ is now used. Further, the mass transfer coefficient is expressed terms of the Stanton number by the following relation:

$$\frac{J_{Aw}}{v_e(C_{A\infty} - C_{As})} = \frac{k_m}{v_e} = St = Sh/(ReSc)$$

Equation 12.52 can now be represented as

$$\frac{\sqrt{C_{fx}/2}}{St} = \int_0^{\delta^+} \frac{1}{1/Sc + (1/Sc_t)(\nu_T^+ - 1)}dy^+$$

Integration can be done if the variation for ν_T^+ with y^+ is specified. For the modified van Driest model, the following equation for ν_T^+ can be derived:

$$\nu_T^+ = 1 + \left(\frac{y^+}{C}\right)^3$$

It is also convenient to use ∞ as the upper limit of integration since the concentration gradients are confined to only small values of y^+ compared to δ^+. The results after integration and some rearrangement can be expressed as

$$\frac{Sh}{ReSc} = St = \frac{1}{17.5}\frac{\sqrt{C_{fx}/2}}{Sc^{2/3}}$$

which fits the data with some adjustment in the numerical constant.

12.8   Turbulent Mass Transfer at Gas–Liquid Interface

An illustrative example of this is mass transfer from a gas to a liquid in a falling film flow. The flow can be laminar at low Reynolds number, become oscillatory, and then become turbulent. We mainly focus on the turbulent regime here since the transport in laminar film was examined earlier in Section 10.4.2. The convection-diffusion equation is now augmented by incorporating an eddy diffusion coefficient D_t for mass transfer:

$$v_z(y)\frac{\partial C_A}{\partial z} = \frac{\partial}{\partial y}\left[(D + D_t)\frac{\partial C_A}{\partial y}\right] \qquad (12.53)$$

Both the concentration and the velocity are to be interpreted as the time-averaged values here. The velocity profiles are steeper in turbulent flow. Further, the mass transfer boundary layer is confined to a narrow region near the interface due to high Sc number values in the liquid phase. Hence a constant value of v equal to v_{max} can be used in the interface region. This approach is similar to that done for the laminar flow case in an earlier chapter. Hence we have

$$v_{max}\frac{\partial C_A}{\partial z} = \frac{\partial}{\partial y}\left[(D + D_t)\frac{\partial C_A}{\partial y}\right] \qquad (12.54)$$

The main problem emerges in the formulation of the expression for the turbulent eddy diffusion coefficient in Equation 12.54. Levich (1962) suggested the following correlation:

$$D_t = a y^n \tag{12.55}$$

Here y is the distance normal to the interface. The value of $n = 2$ seems to fit the data best (which can also be justified by a scaling argument) while the parameter a (units of s^{-1} for $n = 2$) was fitted by the following equation:

$$a = 1.17 \times 10^{-6} (g^2/\nu)^{1/3} Re_L^{1.448} \tag{12.56}$$

Re_L is the Reynolds number for film flow. A number of other correlations are available; see for example Grossman and Heath (1984) or Won and Mills (1982) and the discussion that follows this subsection. A useful review of mass transfer in turbulent film flow is also provided by Bin (1983) and this paper provides additional information on this topic.

One of the first analytical studies based on the above model was that by King (1966), who obtained two asymptotic solutions: the first for short contact time and the second for long contact time.

For very short contact time the eddy diffusivity has no effect on the rate of mass transfer since the penetration distance is small and is within the region where the turbulence is damped. ($D \gg D_t$, which holds for short contact times z/v_{max} or when a is small in the region of interest.) The classical penetration model holds for this case.

For long contact times the concentration profiles get fully established and the steady state mass transfer model holds. The left-hand side of Equation 12.53 is much smaller than the diffusion and the flux can be calculated by direct integration over the (semi-infinite) domain. The solution details are asked as exercise problem 9 and the King expression for mass transfer coefficient is

$$k_L = \frac{n}{\pi} \sin(\pi/n) \, a^{1/n} D^{1-1/n} \tag{12.57}$$

The mass transfer coefficient then depends on the value assigned to the parameter n and its value is assigned in the range of 2 to 4. A brief note on this is shown next.

12.8.1 Damping of Turbulence

The role of surface tension in dampening of the turbulence at the interface does not appear to have been clearly established. Surface tension will damp out smaller eddies, which have less kinetic energy compared to large eddies. On this basis, Tien and Wasan (1963) indicated that the eddy profile should be similar to that near the solid surface. Hence $n = 3$ was indicated in their work.

A value of $n = 4$ has also been suggested. In this case the model for k_L is

$$k_L = 0.90a^{1/4}D^{3/4}$$

Comparison with data suggested that the parameter a should be chosen as

$$a = 0.006\frac{\epsilon\rho}{\mu^2}$$

where ϵ is the energy dissipation per unit liquid volume. Energy dissipation in a falling film is given as $\rho g < v >$ and this provides a means of computing the a parameter.

Some additional references on turbulent mass transfer at a gas–liquid interface are Brumfield et al. (1975), Rashidi and Banerjee, 1988, Takagaki et al. (2016), and Avdeev, (2016).

12.8.2 Marangoni Effect

The Maragoni effect refers to the flow caused by a tangential surface tension gradient. This generates a shear along the interface and affects the motion near the interface and thereby the mass transfer coefficient. Detailed analysis of this effect is outside the scope of this book but qualitative understanding is important and presented briefly in this section.

The displacements are in the direction of increasing surface tension. The equilibrium of the forces at the interface leads to a stress discontinuity at the interface:

$$\tau_1 - \tau_2 = -\frac{d\sigma}{dx}$$

This leads to a reduced rise velocity, for example for a drop or bubble dispersed into a second continuous fluid, when $\frac{d\sigma}{dx}$ is positive; the mass transfer coefficient is then reduced.

On the other hand, if $\frac{d\sigma}{dx}$ is negative, the bubble/drop velocity is increased and the mass transfer coefficient is enhanced. One consequence of this is the directional dependence of the mass transfer coefficient, which has been observed experimentally in some distillation and extraction operations. The transfer coefficient is larger if the mass transfer leads to a decrease of surface tension at the interface and *vice versa*.

Closely related but not so well investigated is the role in changing the transfer area in addition to the mass transfer coefficient. Increase in surface tension as a result of solute mass transfer reduces the coalescence of drops, leading to an increase in interfacial area per unit volume. The opposite effect is observed when the surface tension decreases.

12.8.3 Interfacial Turbulence

A closely related phenomena is the interfacial turbulence. A spontaneous agitation caused by local surface tension gradients is referred to as interfacial turbulence. This is essentially a manifestation of hydrodynamic instability. Any perturbation in convection can cause a small change in concentration and hence cause a local perturbation in the surface tension gradients along the interface. The motion generated by this gradient can either by amplified or dampened. If it is amplified a spontaneous agitation results at the interface. For this reason some systems can be stable if the solute is transferred in one direction and can be unstable if the solute is transferred in the opposite direction. An example was presented by Olander and Reddy (1964) for nitric acid transfer from an aqueous medium to an organic one and *vice versa*. Similar effects have been observed for some systems where a simultaneous reaction is taking place (e.g., Sherwood and Wei [1957] for mass transfer of acetic acid from an organic phase to an aqueous base solution). If one is dealing with some practical systems, one should be aware of these potential complexities and the implication on data interpretation scale-up. Research by Strenling and Scriven (1959) and Berbente and Ruckenstein (1964) provide additional quantifiable information on the onset of flow instability due to surface tension effects.

Summary

- For turbulent mass transport problems the equation of mass transfer is time averaged. The cross-correlation of the velocity and the concentration fluctuations arises as extra terms due to time averaging; these are called turbulent mass flux terms. There is no simple way to predict these terms; this is referred to as the closure problem in turbulent flow.
- For turbulent reactive systems, an additional term arises in the reaction rate term due to the fluctuating components if the rate is a nonlinear function of concentration. For example, for a second-order reaction the

term is $k_2 \overline{C'_A C'_A}$. This term can be viewed as a macromixing effect and is of importance for fast reactions. The closure of this term is a field of considerable research at this moment and is important in many applications, for example, combustion processes. Further details on modeling these systems are not discussed in this chapter.

- An eddy diffusivity parameter is used to model the turbulent mass flux term as though a Fick's law type of model holds. This is only a definition and now the closure problem is transferred to the eddy diffusivity parameter.

- A common way to model the eddy diffusivity is to assume that it is proportional to momentum eddy diffusivity, since data on the latter can be related to a velocity profile. This ratio, the turbulent Schmidt number, is often assigned a constant value of 0.9.

- Turbulent mass transfer is then modeled by adding this extra term to the time-averaged model. Concentration profiles can then be solved for common flows such as boundary layers, channels, and pipes. The profiles are very similar for these three cases and the profiles are sometimes referred to as universal profiles.

- Mass transfer coefficients, usually expressed as a Stanton number, can be extracted from the concentration profiles. The Stanton number is related directly to the friction or the drag coefficient. This is an example of analogy between momentum and mass transfer and the various forms of the analogy follow from the relative contribution of the laminar sublayer, buffer zone, and turbulent cores to mass transfer.

- The molecular Schimidt number is the key dimensionless group that determines the relative thickness of the momentum and mass boundary layers. This dimensionless group is similar to the Prandtl number for heat transfer problems. If this is large, the concentration is mainly confined to the laminar sublayer (and to some extent to the buffer zone). Hence the van Dreist model for turbulent viscosity is more suitable to model the mass transfer since it captures the velocity in the laminar and buffer region more accurately. Correspondingly the expression for the Stanton number is slightly different compared to the low Schmidt case.

- Gas–liquid mass transfer in turbulent flow is modeled in a similar manner with a suitable expression assigned to the eddy diffusivity. The model uses a relation of the form $D_t \propto y^n$ where y is the distance measured from the interface. The nature of the solution and the parametric dependency varies with the exposure time of the gas to the liquid. For short exposure times, the molecular diffusivity can dominate, leading to the classical penetration model. The effect of turbulent flow is not so strong under these conditions. For long exposure times, the eddy diffusion dominates and the convection-diffusion model provides a framework for the analysis of mass transfer.

- Surface tension can vary in a mass transfer system and lead to an additional stress term; this results in Marangoni flow. The instabilities caused by the surface tension variation can amplify in some cases, leading to the phenomena of interfacial turbulence. The direction of mass transfer plays a role in such cases; this is something to watch out in some separation processes, for example, distillation, liquid extraction, and so on.

Review Questions

12.1　What is a "steady on average" for flow or mass transfer problem?

12.2　Cite an example where the flow is not steady on average. Can time averaging be used for such cases?

12.3　What is meant by cross-correlation between two random variables?

12.4　What is meant by the closure problem in the analysis of turbulent systems?

12.5　Define the turbulent mass flux vector and indicate where it comes from.

12.6　Define Reynolds stresses and indicate where they come from.

12.7　What is the RANS equation?

12.8　Define eddy viscosity and turbulent diffusivity.

12.9　Define the turbulent Schmidt number and point out the difference between this and the ordinary (molecular) Schmidt number.

12.10　Briefly explain the mixing length concept introduced by Prandtl.

12.11　How are the velocity and cross-flow distance parameter scaled in turbulent flow?

12.12　What is the relation between wall shear stress and pressure drop in pipe flow?

12.13　Can we assume the turbulent diffusivity to be a constant across the geometry of the system?

12.14　What is the van Dreist model for mixing length?

12.15　If the mass transfer coefficient for gas–liquid mass transfer is found to be proportional to D to the power of 0.75, what is the exponent n in the eddy diffusivity model?

12.16　What is the dependency of the mass transfer coefficient on D if the penetration model holds and if the eddy diffusion model holds with $n = 2$ for gas–liquid mass transfer?

12.17　What is Marangoni flow? State its importance in mass transfer analysis.

12.18　Explain briefly the phenomena of interfacial turbulence.

Problems

12.1 **Cross-correlation terms.** Explain what is meant by the closure problem in turbulence modeling. Explain the significance of the following cross-correlation terms: $\overline{v'_x v'_y}$, $\overline{v'_x C_A'}$, $\overline{C_A' C_A'}$, and $\overline{T' C_A'}$.

12.2 **Expression for the total viscosity.** Start from the basic definition of total shear stress:

$$\tau = \tau_{yx} = (\mu + \mu_t)\frac{d\bar{v}_x}{dy}$$

Use the dimensionless versions of velocity and distance to verify the expression for total eddy velocity given by Equation 12.15.

12.3 **Concentration at the edge of the boundary layer.** Use Equation 12.34 to show that the concentration at the edge of the boundary layer is

$$c_{Ae}^+ = 5Sc + 5\ln(5Sc + 1) + 2.25\ln(\delta^+/30) \tag{12.58}$$

Then using the Prandtl profile show that

$$\ln(\delta^+) = 0.4(v_e - 5.5)$$

Using the definition of v^+ and the drag coefficient relation to wall shear stress, verify the following relation:

$$v_e^+ = \sqrt{2/C_{fx}}$$

Combining all of the preceding, verify Equation 12.37 in the text.

12.4 **Turbulent mass transfer for air flow in a pipe.** Air is flowing at 4 m/s in a 5 cm i.d. pipe. Determine the friction factor, wall shear stress, friction velocity, v_f, v_b^+, and St.

   If the molecular Schmidt number for an evaporating solute is 2, calculate and plot the c_A^+ profile as a function of y^+. Use the Prandtl expression for the velocity here.

12.5 **Turbulent mass transfer for water flow in a pipe.** Water is flowing at 4 m/s in a 5 cm i.d. pipe. The molecular Schmidt number for a species dissolving from the wall is 500. Determine the friction factor, wall shear stress, friction velocity, v_f, v_b^+, and St. Calculate and plot the c_A^+ profile as a function of y^+.

12.6 **Expression for Stanton number for pipe flow.** Starting from the basic definition of the mass transfer coefficient and using the various dimensionless variables defined in the text, verify the relation in Equation 12.42 for the Stanton number for pipe flow mass transfer.

   Also verify Equation 12.43 to calculate the dimensionless cup-mixing concentration.

12.7   **Eddy diffusivity variation near the wall.** Verify that the fluctuating component of velocity (2D assumption) satisfies the following equation:

$$\frac{\partial v'_x}{\partial x} + \frac{\partial v'_y}{\partial y} = 0 \qquad (12.59)$$

The velocity components can be expanded as a Taylor series in y. A cubic approximation to the velocity is then as follows:

$$v'_x = a_0 + a_1 y + a_2 y^2 + a_3 y^3$$

and

$$v'_y = b_0 + b_1 y + b_2 y^2 + b_3 y^3$$

Use the no-slip condition to show that $a_0 = 0$ and $b_0 = 0$. Also since v_x near the wall does not change with x, show that $b_1 = 0$. Hence suggest an expression for the turbulent shear stress near the wall. How does it compare with the Prandtl mixing length theory? Does the linear relation of the turbulent stress with y hold?

12.8   **van Driest model.** Compare the results of exercise problem 5 if the van Driest model is used instead of the Prandtl mixing length model. Plot c_A versus y^+ to compare the two models. Use a large Sc, say 1000, so that the difference in the models can be noticed.

12.9   **Turbulent film: long contact time model of King.** Set the left-hand side of Equation 12.53 to zero and show that k_L is given by the following integral:

$$k_L = \left[\int_0^\infty \frac{dy}{D + ay^n} \right]^{-1}$$

Verify the expression given by Equation 12.57 in the text for long contact times using definite integral tables. Note the integration limit is changed to infinity and the film is assumed to be a semi-infinite domain. Explain the rationale behind this.

12.10   **Gas absorption in a turbulent film.** Gas absorption of CO_2 in water in a turbulent film was studied by Hikita et al. (1979) under the following conditions: liquid flow rate $= 1 \text{ m}^3/\text{sec}$; width of the plate 20 cm, length $= 20$ cm. Calculate the following: (1) the velocity at the surface, (2) the a parameter, and (3) the mass transfer coefficient.

CHAPTER 13

Macroscopic and Compartmental Models

Learning Objectives

After completing this chapter, you will be able to:

- Define the concept of a backmixed unit and the assumptions involved in it.
- Perform steady state and transient analysis of a backmixed unit.
- Show that transient analysis is useful to assess the level of mixing in the system.
- Couple backmixed units to form a compartmental model for a reactor or for a large-scale system.
- Use compartmental models in engineering analysis of complex systems.
- Formulate macroscopic models for two-phase systems.

This chapter elaborates on the principles of macroscopic modeling introduced in Chapter 3 and shows additional applications. As described earlier, in macroscopic analysis, the control volume is sufficiently large; hence the information from the microscale is lost and simple empirical laws or assumptions are needed to "close" the problem. The empirical method of *transport coefficients* for mass transfer problems is commonly used for closure for purely mass transfer applications. Likewise some assumptions on the mixing pattern in the control volume is needed in conjunction with macroscopic mass balance analysis for reacting systems. This is because the average rate of reaction is needed. But this is an integral of the local values and the local values are unknown in the context of macroscopic balances. We will look at some models to capture the mixing in a macroscopic volume.

One commonly used assumption is that the entire system is well-mixed and at uniform concentration. Such a model was described in Section 3.6 and is widely used as a benchmark in reactor and separation process design. For given equipment it may be therefore important to know to what extent this backmixing assumption is valid. This can be ascertained by a transient analysis and we show how this is useful.

If the system is discovered to be not completely backmixed, then a common modeling approach is to split the reactor into subunits connected in series or in parallel. Such a model is called the compartment model or the tanks or cells in series/parallel model. Each compartment is usually modeled as a completely mixed unit. We show how tracer studies are useful to see how many tanks in series should be used.

The compartmental model approach is also useful for modeling larger scale systems. These type of models find application, for instance, in biomedical engineering (e.g., drug distribution in a complex system such as the human body) and in environmental engineering. Here the complex system is viewed as interconnected macroscopic models consisting of several subunits or compartments. The modeling of such systems is therefore a basic extension of macroscopic analysis. They usually require additional *inter-compartmental exchange parameters*. We will study some illustrative problems and cite some key references on this. Computations of compartmental systems are also illustrated using MATLAB using a simple two-compartment model as an example.

Macroscopic models are widely used for two phase systems, for example, mixer-settler for liquid–liquid extraction. Each phase is modeled as separate compartments and the mass exchange between the two phases is included as the interaction term connecting the two phases. One simplification is the equilibrium stage model, which is often used in separation processes as a first level of analysis. The key assumption in this model is that the exit streams are assumed to be in equilibrium. Mass conservation together with phase equilibrium models are sufficient to predict the level of separation that can be achieved in the system; mass transfer effects are not included at the first level of modeling. An efficiency parameter is introduced at the second level and this parameter can be calculated if the mixing pattern and mass transfer parameters are known in the contactor. The mixing pattern is often modeled using the tanks in series approach. This approach is also used to simulate many similar fluid–fluid contactors such as bubble columns, rotating disc extraction equipment, and so on; the conceptual framework for this approach is explained.

13.1   Stirred Reactor: The Backmixing Assumption

The starting point in macroscopic modeling is the species conservation equation shown in Chapter 3; it is repeated here for ease of reference:

$$\dot{M}_{A,i} - \dot{M}_{A,e} - \dot{M}_{A,W} + V \langle R_A \rangle = \frac{d}{dt}(M_A) \tag{13.1}$$

This was applied to a continuous flow reactor in Section 3.6. In this section we revisit this briefly and follow it up by providing more information on the tracer experiments discussed in Section 3.7.

In order to close the reaction term, the backmixing assumption is used at the first level of modeling:

$$\text{Backmixed assumption: } < C_A > = C_{A,e}$$

Note that this alone is not sufficient to close $< R_A >$ for nonlinear kinetics. For example for a second-order reaction we have

$$-<R_A> = \frac{1}{V} \int_V k_2 C_A^2 dV$$

The right-hand side is the average of C_A^2, denoted as $\langle C_A^2 \rangle$. This need not be the same as the square of average, that is, $(<C_A>)^2$. The two will be equal only if there is no variance in the average concentration at every point in the system. The fluid elements have to be mixed on the microscale for this to happen. Hence an additional assumption that the system is micromixed is also used. In particular, this is true only if the fluid elements are mixed at the molecular level, but it is often used as an approximation. It holds unless the reaction time scale is very small compared to the time scale for mixing. The model can then be closed to this level of approximation.

Performance equations for simple power law kinetics are useful in this context and Equations 3.23 and 3.26 from Chapter 3 apply for a first-order and second-order reaction, respectively.

Fractional order kinetics are often encountered and some caution is needed when the model is used for this purpose. For example, for a zero-order reaction, the exit concentration can become negative numerically, but this not possible based on physical considerations. This sets an upper limit on the Damkohler number, which is defined as follows for a zero-order reaction:

$$Da = \frac{V k_0}{Q C_{A,i}} \tag{13.2}$$

The exit concentration can be obtained by solving the mass balance equation and is given as

$$c_{A,e} = 1 - Da$$

If $Da = 1$ then the exit concentration becomes zero and is set to zero for any Da value greater than one since the exit concentration cannot be negative. A numerical or analytical solution will generate unrealistic negative values for the concentration in this range. Hence caution has to be exercised while interpreting numerical results.

For more complex kinetics, a numerical solution is used. For multiple reactions, the balance equations are written for each species, A, B, C, and so on, and the resulting set of algebraic equations can be solved numerically. The MATLAB solver FSOLVE is useful for this purpose.

The assumption of backmixing closure can be tested by transient response analysis. The next section shows how this is done; it is a continuation of the information provided in Section 3.7. It may be useful to review that section before proceeding further.

13.2   Transient Balance: Tracer Studies

The extent of backmixing can be characterized by tracer studies, also known as stimulus-response studies. The stimulus represents a change in the inlet conditions and the change in concentration in the outlet as a function of time is measured and represents the response curve to the system. The exit response predicted from a theoretical model is then matched with the experimental data to test some of the model assumptions, for example, whether the well-mixed assumption holds or not.

The starting point for analysis of any tracer study is the transient species balance equation. The appropriate representation for a macroscopic volume follows from Equation 13.1 with appropriate simplifications:

$$V\frac{d\langle C_A\rangle}{dt} = QC_{A,i} - QC_{A,e}$$

If the system is assumed to be backmixed, then $\langle C_A\rangle = C_{A,e}$ and the model is represented as

$$V\frac{dC_{A,e}}{dt} = QC_{A,i} - QC_{A,e} \qquad (13.3)$$

Two types of tracer injection are common: step injection and pulse or bolus injection. The outlet response for these two cases are presented in the following.

13.2.1 Step Input

A step tracer refers to a case where a feed stream is replaced by a stream containing a tracer at time zero. The initial condition for a step tracer is that at $t = 0$, $< C_A >= 0$, that is, there is no tracer in tank to start. Hence

$C_{A,e}(t=0) = 0$. The inlet concentration is $< C_A > = C_{A,i}$ for all $t > 0$. The solution to Equation 13.3 is then

$$C_{A,e} = C_{A,i}\left[1 - \exp(-tQ/V)\right]$$

This equation is normally written in dimensionless form:

$$c_{A,e} = 1 - \exp(-t^*) \tag{13.4}$$

where c_A is scaled with inlet concentration and t^* is a dimensionless time defined as $t^* = tQ/V$. This is the ratio of the actual time to the mean residence time in the reactor.

The analysis indicates that the tracer concentration should follow an exponential rise in concentration starting at zero and ending at one. If the experimental data indicates such a pattern then we can treat the system as a well-mixed reactor. Analysis of the tracer data for systems that are not well-mixed is treated in Section 13.2.4.

13.2.2 Pulse or Bolus Input

A pulse or bolus tracer refers to a mode where a quantity of tracer is introduced all at once in a very short duration at time zero. No tracer is injected after this. This section looks at the exit response of the tracer for this case.

The inlet concentration now is $C_{A,i} = 0$ since we are dealing with $t > 0$, that is, after the bolus injection. Hence Equation 13.3, the starting mass balance equation, reduces to

$$V\frac{dC_{A,e}}{dt} = -QC_{A,e}$$

The solution for this differential equation is

$$C_{A,e} = C_0 \exp(-Qt/V) = C_0 \exp(-t^*)$$

where C_0 is the integration constant, which has the physical meaning of the initial concentration in the tank at $t = 0^+$, that is, immediately after the tracer injection. The initial concentration is not known *a priori* and depends on the quantity of the tracer injected at time zero; it needs an overall mass balance of A for the total duration of the experiment. Let the quantity of tracer injected be $\mathcal{M}$. The amount of tracer leaving the system over an interval of time Δt is $QC_{A,e}\Delta t$. The total tracer leaving the system is the integral of this from time zero to infinity. By an overall mass balance for A the total tracer leaving the system must be also equal to the quantity of tracer injected, represented as $\mathcal{M}$.

Hence

$$\mathcal{M} = \int_0^\infty QC_{A,e}dt = \int_0^\infty QC_0 \exp(-Qt/V)dt$$

Performing the integration and rearranging we have $C_0 = \mathcal{M}/V$, which is known as the tracer strength.

Hence the response to a bolus injection is

$$C_{A,e} = (\mathcal{M}/V)\exp(-tQ/V)$$

The bolus injection response is therefore an exponentially decreasing function of time and can be expressed in dimensionless form as

$$c_{A,e} = exp(-t^*) \tag{13.5}$$

where c_{Ae} is equal to $C_{A,e}/(\mathcal{M}/V)$. The response curve for a backmixed reactor for a pulse input is shown in Figure 13.1. Here the response is also shown if the system is in plug flow and for an in between case.

For the plug flow case in Figure 13.1b, the tracer moves as a plug in the axial direction without any mixing in the flow direction; all the tracer appears at the exit at the same time, equal to the mean residence time. All the fluid in the exit stream has spent the same amount of time in the reactor, equal to the mean residence time.

The response curve for a case where the reactor mixing does not belong to either of the two ideal cases of plug flow and backmixed flow is presented in Figure 13.1c. It should be noted here that the curve shown with a single peak in 13.1c and a spread around a mean value is only applicable to reactors with no severe maldistribution. Such reactors are often termed non-ideal but well behaved. More complex flow patterns such as severe bypassing or large recirculation will show tracer response with multiple peaks. Similarly reactors with large dead volume with poor exchange between active and dead zones

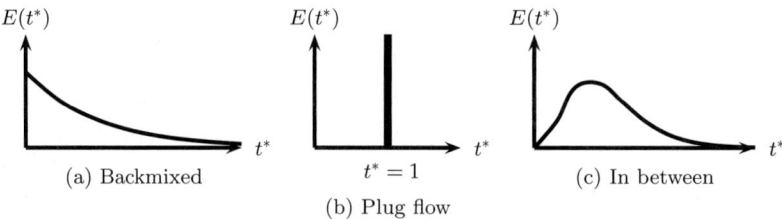

Figure 13.1 Response curve to a pulse tracer for (a) a system that is completely mixed, (b) a plug flow with the same mean residence time, and (c) a case in between the plug and backmixed cases. Note $E(t^*)$ is the dimensionless exit concentration $c_{A,e}$.

(see Figure 3.10) will show a long tail in the tracer response. These cases are sometimes referred to as non-ideal ill-behaved cases.

13.2.3 Age Distribution Functions

In general, the response curves are a measure of time spent by Lagrangian fluid elements in the system; these are characterized by age distribution functions presented in the following subsections. Age means the length of time a fluid has been in the vessel from a Lagrangian perspective. Three age distribution functions are commonly used, as discussed in the following paragraphs.

E-Curve

The dimensionless version of the pulse response is called the E-curve, the exit age distribution curve. The expression for a backmixed reactor follows from Equation 13.5:

$$E(t^*) = \exp(-t^*) = \exp(-tQ/V)$$

The physical meaning is that $E(t^*)dt^*$ represents the fraction of the tracer in the exit stream that has spent time between t^* and $t^* + dt^*$. It represents the fraction of the exit stream that has an age in this interval.

Note that

$$E(t^*)dt^* - E(t)dt$$

Here $E(t)dt$ is the fraction of the tracer in the exit stream that has an age between t and $t + dt$. This E-curve (in dimensional units) for a backmixed reactor is

$$E(t) = \frac{Q}{V} \exp\left(-t\frac{Q}{V}\right) = \frac{1}{\bar{t}} \exp(-t/\bar{t}) \qquad (13.6)$$

Here the $\bar{t}$ is the mean residence time, defined as V/Q.

F-Curve

Another age distribution function is the F-curve, $F(t)$. This represents the fraction of the exit stream, which has an age from 0 to t. From the definition it can be shown that the F-curve is the integral of the E-curve and the E-curve is the derivative of the F-curve:

$$F(t) = \int_0^t E(t)dt$$

and

$$E(t) = \frac{dF}{dt}$$

I-Curve

Another age distribution function is the I-curve, $I(t)$, which is called the internal age distribution. $I(t)dt$ represents the fraction of the fluid inside the reactor, which has an age in the range of t and $t + dt$. The relation between the E-curve and the I-curve follows from the consideration of a fluid of age t either leaving or staying inside the reactor:

$$I(t) = \frac{Q}{V}\left[1 - \int_0^t E(t)dt\right]$$

Hence

$$E(t) = -\frac{V}{Q}\frac{dI(t)}{dt}$$

13.2.4 Tracer Response for Tanks in Series Model

The previous section showed how the tracer response is useful to identify if the reactor or the process equipment is well mixed. If these experiments show that the system is not well mixed, a model for the reactor needs to be developed. One such model is formulated by assuming that the system is made up of a number of tanks or compartments connected in series. The resulting model is called the tanks in series model and is commonly used in chemical reactor analysis and environmental modeling applications. See Figure 3.9 for an illustration of a vessel modeled as two tanks in series. This approach is also useful for multiphase systems as shown in Section 13.10. In this section we derive an expression for the exit response of such a model.

The system is analyzed by writing the mass balance equations sequentially for each tank with the output from tank 1 serving as the input to tank 2, and so forth. For each tank the transient equation of the type given by Equation 13.3 applies:

$$V_t\frac{dC_{A,n}}{dt} = Q(C_{A,n-1} - C_{A,n}) \tag{13.7}$$

where V_t is the volume of each tank; in dimensionless form we have

$$\frac{dc_{A,n}}{dt^*} = c_{A,n-1} - c_{A,n}$$

where the concentration is scaled by the tracer strength, assuming all tracer is dispersed in tank 1, and thereby using, $\mathcal{M}/(V_t)$ as the concentration scale.

The time is scaled as $t^* = tQ/V_t$, that is, for each tank and not that for the overall reactor. At the end of this section we will correct it to base it on the residence time in the whole tank, where t^*, defined as tQ/V, will be used.

A solution to the response curve can be done progressively starting from tank 1 and progressing to tank N. For tank 1 the solution is the same as that for a single tank given earlier (Equation 13.5) with time scaled by the volume of tank 1:

$$c_{A,1} = \exp(-t^*)$$

Hence the equation for tank 2 is

$$\frac{dc_{A,2}}{dt^*} = c_{A,1} - c_{A,2}$$

Substituting the exit from tank 1 as the input to tank 2 we obtain

$$\frac{dc_{A,2}}{dt^*} + c_{A,2} = \exp(-t^*)$$

A solution can be obtained using the integrating factor $\exp(t^*)$. The solution satisfying the initial concentration of zero in tank 2 is

$$c_{A,2} = t^* \exp(-t^*)$$

This becomes the input to tank 3 and the differential equation for this tank is

$$\frac{dc_{A,3}}{dt^*} + c_{A,3} = t^* \exp(-t^*)$$

The integration can be continued to give

$$c_{A,3} = \frac{(t^*)^2}{2} \exp(-t^*)$$

The result can be generalized to N tanks by inspection:

$$c_{A,N} = \frac{(t^*)^{(N-1)}}{(N-1)!} \exp(-t^*)$$

Note that the concentration was scaled by $\mathcal{M}/V_t$. We need to scale it by $\mathcal{M}/V$, the tracer strength based on the reactor volume. Hence we need to multiply the above equation by N to get the new scaled concentration, which is given by

$$c_{A,N} = N \frac{(t^*)^{(N-1)}}{(N-1)!} \exp(-t^*)$$

Further, the results are to be expressed in terms of the overall residence time in the reactor, denoted here as t^*, which is equal to V_{total}/Q. Thus t^* is replaced by Nt^* and the final result for the response of N tanks in series to a pulse of bolus tracer is

$$c_{A,N} = \frac{N^N}{(N-1)!}(t^*)^{N-1} \exp(-Nt^*) \tag{13.8}$$

Note that here $t^* = tQ/V$.

The dimensionless time at which the maximum concentration is obtained is obtained by equating the derivative of the above expression with respect to t^* equal to zero. The result that this occurs at a dimensionless time of $(N-1)/N$ is easily verified. Substituting this value of t^* in the exit concentration, the maximum value of the exit concentration is obtained as

$$c_A,\text{max} = \frac{N(N-1)^{N-1}}{(N-1)!} \exp[-(N-1)]$$

An illustrative response to a system modeled as two and four tanks in series is shown in Figure 13.2 and is compared to a response if the whole tank is assumed to be a single well-mixed tank.

In summary, the tracer studies are a good way to ascertain the degree of mixing in the system and modify the backmixed assumption as needed. If the system is modeled by tanks in series, then the number of tanks needed to represent the system can be obtained by matching the experimental data with the model values given above. This method is called time domain fitting. An alternative method is moment analysis, which is discussed in the following section.

13.3   Moment Analysis of Tracer Data

An important data analysis tool for the interpretation of the tracer experiments is the method of moments where the expressions for the first and second moments are derived and used to fit the parameters. Let the outlet response, $C_{A,e}$, for a pulse input be measured as a function of time. (Note that this data need not be the actual concentration but could be some measure of it,

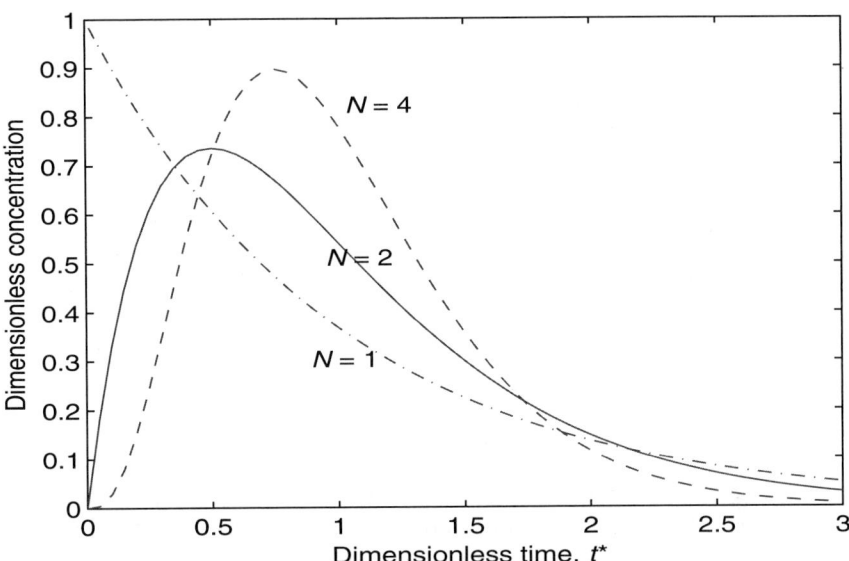

Figure 13.2 Transient concentration profiles in response to a bolus input. Three cases are shown: system modeled as one backmixed tank, system modeled as two tanks in series, four tanks in series.

for example, electrical conductivity of a salt solution, color intensity for a dye tracer, and so on.) Then these curves are normalized by the total area under the tracer response curve:

$$\mu_0 = \int_0^\infty C_{A,e}\, dt$$

This is a measure of the tracer strength and should be equal to $\mathcal{M}/Q$ since all the injected tracer has to appear in the exit stream sooner or later. The comparison of the experimental μ_0 and the amount of tracer injected also provides a check on the mass balance as well. The μ_0 is also called the zeroth moment.

All concentration data can then be scaled by dividing by μ_0, which provides the response in terms of the dimensionless concentration, $c_{A,e}$ verus t. The first moment with units of time is defined as

$$\mu_1 = \int_0^\infty t\, c_{A,e}\, dt$$

This is a measure of the mean passage time, that is, the time when most of the tracer appears in the exit. This will be equal to V/Q for any closed vessel, that is, a vessel with no back-diffusion or back-flow at the flow boundaries.

The second moment with units of time^2 is defined as

$$\mu_2 = \int_0^\infty t^2\, c_{A,e}\, dt$$

A related quantity is the variance σ^2 (s^2), defined as

$$\sigma^2 = \mu_2 - \mu_1^2$$

This is a measure of the spread of the tracer around the mean passage time.

Higher order moments can also be defined, for example the nth moment is

$$\mu_n = \int_0^\infty t^n c_{A,e}\, dt$$

However these are not commonly used in data analysis since these higher order integrals cannot be accurately calculated from the measured data.

In the moments analysis method, the measured values of the moments are compared to theoretical values and the parameters of the model are fitted. Usually the second central moment or the variance is compared. The calculation of theoretical values is discussed next.

13.3.1 Moments from Laplace Transform of Response

Moments are related to the response curve in the Laplace domain and the method of deriving the needed expressions is presented in the following.

The Laplace transform of the tracer response is defined as

$$\bar{c}_{A,e}(s) = \int_0^\infty \exp(-st) c_{A,e}(t)\, dt \tag{13.9}$$

where s is the transform parameter. Note that the integration is with respect to t for various values of the parameter s.

If the exponential function is expanded the integral on the right-hand side of Equation 13.9 can be written as

$$\bar{c}_{A,e}(s) = \int_0^\infty \left(1 - st + \frac{s^2 t^2}{2} + \cdots \right) c_{A,e}\, dt$$

If we expand the left-hand side of Equation 13.9 in a Taylor series, we have

$$\bar{c}_{A,e}(s) = \bar{c}_{A,e}(s=0) + s \left(\frac{d\bar{c}_{A,e}}{ds} \right)_{s=0} + \frac{s^2}{2} \left(\frac{d^2 \bar{c}_{A,e}}{d^2 s} \right)_{s=0} + \cdots$$

Equating terms with the same power of s on either side the following relations are obtained. For $s = 0$ we get

$$\bar{c}_{A,e}(s = 0) = \int_0^\infty \bar{c}_{A,e}\, dt = 1$$

which is the dimensionless tracer strength.

The (scaled) first moment is then related to the first derivative of the response in the Laplace domain:

$$\mu_1 = \int_0^\infty t c_{A,e}\, dt = -\left(\frac{d\bar{c}_{A,e}}{ds}\right)_{s=0}$$

The second moment is

$$\mu_2 = \int_0^\infty t^2 c_{A,e}\, dt = \left(\frac{d^2\bar{c}_{A,e}}{ds^2}\right)_{s=0}$$

Expressions for additional moments can be derived if needed. In general

$$\mu_n = \int_0^\infty t^n c_{A,e}\, dt = (-1)^n \left(\frac{d^n\bar{c}_{A,e}}{ds^n}\right)_{s=0}$$

Thus all the moments can be generated from the solution in the Laplace domain. An illustration of derivation of the Laplace transform and the expression for the second moment is shown in Example 13.1.

Example 13.1   Moments for N Tanks in Series

Derive an expression for the response to N tanks in series in the Laplace domain. From the solution derive expressions for the first and second moment and the variance of the response curve.

Solution

For each tank the transient equation of the type given by Equation 13.7 applies:

$$V_t \frac{dC_{A,n}}{dt} = Q(C_{A,n-1} - C_{A,n})$$

V_t is the volume of each tank. Since $V_t = V/N$, this can be written as

$$\frac{V}{QN}\frac{dc_{A,n}}{dt} = (c_{A,n-1} - c_{A,n})$$

Here the concentration has been normalized by dividing C_A by the tracer strength for the pulse tracer.

Taking the Laplace transform of the previous equation we get

$$\frac{V}{QN} s \bar{c}_{A,n} = \bar{c}_{A,n-1} - \bar{c}_{A,n}$$

Note that the initial concentration is taken as zero here. Rearranging, we get the response of the nth tank in terms of the response from the $n-1$ tank:

$$\bar{c}_{A,n} = \frac{\bar{c}_{A,n-1}}{(1 + sV/QN)}$$

Starting from tank 1 we can get the response of tank 2 and continuing we get the response for N tanks as

$$\bar{c}_{A,e} = \frac{\bar{c}_{A,i}}{(1 + sV/QN)^N}$$

where $\bar{c}_{A,i}$ is the Laplace transform of the inlet concentration. For a pulse response $\bar{c}_{A,i} = 1$ and hence the response curve in the Laplace domain is

$$\bar{c}_{A,e} = \frac{1}{(1 + sV/QN)^N}$$

Taking the derivative and setting $s = 0$ we find $\mu_1 = V/Q$, which is equal to the mean residence time $\bar{t}$.

Similarly μ_2 is found by taking the second derivative and you should verify that

$$\mu_2 = [(N+1)/N](V/Q)^2$$

The variance, σ^2, is calculated as $\mu_2 - \mu_1^2$. This gives us

$$\sigma^2 = \frac{1}{N}\left(\frac{V}{Q}\right)^2$$

This has units of s^2. A dimensionless variance is often defined and used as follows:

$$\boxed{(\sigma^*)^2 = \frac{\sigma^2}{\bar{t}^2} = \frac{1}{N}} \tag{13.10}$$

$\bar{t}$ is the mean passage time equal to V/Q. This equation is a classical result for N tanks in series and can be matched with the experimental variance to find the representation of the reactor in the tanks in series model.

It is interesting to note that the response curve in the time domain can be reconstructed within engineering accuracy from the moments using the Laguerre polynomials. The Laplace inversion is not needed when using this method since the inverse transform calculations can in some cases involve lengthy mathematical manipulations. The full discussion is given by Linek and Dudukovic (1982) and further details are not addressed here.

13.4   Tanks in Series Models: Reactor Performance

In this section, we calculate the reactor performance when the reactor is modeled as N tanks connected in series. For a first-order reaction, the species mass balance for any compartment n leads to

$$Q(C_{A,n-1} - C_{A,n}) = V_t k_1 C_{A,n}$$

where V_t is the volume of each tank. Since $V_t = V/N$, this can be written as

$$c_{A,n-1} - c_{A,n} = \frac{V k_1}{QN} c_{A,n} = \frac{Da}{N} c_{A,n} \qquad (13.11)$$

The concentration has been normalized by the inlet concentration. The Da parameter is the Damkohler number, based on the volume V of the reactor. Equation 13.11 can be rearranged to:

$$c_{A,n} = \frac{c_{A,n-1}}{1 + Da/N}$$

Starting from $n = 1$, the first compartment and progressing to the Nth compartment we find the exit concentration in the reactor:

$$c_{A,N} = \frac{1}{(1 + Da/N)^N} \qquad (13.12)$$

Nonlinear reactions are solved by writing equations similar to 13.11 and solving the set of the resulting set of N algebraic equations numerically. No new concepts are involved in modeling these but closed-form solutions may not be obtained, unlike the first-order case. Note that Damkohler number is a function of the inlet concentration for nonlinear kinetics, unlike the case of a first-order reaction. Multiple species and multiple reactions are handled by a similar approach.

An assumption involved in this modeling approach is that the fluid is mixed all the way to the micro level. If the fluid is mixed only at a macro level the method discussed in the following section is used. For linear kinetics it does not matter whether the fluid is mixed on a macro or micro level.

13.5   Macrofluid Models

Here we show how the exit age distribution is useful to predict the conversion in a reactor assuming the system is well mixed but only on a macro level. The following assumption is made for this analysis: different elements entering the reactor at different times remain segregated and do not mix with elements that have entered earlier or elements that stay behind for a longer time.

The exit stream is visualized as a mixture of different macrofluid parcels that have spent different amounts of time in the reactor. The situation is illustrated in Figure 13.3.

An element spending time t would have reacted as

$$\frac{dC_A}{dt} = R_A(C_A) \tag{13.13}$$

The integration provides us a $C_A(t)$ function for any specified form of R_A.

The exit stream is composed of elements of many ages, each spending a time t. The fraction of the exit stream that has spent time t is $E(t)dt$. Hence the contribution to the exit conversion for the element of age t is $C_A(t)E(t)dt$. The exit concentration, summing for all ages, is thus

$$\boxed{C_{A,e} = \int_0^\infty C_A(t)E(t)dt} \tag{13.14}$$

C_A as a function of time is obtained from the solution of Equation 13.13 and results in the same expression as for a batch reactor. It can be shown that the results for macrofluid models and microfluid models are the same

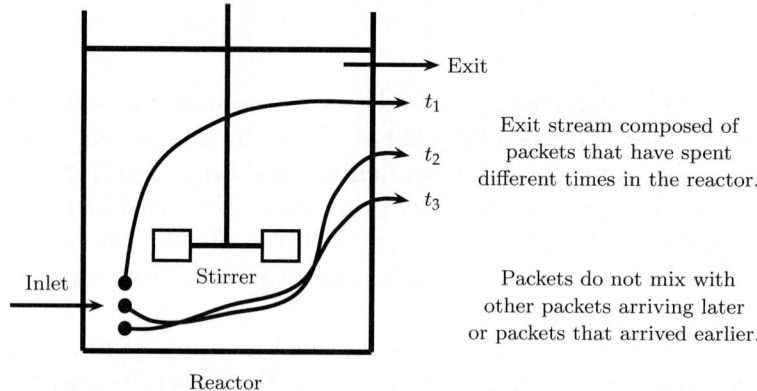

Figure 13.3   The macrofluid representation of a reactor: segregation on an age basis.

for a first-order reaction. An example follows for a second-order reaction for a backmixed reactor.

13.5.1 Second-Order Reaction

For a second-order reaction in a batch reactor we can show that the concentration of a stream that has spent a time t in the reactor is:

$$c_A(t) = \frac{1}{1 + k_2 C_{A,i} t}$$

Combining this with the exit age distribution for a backmixed system (Equation 13.6), we obtain

$$E(t) = \frac{1}{\bar{t}} \exp(-t/\bar{t})$$

Using these results in Equation 13.14 we get

$$c_{A,e} = \int_0^\infty \frac{1}{1 + k_2 C_{A,i} t} \frac{\exp(-t/\bar{t})}{\bar{t}} dt$$

The integral can be evaluated numerically as shown in Listing 13.1. But a closed-form analytical solution can be obtained:

$$c_{A,e} = \frac{C_{A,e}}{C_{A,i}} = \frac{1}{Du} \exp(-1/Da) Ei(1/Da) \tag{13.15}$$

where Ei is an exponential integral, a tabulated mathematical function. This function is readily calculated in MATLAB as *expint*. Table 3.1 in Chapter 3 showed the results comparing the microfluid and microfluid approach for a second-order reaction; we find that the macromixing conditions provide a lower concentration and therefore a higher conversion.

13.5.2 Zero-Order Reaction

The concentration change for a zero-order reaction is given as

$$c_{A,e} = 1 - k_0 t / C_{A,i} \qquad t \le C_{A,i}/k_0$$

and $C_{A,e} = 0$ if $t > C_{A,i}/k_0$. Using this together with the expression for $E(t)$ in Equation 13.14, the result for exit conversion can be determined to be

$$c_{A,e} = \frac{C_{Ae}}{C_{A,i}} = 1 - Da + Da \exp(-1/Da) \tag{13.16}$$

Table 13.1 Comparison of Micro versus Segregation Model: Results for a Backmixed Reactor for Zero-Order Kinetics

Da	$c_{A,e}$, Microfluid	$c_{A,e}$, Macrofluid
0.1	0.9	0.9
0.5	0.5	0.5677
1.0	0.0	0.3679
5.0	0.0	0.0937
10.0	0.0	0.0484

where Da is the Damkohler number defined earlier by Equation 13.2. The derivation is left as an exercise problem. The comparison is shown in Table 13.1. The conversion is lower if the fluid is macromixed, a trend opposite to the second-order case.

The previous two examples provide illustrations of the macrofluid model for second-order and zero-order kinetics. For the general case, it is easier to calculate this by writing the small computational snippet presented in Listing 13.1.

Listing 13.1   Macrofluid Model Calculations with MATLAB+CHEBFUN

```
% conversion in a CSTR assuming a macrofluid.
Da = 10.0;
d = [0, 30];  % domain; t_max = 30 used.
t = chebfun('t',d);        % x variable
L = chebop(d);             % name of operator
L.op = @(c) diff(c,1)+ Da*c.^2; % second order reaction
L.lbc = @(c) [c—1]; % inlet condition
c = L\0;  % batch concentration for time t.
% define exit age curve; CSTR used here
% note t is dimensionless (t/tbar).
E = exp(—t) ; % E—theta curve
% use sum command or quad function to integrate.
c_exit_macro = sum (c.*E)
% analytical solution for comparison (for second order)
c1 = 1/Da * exp(1/Da)* expint (1/Da)
```

The infinte domain for dimensionless time integration was replaced by 30 as the upper limit. Zero-order reactions need small modifications. Only dimensionless time less than $1/Da$ contributes to the exit concentration. Hence the domain should be changed to $1/Da$.

The result for $Da = 10$ for a zero-order reaction should be 0.0408 as in Table 13.1 or could be computed from the analytical solution given by Equation 13.16. Note that even when $Da = 10$, complete conversion is not achieved. Elements exiting earlier have remained nearly unreacted!

13.6   Variance-Based Models for Partial Micromixing

This section may be omitted without loss of continuity.

Here we present some results where the variance in concentration in the reactor is used as an additional variable to predict the effects of deviation from complete micromixing. Such models are developed from a detailed population balance model and details are shown in Froment, Bischoff, and Wilde (2011). The goal of this section is to merely present the key results so that you get an overview of these models. The conceptual basis for the model shown here is useful if you wish to pursue the study and applications of these models further.

The key idea is that the concentration in the reactor can be defined by a probability distribution function f that can then be used to find an average concentration and also its variance. The fraction of the reactor volume having a concentration of C_A in the interval of dC_A is then equal to $f(C_A)dC_A$. The average concentration is then

$$\langle c_A \rangle = \int_0^{C_{A,i}} C_A f(C_A) dC_A$$

This is the same is the exit concentration $C_{A,e}$ for a well-mixed system. The variance of the concentration around the mean is defined as

$$\sigma_R^2 = \int C_A^2 f(C_A) dC_A - [\langle c_A \rangle]^2$$

The variance is non-zero unless the reactor is perfectly mixed.

The population balance model provides a differential equation for the f function. (A number of closure models are needed including the birth and death rate of the population [concentration] in each interval.) The details are not shown here. Using this the following relations can be derived.

For a first-order reaction the following equation can be derived for the dimensionless variance of the concentration in the reactor:

$$\sigma^{*2} = \frac{\sigma_R^2}{C_{A,e}^2} = \frac{Da^2}{2Da + \beta\bar{t} + 1}$$

where β is a micromixing parameter that characterizes the level of micromixing in the system, with β of infinity representing perfect micro-mixing. Of course the variance has no effect on the average concentration for a first-order reaction, which is given as $1/(1 + Da)$ as shown earlier. The estimate of the variance here is mainly useful for experimental verification of mixing effects in the reactor.

For a second-order reaction the average rate of reaction now depends also on the variance and therefore on how close the reactor is to micromixing. For the latter case, the variance is zero and the micromixing model shown in Section 3.6 applies. If the reactor is not completely micromixed then the exit concentration can be given by the following equation:

$$1 - c_{A,e} = Da\, c_{A,e}^2 (1 + \sigma^{*2})$$

Hence an additional equation for the variance is derived and solved. This equation for a second-order reaction can be determined to be

$$(4Da\, c_{A,e} + \beta\bar{t} + 1)\sigma^{*2} = Da\, c_{A.e} \left(\sigma^{*2} + 1\right)^2$$

The above two equations are solved simultaneously for a given value of the micromixing parameter, β. An estimate of this micromixing parameter is needed for these calculations. An approximate value of $\beta = 0.5s^{-1}$ is a useful estimate for low viscosity fluids in standard process vessels (Froment et al., 2011). For viscous fluids β is much smaller and deviation from perfect micromixing can become important. Evangelista (1969) suggests that β scales as

$$\beta = K \left(\frac{\epsilon}{L^2}\right)^{1/3}$$

Here ϵ is the power input per mass of the system and L is a characteristic length parameter of the vessel.

13.7 Compartmental Models

Complex systems such as the human body or the environment are often modeled using compartmental models. Conceptually this is similar to the tanks in series model discussed earlier; the only difference is the interconnectivity of compartments can be quite complex. A complex system is therefore treated as a network of interconnected compartments and macroscopic balances (usually) are applied to each compartment. Thus a complete system is viewed as a network of perfectly mixed cells connected in some suitable manner. The compartments can exchange mass with each other. The schematic basis of the compartment model is shown in Figure 13.4.

The mass balance for each species takes the following form:

$$V_j \frac{d(C_{s,j})}{dt} = Q_{j,i} C_{s,j,i} - Q_{j,e} C_{s,j} + \sum_{m=1}^{NT} K_{m,j}^{(s)} (C_{s,m} - Cs,j) + V_j R_{s,j} \quad (13.17)$$

The notations are pertinent to those shown in Figure 13.4. Here the subscript s refers to species, and j refers to a tank. Total number of tanks is NT; they are all assumed to be interconnected with each tank exchanging mass with other tanks. The summation term on the right-hand side accounts for this interaction. The rate of exchange is $K_{m,j}$, which represents the moles or mass of species under consideration being transferred from compartment m to compartment j. Various species (index s) may have different exchange coefficients. Hence $K_{m,j}$ is tagged as $K_{m,j}^{(s)}$.

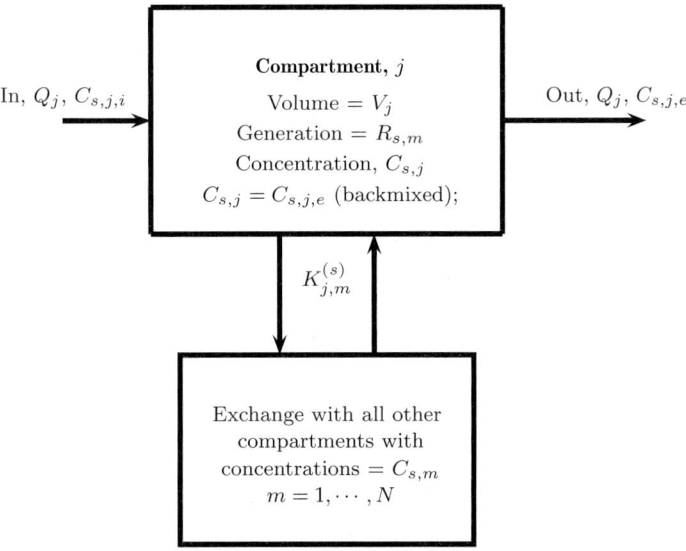

Figure 13.4 Schematic of a compartmental model showing the mass balance for compartment j.

Examples of compartmental models are widespread in biomedical systems, pharmacokinetic analysis, environmental systems, and chemical reactor modeling. We now discuss some computational aspects of the compartmental model with one application. This problem also illustrates the application of the linear algebra technique to solve differential equations. Additional problems at the end of the chapter provide more information.

13.7.1 Matrix Representation

The following assumptions are commonly made to simplify the computation of the model given by Equation 13.17:

- Mixing assumption: Each compartment is assumed to be well mixed and the concentration in the compartment is the same as the exit concentration of that compartment.
- Reaction kinetics: The reaction is assumed to be first order.

These assumptions reduce the model equations into a linear form and the compartmental model can be generalized using matrix-vector representation:

$$\frac{d\boldsymbol{y}}{dt} = \tilde{A}\boldsymbol{y} + \boldsymbol{R} \tag{13.18}$$

Here y is the solution vector consisting of the concentration values in each compartment. You should work out what the coefficient matrix A and the vector R are for the case represented by Equation 13.17. We discuss the solution here for the case where A and R are constants. The formal solution to such problems can be represented using the concepts of eigenvectors and eigenvalues of a matrix and the concept of the matrix exponential. Recall that if y is a scalar variable then the solution to Equation 13.18 is

$$y = B_1 \exp(At) - R/A$$

where B_1 is an integration constant. The solution consists of two parts: a homogeneous solution and a particular solution.

The constant of integration B_1 can be evaluated from the initial condition: if $y(t = 0) = y_0$ then the solution can be written as

$$y = (y_0 + R/A)\exp(At) - R/A$$

The solution for multiple systems of initial value problems (IVP) is represented in exactly same manner as that for a single equation: the solution is

$$\boldsymbol{y} = \exp(\tilde{A}t)[\boldsymbol{y}_0 + \tilde{A}^{-1}\boldsymbol{R}] - \tilde{A}^{-1}\boldsymbol{R} \tag{13.19}$$

Here exp is the exponential of a matrix, which can be readily computed using the function *expm* in MATLAB. y_0 is the vector of initial values. Using this format the MATLAB implementation of the compartmental model is relatively simple and useful.

Note that the exponential of a matrix is defined similarly to an exponential function. This is simply a power series in matrix form and the formal representation is

$$\exp(\tilde{A}) = \tilde{I} + \tilde{A} + \frac{\tilde{A}^2}{2!} + \frac{\tilde{A}^3}{3!} + \cdots$$

However, the numerical implementation uses a method based on the Pade approximations and avoids computing all the powers of the matrix $\tilde{A}$.

The use of MAPLE provides the exponential matrix in a symbolic form rather than the numerical values at a given instant of time. Sample code segments are given by White and Subramanian (2010). Example 13.2 shows an illustrative application.

Example 13.2   A two-compartment model for pharmacokinetic analysis

Set up the model equations for the two-compartment model shown in Figure 1.12 (Chapter 1) for analysis of drug distribution in a body. Solve for the illustrative parameters. Study the effect of the exchange parameter K_{ex} on the response of the system.

Solution

Apply Equation 13.17 for each compartment. The following equations should result.

For compartment 1, the blood compartment:

$$V_1 \frac{dC_1}{dt} = Q_{1,in} C_0 - Q_{1,out} C_1 + K_{ex}(C_2 - C_1) - V_1 k_1 C_1 \qquad (13.20)$$

For compartment 2, the tissue compartment:

$$V_2 \frac{dC_2}{dt} = Q_{2,in} C_{2,in} - Q_{2,out} C_2 + K_{ex}(C_1 - C_2) - V_2 k_2 C_2 \qquad (13.21)$$

We obtain a system of two differential equations that can be expressed in the compact matrix-vector form shown in Equation 13.18. Also the flow terms Q_1 and Q_2 are set as zero; thereby a simpler model is generated. The y vector is composed of C_1 and C_2. The solution can then be computed directly using the *expm* function or can also be implemented numerically using ODE45 or ODE15s if the equations are stiff (defined at the end of this section).

An illustrative result is shown in Figure 13.5 for the following parameter values: $V_1 = 1$, $V_2 = 3$, $k_1 = 0.2$, $k_2 = 2.0$, and $K_{ex} = 5.0$ in arbitrary units. The results for $K_{ex} = 50$ are shown in Figure 13.6.

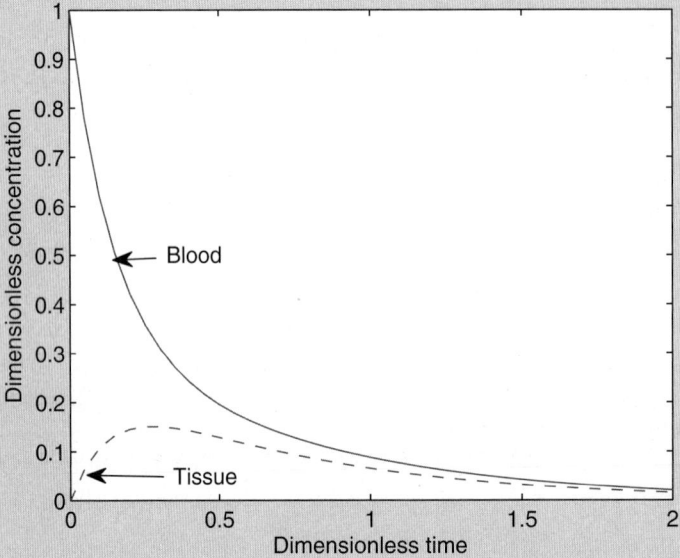

Figure 13.5 Transient concentration profiles in response to a bolus input for a two-compartment model.

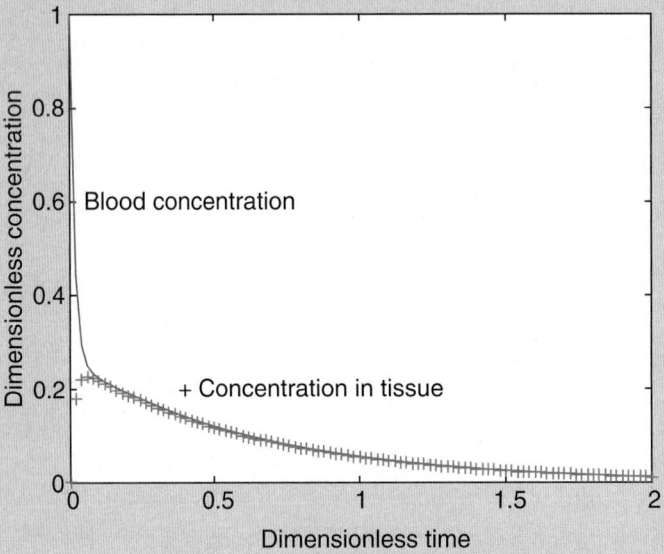

Figure 13.6 Transient concentration profiles in response to a bolus input for a two-compartment model. $K_{ex} = 50$.

> For the second case, since the exchange parameters are large, we find that there is an initial fast decay where the tracer exchanges mass with the tissue. Both blood and tissue reach the same concentration in a short time. This is followed by metabolism in tissue compartments (which is now in equilibrium with the blood compartment) leading to a slow decay in the system. In some cases it may be difficult to experimentally detect the initial fast decay and exercise problem 8 deals with some of these issues.

13.8   Compartmental Models for Environmental Transport

Compartmental models are widely used in environmental engineering for prediction of the environmental fate of chemicals. The goal of this section is to provide a brief overview on this topic. Some useful references are also provided for readers who wish to pursue this topic further.

The fate of chemicals released into the environment is important in predicting the exposure risk of chemicals. This is dependent on two factors:

- Translocation: Will it stay in the same place (e.g., in air) or if not, where will it go?
- Transformation: Will its chemical composition be altered by chemical reaction?

Models to predict this information require defining some type of large-scale compartmental model, sometimes referred to as a megascopic model. The environment can be considered to be composed of four broad compartments, that is air, water, soil, and biota (which includes plant and animals) (see Figure 1.14).

The simplest model considers each compartment a lumped system and the macroscopic level compartmental models described in this chapter are then used for each compartment. Such models are useful for evolution at very large scales (e.g., at the global level). More detailed models with greater levels of segmentation and spatial variation in each compartment are best suited for site-specific analysis problems (e.g., locations near a chemical industry). It should be noted that as the model complexity increases, its resolution and the data needed for prediction also increase.

Mackay et al. (1996a, 1996b, 1996c) and Mackay (2001) suggest modeling at three levels (later extended to four levels) and a brief description of these levels together with the data needed at each level is provided in the following subsections to provide a general overview.

13.8.1 Fugacity of Pollutants in Each Compartment

The fugacity is used as basic variable in these models and the fugacity difference between two compartments is used as the driving force in defining the
inter-compartmental exchange term. The concentration jump across the interface of each compartment is then automatically accounted for. Fugacity is like
a single currency! The concentration and the fugacity in each compartment
are related by defining a fugacity capacity factor. Thus

$$C_{A,j} = f_{A,j} Z_{A,j} \tag{13.22}$$

Here $f_{A,j}$ is the fugacity of A in compartment j, $C_{A,j}$ is its concentration, and
$Z_{A,j}$ is the fugacity capacity factor. The latter will depend on the thermodynamic equilibrium relation for each compartment relative to the air phase and
the relations can be calculated as follows.

For air, using the ideal gas law the fugacity is the same as the partial
pressure of A and hence $Z_A = 1/R_g T$.

For water, using Henry's law, $p_A = H_A C_{AL}$, and equating the fugacities
at equilibrium conditions, it is easy to show that $Z_A = 1/H_A$.

For soil, the following relation (again based on equal fugacity of A in soil
and air) was suggested:

$$Z_A(\text{soil}) = x_{Ac} K_{OC} \rho_s / H_A$$

where x_{Ac} is the mole fraction of organic carbon in the soil, and K_{OC} is the
organic carbon partition coefficient in the soil. This is related to the octanol-
water partition coefficient K_{OW} as $0.41 K_{OW}$.

For biota, the following relation (again based on equal fugacity of A in
soil and air) was suggested by Paterson (1991):

$$Z_A(\text{biota}) = 0.048 K_{OW} \rho_b / H_A$$

13.8.2 Level I or Equilibrium Model

A Level I simulation is the equilibrium distribution of a fixed quantity of conserved (i.e., non-reacting) chemical, in a closed environment at equilibrium,
with no degrading reactions, no advective processes, and no intermedia transport processes. The medium receiving the emission is unimportant because
the chemical is assumed to become instantaneously distributed to an equilibrium condition.

Computations at this level are done using the following approach. The total moles in the system is

$$\mathcal{M}_A(\text{total}) = \sum_j V_j C_{Aj} = f_A \sum_j V_j Z_{Aj} \qquad (13.23)$$

Here V_j is the volume of compartment j. $\mathcal{M}_A$ is the total moles of A released into one or more of the compartments. Note that the fugacity term is outside the summation since it is the same in all the compartments. This is an advantage of the fugacity approach where a single currency is being used. Hence fractional distribution of species A in compartment j is equal to

$$f_{j,A} = \frac{V_j Z_{Aj}}{\sum_j V_j Z_{Aj}}$$

This provides the translocation information. The data needed are physical-chemical properties to find the fugacity capacity factors and the user-defined compartment volumes and densities.

This model is useful for establishing the general features of a new or existing chemical's behavior and provides the likely media into which a chemical will tend to partition.

13.8.3 Level II Model: Advection Effects

This model is similar to Level I, but is a steady state model with a constant input rate, rather than a single dose of chemical. There is both advective in- and outflow of chemicals from the unit world. Chemical losses can also occur through degrading reactions. Equilibrium is assumed in each compartment. Equations similar to Equation 13.17 (with no accumulation and no exchange term) are then written for each compartment and solved simultaneously.

In addition to the data required for Level I the following must be input: emission rate in *lieu* of amount of chemical, advective inflow rates, inflow concentration, and reaction half-lives of the chemical in each medium.

13.8.4 Level III Model: Intermedia Transport Effects

This level does not assume an equilibrium state between the compartments, but only steady state. In addition to the data required for Level II the following must be input: intermedia transfer rates and related parameters such as air-side mass transfer coefficient, water-side mass transfer coefficient, rain rate, aerosol deposition, and so on. The program uses conventional expressions and typical parameters for intermedia transfer by processes such as wet deposition from the air, sediment deposition in the water, and soil runoff. If intermedia transfer rates are not known, then a generic version of the model uses a set of pre-assigned transfer rates.

More details can be found in Paterson (1991) and Mackay (2001). Model formulation is very similar to the compartmental models described earlier in this section. The main difference is that fugacity is used as the primary variable rather than the concentration.

A note of caution is useful here. These models cannot be as such validated in the same sense as some other models in mass transfer, for example, a model for simple equipment such as an absorber. Only an order of magnitude fit can be expected, but even with this limitation, the model predictions are very useful for predicting the likely behavior of new chemicals and their environmental impact.

13.8.5 Level IV Model: Transient Effects

This is an extension of the Level III model and the dynamics of emissions and resulting temporal concentration changes are taken into account. This model is complex and is not widely used; Level III is more common. A thesis by Kilic (2008) is illustrative of the use of this model to predict the pollutant levels in rivers.

13.9   Fluid–Fluid Systems

In this section we show modeling concepts for two-phase flows with the assumption that both phases are backmixed. This is the simplest model and leads to algebraic equations that can be readily solved. The discussion is shown for a gas–liquid case but the method is very similar for the liquid–liquid case as well.

13.9.1 Backmixed–Backmixed Model

A schematic description of the model is shown in Figure 13.7.

Let us consider dilute systems to illustrate the key features of modeling this system. The extension to non-dilute systems is slightly lengthy and generally uses the mole ratio as a variable rather than the mole fraction. But

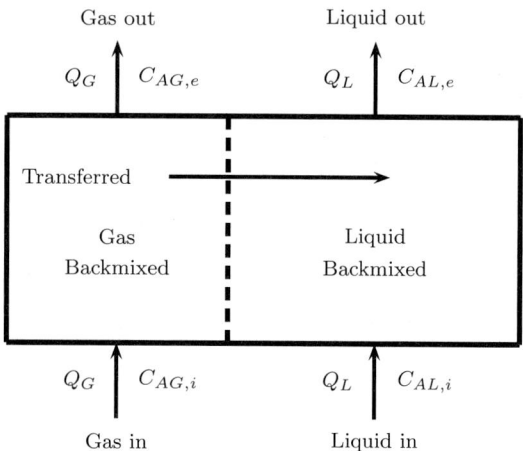

Figure 13.7 Schematic of a two-phase backmixed–backmixed system showing the control volumes for species mass balance.

this does not require any new concepts and the dilute system analysis can be extended to these cases.

The dilute system assumption means that the gas molar flow rate and the liquid molar flow rate are nearly constant in the inlet and outlet of the separator. Further a linear equilibrium is assumed and represented in terms of a Henry parameter:

$$C_{AG} = H_A C_{AL} \text{ at equilibrium}$$

The gas and liquid phase balances for a dilute system are as follows:
Gas phase balance:

$$Q_G(C_{AG,i} - C_{AG,e}) - K_G a_{gl} V_C(C_{AG,e} - H_A C_{AL,e}) = 0$$

Liquid phase balance:

$$Q_L(C_{AL,i} - C_{AL,e}) + K_G a_{gl} V_C(C_{AG,e} - H_A C_{AL,e}) = 0$$

It is more useful to solve these in terms of dimensionless concentrations and dimensionless parameters. The concentration in the gas phase is made dimensionless with respect to the inlet gas concentration. Thus we define

$$c_{AG,e} = \frac{C_{AG.e}}{C_{AG,i}}$$

For a liquid we use the maximum solubility value as the reference concentration. Thus we use $C_{AG,i}/H_A$ as the reference and define

$$c_{AL,e} = \frac{C_{AL,e}}{C_{AG,i}/H_A}$$

Defined in this manner, the maximum concentration in both the liquid and gas phases is equal to one. The concentration jump is implicit in this definition and need not be accounted for separately. The dimensionless version is then as follows for the gas phase:

$$1 - c_{AG,e} - \kappa_{GL}(c_{AG,e} - c_{AL,e}) = 0 \tag{13.24}$$

Here κ_G is a dimensionless mass transfer coefficient defined as

$$\kappa_{GL} = \frac{K_G a_{gl} V_C}{Q_G}$$

For the liquid phase we have

$$c_{Al,i} - c_{AL,e} + \kappa_{GL}\Lambda(c_{AG,e} - c_{AL,e}) = 0 \tag{13.25}$$

where Λ is a flow ratio parameter defined as

$$\Lambda = \frac{Q_G H_A}{Q_L}$$

From Equation 13.24, the exit gas concentration is related to the exit liquid concentration as

$$c_{AG,e} = \frac{1 + \kappa_{GL} c_{AL,e}}{1 + \kappa_{GL}} \tag{13.26}$$

Using this in Equation 13.25, the exit liquid concentration is obtained as

$$c_{AL,e} = \frac{c_{AL,i}(1 + \kappa_{GL})}{1 + \kappa_{GL} + \kappa_{GL}\Lambda} + \frac{\kappa_{GL}\Lambda}{1 + \kappa_{GL} + \kappa_{GL}\Lambda} \tag{13.27}$$

13.9.2 Equilibrium Model

The equilibrium model can be shown to be the limiting case of the preceding equation as κ_{GL} tends to infinity. For the common case of $c_{AL,i}$ equal to zero, the expression for the exit liquid concentration is

$$c_{AL,e}^* = \frac{\Lambda}{1 + \Lambda} \tag{13.28}$$

The dimensionless exit gas concentration will be the same, which can be verified by using the preceding relation in Equation 13.26.

An alternative derivation is obtained by eliminating the κ_{GL} term between the liquid and gas balances. This leads to an overall material balance:

$$c_{Al,i} - c_{AL,e} + \Lambda(1 - c_{AG,e}) = 0$$

Further, the equilibrium condition is $c_{AG,e} = c_{AL,e}$ since both concentrations are scaled with due consideration to the thermodynamic Henry constant. Combining the overall mass balance with the equilibrium condition gives Equation 13.28.

The ratio of the fractional solute absorbed between the two models measures the contacting efficiency or the stage efficiency for zero inlet liquid concentration:

$$\eta_c = \frac{\kappa_{GL}(1+\Lambda)}{1 + \kappa_{GL} + \kappa_{GL}\Lambda}$$

13.9.3 Mixing Cell Model

In order to account for the backmixing effects in many separation columns a mixing cell model is useful. This is an extension of the tanks in series model for a single-phase reactor. For example, consider gas absorption in a tall bubble column. Liquid is close to backmixed in bubble columns but in tall columns this assumption may not be valid. The system is then modeled as two or more tanks (cells) connected in series.

13.10 Models for Multistage Cascades

A similar modeling tool applies to multistage separation equipment. Consider liquid–liquid extraction in a mixer-settler as an example. The extent of separation that can be obtained in a single-stage extractor is usually small and multiple units are needed. Usually the system operates in a countercurrent mode, which is known as a multistage cascade. An illustrative sketch is shown in Figure 13.8. Each stage is modeled assuming a backmix–backmix model similar to that studied in Section 13.9.1. Equations are set up for each stage. For a countercurrent operation all the stages have to be solved simultaneously since feed enters at one end and the solvent enters at the other end. For cocurrent cases the model can be solved in a sequential manner.

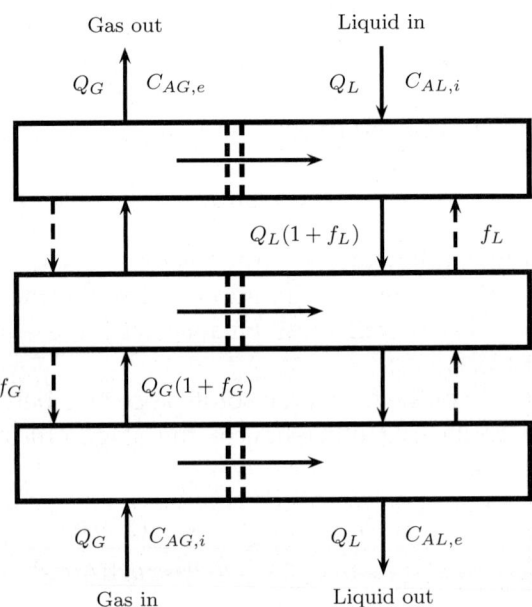

Figure 13.8 Stagewise backmixing model for a two-phase system showing three stages: inter-stage backflow terms f_L and f_G are often added to provide flexibility to characterize the mixing pattern.

The stages need not be discrete units and this could be done even for a single column. For example, column extractors with various internal design arrangements to provide mechanical agitation are common in industrial practice. Common contactors Scheibel columns of different internal configurations, Oldshue-Rushton columns, rotating disk contactors, and Kuhni columns; these are discussed by Seader et al. (2011). These columns can then be modeded as a number of stages or cascades arranged in a countercurrent manner. In some cases some backflow between stages is included, as shown in Figure 13.8. Equations can then be set up for each stage and the resulting set of algebraic equations can be solved simultaneously to get the exit concentrations and also the various inter-stage concentrations.

13.10.1 Equilibrium Model

For preliminary design an equilibrium model where the streams existing any particular stage are assumed to be in equilibrium is used. For linear equilibrium analytical solutions can be obtained; these are summarized in this section.

Single Stage

The single stage was analyzed in Section 3.8.2. The fractional non-extraction is given by Equation 3.33, which can be expressed as

$$\frac{X_e}{X_f} = \frac{1}{1 + E}$$

where E is called the extraction factor, defined as

$$E = \frac{S K_A}{F_C}$$

where S is the solvent rate, F_C is the carrier molar flow rate, and K_A is the equilibrium constant, expressed as mole ratios. This expression assumes that K_A is constant and not a function of composition.

Crosscurrent Cascade

An arrangement used for separations that are not so difficult is the crosscurrent cascade. The schematic is shown in Figure 13.9. Here fresh solvent is added to each stage and the extracts are collected at the exit of each stage. The fraction of A not extracted is given by the following equation for this case:

$$\frac{X_{Ac}}{X_{Af}} = \frac{1}{(1 + E/N)^N} \tag{13.29}$$

In general the countercurrent arrangement provides more extraction than the crosscurent arrangement.

Countercurrent Cascade

The result can be extended to a countercurrent cascade; the following formula (shown without detailed derivation) is useful for practical applications:

$$\frac{X_{Ae}}{X_{Af}} = \frac{E - 1}{E^{N+1} - 1} \tag{13.30}$$

This is known as the Kremser equation.

Absorption and Stripping Factors

Similar equations are used for absorption columns. For an N equilibrium stage absorber, with pure absorbent and the streams in counterflow, the

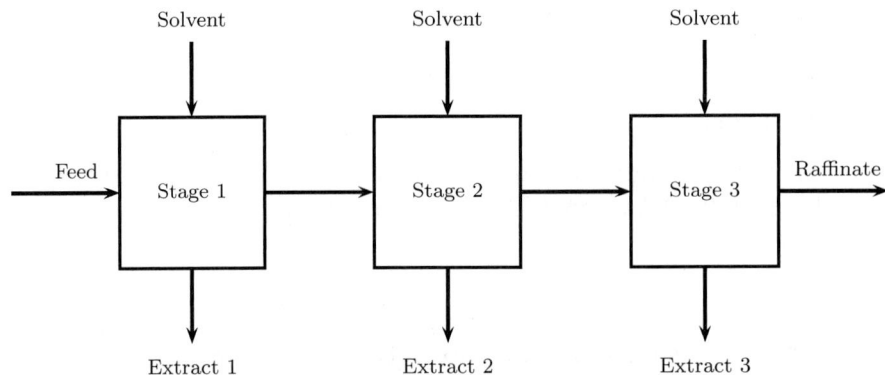

Figure 13.9   Schematic of crosscurrent extraction.

recovery fraction is given by the Kremser equation:

$$\phi_A = \frac{\mathcal{A} - 1}{\mathcal{A}^{N+1} - 1}$$

where $\mathcal{A}$ is known as the absorber factor and defined as

$$\mathcal{A} = \frac{L}{V m_A}$$

Here L is the molar liquid flow rate, V is the gas molar flow rate and m_A is the equilibrium constant defined as $y = mx$.

A similar equation holds for a stripping column. The fraction of species that is not stripped is given as

$$\phi_S = \frac{S - 1}{S^{S+1} - 1}$$

where S is called the stripped factor, defined as

$$S = \frac{V m_A}{L}$$

The stripping factor is the reciprocal of the absorption factor. All three factors, extraction, absorption, and stripping, have similar significance and depend on the linear equilibrium constant times the ratio of the flow rates of the two phases.

Summary

- Macroscopic models are derived by using species balances over a large control volume, often the whole reactor or separator. These models lead to algebraic equations rather than differential equations for the steady state case and are therefore easier to solve. They find wide applications in practical design.

- An assumption about the mixing pattern in the reactor/separator is needed to assign the kinetic rate and mass transfer driving force terms. A common simplification is that the system is backmixed, which is a reasonable assumption for well-mixed tanks with high levels of agitation. The model is then closed by assuming that the reactor concentration is the same as the exit concentration.

- Tracer experiments provide a useful tool to assess the extent of backmixing in the system. For a completely backmixed system for a pulse input of tracer the concentration of the tracer is maximum at time zero and decays as an exponential function. Any deviation of the observed response from this ideal pattern is an indication of less mixing in the system. Such reactors are often modeled as N tanks connected in series. The model parameter to characterize the reactor is N, the number of tanks needed to match the tracer response.

- Tracer data can be interpreted in various ways to find the model parameters. Moment analysis is commonly used, where the dimensionless second central moment of the tracer response is matched to the theoretical value. Theoretical values in turn can be generated using the response in the Laplace domain without the need to find the solution in the time domain.

- The variance of the response curve is an indicator of the extent of mixing and maldistribution in the reactor. The dimensionless variance is equal to one for a backmixed system. It is equal to zero for plug flow. An intermediate degree of mixing takes a value in between these provided there is no gross maldistribution, recycling, or bypassing. The reactor is then referred to as non-ideal but well behaved.

- For reactions with positive order kinetics, plug flow provides the highest conversion and backmixed provides the lowest value. The reactor performance can be bracketed between the high and low values for systems that are well behaved (variance less than one).

- Systems with severe maldistribution or bypassing show a variance larger than one. These are called ill-behaved systems. Reactor performance can be even lower that the backmixed case in such situations. An example is a fluidized bed reactor where part of the gas moves as "bubbles" and does not come in contact with catalyst.

- For non-first-order reactions, more details on the level of mixing is needed in order to predict reactor performance. The reactor may be well mixed but only on a macro scale. Fluid elements arriving at different times remain segregated in such a situation. The exit age distribution can be used to predict the conversion in a macrofluid model and Equation 13.14 can be used. This provides one bound on the conversion. If the reactor is modeled as a microfluid, we obtain a second bound. Population balance–based models are needed for more detailed modeling of these systems since the level of segregation has to be modeled as well.

- Complex systems such as the human body or the environment are often modeled as a network of interconnected compartments and macroscopic balances (usually) are applied to each compartment. Such models find application, for example, in pharmacokinetic modeling to study drug uptake and metabolism. These models are known as compartmental models in general.

- An important application of compartmental models is to predict the distribution and fate of chemicals released in the environment. The environment can be considered to be composed of four broad compartments, that is, air, water, soil, and biota (which includes plant and animals). Models of various levels can be used and the primary difference is the level of spatial and component details in each of the compartments. The simplest is the equilibrium model, which is often referred to as a fugacity Level I model.

- Separation equipment is often operated in multistage mode; this is known as a cascade arrangement. Commonly countercurrent or crosscurrent cascades are used. Each stage is modeled as a mixed compartment.

- The cascades are modeled assuming an equilibrium model at the first level. The mass transfer coefficient and level of mixing is not needed in this model. The extent of separation can then be predicted analytically using the separation factor as a parameter. The number of stages needed to achieve a specified level of separation can be calculated using the expression presented in the text for both the countercurrent and crosscurrent arrangements.

- Equilibrium models can be corrected by using a stage efficiency factor. This depends on the mass transfer coefficient and the mixing pattern in the separator. Analysis for a case where both phases are backmixed is shown in the text.

Review Questions

13.1 If the tracer response is an exponentially decaying function of time, what is the mixing pattern in the reactor?

13.2 If the tracer response is close to a Dirac delta function, what is the mixing pattern in the reactor?

13.3 Can the tracer response to a pulse input show multiple peaks? If so, when?

13.4 Define the E-curve and indicate its physical meaning.

13.5 Define the F-curve and indicate its physical meaning.

13.6 Define the I-curve and indicate its physical meaning.

13.7 What is the expression for the I-curve for a completely mixed reactor?

13.8 A reactor is modeled as four tanks in series. What is the time at which the maximum concentration in the exit for an input of a pulse tracer is observed?

13.9 What is meant by the stiffness factor for a set of first-order differential equations?

13.10 Why is fugacity preferred as the variable in environmental models?

13.11 Which mode gives higher separation: countercurrent or crosscurrent?

13.12 What is the Kremser equation?

13.13 Define absorption, extraction, and stripping factors.

Problems

13.1 **Limit for large N for the tanks in series model.** What is the limit of expression 13.12 for the concentration in the tanks in series model for N tending to ∞? Show that the plug flow model is approached.

13.2 **Reactor performance for the tanks in series model.** Calculate and plot the conversion if a reactor is modeled as tanks in series for $Da = 3$ if the reaction is first order and second order. Take N from 1 to 10. Plot on the same graph the conversion for a plug flow and a backmixed model.

13.3 **Response of a tank with a deadzone.** A tank is assumed to have a dead zone that exchanges mass with the main zone as shown in Figure 3.10 in Chapter 3. Derive an expression for the response curve in the time domain as well as in the Laplace domain. From the solution derive an expression for the first and second moment and the variance of the response curve.

13.4 **Internal age distribution.** Show that for a completely backmixed system the I-curve is the same as the E-curve. Thus the probability of an age t in the exit stream is the same as the probability of an age t within the reactor.

13.5 **Bolus injection with a first-order reaction.** The bolus injection is a very useful and important tool in pharmacokinetic analysis. Repeat the analysis for a bolus injection of tracer that undergoes a first order reaction. Sketch typical exit concentration versus time plots for various values of rate constant (expressed as Da). How is the time constant of the response affected by the rate constant of the reaction?

13.6 **Macro- and micro-models for a first-order reaction in a CSTR.** For a first-order reaction we have

$$c_A(t) = \exp(-k_1 t)$$

while

$$E(t) = \frac{1}{\bar{t}} \exp(-t/\bar{t})$$

where $\bar{t}$ is the mean residence time. Substitute and integrate the macrofluid model given by Equation 13.14. Show that the result is

$$\langle c_{A,e} \rangle = \frac{1}{1 + k_1 \bar{t}}$$

This is the same as that for a micromixed system. Hence show that for a first-order reaction the level of segregation is not important.

13.7 **Macro- and micro-models for a half-order reaction in a CSTR.** Compare the macro- and micro-models for a half-order reaction in a CSTR. Use a Damkohler number based on an inlet concentration equal to two.

13.8 **Macrofluid model for second-order reaction.** For a macro-model the integral given by Equation 13.14 can be integrated analytically, leading to Equation 13.15 in the text. Verify the result.

13.9 **Two-compartment versus one-compartment model for drug distribution.** The following data was found as a response to a drug that was injected as a pulse. The concentration distribution in the blood was obtained as a function of time. Concentration in the tissue compartment was not measured.

Time	Concentration
0	1.0000
0.3000	0.4867
0.6000	0.2748
0.9000	0.1700
1.2000	0.1101
1.8000	0.0487
2.4000	0.0219

Fit a one-compartment model and determine the time constant. Fit a two-compartment model and find the time constants and the exchange parameter.

Comment on the accuracy of the results in view of the fact that the data are limited and no explicit measurement of the tissue concentration was made.

13.10 **Pollutant distribution with Level I model.** Mackay (2001) studied the distribution of hexachlorobiphenyl (6-CB) in the environment. The molecular weight of this compound is 350 g/gmol, vapor pressure is 0.0033 Pa, and solubility in water is 0.035 g/cm^3. The octanol water partition coefficient is $10^{6.8}$. Assume an amount of 5×10^5 mol is dispersed (on an annual basis) in a particular region. Find the distribution in the various compartments. The following values were used for the volumes of each compartment in Mackay's work: air = 4×10^{14} m^3; water = 2×10^9 m^3; soil = 1.2×10^{10}; biota = 8×10^8 m^3. The organic carbon in solids is taken as 0.02.

13.11 **Extraction efficiency equations.** Starting from the basic mass balances coupled with the assumption that the exit streams are in equilibrium, derive Equations 13.30 and 13.29 relating the extent of extraction to the extraction factor. Make a comparison plot between the crosscurrent and countercurrent for extraction factors of 2 and 5.

13.12 **Extraction efficiecy of a countercurrent cascade.** Water and p-dioxane formed by catalytic dehydration of ethylene glycol are to be separated by liquid–liquid extraction. Since the boiling points of these compounds are very close (100 and 101.1 °C (Seader et al.), distillation is not feasible. Liquid extraction with benzene as a solvent is used. Consider a feed of 4530 kg/hour with a 25% solution of p-dioxane. Benzene and water are mutually insoluble and the distribution coefficient of dioxane (mole ratio in benzene to water) is near 1.2. The flow rate of benzene is 6800 kg/hour. Find the fractional extraction for single-stage and two-stage countercurrent extraction, and two stage crosscurrent extraction. Assume an equilibrium model. If 99% extraction is needed, find the number of countercurrent stages that need to be used.

CHAPTER 14

Mesoscopic Models and the Concept of Dispersion

Learning Objectives

After completing this chapter, you will be able to:

- Understand how mesoscopic models result from a cross-sectional averaging of differential models.
- Identify additional parameters that result upon averaging.
- Explain the meaning of the dispersion coefficient.
- Model and solve reactor models using the dispersion concept.
- Understand the usefulness of the transient response data to determine the dispersion coefficient.
- Understand the classical work of the Taylor model for axial dispersion in laminar flows.
- Model and solve cocurrent and countercurrent mass exchangers using the dispersion coefficient for each of the phases.

Mesoscopic models use one spatial coordinate oriented in the main flow direction as the primary distance variable. Any concentration variation in the cross-flow directions are represented by average values. We have seen some examples of such models in Chapter 4 and this chapter provides additional details.

A differential model averaged over a cross-section gives us the meso-model. It is useful to understand the process of averaging and how meso-models arise naturally. This also helps us to identify the missing information, the so-called closures, which are needed when using the meso-models. This is the first topic addressed in this chapter.

The main closure relation required is a relation between the cup-mixing average and the cross-sectional average. Commonly this is done using a parameter called the dispersion coefficient. This was introduced in Chapter 4 and simple applications to a homogeneous reactor were demonstrated. Here we present additional information and generally useful MATLAB code to execute the performance calculations. We also show how tracer data can be interpreted to evaluate the dispersion coefficient experimentally.

The concept of dispersion followed from the classic work of Taylor (1953), who developed a theoretical model for this coefficient for some well-defined flow fields. It is useful to understand the details leading to this development; this is presented in this chapter as well.

The mesocopic modeling approach from Chapter 4 for mass exchangers (two-phase systems) is then reviewed and a model for plug-backmixed flow is presented. Further, the plug-plug model shown in Section 4.3.2 is extended to include dispersion. Such models are useful in modeling many separation processes. Both the mass transfer coefficient and the dispersion coefficient for each of the phases are needed as input parameters for such models. Illustrative results are presented for the effect of dispersion on the concentration profiles in mass exchangers.

14.1   Plug Flow Idealization

Meso-models are simple in form and structure and can be formulated in two ways: by the direct use of the conservation for meso control volume, and by averaging of the differential models. In both cases, the convection term depends on the cup-mixed average value while the reaction term depends on the cross-sectional average. The resulting model equation (Equation 4.12) for a first-order reaction (for a constant density system) is revisited here for ease of reading:

$$- \langle v \rangle \frac{dC_{Ab}}{dz} = k_1 \langle C_A \rangle \tag{14.1}$$

The left-hand side represents the convection term; the cup-mixing average is the proper concentration to use in defining this term. The right-hand side is the rate of generation term and the cross-sectional average should be used as the measure of concentration here. Thus two types of averages are involved in the same model and there is already a closure problem. Section 4.2 explained this in some detail and should be reviewed at this point before proceeding further. In this chapter we further elaborate on these concepts and in particular discuss the dispersion models in more detail. The plug flow assumption is reviewed first. It is useful to review Section 4.2.1 at this point.

The concept of plug flow is an idealized representation of the cross-direction concentration profile in the reactor. The assumption is that there is no concentration gradient in this direction. that is, the concentration is constant across any cross-section. The velocity profile has to be uniform for this to occur. In addition there should be no other mechanism that can cause a concentration gradient. For example, if a heterogeneous reaction occurs at the wall, then the concentration at the wall will be smaller (or even near zero for a rapid reaction) causing a deviation from plug flow. Similarly if a temperature profile exists then the change in rate due to this causes the concentration to be different at different radial positions, causing the assumption to be invalid.

The plug flow approximation then uses the idealization that $C_{Ab} = \langle C_A \rangle$. For power law kinetics at constant velocity the plug flow model is

$$ -\langle v \rangle \frac{d\langle C_A \rangle}{dz} = k_n \langle C_A \rangle^n $$

The performance equations for a simple single reaction with no volumetric change (constant density case) are shown in Table 14.1 for common power law kinetics and are useful for a first level of calculations for many applications. Results are presented for the dimensionless exit concentration, $\langle c_{A,e} \rangle$, in terms of the Damkohler number, which is the key dimensionless groups for such problems. This is defined as

$$ Da = k_n C_{A,i}^{n-1} \frac{L}{\langle v \rangle} \tag{14.2} $$

Note that this represents the ratio of the mean residence time in the reactor to the reaction time. Also note this is a function of the inlet concentration, $C_{A,i}$, unless the reaction is first order.

Note that the results in Table 14.1 are similar to the batch reactor with the mean passage time, defined as $L/\langle v \rangle$, replacing the batch time. The equations are applicable for a constant density system, that is, $\langle v \rangle$ is treated as a constant.

Extension to systems where the average velocity is changing in the flow direction due to changes in density of the system can be readily accomplished as well. Additional relations for change in average velocity can be computed based on an overall mass balance and included in the model.

The dispersion model introduced in Section 4.2.2 and continued in the following section is an attempt to correct for the deviation from plug flow. More details are presented next.

Table 14.1   Plug Flow Exit Concentration for Power Law Kinetics

First-order reaction:

$$\langle c_{A,e} \rangle = \exp(-Da)$$

Second-order reaction:

$$\langle c_{A,e} \rangle = \frac{1}{1 + Da}$$

Zero-order reaction:

$$\langle c_{A,e} \rangle = \begin{cases} 1 - Da & \text{if } Da \leq 1 \\ 0 & \text{if } Da > 1 \end{cases}$$

Half-order reaction:

$$\langle c_{A,e} \rangle = \begin{cases} (1 - Da/2)^2 & \text{if } Da \leq 2 \\ 0 & \text{if } Da > 2 \end{cases}$$

14.2   Dispersion Model

Plug flow assumes no mixing in the flow direction. We can account for the deviation by assuming there is a mixing component superimposed on the convective flow; this is as though some diffusion is taking place in the flow direction. A dispersion parameter is then imposed and the *in* and *out* terms are modified. Equation 4.15 shows how the additional terms get added to the plug flow model and it may be useful to review that at this stage. The following second-order differential equation for the cross-sectional average concentration applies:

$$D_E \frac{d^2 \langle C_A \rangle}{dz^2} - \langle v \rangle \frac{d \langle C_A \rangle}{dz} + \langle R_A \rangle = 0 \tag{14.3}$$

This can be obtained by a differential mass balance or the use of the relation in 1.28 in Chapter 1 to close the cup mixing average terms on the left-hand side of Equation 14.1. Here $\langle R_A \rangle$ is based on the cross-sectional averaged rate of reaction. For example, for a first-order reaction, $\langle R_A \rangle = -k_1 \langle C_A \rangle$. Now all terms are in terms of the cross-sectional average and the model is closed. This dimensionless form of the model and the dimensionless parameters needed to calculate the reactor performance is presented next.

Dimensionless Parameters

The distance is scaled by the reactor length and a dimensionless distance $\eta = z/L$ is used. The concentration is scaled by the inlet concentration. The notation $\langle c \rangle$ is used for the dimensionless concentration of A; that is, the subscript A is dropped for brevity.

The resulting equation for a first-order reaction is

$$D_E^* \frac{d^2 \langle c \rangle}{d\eta^2} - \frac{d \langle c \rangle}{d\eta} - Da <c> = 0 \qquad (14.4)$$

The three terms can be identified as disperion, convection, and reaction. The dimensionless parameter appearing on the first term is referred to as the dispersion number. This is defined as

$$D_E^* = \frac{D_E}{\langle v \rangle L} \qquad (14.5)$$

The second parameter appearing in the reaction term is the Damkohler number, Da, which equals $k_1 L / \langle v \rangle$ for first-order reaction and is defined by Equation 14.2 for a power-law kinetics.

Thus two dimensionless parameters are needed in the disperion model; plug flow needs only one. Also note that the plug flow model is recovered if D_E^* approaches zero. It can also be shown that the backmix model is recovered if D_E^* approaches infinity.

The reciprocal of D_E^* is defined as the dispersion Peclet number:

$$Pe^* = \frac{1}{D_E^*} = \frac{\langle v \rangle L}{D_E}$$

Note: This should not be confused with the Peclet number, Pe, introduced in 9.3.1. Both definitions have a similar grouping format but in one case (Pe^*) we use the axial dispersion coefficient while in the other case (Pe) we use the molecular diffusion coefficient. In some books Pe^* is referred to as the Bodenstein number. In terms of the (dispersion) Peclet number the model is written as

$$\frac{1}{Pe^*} \frac{d^2 \langle c \rangle}{d\eta^2} - \frac{d \langle c \rangle}{d\eta} - Da \, \langle c \rangle = 0 \qquad (14.6)$$

Hence two parameters are needed to describe the reactor behavior, unlike the plug flow model where Da alone is sufficient. Note that the plug flow model is recovered if Pe^* is large.

The application of the dispersion model requires the following considerations:

- How do we know what the value of D_E is? Can we predict it from theory? Can we measure it for a real reactor?
- What boundary conditions should be used in the solution of the dispersion model? Note that two boundary conditions are now needed.
- How do we solve this for multiple reaction case and complex nonlinear kinetics?

The answers to these questions will be addressed as we progress through this chapter. First we start with the boundary conditions and the solution to the dispersion model.

14.2.1 Boundary Conditions

The boundary conditions to be used have been the subject of considerable discussion and many research papers. See Froment et al. (2011) for key references. The boundary conditions depend on what assumptions are used in the inlet and outlet. At both of these points, the reactor is characterized as "closed" or "open." A closed inlet boundary is considered one where plug flow exists prior to the inlet section. A open boundary is one where the flow has the same characteristics within and adjacent to the test section, that is, there is dispersion across the plane under consideration. Similar definitions are used for the exit boundary. Thus there can be four different combinations for the boundary conditions. Most commonly closed-closed or open-open conditions at the inlet and exit respectively are used.

The Danckwerts boundary conditions are commonly used in the context of the dispersion model, assuming a closed-closed system. A material balance at the inlet leads to

$$\text{At } \eta = 0, \quad \langle c \rangle - \frac{1}{Pe^*} \frac{d \langle c \rangle}{d\eta} = 1 \tag{14.7}$$

This condition assumes some drop in concentration due to dispersion as we enter the reactor, that is, at $\eta = 0^+$, where the concentration is not the same as the concentration at $\eta = 0^-$, that is, just before the entry.

The schematic of the assumptions in the closed-closed approximation is shown in Figure 14.1. The figure also illustrates the balance across the inlet plane, $\eta = 0$, which leads to the above boundary condition. The concentration is discontinuous when crossing this plane.

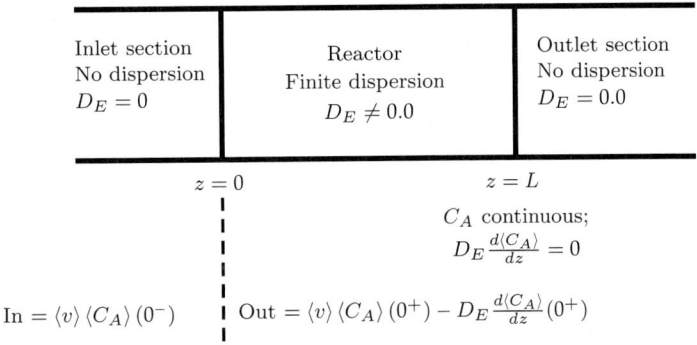

Figure 14.1 Schematic of conditions used for closed-closed boundary conditions.

At the exit the stream is allowed to blend smoothly with the outlet fluid, that is, the concentration is assumed to be a continuous function. This leads to

$$\text{At } \eta = 1, \quad \frac{d\langle c\rangle}{d\eta} = 0 \tag{14.8}$$

14.2.2 Solution for a First-Order Reaction

The solution can be obtained analytically for a first-order reaction and numerically for other cases. The result for a first-order reaction using the boundary conditions stated in the previous section is as follows for the dimensionless exit concentration:

$$\langle c_{A,e}\rangle = \frac{4\alpha \exp(Pe^*/2)}{(1+\alpha)^2 \exp(\alpha Pe^*/2) - (1-\alpha)^2 \exp(-\alpha Pe^*/2)} \tag{14.9}$$

where

$$\alpha = \sqrt{(1 + 4Da/Pe^*)}$$

The effect of dispersion was shown qualitatively in Chapter 4 and Figure 4.3 should be reviewed again at this stage. The results were for a first-order reaction with $Da = 3$ and the effect of D_E^* (the reciprocal of Pe^*) was shown. It was also shown that the results of the two ideal cases, backmix and plug flow, are recovered as $1/(1 + Da)$ and $\exp(-Da)$, respectively. Thus we observe that the reactor behavior is bracketed between these limits.

14.2.3 Nonlinear Reactions

For nonlinear reactions one can use cross-sectional averaging as explained earlier. However, one obtains the cross-sectional average reaction rate as a term. The cross-section average of the rate of reaction is not the same as $r(\langle c \rangle)$, the rate based on the average concentration, unless the reaction is first order. For example, the average of c^2 is not the same as the square of the average of c. Note that these two averages are the same only if the variance of the radial concentration profile is zero. The unknown variance introduces yet another uncertainty in the dispersion model. One has to assume that

$$\langle r(c) \rangle \approx r(\langle c \rangle)$$

in order to close the model as a first approximation. This assumption is similar to the micromixed model, now at each local axial location level. With this assumption the dispersion model for power-law kinetics is

$$\frac{1}{Pe^*} \frac{d^2 \langle c \rangle}{d\eta^2} - \frac{d \langle c \rangle}{d\eta} - Da \, \langle c \rangle^n = 0 \qquad (14.10)$$

The Danckwerts boundary conditions are commonly used. The equation is solved numerically and a CHEBFUN-based solution appears to do the trick. We therefore provide sample code that is written for a series reaction, but it can be used generally.

14.2.4 Dispersion Model: Numerical Code Using CHEBFUN

The problem considered is the simultaneous solution of the following equations:

$$D_E^* \frac{d^2 u}{d\eta^2} - \frac{du}{d\eta} - Da1 \, u = 0$$

$$D_E^* \frac{d^2 v}{d\eta^2} - \frac{dv}{d\eta} + Da1 \, u - Da2 \, v = 0$$

which represents consecutive reactions in a tubular reactor with axial dispersion. Equations are in dimensionless form.

   Here u is the concentration (cross-section average) of species A, which reacts to form B with a Damkohler number of $Da1$. Species B, denoted as v, undergoes an additional reaction to form C. The Damkohler number for this is $Da2$. Both reactions are taken as first order. Due to stoichiometric constraints only A and B need to be tracked and the concentration of the final product C can be found by material balance. The reactor is modeled with a dispersion number D_E^*. Danckwerts conditions are used here for both components.

Listing 14.1   Numerical Solution of Dispersion Model

```
% Parameter values used are:
da1=  0.5; DE_star = 0.25; da2 = 1.0; Pe1 = 1/DE_Star;
   Pe2= 1/DE_Star;
% Inlet concentrations are
cin1= 1.; cin2= 0.;
% Set up a CHEBFUN operator
x = chebfun ('x', [0,1] ); % distance
u = chebfun ('u', [0,1]);  % species A
v= chebfun ('v', [ 0,1 ] ) % species B
N = chebop (0,1);
 % Define diffusion—convection—reaction operator
N.op =  @(x,u,v) [DE_star* diff(u,2)—diff(u,1)—da1*u,...
     DE_Star* diff(v,2)—diff(v,1)+ da1*u—da2*v ]
% Danckwerts boundary conditions at exit
N.rbc = @(u,v)[ diff(u,1), diff(v,1) ];
% Inlet boundary condition; some dispersion allowed
N.lbc = @(u,v)[ diff(u,1)—Pe1.*u+Pe1*cin1 ,...
     diff(v,1)—Pe2*v+Pe2*cin2]
C= N\0   % solution found by overloaded \ operator
   C = chebfun(C)
 plot(C) % plot of solution.
```

14.2.5 Criteria for Negligible Dispersion

It is useful to know when dispersion effects can be neglected and the plug flow approximation is adequate. The following guidelines are useful:

- For slow reactions, dispersion effects are not important. If the exit conversion is less than 30% or so, the mixing pattern in the reactor is not important. The plug model, the backmixed model, and the dispersion model provide nearly the same conversion.
- For turbulent flow, dispersion effects are not so significant, unless the reaction is fast. The velocity profiles are nearly flat except in the wall region. Hence the tube behaves almost like a plug. Dispersion effects may need to be considered if the conversion is above 90% or more. The plug flow assumption will provide an underdesign for reactor volume while

the backmixed model will provide an overdesign for the same level of conversion.

- In long reactors plug flow will be generally achieved. This is because the dispersion number depends on the length of the reactor and becomes small for long reactors; hence the model predictions become close to that for plug flow.

One method to evaluate the dispersion parameter is the tracer method, which we will discuss now.

14.3   Dispersion Coefficient: Tracer Response Method

In this section we discuss the tracer response to a system modeled by a dispersion model. This is an useful technique to experimentally determine the value of the dispersion coefficient in the system. The methodology was introduced in Chapter 13 for the tanks in series model and is similar for the dispersion model. All of the following methods can be used to fit the tracer response curve and the value of the dispersion coefficient can be evaluated: Laplace domain fitting, moment analysis, and time domain fitting.

The calculation of the response in the Laplace domain is presented next. From the Laplace domain solution the moments can be extracted and the variance, which is the necessary equation for the moment method, is calculated. The solution in the time domain is needed for time domain fitting and involves some additional mathematical manipulations. Only the final solution is shown for this case.

The starting point is the dispersion model with the time dependency included but no reaction. In terms of the dispersion Peclet number the model to be solved is

$$\frac{\partial \langle c \rangle}{\partial t^*} = \frac{1}{Pe^*} \frac{\partial^2 \langle c \rangle}{\partial \eta^2} - \frac{\partial \langle c \rangle}{\partial \eta} \tag{14.11}$$

Time is dimensionalized as $t^* = t/[L/\langle v \rangle]$, that is, scaled by the mean residence time.

Danckwerts boundary conditions are commonly used in the context of the dispersion model. At the inlet we allow some drop in concentration due to dispersion:

$$\text{At } \eta = 0, \quad \langle c \rangle - \frac{1}{Pe^*} \frac{\partial \langle c \rangle}{\partial \eta} = \langle c \rangle_{\text{in}} (t^*) \tag{14.12}$$

where $\langle c \rangle_{in}(t)$ is the inlet concentration of the tracer. This will depend on the mode of tracer injection. Commonly the step and pulse functions are analyzed.

At the exit, the stream is allowed to blend smoothly with the outlet fluid:

$$\text{At } \eta = 1, \quad \frac{\partial \langle c \rangle}{\partial \eta} = 0 \tag{14.13}$$

14.3.1 Laplace Domain Solution

The Laplace transform of Equation 14.11 is

$$\frac{1}{Pe^*} \frac{d^2 \bar{c}}{d\eta^2} - \frac{d\bar{c}}{d\eta} - s\bar{c} = 0 \tag{14.14}$$

Here $\bar{c}$ is used as the Laplace transform of $\langle c \rangle$. The initial tracer concentration in the reactor is normalized to zero.

The equation has the same form as that for the steady state case with reaction with Da now being replaced by s. Equation 14.9 can be used with this substitution. Hence the solution for exit concentration in the Laplace domain can be written (for a pulse input) as

$$\bar{c}_e = \frac{4\alpha \exp(Pe^*/2)}{(1+\alpha)^2 \exp(\alpha Pe^*/2) - (1-\alpha)^2 \exp(-\alpha Pe^*/2)} \tag{14.15}$$

where

$$\alpha = \sqrt{(1 + 4s/Pe^*)}$$

The inversion, which is mathematically challenging, provides the time response, but it is easier to work with the moments, which are shown in the following paragraph. Of course one can use MATHEMATICA or MAPLE directly to generate the time domain results numerically as well.

14.3.2 Moments of the Response Curve

The first and second moment can be computed from the Laplace domain using the formulae shown in Section 13.4. The following results may be verified:

$$\mu_1^* = 1 \tag{14.16}$$

$$\sigma^{*2} = \mu_2^* - \mu_1^{*2} = \frac{2}{Pe^*} - \frac{2}{Pe^{*2}}(1 - \exp(-Pe^*)) \tag{14.17}$$

The dimensionless first moment is equal to one. Therefore the actual first moment is equal to the mean passage time.

The second central moment (scaled by the square of the mean passage time) is indicative of the dispersion coefficient. For large values of Pe^* (reactors closer to plug flow) this reduces to

$$\sigma^{*2} = \frac{2}{Pe^*}$$

If the reactor were modeled as a series of backmixed reactors (tanks in series model) instead of using the dispersion model, the dimensionless second moment would be

$$\sigma^{*2} = \frac{1}{N}$$

This was shown in Example 13.1 in the last chapter.

Hence we find an equivalence between the two models. The number of tanks needed to model the system is approximately equal to $Pe^*/2$. Hence either of the two models can be used and the prediction will be close if the second moments are matched.

Open-Open Boundary Conditions

The open-open boundary conditions are also used to analyze the tracer data. This type of approach is useful, for example, when the tracer is injected at some distance into the reactor. For example in a packed column, the tracer may be injected at a length equal to three or four packing diameters rather than right at the inlet. Similarly the tracer may be measured at some intermediate point near the exit rather than the actual exit. In this case there is dispersion on both the entrance side and the exit side and open-open boundary conditions are more suitable. A schematic of an open-open system is shown in Figure 14.2.

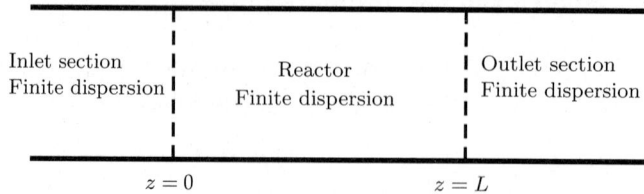

Figure 14.2   Schematic of open-open boundary conditions.

The moments are given by the following expressions for the open-open case:

$$\mu_1^* = 1 + \frac{2}{Pe^*}$$

$$\sigma^{*2} = \frac{2}{Pe^*} + \frac{8}{Pe^{*2}}$$

14.3.3 Time Domain Solution

Time domain solutions are lengthy and can be evaluated numerically. The analytical solution (Wen and Fan, 1975) for the (cross-sectional averaged) exit concentration is as follows for a pulse injection.

$$\langle c_{A,e} \rangle (t^*) = \sum_{i=1}^{\infty} \frac{2\lambda_n (Pe^* \sin(\lambda_n) + \lambda_n \cos(\lambda_n))}{(Pe^{*2}/4 + Pe^* + \lambda_n^2)} \exp(Pe/2)$$

$$\times \exp\left[-t^* \left(\frac{Pe^{*2} + \lambda_n^2}{Pe^*}\right)\right]$$

(14.18)

The λ_n are the eigenvalues, given as the roots of

$$2\cot(\lambda) = \frac{\lambda}{Pe^*} - \frac{Pe^*}{4\lambda}$$

It is interesting to note that for small deviations from plug flow (large Pe^* values) the E-curve can be reduced to

$$\langle c_{A,e} \rangle (t^*) = E(t^*) = \frac{1}{\sqrt{4\pi D_E^*}} \exp\left[\left(\frac{-(1-t^*)^2}{4D_E^*}\right)\right]$$

(14.19)

This has the same form as the Gaussian (normal) distribution.

Illustrative results for the exit concentration for a step tracer are shown in Figure 14.3. The response is shown for a backmixed system, for Pe^* values of 2 and 10, and for the case of plug flow ($Pe^* = \infty$). The plug flow is a sharp step function placed at the mean residence time ($t^* = 1$ in dimensionless units). For the case of $Pe = 0$, the response curve approaches the backmixed system. For other Pe values the response is spread on either side of the plug flow step function, approaching one for large values of time.

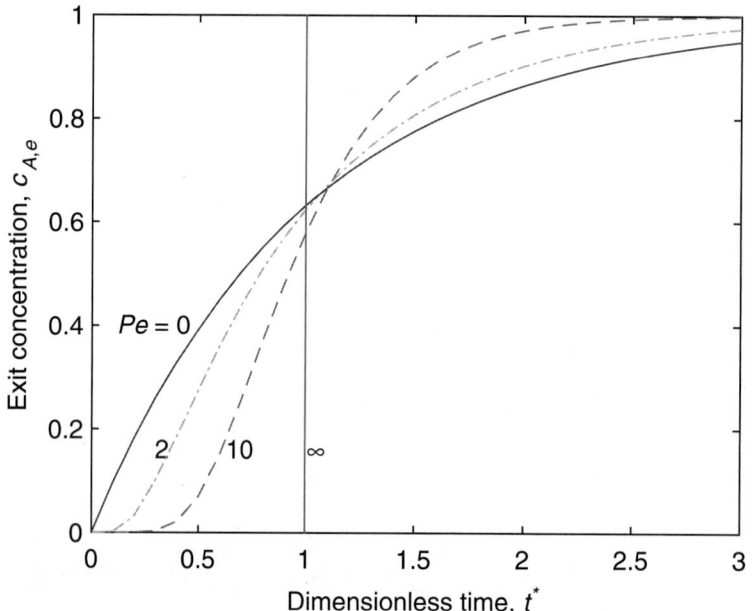

Figure 14.3 Tracer exit concentration versus time based on the dispersion model computed using the MATLAB PDEPE solver for a step input case.

14.4   Taylor Model for Dispersion in Laminar Flow

Taylor (1953) derived the classical result given by Equation 4.6 for the dispersion coefficient, D_E, of a solute for pipe flow under laminar flow conditions. This expression is derived here following Taylor's analysis.

Consider a pulse of dye introduced into a liquid laminar flow as shown in Figure 14.4.

This pulse is sheared due to convection due to the variation of the velocity as a function of radial position. This creates a concentration gradient in the radial direction and causes the spread of the dye by diffusion. After some entry region, the dye is smeared into a slug and further on it appears that the dye is spreading on either side of the slug through a diffusion-type process. The process is sketched in Figure 14.4. The spreading of the pulse is called dispersion and is a combined effect of convection (due to a non-uniform velocity profile) and radial diffusion. The goal of the Taylor dispersion analysis is to derive an expression for the spread of the dye using a Fick's law type of model for the dispersion process.

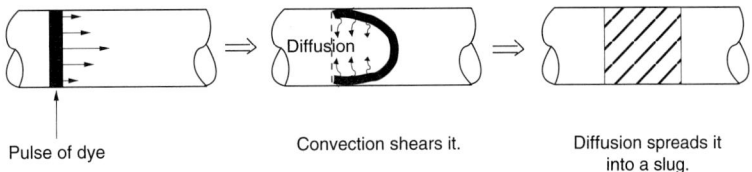

Pulse of dye   Convection shears it.   Diffusion spreads it
                                             into a slug.

Figure 14.4 Illustration of the concept of Taylor diffusion: combined effect of axial convection and radial diffusion caues the pulse
of dye to be spread with a mechanism similar to diffusion.

The starting point is the convection-diffusion model for Newtonian laminar flow studied in Chapter 10. We now add the time derivative in order to
investigate the transient response. The resulting model for laminar flow is

$$\frac{\partial C_A}{\partial t} + 2 \langle v \rangle (1 - \xi^2) \frac{\partial C_A}{\partial z} = \frac{D}{R^2} \left[\frac{1}{\xi} \frac{\partial}{\partial \xi} \left(\xi \frac{\partial C_A}{\partial \xi} \right) \right] \qquad (14.20)$$

Dimensional (actual) variables are used here except for the radial coordinate, which is represented by ξ, equal to r/R. The coordinate system can be
transformed to moving coordinates as

$$z^* = z - \langle v \rangle t$$

and

$$t^* = t$$

Here z^* is the axial position in a coordinate system moving with the average velocity. The variable t^* is the same time variable t but a different symbol
is used to distinguish the two sets of independent variables in the two systems
(stationary vs. moving). Also note that t^* and z^* here have the units of time
and distance, respectively, and are not dimensionless in this section.

The partial derivatives in Equation 14.20 have to be transformed using
the chain rule. The following results can be derived:

$$\frac{\partial C_A}{\partial t} = \frac{\partial C_A}{\partial t^*} - \langle v \rangle \frac{\partial C_A}{\partial z^*}$$

Using similar mathematical manipulations, you should verify that

$$\frac{\partial C_A}{\partial z} = \frac{\partial C_A}{\partial t^*} \frac{\partial t^*}{\partial z^*} + \frac{\partial C_A}{\partial z^*} \frac{\partial z^*}{\partial z} = \frac{\partial C_A}{\partial z^*}$$

since the term $\frac{\partial t^*}{\partial z^*}$ is zero and the term $\frac{\partial z^*}{\partial z}$ is one.

Hence the model in the moving reference frame is

$$\frac{R^2}{D}\frac{\partial C_A}{\partial t^*} + \frac{\langle v \rangle R^2}{D}[2(1-\xi^2)-1]\frac{\partial C_A}{\partial z^*} = \frac{1}{\xi}\frac{\partial}{\partial \xi}\left(\xi\frac{\partial C_A}{\partial \xi}\right) \tag{14.21}$$

The time derivative term can be dropped if $t^* > R^2/D$ (the radial diffusion time). The interpretation is that after the time is greater than the radial diffusion time in a moving coordinate system, the system reaches a pseudo-steady state case with the convection term balancing the diffusion term. Referring to Figure 14.4, this means that the smearing of the tracer due to diffusion has already taken place (stage 3 of the figure). The corresponding distance will be $\langle v \rangle R^2/D$ and the analysis will be valid if the observation point is larger than this length. Hence the Taylor analysis will hold if L/R is larger than $\langle v \rangle R/D$.

In a moving reference frame $\frac{\partial C_A}{\partial z^*}$ is not expected to vary significantly and Taylor assumed this to be a constant. Let this constant be called $\mathcal{A}$:

$$\mathcal{A} = \frac{\partial C_A}{\partial z^*} = \frac{\partial < C_A >}{\partial z^*} \tag{14.22}$$

$< C_A >$ is the cross-sectional average concentration. Hence Equation 14.21 reduces to

$$(1-2\xi^2)\frac{\mathcal{A} < v > R^2}{D} = \frac{1}{\xi}\frac{\partial}{\partial \xi}\left(\xi\frac{\partial C_A}{\partial \xi}\right) \tag{14.23}$$

This can be integrated twice with respect to ξ to obtain an approximate expression for the variation of the concentration as a function of dimensionless radial position. The result after applying the two boundary conditions of no flux at both the center and wall is

$$C_A = \left(\frac{\xi^2}{4} - \frac{\xi^4}{8}\right)\frac{\mathcal{A} < v > R^2}{D} + C_0 \tag{14.24}$$

where C_0 is the unknown center line concentration.

Taking the cross-sectional average we have

$$< C_A >= \frac{1}{12}\frac{\mathcal{A} < v > R^2}{D} + C_0 \tag{14.25}$$

Taking the flow average (cup-mixing average), we obtain

$$C_{Ab} = \frac{1}{16}\frac{\mathcal{A} < v > R^2}{D} + C_0 \tag{14.26}$$

The difference may be expressed as

$$C_{Ab} = <C_A> - \frac{\mathcal{A} <v> R^2}{48D} \tag{14.27}$$

Multiplying by $\langle v \rangle$ and also substituting for $\mathcal{A}$ as the derivative of the concentration $\partial <C_A> /\partial z^*$ as per Equation 14.22, we get the following result:

$$<v> C_{Ab} = <v><C_A> - \left(\frac{1}{48} \frac{\langle v \rangle^2 R^2}{D} \right) \frac{\partial <C_A>}{\partial z^*} \tag{14.28}$$

The first term can be viewed as the mass of A crossing any axial position (per unit cross-sectional area). The second term is the mass of A crossing computed using the mean (area-averaged) concentration. Hence the last term is the extra term and can be interpreted as a correction term. This term has the appearance of Fick's law. The dispersion representation is

$$<v> C_{Ab} = <v><C_A> - D_E \frac{\partial <C_A>}{\partial z^*} \tag{14.29}$$

Comparing the last two equations, the following expression for D_E can be obtained:

$$\boxed{D_E = \frac{1}{48} \frac{\langle v \rangle^2 R^2}{D}} \tag{14.30}$$

This is the classical result obtained by Taylor for a dispersion coefficient in laminar flow. In a follow-up paper, Taylor (1954a) used the theory to measure molecular diffusivity of solutes present in a flowing stream and demarcated the conditions under which the diffusivity can be obtained. Although derived originally for laminar flow, the concept has been used in many applications, including chemical reactor analysis, and we have already seen an example of this for reactors; additional examples in separation processes, chromatography columns, and so on, are discussed later in the text.

14.5 Segregated Flow Model

Note that the Taylor dispersion model does not apply to short columns. The condition for validity of the Taylor model can be stated as

$$\frac{L}{d_t} >> \frac{ReSc}{4}$$

Normally this condition is difficult to fulfill for liquids where $ReSc$ can be quite large. Very long pipes are needed for Taylor dispersion to be observed in such pipes. For gases this can be satisfied more easily since Sc is comparable to one.

For shorter pipes, the process is dominated by convection effects caused by variation of the axial velocity in the radial direction. The tracer response can be modeled in such cases by what is called the segregated flow model. The diffusion terms are dropped in this case and the convection balances the accumulation. The model equation (Equation 14.20) is now simplified as

$$\frac{\partial C_A}{\partial t} = -2 \langle v \rangle (1 - \xi^2) \frac{\partial C_A}{\partial z} \tag{14.31}$$

The exit age distribution based on the solution of this equation can be derived and shown to be:

$$E(t) = \begin{cases} 0 & \text{if } t \leq \bar{t}/2 \\ \bar{t}^2/t^3 & \text{if } t \geq \bar{t}/2 \end{cases} \text{ where } \bar{t} = L/\langle v \rangle$$

The model applies when the diffusion coefficient is small; that is, tracer spread is mainly caused by the difference in velocity at various radial locations. The tracer at the center moves faster and arrives at the exit at time equal to half the mean passage time, since the fluid moves twice the mean velocity at the center. No tracer will appear before that. Subsequently tracer at each location will spend a time equal to the local residence time at that radial position and arrive at the exit at that tine. The response curve is shown in Figure 14.5.

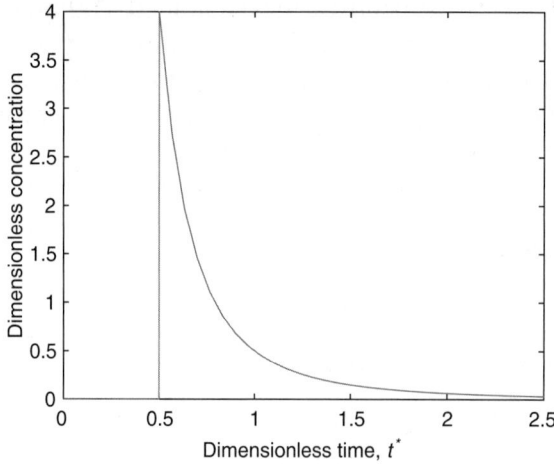

Figure 14.5 *E*-curve for laminar flow reactor under segregation. This model applies when the diffusion coefficient is small.

Both the dispersion and segregation models have their own range of applications. The model predictions for a reacting system are shown in Chapter 17 where a detailed analysis of a laminar flow reactor is provided. In that chapter the segregated model and the dispersion model are compared with a full 2-D differential model and the range of validity of each of these models is presented.

14.6   Dispersion Coefficient Values for Some Common Cases

Dispersion coefficient values for some common cases are presented in this section.

Pipe Flow of a Non-Newtonian Liquid. If the fluid behavior is given by a power law model, the following equation derived by Fan and Hwang (1965) can be used for the dispersion Peclet number:

$$Pe^* = \frac{2(3n+1)(5n+1)}{n^2} \frac{\bar{t}D_A}{R^2}$$

Here n is the power law index. For $n = 1$, the coefficient above is 48, consistent with the Taylor model for a Newtonian fluid.

Axial Dispersion in Channel Flow. Consider pressure-driven laminar flow in a channel of height $2h$. The following formula from which the axial dispersion coefficient can be derived closely follows the Taylor approach. Details are left as an exercise problem (14.16).

$$D_E = \frac{2h^2 \langle v \rangle^2}{105 D_A}$$

where $\langle v \rangle$ is the averaged velocity and D_A is the diffusivity of the species under consideration.

Axial Dispersion in Turbulent Flow. Taylor (1954b) showed that the following expression is suitable for the calculation of the axial dispersion coefficient for turbulent flow in a pipe:

$$D_E = 5d_t v_f$$

Here d_t is the pipe radius and v_f is the friction velocity. This is related to the friction factor by the following formula:

$$v_f = \langle v \rangle \sqrt{f/2}$$

Using the Blasius equation for the friction factor, $f = 0.0791 Re^{-1/4}$ for turbulent flow, we find

$$D_E^* \propto Re^{-1/8}$$

The above correlation predicts somewhat lower values than the experimental values. The following correlation proposed by Wen and Fan (1975) fits the data more closely:

$$D_E^* = \frac{3 \times 10^7}{Re^{2.1}} + \frac{1.35}{Re^{1/8}}$$

The first term is the correction due to transitional flow and at high Reynolds numbers only the second term is important. Note that the Re to the power of $-1/8$ proportionality predicted by the Taylor model is retained.

Dispersion Coefficient in Packed Beds. Knowledge of axial dispersion in packed beds is of importance in many applications such as adsorber design, chromatographic separations, and so on, in addition to packed reactors. The axial dispersion coefficient is expressed in terms of a dimensionless parameter, the dispersion Peclet number, defined as

$$Pe_d^* = \frac{\epsilon_B v_z d_p}{D_E}$$

Note that the particle diameter is used as the length scale here. v_z is the superficial fluid velocity.

Two regimes can be identified: the molecular diffusion controlled regime and the hydrodynamic regime.

The molecular control holds for low values of $ReSc$. The dispersion coefficient is modeled using a torutosity factor:

$$D_E = D_A/\tau$$

where the torutosity τ is taken as $\sqrt{2}$ for packed beds; hence $D_E = D_A/\sqrt{2} = 0.707 D_A$ is suggested for this regime.

The hydrodynamic regime holds for high values of $ReSc$, a molecular Peclet number greater than two or so. Aris and Amundson (1957) showed that the dispersion Peclet number based on the particle diameter can be represented as

$$Pe_d^* = \frac{2}{\eta}$$

where η is a geometric parameter that is representative of distance between successive particle layers. For different arrangements of spherical particles, the range of values of η is between 0.817 and 1.0. Hence $D_E^* = 2u_g d_p$ may be used as an approximate formula in this range.

The equation was modified to include the Reynolds number dependency. The factor η was replaced by $(1-p)/p$ and the equation was modified to

$$Pe_d^* = \frac{2p}{1-p}$$

where p is defined as

$$p = 0.17 + 0.33\exp(-24/Re)$$

For an intermediate regime, a combined correlation of the following form is useful for spherical particles:

$$\frac{1}{Pe_d^*} = \frac{1-p}{p}[Y + Y^2(\exp[-Y^{-1}] - 1)] + \frac{\epsilon_B}{\tau ReSc} \qquad (14.32)$$

where Y is defined as

$$Y = \frac{p(1-p)ReSc}{23.16(1 - \epsilon_D)}$$

A summary and a critical review of dispersion in packed beds is presented in Delgado (2006) and is useful for further evaluation of this parameter. The parametric effects of column to particle diameter, column length to particle diameter, fluid properties, and so on, are discussed in detail in this reference.

14.7   Two-Phase Flow: Models Based on Ideal Flow Patterns

Mesoscopic models for two-phase flow systems follow the developments shown in Section 4.3. The first modeling approach uses the ideal flow patterns, now for both phases. Then non-idealities to account for the deviation can be added. In the ideal contacting assumption, each phase may be assumed to be in either in plug flow or completely mixed in the ideal contacting assumption. Thus we get four combinations:

- Gas and liquid both in plug flow
- Gas and liquid both backmixed
- Gas in plug flow and liquid backmixed
- Gas backmixed and liquid in plug flow

The last case (gas backmixed but liquid in plug flow) is not very common in process equipment. Systems with batch liquid and continuous gas flow may be modeled using this contacting pattern. In any case the analysis of the fourth case is mathematically the same as the third with the fluids switched. The first two cases have been analyzed in earlier chapters. Here we discuss the third case and show how models for such systems can be formulated.

14.7.1 Plug-Backmixed Model

The case where gas is in plug flow and the liquid is backmixed is analyzed in the following section. Note that the discussion is not specific to gas–liquid systems and applies to any fluid–fluid system with one of the phases backmixed and the other phase in plug flow. The schematic of the control volumes used in the analysis is shown in Figure 14.6.

The control volume for the gas is mesoscopic and occupies a section Δz of the equipment in the flow direction. The control volume for the liquid is macroscopic and occupies the entire liquid volume in the system. Thus both meso and macro balances have to be used for the same problem.

The species mass balance in the gas phase balance leads to

$$-u_G \frac{dC_{AG}}{dz} - K_L a_{gl}\left(\frac{C_{AG}}{H_A} - C_{AL}\right) = 0$$

u_G is the superficial gas velocity equal to Q_G/A. $K_L a_{gl}$ is the overall mass transfer coefficient based on the liquid driving force and H_A is the Henry constant expressed in concentration ratio.

The equation is made dimensionless by defining the dimensionless concentration as

$$c_{AG}(\eta) = \frac{C_{AG}(z)}{C_{AG,i}}$$

η is the dimensionless location along the column height, defined as z/L.

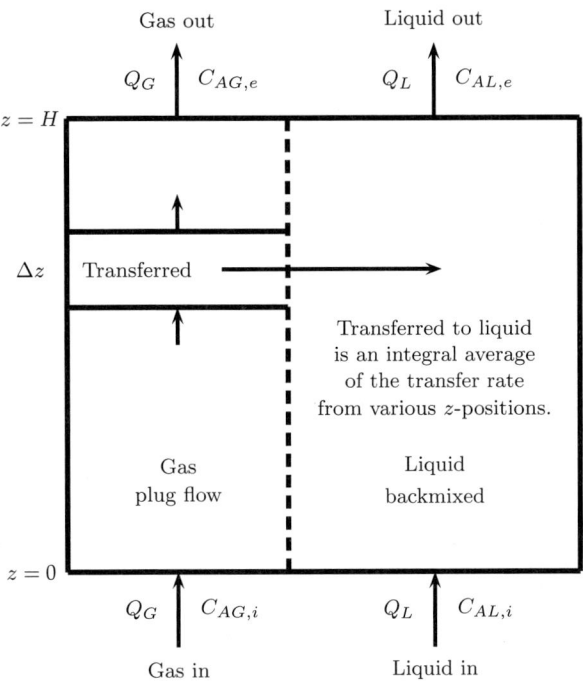

Figure 14.6 Schematic of a two-phase plug-backmixed system showing the control volumes. For gas a differential volume is used while for liquid the entire liquid volume is used.

For liquid we use the maximum solubility value as the reference concentration. Thus we use $C_{AG,i}/H_A$ as the reference and define the dimensionless exit liquid concentration as

$$c_{AL,e} = \frac{C_{AL,e}}{C_{AG,i}/H_A}$$

The gas phase balance is then

$$-\frac{dc_{AG}}{d\eta} - \frac{\kappa_L}{\Lambda_{gl}}(c_{AG} - c_{AL,e}) = 0 \tag{14.33}$$

Here κ_L is a dimensionless mass transfer coefficient:

$$\kappa_L = \frac{K_L a_{gl} V_R}{Q_L}$$

Λ_{gl} is a flow ratio parameter (corrected by the Henry coefficient):

$$\Lambda_{gl} = \frac{H_A Q_G}{Q_L}$$

The liquid phase balance can be formulated as

$$Q_L(C_{AL,i} - C_{AL,e}) + K_L a_{gl} \int_0^H (\frac{C_{AG}}{H_A} - C_{AL,e}) A dz = 0$$

Note how the mass transfer rate term from gas to liquid is written as an integral. Locally the driving force is therefore $\frac{C_{AG}}{H_A} - C_{AL,e}$. Here $C_{AL,e}$ is the exit liquid concentration, which is a constant, but C_{AG}, the gas concentration, is a function of z. Hence a local differential volume $A\Delta z$ is used to find the local rate of transfer. The integral of this provides the total transferred from the gas to the liquid.

The liquid phase balance in dimensionless variables is

$$c_{AL,i} - c_{AL,e} + \kappa_L \int_0^1 (c_{AG}(\eta) - c_{AL,e}) d\eta = 0 \qquad (14.34)$$

The solution is straightforward but consists of several algebraic steps. Equation 14.33 can be integrated to get the profiles in the gas phase:

$$c_{AG} - c_{AL,e} = (1 - c_{AL,e}) \exp\left(-\eta \frac{\kappa_L}{\Lambda_{gl}}\right) \qquad (14.35)$$

This can then be used to find the integral term in the liquid mass balance:

$$\int_0^1 (c_{AG}(\eta) - c_{AL,e}) d\eta = (1 - c_{AL,e}) \frac{\Lambda_{gl}}{\kappa_L} \mathcal{E}$$

The term $\mathcal{E}$ is defined as

$$\mathcal{E} = 1 - \exp(-\kappa_L / \Lambda_{gl})$$

Using this in Equation 14.34 and rearranging, we obtain the exit liquid concentration in an explicit form:

$$c_{AL,e} = \frac{c_{AL,i} + \Lambda_{gl}\mathcal{E}}{1 + \Lambda_{gl}\mathcal{E}}$$

The corresponding exit gas concentration is given as

$$c_{AG,e} = \frac{\mathcal{E}c_{AL,i} + 1 + \Lambda_{gl}\mathcal{E} - \mathcal{E}}{1 + \Lambda_{gl}\mathcal{E}}$$

For large mass transfer coefficients, $\mathcal{E}$ tends to one and the model reduces to the equilibrium model:

$$c^*_{AL,e} = \frac{c_{AL,i} + \Lambda_{gl}}{1 + \Lambda_{gl}}$$

The model finds application in many areas, for example, in the calculation of tray efficiency in a distillation column, for simulation of fluid bed reactors, and for gas absorption in well-mixed continuous flow bubble columns. Some of these are illustrated as exercise problems at the end of this chapter.

Having considered three ideal contacting patterns, we now look at how non-idealities can be modeled using the dispersion concept.

14.7.2 Non-Idealities in Two-Phase Flow

In this section we show how deviations from the ideal contacting can be modeled using the dispersion concept. First we show some cases where it is required to consider these deviations from ideal flow patterns. The following guidelines are useful in this regard:

- If the solute concentration changes by large amount, say 99% removal, then dispersion effects need to be considered.
- If the system has a relativly low HTU, which means a relatively short contactor can achieve significant separation, then dispersion effects are important. Since the contactor is short, the dispersion Peclet number is small and plug flow may not be a good assumption.
- When large eddies or circulation patterns develop in the system, for example, a bubble column absorber, the system can be modeled using the dispersion model. But the tanks in series model (the same as the mixing cell model) is closer to the physical picture and is preferred.
- When there is a wide range of drop size in a gravity-driven liquid–liquid separation unit, the plug flow may not be suitable due to varying velocity of the drops and some drop phase backflow in the separator.
- When there is a very large or very small flow ratio there is likely flow segregation and non-ideal contacting models should be used.
- When there is significant channeling, e.g., due to inhomogeneity in packing near the wall, the flow pattern is again non-ideal. This is often important in small laboratory reactors. Here the wall region is comparable to the size of the reactor and bypassing can occur. A dispersion model may not be a good representation when bypassing occurs. Data in such systems should be used with caution when scaling up the reactor.

The model in the following section is based on the assumption that both phases are in dispersed plug flow. This covers a wide range of mixing and the four ideal cases can be derived as special cases by putting either a very small or very large value for the dispersion coefficient for each of the phases.

We also use dilute systems for simplicity, that is, the gas and the liquid flow rates are nearly constant. A linear equilibrium model is also used. Formulation is shown for both cocurrent and countercurrent flow although the models are similar with some minor switches.

Concurrent Flow

The model equations consist of the gas phase balance and the liquid phase balance for the solute in each phase.

For the gas phase, the following equation can be derived starting from the species mass balance:

$$D_{EG}\frac{d^2 C_{AG}}{dx^2} - u_G\frac{dC_{AG}}{dz} - K_L a_{gl}(\frac{C_{AG}}{H_A} - C_{AL}) = 0 \qquad (14.36)$$

The three terms represent the dispersion, convection, and mass transfer from gas to liquid in that order. D_{EG} is the dispersion coefficient in the gas phase and u_G is the superficial gas velocity.

Liquid phase mass balance is given by the following equation:

$$D_{EL}\frac{d^2 C_{AL}}{dx^2} - u_L\frac{dC_{AL}}{dz} + K_L a_{gl}(\frac{C_{AG}}{H_A} - C_{AL}) = 0 \qquad (14.37)$$

The three terms represent the dispersion, convection, and transfer from gas to liquid in that order. The transfer term is common to the gas phase balance but now appears with the opposite sign. D_{EL} is the dispersion coefficient in the gas phase and u_L is the superficial gas velocity.

Boundary conditions, usually of the Danckwerts type, are applied to each of the phases. If the dispersion terms are dropped we get the model for plug flow of both phases.

The equations can be made dimensionless by introduction of the following parameters:

- Dimensionless distance, $\eta = \frac{z}{L}$
- Dimensionless gas phase concentration, $c_{AG} = \frac{C_{AG}}{C_{AG,i}}$
- Dimensionless liquid phase concentration, $c_{AL} = \frac{C_{AL}}{(C_{Ag,i}/H_A)}$

Note that the liquid concentration is scaled by the saturation value corresponding to the inlet gas concentration. The liquid concentration then varies from 0 to 1. Gas concentration is scaled by its inlet value and goes from 0 to 1 as well. These normalizations lead to the dimensionless version of the model:

$$\frac{1}{Pe_G}\frac{d^2 c_{AG}}{d\eta^2} - \frac{dc_{AG}}{d\eta} - \frac{\kappa_{gl}}{\Lambda_{gl}}(c_{AG} - c_{AL}) = 0 \tag{14.38}$$

$$\frac{1}{Pe_L}\frac{d^2 c_{AL}}{dx^2} - \frac{dc_{AL}}{dz} + \kappa_{gl}(c_{AG} - c_{AL}) = 0 \tag{14.39}$$

The dimensionless parameters appearing in this equation are as follows:

- Dispersion Peclet number for the gas phase, $Pe_G = \frac{u_G L}{D_{EG}}$
- Dispersion Peclet number for the liquid phase $Pe_L = \frac{u_L L}{D_{EL}}$
- Dimensionless mass transfer coefficient, $\kappa_{gl} = \frac{K_L a_{gl} L}{u_L}$
- Flow ratio parameter, $\Lambda_{gl} = \frac{u_G H_A}{u_L}$

Thus four parameters are needed to characterize the separator performance.

The Danckwerts boundary conditions in dimensionless form are used at the entrance and exit and these are as follows.

For the gas phase, the boundary conditions are

$$\eta = 0, \ \text{gas inlet,} \ c_{AG} - \frac{1}{Pe_G}\frac{dc_{AG}}{d\eta} = 1 \tag{14.40}$$

and

$$\eta = 1, \ \text{gas exit,} \ \frac{dc_{AG}}{d\eta} = 0 \tag{14.41}$$

For the liquid phase the boundary conditions are

$$\eta = 0, \ \text{liquid inlet,} \ c_{AL} - \frac{1}{Pe^*}\frac{dc_{AL}}{d\eta} = c_{AL,i} \tag{14.42}$$

and

$$\eta = 1, \ \text{liquid exit,} \ \frac{dc_{AL}}{d\eta} = 0 \tag{14.43}$$

Model equations can be solved numerically with minor changes in the code given in Listing 14.1. The model is useful to predict the effect of non-ideality in separator. It can also be used for two-phase reactor models by adding the reaction term.

Countercurrent Flow

The dispersion model for countercurrent flow is similar to cocurrent flow with some minor modifications. The gas phase equation and the boundary conditions are the same.

But for the liquid phase a change in sign in the convection term is needed as the flow direction is now opposite to that of the gas. The corresponding equations in dimensionless form are as follows:

$$\frac{1}{Pe_L}\frac{d^2 c_{AL}}{dx^2} + \frac{dc_{AL}}{dz} + \kappa_{gl}(c_{AG} - c_{AL}) = 0 \tag{14.44}$$

The boundary conditions are also switched and now read as

$$\eta = 1, \quad \text{liquid inlet, } c_{AL} + \frac{1}{Pc_L}\frac{dc_{AL}}{d\eta} = c_{Al,i} \tag{14.45}$$

and

$$\eta = 0, \quad \text{liquid exit, } \frac{dc_{AL}}{d\eta} = 0 \tag{14.46}$$

Since the problem is linear it can be solved by analytical methods. Pratt (1975) solved the problem analytically and presented results for the concentration profiles. Nowadays it is easier and more convenient to solve this numerically. Listing 14.1 is applicable simply by changing some terms. An illustrative result is shown in Figure 14.7.

The system behavior for the plug-plug flow case is generated by simply using large enough values of the Peclet parameters. Similarly the results for a backmix-backmix case can be simulated by setting Pe to a small value. Hence the dispersion model is flexible to handle the limiting cases. These limiting cases are shown in Table 14.2 for high and low values of the disperison coeficients in the gas and liquid phases.

Tanks in Series Model

The tanks in series model presented in Chapter 13 is an alternative method to handle non-idealities in two-phase flow. Each phase is represented by N tanks connected in series. If $N = 1$, a backmixed model is obtained. Similarly a large value of N simulates a plug flow case. Both dispersion and the tanks in series models show similar trends provided the parameters are matched

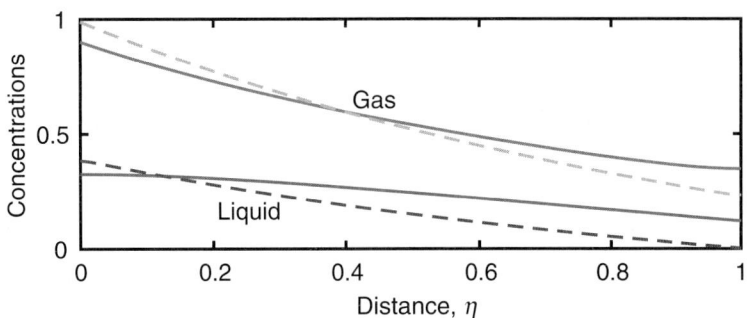

Figure 14.7   Effect of dispersion on the concentration profiles in a gas–liquid absorber; the dotted line is plug flow. The parameters used for the dispersion case are $\kappa_{gl} = 1$, $\Lambda_{gl} = 0.5$, $Pe_G = 10$, and $Pe_L = 5$.

Table 14.2   Ideal Contacting Patterns for Fluid–Fluid Systems

	Small D_{EL}	Large D_{EL}
Small D_{EG}	plug-plug	plug-backmixed
Large D_{EG}	backmixed-plug	backmixed-backmixed

to maintain similar second central moments. Often the choice is therefore the preference of the modeler.

The tanks in series model is preferable if there is a large extent of back-mixing, for example, a bubble column type of reactor. The dispersion is more useful for tall columns that are closer to plug flow, such as a packed bed absorber. One minor difficulty with the tanks in series model is that the number of tanks to be used for the liquid phase can be different from that for the gas phase. This can create some problems in coupling the two balances.

14.8   Tracer Response in Two-Phase Systems

In order to use the dispersion model shown in the last section, the dispersion coefficients for both phases have to be experimentally determined and the tracer methods used for single-phase systems are again useful. However, there are some caveats that have to be kept in mind and properly addressed in the interpretation of the tracer data.

The tracer may not stay in the same phase where it was introduced and may be transferred to the second phase. The overall response is then modulated by transport to the other phase. The observed tracer response can in such case show a dispersion effect even if the phase into which the tracer was introduced was in plug flow. Hence the tracer data cannot be interpreted as a single-phase system model and the effect of interphase mass transfer may need to be considered. Only if the tracer stays in the same phase where it was introduced will the response curve be a true representation of the backmixing in that phase. We will show some examples to illustrate the effects qualitatively. Key references are also cited for those interested in following this industrially important and relevant topic further.

First we consider a two-phase system with only one flowing phase, that is, gas–solid or liquid–solid systems. The effect of mass transfer on the tracer response is more readily visualized by looking at this simple example. Then we show the considerations needed for systems with two flowing phases.

14.8.1 Single Flowing Phase

Consider a single flowing phase in contact with a solid that can adsorb some gas. Models for such systems are important in the design of packed bed adsorption columns and are treated in more detail in Chapter 29. Here we show how the tracer response is modified by the processes taking place in the solid phase. A schematic of the processes taking place in the gas and solid phases is shown in Figure 14.8.

A simplified analysis is presented that assumes the gas is in plug flow and the process is in equilibrium at each point in the column. Thus the balance equation is

in at z − out at $z + \Delta z$ − transferred to solid = accumulation

In mathematical form the gas phase balance is

$$u_G \frac{dC_A}{dz} = -k_{ex}(C_A - q_A/K_A) + \epsilon_B \frac{\partial C_A}{\partial t}$$

where k_{ex} is the volumetric transfer coefficient from gas to the solid. Here q_A is the concentration in the second phase and K_A is the adsorption coefficient of the gas in the solid.

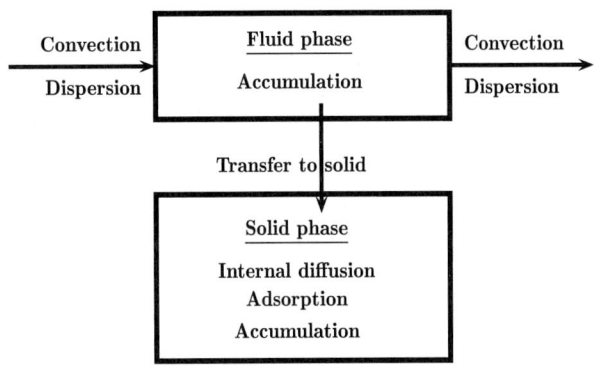

Figure 14.8 Schematic of processes in a mescoscopic control volume of a gas–solid adsorption column.

The balance on the second phase (solid) leads to

$$\text{Accumulation} = \text{transfer from gas}$$

The mathematical formulation of this is

$$(1 - \epsilon_B)\frac{dq_A}{dt} = k_{ex}(C_A - q_A/K_A)$$

The two equations can be combined to give

$$\left[\epsilon_B\frac{\partial C_A}{\partial t} + (1 - \epsilon_B)\right]\frac{\partial q_A}{\partial t} = -u_G\frac{\partial C_A}{\partial x} \tag{14.47}$$

and this eliminates the transfer term.

If the mass transfer is rapid and equilibrium exists at every location in the column then

$$q_A = K_A C_A$$

Hence Equation 14.47 can be reduced to a single equation for the gas phase response:

$$[\epsilon_B + K_A(1 - \epsilon_B)]\frac{\partial c_A}{\partial t} = u_G\frac{\partial C_A}{\partial z} \tag{14.48}$$

This is the equation for the tracer response. The model is similar to a single-phase plug flow except for the coefficient term in the time derivative, known as the capacity term. This term now includes the adsorption constant. It is now larger by a factor of $K_A(1 - \epsilon_B)$. The tracer will exist after a time L/u_G times $\epsilon_B + K_A(1 - \epsilon_B)$ rather than at the mean passage time in the gas phase. The first moment can be readily found from Equation 14.48 as

$$\mu_1 = \frac{L}{v_G}\left[1 + \frac{1 - \epsilon_B}{\epsilon_B}K_A\right]$$

where $v_G = u_G/\epsilon_G$, the interstitial velocity. The second moment is zero.

Effect of Internal Diffusion

If the solid is porous, part of the tracer will diffuse into the solid and the internal diffusion will add a further time delay. Similarly any axial dispersion in the gas phase will again add to the tracer spread. The second moment will no longer be zero. Including all these factors the second moment can be shown to be

$$\mu_2 = \left[\frac{2}{Pe^*} - \frac{2}{Pe^{*\,2}}(1 - \exp(-Pe^*))\right]\mu_1^2 +$$
$$\frac{6(1 - \epsilon_B)R^2 L}{u_G}(\epsilon_p + K_A)^2\left(\frac{1}{45D_e} + \frac{1}{9k_{sl}R}\right) \tag{14.49}$$

which was derived by Schneider and Smith (1968). We find that the tracer response is affected by a number of factors in addition to the backmixing in the gas phase (Pe^* parameter). Hence interpretation of tracer data has to include the effect of the additional contribution to the tracer spread.

14.8.2 Two Flowing Phases

If there are two flowing phases, the tracer can be introduced in either the gas phase or the liquid phase. The tracer can appear in the outlet of both phases as a result of mass transfer, even when it is introduced in only one of the phases. This can cause problems in data interpretation since the outlet response is not entirely due to dispersion. Tracer spread is due to dispersion only if the tracer stays in the same phase where it was introduced, for example. a liquid phase tracer such as NaCl that stays entirely in the liquid or a relatively non-volatile tracer in general. Similarly for a tracer introduced in the gas phase, the solubility of the gas has to be rather low in the liquid so that it stays mostly in the gas. Helium is often used as a tracer for such cases. Other gases have reasonable solubility and the interpretation of the tracer data is difficult since it is a function of not only the dispersion coefficient but also the mass transfer coefficient. It is also a function of the mixing pattern in the liquid.

The effect of tracer mass transfer effects can be examined by looking at the zeroth moment of the tracer response. For a tracer introduced in the gas phase we can define the zeroth moment for both the gas and the liquid phase:

$$\mu_{0G} = \int_0^\infty C_{AG,e}(t)dt$$

and

$$\mu_{0L} = \int_0^\infty C_{AL,e}(t)dt$$

Thus the zeroth moment of the tracer in the liquid phase can be non-zero although the tracer was introduced in the gas phase. Some tracer may undergo mass transfer to liquid and appear in the liquid exit stream. These considerations have to be included in the analysis of tracer response in two-phase systems.

The two zeroth moments are related by an overall mass balance requirement, which then leads to

$$Q_G\mu_{0G} + Q_L\mu_{0L} = \mathcal{M}$$

where $\mathcal{M}$ is the quantity of tracer introduced in the inlet stream of the gas phase.

Detailed analysis was undertaken by Ramachandran and Smith (1979) who showed that the zeroth moment is a function of the dimensionless parameters κ_{gl} and Λ_{gl}. It also depends on whether two phases are flowing in a concurrent or countercurrent mode and what the level of mixing is in the liquid phase. Hence interpretation of tracer data in two-phase systems is prone to a number of assumptions and has to be handled with care. This could also possibly be the reason for the wide discrepancy seen in the literature for the reported dispersion coefficients even when experiments were done under very similar conditions.

Summary

- Mesoscopic models are useful when there is significant variation in concentration in one direction, the flow direction, while the variation in the direction normal to flow is relatively small. Examples where such models are used are tubular reactors, packed bed catalytic reactors, packed bed absorption columns, countercurrent extraction columns, chromatographic columns, and so on.

- In the mesoscopic model cross-flow concentration variations are not modeled and represented by average values. Note that two types of average concentrations arise upon the cross-sectional average of differential models: the flow-weighted average or the cup-mixing average and the cross-sectional average.

- The convection terms are dependent on the flow-weighted average while the generation (and accumulation) term is based on the cross-sectional average. This leads to a closure problem unless the two are nearly equal.

- The concept of plug flow is an idealized representation of the cross-direction concentration profile in the reactor. The assumption is that there is no concentration gradient in this direction, that is, the concentration is constant across any cross-section. Hence the cup-mixed value is the same as the cross-sectional average and no additional closure is needed.

- Plug flow is a reasonable assumption for a nearly flat velocity profile, moderate or slow rate of reaction, no significant radial temperature profile, and no mass transfer to the wall.

- Plug flow models can be corrected by introducing some dispersion leading to the axial-dispersed plug flow model. The concept of dispersion arises by connecting the cup mixing and the cross-sectional flux through a Fick's law type of model.

- The dispersion Peclet number is the important dimensionless group that determines the reactor performance in this model. This can be predicted from theory for simple flows such as laminar flow in long tubes as shown in the pioneering work of Taylor. In general this is a fitted parameter based on experimental data.

- Dispersion effects are important in many applications other than separator and reactor modeling, for example, transport of pollutants in the environment, pharmacokinetics, chromatography, and so on; dispersion effects are used in modeling such processes.

- Tracer experiments are widely used to characterize the non-ideal flow pattern in the reactors and the dispersion coefficient can be determined by matching the experimental tracer response with theory. The moments method is a simple and useful way of interpreting the data. In particular the dimensionless variance of the tracer (second central moment) can be related to the Peclet number.

- A significant extent of non-idealities can be caused by bypassing dead zones and recirculation zones in some types of reactors. For such cases, the dispersion model may not be suitable for interpretation of data. The variance can be larger than one in such cases and these are referred to as non-ideal ill-behaved systems. Phenomenological models linking the various flow regions (similar to compartmental models discussed in the last chapter) are more useful in such cases. The tracer response can be used to find the parameters needed in such models.

- Reactor behavior can be bracketed between plug flow and backmixed flow when the variance is less than one (non-ideal but well-behaved systems). The tanks in series model or the dispersion model can be used and provide similar predictions if the model parameter (number of tanks in one case, dispersion Peclet number in the other cases) is matched with the variance of the tracer response curve. However, reactor behavior can be smaller than even the backmix reactor for the non-ideal ill-behaved systems and therefore it cannot be bracketed between the two ideal flow patterns in such cases. An example is the fluid bed reactor.

- Two-phase systems can be modeled by using a dispersion parameter for each of these phases. An alternative way of modeling such systems is the tanks in series model. The dispersion effects in separators can be assessed by the MATLAB code shown in the text. For low dispersion values the phase may be assumed to be in plug flow and vice versa. This leads to the four ideal contacting patterns shown in Table 14.2 for two-phase systems.

- Tracer experiments in fluid–solid systems need to be interpreted by consideration of what is happening to the tracer in the solid phase. The internal diffusion within the particle can cause a spread in the tracer although the fluid phase itself may be in plug flow. Similarly the mean residence time is increased if the tracer adsorbs significantly in the particle. Hence the second moment of the tracer depends on many parameters (see Equation 14.49) in addition to the gas dispersion. The data interpretation needs an estimate of the adsorption constant, internal diffusion coefficient, and in some cases the fluid–solid mass transfer coefficient as well. Hence the measured data may not be representative of the true gas phase non-idealities.

- Tracer methods are still useful for systems with two flowing phases for determining the extent of backmixing in each of the phases. However, the nature of the tracer becomes important and the true extent of mixing can be determined only if the tracer stays in the same phase where it was injected. In other cases, the tracer response needs to be corrected for the mass transfer term to account for the interfacial mass transfer. Thus a part of the tracer introduced in the gas phase may appear in the liquid phase and vice versa. Interpretation of such tracer data requires an estimate of the mass transfer rate between the two phases and is therefore prone to inaccuracies.

Review Questions

14.1   What is meant by the plug flow assumption?

14.2   What closure is accomplished by using the concept of dispersion?

14.3   State conditions when the plug flow model may be reasonably used.

14.4   Can the conversion in a reactor be less than that predicted by the backmixed assumption?

14.5   What are Danckwerts boundary conditions?

14.6   State the difference between closed-closed and open-open boundaries.

14.7   Is the mean residence time equal to the first moment for an open-open boundary?

14.8   For low values of D_E state the expression for the dimensionless variance of a tracer curve.

14.9   Explain why the dispersion coefficient is inversely proportional to the molecular diffusion coefficient.

14.10   State an approximate formula that relates the dispersion Peclet number to the number of tanks in series needed to model the same reactor.

14.11   List different models for accounting for deviation from plug flow in a two-phase contactor.

14.12   Define the four dimensionless parameters needed to model a separator using a dispersion model for both phases.

14.13   Why is the interpretation data in a packed bed with an adsorbing tracer more difficult than that for a non-adsorbing tracer?

14.14   What are the additional problems in interpreting tracer data in systems with two flowing phases compared to single-phase flow?

Problems

14.1   **Plug flow reactor with variable velocity.** When the velocity is changing in the system, show that the plug model can be written as

$$-\frac{dN_{Az}}{dz} = k_1 < C_A >$$

Here N_A is the moles of A crossing at z per unit area per unit time. The concentration terms can be related to the total molar flux, $\tilde{V}_M N_A / N_t$, where $\tilde{V}_M$ is the local molar volume of the mixture. This sets up all the equations in terms of N_A, N_B, and so on, and accounts for the variation in velocity along the length. A thermodynamic equation of state is needed to calculate the molar volume given the mixture composition.

Apply this to the following problem of acetaldehyde decomposing to methane and CO in a plug flow reactor operated at 793 K and 101 kPa pressure. The reaction is irreversible but second order with a rate constant of 0.43 m^3/mol s; the feed to the reactor is 0.1 kg/sec. Simulate using ODE45 and plot the concentration profile of acetaldehyde along the reactor length.

14.2 **Analytical solution for a first-order reaction with dispersion.** Verify Equation 14.9 by finding the roots of the characteristic polynomials of the governing differential equation (Equation 14.6) and then using the Danckwerts boundary conditions.

14.3 **Conversion for a first-order reaction based on dispersion model.** A first-order reaction has rate constant of 3.33×10^{-3} s^{-1} and is carried out in a tubular reactor of length 3 m and a velocity of 1 m/s. The dispersion coefficient was measured by tracer experiments and found to have a value of 0.2 m^2/s. Find the conversion in the reactor. Determine the deviation from the plug flow value.

14.4 **From meso to macro.** Macroscopic models result from mesoscopic models by averaging over the reactor length. Start with the meso-model given by Equation 14.1, which is repeated here for ease of reference:

$$ -\langle v \rangle \frac{dC_{Ab}}{dz} = k_1 \langle C_A \rangle \tag{14.50} $$

Integrate both sides from 0 to L for z and show that a macroscopic model is obtained. State the definition of the rate term to be used in this (macro) model.

14.5 **Dispersion coefficient from tracer data.** Levenspiel and Smith (1957) injected a pulse of KMnO$_4$ into a tubular reactor 2.72 m long with a flow velocity of 35.7 cm/s. The tracer concentration measured at the exit was as follows:

Time	0	2	4	6	8	10	12	14	16	18	20
Concentration	0	11	53	64	58	48	39	29	22	16	11

Time	22	24	26	28	30	32	34	36	38	40
Concentration	9	7	5	4	2	2	2	1	1	1

Evaluate the moments of the above tracer data. From the second moment calculate the dispersion coefficient in the system.

14.6 **Dispersion model for laminar flow reactor.** Evaluate the performance of a laminar flow reactor using the dispersion model for the following conditions: first-order reaction with a rate constant of 0.1 s^{-1}, $L = 100$ cm, tube diameter = 1 cm, flow velocity = 0.2 cm/sec, and diffusivity = 2×10^{-9} m^2/s.

14.7 **Dispersion model for turbulent flow reactor.** Evaluate the performance of a reactor in turbulent flow and compare it with a plug flow reactor for the following conditions: first-order reaction with a rate constant of $5\,\mathrm{s}^{-1}$, L = 100 cm, tube diameter = 1 cm, flow velocity = 10 cm/sec, diffusivity = $2 \times 10^{-9}\,\mathrm{m}^2/\mathrm{s}$.

14.8 **Mesoscopic model for a heterogeneous reaction.** Consider a case where there is no homogeneous reaction and species A diffuses to the wall and undergoes a heterogeneous reaction. State the convection-diffusion model together with the appropriate boundary conditions. Determine a cross-sectional average of this model to provide a mesoscopic model for the system. State the closure relations needed for this case. Extend the analysis where both homogeneous and heterogeneous reactions of species A are taking place.

14.9 **Numerical solution of the transient dispersion model.** Determine the time domain solution for the transient dispersion model (Equation 14.11) using PDEPE. Note that the right-hand side of Equation 14.11 is to be rearranged to the derivative of a (combined) flux term. You may also want to compare your numerical solution with the analytical results shown in Equation 14.18.

14.10 **Laplace domain response.** Show all steps leading to Equation 14.15 and derive this equation. State the formula to find the moments of the response directly from the Laplace domain. Verify that the first moment is one and that the variance is given by Equation 14.17.

14.11 **Laplace domain response for an open-open system.** Consider the open-open case discussed in the text. State the proper boundary conditions to be used in this situation. Derive an expression for the response at $z = L$ in the Laplace domain: find the moments of the response and verify the results shown in the text.

14.12 **Time domain response for an open-open system.** Find the time domain solution for the open-open system and verify the following result by Levenspiel and Smith (1957):

$$\langle c \rangle_e = \frac{1}{2\sqrt{\pi t^*/Pe^*}} \exp\left[\frac{-(1 - t^*)^2}{4t^*/Pe^*}\right]$$

14.13 **Comparison with ideal reactors.** A first-order reaction is taking place in a reactor 2 m long with a rate constant of $2\,\mathrm{s}^{-1}$. The flow velocity is 5 cm/sec. A tracer experiment was performed and the variance was obtained as 40 sec². Find the following: dispersion coefficient in the reactor; the corresponding exit conversion; the conversion if the reactor were modeled as plug flow; the conversion if the reactor were modeled as a backmixed flow; the number of tanks that can be used to represent the system; the conversion based on the tanks in series model.

14.14 **Comparison of the mixing cell and the dispersion model.** Tracer experiments for a reactor showed that the dimensionless variance is equal to 4. A first-order reaction is taking place in this reactor with the Da parameter equal to two. If you were to model this by the dispersion model, determine the value of the dispersion Peclet number to be used. Find the conversion in the reactor. If the same reactor is to modeled by the tanks in series model, determine the number of tanks needed to provide the same exit age distribution. Find the conversion in the reactor based on the tanks in series model.

14.15 **Taylor dispersion in non-Newtonian flow.** The following equation can be derived for the velocity profile for a power law fluid under laminar flow conditions:

$$v_z = \frac{3n + 1}{n + 1} \langle v \rangle (1 - \xi^2)$$

with n being the power law index. Follow the approach given in Section 14.3 and derive an expression for the Taylor dispersion coefficient for a power law fluid.

14.16 **Taylor dispersion in channel flow.** The following equation can be derived for the velocity profile for a Newtonian law fluid under laminar flow conditions:

$$v_x = \frac{3}{2} \langle v \rangle (1 - \xi^2)$$

Here ξ is the dimensionless distance starting from the center of the channel. This is defined as y/H with H being half the channel width. Follow the approach given in Section 14.3 and derive an expression for the Taylor dispersion coefficient for channel flow given in the text.

14.17 **Dispersion coefficient in packed bed.** A packed bed contains 3 mm particled in a 2 cm diameter tube that is 1 m long. The bed porosity is estimated as 0.6. Find the dispersion coefficient for a superficial liquid velocity of 5 cm/s. Properties of the fluid may be assumed to be the same as water.

14.18 **Murphree efficiency for tray columns.** An application of the plug-backmixed model shown in Section 14.7.1 is to find the vapor tray efficiency in a distillation column. The schematic of this is based on Figure 14.9. Here a vapor is bubbling

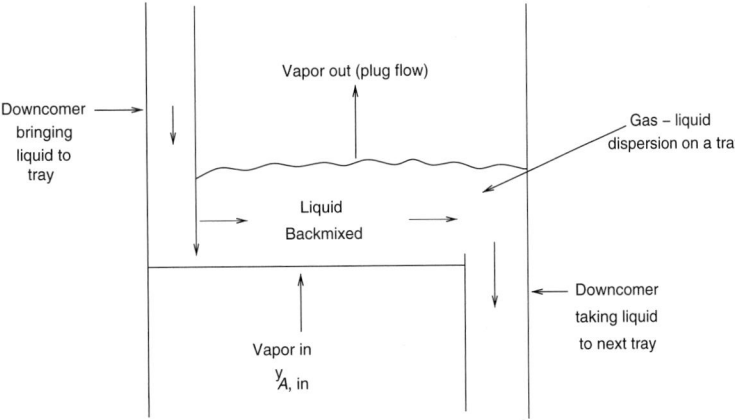

Figure 14.9 Schematic of a distillation tray for derivation of the Murphee vapor efficiency, from Seader et al. (2011).

through a pool of liquid and the vapor flow is assumed to be in plug flow. The liquid on the tray is assumed to be well mixed. The Murphree vapor efficiency is defined as

$$E_{MV} = \frac{y_{A,in} - y_{A,out}}{y_{A,in} - y_{A^*}}$$

where y_{A^*} is the vapor composition if the vapor were to achieve equilibrium with the exit liquid. Show that this can be expressed as

$$E_{MV} = \frac{1 - c_{AG,e}}{1 - c_{AL^*,e}}$$

in terms of the variables shown in Section 14.7.1.

Show that the Murphree efficiency in a tray column is equal to $\mathcal{E}$ directly by using Equation 14.35.

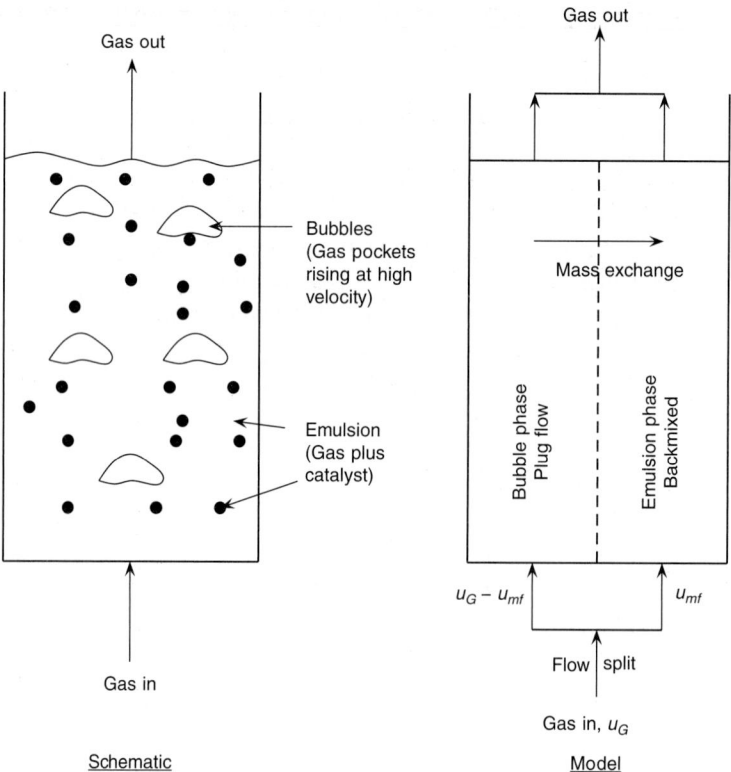

Figure 14.10   Schematic of a fluid bed reactor and its model representation as a two-phase plug-backmixed system.

The assumption of a liquid being backmixed is suitable for small-diameter columns. Corrections to account for the deviation of the liquid flow from complete backmixing have been also proposed in the distillation literature.

14.19 **A model for a fluid bed reactor.** A fluid bed reactor is modeled assuming that the bed consists of two phases: a bubble phase in plug flow and an emulsion phase in backmixed flow. Reaction takes place only in the emulsion phase since this is where most of the catalyst is present. Develop a model along the lines of the plug-backmixed two-phase flow shown in the text. You need to add a reaction term to the liquid (in this case the emulsion) balance. Find the exit conversion. Show that this can be lower than that in a CSTR. A schematic of the model is shown in Figure 14.10.

14.20 **Breakthrough time in an adsorption column.** An adsorption column is operating at a gas superficial velocity of 30 cm/s in a packed bed with a bed porosity of 0.5. The adsorption equilibrium constant is 5000. If a step tracer of an adsorbing gas is introduced into the system inlet, at what time will the tracer start to appear at the exit? Assume plug flow.

CHAPTER 15

Mass Transfer: Multicomponent Systems

Learning Objectives

After completing this chapter, you will be able to:

- State the Stefan-Maxwell constitutive model for multicomponent mass transfer.
- Develop a MATLAB implementation for multicomponent diffusion with reaction.
- Apply the model to heterogeneous reaction based on the film model.
- Use the model with appropriate determinacy condition for non-reacting systems.
- Understand the concept of the diffusivity matrix for multicomponent systems.

The goal of this chapter is to revisit the transport laws for multicomponent systems and to demonstrate tools by which multicomponent models are solved. In particular, the numerical solution of Stefan-Maxwell model is introduced and applied to many illustrative problems.

The set of problems examined is similar to those in Chapter 6 and later in 18 and show the multicomponent extension to these problems. First we show somewhat general code that couples and solves the Stefan-Maxwell model with the conservation equation. This code can be used for a variety of problems in both reacting and non-reacting systems. The use of this code for the analysis of diffusion with reaction in a porous catalyst is demonstrated and the results are presented for the concentration profile and the effectiveness factor for a four-component system with a simple reversible bimolecular reaction. This is an extension of the Fick's law–based approach to such problems studied in Section 18.3.

518 Chapter 15 Mass Transfer: Multicomponent Systems

The transport effects in a heterogeneous reaction at a catalytic surface are shown next. The film model is used, which is similar to that in Section 6.3 where the simpler concept of mass transfer coefficient was used. Here we demonstrate how the Stefan-Maxwell model can be used for transport in the film. A three-component system with a surface reaction is studied.

Examples of non-reacting systems are then discussed. The first example is evaporation of a volatile species in a gas mixture of two inert components, which makes it a ternary system. Analytical solutions can be derived for this case. We also demonstrate here that the Stefan-Maxwell equation can be approximated by the Wilke (1950) equation. Evaporation of two liquid components into an inert gas is studied next, followed by mass transfer in equimolar counterflow in a diffusion film; examples are shown where reverse diffusion and osmotic diffusion take place.

The last section introduces the concept of a generalized Fick's law and the use of a matrix of diffusion coefficients to characterize a multicomponent system. The inverted form of the Stefan-Maxwell equation leads to the definition of a multicomponent diffusion matrix. The advantages and disadvantages of this approach are pointed out.

15.1 Constitutive Model for Multicomponent Transport

In this section we revisit the Stefan-Maxwell model as a constitutive model for multicomponent mass transfer. We start with a binary diffusion first to show the connection of the inverted form of the Fick's law with the Stefan-Maxwell model for multicomponent diffusion.

15.1.1 Binary Revisited

It is useful to recall some relations for binary diffusion that follow directly from Fick's law. In particular, the difference in velocity between species A and B can be shown to be equal to

$$\boldsymbol{v}_A - \boldsymbol{v}_B = -D_{AB}\nabla \ln(y_A/y_B) \tag{15.1}$$

The velocity can be either with reference to a stationary frame or a moving frame since the difference of two velocity values rather than individual species velocity appears on the left-hand side of the above equation. The y is the mole fraction of the subscripted species.

The flux of A (in a stationary frame of reference) is related to the velocity of A by the relation $N_A = Cy_Av_A$; a similar relation for N_B holds as well. Equation 15.1 now reduces to the following equation with minor rearrangement:

$$-\nabla y_A = \frac{y_B N_A - y_A N_B}{C D_{AB}} \tag{15.2}$$

This is the same as Fick's law but the form is inverted; that is, the mole fraction of a species (A or B) is now given as a combination of the fluxes of the two species.

The following equation (in terms of the diffusive fluxes) holds as well since the velocity difference is the same with respect to a moving frame as well:

$$-\nabla y_A = \frac{y_B J_A - y_A J_B}{C D_{AB}} \tag{15.3}$$

This is again an inverted form with the mole fraction term appearing explicitly.

15.1.2 Generalization: The Stefan-Maxwell Model

Generalization of the above equation of writing the diffusion model for a binary diffusion in inverted form leads to the Stefan-Maxwell model for multicomponent systems:

$$\boxed{-\nabla y_A = \sum_{j=1}^{ns} \frac{y_j \mathbf{N}_A - y_A \mathbf{N}_j}{C D_{Aj}}} \tag{15.4}$$

where ns is the number of components and D_{ij} is the binary pair diffusivity for the i-j pair. The summation term is zero if $j = A$ and is not needed.

Note that the equation can also be written in terms of the diffusive fluxes as

$$-\nabla y_A = \sum_{j=1}^{ns} \frac{y_j \mathbf{J}_A - y_A \mathbf{J}_j}{C D_{Aj}} \tag{15.5}$$

You should verify that both forms are equivalent. For problem-solving purposes, the form in terms of the combined flux (Equation 15.4) is more useful. The constitutive equations are supplemented with the species mass balance equations, which have the same form as in Section 5.1. Thus for a steady state case we have

$$\boxed{\nabla \cdot \mathbf{N}_A = R_A} \tag{15.6}$$

The simultaneous solution to the two boxed equations, 15.4 and 15.6, will be our focus now and we shall present the application to illustrative multicomponent mass transfer problems.

15.2   Computations for a Reacting System

The principle difficulty, compared to binary problems, is that the equations are flux implicit and hence the species balance equation (Equation 15.6) cannot be directly expressed in terms of species mole fractions (or species concentrations). Hence Equation 15.6 must be solved simultaneously with Equation 15.4. Many methods have been developed for this mainly based on matrix manipulations and Taylor and Krishna (1993) provide considerable detail on this. Additional information is also presented by Ramachandran (2013). Here we follow simple numerically based tools and focus on applications to practical problems. The BVP4C method is used directly to solve these equations. MATLAB code is shown and used in several applications.

Consider the reaction scheme taking place in a porous catalyst:

$$A + B \rightleftharpoons C + D$$

Let the rate model be given as

$$-R_A = k_f C_A C_B - k_b C_C C_D$$

This can be used in the conservation law (Equation 15.6) for species A to D. For example, using this for A we have

$$\frac{dN_A}{dz} = -(k_f C_A C_B - k_b C_C C_D) \tag{15.7}$$

This generates four equations. Note that there is an invariance property due to the fact that the rate term is common to all four equations. Hence the fluxes are tied together by this invariant relation. (Hence there is only one independent value for the flux.) However, for numerical solution and for ease of code setup, we retain all four equations.

Additional equations arise by using the Stefan-Maxwell flux expressions (Equation 15.4) for each of the components. For example A we have

$$\frac{dy_A}{dz} = \sum_{j=1}^{ns} \frac{y_A \mathbf{N}_j - y_j \mathbf{N}_A}{C D_{Aj}}$$

Similar equations apply for species B, C, and D. Thus we have $2 * ns$ equations, eight for a four species system. Eight boundary conditions are to be specified at the two end points of the domain, 0 to L. These are problem specific and the examples provided illustrate how these may be specified. For example, if this reaction is taking place in a porous catalyst that is modeled as a slab of half-thickness L then the following boundary conditions can be applied:

- The flux is zero at $x = 0$ and hence N_A, N_B, and so on, are set as zero.
- The concentrations are specified at $x = L$ and hence the mole fractions y_A and so on are specified.

Hence we generate eight boundary conditions by specifying either the flux or mole fractions.

The code in Listing 15.1 is somewhat general and useful for a variety of multicomponent diffusion problems with and without reaction. The main driver is shown first. The code uses CHEBFUN as a wrapper for the BVP4C code.

Listing 15.1   Program to Solve Multicomponent Diffusion with Simultaneous Chemical Reactions

```
global ns D Da K
ns = 4;
D =[  1.0000     0.7319     0.7188     0.7141
      0.7319     1.0000     0.2289     0.2138
      0.7188     0.2289     1.0000     0.1633
      0.7141     0.2138     0.1633     1.0000];
Da = 9; K = 5;
global db
db = D;
global ys
%% solver
length = 1. ; % dimensionless
ys = [ 0.33333  0.66667     0 0 ] % surface concentrations
d = [0,length];
one = chebfun(1,d);
 eta = chebfun ( 'eta', d) ;
  % initial guess of rate
  rate = reactionrate (ys);
initguess = [ys(1)*one ys(2)*one ys(3)*one ys(4)*one ...
   rate*one rate*one —rate*one —rate*one];
%%%%% call bvp4c now
sol = bvp4c (@odes, @bcs, initguess) % bvp solved
y = sol(:,1); % mole fraction of species one
plot(sol(:,1:2))
% effectiveness factor
```

```
rate_s = reactionrate (ys);
eta1 = sol(0,5)/rate_s  % effectineess factor
%%%%
```

The dimensionless representation is used here with the distance being scaled by L. The flux is scaled as

$$N_A^* = \frac{N_A}{D_{ref}C/L}$$

A similar definition holds for B, C, and D. Here C (in the previous equation) is the total concentration in the bulk gas and D_{ref} is the diffusivity of a chosen species A. All other binary diffusivity is scaled by this value entered as a matrix D.

The set of functions to be solved is then specified in an m file denoted as *odes* here (Listing 15.2). The first four derivatives are for $dy_i/d\xi$ and the next four are for $dN_i^*/d\xi$ with i varying from 1 to 4 to cover the four components.

Listing 15.2   Code Defining the Differential Equations

```
%
function dydx = odes ( x, y )
global dk db
global ns
rate = reactionrate(y);
    n1 = y(5); n2= y(6) ; n3= y(7); n4 = y(8);
    % na = y(3) ; nb= y(4) ; nc= -(na); y3= 1.-y(1) - y(2);
    bulkterm = zeros (ns,1);
    % species 1.
    T12   = y(2) * n1 /db(1,2) - y(1) * n2 /db(1,2) ;
    T13   = y(3) * n1 /db(1,3) - y(1) * n3 /db(1,3) ;
    T14   = y(4) * n1 /db(1,4) - y(1) * n4 /db(1,4) ;
    % species 2
    T21   = y(1) * n2 /db(2,1) - y(2) * n1 /db(2,1) ;
    T23   = y(3) * n2 /db(2,3) - y(2) * n3 /db(2,1) ;
    T24   = y(4) * n2 /db(2,4) - y(2) * n4 /db(2,4) ;
    % T2% species 3
    T31   = y(1) * n3 /db(3,1) - y(3) * n1 /db(3,1) ;
    T32   = y(2) * n3 /db(3,2) - y(3) * n2 /db(3,2) ;
    T34   = y(3) * n3 /db(3,4) - y(3) * n4 /db(3,4) ;
    % species 4
    T41   = y(1) * n4 /db(4,1) - y(4) * n1 /db(4,1) ;
    T42   = y(2) * n4 /db(4,2) - y(4) * n2 /db(4,2) ;
    T43   = y(3) * n4 /db(4,3) - y(4) * n3 /db(4,3) ;
    % dk is the Knudsen diffusion term.
```

```
dydx = [− y(5) /dk(1)−T12−T13−T14
          − y(6)/dk(2)−T21−T23−T24
           −y(7)/dk(3)−T31−T32−T34
           −y(8)/dk(4)−T41−T42−T43
        −rate
        −rate
        rate
        rate ] ;
```

Boundary Conditions

The boundary conditions are set as the mole fraction values at the surface and zero flux at the pore end. Dirichlet is applied at $\xi = 1$, the pore mouth, and Neumann is applied at the pore end, $\xi = 0$. The code for the boundary conditions is as follows:

Listing 15.3   Code Defining the Boundary Conditions

```
function res = bcs ( ya , yb)
global ys
res = [ ya(5)
        ya(6)
        ya(7)
        ya(8)
        yb(1)− ys(1)
        yb(2)− ys(2)
        yb(3)− ys(3)
        yb(4)− ys(4) ] ;
```

The kinetic model to define the rate term is set as a separate subroutine to provide code flexibility. This will be zero for the non-reacting systems considered in Section 15.4. The code for simple bimolecular reverse reaction model is presented in the following:

Listing 15.4   Code Defining Simple Bimolecular Reverse Reaction Model

```
%
function ratemodel = reactionrate(y)
global ys
global ns D Da K
 ratemodel = Da * (y(1)*y(2) − y(3)* y(4)/K)
```

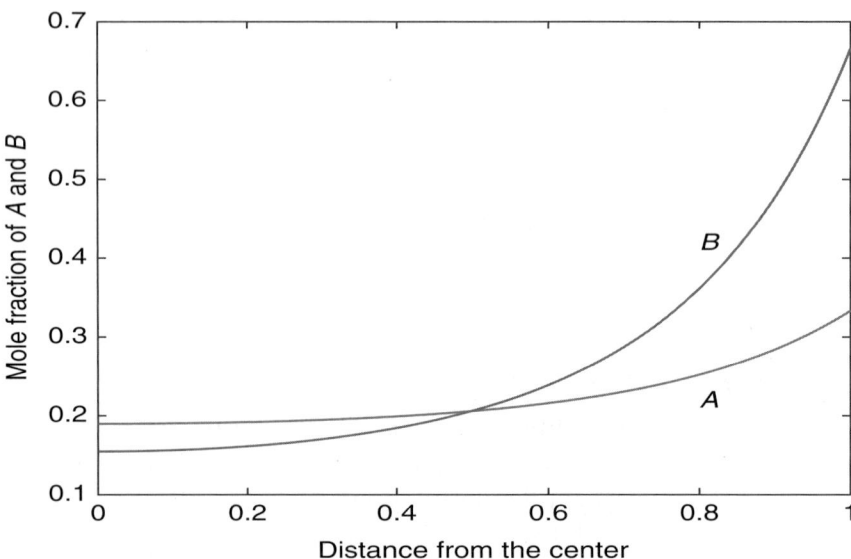

Figure 15.1 Composition profiles of the example studied in this section. The parameters align with those in Listing 15.1.

Illustrative results are shown in Figure 15.1. The dimensionless parameters are $k_f^* = 5$ and $K = k_f/k_b = 5$. The bulk mole fractions are $y_A = 1/3$, $y_B = 2/3$. The effectiveness factor is calculated as 0.21. Species A is taken as the faster diffusing species here. We note that although the surface mole fraction of B is higher than A, the drop in concentration of B is much more than that of A, since B diffuses much slower than A.

It is fairly easy to modify and apply this code for multiple reactions. The rate of reaction can be calculated by using the stoichiometric matrix and a rate function for each reaction. Thus the rate of production for species i is given by the summation of its rate of production from all the reactions (subscript r) as

$$R_i = \sum_r^{nr} \nu_{ri} \mathcal{R}_r$$

where nr is the number of reactions, ν_{ri} is the stoichiometry of the ith species in the rth reaction and $\mathcal{R}_r$ is the rate term for the rth reaction (which represents the molar rate of production of any species with unit stoichiometry). The rate term is now introduced into the species balance equations:

$$\frac{dN_i}{dx} = \sum_r^{nr} \nu_{ri} \mathcal{R}_r \tag{15.8}$$

Equations 15.4, the constitutive model and Equation 15.8, the conservation model, have to be integrated together now. There are two equations for each species (one for the conservation law and the other for the constitutive model) and hence there are $2 * ns$-first order differential equations to be solved simultaneously. These are boundary value types of problems.

Note that the number of equations can be simplified using the rank of the stoichiometric matrix; this is left as an exercise. For example, if there is a single reaction then there is only one independent flux, usually the flux based on a key component. The flux of all other species is determined simply by the stoichiometry of the reaction. Here we keep all $2 * ns$ equations for simplicity. This takes up slightly more computational time but keeps the implementation to new problems somewhat simpler.

For porous catalysts both Knudsen and bulk diffusion may be important. In such cases the contributions are added together and the Stefan-Maxwell model is modified as

$$-\frac{dC_i}{dx} = \frac{N_i}{D_{Ke,i}} + \sum_{j=1}^{n} \frac{y_j N_i - y_i N_j}{D_{ij,e}} \tag{15.9}$$

Here $D_{Ke,i}$ is the effective Knudsen diffusivity of i while $D_{ij,e}$ is the effective binary pair gas phase difficulty (D corrected with the porosity and tortuosity factors).

The relative magnitude of the two terms in Equation 15.9 determines the nature of the solution method. If Knudsen dominates the fluxes are uncoupled while if the bulk gas diffusion dominates we expect multicomponent coupling effects, especially if the molecular weights of the diffusing species are vastly different. Uncoupled problems can be reduced to a set of second-order differential equations in concentrations and solved directly. Coupled problems are best solved by simultaneous computations of fluxes and concentrations by the code shown in Listing 15.1.

15.3 Heterogeneous Reactions

Illustrations of computations to a heterogeneous reaction are presented in this section. Consider a case where the reaction

$$A \rightarrow B + C$$

is taking place over a catalyst surface.

Then by stoichiometry the following conditions apply for the fluxes:

$$N_B = -N_A \text{ and } N_C = -N_A$$

Thus N_B and N_C can be eliminated from the Stefan-Maxwell equations. In addition the mole fraction of one of the species, say, C, can also be eliminated using

$$y_C = 1 - y_A - y_B$$

The Stefan-Maxwell equation for A reduces to the following equation when these conditions are applied:

$$\frac{dy_A}{dz} = -\frac{y_A N_A}{CD_{AB}} + y_B N_A \left(\frac{1}{CD_{AC}} - \frac{1}{CD_{AB}} \right) - \frac{N_A}{CD_{AC}} \qquad (15.10)$$

The equation for B becomes

$$\frac{dy_B}{dz} = y_A N_A \left(\frac{1}{CD_{AB}} - \frac{1}{CD_{BC}} \right)$$
$$+ y_B N_A \left(\frac{1}{CD_{BA}} - \frac{2}{CD_{BC}} \right) + \frac{N_A}{CD_{BC}} \qquad (15.11)$$

The condition that no homogeneous reaction takes place provides the third equation:

$$\frac{dN_A}{dz} = 0 \qquad (15.12)$$

Thus we track three variables simultaneously: y_A, y_B, and N_A. Three boundary conditions are needed to solve the problem. The gas composition in the bulk gas is specified, which provides two conditions. The additional condition to be imposed is based on the kinetics of the heterogeneous surface taking place at the surface and is discussed in the following.

- For fast reactions the Dirichlet condition is applied at the reacting surface. The mole fraction of A at the surface is now set as zero, which provides the third boundary condition. These equations can be used to solve for N_A and also for $y_{B,L}$, the mole fraction of B at the reacting surface.
- For a finite reaction case, the balance of flux of A at the reacting surface leads to the following boundary condition at the surface:

$$N_A(z = L) = -R_A = k_s C y_{As}$$

Note that the last term is specific for a first-order reaction at the catalyst surface. ($z = L$ is the location of the reacting surface.) Thus a Robin type of boundary condition is now specified at the reaction surface.

This completes the problem specification; the system equations can be implemented in BVP4C and solved. Example 15.1 provides an illustration.

Example 15.1   Ternary Diffusion with a Heteogeneous Reaction

Vapor phase dehydrogenation of ethanol (designated as A or 1) to produce acetaldehyde (2) and hydrogen (3) was studied by Froment, Bischoff, and de Wilde (2011) and is an example of a system of the preceding type where a heterogeneous catalytic reaction takes place on a surface:

$$\text{ethanol} \rightarrow \text{acetaldehyde} + \text{hydrogen}$$

The kinetics of the heterogeneous reaction was modeled as

$$\text{Rate} = -k_s y_A \text{ with } k_s \text{ equal to } 10.0 \text{ mol/m}^2 \text{ s mole fraction}$$

The bulk gas phase concentration was $y_1 = 0.6$, $y_2 = 0.2$, and $y_3 = 0.2$. Find the rate of reaction assuming a mass transfer film thickness of 1 mm. Temperature = 548 K and pressure = 101.3 kPa. The binary diffusion coefficients are $D_{12} = 7.2 \times 10^{-5}$ m^2/s, $D_{13} = 23.0 \times 10^{-5}$ m^2/s, and $D_{23} = 23.0 \times 10^{-5}$ m^2/s.

Solution

The Stefan-Maxwell model is written for each of the species, resulting in three differential equations. These are simplified using the determinancy condition and the constraint on the mole fraction to two equations. The resulting equations are linear and hence they can be set up in matrix form so that the solution can be written, *in principle,* in terms of the *expm* function. Finally the boundary conditions at the surface are applied to close the problem and the resulting nonlinear algebraic equations are solved by FSOLVE or similar routines. The unknown variables are the mole fractions of species 1 and 2 at the catalytic surface.

An alternative is to integrate the system of equations by using the BVP4C solver in MATLAB. The code prsented in Listing 15.1 can be easily adapted for this case.

The following values of the mole fraction at the catalytic surface should be obtained upon using either of these procedures:

$$y_1 = 0.0592, \ y_2 = 0.6055, \text{ and } y_3 = 0.3353$$

Note the low value of y_1 (compared to bulk) indicates significant mass transport resistance. The flux of A is computed from the code as 0.9473 mol/m^{2}s.

15.4 Non-Reacting Systems

Application to non-reacting systems based on the film theory of mass transfer is discussed in this section. The first example is the multicomponent version of the evaporation of liquid in a container. The binary version was studied in Section 6.2.

15.4.1 Evaporation of a Liquid in a Ternary Mixture

Consider a liquid evaporating into an open space in a tube. The system analyzed in Section 6.2 was a binary mixture; A was the evaporating species while B was the inert gas into which A was evaporating. The use of a determinancy condition led to the following equation for the flux of A:

$$N_A = -\frac{D_{AB}C}{1 - y_A}\nabla y_A \tag{15.13}$$

This, integrated across the system, provided an expression for the flux (Equation 6.18). In this section we extend the model to a ternary system with A evaporating into an inert gas mixture of B and C.

One approach is to use the pseudo-binary diffusivity for A, D_{A-m}, which can be calculated from the Wilke equation shown in Section 5.3. The expression for the flux then has a similar form as that for a binary:

$$N_A = -\frac{D_{A-m}C}{1 - y_A}\nabla y_A \tag{15.14}$$

Let us explore this in connection with the Stefan-Maxwell model first, where we show that for this specific problem it is an exact result that can be derived from the Stefan-Maxwell model. However, we will find that a unique value for D_{A-m} cannot be assigned immediately. A value based on some average composition has to be calculated and used.

Here we solve the Stefan-Maxwell model directly without having to use the pseudo-binary approach. First we demonstrate that the Wilke equation results from simplification of the Stefan-Maxwell equation.

Basis for Wilke Equation

The Stefan-Maxwell model transport of species A in a ternary mixture is

$$-\nabla y_A = \frac{y_B \boldsymbol{N}_A - y_A \boldsymbol{N}_B}{CD_{AB}} + \frac{y_C \boldsymbol{N}_A - y_A \boldsymbol{N}_c}{CD_{AC}} \tag{15.15}$$

If we have UMD then N_B and N_C are zero, which provides the determinacy condition. These are now used in the above equation, which simplifies to

$$-\nabla y_A = \frac{y_B \mathbf{N}_A}{C D_{AB}} + \frac{y_C \mathbf{N}_A}{C D_{AC}} \tag{15.16}$$

Rearranging this we have the following expression for the flux:

$$N_A = -C \nabla y_A \left[\frac{y_B}{D_{AB}} + \frac{y_C}{D_{AC}} \right]^{-1} \tag{15.17}$$

The combined flux above is the sum of the diffusion and convection flux:

$$N_A = J_A + y_A N_t$$

For the system being studied, $N_t = N_A$. Also, expressing the diffusion in terms of a pseudo-binary diffusivity of A in the mixture, D_{A-m}, we have

$$N_A = -C D_{A-m} \nabla y_A + y_A N_t \tag{15.18}$$

Comparing this expression with that defined using the pseudo-binary diffusion coefficient (Equation 5.23 in Chapter 5), we find the pseudo-binary is defined according to the Wilke equation as

$$\frac{1 - y_A}{D_{A-m}} = \frac{y_B}{D_{AB}} + \frac{y_C}{D_{AC}} \tag{15.19}$$

This is the basis for the Wilke equation. It therefore applies strictly speaking when only one species (A) is diffusing in a ternary mixture. The other two species have zero fluxes. Note that this is used as an approximation often even when these conditions do not hold. Hence the flux is given as

$$N_A = -\frac{D_{A-m} C}{1 - y_A} \nabla y_A \tag{15.20}$$

The advantage of this over the Stefan-Maxwell model is that it is flux explicit and similar to that for a binary case. The disadvantage is that D_{A-m} is mole fraction dependent (as can be seen from Equation 15.19) and hence varies along the diffusion or reaction path and has to be based on some representative mole fraction in the system. We do not know the mole fraction profiles *a priori*. The mole fractions are known in the bulk gas and can be used to find the pseudo-binary diffusivity as an approximation.

As an alternative we can track the concentration profiles for B and C, in other words solve the Stefan-Maxwell formulation directly. This approach is rather general and explored further in this section.

Evaporation in Ternary: Stefan-Maxwell Model

Let us formulate two differential equations for B and C using the Stefan-Maxwell formulation. We can use these conditions as simplifications:

$$y_A = 1 - y_B - y_C$$

Also $N_B = 0$ and $N_C = 0$ (UMD).

With these substitutions the flux expressions for B and C are as follows:

$$\frac{dy_B}{dz} = \frac{N_A}{CD_{AB}} y_B$$

and

$$\frac{dy_C}{dz} = \frac{N_A}{CD_{AC}} y_C$$

Each equation can be integrated separately here since N_A is not a function of z. Note that this is not always true, for example, if a simultaneous homogeneous reaction of A is also taking place.

Integration with the boundary condition at $z = L$ (the bulk gas) gives

$$y_B = y_{BL} \exp[-N^*(1 - z/L)]$$

and

$$y_C = y_{CL} \exp[-N^*\beta(1 - z/L)]$$

where N^* is a dimensionless flux defined as

$$N^* = \frac{N_A}{(CD_{AB}/L)}$$

and β is the diffusivity ratio defined as D_{AB}/D_{AC}.

Finally, applying the boundary condition at the surface, $z = 0$, (where the mole fraction of A is known by equilibrium considerations) we have

$$y_{As} = 1 - y_{BL} \exp(-N^*) - y_{CL} \exp(-N^*\beta) \tag{15.21}$$

which provides an implicit equation for find N^*.

We also note that the binary expression is recovered if $\beta = 1$, that is, when the binary diffusivity of A in B is the same as that for A in C, $D_{AB} = D_{AC}$. The effect of multicomponent interaction therefore depends on the β parameter.

15.4.2 Evaporation of a Binary Liquid Mixture

Here we consider a similar problem with now the liquid containing a mixture of two species, A and B. This mixture is evaporating into a inert gas C. The problem is now to compute both N_A and N_B.

Since C is inert we can set $N_C = 0$, which is the required determinacy condition. Also, since the mole fractions add up to unity, we have $y_C = 1 - y_A - y_B$.

When these simplifications are introduced into the Stefan-Maxwell equations, we obtain the following equations for the mole fraction profiles of A and B:

$$\frac{dy_A}{dz} = y_A \left(\frac{N_B}{CD_{AB}} + \frac{N_A}{CD_{AC}} \right) + y_B \left(\frac{N_A}{CD_{AC}} - \frac{N_A}{CD_{AB}} \right) - \frac{N_A}{CD_{AC}} \quad (15.22)$$

and

$$\frac{dy_B}{dz} = y_A \left(\frac{N_B}{CD_{BC}} - \frac{N_B}{CD_{AB}} \right) + y_B \left(\frac{N_A}{CD_{AB}} + \frac{N_B}{CD_{BC}} \right) - \frac{N_B}{CD_{BC}} \quad (15.23)$$

The boundary conditions needed are as follows: at the evaporating surface, $z = 0$, the mole fractions are related to the vapor pressure values (assuming an ideal liquid), while at the bulk gas, $z = L$, the mole fractions are specified from the bulk gas values. Often a value of zero is used as a simplification.

The above differential equations are first order and look like initial value problems (IVPs); they are candidates for explicit marching in z or solvers such as ODE45. However, the fluxes are unknown and have to be computed by imposing the boundary condition at $x = L$. Hence this is a boundary value problem. We discuss two methods for solution of this problem. First is the direct use of BVP4C, which is similar to that discussed in Section 15.2. The second is based on matrix algebra, which is discussed next.

Matrix Method

Equations 15.22 and 15.23 can be expressed in matrix form as

$$\frac{d\boldsymbol{y}}{dz} = \tilde{A}\boldsymbol{y} + \boldsymbol{R}$$

where y is the solution vector; in the present case the solution vector has a dimension of two, representing y_A and y_B.

The term $\tilde{A}$ above is the coefficient matrix and $\boldsymbol{R}$ is the non-homogeneous term. Since both the coefficient matrix and non-homogeneous term are constants, the solution can be represented in matrix form using the exponential matrix concept:

$$\boldsymbol{y} = (\boldsymbol{y}_0 + \tilde{A}^{-1}\boldsymbol{R})\mathrm{expm}(Az) - \tilde{A}^{-1}\boldsymbol{R}$$

where $\boldsymbol{y}_0$ is the mole fraction at the evaporating surface.

Now using the conditions at the bulk gas the following algebraic relation is obtained:

$$\boldsymbol{y}_L = (\boldsymbol{y}_0 + \tilde{A}^{-1}\boldsymbol{R})\mathrm{expm}(\tilde{A}L) - \tilde{A}^{-1}\boldsymbol{R} \qquad (15.24)$$

This equation contains the unknowns N_A and N_B, which are buried in the $\tilde{A}$ matrix. The solution of these "implicit" algebraic equations gives the fluxes. This can be accomplished, for instance, by using a nonlinear algebraic solver such as FSOLVE in MATLAB. Example 15.2 provides an illustration for evaporation rate in a binary mixture.

Example 15.2   Evaporation Rate in a Liquid Mixture

The problem analyzed here is an Arnold cell containing a binary mixture of acetone (1) and methanol (2). The mixture is evaporating into an inert gas, say air (3). The pressure and temperature are maintained at 99.4 kPa and 328.5 K respectively. The gas composition at the liquid surface is $y_1 = 0.319$ and $y_2 = 0.528$. The height of the Arnold cell was 0.238 m and composition of both vapor species (1) and (2) at this end was zero. Find the evaporation rate and sketch the mole fraction profiles in the cell. This problem is adapted from Taylor and Krishna (1993). The binary diffusion coefficients are $D_{12} = 8.48 \times 10^{-6}$ m^2/s, $D_{13} = 13.72 \times 10^{-6}$ m^2/s, and $D_{23} = 19.91 \times 10^{-6}$ m^2/s.

Solution

The problem was solved by both the matrix method using the MATLAB FSOLVE function for the solution of the algebric equation and by the boundary value method using the BVP4C solver in MATLAB. The results for both cases agree and are as follows:

$$N_1 = 1.78 \times 10^{-3} \mathrm{mol/m}^2.\mathrm{s}$$
$$N_2 = 3.13 \times 10^{-3} \mathrm{mol/m}^2.\mathrm{s}$$

The concentration profiles are sketched in Figure 15.2. These profiles are in good agreement with the experimental data shown in Taylor and Krishna as well. Students should verify these results by using the code in Listing 15.1.

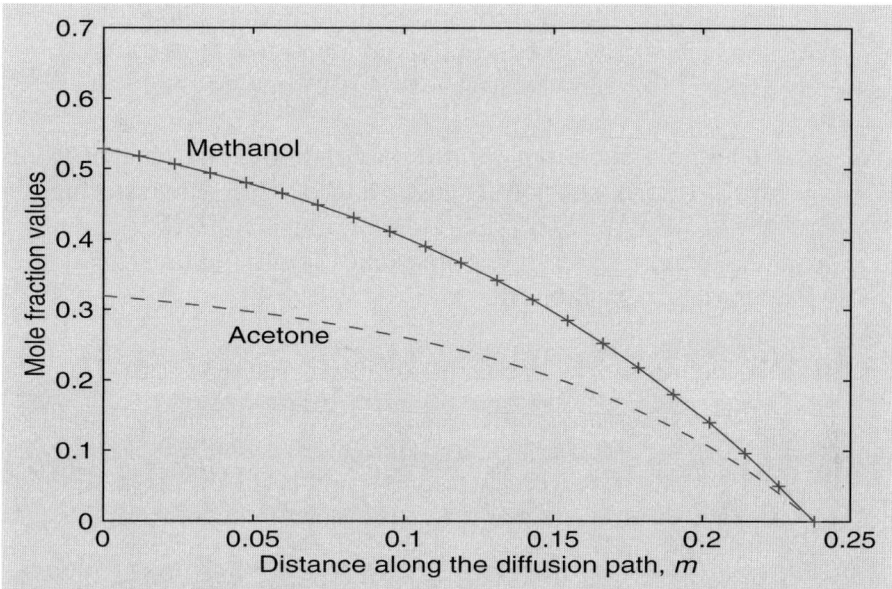

Figure 15.2 Composition profiles in an Arnold type of cell for evaporation of a binary liquid mixture.

15.4.3 Equimolar Counter-Diffusion

This section discuses the use of the equimolar counter-diffusion condition as the determinancy condition to relate the fluxes.

An example of equimolar counter-diffusion is found in the modeling of diffusion in a porous plug separating two regions of different gas compositions. The binary case was examined in Example 5.2 and exercise problem 4 in Chapter 6. The compositions at the two ends are specified and the fluxes of the various species are to computed. The determinacy condition used here is that there is equimolar counter-diffusion.

A second example is in distillation of a ternary mixture. Since distillation is usually under adiabatic conditions, we assume equimolar diffusion. This assumes that the molar heat of vaporization of the three species is nearly equal. Then the total flux across the system is zero in order to maintain the heat balance and the problem reduces to equimolaar counter-diffusion.

The Stefan-Maxwell model is simplified by using the following conditions. The total flux is zero, which provides the determinancy $N_t = 0$, or

$$N_C = -N_A - N_B$$

The second simplification to be used in the Stefan-Maxwell model is that mole fractions have to add up to unity and hence

$$y_C = 1 - y_A - y_B$$

The mole fraction y_C and flux N_C are then eliminated from the Stefan-Maxwell equations for both dy_A/dz and dy_B/dz. The resulting equations can be solved by the matrix method (followed by FSOLVE) or by the BVP4C routine. Example 15.3 shows an application to the flux calculation in the distillation of ternary mixture.

Example 15.3   Distillation of a Ternary Mixture

Estimate the mass transfer rates during distillation of ethanol (species 1), t-butyl alcohol (species 2), and water (species 3) at a point in the column where the following conditions are specified: bulk vapor composition: $y_{10} = 0.6$, $y_{20} = 0.13$; interface vapor composition: $y_{1\delta} = 0.5$, $y_{2\delta} = 0.14$; film thickness $\delta = 1$ mm; The following values may be used for the three binary pair diffusion coefficients: $D_{12} = 8.0 \times 10^{-6}$ m^2/s $D_{13} = 21.0 \times 10^{-6}$ m^2/s, and $D_{23} = 17.0 \times 10^{-6}$ m^2/s.

Solution

The number of species is three and hence six differential equations can be set up, three from conservation law and three from constitutive equations. The mole fractions are specified at both ends and therefore we have six conditions. The problem specification is complete and the BVP4C code (Listing 15.1) can be modified and used. The following values were obtained for the fluxes:

$$N_1 = 8.09 \times 10^{-2} \text{mol/m}^2.\text{s}$$
$$N_2 = 0.434 \times 10^{-2} \text{mol/m}^2.\text{s}$$
$$N_3 = -8.52 \times 10^{-2} \text{mol/m}^2.\text{s}$$

It may be noted that the number of equations to be solved can be reduced to four by using the sum of the mole fraction requirement and the determinacy condition. Computations are not intense even with six equations and hence this simplification is optional.

Some interesting results pertinent to diffusional interaction effects are now available. Note that component 2 diffuses against its concentration gradient, an example of reverse diffusion. The mole fraction is 0.13 on one side and 0.14 on the other side, as specified in the problem, but the flux is positive!

The conditions simulated leading to reverse diffusion are shown in Figure 15.3 for t-butyl alcohol (species 2). Species 2 moves from left to right while Fick's law would indicate it should move the other way.

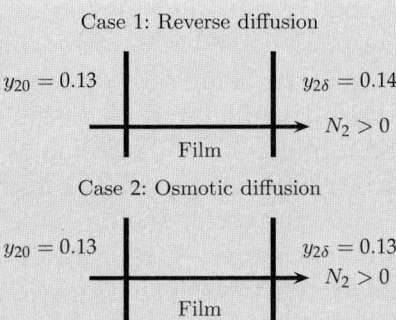

Figure 15.3 Illustration of reverse and osmotic diffusion for t-butyl alcohol (species 2) for the conditions studied in Example 15.3.

We now change the problem specification slightly to indicate the phenomena of osmotic diffusion (Case 2). We set $y_{2\delta} = 0.13$, adjusting the mole fraction of C. The following values can now be computed and students should verify these by running the code:

$$N_1 = 8.30 \times 10^{-2} \text{mol/m}^2.\text{s}$$
$$N_2 = 0.79 \times 10^{-2} \text{mol/m}^2.\text{s}$$
$$N_3 = 9.10 \times 10^{-2} \text{mol/m}^2 \text{ s}$$

Species 2 diffuses to the liquid even though its own mole fraction is the same both in the vapor and the interface, which is an example of "osmotic" diffusion. The effect is due to the multicomponent diffusion coupling of the transport of three species.

The conditions simulated to leading to osmotic diffusion are also shown in Figure 15.3. Species 2 moves from left to right while Fick's law would indicate it should have no flux.

The major flux components are due to ethanol (1) and water (3) and these are in opposite directions. These are much larger (in magnitude) than the t-butyl alcohol flux. The flux of t-butyl alcohol varies to satisfy the equimolar constraints and therefore shows unusual effects.

15.5 Multicomponent Diffusivity Matrix

In this section we discuss an inverted formulation of the Stefan-Maxwell model that has been used in many earlier studies (e.g., Toor, 1957; Stewart and

Prober, 1964). Although the direct numerical solution of the Stefan-Maxwell model is efficient these days due to many computational tools, the inverted form has many useful pedagogical values and is also useful if one wishes to follow some of the earlier literature in this field.

An inverted formulation of the Stefan-Maxwell model leads to a generalized Fick's law for diffusion fluxes. The form of the equation is similar to a binary mixture and is the vector analog of the law. A matrix is introduced for the diffusivities in this model.

We consider a ternary system for simplicity. Extension to quaternary systems and so on is straightforward but lengthy.

The idea is to write the diffusion flux explicitly. Since the diffusion flux of species A (denoted as 1 here) may depend on the mole fraction gradient of both A (1) and B (2), it is written as

$$J_1 = -C\mathcal{D}_{11}\nabla y_1 - C\mathcal{D}_{12}\nabla y_2$$

Here $\mathcal{D}$ is used as the coefficient matrix and is a measure of the diffusion coefficients. Note that these are not binary pair diffusion coefficients and the relation between the two will be shown soon. Hence the symbol $\tilde{\mathcal{D}}$ is used for this matrix. Note that ∇y_3 is not independent and therefore not used in the flux expression.

A similar equation is written for species 2:

$$J_2 = -C\mathcal{D}_{21}\nabla y_1 - C\mathcal{D}_{22}\nabla y_2$$

Thus we use four constants to model the diffusion. These four constants can be put into a matrix, which is called the generalized Fick's law diffusion matrix. Also note that the equation for species 3 is redundant since the diffusion fluxes have to add up to zero.

Note: Although four values for $\tilde{\mathcal{D}}$ are needed for a ternary case (and $(ns-1)^2$ in general for a ns component system), there are some restraints between these values, which reduces it to three independent values or $ns(ns-1)/2$ in general. These are shown by De Groot and Mazur (1968). The restraint arises from what are known as Onsager reciprocal relations. But in practice these restraints are difficult or even impossible in most cases to apply and the four for binary or $(ns-1)^2$ values in general are commonly used.

15.5.1 $\tilde{\mathcal{D}}$ Matrix Relation to Binary Pair Diffusivity

The expression for the $\tilde{\mathcal{D}}$ matrix can be evaluated by matrix inversion of the Stefan-Maxwell model and the procedure follows.

The Stefan-Maxwell model in diffusion flux form (Equation 15.5) is used here, and is repeated for ease of reading:

$$-\nabla y_A = \sum_{j=1}^{ns} \frac{y_j \boldsymbol{J}_A - y_A \boldsymbol{J}_j}{CD_{Aj}} \tag{15.25}$$

A similar equation holds for species 2. We take $ns = 3$ here, that is, a ternary system.

Two substitutions are made here: (1) $J_3 = -(J_1 + J_2)$ since the diffusion fluxes have to add up to zero and (2) $y_3 = 1 - (y_1 + y_2)$, as the summation of mole fractions should be one.

This permits the flux and the mole fraction of the species to be eliminated from the Stefan-Maxwell equation. The eliminated species is sometimes called the solvent. The resulting equation after some simple algebra is

$$\begin{aligned}
\frac{dy_i}{dz} = &-J_i \sum_{j=1}^{ns} \left(\frac{y_j}{CD_{ij}} \right) \\
&+ y_i \sum_{j=1}^{ns-1} \left(\frac{J_j}{CD_{ij}} - \frac{J_j}{CD_{i,ns}} \right) \text{ for } i = 1, (ns-1)
\end{aligned} \tag{15.26}$$

where ns represents the species that was eliminated, species 3 for the ternary case considered above. The summation is not applied if i equals j.

Equation 15.26 can be represented in compact vector-matrix form as

$$C\nabla y = -\tilde{B}J \tag{15.27}$$

Here $\tilde{B}$ is a 2 by 2 coefficient matrix for a ternary system. This matrix is obtained by gathering all the coefficients of J_i on the right-hand side of Equation 15.26.

The matrix representation can be inverted to obtain J in terms of the mole fraction gradients and the resulting equation can be represented as

$$\boxed{J = -C\tilde{D}\nabla y} \tag{15.28}$$

where $\tilde{D}$ is equal to

$$\tilde{D} = [\tilde{B}]^{-1}$$

Equation 15.28 defines the diffusivity matrix $\tilde{D}$ and has a form similar to Fick's law except that it is now in a vector matrix form. The matrix $\tilde{D}$ is the

Table 15.1   Explicit Expressions for the
$\tilde{D}$ Matrix for a Ternary System

$$\mathcal{D}_{11} = \left[\frac{y_1}{D_{12}} + \frac{y_2 + y_3}{D_{23}} \right] / \text{DENO}$$

$$\mathcal{D}_{12} = \left[y_1 \left(\frac{1}{D_{12}} - \frac{1}{D_{13}} \right) \right] / \text{DENO}$$

$$\mathcal{D}_{21} = \left[y_2 \left(\frac{1}{D_{12}} - \frac{1}{D_{23}} \right) \right] / \text{DENO}$$

$$\mathcal{D}_{22} = \left[\frac{y_2}{D_{12}} + \frac{y_1 + y_2}{D_{13}} \right] / \text{DENO}$$

$$\text{DENO} = \frac{y_1}{D_{12} D_{13}} + \frac{y_2}{D_{12} D_{23}} + \frac{y_3}{D_{13} D_{23}}$$

tensor equivalent to the binary diffusivity. The form above is often referred to as the generalized Fick's law.

The explicit form for the $\tilde{D}$ matrix can be obtained for a ternary system analytically and are presented in Table 15.1.

An often quoted and well studied problem is the hydrogen(1)-methane(2)-argon(3) system and we calculate the values of $\tilde{D}$ for this case from the equation shown previously for a composition of $y_1 = 0.2$, $y_2 = 0.2$, and $y_3 = 0.6$; the results are as follows:

$$\mathcal{D}_{11} = 0.76 \times 10^{-4} \text{ m}^2/\text{s}$$
$$\mathcal{D}_{21} = -0.01 \times 10^{-4} \text{ m}^2/\text{s}$$
$$\mathcal{D}_{21} = -0.12 \times 10^{-4} \text{ m}^2/\text{s}$$
$$\mathcal{D}_{22} = 0.25 \times 10^{-4} \text{ m}^2/\text{s}$$

Note the reasonably large value for the cross-diagonal term $\mathcal{D}_{21}$. Complex diffusion effects are signaled by this term. Also note that the coefficients are concentration dependent, unlike the binary pair values. We have calculated this for one composition but if you do this for another composition you will get a different set of values. The diffusivity matrix based on an average composition along the diffusion path is often used as the representative values and the composition dependency is ignored as an approximation.

Another point to note is the choice of the third component, that is, the species that is eliminated in the Stefan-Maxwell model. In the above case argon was eliminated and equations for the diffusion fluxes of the other two species are obtained from the diffusivity matrix. The coefficients will be different if another species, say hydrogen, is eliminated.

Multicomponent diffusion in liquids is often modeled by this approach and the $\tilde{\mathcal{D}}$ matrix is fitted to experimental data. This is therefore more of an empirical approach. Often the non-ideal effects are not considered explicitly and lumped into the $\tilde{\mathcal{D}}$ matrix. Similarly, diffusion in ternary solid alloys are often fitted in this manner.

The advantage of diffusion matrix formulation is that the models are analogous to a matrix version of Fick's law. Hence linear algebra–based methods can be used to get a quick analytical solution. However the $\tilde{\mathcal{D}}$ matrix is mole fraction dependent. Hence it has to be evaluated at some average mole fraction in the domain and further a constant value for this matrix has to be assigned. Such a model is called the linearized model, first introduced in the field by Toor (1957). On the other hand, in the direct solution method no such assumption is needed, but the problem has to be solved by numerial methods as shown in Section 15.2.

Summary

- Stefan-Maxwell equations provide a convenient constitutive model for multicomponent systems. The model is shown to be accurate for ideal gases at low pressures. The model equation is given by Equation 15.4 and is worth memorizing. The values of only binary pair diffusion coefficients, one for each pair in the mixture, are needed. Thus for a ternary mixture we need three diffusion coefficients: D_{12}, D_{21}, and D_{23}.

- Stefan-Maxwell equations need to be supplemented with the species conservation laws and some determinancy condition relating the fluxes of various species, similar to that done for a binary mixture. Common conditions useful in many problems are only one non-diffusing gas (uni-molecule diffusion or UMD), equimolar counter-diffusion (EMD) and reacting systems where the fluxes are related by reaction stoichiometry.

- The main difficulty in computations with the Stefan-Maxwell model is that the equations are implicit in fluxes. We often wish to compute the fluxes given the mole fraction gradients but Stefan-Maxwell gives the mole fraction gradients in terms of all the component fluxes.

- One solution method, applicable when there are no homogeneous reactions, is the matrix method. A matrix-vector solution can be written formally in terms of the exponential matrix. The fluxes are implicit in the coefficients of this exponential matrix and can be solved by a nonlinear algebraic equation solver. Once the fluxes are computed the mole fraction profiles can be calculated using the formal matrix-vector solution.

- A second method is to augment the Stefan-Maxwell equations with species mass balance equations. The resulting problem is of a boundary value type and can be solved using any standard package for the solution of boundary value problems, for example BVP4C. Code based on this is provided in Listing 15.1.

- An inverted formulation of Stefan-Maxwell models leads to a generalized Fick's law for diffusion fluxes. The form of the equation is similar to a binary mixture and is the vector analog of the law. The matrix appearing in this expression, the $\tilde{D}$ matrix, can be calculated from the binary pair values at an average concentration in the system. Once this matrix is evaluated, solution methods for binary systems have a direct vector-matrix analog that may be useful for solution of many problems.

- The $\tilde{D}$ matrix is concentration dependent. An often used simplifying assumption is that the concentration dependency of $\tilde{D}$ is ignored. This is often referred to as a linear model for multicomponent diffusion. Linear algebra tools can then be used to solve the model equations.

- Complexities associated with multicomponent diffusion can be associated with the off-diagonal terms of the $\tilde{D}$ matrix. If these terms are small or zero then each species diffuses in proportion to its own gradient. Complexities such as reverse diffusion, osmotic diffusion, and so on, are absent in such cases.

- The $\tilde{D}$ matrix can also be viewed as a fitted parameter relating the fluxes to the mole fraction gradients and used as such in an empirical setting. Such an approach for instance is useful for non-ideal systems.

Review Questions

15.1 What is the molecular basis for the Stefan-Maxwell model?

15.2 What is the basis for the Wilke equation for the pseudo-binary diffusion coefficient?

15.3 Under what conditions is the Wilke equation an exact representation of the Stefan-Maxwell model?

15.4 When is the exponential matrix-based method not suitable for solution of multicomponent diffusion problems?

15.5 What are reverse diffusion and osmotic diffusion?

15.6 If two independent reactions take place and there are five species, how many of the species fluxes are independent?

15.7 Define a diffusivity matrix and distinguish its components from binary pair diffusion coefficients.

15.8 What is the difference between D_{12} and $\mathcal{D}_{12}$?

15.9 Can D_{11} be non-zero? Can $\mathcal{D}_{11}$ be non-zero?

15.10 What is the implication of $\mathcal{D}_{12}$? Can it be zero?

Problems

15.1 **Evaporation into a mixture of two inert gases.** Acetone is evaporating in a mixture of nitrogen and helium. Find the rate of evaporation and compare it to the rate in pure nitrogen and pure helium.

15.2 **Comparison of Stefan-Maxwell model with Wilke model for ternary evaporation.** Compare the results in exercise problem 1 with the model using a pseudo-binary diffusivity value for acetone. The pseudo-binary diffusivity can be calculated from the correlation given by Wilke.

15.3 **Evaporation of a binary mixture: *expm*-based code.** Write MATLAB code to solve Equations 15.22 and 23 for the unknowns N_A and N_B. You will need to write a function m-file to be used in FSOLVE, which takes in trial values of N_A and N_B, calculates the A and R matrices based on these values, and returns two equations based on Equations 15.22 and 15.23 as the function values. FSOLVE will then iterate using this function program to find N_A and N_B.

Test the results for the data given in Example 15.2.

15.4 **Arnold cell analysis for a binary liquid.** A binary mixture of A and B is contained in a tube with a starting level of H_0. Over time evaporation proceeds and the level changes with time. Develop a model to calculate the level change in the tube. Note that the mass balances for A and B in the liquid are needed since they change in different manners with time. Thus the liquid will get enriched with the less volatile compound.

15.5 **CO diffusion to a catalytic surface followed by a heterogeneous reaction.** Catalytic oxidation of CO is an important reaction in pollution prevention. The reaction scheme is

$$O_2(A) + 2CO(B) \rightarrow 2CO_2(C)$$

Set up the Stefan-Maxwell model for this problem. State the determinacy condition and simplify the equations into two differential equations for the mole fractions of any two of these species. Calculate the solution for the mole fraction profiles and find the values of the fluxes using the matrix method and the BVP method. Assume fast reaction at the catalyst surface.

15.6 **Rank of a stoichiometric matrix to find the invariant relations between the component fluxes.** Examine how the rank of the stoichiometric matrix can be used to reduce the number of equations to be solved. For example, if there are ns components and nr equations we can show that only $ns - nr$ independent mass balance equations are needed. The remaining variables form an invariant that connects the flux of the selected key components to those of the non-key components. Write MATLAB code to find the invariant given a stoichiometric matrix.

15.7 **Methane reforming in a porous catalyst.** Extend the analysis in Section 15.2 to two reactions that are applicable for steam reforming of methane. The first reaction is

$$CH_4 + H_2O \rightleftharpoons CO + 3H_2$$

followed by the water-gas shift reaction. Kinetic and diffusion parameters can be found in the work by Haynes (1984) or other sources.

15.8 **$\tilde{\mathcal{D}}$ matrix values for hydrogen-methane-argon mixture.** Consider the calculation of a generalized diffusion coefficient ($\mathcal{D}$ matrix) for a gas mixture of hydrogen (1), methane (2), and argon (3). The following coefficient values are reported in Cussler (2009) at 298 K:

$$\tilde{\mathcal{D}} = \begin{pmatrix} 0.76 & -0.01 \\ 0.12 & 0.25 \end{pmatrix}$$

These values are for a composition of $x_1 = 0.2$ and $x_2 = 0.2$. The units are cm^2/s. Argon was used as the species that was eliminated from the Stefan-Maxwell equation.

Verify the results if the binary pair values are $D_{12} = 0.726$; $D_{13} = 0.902$ and $D_{23} = 0.218$ at 307 K temperature. Units are cm^2/s.

15.9 **Effect of composition numbering on the Fick matrix $\tilde{\mathcal{D}}$.** In the matrix calculation of $\tilde{\mathcal{D}}$, a particular species n chosen for elimination is usually referred to as the solvent. For example, in the above example argon was chosen as the reference. The choice of "solvent" species is arbitrary but it can have an effect on the coefficient and the structure of the resulting matrix. For this exercise use argon (1), methane (2), and hydrogen (3). Eliminate hydrogen rather than argon, which was done in the previous exercise. Use the same composition as earlier. Show that the $\tilde{\mathcal{D}}$ matrix is now

$$\tilde{\mathcal{D}} = \begin{pmatrix} 0.64 & -0.39 \\ -0.12 & 0.37 \end{pmatrix}$$

15.10 **Transient profiles in a two bulb apparatus.** Derive the solutions for transient concentration profiles in the two bulb apparatus (Example 6.4 in the text) for a binary case and show that the multicomponent case can be derived as an extension of this. What assumption is implicit in extending the binary case to the multicomponent case?

CHAPTER 16

Mass Transport in Electrolytic Systems

Learning Objectives

After completing this chapter, you will be able to:

- Understand the role of the electric field in the transport of ions, and explain an important equation for electrochemical transport, the Nernst-Planck equation.
- Understand how the electric field modifies the transport rate and the apparent diffusivity of ions.
- Explain how the field in turn is dependent on the diffusion rate.
- Understand the concept of charge neutrality and how this concept can be used to simplify the models to calculate the electric potential.
- Apply the modeling concepts to three prototypical problems in electrochemical transport: transport across an uncharged membrane, transport across a membrane carrying a fixed charge, and transport in film near an electrode.

Transport of ions or charged species is encountered in a wide variety of processes. For example, sodium chloride is ionized in water and exists as sodium and chloride ion pair. Thus the diffusion of sodium chloride in reality involves the diffusion of positively charged sodium ions and negatively charged chloride ions. Such systems are of importance in electrochemical reaction engineering and in electric field–assisted separations. Examples can be found, for example, in electro-winning of metals, fuel cells, batteries, and so on. Equally important is the transport of charged species in charged membranes or solids carrying a net surface charge. An example of such a system is the ion-exchange membrane, which is widely used in water purification, metal recovery by electrodialysis, and selectivity modulation of chemical reactors. Transport of charged species in biological membranes is another example of transport across charged membranes. In addition, a number of techniques to separate large molecules such as DNA or protein involve application of an electric field. Electrophoresis is such an example. The ionic transport effects are also important in the study of surface properties of colloids.

The key feature in the model for transport in such systems is the charge migration due to the electric field. The constitutive equation for diffusion in such systems the classical Nernst-Planck equation, which is discussed in this chapter together with some application examples. Further complexity in models for these systems arise due to the requirement that the electric field has to be computed simultaneously with the concentration field. We explore these situations in detail in this chapter. This chapter therefore provides the background for modeling electrochemical reactors and electric field–based separation processes. These are taken up in some additional details in later chapters (24 and 30).

16.1   Transport of Charged Species: Preliminaries

Basic definitions and concepts needed for modeling of transport processes in ionic systems are reviewed here to provide the basic foundation of electrochemical engineering. First we define the mobility of an ion and its relation to diffusivity of the ion.

16.1.1 Mobility and Diffusivity

Consider a positive charge placed in an electric field. The field produces an acceleration on the particle but the frictional resistance of the neighboring molecules retard the motion. The net result is that the charge acquires a terminal velocity that is proportional to the electric field. The relationship can be related by the use of a mobility parameter, μ_i. This is defined as the velocity acquired by a unit charge of 1 Coulomb divided by the electric field strength that moves the charge. Hence the following relation holds:

$$v_i = \mu_i E$$

Note that the velocity v_i is defined for a system with zero total velocity, that is, in a frame of reference moving with a mixture velocity. The previous equation also applies to a positive charge. A negative charge moves in the opposite direction to the field. The two cases can be reconciled by introducing z_i, the valency of the ion. Hence the previous equation can be written as

$$v_i = z_i \mu_i E \tag{16.1}$$

Here z_i is positive for cations and negative for anions.

The unit for electric field is N/C as it is force on a Coulomb of charge and hence the unit of mobility is m.C/N s or, equivalently m^2/Vs.

The electric field can be represented as the negative of the gradient of a scalar potential:

$$E = -\nabla \phi \tag{16.2}$$

The ϕ is the electric potential with units of J/C same as 1 V.

The mobility is related to diffusivity by the Einstein equation:

$$\boxed{\mu_i = D_i \frac{F}{R_g T}} \tag{16.3}$$

where D_i is the diffusion coefficient for the ion. Here F is the Faraday constant, which has a numerical value of 96485 C/mole and represents the charge on one mole of electrons.

Typical values of diffusion coefficients of the various common ions are shown in Table 16.1. Note the relatively high value of H^+ (protons), which is somewhat inconsistent with its size. The higher value is due to molecular interaction with the water molecules, the so-called Grotthuss effect.

The mobility relation can be combined with the Fick's law of diffusion, leading to the Nernst-Planck equation discussed next.

16.1.2 Nernst-Planck Equation

The electric field causes the movement of ions and the flux of a charged species caused by the electric field is called the migration flux. It is similar to a convection flux and is therefore equal to $C_i v_i$. Using Equation 16.1 for the migration velocity, the flux is equal to $C_i z_i \mu_i E$. Equation 16.3 can be used to relate the mobility to diffusivity. Hence the flux can be expressed as $C_i z_i D_i (F/R_g T) E$.

Table 16.1 Diffusion Coefficients of Common Ions in Water at 25°C

Ion	H^+	Na^+	K^{++}	Ca^{++}	OH^-	Cl^-	$SO4^{--}$
$D \times 10^9 m^2/s$	9.313	1.334	1.957	0.7920	5.860	2.032	1.065

Finally, expressing the electric field in terms of the potential gradient as per Equation 16.2 we have

$$\text{Migration flux due to electric field} = -D_i z_i C_i (F/R_g T) \nabla \phi$$

The total flux of a charged species is then obtained by combining Fick's law, which gives the diffusion due to the concentration gradient, and the flux due to mobility under the electric field. The x-component of this flux, for example, is therefore

$$J_{i,x} = -D_i \frac{dC_i}{dx} - D_i z_i C_i \frac{F}{R_g T} \frac{d\phi}{dx} \tag{16.4}$$

If there is a net superimposed system velocity of v, the contribution of the convective term is also added to obtain the combined flux:

$$N_{i,x} = -D_i \frac{dC_i}{dx} - D_i z_i C_i \frac{F}{R_g T} \frac{d\phi}{dx} + v_x C_i \tag{16.5}$$

This equation is known as the Nernst-Planck equation. The vector representation is

$$\boldsymbol{N}_i = -D_i \nabla C_i - D_i z_i C_i \frac{F}{R_g T} \nabla \phi + \boldsymbol{v} C_i \tag{16.6}$$

The three terms on the right-hand side can be identified as diffusion, migration, and convection, respectively. The migration term is the key to transport in charged systems and does not appear in the flux model for uncharged species.

It is common to express the fluxes in terms of electric current in electrochemistry. Flux of a charged species produces a current. The total current density in an electrolyte solution is related to the fluxes by the following equation:

$$\boldsymbol{i} = F \sum_i z_i \boldsymbol{N}_i \tag{16.7}$$

Also, due to conservation requirements the divergence of the current density is zero:

$$\nabla \cdot \boldsymbol{i} = 0 \tag{16.8}$$

This provides an additional relation between the fluxes and is similar in status to the determinancy condition used in diffusion type of problems.

Note that both the concentration field and the electric potential field have to be computed simultaneously to calculate the flux of each ionic species. The additional equations needed to compute the potential field are presented in Section 16.3.

16.2   Charge Neutrality

The charge neutrality condition states that the net charge at any point is zero:

$$\sum_{i}^{N} z_i C_i = 0 \qquad (16.9)$$

The following discussion shows why the charge neutrality holds in general and the region where it does not hold.

The electric field is related to charge density ρ_c by Maxwell's equation:

$$\nabla \cdot \boldsymbol{E} = \frac{\rho_c}{\epsilon}$$

The ϵ is the dielectric permittivity of the medium. The electric field can be expressed as the gradient of the electric potential (Equation 16.2). Hence the equation for the potential field is Poisson's equation, given as

$$\nabla^2 \phi = -\frac{\rho_c}{\epsilon} \qquad (16.10)$$

The electrial charge density ρ_c is given as

$$\rho_c = F \sum_{i}^{N} z_i C_i \qquad (16.11)$$

Using these definitions in Equation 16.10 we have

$$\nabla^2 \phi = -\frac{F}{\epsilon} \left(\sum_{i}^{N} z_i C_i \right) \qquad (16.12)$$

The permittivity is the product of the free space permittivity and the dielectric constant of the system. The value of ϵ_0 (the free space permittivity) is 8.8×10^{-12} C/V.m and the dielectric constant for water is 80.2. Hence the value of the parameter F/ϵ is equal to 1.3592×10^{16} V.cm/gmol.eq for water. Hence the coefficient on the right-hand side of Equation 16.12 is large. This implies that the left-hand side is nearly zero and hence that the charge density (Equation 16.12) will be nearly zero in most cases. This provides the basis for the charge neutrality relation widely used in electrochemical reaction engineering.

The charge neutrality condition is valid everywhere except near the solid (electrode) surface where there is a thin layer (of the order of 1 to 10 nm) over which the charge density is non-zero. This thin region is known as the electrical double layer or the Debye layer. The thickness of the Debye layer, denoted as λ, can be calculated by the following equation:

$$\lambda = \sqrt{\frac{\epsilon R_g T}{2 z_i^2 F^2 C_\infty}} \tag{16.13}$$

where C_∞ is the concentration of either the cation or anion just outside the Debye layer where electroneutrality holds.

Electroneutrality does not hold in the Debye region but holds everywhere else to a first approximation. However, often electroneutrality is assumed to hold all the way to the electrode surface and the effects due to a double layer are incorporated indirectly into the boundary condition at the electrode.

16.3   General Expression for the Electric Field

The following expression can be derived starting from Equation 16.7 for the current and the Nernst-Planck equation for the fluxes:

$$\boldsymbol{i} = -F \sum \left(z_i D_i \nabla C_i - D_i z_i^2 C_i \frac{F}{R_g T} \nabla \phi + z_i \boldsymbol{v} C_i \right) \tag{16.14}$$

The last term is taken as zero due to electroneutrality and is dropped from further discussion. We define the conductivity of the solution by the following equation:

$$\kappa = F \sum_i z_i^2 \mu_i C_i$$

We can also write this in terms of the diffusivity using the Einstein relation as

$$\kappa = \frac{F^2}{R_g T} \left(\sum_i z_i^2 D_i C_i \right)$$

Both expressions are completely equivalent. Using this expression for the conductivity, Equation 16.14 can be written as

$$i = -\kappa \nabla \phi - F \sum z_i D_i \nabla C_i + z_i v C_i \qquad (16.15)$$

Rearranging this we get the following general expression for the electric field:

$$\boxed{\nabla \phi = -\frac{i}{\kappa} - \frac{F}{\kappa} \left(\sum_i z_i D_i \nabla c_i \right) + \frac{Fv}{\kappa} \left(\sum_i z_i c_i \right)} \qquad (16.16)$$

The three terms on the right-hand side can be interpreted as an Ohmic term with the actual current, the potential caused by diffusion, and the streaming potential caused by the bulk flow of the charges. The last term is generally zero whenever electroneutrality holds but may not be zero in some cases, for example, transport in charged membranes. Here the electroneutrality applies only if the immobile charge of the membrane is also included. Thus the term $\sum z_i c_i$ is not zero for mobile charges (charges transported across the charged membranes) and this term contributes to the streaming potential.

16.3.1 Laplace Equation for the Potential

Often the Laplace equation is used to calculate the potential. In this section we show why and also show when, that is, cases where it is not applicable.

Consider Equation 16.15. If there are no concentration gradients in the system, the current is simply related to the electric potential gradient and Ohm's law is applicable:

$$i = -\kappa \nabla \phi \qquad (16.17)$$

Using the current continuity condition in Equation 16.7 we find that

$$\nabla^2 \phi = 0$$

Hence the Laplace equation holds for the potential field. As seen from the previous discussion, this equation for electric potential is valid only under certain conditions such as constant electrical conductivity of the medium and no concentration gradients in the system (e.g., bulk liquid outside the diffusion boundary layers). In such cases, it is appropriate to use the Laplace equation to calculate the potential field, but such calculations may be inaccurate near the concentration boundary layer (diffusion film) near the electrodes. In general Equation 16.16 is appliable for finding the potential.

16.3.2 Transference Number

The fraction of the current carried by species i is called the transference number; it is defined by the following equation:

$$t_i = \frac{z_i^2 \mu_i c_i}{\sum_i z_i^2 \mu_i c_i}$$

Differences in transport number arise from differences in electrical mobility. For example, in a solution of sodium chloride, less than half of the current is carried by the positively charged sodium cations and more than half is carried by the negatively charged chloride anions because the chloride ions are able to move faster, that is, chloride ions have higher mobility/diffusivity than sodium cations (see Table 16.1 for values). The sum of the transport numbers for all of the ions in a solution equals unity.

16.3.3 Mass Balance for Reacting Systems

The flux expression N_i is coupled with species mass balance in the usual manner as described in Chapter 5. Thus

$$-\nabla \cdot \boldsymbol{N}_i + R_i = 0$$

where R_i is the rate of production of species i by homogeneous reaction in the liquid. Note that the reaction on the electrode surface is not included here and will appear as the wall boundary condition.

The reaction at the electrode surface is modeled as a Robin boundary condition at the wall. Simply stated this is the balance of the flux to the surface to the reaction rate for each species. The surface concentration of a reacting species at the electrode is determined by the kinetics of the reaction. There is discussion on the kinetic models for electrode reactions in Chapter 24 where we show that the kinetics of the reaction is usually modeled by the Butler-Volmer kinetics or by a simple Tafel model. In general, rate is a function of the overpotential (potential at the surface minus the equilibrium value). Thus rate can be varied by changing the potential of the electrode. Rate is also a function of the concentration and temperature.

If the surface reaction is rapid, the concentration of the reacting species is zero at the electrode. This condition determines the maximum current in the system, also known as the limiting current. This concept is again similar to a catalytic heterogeneous reaction; the concentration of a limiting reactant is nearly zero for a fast reaction at a surface.

With this background material in hand, we show some illustrative examples for some prototypical problems.

16.4   Electrolyte Transport across Uncharged Membrane

Consider an binary electrolyte, MX, diffusing across a membrane due to a concentration gradient across the membrane. Here we derive an expression for the flux and also for the potential gradient developed across the membrane.

We assume that the electrolyte is fully ionized and diffuses as $M+$ and $X-$ ions. The fluxes are given by the Nernst-Planck equation:

$$J(M+) = -D_+ \frac{dC_M}{dx} - z_+ D_+ (F/R_g T) C_M \frac{d\phi}{dx}$$

A similar equation holds for $X-$:

$$J(X-) = -D_- \frac{dC_X}{dx} - z_- D_- (F/R_g T) C_X \frac{d\phi}{dx}$$

A univalent electrolyte is considered for further discussion; $z_+ = 1$ and $z_- = -1$ is used further on. The electroneutrality implies that

$$C_M = C_X = \text{ say}, C$$

where C is either C_M or C_X.

Also, if no net current is drawn across the membrane the two fluxes are equal:

$$J(M+) = J(X-) = J_{\text{salt}}$$

This helps us to eliminate the potential term. Hence the following expression for $d\phi/dx$ can be obtained by combining the two flux equations with the requirements of electroneutrality and equal fluxes for both cations and anions:

$$\frac{d\phi}{dx} = -\frac{R_g T}{F} \left(\frac{D_+ - D_-}{z_+ D_+ - z_- D_-} \right) \frac{1}{C} \frac{dC}{dx} \qquad (16.18)$$

Integrating across the system gives an expression for the potential:

$$\phi(L) - \phi(0) = \left(\frac{D_+ - D_-}{z_+ D_+ - z_- D_-} \right) \frac{R_g T}{F} \ln \left(\frac{C_0}{C_L} \right) \qquad (16.19)$$

This potential is known as the diffusion potential and arises whenever the diffusion coefficients of the cation and anion are not equal.

Using this back in the flux expression, we can now calculate the flux of either of the species. The result can be expressed in a Fick's law type of equation as

$$J(M+) = J(X-) = J(\text{salt}) = -D_{eff}\frac{dC}{dx}$$

where the effective diffusion coefficient turns out to be

$$D_{eff} = \left(\frac{(z_+ - z_-)D_+D_-}{z_+D_+ - z_-D_-}\right) \tag{16.20}$$

which is known as the ambipolar diffusion coefficient. The expression can be written in a simple form for the case of equal charges, $z_+ = -z_-$, as

$$D_{eff} = \frac{2D_+D_-}{D_+ + D_-}$$

or

$$\frac{2}{D_{eff}} = \frac{1}{D_+} + \frac{1}{D_-}$$

Thus the effective diffusivity is a harmonic mean of the positive and negative ion diffusivities. The concept of ambipolar diffusion is illustrated in Figure 16.1.

In Figure 16.1 the direction of the transport is from left to right and both H^+ and Cl^- move to the right. The faster moving H^+ ions set up an electric field in the opposite direction to its motion (to the left). The electric field slows the hydrogen ions down and these ions are pushed back somewhat by the

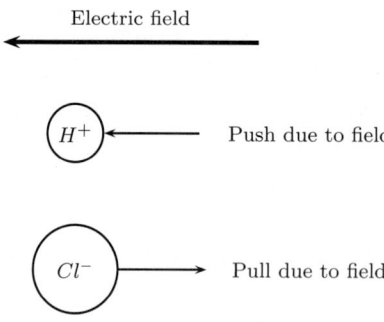

Figure 16.1 Illustration of the concept of ambipolar diffusion. Diffusion of HCl across an uncharged membrane or barrier.

field. The overall result is that H^+ and Cl^- diffuse at the same rate and have the same (ambipolar) diffusivity.

16.5   Transport across a Charged Membrane

In this section we analyze a situation where there is a fixed charge in the membrane. An example would be a hydrogel, which typically consists of >90% water with a crosslinked polymer network. Such a network may have an acidic group that produces a net charge on the network. Many examples are also found in biological systems.

The problem analyzed is that of a membrane exposed to two different salt concentrations on either side ($x = 0$ and $x = L$). The transport of the salt in ionized form as cations and anions take place across the membrane. Also we assume there is no current in the system. (The current will be non-zero if some electrode reactions are taking place downstream and upstream of the membrane, for example, in electrodialysis; a modified analysis is needed for such cases.) The flux of the salt across the system is to be calculated. External mass transport resistance near the diffusion film on either side of the membrane is neglected for simplicity since we want to focus on the transport rate in the membrane itself. A univalent electrolyte is considered for simplicity.

The concentrations at the interfaces ($x = 0$ and L) are different at the liquid side and the membrane side. This is first calculated from thermodynamic considerations.

16.5.1   Interfacial Jump: Donnan Equation

Usually the concentration values in the bulk solutions are known. In the absence of external resistance in the diffusion film on the liquid side of the membrane, we can use the bulk values as the values at the liquid side of the membrane as well. The concentration at the membrane side of the interface needs to be used for transport calculations across the membrane and such a calculation is addressed here.

Since the charge of the membrane is fixed and does not penetrate into the solution there is a concentration jump/fall at the solution–membrane interface. Hence the concentrations at the membrane surface are not equal to the bulk concentrations and are to be calculated by the following procedure. This discussion uses a univalent electrolyte of type MX with $z_+ = 1$ and $z_- = -1$, but the concepts can be extended to other types of electrolytes.

The ionic products of the cations and anions are fixed. They should be the same on the bulk side and the membrane side of the interface. Hence

$$C_{+,b} \times C_{-,b} = C_{+,0} \times C_{-,0}$$

where b represents the bulk values and 0 refers to values at $x = 0^+$, the membrane side values.

Electroneutrality on the membrane side requires that $C_{+,0} = C_{-,0} + C_{-,m}$ for a negatively charged membrane carrying a charge of $C_{-,m}$. Electroneutrality in the bulk requires $C_{+,b} = C_{-,b}$. This is illustrated in Figure 16.2.

Now, equating the ionic products on either side we get a quadratic equation for $C_{-,0}$, the solution of which results in the boxed equation for $C_{-,0}$ shown in Figure 16.2. This equation is often called the Donnan equilibrium conditions. The value of $C_{+,0}$ can then be computed using the electroneutrality on the membrane side of the interface. A similar equation can be derived for $C_{-,L}$ as well.

Having obtained the concentrations on the membrane side of each of the boundaries, we now look at the transport model.

16.5.2 Transport Rate

The Nernst-Planck equation is applied for both the positive and negative ions. Hence

$$N_+ = -D_+ \frac{dC_+}{dx} - D_+ C_+ \frac{F}{R_g T} \frac{d\phi}{dx} \tag{16.21}$$

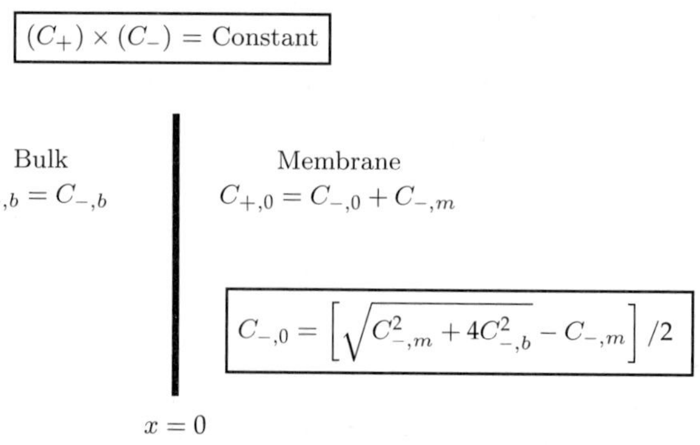

$$\boxed{(C_+) \times (C_-) = \text{Constant}}$$

Bulk
$$C_{+,b} = C_{-,b}$$

Membrane
$$C_{+,0} = C_{-,0} + C_{-,m}$$

$$\boxed{C_{-,0} = \left[\sqrt{C_{-,m}^2 + 4C_{-,b}^2} - C_{-,m} \right] / 2}$$

$$x = 0$$

Figure 16.2 Sketch for the derivation of the Donnan equilibrium equation. The membrane is assumed to have a negative charge here.

and

$$N_- = -D_- \frac{dC_-}{dx} + D_- C_- \frac{F}{R_g T} \frac{d\phi}{dx} \tag{16.22}$$

The electroneutrality condition, as stated earlier, is

$$C_+ = C_- + C_{-m}$$

Differentiating this with respect to x we get

$$\frac{dC_+}{dx} = \frac{dC_-}{dx}$$

Here C_{-m} is taken as a constant, that is, there is constant fixed charge across the membrane.

The current in the system is given as $F(N_+ - N_-)$ for a univalent binary electrolyte. We analyze the case here where the net current is zero. This means that the fluxes are equal. Hence

$$N_+ = N_- = N_{salt}$$

Subtracting the two Nernst-Planck equations and using these two conditions you will be able to show that the potential field is given by the following equation:

$$\frac{F}{R_g T} \frac{d\phi}{dx} = \left[\frac{D_- - D_+}{D_- C_- + D_+ (C_- + C_{-m})} \right] \frac{dC}{dx} \tag{16.23}$$

This can be now backsubstituted into either of the two Nernst-Plank equations to get the flux of each ion, which is also the flux of the salt transported across the system. This results in the following expression for the salt flux in terms of the concentration gradient of the anions:

$$N_{salt} = -D_- \frac{dC_-}{dx} \left[1 + \frac{C_-(D_+ - D_-)}{(D_+ + D_-)C_- + D_+ C_{-m}} \right] \tag{16.24}$$

Integrating this across the system, keeping N_{salt} constant, yields an expression for the flux across the membrane:

$$N_{salt} = \frac{2D_+ D_-}{D_+ + D_-} \left(\frac{C_{-0} - C_{-L}}{L} \right)$$

$$+ \frac{D_+ D_- (D_- - D_+)}{(D_+ + D_-)^2} \frac{C_{-m}}{L} \ln \left[\frac{D_+ C_{-m} + (D_+ + D_-)C_{-0}}{D_+ C_{-m} + (D_+ + D_-)C_{-L}} \right] \tag{16.25}$$

The following limiting cases can be identified:

- No charge on the membrane: In this case $C_{-m} = 0$ and the model reverts to the case of an uncharged membrane. Only the first term on the right-hand side of Equation 16.25 remains and the term with all the diffusion coefficients can be identified as the ambipolar diffusion coefficient.
- Cations and anions have the same diffusion coefficient: In this case again the uncharged membrane model applies (with Donnan correction at the end points) since the potential across the membrane can now be shown to be equal to zero.
- Highly charged membrane: The limit of C_{-m} tending to infinity is now taken. It can be shown that the mass transfer of anions is the rate limiting step:

$$N_{\text{salt}} = -D_- \frac{dC_-}{dx}$$

The Donnan relation simplifies to C_- equals $C_{0,b}^2/C_{-m}$ for this case. Hence the previous equation can be expressed in terms of the bulk concentrations of the anions as

$$N_{\text{salt}} = \frac{D_-}{L} \frac{\left(C_{0,b}^2 - C_{L,b}^2\right)}{C_{-m}} \tag{16.26}$$

16.6   Transfer Rate in Diffusion Film near an Electrode

In this section, we consider transport to an electrode where an electrochemical reaction is taking place. For example, a metal ion M^+ gets transported to a cathode where a reduction reaction is taking place. The process is similar to mass transfer with a surface reaction and the film model is useful to describe mass transfer from the bulk liquid to the electrode surface.

We assume quasi-steady state conditions and use the concept of a diffusion film near the cathode. We also neglect convection effects and use the low mass flux approximation: $J_A = N_A$. The concentration of the salt is denoted as C_{Mb} in the bulk liquid and C_{Ms} at the cathode surface. Recall that the film model flux is described by the following relation in the absence of migration effects:

$$J(M^+) = \frac{D_M}{\delta_f}(C_{Mb} - C_{Ms}) = k_L(C_{Mb} - C_{Ms})$$

Here δ_f is the film thickness. Since M is a charged species, the migration contribution to the flux has to be included and the goal of this section is to study the effect of migration and reexamine the enhancement in transport rate due to migration. A schematic for this problem is presented in Figure 16.3.

The model starts with the Nernst-Planck equation, which is applied for both species X^- and M^+. The procedure is similar to that presented in the previous two sections with the difference that there is a net current in the system. Thus the determinancy condition has to be expressed differently here.

Since there is no flux of species X^- (denoted by subscript X) we can equate J_X to zero:

$$J_X = -D_X \frac{dC_X}{dx} + D_X \frac{F}{R_g T} C_X \frac{d\phi}{dx} = 0 \tag{16.27}$$

Note that z_i is taken as -1 in the Nernst-Planck equation. The electric field can be solved from the previous equation:

$$\frac{d\phi}{dx} = \frac{R_g T}{F} \frac{1}{C_X} \frac{dC_X}{dx} \tag{16.28}$$

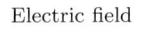

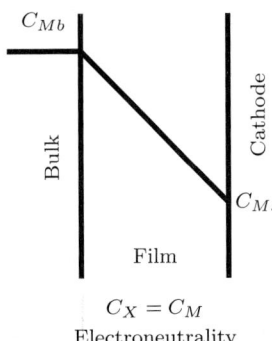

Figure 16.3 Concentration profile of cation M in the diffusion film near a cathode.

Charge equality means $C_M = C$. (M here means the species M^+.) Hence the electric field can also be expressed as

$$\frac{d\phi}{dx} = \frac{R_g T}{F}\frac{1}{C_M}\frac{dC_M}{dx}$$

For M^+ we have from the Nernst-Planck equation

$$J_M = -D_M\frac{dC_M}{dx} - \frac{D_M F}{R_g T}C_M\frac{d\phi}{dx}$$

Note the minus sign on the second term of the right-hand side since $z_i = +1$ now.

Using the expression for the electric field given by Equation 16.28 and simplifying we find

$$J_M = -2D_M\frac{dC_M}{dx} \qquad\qquad (16.29)$$

The simple Fick's law would have produced the above result but without the factor of two. Hence the effect of the electric field is to enhance the transport by a factor of two over that given by the simple Fick's law.

Since J_M is constant in the diffusion film, we find the concentration profile is linear. Also, the concentration profiles for both species M and X are the same. Thus there is a concentration gradient for species X in the film as well, but this does not imply that X is diffusing toward the cathode. The diffusive flux of X caused by the favorable concentration gradient is balanced by the migration flux of X in the opposite direction. The net flux of X is therefore zero! The transport processes of diffusion and migration for both cations and anions near a cathode are schematically explained in Figure 16.4.

The electric potential field can be computed by integration of Equation 16.28, keeping the concentration gradient constant. The result is

$$\phi(0) - \phi(\delta_f) = \frac{R_g T}{F}\ln\left(\frac{C_X(0)}{C_X(\delta_f)}\right)$$

The surface concentration of M will depend on the rate of electrode reaction. Hence the transport rate of M in the film is coupled with a kinetic model for the electrode reaction. Commonly used kinetic models are shown in Chapter 24 where we demonstrate examples of the combined model to find the rate of electrode reaction. The emphasis here is to show the analogy with mass transfer followed by the heterogeneous reaction studied in Section 6.4. A limiting case is examined next.

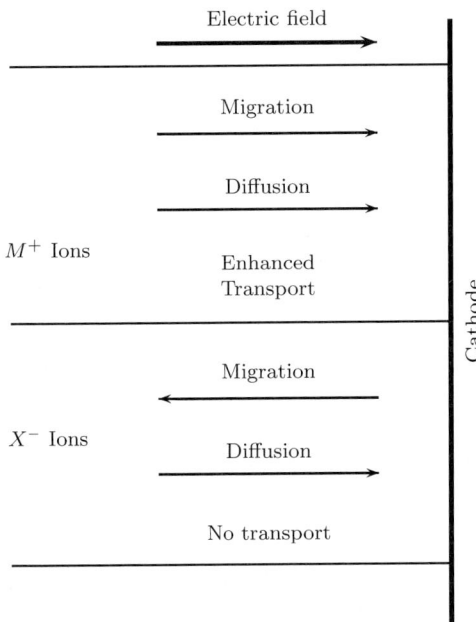

Figure 16.4 Transport processes near a cathode for cations and anions. Cations are deposited at the cathode in this example and have a non-zero net flux, while anions have no net flux.

For a fast reaction the concentration of M at the electrode is nearly zero. Hence C_{Ms} is set to zero and the process is limited by mass transfer rate to the electrode. The corresponding current in the system can be calculated as

$$i_L = 2F \frac{D_M}{\delta_f} C_{Mb}$$

and is called the limiting current.

Note that the factor of 2 arises due to the enhancement in diffusional mass transfer due to migration. This factor is applicable when there are no other electrolytes in the system. The role of a second electrolyte is discussed briefly in the following paragraph.

The enhancement factor of two (over Fick's value) in transport rate is valid if there are no additional ions in the system. If there is another salt of composition NX then the effect of the migration term will be smaller than one and in the limiting case there will be no enhancement and Fick's law simply holds. This happens because the electroneutrality now requires $X = M + N$. Since N is not reacting, the overall contribution of migration is reduced. Thus the rate of mass transport can be calculated based on the diffusional considerations alone for cases where there is an excess of a "supporting" electrolyte.

Summary

- Mass transport of charged species is important in many applications. The transport rate is now also affected by the electric field. The Nernst-Planck equation (Equation 16.5) is commonly used as the transport law for such systems.

- The electric field (or equivalently the negative gradient of the electric potential) has to be either specified or simultaneously calculated to find the transport rates of charged species. In general the electric potential is given by the Poisson equation (Equation 16.12). However, the electroneutrality condition (positive and negative charges are equal) holds for most of the region and is used as an implicit condition to find the electric field.

- Electroneutrality does not hold in a thin region near a charged surface. This layer is called the Debye layer and equations to calculate the thickness of the Debye layer are presented in the text. Ions with charge opposite to that of the surface (counterions) accumulate near the surface. A number of important electrokinetic effects are dependent on the charge distribution in the Debye region but these are outside the scope of this chapter.

- The current in an electrochemical system cannot be computed simply by using Ohm's law based on the applied potential. The potential gradient is composed of three components as indicated by Equation 16.16: the Ohmic term with the actual current, a diffusion potential due to concentration gradient, and a streaming potential due to bulk flow of the liquid.

- The key components in modeling of an electrochemcical processes are the Nernst-Planck equation for each ion present in the system, the current continuity condition, and the electroneutrality condition. In general the concentration field and the potential field can be computed simultaneously. Often numerical simulation is needed. Simpler examples where analytical solutions can be obtained are shown in the chapter.

- For transport of a single electrolyte across an uncharged membrane, an electric field develops across the membrane whenever the diffusion coefficient of the cation is not equal to that of the anion. The result is that the two ions diffuse with an apparent diffusivity that is in between the true diffusivity of the individual ions. This diffusion coefficient is called the ambipolar diffusion coefficient.

- For transport across a charged membrane, there is a concentration discontinuity at the liquid–membrane interface. The concentrations on the membrane side of the interface can be calculated by the Donnan model given the concentrations on the liquid side of the interface. A transport model for the flux across the membrane can then developed that includes this concentration discontinuity.

- For a binary electrolyte MX reacting on an electrode, the flux of only one of the species is non-zero (e.g., cations diffusing and depositing on the cathode surface). The transport rate of this ion is enhanced due to migration and the migration enhancement factor is two in the absence of a supporting electrolyte. The presence of the supporting electrolyte reduces the migration enhancement in film transport due to migration.

- For fast reactions the flux or equivalently the current is limited by film diffusion (and enhanced by the migration factor as appropriate) and represents a transport limited situation. The current in the system is called the limiting current and is the maximum current that can be produced due to the electrode reaction.

Review Questions

16.1 State the values with proper units for the following fundamental constants in physics: (1) charge on an electron, (2) Boltzmann constant, (3) Faraday constant, (4) permittivity of free space, and (5) dielectric constant of water.

16.2 What is the relation between $R_g T/F$ and $k_B T/e$?

16.3 Calculate the value of $R_g T/F$ at 298 K.

16.4 Verify that the Einstein relation given by Equation 16.3 is dimensionally consistent.

16.5 Find the mobility of H^+, OH^- ions from the diffusivity data given in Table 16.1.

16.6 Define the conductivity of an electrolyte solution and the transference number and give equations to calculate these.

16.7 When does Ohm's law apply in an electrolyte solution? When does it not apply?

16.8 What is the Debye length and what is its physical significance?

16.9 What is meant by the term limiting current in an electrochemical reactor?

16.10 Explain briefly the concept of ambipolar diffusion.

16.11 When does an electric potential develop for transport of ions across an uncharged membrane?

16.12 What is the direction of the electric field in an uncharged membrane?

16.13 State the basis of the Donnan equilibrium relation.

16.14 When is the potential across a charged membrane equal to zero?

16.15 What is the enhancement in mass transfer for a cation diffusing and reacting at a cathode?

16.16 When should the enhancement due to migration effects be applied for film diffusion toward an electrode?

Problems

16.1 **Conductivity of pure water.** Estimate the electrical conductivity of pure water. You need the ionic product of water to find how much water is ionized.

16.2 **Conductivity of a salt solution.** Calculate the value conductivity of 0.1 M and 1 M solutions of NaCl. Also calculate the transference numbers and indicate the current carried by the cation and anion separately.

16.3 **Conditions for Laplace equation to hold.** Verify that the divergence of the current is equal to zero. Hence show that the Laplace equation holds for the potential field for cases with no concentration gradients.

16.4 **Poisson equation for the electric potential.** When there is a concentration gradient in the system show that the potential gradient is composed of two terms: an Ohm's law contribution and a diffusional contribution. Here we assume there is no contribution of the superimposed velocity v. State the equation for the current. Now take the divergence of the current and show that the following Poisson equation holds for the potential field:

$$\nabla \cdot (\kappa \nabla \phi) = -F \sum_i z_i \nabla \cdot (D_i \nabla c_i)$$

Show the similarity to mass transport with generation with variable diffusivity. What assumption is implicit in the above equation?

16.5 **Induced potential due to difference in mobilities.** A membrane separates two bulk solutions of 0.5 M NaCl and 0.1 M NaCl. Find the potential difference developed across the membrane and the flux of NaCl across the membrane.

16.6 **Diffusion potential calculations.** Calculate the diffusion potential for an uncharged membrane with a concentration of 0.5 M on one side and 0.1 M on the other side for the solutions of (1) $CuSO_4$, (2) $MgCl_2$, and (3) KCl. Also calculate the ambipolar diffusion coefficient for these systems.

16.7 **Transport rate across an uncharged membrane.** Consider an uncharged membrane of thickness 100 μm with concentrations of 1 M HCl on one side and 0.1 M

HCl on the other side. Assume that the HCl is completely ionized. (1) Find the flux of the HCl in the system. (2) Find the diffusion potential generated in the system. (3) Which side of the membrane is at a higher potential? (4) Find the diffusion flux and migration flux of Cl^- and H^+ ions and tabulate these. Comment on the results.

16.8 **Diffusion across a disk for a mixture of two salts.** Two solutions are separated by a porous sintered disk (1 mm thick) that permits diffusion across the disk. On one side we have a mixture of 1 M HCl and 1 M BaCl$_2$ while on the other side we have pure water. Both salts are completely ionized and diffuse as H^+, Cl^-, and Ba^{++} across the disk. It is required to find the flux across the system. Assume zero current flow across the membrane. Set up the model to compute the fluxes. Based on the fluxes find the effective diffusivity of these ions across the disk. Use the following value of ionic diffusivity for Ba ions: $0.85 \times 10^{-9} \mathrm{m^2/s}$. For the other ions use the values in Table 16.1.

16.9 **Diffusivity of a weakly ionizing acid solution.** Diffusion of weakly ionized acids such as acetic acid is an interesting and complex problem in diffusion. The diffusion occurs by transport of the ionized species CH$_3$COO$^-$ as well as by the unionized acid CH$_3$COOH. The diffusion of acetate ion is affected by H-ions due to electroneutrality while that for unionized is constant. For acetate ions an ambipolar diffusion coefficient may be used. Set up a model to compute the flux based on this model. Express the results in terms of the total concentration of acetic acid on either side of a region. What is the dependency of flux on concentration? Is flux linear in acetic acid concentration?

16.10 **Concentration jump at a charged membrane interface.** A surface of a charged membrane is in contact with a 0.25 M NaCl solution. The membrane carries a fixed anion concentration of 0.05 M. Find the concentrations of sodium and chloride ions on the membrane side of the interface.

16.11 **Transport rate across a charged membrane.** Consider a charged membrane of thickness 100 μm with concentrations of 1 M HCl on one side and 0.1 M HCl on the other side. The membrane has a negative charge. Assume this generates a constant electric field of 1 N/C. Find the flux of HCl in the system using the model for transport across a charged membrane given in the text.

16.12 **Transport across a charged membrane for a mixture of two electrolytes.** Extend the analysis for a system for transport in a charged membrane for the case of a mixture of two electrolyte salts with a common anion (a mixture of NaCl and KCl for example). Derive a formula for the membrane potential when one side of the membrane is exposed to concentrations of $C_1(0)$ for KCl and $C_2(0)$ for NaCl while the other side at L is exposed to $C_1(L)$ and $C_2(L)$. Calculate the potential for membrane that is 100 μm thick with the concentrations of 0.1 and 0.01 M on either side.

16.13 **Copper deposition at an electrode and the limiting current.** Copper is deposited at a cathode from solution with a bulk concentration of 0.5 M at a rate of 3.0 g/m^{2}s. Find the surface concentration of Cu^{++} at the cathode if the mass transfer coefficient from the bulk to the surface is 1×10^{-4} m/s. Find the current density. Determine the maximum rate of deposition.

16.14 **Effect of supporting electrolyte in copper deposition.** Consider a system with a "supporting" electrolyte, for example, $CuSO_4$ and a second salt Na_2SO_4 that serves as the supporting electrolyte. The system now consists of Cu^{++}, SO_4^{--}, and Na^+. Assume that only Cu^{++} can react at the cathode. Hence the flux of Cu^{++} should be assumed to be non-zero while that of the other species is zero. Develop an equation to find the diffusion potential developed in this system and the factor by which Cu^{++} flux to the cathode is enhanced.

PART II

Reacting Systems

CHAPTER 17

Laminar Flow Reactor

Learning Objectives

After completing this chapter, you will be able to:

- Apply differential equations of mass transfer to laminar flow with homogeneous and heterogeneous reactions.
- Render the equation to a dimensionless form and identify the key dimensionless groups.
- Solve the equation by a number of mathematical and computational tools.
- Identify two limiting cases of the model, that is, the segregated flow and the plug flow model.
- Compare the mesoscopic approach based on the dispersion model with the exact 2-D model, segregated model, and plug flow model.
- Examine briefly the temperature effect in a laminar flow reactor and modeling of a turbulent flow reactor.

Reactors under laminar flow conditions are often encountered in industrial practice, especially for viscous liquids. The modeling of such systems involves the convection-diffusion equation for mass transfer together with homogeneous and, in some cases, wall catalyzed heterogeneous reactions. Study of such systems is thus important not only in reaction engineering but also in mass transfer analysis. The chapter can be viewed as a case study example of this analysis. We first formulate a full 2-D model, identify the key dimensionless groups, and show some numerical solutions. These solutions will then be used as benchmarks to examine some approximate limiting cases.

Two simplified limiting models are often used. These are a pure convection model and a plug flow model. In the convection model it is assumed that there is no radial diffusion (which is relatively small and therefore neglected) and each fluid element slides past each other with no interaction by molecular diffusion. In the second model, the radial diffusion is rapid and the concentration variation across the radial direction is not assumed to exist; therefore the system is close to plug flow conditions. From this qualitative discussion, you can guess that the convection model applies when the diffusion coefficient is small.

Another approach is to use the dispersion model in conjunction with a mesoscopic model. We compare these various models and this enables you to get a good conceptual understanding of the interplay of the various transport mechanisms involved in the system.

The focus is mainly on isothermal systems with Newtonian flow. This is the simplest example of a laminar flow reactor, enables you to understand the modeling approach, and permits you to extend to other cases. A few examples of complex cases are mentioned in this chapter but not studied in detail. These are non-Newtonian fluids, turbulent flow, and non-isothermal systems. Some additional references are provided so that the interested reader can pursue further study of these topics.

17.1   Model Equations and Key Dimensionless Groups

Consider a fully developed flow of a Newtonian fluid in a circular pipe. The fluid is a mixture containing a compound A at a concentration of $C_{A,i}$ at the inlet. This compound undergoes a chemical reaction in the system with a rate constant of k. The reaction is assumed to be first order. Also the walls are coated with a catalyst and the compound A reacts at the wall with a surface reaction constant of k_S.

The goal is to examine and solve the model differential equation and to obtain the concentration distribution of A as a function of both z, the axial distance from the entrance, and r, the radial coordinate. We assume the concentration is symmetric in the θ direction in cylindrical coordinates.

17.1.1   Dimensionless Model Equations

The differential equation for concentration distribution of species A as a function of r and z given by the convection-diffusion-reaction (CDR) model. The model is given by the following equation:

$$2 \langle v \rangle \left(1 - (r/R)^2\right) \frac{\partial C_A}{\partial z} = D \left[\frac{\partial^2 C_A}{\partial z^2} + \frac{1}{r} \frac{\partial}{\partial r} \left(r \frac{\partial C_A}{\partial r} \right) \right] - k C_A \qquad (17.1)$$

In this equation a parabolic velocity profile has been used, which is valid for a Newtonian fluid. D is the diffusion coefficient of A. The reaction is assumed to be first order. If the reaction is not first order the rate term kC_A is replaced by some "rate" function of C_A and the differential equation becomes nonlinear. We consider only the linear case and the nonlinear case is left as a computational exercise.

The following dimensionless variables are introduced:

- Dimensionless radial position $\xi = r/R$.
- Dimensionless concentration $c_A = C_A/C_{A,i}$.
- Dimensionless axial length $\eta = z/L$; here L is some specified length of the reactor and the variable η goes from 0 to 1 (entrance and exit of the reactor, respectively).

The axial diffusion term in Equation 17.1 (the first term in the square bracket on the right-hand side) can usually be neglected. This is similar to the case of mass transfer with no reaction examined in Section 10.1 and is justified if the Peclet parameter is greater than, say, 10. In dimensionless form the following equation will then be obtained:

$$2(1 - \xi^2)\frac{\partial c_A}{\partial \eta} = \frac{L}{R}\frac{1}{Pe_R}\left[\frac{1}{\xi}\frac{\partial}{\partial \xi}\left(\xi\frac{\partial c_A}{\partial \xi}\right)\right] - Da\, c_A \qquad (17.2)$$

The dimensionless numbers appearing here are the Damkohler number, the Peclet number, and the length to radius ratio. The Damkohler number Da is defined as

$$Da = kL/\langle v \rangle$$

and the Peclet number Pe_R (using radius as a length scale) is defined as

$$Pe_R = \langle v \rangle R/D$$

The length to tube radius ratio, which is the third parameter, can be combined with Pe_R to get a composite dimensionless group B defined as $(L/R)/Pe_R$.

Note: We use the tube radius as the representative length to define the Peclet number while the diameter was used in Chapter 10. The subscript R on the Peclet number is used to distinguish this defnition.

Time Scales

Four time scales can be defined as indicated in a seminal work by Chakraborty and Balakotaiah (2002). These time scales are:

- Mean residence time = $\frac{L}{<v>}$
- Reaction time = $\frac{1}{k_1}$ for a first-order reaction or $\frac{1}{k_n C_{Ai}^{n-1}}$ for a nth-order reaction.
- Radial diffusion time = $\frac{R^2}{D_A}$
- Axial diffusion time = $\frac{L^2}{D_A}$

One should estimate the values of these time constants and the reactor performance will not depend on the phenomena having large time constants. (The effect of axial diffusion is usually neglected since the axial diffusion time is relativity larger compared to other time scales in this system.)

Note that due to low values of the diffusion coefficient in the liquid phase (compared to the gas phase), the radial diffusion time is quite large in liquid. Hence the effect of radial diffusion can often be neglected for liquids, but this may not be so for gases in small diameter tubes as shown in an early and seminal work by Cleland and Wilhelm (1956).

Relation of the Dimensionless Groups to the Time Constants

It is also useful to write the dimensionless groups as the ratio of time constants.

The dimensionless group Pe_R has the following significance:

$$Pe_R = \frac{\text{Mass transport by convection}}{\text{Mass transport by diffusion}}$$

It can be shown to be propotional to the ratio of the radial diffusion time to the convection time.

Likewise Da can be shown to be the ratio of the residence time to the reaction time:

$$Da = \frac{L/\langle v \rangle}{1/k} = \frac{\text{Residence time}}{\text{Reaction time}}$$

Finally the group appearing at the front of the right-hand side of Equation 17.2, denoted as B, can be shown to be

$$B = \frac{L}{R}\frac{1}{Pe_R} = \frac{L/\langle v \rangle}{R^2/D} = \frac{\text{mean residence time}}{\text{radial diffusion time}}$$

This group plays an important role in determining the importance of the radial diffusion. It is the ratio of time spent by fluid in the reactor (on average) to the time it takes for the diffusion to equalize the radial variation in concentration. If this parameter is large, radial diffusion time is small and the concentration is likely to be equalized in the radial direction. Plug flow behavior can be expected. Results shown later will verify this effect.

On the contrary if the B parameter is small, the radial diffusion time is large and the radial diffusion is not likely to start to click in during the mean passage time of the fluid. Fluid elements at each radial position will behave as though they are flowing independently of each other and the segregated flow model is expected to be applicable.

17.1.2 Boundary Conditions

At the inlet $\eta = 0$ we set $c_A = 1$. At the center $\xi = 0$ we have symmetry and therefore $\partial c_A / \partial \xi = 0$.

The boundary condition at the wall in the presence of a heterogeneous wall reaction is of the Robin type and can be expressed as

$$\left(\frac{\partial c_A}{\partial \xi} \right)_{\xi=1} = -Bi_w c_A \text{ at } \xi = 1 \tag{17.3}$$

Thus an additional dimensional parameter, wall reaction Biot number, $Bi_w = k_w R / D_{A-m}$, arises for this problem. This group can be shown to signify the relative ratio of the radial diffusion time to the surface reaction time.

No flux condition can be used for the case of no wall reaction, leading to a Neumann condition. Further if the surface reaction is extremely rapid, the concentration at the wall can be set to zero, leading to a Dirichlet problem. Hence it is interesting to note that all three common conditions arise for this problem depending on the wall condition.

The result of dimensionless formulation thus leads to the following parametric representation of the problem:

$$c_A = c_A(\eta, \xi, ; B, Da, Bi_w)$$

A numerical solution can be readily obtained using the PDEPE software.

The results are usually represented in terms of a cup-mixing average concentration, which is defined as

$$c_{Ab}(\eta) = 4 \int_0^1 \xi(1 - \xi^2) c_A(\xi) d\xi \tag{17.4}$$

Usually the value of the cup-mixing concentration at the exit ($\eta = 1$) is the quantity of design interest. Note conversion equals $1 - c_{Ab}$ at $\eta = 1$.

Limiting cases of the model are now examined.

17.2   Two Limiting Cases

This section examines two limiting cases that depend on the value of parameter B and develops a simplified solution for these limiting cases.

Parameter B, which appears in the front of the radial diffusion term, can be written as

$$B = \frac{L}{<v>} \frac{D}{R^2} = \bar{t}(D/R^2)$$

Observe that this is the ratio of mean residence time to the diffusion time in the radial direction. Equation 17.2 can be written as

$$2(1 - \xi^2)\frac{\partial c_A}{\partial \eta} = B \left[\frac{1}{\xi} \frac{\partial}{\partial \xi} \left(\xi \frac{\partial c_A}{\partial \xi} \right) \right] - Da\, c_A \qquad (17.5)$$

Some limiting cases will now be analyzed depending on the value of the B parameter. It is useful to have a full 2-D solution at hand to compare the limiting cases, which can be simulated using the MATLAB tool PDEPE. An illustrative result is shown in Figure 17.1 as a plot of $c_{Ab}(exit)$ versus B for a fixed $Da = 3$. These results are useful in understanding the role of radial diffusion. We discuss the limiting cases of small B and large B now.

17.2.1   Small B: Pure Convection Model

If the B parameter is small then the radial diffusion term can also be dropped in Equation 17.5, leading to a model called a pure convection model, also known as the segregated flow model. A value of $B < 5 \times 10^{-3}$ was suggested by Merrill and Hamrin (1970) for liquid phase systems as an approximate criteria.

The segregated model has a simple representation by dropping the B term in Equation 17.5:

$$2(1 - \xi^2)\frac{\partial c_A}{\partial \eta} = -Da\, c_A \qquad (17.6)$$

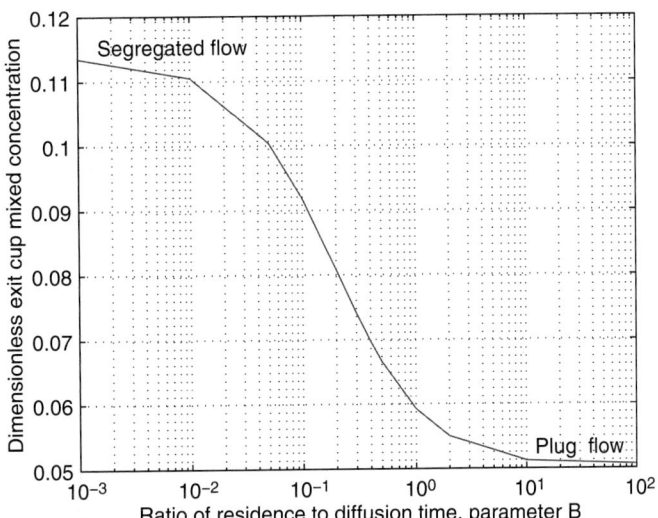

Figure 17.1   Effect of radial diffusion on performance of a laminar flow reactor; results are for $Da = 3.0$.

The model can be readily solved and integrated over the cross-section to model reactor performance. The concentration at any radial position is obtained as

$$c_A(\xi) = \exp\left(-\frac{Da\,\eta}{2(1 - \xi^2)}\right)$$

The reactor exit is at $\eta = 1$ and the exit concentration profile can be computed. The cup-mixing exit concentration can be then evaluated. An illustrative exit profile is shown in Figure 17.2.

Note that the center has higher concentration since the fluid elements at the center have spent only half the mean residence time. For example, for $Da = 1$, the center concentration is $\exp(-Da/2) = 0.6035$. The wall concentration is zero. Note that the diffusion will act as an equalizer and spread these concentration profiles a bit, but in the segregated flow model the radial diffusion is assumed to be absent.

Knowing the radial distribution of the concentration it remains to find the cup-mixing concentration by radial integration as per Equation 17.4. The final result can be expressed as an exponential integral:

$$c_{Ab}(\text{exit}) = \frac{Da^2}{4}\text{expint}(Da/2) + (1 - Da/2)\exp(-Da/2) \tag{17.7}$$

where *expint* denotes the exponential integral.

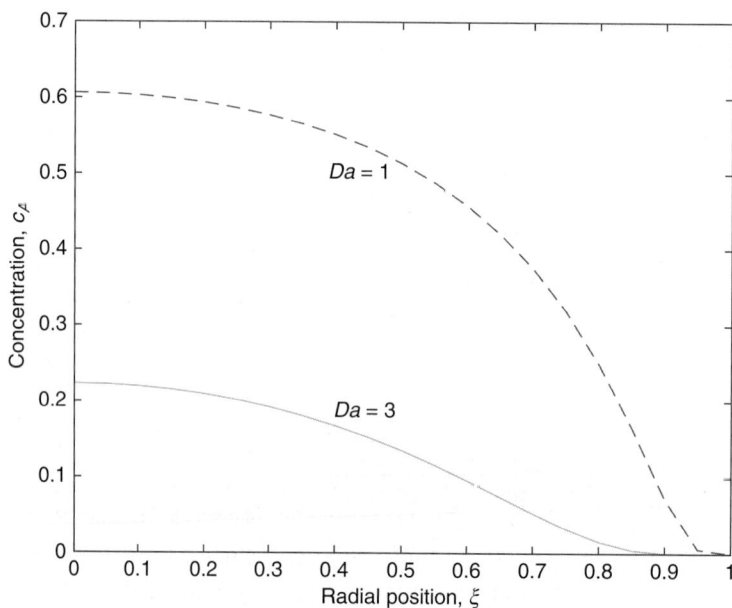

Figure 17.2 Exit radial concentration profile in a laminar flow reactor on the basis of the segregated flow model; results for $Da = 1.0$ and $Da = 3.0$ for first-order reaction.

17.2.2 Large B: Plug Flow Model

A second limiting case is the plug flow model for large B, which results after some cross-sectional averaging of Equation 17.5. The exit concentration is given as

$$c_{Ab}(\text{exit}) = \exp(-Da) \tag{17.8}$$

Figure 17.3 shows a comparison of these limiting case for various values of Da. The difference starts to appear at larger values of Da, that is, for large reactor conversions. The reactor performance can be bracketed between the limits of the segregated model and plug flow model. For example, if $Da = 3$ the segregated model gives a conversion of 89% while the plug flow model gives a conversion of 95%.

The dispersion model based on a mesoscopic average is also used and we now present and show some results for this model.

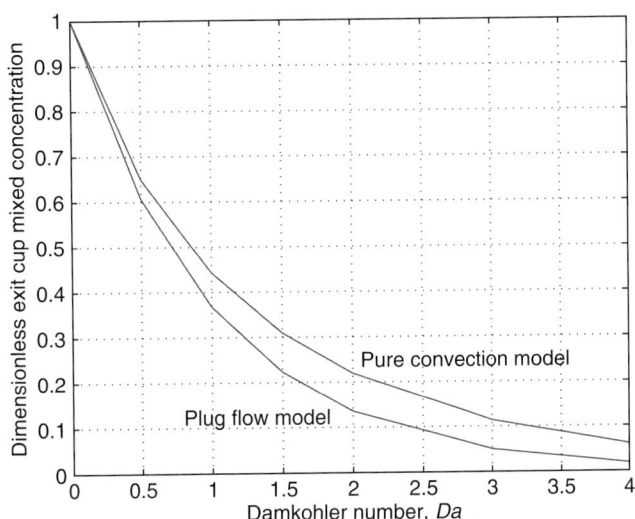

Figure 17.3   Comparison of pure convection and the plug flow model for laminar flow reactor.

17.3   **Mesoscopic Dispersion Model**

Sections 14.2 and 14.3 should be reviewed at this stage. A few concepts are, however, repeated here for ease of understanding.

If the B parameter is in the intermediate range, neither the segregated model nor the plug flow model will be reasonable. There is some smearing of concentration in the radial direction. The full 2-D model should be used for such cases. However, the model can be reduced to a 1-D or mesoscopic model by radial averaging, which is taken up in the following section. The radial averaging needs additional closure terms, which are determined using the Taylor dispersion model. The resulting 1-D model for the reactor is known as the axial dispersion model. These concepts were already introduced in Section 4.2.2 and Section 14.2 and these sections should be reviewed at this stage.

For laminar flow the dispersion coefficient can be obtained from the theory derived by Taylor. The equation for the dispersion coefficient was presented in Section 14.4 and with some rearrangement of the equation, we have the following relation for the dispersion number:

$$D_E^* = \frac{1}{48B} \qquad (17.9)$$

The reciprocal of D_E^* is defined as the dispersion Peclet number:

$$Pe^* = \frac{1}{D_E^*} = 48B$$

In terms of the dispersion Peclet number the dispersion model is written as

$$\frac{1}{Pe^*}\frac{d^2 <c>}{d\eta^2} - \frac{d<c>}{d\eta} - Da <c> = 0 \qquad (17.10)$$

Danckwerts boundary conditions are commonly used in the context of the dispersion model. In the previous equation, $<c>$ is the cross-sectional average concentration of species A.

The result for the average concentration for a first-order reaction using the above-stated boundary condition was given in Section 14.2.1. Please review this section at this stage. The results are summarized in the following for easy reference:

$$<c> (\eta = 1) = \frac{4\alpha \exp(Pe^*/2)}{(1+a)^2 \exp(\alpha Pe^*/2) - (1-a)^2 \exp(-\alpha Pe^*/2)} \qquad (17.11)$$

where

$$\alpha = \sqrt{(1 + 4Da/Pe^*)}$$

The effect of the B parameter is examined now based on the dispersion model. For each B parameter the value of Pe^* is equal to 48 B. The exit concentration is then calculated using Equation 17.11. The results are compared in Table 17.1. Results for the exit concentration are compared for various values of the relative diffusion parameter, B, for a fixed value of Da equal to 3.

The following conclusions are useful to note. The dispersion model should not be used if the B parameter is less than 0.1. For such cases the dispersion model results are closer to the backmixed reactor model rather than to the more realistic segregated flow model. For large values of $B > 1$ there is hardly any difference between the dispersion model and the full-D

Table 17.1   Comparison of the Dispersion Model with the Detailed 2-D Model for $Da = 3$

B values	0.01	0.1	1.0	10.0
Pe^* values	0.48	4.8	48.0	480.0
$\langle c_A \rangle$ (1) 2-D model	0.1102	0.0915	0.0587	0.0507
$\langle c_A \rangle$ (1) Dispersion model	0.2134	0.1123	0.0587	0.0507

model. Plug flow is nearly achieved as well. Thus the dispersion model is useful for the intermediate range of the relative diffusion parameter, that is, for $0.1 < B < 1.0$.

17.4 Other Examples of Flow Reactors

The previous sections provided a complete treatment of a laminar reactor in a pipe flow for a Newtonian fluid. A full 2-D model, dispersion model, segregated flow model, and plug flow model were compared and the range of validity of each model was discussed. In this section, we present some discussion on flow reactors for some other useful cases.

17.4.1 Channel Flow

Channel flow is encountered in some applications, for example, microchannel reactors, monoliths with catalytic walls, and so on. The system is simulated using the appropriate velocity profile and Laplacian in Cartesian coordinates. The governing differential equation is

$$\frac{3}{2} \langle v \rangle \left[1 - (y/H)^2 \right] \frac{\partial C_A}{\partial x} = D \frac{\partial^2 C_A}{\partial y^2} - kC_A \qquad (17.12)$$

A parabolic velocity profile has been used here that is valid for a Newtonian fluid. The wall boundary conditions can be appropriately defined depending on whether there is a wall reaction or not. The PDEPE solver is useful to compute this with only minor changes in the various definitions. If the dispersion model is used (which again is useful in the intermediate range of the B parameter) the dispersion coefficient to be used is given in Section 14.6.

17.4.2 Non-Newtonian Fluids

Many important practical systems (e.g., polymerization reactions) obey the non-Newtonian model for fluid behavior. A power law model is often used for the stress-strain rate relation. In such cases the velocity profile is given as

$$v_z = \frac{3n + 1}{n + 1} \langle v \rangle \left[1 - \left(\frac{r}{R} \right)^{(n+1)/n} \right] \qquad (17.13)$$

Table 17.2   Effect of Power Law Index on the Performance of a Laminar Flow Reactor

B values	0.01	0.1	1.0	10.0
$n = 1$	0.1102	0.0915	0.0587	0.0507
$n = 0.5$	0.0948	0.0791	0.0559	0.0504

Numerical solutions can be obtained by PDEPE using this velocity profile. The effect of the power law index on reactor performance can then be examined. Illustrative results are presented in Table 17.2.

In Table 17.2 the results for the exit concentration are compared for various values of the relative diffusion parameter, B. The value of Du was set as 3. For $B > 1$, the power law index parameter does not appear to be important since the system is moving toward plug flow. For low values of B, the segregated model can be used and Novosad and Ulbrecht (1966) provide useful results for this case. Hence the modeling approach and the observations for Newtonian flow are still valid and only minor modifications are needed to model the non-Newtonian case. Pure convection models for laminar flow are studied by Osborne (1975), which is a useful additional reference.

An alternative is the dispersion model. This will be useful in the intermediate range of the B parameter. The dispersion coefficient is given by the equation of Fan and Huang (1975):

$$Pe^* = \frac{2(3n + 1)(5n + 1)}{n^2} B$$

Equation 17.11 can then be used with this value for Pe^* for a first-order reaction.

17.4.3   Heat Transfer Effects

Temperature effects are handled by using the analogous heat transfer equation:

$$2 \langle v \rangle \rho c_p \left(1 - (r/R)^2\right) \frac{\partial T}{\partial z} = k \frac{1}{r} \frac{\partial}{\partial r} \left(r \frac{\partial T}{\partial r}\right) + k_1(T)C_A(-\Delta H_R) \qquad (17.14)$$

Here k_1, the rate constant, is a function of temperature according to the Arrhenius relation. k is the thermal conductivity of the fluid.

Simultaneous solution of the concentration equation and the temperature equation is needed. The commonly used assumptions are no axial conduction of heat and no change in viscosity, The coupled equations can be easily set up in MATLAB with PDEPE. Churchill and Yu (2006) also provide some numerical solutions and indicate that the computer implementation and solution of such coupled models are within the capability of ordinary desktop computers.

A more general case where the variation of viscosity is important as well as multiple reactions, and so on, requires a detailed solution. Fluid velocity has to be computed simultaneously due to visocity variation with temperature. Hence this is an example of a problem where all three transports (momentum, heat, and mass) are involved. COMSOL-based tutorials are available on the web and these may be useful for such complex situations. A detailed and useful study of the industrially important styrene polymerization reaction is presented in Wyman and Carter (1976). This is a system with large heat effects and significant change in viscosity as one goes from the monomer at the entrance to a polymeric fluid at the exit of the reactor, if all goes well. The system can become unstable for larger tube diameters and for large inlet temperatures and the mathematical models provide us the tools to identify the stable regions of operation. Simulation models have proven to be of great value in safe operation of such reactors in addition to fixing the optimum operating conditions for the reactor.

17.4.4 Turbulent Flow Reactor: 2-D Model

2-D models for turbulent flow can be set up in a similar manner. The key points to note are that the velocity and the concentration should be interpreted as time-averaged values. The diffusion coefficient is also augmented by adding the eddy diffusivity, D_t. The model representation in terms of the time-averaged quantities is

$$\bar{v}_z(r)\frac{\partial \overline{C}_A}{\partial z} = \frac{1}{r}\frac{\partial}{\partial r}\left(r\,(D + D_t)\frac{\partial \overline{C}_A}{\partial r} \right) - k\overline{C}_A \tag{17.15}$$

The velocity profile can be approximated by the 1/7th power law as a simplification:

$$v_z(r) = \frac{60}{49}\langle v \rangle \left[1 - (r/R) \right]^{1/7}$$

An alternative is to use the Prandlt universal velocity profile. Note that the velocity in turbulent flow is a steep function of radius in contrast to the smooth parabolic function characteristic of laminar flow of Newtonian fluid.

The eddy diffusion coefficient is also a function of position and hence is retained inside the derivative sign in Equation 17.15. The following relation for turbulent stress and eddy viscosity proposed by Pai (1953) is useful:

$$\tau_{rz}^{(t)} = \mu_t \frac{\partial v_z}{\partial r} = 0.9835 \rho v_f^2 \left(1 - \frac{y^+}{H^+}\right) \left[1 - \left(1 - \frac{y^+}{H^+}\right)^{30}\right] \tag{17.16}$$

where v_f is the friction velocity. y^+ is the dimensionless distance measured from the wall. The turbulent diffusivity is then calculated as ν_t/Sc_t with a value of 0.9 assigned for Sc_t, the turbulent Schmidt number.

Further complexity arises for non-first-order reactions. Additional contribution due to the fluctuating concentration results in the rate term. For example, for a second-order reaction an additional rate term $k_2 < C_A' >^2$ is needed. However, these effects can usually be neglected, except for gas phase reactions at high temperature, as pointed out by Glassman (1966).

Detailed computations of tubular reactors under turbulent flow conditions have been done by Yu and Churchill (2006). They provide useful benchmark results for comparison of the numerical tools, which may be useful for solution of the governing equations. They found that the central finite difference is fairly accurate and also indicate that these days, the numerical study of these type of problems is entirely within the capacity of undergraduate students.

Most CFD codes have provision for simultaneous solution of the velocity profile and the concentration profile. COMSOL-based simulation modules are also available.

17.4.5 Axial Dispersion Model for the Turbulent Case

A simplified model is obtained by using the axial dispersion model. The correlation for an axial dispersion coefficient in turbulent flow is discussed in Section 14.6, and it can be used to predict conversion in the reactor (see Ramachandran and Mashelkar (1976), for detailed numerical results on the effect of axial dispersion on the performance of a turbulent flow reactor).

In general the departure from plug flow is small unless the reactor is operated at high conversion levels. The following criteria was suggested by

Ramachandran and Mashelkar to estimate the deviation from plug flow:

$$\frac{c_A}{c_{A,p}} = 1 + Re^{-1/8} \frac{d_t}{\langle v \rangle} \frac{R_A(C_{A,p})}{C_{A,p}} \ln\left[\frac{R_A(C_{A,p})}{R_A(C_{A,i})}\right] \quad (17.17)$$

Here $c_{A,p}$ is the concentration if plug flow existed. Equation 17.17 is based on the Blasius correlation for the friction factor, the Taylor model for dispersion in turbulent flow, and the perturbation analysis of the dispersion model done by Horn and Parish (1967) to study small deviation effects from plug flow.

For a first-order reaction this reduces to

$$\frac{c_A}{c_{A,p}} = 1 - Re^{-1/8}\frac{d_t}{L}Da^2$$

For example if $Da = 3$ and $Re = 5000$, the ratio is 0.969 for an L/d_t ratio of 100. The conversion values would be only 3% lower, showing that the deviation from plug flow is not significant even at this conversion level.

Summary

- An important application of the convective transport equations is in the simulation of the laminar flow reactor. A 2-D model shows that the Damkohler number and modified Peclet number (B parameter) are the key dimensionless groups that influence reactor performance.

- A laminar flow model can be simplified in some cases where the radial diffusion term is small. This holds when the B parameter is small and leads to a model called the pure convection model. If the velocity variation is neglected, the reactor can be modeled as a plug flow model. This will hold for large B. These two models provide two limiting cases for the complete 2-D laminar reactor model.

- The dispersion model can also be used to model the performance of the laminar flow reactor. This model predicts the laminar reactor performance in the intermediate range of values for B. The dispersion model should not be used if the B parameter is small, say less than 0.1. In such cases the segregated flow model should be used.

- A detailed 2-D model can be solved numerically and code based on the PDEPE solver given in the text in Section 8.11 can be adapted to simulate the reactor. This is useful for a variety of similar applications, for example, channel flow, non-Newtonian fluids, turbulent flow, and exothermic reactions (where simultaneous solution of the heat equation is needed).

Both homogeneous reaction in the bulk fluid and heterogeneous reaction on the wall coated with a catalyst can be simulated by this numerical tool.

- Heat effects can be included by adding an additional heat equation for the temperature profile and incorporating the Arrhenius law for the effect of temperature on the rate constant. The coupling can be strong for highly exothermic reactions and can lead to numerical difficulties in the solution. The reactor can show instability and temperature run-away. The identification of safe regions of operation using model-based computations is extremely important in control of such exothermic reactors.

- Turbulent flow reactors need additional model closures. The velocity profile needs to be modified and a profile for turbulent diffusivity must be included in the species mass conservation equation. In general turbulent reactors are close to plug flow unless high conversions (>95%) are encountered.

Review Questions

17.1 Define the relative diffusion parameter, B, and express this as a ratio of two time constants.

17.2 What is the role of the B parameter in determining which type of model to use for the laminar flow reactor?

17.3 When will the plug flow model apply to a laminar flow reactor?

17.4 When will the segregated flow model (pure convection model) apply for a laminar flow reactor?

17.5 When would you use the dispersion model for laminar flow?

17.6 Under what conditions would you not use the dispersion model to simulate a laminar flow reactor?

17.7 What additional parameters are needed to model a turbulent flow reactor?

Problems

17.1 **Conversion in laminar flow reactor.** An aqueous solution containing a reactant A is flowing in laminar flow in a reactor and undegoes a first-order reaction with a rate constant of 0.5 s^{-1}. The physical properties are the same as water and the diffusion coefficient is estimated as $2 \times 10^{-9} \text{ m}^2/\text{s}$. The tube diameter is 50 mm, the length is 3 m, and the average velocity is 0.1 m/s. Find the conversion based on a full 2-D simulation. PDEPE is directly useful here.

17.2 **Time constants and model to use.** In the previous problem estimate the various time constants and state what type of simpler 1-D model would be most appropriate rather than a full 2-D numerical simulation.

17.3 **Second-order reaction.** For the same conditions find the conversion if the reaction is second order with a rate constant of $2.5 \times 10^{-4} \, \text{m}^3/\text{mol s}$.

17.4 **Comparision of plug flow and laminar flow.** A reactor gives a conversion of 95% if operated as a plug flow. But the fluid is highly viscous and laminar flow is applicable. What would be the conversion in the reactor?

17.5 **Pure convection model: zero-order reaction.** Extend the analysis of the pure convection model shown in Section 17.1.1 to a zero-order reaction. Verify the following equation for the conversion given by Levenspiel (1999):

$$c_A(\text{ exit }) = \left(1 - \frac{Da}{2}\right)^2$$

where Da is the Damkohler number for a zero-order reaction. How is it defined? Find the length for complete conversion for a zero-order reaction based on the above model if the length needed is 3 m for a plug flow situation.

17.6 **Pure convection model: second-order reaction.** Extend the analysis in Section 17.1.1 to a second-order reaction. Verify the following equation for the conversion given by Levenspiel (1999):

$$c_A(\text{ exit }) - 1 - Da \left[1 - \frac{Da}{2} \ln\left(1 + \frac{2}{Da}\right)\right]$$

where Da is the Damkholer number for a second-order reaction. How is it defined?

17.7 **Multiple reactions.** Set up the model for a series reaction taking place in a pipe under laminar flow conditions:

$$A \rightarrow B \rightarrow C$$

Solve with PDEPE and compare the limiting cases of the convection model and the plug flow model. Because of the velocity profile, the maximum concentration of the intermediate species, B, will be less than that for a plug flow reactor. Verify this observation.

17.8 **Non-Newtonian fluids.** Consider a fluid with a power law index of 0.5. The value of the relative diffusion parameter, B, is 0.6. The Damkohler number is 3. Compare the 2-D model with the axial dispersion model.

17.9 **Laminar flow in a channel with wall reaction.** Often wall-coated channels are used in pollution abatement. In such cases there is no homogeneous reaction and the wall reaction is the dominant term. Set up and solve the 2-D model for the following conditions: channel width = 5 mm; channel length = 2 m; average velocity = 5 cm/s, wall rate constant = 10^{-4} m/s; $D = 2 \times 10^{-4} \, \text{m}^2/\text{s}$. An alternative is to use the mesoscopic model. Set up this model as well and compare the predictions.

17.10 **Homogeneous and heterogeneous reactions.** Solve the laminar flow reactor where both homogeneous and heterogeneous reactions are taking place for the following values of the dimensionless parameters: $B = 1.0$; $Da = 4$; $Bi_w = 2.0$. Find the cup-mixing concentration at the exit of the reactor. Also model this as a plug flow and compare the 2-D model with the plug flow model.

17.11 **Axial dispersion model for turbulent flow.** A reactor is operated at a Reynolds number of 10^5. The tube to pipe diameter is 100 and the Damkohler number is 6. Estimate the conversion and the deviation from plug flow.

CHAPTER 18

Mass Transfer with Reaction: Porous Catalysts

Learning Objectives

After completing this chapter, you will be able to:

- Use the equation of mass transfer to set up models for porous catalyst a where diffusion and reaction are occurring in parallel.
- Solve the equations for simple 1-D problems analytically for first-order and zero-order reactions.
- Use semi-analytic tools to solve problems involving nonlinear reactions.
- Understand the concept of effectiveness factor and how diffusion masks the true kinetics.
- Learn and use the method of orthogonal collocation to solve nonlinear diffusion-reaction problems in 1-D.
- Learn simple finite difference methods to solve 2-D problems in regular geometry.
- Model diffusion with reaction with heat effects and evaluate some of the complexities associated with such systems.
- Solve a simple plug model for a packed bed reactor incorporating local internal diffusional effects.

In industrial practice, reactions where a fluid species reacts over a porous solid catalyst is very common. The concentration of the reacting species may vary in the interior of the catalyst especially if the diffusion rate is low or if the reaction rate is rapid compared to diffusion. Thus the rate calculated on the basis of the external concentration may not be representative and an average rate that accounts for the concentration variation in the pores of the catalyst is needed in order to design industrial gas–solid catalyzed reactions. Hence mass transfer effects are of importance in design of catalytic gas–solid reactions. It is the goal of this chapter to examine the role of mass transfer in porous catalysts.

The chapter is organized in the following manner. First we show some industrially relevant examples and properties of common catalysts. Then the diffusion-reaction problem is analyzed for simple cases where diffusion is only in one coordinate direction. First-order and zero-order reactions are studied where analytical solutions are possible. Then we show how semi-analytical solutions can be obtained for nth-order reactions. The important concept of effectiveness factor, which is a measure of pore diffusional effects is introduced, and its relation to a key dimensionless parameter, the Thiele modulus, is discussed. Application of this relationship results in the correct interpretation of kinetic data measured from laboratory reactors and examples of this are presented.

Two computational tools for studying the effect of internal diffusion for nonlinear and multistep reactions are then discussed. Finally the effect of heat generation due to reaction is studied based on the analysis of simultaneous heat and mass transfer within the catalyst. Many complex effects such as multiple steady states, effectiveness factor larger than one, and so on, are observed in such systems and we show some examples of this behavior. Linking of the local model to the mesoscopic models for the gas phase is then shown with some examples. Overall, the chapter provides a solid fundamental background for further study of catalytic reaction engineering in detail.

18.1　Catalyst Properties and Applications

Some examples of industrial scale gas–solid catalyzed reactions are shown in this section followed by properties useful to characterize the catalyst. Applications are widespread and cover a wide range of industries. We cite only a few large-scale applications in the following. Smaller scale applications are also important, especially in the fine chemicals and pharmaceutical industries.

- Sulfuric acid manufacture: The first step is in this process is the oxidation of sulfur dioxide over a vanadium pentaoxide catalyst. Worldwide production is on the order of 140 million tons.
- Ammonia production: Reaction of nitogen and hydrogen over iron gauze catalyst at high pressures is used to produce ammonia. This is the starting material for the fertilizer industry. This is also a large-scale process.
- Ethylene oxide production from ethylene: Ethylene and oxygen (air with 95 mole % of oxygen) are mixed in a ratio of 1:10 by weight and passed

over a catalyst consisting of silver oxide deposited on an inert carrier such as corundum. Commercial processes operate under recycling conditions in a packed bed multitubular reactor. The reaction is highly exothermic and heat management and heat recovery are important (not addressed in this book). It can also lead to a side reaction of complete oxidation that needs to be minimized as well.

- Methanol synthesis from a CO and H_2 gas mixture over a Cu-based catalyst. Packed bed tubular reactors are commonly used.
- Maleic anhydride production by butene oxidation. Fluidized bed reactors are used since they provide a uniform temperature environment and facilitate control of the reactor temperature.

The catalysts are generally porous with a large internal area and the species diffuses into the pores and simultaneously interacts with the pore walls and undergoes a chemical reaction. The products counter-diffuse back to the bulk gas. The diffusion in the pores sets up a concentration gradient within the pellet and mass transfer models are used to calculate this. The rate of reaction is lowered (generally) due to this concentration gradient and a primary goal is to calculate this reduction factor, which is usually calculated as an effectiveness factor. The methodology and the results will now be studied. We first review some properties needed to characterize a given catalyst.

18.1.1 Catalyst Properties

The bulk density is defined as the mass of catalyst per total volume, which includes the solid matrix and the pores. The porosity ϵ_p is defined as the volume of the pores to the total volume. Hence the bulk density and the solid density are related as $\rho_b = (1 - \epsilon_p)\rho_s$.

The internal surface area of the catalyst, S_i, is another important property and usually expressed on the basis of the mass of the catalyst. Most of the surface area is in the internal pores and hence surface area is a usually large, on the order of 1000 m^2/g. This should not be confused with the external surface area, denoted as S_p, which for example is equal to $3/R\rho_b$. This is external area per unit mass of the catalyst for a spherical particle of radius R.

The average pore radius of a catalyst is defined as

$$\bar{r}_{pore} = \frac{2V_{pore}}{S_i}$$

Here V_{pore} is the pore volume per unit mass of the catalyst. There is usually a pore size distribution within the catalyst, which is characterized by a distribution function, f. Thus $f(r)dr$ denotes the fraction of the pores in the pore radius range r and $r + dr$. The distribution can be unimodal, which represents a case where there are only pores near one size range. An average pore radius can be defined as the first moment of the distribution:

$$\bar{r}_p = \int_0^\infty r f(r) dr$$

The distribution of pore sizes can be measured by mercury porosimetry, where Hg is forced into the pores under pressure. The pressure required is a measure of the radius of the pore, with the smaller pores requiring higher pressures for the mercury to penetrate.

Many catalysts have a bimodal distribution and consist of macropores and micropores. In such cases the distribution function shows two peaks, one at the average micropore diameter and one at the average macropore diameter. Correspondingly a micropore porosity and a macropore porosity can be defined. These properties are useful, for example, in the Wakao-Smith model to predict the internal diffusivity (see Chapter 7).

18.2   Diffusion-Reaction Model

Diffusion sets up an internal concentration gradient in the catalyst as noted in Section 2.1.2 and it is necessary to calculate the concentration drop and the average reactant concentration in the catalyst so that the rate of reaction in the presence of diffusional gradients can be calculated. This is the goal of this section, followed by examples for simple power-law type reactions.

The model equations are simply truncated versions of the general equations of mass transfer and are known as the diffusion-reaction equations. For the steady state case the rate of diffusion balances the local rate of reaction and the model equation is of the form

$$D_e \nabla^2 C_A + R_A = 0 \tag{18.1}$$

where C_A is the concentration of the diffusing species in the pores of the catalyst and D_e is an effective diffusivity of the species A in the pores of the catalyst. One assumption here is that diffusion-induced convective flux term is neglected here. Also, the diffusivity is taken as a constant. The effective diffusivity can be calculated by the models described in Chapter 7.

A note on the basis used in the definition of the local rate of reaction R_A is in order. The local rate at any point within the catalyst is defined based on the local volume of the catalyst. This includes the solid and the pores and is treated as a volumetric or homogeneous source of mass. The actual reaction takes place by adsorption followed by reaction on the pore surface and these are lumped into a composite "homogeneous" type of rate based on the catalyst volume.

Rate is also defined in some sources based on the unit mass of the catalyst, per surface area of the catalyst, on the mass of active metal loaded on the surface of an inert support, and so on. All these are valid definitions as long as the basis is clearly stated. A proper conversion factor needs to be applied in order to base the rate based on these other definitions on the unit catalyst volume basis used in this chapter.

Solutions to the diffusion-reaction equation in complex shapes with 3-D models and mixed boundary conditions (Robin in some parts of the control surface and Dirichlet over the other area) are difficult and generally require numerical solutions. We examine here simple 1-D cases by taking simple geometries. These geometries are slab, long solid cylinder, and solid sphere, where the Laplacian can be represented as a function of one spatial coordinate. Examples of these were already introduced in Chapter 2 and should be revisited.

The boundary condition at the surface can be of the Dirichlet type, or the Robin type as discussed later. Also, the method of solution depends on the kinetics of the reaction. For zero and first-order kinetics, analytic solutions can be obtained. Zero-order reactions have to be given special consideration as shown later since the concentration can become zero at some point in the interior of the catalyst. Nonlinear kinetics can be solved semi-analytically by a method of p-substitution for simple slab geometry. More general cases involving multiple species or multiple reactions or simultaneous heat transfer effects are best handled with numerical methods. We will study some numerical tools in this chapter as well for some simple cases; these form the foundation for the solution of other complex cases.

18.2.1 First-Order Reaction

First we analyze a first-order reaction in simple 1-D geometries, starting with the slab geometry first. Note that this section builds on the models shown in Section 2.1.2, which should be revisited at this point.

For a first-order reaction $R_A = -k_1 C_A$ and the differential equation in Equation 18.1 can be represented for slab geometry as

$$D_e \frac{d^2 C_A}{dx^2} = k_1 C_A \qquad (18.2)$$

The scaled concentration c_A is $C_A/C_{A,b}$, where $C_{A,b}$ is the concentration in the bulk gas outside the catalyst while ξ is a dimensionless spatial location equal to x/L. The L is taken as half the thickness of the slab in order to take advantage of the symmetry for ease of solution. The distance x starts now from the center of the slab.

Using these variables the dimensionless representation is

$$\frac{d^2 c_A}{d\xi^2} = \phi^2 C_A \tag{18.3}$$

The key dimensionless parameter appearing in the above dimensionless formulation is ϕ and is called the Thiele modulus. Thus the square of the Thiele modulus is defined as

$$\phi^2 = k_1 L^2 / D \tag{18.4}$$

The Thiele modulus (squared) can be interpreted as the ratio of the characteristic time for diffusion L^2/D to the time for reaction $1/k$. It can also be interpreted as the ratio of the relative rate of reaction to the rate of diffusion within the pores:

$$\text{Thiele squared} = \frac{\text{diffusion time}}{\text{reaction time}} = \frac{\text{reaction rate}}{\text{diffusion rate}}$$

The boundary condition at the plane of symmetry for slab is $dc_A/d\xi = 0$. The plane of symmetry is taken at ξ equal to 0. This boundary condition is also applied at the center for the long cylinder and sphere as well.

The boundary condition at the surface $\xi = 1$ depends on whether external mass transfer effects are accounted for or not. We postulate the existence of a thin film or a boundary layer near the surface in accordance to film theory. This thin film offers all the resistance for gas transport from the bulk to the surface of the catalyst. A balance of fluxes at the surface can then be represented as

$$k_m(C_{A,b} - C_{A,s}) = D_e \left(\frac{dC_A}{dx}\right)_{x=L}$$

where k_m is the external (gas film) mass transfer coefficient and the left-hand side above represents the mole of A transported from the gas to the surface. The right-hand side is the rate of diffusion into the catalyst surface. The dimensionless version of the condition is as follows and is of the third kind (Robin):

$$\left(\frac{dc_A}{d\xi}\right)_{\xi=1} = Bi\left[1 - (c_A)_{\xi=1}\right]$$

where Bi for external mass transfer is defined as

$$Bi = \frac{k_m L}{D_e}$$

The Biot number represents the ratio of the external transport rate to the internal transport rate and has the same significance as in heat transfer:

$$\text{Biot number} = \frac{\text{externl mass transfer rate}}{\text{internal diffusional rate}}$$

For large values of Bi we would expect the problem to reduce to the Dirichlet type (external transport rate would be large and therefore there is no external transport resistance). Hence if there are no transport resistances on the film near the catalyst then the Dirichlet condition $c_A = 1$ can be directly applied at $\xi = 1$, which simplifies the math. The parametric representation of the problem is then simpler and is represented as: $c_A = c_A(\xi; \phi)$.

Slab Solution for Dirichlet Condition

The solution of Equation 18.3 for large Bi is as follows:

$$c_A = \frac{\cosh(\phi \xi)}{\cosh(\phi)}$$

This was already shown in Section 2.4 for a similar problem of oxygen diffusion with reaction in a pool of liquid.

Effectiveness Factor

The effectiveness factor is the quantity of interest in the design of catalytic reactor and is defined in words as

$$\text{Effectiveness factor} = \frac{\text{actual rate}}{\text{maximum rate}}$$

The maximum rate is the rate based on the external gas concentration. This is, of course, the same as the rate based on the surface concentration for large Bi. It is equal in this case to $k_1 C_{As}$.

The actual rate can be calculated in two ways: by taking the local rate and integrating over the whole catalyst or by calculating the rate of transport into the catalyst at the surface by using Fick's law.

In the first case we have to integrate the concentration profile while in the second case we need the derivative of the concentration profile at the surface (in order to apply Fick's law here). Both cases lead to the same answer.

The effectiveness factor can be shown to be the same as the average rate of reaction in dimensionless units. The following expression can be easily derived:

$$\eta = \int_0^1 c_A \, d\xi = \frac{\tanh \phi}{\phi} \tag{18.5}$$

The concentration profiles are shown in Figure 18.1 together with illustrative values for the effectiveness factor. Note that the concentration profiles get steeper as the Thiele modulus is increased with a corresponding drop in the effectiveness factor. For example for $\phi = 10$, it is hard for the reactant A to reach the interior of the catalyst, leading to an effectiveness of only 0.1.

Slab: External Transport Effects

The solution for any general case where a Robin condition applies is slightly lengthy:

$$c_A = \frac{Bi \, \cosh(\phi \xi)}{\phi \sinh(\phi) + Bi \cosh(\phi)} \tag{18.6}$$

The reader may wish to verify the algebra and also show that the Dirichlet limit is approached as $Bi \rightarrow \infty$. Such limiting cases should always be checked in general.

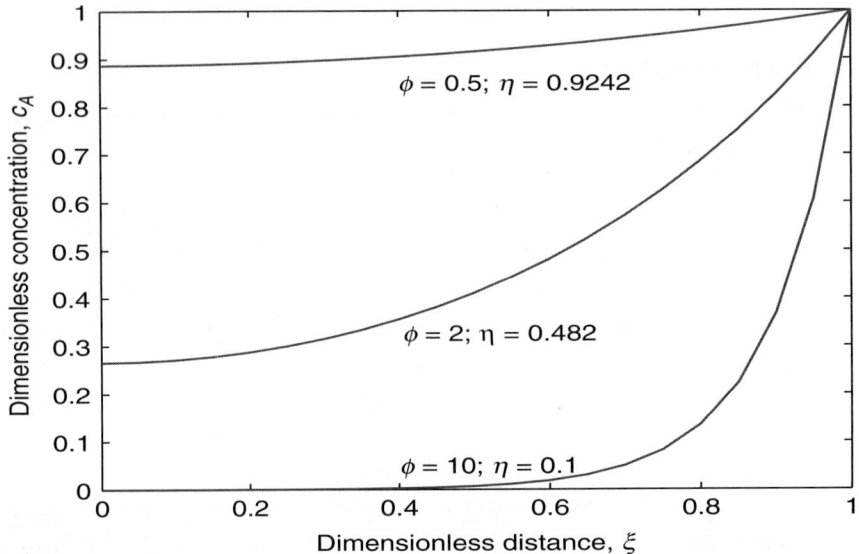

Figure 18.1 Concentration profiles for a first-order reaction in a catalyst in a slab shape. Dirichlet condition used at $\xi = 1$.

The effectiveness factor expression for the above case can be derived as

$$\eta = \frac{\tanh \phi}{\phi} \left(\frac{Bi}{Bi + \phi \tanh(\phi)} \right) \tag{18.7}$$

Long Cylinder: Dirichlet Problem

The governing equation for a cylinder in dimensionless form is

$$\frac{1}{\xi^s} \frac{d}{d\xi} \left(\xi^s \frac{dc_A}{d\xi} \right) = \phi^2 c_A \tag{18.8}$$

Here $s = 1$ the geometry parameter and ξ is the dimensionless parameter, r/R. Note that if s is set as zero the slab equation is recovered. Similarly if $s = 2$ the model for a sphere is obtained.

The previous differential equation for a cylinder is now the modified Bessel equation. This can be seen by expanding the Laplacian and rewriting the equation as

$$\xi^2 \frac{d^2 c_A}{d\xi^2} + \xi \frac{dc_A}{d\xi} - \xi^2 \phi^2 c_A = 0$$

The general solution is

$$c_A = A_1 I_0(\phi \xi) + A_2 K_0(\phi \xi)$$

Students may wish to verify the general solution using MAPLE or the information given in advanced engineering mathematics textbooks such as Kreyzig (2010).

The constants of integration are readily found for a solid cylinder. The function K_0 goes to infinity as $r \to 0$ and hence A_2 is set as zero so that the concentration remains finite at the center. The second constant is then fitted from the boundary condition at $\xi = 1$, which may be of the Dirichlet or Robin type. For the Dirichlet type (infinite Biot number) the solution is

$$c_A = \frac{I_0(\phi \xi)}{I_0(\phi)}$$

The effectiveness factor is the ratio of the actual rate to maximum rate. The actual rate is the integral of the local rate multiplied by the local volume:

$$\text{Actual rate} = \int_0^R 2\pi r L k_1 C_A dr$$

The maximum rate is

$$\text{Maximum rate} = \pi R^2 L k_1 C_{A,s}$$

L here is the length of the cylinder.

The ratio the two rates leads to the following expression in dimensionless version:

$$\eta = 2 \int_0^1 \xi c_A d\xi$$

An alternative expression for the effectiveness factor is found from the flux crossing into the cylinder at the surface:

$$\eta = \frac{2}{\phi^2} \left(\frac{dc_A}{d\xi} \right)_{\xi=1}$$

Students should verify that this will be the expression from flux considerations at the surface; both expressions will lead to the same final answer, which is

$$\eta = \frac{2}{\phi} \frac{I_1(\phi)}{I_0(\phi)} \tag{18.9}$$

This is valid for large Biot numbers; that is, when the gas film transport is not rate limiting. The solution to the case of a finite Biot number is left as an exercise.

Sphere

Here Equation 18.8 applies with $s = 2$. A variable transformation $f = \xi c_A$ reduces the equation to a form that can be integrated more easily (see exercise problem 1). The solution is

$$c_A = \frac{\sinh(\phi\xi)}{\xi \sinh(\phi)} \tag{18.10}$$

The effectiveness factor is given as

$$\eta = 3 \int_0^1 \xi^2 c_A d\xi$$

and the final result is

$$\boxed{\eta = \frac{3}{\phi^2} [\phi \coth(\phi) - 1]} \tag{18.11}$$

Example 18.1 shows an illustration of the drop in the concentration profile due to pore diffusion.

Example 18.1   Effectiveness Factor and Center Concentration in a Spherical Catalyst

A first-order reaction with a rate constant of 10 sec^{-1} is taking place in a porous spherical catalyst. The effective diffusion coefficient is estimated as 10^{-6} m^2/s. Estimate the rate of reaction for a 1 mm and 5 mm diameter pellets. Also evaluate the center concentration for these cases to get a feel of the drop in the concentration in the pellet. Gas phase concentration is estimated as 30 mol/m^3.

Solution

For the 1 mm case the Thiele modulus is calculated as 1.5811 and the corresponding effectiveness factor is 0.865. The rate is calculated as $k_1 C_{Ab} \eta = 260$ mol/m^{3}s. The center concentration calculation requires taking the limit of Equation 18.10 and is equal to $\phi/\sinh(\phi)$. The value for a ϕ of 1.58 is 0.68, a 32% drop.

For the second case we have a Thiele modulus of 7.9057 and η of 0.3315. The rate is only 99 mol/m^{3}s now and the center concentration is 0.0058, a significant drop. The reactant is not able to reach the interior of the sphere.

Shape Normalization

A useful definition is the shape-normalized Thiele modulus. This definition can be used as an approximation for complex shapes. A full 3-D solution to the diffusion-reaction model may often not be needed.

We define a characteristic length parameter (equivalent half-length of a slab catalyst) as

$$L_c = V_p/S_e$$

where V_P is the volume of the particle and S_e is the external surface for diffusion of the gas into the porous solid.

A Thiele modulus is then defined based on this length scale as

$$\phi^* = L_c \sqrt{\frac{k_1}{D_A}}$$

irrespective of the actual shape of the catalyst; the slab equation is used for all the geometries as an approximation with the above-defined Thiele modulus:

$$\eta = \frac{\tanh(\phi^*)}{\phi^*}$$

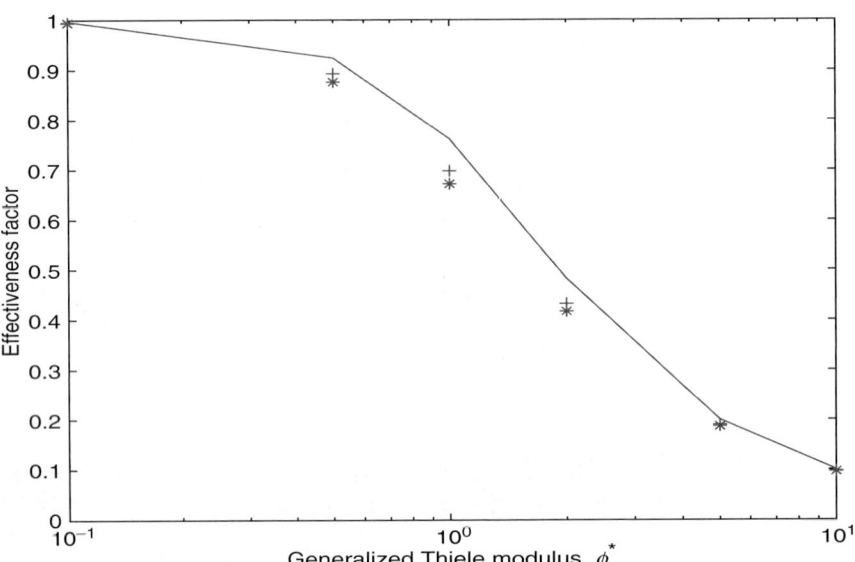

Figure 18.2 Effectiveness factor for all three geometries plotted in terms of the shape-normalized Thiele modulus. + is a cylinder and * is a sphere from analytical solutions; the solid solid line is for a slab.

Note that ϕ^* for a cylinder is $\phi/2$ (the length parameter is now $R/2$). Similarly it is $\phi/3$ for sphere. The length parameter is $R/3$. A plot of the effectiveness factor for all three geometries is shown in Figure 18.2.

We find from this figure that the slab approximation is reasonable for all three geometries. The results for all three geometries get compacted with this modified (shape-normalized) Thiele modulus. The maximum error is on the order of 10% and is observed at an intermediate value of the Thiele modulus. The concept can be used for complex shapes; Example 18.2 shows one application.

Example 18.2   Annular Ring Catalyst

As an example, consider a catalyst shaped as an annular ring, with three dimensions: inside radius = R_i; outside radius = R_o; length = L.

The volume of the catalyst is $\pi(R_o^2 - R_i^2)L$, while the external surface area is the sum of the area at the top and the bottom of rings, the outer surface area, and the inner surface area:

$$S_e = 2\pi(R_o^2 - R_i^2) + 2\pi R_o L + 2\pi R_i L$$

Hence the equivalent length of the "slab" catalyst for diffusion is the volume of the catalyst divided by the external area:

$$L_c = \frac{R_o - R_i}{2} \frac{1}{1 + (R_o - R_i)/L}$$

Thiele modulus based on this length can be used and the slab model can be used as an approximation.

For a solid cylinder $R_i = 0$ and $R = R_o$; the length parameter is $R/[2(1 + R/L)]$. If $L/R >> 1$ this further reduces to $R/2$, the value of the characteristic length for the long cylinder case.

External Resistance

It is often useful to know if external resistance is important or whether it can be neglected and the simple Dirichlet condition used as the boundary condition.

At steady state the rate of external mass transfer is $k_m S_e (C_{A,b} - C_{As})$, which is also equal to the rate of reaction in the interior of the catalyst, $V_p k_1 C_{A,s}\eta$. The latter can be also represented as $V_p \bar{R}_A$, which is the measured rate. Hence

$$C_{A,b} - C_{A,s} = \frac{V_p \bar{R}_A}{k_m S_e}$$

If the rate is measured and the external mass transfer coefficient is measured independently or estimated from correlations we can find the concentration drop in the film from the previous equation. If the drop in the external film should be small relative to the bulk concentration, then the external mass transfer resistance can be neglected.

Apparent Rate Constant

The significance of η and the effect of Thiele modulus on the measured rate of reaction is discussed here.

Note that the local rate of reaction is $k_1 C_A(x)$ per unit catalyst volume. As the concentration is changing within the slab, the rate is different at different points in the slab. What one would observe is an average rate, which can be calculated using the definition of η. Hence

$$-\bar{R}_A = k_1 C_{A,b}\eta$$

Using the expression for η (for the slab model with no external resistance)

$$-\bar{R}_A = k_1 C_{A,b} \frac{\tanh \phi}{\phi}$$

an apparent rate constant can be defined as the observed rate divided by the bulk concentration. This is observed to be

$$k_{app} = k_1 \frac{\tanh \phi}{\phi}$$

which is not in general equal to the true rate constant k_1. This indicates that the measured value should be appropriately corrected for diffusional effects. Example 18.3 illustrates this.

Two limiting cases are worth noting:

- No pore resistance: This holds for low ϕ. The effectiveness factor $\eta = 1$. The average rate, which is also the observed rate, is

$$-\bar{R}_A = k_1 C_{A,b}$$

The apparent rate constant and true rate constants are the same. The true rate constant can therefore be directly estimated from the measured data in this case. Also note that the particle size has no effect. One way to check this is to take two relatively small particle sizes. The measured rate will be the same if there are no pore resistances due to diffusion.

- Strong diffusional resistance, which holds for high ϕ; Usually $\phi > 3$ is taken as the criteria. Here we can show that $\eta = 1/\phi$ since the function $\tanh\phi$ will now be one. Hence the measured rate would be

$$-\bar{R}_A = C_{A,b} \left(\frac{\sqrt{D_{eA} k_1}}{L} \right) \tag{18.12}$$

The apparent rate constant is now the bracketed term on the right-hand side. The rate is inversely proportional to the size of the catalyst. This is called the strong diffusional regime. A true rate constant cannot be directly obtained here but can be back-calculated if the effective diffusion coefficient is known or estimated. Example 18.3 illustrates the procedure.

Example 18.3   Correcting the Rate Constant for Diffusional Effects

Consider the problem examined in Example 18.1. Assume now that the rate constant is not known and the reaction rate is measured. For a 5 mm diameter particle, let the measured rate be 99 mol/s.m^3. Find the true rate constant. To what extent is the internal diffusion resistance controlling? Assume a first-order reaction.

Solution

The gas phase concentration of A is fixed here as 30 mol/m^3. Dividing the rate by the concentration, the apparent rate constant is known:

$$k_{app} = \frac{\text{rate}}{\text{concentration}} = 99/30 = 3.3 \text{ s}^{-1}$$

To find the true rate constant the extent of the internal diffusion coefficient in the pores must be measured or estimated. We use the value of $10^{-6} \text{ m}^2/\text{s}$ (same as in Example 18.1 for comparison purposes).

Based on the apparent rate constant, the Thiele modulus is calculated as 5.63. Hence $\eta = 0.5331$, which is only an approximate value. The reaction rate is calculated as $k_A C_{Ab} \eta$, $33 \text{ mol/m}^3\text{s}$, which is far lower than the measured value.

If the rate constant is changed to 6, the rate can be recalculated. We calculate the rate to be 73 mol/s.m^3. This is still lower than the measured rate and an increased value needs to be tested. A few trial and error calculations will zero in on the value of the rate constant to 10, which is the true value. The effectiveness factor is 0.3315 and the internal diffusion is strongly rate limiting.

Note: The reaction order is assumed to be first order. This may not be known *a priori*. In practice the rate will have to be measured at different concentrations to ascertain the reaction order.

18.2.2 Zero-Order Reaction

This is often important, especially in biological systems, as many metabolic reactions can often be approximated as a zero-order process. The zero-order kinetics is also observed in some hydrogenation reactions. Although the model equation is straightforward, special considerations are needed since the concentration value can actually become zero at some position in the diffusion path. This is unlike a first-order reaction where the concentration decays as an exponential function (which can of course become very small but never actually becomes zero). For zero-order reactions the concentration (a parabolic function as shown in the following) can actually become zero at some point within the porous catalyst, especially if diffusion is slow compared to the reaction rate. Hence we have to demarcate the position where the concentration actually becomes zero and analyze the problem separately in two regions. A simple example for a slab geometry is illustrated and other cases are left as exercises. An important problem application is in biomedical systems, that is, oxygen transport in tissues, which is also known as the Krogh cylinder problem. This is discussed in Chapter 23.

The slab analysis proceeds as follows with the basic diffusion-reaction equation:

$$D_e \frac{d^2 C_A}{dx^2} - k_0 = 0$$

where the production rate of A, R_A, has been replaced by minus k_0, the zero-order rate constant for the metabolic process. The boundary conditions are set as follows:

- At $x = L$, $C_A = C_{As}$, a prescribed or known surface concentration (Dirichlet case).
- At $x = 0$, the center of the slab, we use the no flux condition $dC_A/dx = 0$.

The dimensionless version of the previous equation can be easily derived as

$$\frac{d^2 c_A}{d\xi^2} = \phi_0^2 \tag{18.13}$$

Here ϕ_0^2 is the square of the Thiele modulus for a zero-order reaction. This should be defined as

$$\phi_0^2 = \frac{k_0 L^2}{D_e C_{As}} \tag{18.14}$$

Note that the Thiele modulus depends on the surface concentration, unlike the case of a first-order reaction. This is true in general for any nonlinear kinetics.

The solution for the concentration profile (in dimensionless form) is very simple:

$$c_A = 1 - \frac{\phi_0^2}{2} + \frac{\phi_0^2}{2} \xi^2 \tag{18.15}$$

The center concentration is assumed to be above zero so that this equation can be applied.

The condition at which the center becomes zero or remains positive is then obtained from Equation 18.15 by setting $\xi = 0$ as

$$\phi_0 \leq \sqrt{2}$$

For this case, there is no reactant starved zone in the interior of the catalyst. The rate of consumption of A per unit volume is k_0; this is independent of the concentration, as expected for a zero-order reaction. Hence the effectiveness factor is equal to one since the entire catalyst is exposed to the reactant as long as this condition is satisfied.

The solution for a larger Thiele modulus is studied next where a reactant starved region develops near the center of the catalyst.

Solution for Larger Thiele Modulus

For $\phi_0 > \sqrt{2}$, the concentration becomes zero at a dimensionless location λ. The no flux condition is now imposed at $\xi = \lambda$ since no mass of A crosses this point, rather than at $\xi = 0$. This leads to the following equation for the concentration profile:

$$c_A = 1 - \frac{\phi_0^2}{2}(1 - 2\lambda) + \frac{\phi_0^2}{2}(\xi^2 - 2\xi\lambda)$$

The position λ is not known. Hence we impose an additional condition that the concentration c_A is also zero at $x = \lambda$. This provides the following condition for the calculation of λ:

$$\lambda = 1 - \sqrt{\frac{2}{\phi_0^2}}$$

Note that only the region λ to 1 is effective for a reaction. Hence the effectiveness factor is simply the length of the region and can be calculated as

$$\eta = \sqrt{\frac{2}{\phi_0^2}} \quad \text{for } \phi_0 > \sqrt{2} \tag{18.16}$$

The rate of consumption of A per unit volume is now $k_0\eta$ and turns out to be

$$R_A = \sqrt{2 D_e k_0 C_{A,s}}/L \tag{18.17}$$

This shows a square root dependency on the surface concentration. Thus a zero-order reaction appears as a half-order reaction due to the presence of diffusional effects.

This is a general effect and can be stated as follows:

A reaction of true order n appears as a reaction of $(n + 1)/2$ when strong pore diffusional resistances are present.

The concentration profiles for various ranges of the Thiele parameter are shown in Figure 18.3.

In Figure 18.3 three cases are shown. The first case is for $\phi_0 = 1$ and represents the no starvation case; the concentration remains finite everywhere. The second case is for $\phi_0 = \sqrt{2}$. The concentration becomes zero at the center; this represents the onset of starvation. The effectiveness factor is one for both these cases. The third case is for $\phi_0 = 3$ and represents a case of strong diffusional gradients. The concentration becomes zero at $\xi = 0.53$. The region from 0 to 0.53 is not effective for the reaction. The effectiveness factor is equal to 0.47, the width of the zone where the reactant concentration is non-zero.

Similar to a zero-order reaction, reactions with power law order less than one can exhibit a depleted zone near the center. The analysis of this problem was undertaken by Mehta and Aris in 1971 following the work of another earlier researcher and resurrected in 2011 by York et al. (2011). The mathematical method to find the critical Thiele modulus is examined in exercise problem 10.

We now move to a nonlinear nth-order reaction and illustrate a mathematical solution method for such problems for large values of the Thiele modulus. The corresponding result for the effectiveness factor is known as the asymptotic solution.

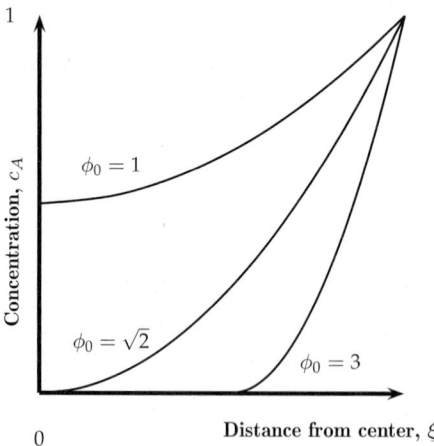

Figure 18.3 Concentration profiles for a zero-order reaction in a porous catalyst for three values of the Thiele parameter.

18.2.3 nth-Order Reaction

In this section we show a variable substitution method for a nonlinear diffusion reaction problem and derive a solution for large values of the Thiele modulus. In order to illustrate the key points, we take an nth-order reaction and show the derivation directly in dimensional form for a slab geometry. The Dirichlet condition is also used at the surface.

Expression for the Surface Flux

Consider diffusion with an nth-order reaction in a porous catalyst, which can be modeled for a slab geometry as

$$D_e \frac{d^2 C_A}{dx^2} = k_n C_A^n \tag{18.18}$$

The order of the differential equation can be reduced by one by using the transformation $p = dC_A/dx$. (Hence the method of solution is called the p-substitution.) Noting that

$$\frac{d^2 C_A}{dx^2} = p \frac{d\,p}{dC_A} \tag{18.19}$$

we find that Equation 18.18 can be written as

$$D_e\, p\, \frac{d\,p}{dC_A} = k_n C_A^n \tag{18.20}$$

This equation can be solved by separation of variables to get an expression for p, the concentration gradient:

$$\int_0^{p_s} p\,dp = \frac{k_n}{D_e} \int_{C_{Ac}}^{C_{As}} C_A^n dC_A \tag{18.21}$$

or

$$p_s^2/2 = \frac{1}{n+1} \frac{k_n}{D_e} \left[C_{As}^{n+1} - C_{Ac}^{n+1} \right] \tag{18.22}$$

The limits are set as follows. For flux the limits are set as 0 at the center and the unknown surface gradient p_s. For concentration the unknown center concentration and the known surface concentration are used.

Note that the center concentration is unknown but an expression can be derived in integral form upon a second integration. However, these details are not important in the asymptotic region, which applies for large values of the

Thiele modulus. Here the concentration drops to nearly zero at some point in the interior of the catalyst. Thus we can set $C_{Ac} \approx 0$ and use this in the above expression. The flux at the surface is then given as

$$p_s^2/2 = \frac{1}{n+1} \frac{k_n}{D_e} C_{As}^{n+1} \tag{18.23}$$

Taking the square root with the positive sign (since the concentration is increasing with increase in x), we obtain an expression for the surface gradient:

$$p_s = \sqrt{\frac{2}{n+1}} \sqrt{\frac{k_n}{D_e}} C_{As}^{(n+1)/2} \tag{18.24}$$

The quantity entering the catalyst $\dot{\mathcal{M}}_A$ is given as

$$\dot{\mathcal{M}}_A = AD_e \left(\frac{dC_A}{dx}\right)_{x=L} = AD_e p_s \tag{18.25}$$

Correspondingly the average rate of reaction is obtained as

$$-\bar{R}_A = \frac{\dot{\mathcal{M}}_A}{AL} = D_e p_s L \tag{18.26}$$

Using Equation 18.24 for p_s, the following expression is obtained for the average rate:

$$\boxed{-\bar{R}_A = \sqrt{\frac{2}{n+1}} \frac{\sqrt{D_e k_n}}{L} C_{As}^{(n+1)/2}} \tag{18.27}$$

Note that this reduces to Equations 18.12 and 18.17 for first-order and zero-order reactions with n equal to 1 and 0, respectively.

Effectiveness Factor

The expression for the average rate can be put in terms of an effectiveness factor. This is given by the ratio of the actual rate to that based on surface concentration (only internal gradients considered):

$$\eta = \frac{-\bar{R}_A}{k_n (C_{As})^n}$$

Therefore the following expression holds for the effectiveness factor upon using Equation 18.27 in the asymptotic region:

$$\eta = \sqrt{\frac{2}{n+1}}\frac{1}{L}\sqrt{\frac{D_e}{k_n C_{As}^{(n-1)}}}$$

Note that $\eta = 1/\phi$ for a first-order reaction in the asymptotic region. Now if we wish to write the previous result for an nth-order reaction in a similar format as: $\eta = 1/\Lambda$, the Thiele modulus Λ should be defined as

$$\Lambda = L\sqrt{\frac{(n+1)k_n C_{As}^{(n-1)}}{2D_e}} \tag{18.28}$$

This is known as the kinetic generalized Thiele modulus. Now we have $\eta = 1/\Lambda$ for any kinetics, but only for the asymptotic case.

The expression for η is then generalized as

$$\eta = \frac{\tanh(\Lambda)}{\Lambda} \tag{18.29}$$

This expression is strictly valid only in the asymptotic region where $\tanh(\Lambda)$ tends to one but is used for the entire range of Λ values as an approximate solution.

Further, shape normalization can be introduced in order to apply this to other shapes as well. The length parameter is now changed to V_p/S_e and the Thiele modulus is further generalized as

$$\boxed{\Lambda = \frac{V_p}{S_e}\sqrt{\frac{(n+1)k_n C_{As}^{(n-1)}}{2D_e}}} \tag{18.30}$$

The expression for η given by Equation 18.29 is then used for the nth-order reaction as an approximation for the entire range of Thiele values and for any shape of catalyst. This approximate method of calculation of the effectiveness factor is known as the Bischoff (1965) approximation.

18.3   Multiple Species

Here we examine two species reacting together. Diffusion equations for both species are needed and must be solved simultaneously. Multiple reactions can be handled in the same manner.

The following reaction scheme is used in the the analysis presented in this section:

$$A + \nu B \rightarrow \text{Products}$$

The diffusion equations for species A and B both have to be solved together here. Both of these concentrations are lower in the pellet compared to the surface. We illustrate the method by taking a slab geometry and a bimolecular (1,1) order reaction. Note that a product concentration increase in the pellet can also be computed by a similar procedure.

The model equations to be solved are as follows for components A and B:

$$D_{eA} \frac{d^2 C_A}{dx^2} - k_2 C_A C_B = 0 \tag{18.31}$$

and

$$D_{eB} \frac{d^2 C_B}{dx^2} - \nu k_2 C_A C_B = 0 \tag{18.32}$$

The concept of a limiting reactant is very useful here. The reactant whose value of its surface concentration times the diffusivity is lower than the other is usually the limiting reactant. Species A is limiting if

$$q = \frac{D_{eB} C_{Bs}}{\nu D_{eA} C_{As}} > 1$$

and vice versa. The second species is called the excess reactant. For further discussion A can be assumed to be limiting without loss of generality. Thus if B is limiting, the notations can be switched.

One of the important relations that is useful here and also in a general context is the invariant property of the equations. Thus by combining the equations for A and B where we have the common rate term we find the following relation holds between the concentrations of species A and B:

$$D_{eB} \frac{d^2 C_B}{dx^2} = \nu D_{eA} \frac{d^2 C_A}{dx^2}$$

The integrated form suggests that

$$C_B = C_{Bs} - \frac{\nu D_{eA}}{D_{eB}} (C_{As} - C_A)$$

Hence C_B (the excess reactant) can be eliminated in Equation 18.31 and a single differential equation can be solved for A only. The number of differential equations to be solved is thus reduced to one. In the asymptotic region an

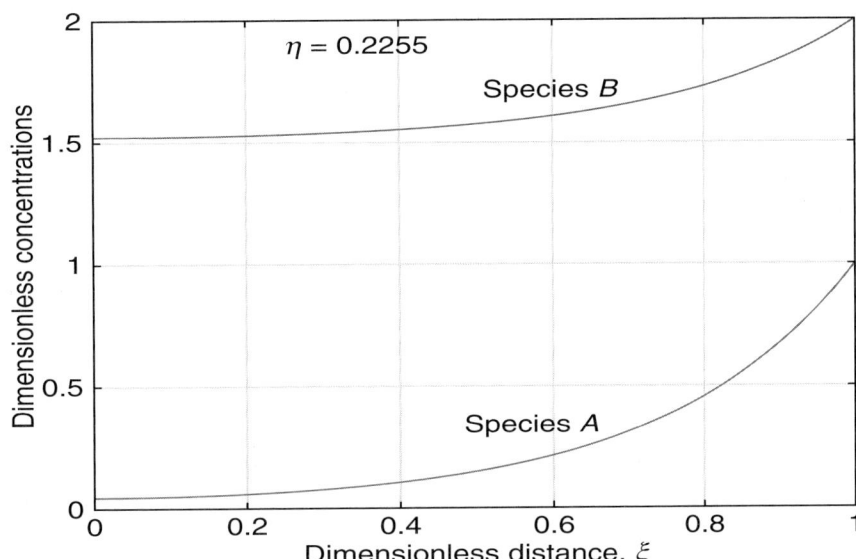

Figure 18.4 Concentration profiles for bimolecular reaction in a porous catalyst.

analytical value can also be obtained by the p-substitution method since we have only one differential equation to handle. An illustrative result is shown in Figure 18.4.

The results shown in this figure are for a slab geometry with $\phi = 3$ and $q = 2.0$. Here species B is the excess reactant since $q > 1$. Note that the concentration of the excess reactant (B) drops only marginally in the catalyst while that of the limiting reactant drops much more significantly here. The effectiveness factor can be calculated as 0.2255 for this case. Analysis for multiple reactions follows a similar methodology.

18.4   Three-Phase Catalytic Reactions

A large class of industrially important examples include three phases: gas, liquid, and a solid catalyst. Transport of a species from the gas phase and transport of a second species from the liquid are often the rate determining steps. This section provides a brief introduction to mass transport effects in three-phase systems. A simple first-order reaction is considered and transport effects are accounted for by a resistance in series approach. For more details a complete textbook (Ramachandran and Chaudhari, 1983) and many other monographs are available. We first cite some industrial examples and some common reactor types.

18.4.1 Application Examples

Some examples of industrial-scale gas–liquid–solid catalyzed reactions are the following:

- Hydrogenation reactions: Examples include hydrogenation of glucose to sorbitol, unsaturated oils, nitrocompound, aldehydes and esters, and so on.
- Hydrodesulfurization of diesel to remove sulfur compounds.
- Hydrocracking of petroleum fractions using supported metal catalysts to produce lower molecular compounds, such as the gasoline fraction from heavy oil.
- Methanol synthesis from a CO and H_2 gas mixture in the liquid phase with suspended solid catalyst.
- Fisher-Tropsh synthesis from syngas to produce fuels where synthesis gas (CO, H_2 mixture) is converted to heavy hydrocarbon fractions.
- Oxidation reactions; these include oxidation of cumene, phenol, p-xylene, and so on, using oxides of Cr, Cu Co, and so on, as solid catalysts. Note that use of soluble homogeneous catalysts is more common for oxidation and such reactors belong to the category of two-phase reactors.

Reactor Types

The common industrial reactors can be classified into the following types:

- Slurry type: Here the catalyst particles are suspended in a liquid with gas being simultaneously sparged into the liquid. The dispersion of the gas and the suspension of the solid catalyst can be done simply by bubbling the gas (bubble column reactor) or by mechanical agitation.
- Fixed bed type: Here the catalyst is stationary and the gas and liquid flow over the catalyst bed similar to a packed bed absorption column. The flow may be cocurrent downward (e.g., trickle bed reactor shown in Figure 1.4) or concurrent upflow (packed bubble column reactor) or countercurrent flow over a packed catalyst bed.

Relative comparisons and the choice of reactors are discussed by Ramachandran and Chaudhari (1983), Mills et al. (1992), and other references;

these should be consulted if you are designing these types of reactors. The focus here is mainly on mass transfer effects and how the effectiveness factor concept is useful to calculate the rate of reaction for a simple first-order reaction.

18.4.2 Mass Transfer Effects

For the simple case shown in Figure 18.5, a number of steps have to occur before species A can be converted on the active sites for production:

1. Transport of gas from bulk gas to the bulk liquid. This can be modeled using an overall mass transfer coefficient from gas to liquid:

$$-R_A^v = K_L a_{gl}(C_{Ag}^* - C_{AL})$$

   C_{Ag}^* is the saturation solubility of the gas in the liquid based on the bulk gas concentration of gas A. C_{AL} is the concentration in the bulk liquid. The R_A^v is the volumetric absorption of A in the system which equals the rate of reaction of A per unit reactor volume. This step has a resistance of $1/K_L a_{gl}$.

2. Transport of dissolved gas from the bulk liquid to the catalyst surface. This can be modeled using a solid–liquid mass transfer coefficient:

$$-R_A^v = k_{ls} a_{ls}(C_{AL} - C_{As})$$

   Here k_{ls} is the intrinsic liquid–solid mass transfer coefficient and a_{ls} is the external surface area of the catalyst per unit reactor volume. This step has a resistance of $1/k_{ls} a_{ls}$.

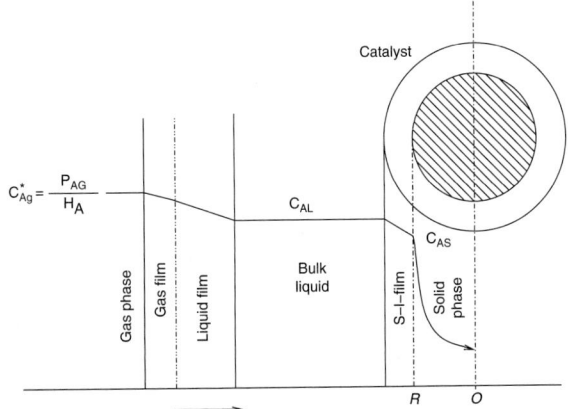

Figure 18.5 Mass transfer effects in three-phase catalytic reactors S-L means solid liquid; R = radius of the catalyst particle.

3. **Internal diffusion with reaction.** This can be modeled using $k_1\eta$ as the rate coefficient:

$$-R_A^v = wk_1\eta C_{As}$$

A simple first-order kinetics is used here. w is the mass of the catalyst per unit reactor volume. This step has a resistance of $1/wk_1\eta$. The rate constant k_1 here is defined on the mass of the catalyst.

Usually species A will react with a second liquid phase B, which gets transported to the catalyst as well. However if the second species is in excess, the reaction may be approximated as pseudo-first-order case shown here. The details of surface interactions such as adsorption on the active surface followed by reaction are also lumped into the rate constant. The effect of pore diffusion is captured by using the effectiveness parameter, which in turn depends on the Thiele modulus.

Each of the transport steps contribute to a corresponding resistance. For a linear process the resistances can be added. The overall rate can be then calculated as

$$-R_A^v = C_{Ag}^* \left[\frac{1}{K_L a_{gl}} + \frac{1}{k_{ls} a_{ls}} + \frac{1}{wk_1\eta_c} \right]^{-1}$$

where η_c in turn can be calculated based on the Thiele parameter for the catalyst.

18.5   Temperature Effects in a Porous Catalyst

In this section we show the transport effect where an exothermic reaction is taking place in the catalyst. Due to the release of heat due to the reaction, a temperature profile develops in the catalyst. This in turn affects the rate of reaction via the Arrhenius dependency of the rate constant on the temperature.

18.5.1   Equations for Heat and Mass Transport

The simultaneous transport of heat and mass in a porous catalyst is needed for analysis and can be represented by the following set of equations for a slab geometry:

$$D_e \frac{d^2 C_A}{dx^2} = f(C_A, T) \tag{18.33}$$

$$k_e \frac{d^2T}{dx^2} = (\Delta H)f(C_A, T) \tag{18.34}$$

where k_e is the effective conductivity of the catalyst and ΔH is the heat of reaction. The local rate of disappearance $f(C_A, T)$ is now a function of both the pore concentration and the temperature at any point in the pellet.

The boundary conditions are specified as follows: no flux condition is imposed at the pore end ($x = 0$) for both heat and mass:

$$\text{at } x = 0 \quad dC_A/dx = 0; \quad \text{and } dT/dx = 0$$

At the pore mouth ($x = L$) the flux into the pellet must match the transport through the gas film. This leads to the following boundary conditions:

For mass transport:

$$D_e \left(\frac{dC_A}{dx} \right)_{x=L} = k_m (C_{Ag} - C_{A,x=L})$$

For heat transport:

$$k_e \left(\frac{dT}{dx} \right)_{x=L} = h(T_g - T_{x=L})$$

18.5.2 Dimensionless Representation

The equations can be put in terms of dimensionless variables by scaling the concentration and temperature by reference values. The bulk concentration and temperature can be used as reference values if the particle-scale model is being analyzed. For reactor-scale models, the inlet feed conditions can be used as a reference.

The dimensionless mass and heat transport equations are then as follows: the mass balance equation is

$$\frac{d^2 c_A}{d\xi^2} = \phi^2 f^*(c_A, \theta) \tag{18.35}$$

Here f^* is a scaled rate of reaction, defined as

$$f^* = f(C_A, T)/[f(C_{A_{ref}}, T_{ref})]$$

ϕ^2 is a Thiele modulus based on reference conditions:

$$\phi^2 = \frac{f(C_{ref}, T_{ref})L^2}{D_e C_{ref}}$$

The dimensionless heat balance equation is

$$\frac{d^2\theta}{d\xi^2} = -\beta\phi^2 f^*(C_A, \theta) \tag{18.36}$$

where β is the dimensionless parameter (thermicity group), defined as

$$\beta = \frac{(-\Delta H)D_e C_{ref}}{k_e T_{ref}} \tag{18.37}$$

Note that β takes a positive value for exothermic reactions and a negative value for endothermic reactions.

The dimensionless rate has to be appropriately defined, including the temperature effect on the rate constant. For example, using the Arrhenius equation, the dimensionless rate for a first-order reaction can be expressed as

$$f^* = c_A \exp[\gamma(1 - 1/\theta)] \tag{18.38}$$

where $\gamma = E/R_g T_{ref}$. E is the activation energy for the reaction. Parameter γ is a dimensionless group sometimes referred to as the Arrhenius number. The governing equations and the boundary conditions shown in Section 18.5.1 to be solved are then calculated as follows. The species mass balance is:

$$\frac{d^2 c_A}{d\xi^2} = \phi^2 c_A \exp[\gamma(1 - 1/\theta) \tag{18.39}$$

and the heat balance is

$$\frac{d^2\theta}{d\xi^2} = -\beta\,\phi^2\,c_A \exp[\gamma(1 - 1/\theta)] \tag{18.40}$$

18.5.3 Dimensionless Boundary Conditions

The normalized boundary conditions can be stated as follows.

At the center the no flux condition is used for both the concentration and the temperature:

$$\text{At } \xi = 0, \quad dc_A/d\xi = 0; \quad d\theta/d\xi = 0$$

The conditions at $\xi = 1$ can be related to the Biot numbers for heat and mass:

$$\left(\frac{dc_A}{d\xi}\right)_s = Bi_m(c_{Ag} - c_{As})$$

$$\left(\frac{d\theta}{d\xi}\right)_s = Bi_h(\theta_g - \theta_s)$$

Here the subscript s is used for $\xi = 1$, that is, the surface values. The subscript g is used for the bulk values. For the particle problem (local particle-scale model) we can set c_g and θ_g as one since the reference concentration and temperature are the bulk values.

The Biot numbers for mass and heat are defined as

$$Bi_m = k_m L/D_e$$

and

$$Bi_h = hL/k_e$$

This completes the problem formulation. The problem requires a numerical solution in general.

The results are often shown in terms of an effectiveness factor as a function of Thiele modulus, defined as

$$\eta = \int_0^1 c_A \exp[\gamma(1 - 1/\theta)] \; d\xi$$

Even for this simple first-order reaction, the overall effectiveness factor is a function of five parameters shown previously. It is a numerical problem with a daunting solution and has been the subject of many research papers. However, certain features can be examined even without solving the full set of equations. Let us now examine if we can calculate the maximum temperature rise (exothermic case).

18.5.4 Estimate of the Temperature Gradients

The magnitude of the internal temperature gradient can be assessed by combining the two differential equations, Equations 18.39 and 18.40, without even actually solving for the full temperature profiles. These mass and heat transport equations can be combined as

$$\frac{d^2\theta}{d\xi^2} + \beta\frac{d^2 c_A}{d\xi^2} = 0 \qquad (18.41)$$

Hence the integration of this equation twice leads to

$$\theta - \theta_s = \beta(c_{As} - c_A) \qquad (18.42)$$

This is an invariant of the system.

The maximum value of the temperature occurs for an exothermic reaction when the center concentration drops to zero. This occurs in the strong pore diffusion resistance for mass transfer. Hence an estimate of the internal gradients is

$$\text{Max } \theta_c - \theta_s = \beta c_{As} = O(\beta) \qquad (18.43)$$

Here θ_c is the center temperature. Further the maximum value of c_{As} is one. Hence the dimensionless maximum temperature difference between the center and surface is β. Using dimensional values we find the following expression:

$$T_c - T_s = \frac{(-\Delta H)D_e}{k_e}(C_{As} - C_{Ac}) \qquad (18.44)$$

The maximum difference occurs when C_{Ac} is nearly zero; hence this analysis provides a quick method of calculating the maximum internal temperature that can develop in the catalyst.

The temperature rise across the gas film can be related to the measured rate of reaction if an estimate of the film heat transfer coefficient is made. This is simply a heat balance across the gas film, which is observed by rewriting this as

$$hS_e(T_s - T_g) = V_p \bar{R}_A \Delta H \qquad (18.45)$$

The left-hand side is heat transferred from the solid while the right-hand side is the heat generated due to reaction in the pellet. Note that V_p/S_e is the characteristic length parameter, L_c.

The above relations enable a rapid determination of the relative importance of external and internal temperature gradients. It can be, for instance, used to determine whether the pellet is operating under near isothermal conditions or whether a temperature correction to the data is necessary.

Detailed numerical results for various combination of parameters can be generated, for example, using BVP4C with MATLAB. The solution can show multiple steady states for a certain range of parameters and the effectiveness factors can be larger than one. This happens only for an exothermic reaction where the interior of the catalyst is hotter than the bulk gas. The rate increase due to temperature rise can often compensate for the rate decrease due to concentration increase. This can cause effectiveness factors to be greater than one.

In the next two sections, we discuss two numerical tools to solve diffusion-reaction problems. The first tool is based on orthogonal collocation and is suitable for 1-D problems in the three standard geometries. The second tool is a finite difference method and is suitable for 2-D and 3-D problems.

18.6   Orthogonal Collocation Method

Collocation methods are powerful and accurate tools for the solution of diffusion-reaction problems. The goal of this section is to provide a brief introduction to the method together with a worked example. The detailed theory behind the collocation method is beyond the scope of this book. The interested reader should refer to Villadsen and Michelson (1978).

18.6.1   Basis of the Method

The method is based on expansion of the solution variable, denoted here as c, in terms of orthogonal polynomials in the distance variable, ξ. We select a number of collocation points, which are the roots of the Jacobi polynomial defined in an optimum manner. It may also be noted here that for problems with symmetry at $\xi = 0$, it is not required to select the point $\xi = 0$ as a collocation point. This is because the expansion uses polynomials with even powers in ξ and hence the symmetry boundary condition is automatically satisfied.

Collocation points are chosen differently for each of the three geometries so as to minimize the error in the calculation of the average rate of reaction or equivalently the effectiveness factor. Again these details are not addressed here as the main theme is to demonstrate the numerical implementation of the method and to provide useful code. The key result needed from the user point of view is that the Laplacian at each interior collocation point (or node) is approximated as an algebraic equation as shown in Figure 18.6.

This equation, known as the discretization formula, is applied at all interior nodes, that is, $i = 1$ to $nt - 1$, generating a set of algebraic equations. The

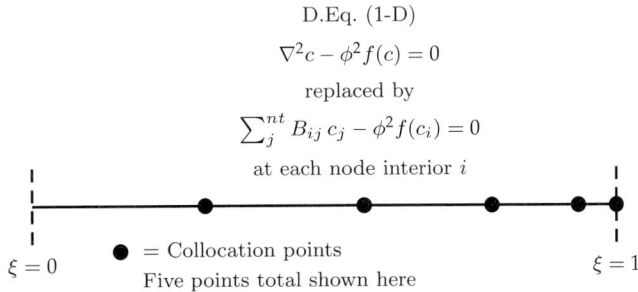

D.Eq. (1-D)

$$\nabla^2 c - \phi^2 f(c) = 0$$

replaced by

$$\sum_j^{nt} B_{ij}\, c_j - \phi^2 f(c_i) = 0$$

at each node interior i

$\xi = 0$ ● = Collocation points $\xi = 1$
Five points total shown here

Figure 18.6   Representation of the collocation approximation to the diffusion-reaction problem. **Note:** c_A is abbreviated as c; f is the dimensionless form of the rate of reaction term.

discretization is not applied at the boundary node at $\xi = 1$. Instead the set of equations is augmented by applying the boundary condition at the node designated as nt. This may be a simple Dirichlet for concentration ($c = 1$) or a Robin boundary. A Robin boundary needs an approximation of the first derivative, which is given by the A matrix:

$$\left[\frac{dc}{d\xi}\right]_i = \sum_j^{nt} A_{ij}\, c_j \tag{18.46}$$

Thus nt equations for nt variables are obtained. This set of nonlinear algebraic equations can be solved in an iterative manner by the Newton-Raphson method, which requires the linearization of the rate term $f(c)$ at each level of iteration.

18.6.2 Two-Point Collocation

Often when the concentration profiles are not too steep, a single interior point collocation is sufficient. Together with the boundary, this is known as a two-point collocation and provides a quick method for calculation of the effectiveness factor. Example 18.4 demonstrates this for a spherical geometry with a simple first-order example for which an analytical solution is also available for comparison.

For a sphere the values of the A and B matrices for a two-point collocation are listed in the following. In addition, the quadrature weights w to approximate the integral of the concentration and the location of the points ξ are also given.

Example 18.4   Two-Point Approximation

Solve the diffusion-reaction problem for a first-order reaction in a spherical geometry with one-point collocation in the interior of the pellet and compare the results with the analytic solution.

Solution

We apply the collocation at one interior point:

$$B_{11}c_1 + B_{12}c_2 = \phi^2 c_1$$

If the Dirichlet condition is applied at the surface, $c_2 = 1$. Hence we get the concentration at the interior point:

$$c_1 = \frac{-B_{12}}{B_{11} - \phi^2}$$

Table 18.1   **Collocation Matrices for Two-Point Collocation for Sphere**

```
A =[ -2.2913  2.2913
      -3.5000  3.5000]
B = [ -10.5000 10.5000
      -10.5000 10.5000 ]
w = 0.2333 0.1000
xi = 0.6547 1.0000
```

The effectiveness factor is then calculated using the quadrature weights as

$$\eta = (s+1)[w_1 c_1 + w_2 c_2]$$

For a specific value of $\phi = 2$ we get the following results using the matrix coefficient values reported in Table 18.1:

$$c_1 = 0.7241 \text{ and } \eta = 0.8067$$

These can be compared with the analytical values, which are 0.7231 and 0.8069. We see a pretty close match.

If $\phi = 5$ we get $\eta = 0.5070$, which is also close to the analytical value of 0.4801. The concentration at the interior point is now 0.2958 while the analytical value is 0.2713.

For a second order reaction with the Dirichlet condition at the surface, the one point representation would be as follows:

$$B_{11} c_1 + B_{12} = \phi^2 c_1^2$$

This can be solved as a quadratic. The effectiveness factor can then be calculated using the quadrature weights.

Note: Often reactors are operated in the intermediate range of the Thiele modulus and one (interior) point is sufficient for many practical purposes.

18.7 Finite Difference Methods

Finite difference methods are simple to apply for regular geometries, which are defined as geometries where the sides are parallel to the coordinate axis, that is, there are no curved boundaries. These include square- or rectangular-shaped domains or L-shaped or similar domains in 3-D. Domains with curved boundaries (e.g., an ellipse) need special treatment for the second derivative near the boundaries and are not considered here. There are many sources for the interested reader to pursue this topic further.

Consider the solution of the following problem:

$$\frac{\partial^2 c_A}{\partial x^2} + \frac{\partial^2 c_A}{\partial y^2} = \phi^2 c_A \tag{18.47}$$

Here x and y are treated as dimensionless variables. c_A is a dimensionless concentration and ϕ is the Thiele modulus.

The domain is divided into meshes along the x- and y-directions with a chosen scale Δx and Δy in each direction. We now abbreviate c_A by c.

18.7.1 Central Difference Equations

The second derivatives are approximated by central difference approximations presented in the following.

In the x-direction the approximation at node i is

$$\left(\frac{\partial^2 c}{\partial x^2}\right)_i = \frac{c_{i-1,j} - 2c_{i,j} + c_{i+1,j}}{\Delta x^2} = O(\Delta x^2)$$

Similarly in the y-direction we use

$$\left(\frac{\partial^2 c}{\partial y^2}\right)_j = \frac{c_{i,j-1} - 2c_{i,j} + c_{i,j+1}}{\Delta y^2} = O(\Delta y^2)$$

Hence the finite difference discretization applied at node (i, j) leads to

$$\frac{c_{i-1,j} - 2c_{i,j} + c_{i+1,j}}{\Delta x^2} + \frac{c_{i,j-1} - 2c_{i,j} + c_{i,j+1}}{\Delta y^2} = \phi^2 c_{(i,j)} \tag{18.48}$$

For direct iteration purposes, the equation is rearranged with $c(i, j)$ as the term on the left-hand side and with all the other terms moved to the right-hand side:

$$\text{c NEW}_{i,j} = \frac{[c_{i-1,j} + c_{i+1,j}]\Delta y^2 + [c_{i,j-1} + c_{i,j+1}]\Delta x^2}{2(\Delta x^2 + \Delta y^2) + \phi^2 * \Delta x^2 * \Delta y^2}$$

Here the old values (values at previous iteration) are used on the right-hand side and these are known. This permits the calculation of c at node (i, j) (the new value). By sweeping over all the nodes one round of iteration is completed and all the old values are replaced by the new values calculated using the left-hand side of the preceding equation. This sets up an iteration scheme that converges to the solution after a few rounds of updating the concentration values. Illustrative results are shown in Figure 18.7.

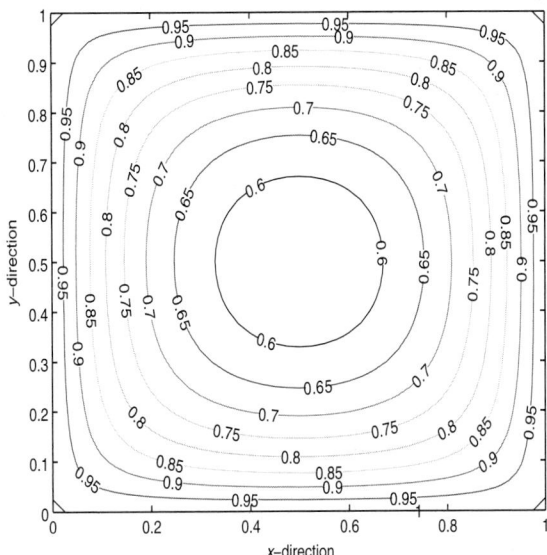

Figure 18.7 Concentration profiles for a first-order reaction in a square catalyst shown as contour plots; $\phi = 3$. Dirichlet condition: $c = 1$ is applied along the boundaries.

The contour plots of the results in Figure 18.7 are for $\phi = 3$. The center concentration is found to be equal to 0.5614. The effectiveness factor can be computed either from the average concentration or by calculation of the flux values at the boundary. Both approaches will give (nearly) the same result. The value is 0.7899. A shape-normalized approach gives a value of 0.8469.

18.7.2 Zero-Order Reaction

The simulation for a zero-order reaction is done in a similar manner. Here $f(c) = 1$. Also the Thiele parameter is renamed ϕ_0.

The iteration scheme now is

$$\text{cNEW}_{i,j} = \frac{[c_{i-1,j} + c_{i+1,j}]\Delta y^2 + [c_{i,j-1} + c_{i,j+1}]\Delta x^2 - \phi_0^2 * \Delta x^2 * \Delta y^2}{2(\Delta x^2 + \Delta y^2)}$$

The results are shown in Figure 18.8 for $\phi_0 = 3$. If this parameter is increased you will find a region of negative concentration in the center of the

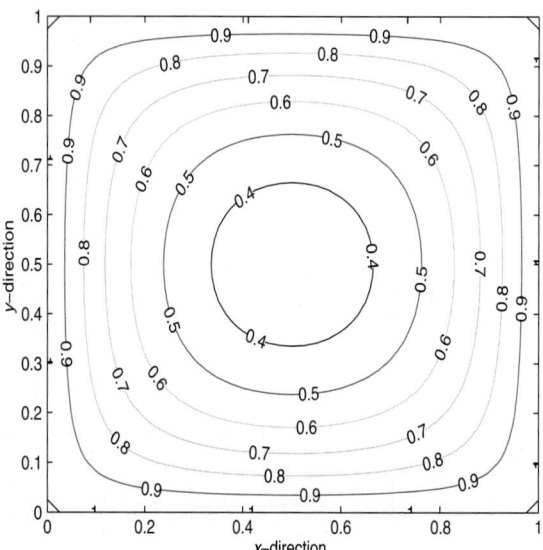

Figure 18.8   Concentration profiles for a zero-order reaction in a square catalyst shown as contour plots; $\phi = 3$. Dirichlet condition: $c_A = 1$ is applied along the boundaries.

system. This indicates the onset of starvation. It can be shown that a Thiele modulus of 3.685 leads to starvation.

18.7.3 Nonlinear Kinetics

Nonlinear kinetics can be determined by local linearization. Consider the rate term $\phi^2 f(c)$. This is linearized based on the current iteration values as

$$f(c) \approx \kappa_0 + \kappa_1 c \tag{18.49}$$

The two coefficients κ_0 and κ_1 are obtained by Taylor series approximation at the current value of the nodal concentration, denoted as c_A^* here:

$$\kappa_0 = f(c_A^*) - c_A^* \left(\frac{df}{dc_A} \right)_{c_A^*} \tag{18.50}$$

and

$$\kappa_1 = \left(\frac{df}{dc_A} \right)_{c_A^*} \tag{18.51}$$

In the iteration scheme κ_0 is treated as a zero-order reaction term and κ_1 as a first-order term. These terms change values during each iteration as they depend on the current value of the nodal concentration. Hence the iteration scheme is a combination of a first-order and a zero-order reaction.

For, example for a second-order reaction $f(c_A) = c_A^2$ and we have $\kappa_0 = -c_A^{*2}$ and $\kappa_1 = 2c_A^*$ where * denotes the old values, that is, the values at the previous iteration. The κ_0 and κ_1 are adjusted at each iteration until the iteration values at all the nodes converge.

18.7.4 Neumann and Robin Conditions

We now show how other types of boundary conditions can be used. Along the boundaries perpendicular to the x-direction we approximate the normal gradient as

$$\frac{dc}{dn} = \frac{3c_s - 4c_{s-1} + c_{s-2}}{2\Delta x} \tag{18.52}$$

Here s is the surface node, $s - 1$ is the next node adjacent to it, and $s - 2$ the third node from the surface.

If a no flux condition is applied at the Neumann boundary node, the normal flux is set to zero and concentration values at the surface node are given as

$$c_s = \frac{4}{3}c_{s-1} - \frac{1}{3}c_{s-2}$$

Thus the surface values for these nodes are updated using this equation at each iteration along with the interior values.

Along a Robin boundary the boundary condition is

$$\left(\frac{dc}{dn}\right)_s = -Bi(c_s - c_b)$$

Using Equation 18.52 as the approximation for the normal derivative and rearranging, the following expression is obtained for the surface concentration:

$$c_s = \frac{4c_{s-1} - c_{s-2} + 2\Delta s\, Bi\, c_b}{2\Delta s\, Bi + 3}$$

where Δs is the mesh spacing at these nodes. Thus along a Robin boundary the iteration is adjusted to maintain this value for c_s.

18.8   Linking with Reactor Models

This section provides a simple example where the local model for the porous catalyst is linked to a reactor model to demonstrate how models at two scales are linked. More complex cases (not shown here) follow simply by extension of the method.

A model for a plug flow case is considered here for an isothermal system with no appreciable change in gas velocity. The species mass balance equation based on a mesoscopic control volume is

$$In - out + generation = zero$$

The generation is calculated as the rate of reaction based on the local gas concentration (and temperature for non-isothermal reactions) times the local effectiveness factor. The effectiveness factor accounts for the effect of concentration drop within the catalyst and in the gas film as well. The rate of reaction is based on the volume of the catalyst. Since the mass balance is based on the total control volume (gas+solid) the rate should be multiplied by the catalyst volume per unit bed volume $(1 - \epsilon_B)$ with ϵ_B being the bed porosity, that is, the volume of the gas per unit volume of the bed. Putting all this together the generation term is

$$Generation = (1 - \epsilon_B)R_A(\text{gas})\eta$$

$R_A(\text{gas})$ is the rate of reaction based on the local gas concentration at a given point in the reactor. The effectiveness factor thereby couples the reactor-level and particle-level models. The mass balance for the concentration C_A in the gas phase is therefore

$$u_G \frac{dC_A}{dz} = (1 - \epsilon_B)R_A(\text{gas})\eta$$

This can be integrated by a finite difference marching scheme in z or by any solver such as ODE45. The point to note is that η is a function of local concentration C_A (unless the reaction is first order at isothermal conditions). Hence at each point of the integration scheme, the value of η has to be calculated and incorporated in the marching scheme. For a non-isothermal reaction both mass and heat balance for the gas phase has to be solved together with the effectiveness factor computed from the particle scale model shown in Section 18.5 locally at each axial location. This is a challenging computational problem.

A simple illustration for a power-law kinetics under isothermal conditions is provided in the following. The kinetics is represented as $-R_A = k_n C_A^n$. The gas phase balance is

$$u_G \frac{dC_A}{dz} = -(1 - \epsilon_B) k_n C_A^n \eta \tag{18.53}$$

This is written is dimensionless form as

$$\frac{dc_A}{d\zeta} = -Da \, c_A^n \eta \tag{18.54}$$

where ζ is z/L, the dimensionless distance. The parameter Da is the Damkohler number, defined as

$$Da = \frac{(1 - \epsilon_B) k_n C_{A,i}^n L}{u_G}$$

The calculation of η can be simplified by using the generalized Thiele modulus. For a nth-order reaction the Thiele modulus is proportional to $c_A^{(n-1)/2}$ as can be garnered from Equation 18.28. Hence the local Thiele modulus can be written in terms of the inlet Thiele modulus Λ_i as

$$\Lambda = \Lambda_i c_A^{(n-1)/2}$$

The effectiveness factor can be calculated as $\tanh(\Lambda)/\Lambda$ at each location. Using these in Equation 18.54, the concentration change in the gas phase is given by the following differential equation:

$$\frac{dc_A}{d\zeta} = -Da \, c_A^n \frac{\tanh[\Lambda_i c_A^{(n-1)/2}]}{\Lambda_i c_A^{(n-1)/2}} \tag{18.55}$$

This is now in a form ready for the ODE45 solver. The reactor exit conversion is now dependent on two dimensionless parameters: Damkohler number and the inlet Thiele modulus.

18.8.1 First-Order Reaction

The computation is easy for first order since the Thiele modulus is not a function of the local gas concentration, C_A. Hence η calculated based on the inlet Thiele modulus is valid throughout the reactor. Hence Equation 18.53 can be directly integrated analytically:

$$c_{A,e} = \exp[-(1 - \epsilon_B) k_1 \eta L/u_G]$$

The equation resembles a plug flow reactor and is often represented as

$$c_{A,e} = \exp[-k_{app}L/u_G] \qquad (18.56)$$

Here k_{app} is an apparent rate constant:

$$k_{app} = (1 - \epsilon_B)k_1 \eta$$

Equation 18.56 can be used to fit the reactor data, but the fitted value of apparent rate constant is not a true rate constant and will in this case vary with particle size if internal gradients are rate limiting. Hence corrections for the internal diffusional effects have to be applied to get the value of the true rate constant.

18.8.2 Second-Order Reaction

An illustrative result for the exit concentration for a second-order reaction is shown in Table 18.2 using numerical integration of Equation 18.55 with ODE45.

The results are for $Da = 3$ for various values of the inlet Thiele modulus. The effectiveness factor varies along the reactor for a second-order reaction, is lower near the inlet, and increases toward the exit of the reactor for a second-order reaction. This is accounted for in the numerical scheme. For low values of the inlet Thiele modulus, the effectiveness factor is one throughout and an exit concentration is $1/(1 + Da)$ or 0.25 is observed. The corresponding conversion is 75% At high values of the Thiele modulus, internal diffusion limits the rate and the conversion is only 25% at the same gas residence time.

18.8.3 Zero-Order Reaction

The case of the zero-order reaction is rather interesting. In the absence of internal gradients, a value of $Da = 1$ is sufficient for complete conversion. This is

Table 18.2 Exit Concentration in a Packed Bed Reactor for Power-Law Kinetics

Λ_i	$n = 1$	$n = 2$
0.1	0.0503	0.2503
1.0	0.1017	0.2786
10.0	0.7408	0.7561

not so in the presence of internal diffusion. At some point in the reactor a critical value of the Thiele modulus of $\sqrt{2}$ (for slab) is reached. The point where it is reached depends on the inlet value of the Thiele modulus. If the inlet Thiele is 1 then the critical value is reached at a point where the concentration is 0.5. Diffusional effects start to play a role beyond this point. The exit concentration of 0.22 is reached for this and complete conversion is not achieved. It may also be noted that beyond the critical concentration, the reaction would exhibit an apparent order of one half rather than the intrinsic order of zero.

Summary

- Mass transfer with reaction in a porous catalyst is an important problem in reaction engineering. The model (when applied to catalysts of simple shapes) leads to second-order ordinary differential equations of the boundary value type.

- The concentration profile within the porous catalyst depends on a dimensionless parameter, the Thiele modulus. Small Thiele modulus values lead to an almost uniform profile in the catalyst, leading to a complete utilization of all of the catalyst. For large Thiele modulus values, the concentration profile is confined to a thin region near the surface and only this part of the catalyst is being used for reaction. An effectiveness factor can be defined as the rate of reaction divided by the rate based on the bulk concentration. This is a measure of the effect of the internal gradients on the rate of reaction.

- Mass transfer accompanied by a zero-order reaction has some special features. Here the rate is not affected by diffusion and remains constant up to a critical value of the Thiele parameter. Beyond this the rate is affected by diffusional gradients. The problem has important applications in oxygen transport in tissues, some hydrogenation reactions, and growth of microbial cells, which often follows a zero-order metabolism.

- Catalysts of different shapes (rings, trilobes, quadrilobes, wagon wheels, and so on) are commonly used in industrial reactors. The motivation is to reduce the pressure drop using larger sizes but at the same time keep the interior accessible to the reactant. For these complex shapes, a slab type ratio of model is used as an approximation with an effective length parameter defined as the volume to the external surface area of the catalyst.

- For nonlinear reactions a generalized Thiele modulus can be defined and used to find the effectiveness factor. This provides an approximate way of executing the calculation without needing to solve nonlinear differential equations. The procedure is especially useful to couple the particle-scale model to a reactor-level model.

- Multiple reactions are handled by solving the diffusion-reaction model for each species. Simplifications based on the invariance property of the system of equations can be used. The number of species balances to be solved is equal to the number of independent reactions. The species balances for the other species simply follows from the stoichiometry of the set of reactions.

- Systems with three phases (gas, liquid, and solid) are common in industrial practice. Additional transport effects (gas–liquid and liquid–solid mass transfer) are included in the analysis in addition to intraparticle diffusion and reaction. For a first-order reaction, the resistances of each transport step can be added and the rate can be calculated by dividing the driving force by the total resistance.

- Reactions in porous catalysts can be accompanied by temperature changes and lead to a coupled problem in heat and mass transfer. Temperature rise for exothermic reactions causes an exponential rise in the rate of reactions, leading to many complexities such as multiple steady states, large sensitivity to small changes in parameters, and so on. Even for a simple first-order reaction, five dimensionless parameters are needed in the model formulation.

- The orthogonal collocation method is a very fast and simple method for diffusion-reaction problems posed in slab, cylinder, or sphere geometries. The collocation points are chosen in such a manner that the result for the effectiveness factor is as accurate as possible for the given approximation. A symmetry condition is implicit in this formulation and need not be applied as a boundary condition. This is a numerical tool well worth learning. Often one (interior) point collocation is sufficient to give accurate results for moderate values of the Thiele modulus.

- The finite difference method is another useful tool especially for 2-D and 3-D geometries. Due to the elliptic nature of the differential equation, a method based on successive relaxation is useful and provides a simple computational tool for these problems. These can be implemented easily on the MATLAB platform.

- Particle-scale models provide a submodel for linking with reactor-level macro- or meso-level models. Mixing at the reactor scale needs to be assumed. Thus the gas phase can be in backmixed or plug flow for the two ideal cases. Reactor performance is usually bracketed between these values. The variation of the effectiveness factor along the reactor position needs to incorporated into the reactor model unless the reaction is first order and isothermal. If the reactor is assumed to be well mixed, the effectiveness factor should be calculated using the Thiele modulus based on the the exit concentration in the reactor.

Review Questions

18.1 What are the units for first-order, zero-order, and second-order rate constants if the rate is based on volume of catalyst, mass of catalyst, and surface area per unit volume of the catalyst?

18.2 How is the Thiele modulus defined and what is its significance?

18.3 If the concentration profile is known, state two ways of calculating the average rate of reaction for the catalyst.

18.4 Define the Thiele modulus for zero- and second-order reactions and verify dimensional consistency.

18.5 What is the Biot number and what role does it play in affecting the reaction rate in a porous catalyst?

18.6 Under what conditions does the rate of reaction become dependent on the catalyst size? Does it increase or decrease with size?

18.7 If the rate of reaction decreases linearly with catalyst size, what is the controlling resistance?

18.8 If the rate of reaction does not change with catalyst size, what is the controlling resistance?

18.9 What is the definition of the shape-normalized Thiele modulus? Why is it useful?

18.10 Can a zero-order reaction exhibit a dependency on the bulk gas concentration? If so, when?

18.11 Define the generalized Thiele modulus for an nth-order reaction.

18.12 A second-order reaction is taking place under strong pore diffusional resistance, that is, $\phi > 3$. If the pressure is doubled by what factor would the rate change?

18.13 What dimensionless parameter is indicative of the maximum temperature difference between the center and the surface of a catalyst?

18.14 Can the effectiveness factor be larger than one? When?

18.15 A second-order reaction is taking place in a packed bed reactor. How would the effectiveness factor change along the axial length in the reactor?

18.16 A half-order reaction is taking place in a packed bed reactor. How would the effectiveness factor change along the axial length in the reactor?

Problems

18.1   **Comparison of the three 1-D geometries.** Consider the diffusion reaction problem represented in the three geometries. Verify the analytical solutions shown in the text for the three geometries with the Dirichlet condition of $c_A = 1$ at $\xi = 1$ and the symmetry condition (Neumann) at $\xi = 0$. Note that the solution for a sphere needs a small coordinate transformation. ($c_A = f(\xi)/\xi$, which reduces the governing equation to a simpler one in f.) Find the average concentration in the system for the three cases, which represents the effectiveness factor. Make a plot of the effectiveness factor versus ϕ^* for all three cases where ϕ^* is defined as a shape-normalized Thiele modulus defined as

$$\phi^* = \frac{\phi}{s+1}$$

Show that the results for the three geometries are quite close when η is plotted as a function of ϕ^*, which is referred to as the generalized Thiele modulus.

18.2   **Robin boundary condition.** Consider the same problem as above but now use a Robin condition at the surface. Derive an expression for the effectiveness factor as a function of Bi in addition to the ϕ parameter. Do the analysis for all three geometries.

18.3   **Choosing an operating catalyst size.** The rate of reaction was measured in a powdered catalyst of fine size and was found to be 200 mol/m^{3}s. Assume the reaction to be first order and there are no internal gradients. The gas is at 1 atm and 400 K with a mole fraction of 0.1 for the reactant. Now we wish to change to a spherical pellet catalyst and we have an estimate of the effective diffusivity of the catalyst. $D_e = 2 \times 10^{-6}$ m^2/s. From pressure drop considerations we wish to operate at a Thiele modulus of 0.4 or so. What should be the radius of the catalyst? How much will the catalyst effectiveness factor drop?

18.4   **First-order catalytic reaction: effect of particle size.** A first-order catalytic reaction was carried out with two different-sized spherical pellets and the following data was reported. For run 1: particle radius = 10 mm; observed rate = 3×10^{-5} mol reacted per cm^3 catalyst per sec. For run 2: particle size = 1 mm; observed rate = 15×10^{-5} mol reacted per cm^3 catalyst per sec. Estimate the Thiele modulus and the effectiveness factor for each of the above runs.

Estimate the rate constant and the effective diffusion coefficient for this system. Use a bulk gas phase concentration of A as 100 mol/m^3. Neglect gas side resistance.

Hint: Note that from the definitions the following relations can be deduced:

$$\frac{\text{Rate } 1}{\text{Rate } 2} = \frac{\eta_1}{\eta_2}$$

and

$$\frac{\phi_1}{\phi_2} = \frac{R_1}{R_2}$$

With these relations and the equation relating η to ϕ you will be able to derive an equation for ϕ_1 (or ϕ_2); solve this and get the required results.

18.5 First-order catalytic reaction: strong pore resistance case. A first-order catalytic reaction was carried out with two different-sized spherical pellets and the following data was reported. For run 1: particle radius = 10 mm; observed rate = 3×10^{-5} mol reacted per cm^3 catalyst per sec. For run 2: particle size = 5 mm; observed rate = 6×10^{-5} mol reacted per cm^3 catalyst per sec. Explain why it is not possible to estimate both the rate constant and pore diffusivity from this data. What can be estimated? Assume the bulk concentration is known.

18.6 Shape normalization. A first-order reaction has a rate constant of 5 sec^{-1} and is taking place in a catalyst with an effective diffusion coefficient of 10^{-6} m^2/s. Use shape normalization to estimate the approximate value of the effectiveness factor for the following cases: (1) Long cylinder with a radius of 4 mm; (2) a short cylinder with a radius of 4 mm and a length of 4 mm; (3) a long annular cylinder with an outer radius of 4 mm and an inner radius of 2 mm; (4) a short annular cylinder with an outer radius of 4 mm, an inner radius of 2 mm, and a length of 4 mm. For case (1), compare the value with the exact answer.

18.7 Weisz modulus. For diagnostic purposes a dimensionless quantity called the Weisz modulus is useful. This is a combination of the Thiele modulus and the effectiveness factor and is defined as

$$\Phi_W = \phi^2 \eta$$

Show that the Weisz modulus is related to the measured rate of reaction as

$$\Phi_W = L_c^2 \frac{-\langle R_A \rangle}{D_{eA} C_{Ab}}$$

Hence conclude that it can be directly calculated from the measured rate. A measured value or estimated value of the internal diffusion in the pore is needed for this. Show that if $\Phi_W < 0.4$ or so the internal gradients are negligible and the measured rate is representative of the true kinetic rate constant. Replot the η versus ϕ plot as a η versus Φ_W plot. What is the usefulness of this new plot?

18.8 Zero-order reaction in a sphere. Many hydrogenation reactions over Pd-supported porous catalyst show a zero-order kinetics. Consider such a reaction in a spherical catalyst. Show that the concentration at the center of the catalyst drops to zero if the Thiele parameter, defined as $\sqrt{k_0 R^2 / (D_e C_{As})}$, is equal to $\sqrt{6}$. Here C_{As} is the concentration of the reactant (hydrogen) at the surface of the catalyst. Based on this conclude that the pore diffusional resistances are unimportant and the effectiveness factor is equal to one if the surface concentration is maintained above $k_0 R^2 / 6 D_e$.

18.9   **Second-order reaction: sulfur removal from diesel.** Sulfur compounds present petroleum fractions such as diesel that can be removed by contact with a porous catalyst containing active metals such as Mo in the presence of hydrogen. If the catalyst size is a sphere of radius 3 mm and the concentration of sulfur in the liquid surrounding the catalyst is 2 mol/m³ find the rate of reaction. Assume the reaction is second order in sulfur concentration with a volumetric rate constant of 0.5 m³/mol.s. The effective diffusion coefficient of sulfur compounds in the liquid-filled pores is estimated as 6×10^{-10} m²/s. Use a slab geometry as an approximation and find the characteristic length parameter L. Define and calculate the generalized Thiele modulus for this second-order reaction. Calculate the effectiveness factor based on the generalized Thiele modulus.

18.10   **Fractional order: dead zones.** Similar to the zero-order reaction, the fractional order reactions exhibit the existence of dead zones beyond a critical value of the Thiele modulus. The problem examines the derivation of this value.

Start with the dimensionless version of Equation 18.18 and use the p-substitution to eliminate ξ. Use the condition that if $c_A = 0$ then $p = 0$ (i.e., a dead zone exists near the center) and derive an expression to get the following expression for the concentration gradient:

$$\frac{dc_A}{d\xi} = \phi \sqrt{\frac{2}{n+1}} \; c_A^{(n+1)/2}$$

At the critical Thiele modulus ϕ_c, c_A is just equal to zero at $\xi = 0$. Use this as the boundary condition, integrate the above equation by separation of variables and then show that the value of this modulus is

$$\phi_c = \frac{\sqrt{2(n+1)}}{1-n}$$

18.11   **Porous catalyst with series reactions.** Consider a porous catalyst with a series reaction represented as

$$A \rightarrow B \rightarrow C$$

Write governing equations for A and **B**. Express these in dimensionless form. Solve the equations for a case where the dimensionless concentration at the catalyst surface is one and zero for A and B respectively. **Note:** Both concentrations are normalized with the concentration of A at the surface. Calculate the dimensionless gradient for these two species at the surface. From these expressions calculate the selectivity parameter, defined as

$$S = -(dc_B/d\xi)/(dc_A d\xi) \text{ at } \xi = 1$$

State the significance of this parameter. Plot S as a function of ϕ and show how the high selectivity changes with the Thiele modulus.

18.12 Bimolecular reaction: asymptotic solution. Derive an expression for the surface flux modulus for a bimolecular second-order reaction using the p-substitution approach. Show that the generalized Thiele modulus should be defined as follows:

$$\Lambda = \frac{\phi^*}{\sqrt{1 - 1/3q}}$$

Here ϕ^* is the shape-normalized Thiele modulus, assuming the reaction to be pseudo-first order in species B, and q is the effective concentration ratio of B to A in the bulk gas.

18.13 Bimolecular reaction in a porous catalyst. Selective catalytic reduction of NO_x with NH_3 is studied in a porous catalyst of 4 mm diameter. The internal diffusivity was estimated independently as $D_A = 0.03 \times 10^{-4}$ m^2/s and $D_A = 0.05 \times 10^{-4}$ m^2/s. The reaction is second order with a rate constant of 2 m^3/mol sec. The temperature was 600 K and the total pressure was 1 atm; the partial pressures of NO and NH_3 were 500 kPa and 1000 kPa, respectively. Which reactant is limiting? Simulate the profiles in the interior of the catalyst. Find the effectiveness factor. To what extent is the diffusion limiting?

18.14 Hydrogenation of butynediol. Hydrogenation of butynediol was carried out in a slurry reactor with a Pd-based catalyst of size 0.01 cm with a catalyst loading of 0.2 kg/m^3. The density of the catalyst is 1500 kg/m^3. The reaction is bimolecular (first order with respect to both A and B) with a rate constant of 0.048 m^6/kg kmol s based on the mass of the catalyst. The mass transfer parameters are as follows: $k_L a_{ql} = 0.3$ s^{-1} and $k_{ls} = 0.005$ cm/sec for hydrogen and 0.009 m/s for butynediol. The diffusion coefficient is 0^{-9} m^2/s for hydrogen. The Henry constant for hydrogen is 0.01 mol/m^3 atm. The initial concentration of butynediol is 2.5 M. Find the time needed for 95% conversion of butynediol.

18.15 Internal and external temperature gradients. Catalytic cracking of gas oil is done at 903 K and 1 atm with a silica-alumina catalyst of 1.62 mm diameter. The effective diffusivity of the catalyst is 8×10^{-8} m^2/s and the effective thermal conductivity of the catalyst is 0.36 W/m.K. The heat of the reaction is -167 kJ/mol and the activation energy is 176 kJ/mol. The mass transfer coefficient is 1 m/s and the heat transfer coefficient is 400 W/m^{2}K. The measured rate is 0.4 mol/m^{3}s. Estimate the internal and external temperature differences.

18.16 Second-order reaction in a packed bed. A second-order reaction is taking place in a packed column with 3 mm radius particles. The internal diffusivity is 10^{-2} cm^2/s. The porosity of the bed is 0.4 and the rate constant is 0.75 m^3/mol.s. The concentration of A in the feed is 12 mol/m^3 and the volumetric flow rate is 0.5 m^3/s. Find the reactor volume needed to get 90% conversion. What are the values of the effectiveness factor at the inlet and the exit of the reactor?

18.17 Two-point collocation for three geometries. The matrices needed for a slab and cylinder are shown in the following. The values for a sphere were shown earlier.

For slab geometry the values are as follows:

```
A  =  [   -1.1180       1.1180
          -2.5000       2.5000]
B  =  [   -2.5000       2.5000
          -2.5000       2.5000]
x  =      0.4472     1.0000
w  =      0.8333     0.1667
```

The values for the cylindrical geometry are as follows:

```
A  =  [    -1.7321       1.7321
        -3.0000       3.0000]
  B  =  [     -6         6
          -6       6]
  w  =      0.3750       0.1250
  x  =      0.5774     1.0000
```

Use these to find the effectiveness factor for a second-order reaction with $\phi = 2$ for all three geometries. Compare the results with the Bischoff approximation.

18.18   **Collocation method for a reaction with substrate inhibition.** A reaction with substrate inhibition can exhibit multiple steady states. As an example consider a diffusion-reaction model in a slab geometry with the following form: $\phi^2 c_A/(1 + \alpha c_A + \beta c_A^2)$. Evaluate the effect of the Thiele modulus, keeping $\alpha = 110$ and $\beta = 1000$. Compare simulation with two point and five point collocation method. Plot η versus ϕ. Show that multiple steady states are likely for a Thiele modulus in the range of 23 to 26.8.

18.19   **Collocation method with larger number of nodes.** In Example 18.4 we showed the collocation method using only one interior point. This was found to be accurate for moderate values of the Thiele modulus, showing the power of the method. Here we wish to examine the solution for a larger number of points. The collocation matrices for $nt = 5$ for a slab geometry are presented in Table 18.3. Set up and solve the equations for various values of ϕ for a first-order reaction. Calculate η as a function of ϕ with an analytical solution. Repeat for a second-order reaction. Compare the effectiveness factor to that calculated using the shape-normalized Thiele modulus. Verify that the five-point collocation is accurate for most practical applications, unless the Thiele modulus is extremely large. In latter case the asymptotic solution is good enough.

Table 18.3 Collocation Matrices for Five-Point Total Collocation for Slab Geometry.

```
A =  -5.0717     8.6586    -6.3671      4.6458    -1.8656
     -1.4346    -2.6538     6.3920     -3.7114     1.4077
      0.5230    -3.1687    -1.9121      6.6239    -2.0661
     -0.3861     1.8615    -6.7019     -1.6060     6.8325
      0.7615    -3.4682    10.2684    -33.5618    26.0000
B =  -34.4654    44.1493   -14.2677      7.0025    -2.4186
      13.9798   -52.3162    48.8370    -15.1776     4.6771
      -3.1083    33.6001   -94.5282     80.8944   -16.8579
       1.8377   -12.5790    97.4469   -286.6901   199.9845
      38.5462  -170.3650   457.9635   -794.1446   468.0000
w =       0.0251     0.0792     0.1152     0.0956      0.0182
x =       0.2958     0.5652     0.7845     0.9340      1.0000
```

CHAPTER 19

Reacting Solids

Learning Objectives

After completing this chapter, you will be able to:

- Show how mass transfer affects the rate at which a gas reacts with a solid.
- Calculate the rate of reaction in a relatively non-porous solid based on the so-called shrinking core model.
- Calculate the rate of reaction of a porous solid on the basis of a diffusion-reaction model.
- Derive expressions for the conversion of the solid as a function of time.
- Characterize the structural changes of the solid due to a reaction and see how it affects transport properties and the rate of conversion of the solid.
- Examine the effects of structural changes such as pore plugging/opening and in some cases, an incomplete conversion of the solid reactant.
- Examine diffusional effects in solid–solid reactions.

Reactions where a solid reacts with a gas are very common in chemical, metallurgical, environmental, and electronic processes. Combustion of a coal particle is an example in the energy production sector as well as the process of gasification of coal where coal is reacted with steam to produce a mixture of CO and H_2 (syngas). In metallurgical industries, the reduction of metal ores (e.g., NiO) with hydrogen to produce metals (e.g., Ni) is an important application where a gas–solid reaction is encountered. Direct reduction of iron ores with hydrogen is another example. Oxidation of silicon to form an insulating film is an important application in the semiconductor fabrication of MOS (metal oxide semiconductor) devices. Similarly, in the production of polysilicon for solar cells, the precursor production involves many steps of gas–solid reactions such as chlorination of silicon, the deposition reaction of silane on a solid substrate, hydro-chlorination of silicon, and so on. In the pollution treatment field, CO_2 removal is an important problem and CO_2 capture by reaction with calcined lime (CaO) is an example of reacting solids.

In all these cases and other similar processes mass transfer effects play an important role as these are heterogeneous systems. Thus external (gas film) diffusion and pore diffusion in the interior of the solid affect the rate at which the solids react at the gas–solid interface. The scope of this chapter is therefore to study mass transfer effects in these systems.

The starting reacting solid can be relatively non-porous or highly porous and the modeling of the systems for these cases takes different approaches. For non-porous systems the reaction occurs at a sharp gas–solid interface. External or film mass transfer followed by diffusion through the product layer (in some cases) and the surface reaction are the factors to be considered. Analysis of such systems is discussed first.

For porous systems, the reaction occurs on the interior surface of the solid and hence simultaneous reaction during pore diffusion plays a role. The mass transfer steps here are similar to those for a porous catalyst analyzed in Chapter 18. However, an important difference is that the solid properties change as the solids react. This changes the porosity and in turn the effective diffusivity in the solid. We consider the various ways to represent how the pore structure affects the rate of reaction.

Diffusional effects in scenarios of a solid reacting with a second solid are also briefly discussed.

19.1 Shrinking Core Model

The reaction scheme considered is that of a gaseous reactant A reacting with a solid B in the following reaction scheme:

$$A + \nu_B B \rightarrow \text{Products}$$

One widely used model for this reaction is the shrinking core model. This model can be applied to the case where there is no solid product and also to a case where there is solid product formed. The analyses for the two cases are slightly different and therefore treated separately in the following subsections. In both cases the solid reactant B is assumed to be non-porous so that the reaction occurs at a sharp interface.

19.1.1 No Solid Product

Figure 19.1 illustrates the transport steps leading to the reaction. The reaction is assumed to follow two steps: external diffusion of the gaseous reactant, A, through the gas film from bulk gas to the surface of the solid followed by a reaction at the solid surface. The combination of these two steps leads to an overall rate constant, which is discussed next.

Overall Rate Constant

The pseudo-steady state model is used for the gaseous transport and reaction, which is then coupled with the transient macroscopic model used for

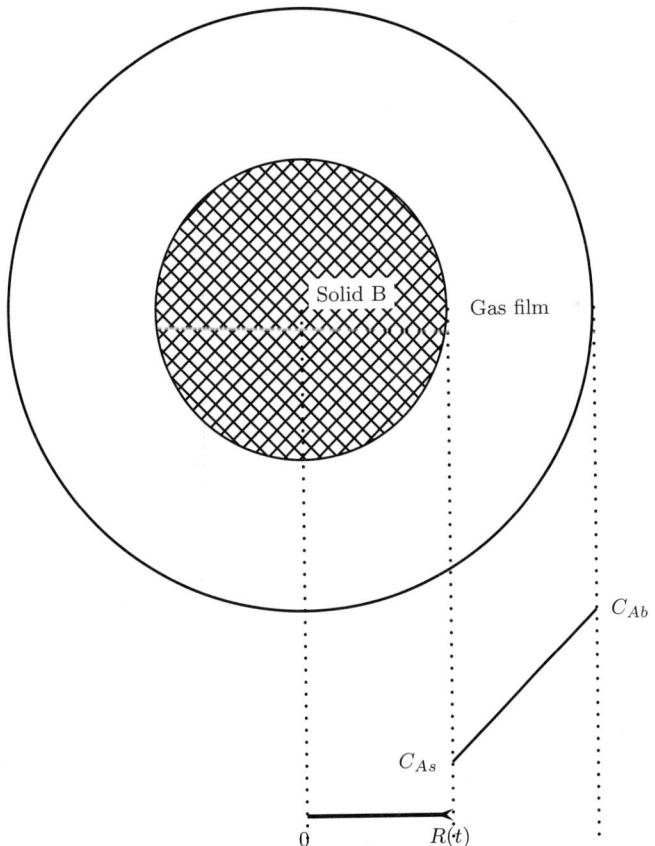

Figure 19.1 Schematic of a solid reacting to form a gaseous product. **Note:** The gas film thickness is exaggerated.

the solid. The modeling goal is to see the rate at which the solids reacts and to examine the effect of operating parameters on the rate:

Rate of mass transfer from gas to the surface $= k_m(C_{Ab} - C_{As})$

This is balanced by the surface reaction assuming a pseudo-steady state:

Rate of reaction per unit surface area of the solid $= k_s C_{As}$

A first-order surface reaction is considered here. The two rates are equal due to the pseudo-steady state assumption and hence C_{As} can be eliminated and a rate expression based on bulk gas concentration can be derived.

This rate equation combining the two transport effects in turn leads to an apparent or overall rate constant, which is defined as

Rate per unit area of B $= k_0 C_{Ab}$

This overall rate constant can be shown to be

$$k_0 = \left[\frac{1}{k_s} + \frac{1}{k_m} \right]^{-1} \tag{19.1}$$

Note that this can also be expressed as $k_0 = k_s k_m/(k_s + k_m)$. Also note that depending on the relative values of the mass transfer coefficient and the surface reaction constant two limiting steps can be identified:

- The process can be mass transfer limiting if $k_m \ll k_s$ and in this case $k_0 = k_m$.
- The chemical kinetics can be rate limiting if $k_s \ll k_m$ and in this case $k_0 = k_s$.

The rate of reaction can now be incorporated into the transient mass balance for the solid to study the conversion of the solid as a function of time, which is presented next. It is useful to note that this part of the model is similar to the model for sublimation of a solid considered in Section 3.2.

Solid Reactant Mass Balance

The rate of shrinkage of the particles (assuming a spherical shape) is calculated by writing a mass balance for the solid phase reactant:

$$\frac{d}{dt} \left([4/3]\pi R^3 C_{B0} \right) = -4\pi R^2 \nu_B k_0 C_{Ab}$$

where ν_B is the stoichiometric coefficient of B and C_{B0} is the concentration of solid B, which is equal to solid density divided by the molecular weight of B, that is, ρ_B/M_B. The previous equation can be simplified to

$$\frac{dR}{dt} = -\frac{\nu_B C_{Ab}}{C_{B0}} k_0$$

The variables can be separated and the solution for time needed to reach a particular radius $R(t)$ can be formally written as

$$t = \frac{C_{B0}}{\nu_B C_{Ab}} \left[\int_R^{R_0} \left(\frac{1}{k_s} \right) dR + \int_R^{R_0} \left(\frac{1}{k_m(R)} \right) dR \right] \qquad (19.2)$$

Here we have also used Equation 19.1 for k_0 in terms of k_s and k_m. We also note that k_m depends on R and hence the explicit dependency of this term on R is also noted in the second integral on the right-hand side. The two integrals on the right-hand side can be written separately and the time needed for the solid to shrink to any R is given as

$$t = t_R + t_M \qquad (19.3)$$

where t_R is the first term on the right-hand side of Equation 19.2:

$$t_R = \frac{C_{B0}}{\nu_B C_{Ab}} \int_R^{R_0} \left(\frac{1}{k_s} \right) dR \qquad (19.4)$$

and t_M is the second term:

$$t_M = \frac{C_{B0}}{\nu_B C_{Ab}} \int_R^{R_0} \left(\frac{1}{k_m(R)} \right) dR \qquad (19.5)$$

The first term on the right-hand side can be integrated directly since k_s is not a function of the radius. The result is usually expressed in terms of the conversion of the solid B, denoted as X_B. This is related to R as

$$X_B = 1 - \left(\frac{R}{R_0} \right)^3$$

Hence the integrated form of the Equation 19.4 is

$$t_R = \frac{C_{B0} R_0}{\nu_B C_{Ab} k_s} \left[1 - (1 - X_B)^{1/3} \right] \qquad (19.6)$$

In order to integrate the second term in Equation 19.2 the functional dependency of k_m on R should be taken into account. In accordance with the

Ranz-Marshall equation shown in Chapter 3, the dependency of mass transfer coefficient on particle radius is as follows:

- For small particles and/or small gas velocity $k_m \propto R^{-1}$.
- For large particles and/or large external gas flow rates $k_m \propto R^{-0.5}$.

Hence the radius dependency may change with time. At the start if we have large particles the dependency is -0.5 and will keep changing with time as the solid radius keeps decreasing; it may switch to the small particle regime where the dependency is -1.

Consider the case for small particles with low gas flow; here the mass transfer coefficient is given as

$$k_m = \frac{D_A}{R}$$

Hence the integrated form of Equaton 19.5 is

$$t_M = \frac{C_{B0}R_0^2}{2\nu_B C_{Ab} D_A} \left[1 - (1 - X_B)^{2/3}\right] \tag{19.7}$$

If $k_s \ll k_m$ right from the beginning, the process is under a kinetic regime and the film mass transfer resistance need not be accounted for.

A few remarks on the kinetics and mechanism of solid–gas reactions may be in order:

- The order of the reaction with respect to the solid concentration is often a fitted parameter rather than that result due to mechanistic considerations. This is unlike gas–gas reactions. The effect of concentration of a solid cannot be easily studied unlike a gaseous system since the activity of the solid reactant is one.
- The surface reaction can proceed in two ways. The first is an adsorption + surface reaction mechanism. Gas first adsorbs on the surface and the adsorbed species then reacts with B on the surface of the solid. Models similar to gas–solid catalytic reactions, that is, Langmuir-Hinshelwood type of models, have been proposed for these cases. The second is the direct attack of gas on individual sites of a solid that are the most reactive (active sites). Crystal defects and orientation of the surface will have a strong effect in this case.
- In some cases the reaction may proceed at only some preferential directions on the solid surface, the so called wormhole effect. Modeling and analysis of wormhole formation in reactive dissolution of carbonate rocks has been studied by Kalia and Balakotaiah (2007, 2009) and Fredd and Fogler (1998) and other references cited within these sources.

19.1.2 Solid Product: Ash Layer Effects

If the product is a solid it forms a layer near the external surface of the solid. This is known as the ash layer. The layer is assumed to be porous and a diffusion from the surface through the ash layer needs to be considered as an additional resistance.

The concentration profiles when the product layer forms are shown in Figure 19.2. The ash layer is assumed to occupy the region from λ to R at any

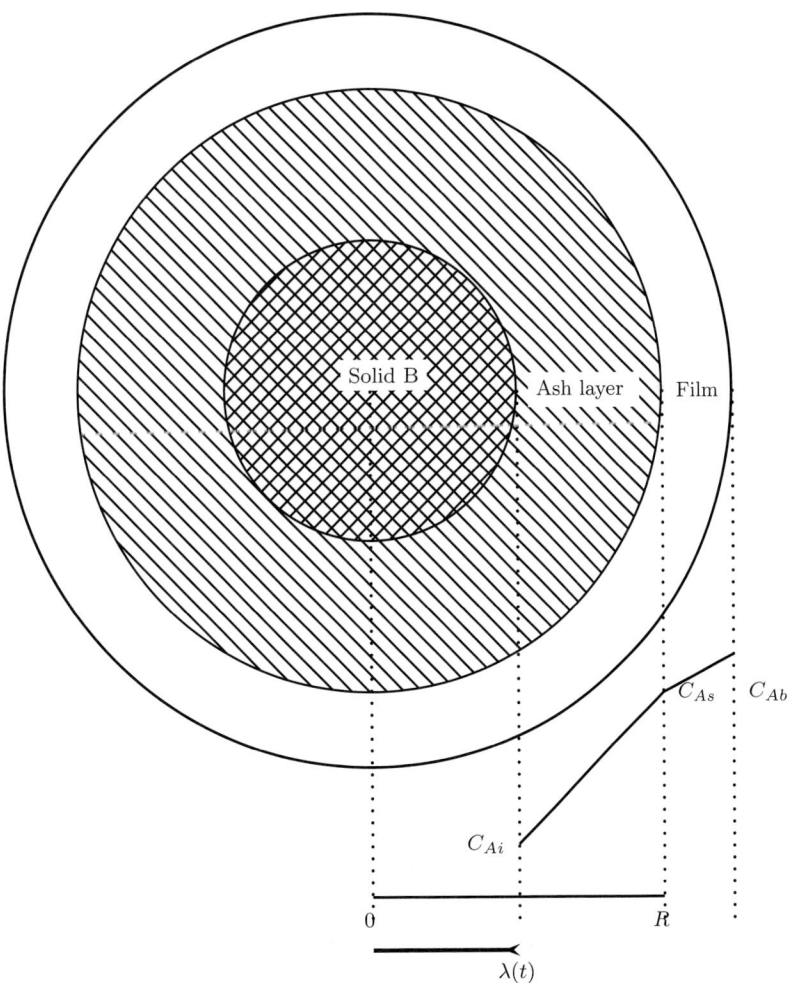

Figure 19.2 Representation of steps in reaction of a non-porous solid B with a gas leading to a solid product. The solid product is called the "ash" layer.

instant of time. The reaction plane is now located at λ, which moves inward with time.

The pseudo-steady state assumption is made for the solid. Let λ be the position of the ash layer at any instant of time as shown in Figure 19.2.

The reaction takes place due to three steps in series, noted in the following together with the corresponding resistance.

The first step is the gas film diffusion; the mole/s transported are given as

$$\dot{M}_A = 4\pi R^2 k_m (C_{Ab} - C_{As}) \tag{19.8}$$

The resistance for this step is $1/(4\pi R^2 k_m)$.

The second step is ash layer diffusion with the following rate:

$$\dot{M}_A = \frac{4\pi R \lambda D_{eA}}{R - \lambda}(C_{As} - C_{Ai}) \tag{19.9}$$

The resistance is $(R - \lambda)/(4\pi R \lambda D_{eA})$.

Finally the surface reaction takes place at a sharp interface located at λ:

$$\dot{M}_A = 4\pi \lambda^2 k_s C_{Ai} \tag{19.10}$$

The resistance is $1/(4\pi \lambda^2 k_s)$.

The overall resistance is the sum of these resistances, and the rate at which the gaseous species is consumed by reaction is

$$\dot{M}_A = 4\pi R^2 k_0 C_{Ab} \tag{19.11}$$

Here k_0 is the overall rate constant, defined as

$$\frac{1}{k_0} = \left[\frac{1}{k_m} + \frac{R}{D_{eA}}\frac{1 - \lambda^*}{\lambda^*} + \frac{1}{k_s \lambda^{*2}}\right] \tag{19.12}$$

The three terms on the right-hand side are due to the three resistances indicated earlier. The dimensionless location of reaction plane λ^*, defined as λ/R, is used for the convenience of later algebra.

Now we are in position to calculate the rate at which the solid phase reactant is consumed by writing a transient mass balance for the solid reactant. The method is similar to the solid balance discussed in the previous section and therefore not shown in detail. The following equation can be derived for the change in the reaction plane location with time:

$$\lambda^{*2}\frac{d\lambda^*}{dt} = -\frac{\nu_B C_{Ag}}{R C_{B0}}k_0 \tag{19.13}$$

Using Equation 19.12 for k_0 and separating the variables the result for the time needed to achieve a given conversion can be expressed as a sum of three terms, each corresponding to one of the resistances:

$$t = t_m + t_D + t_R \qquad (19.14)$$

Here t_m is the contribution of the mass transfer resistance:

$$t_m = \frac{RC_{B0}}{\nu_B C_{Ab} k_m} \int_{\lambda^*}^{1} \lambda^{*2} d\lambda^* \qquad (19.15)$$

Note that the external mass transfer coefficient does not change as a function of time now since the overall radius of the particle remains constant. Particle size may change slightly due to shrinkage or expansion, which can be accounted for by a volume balance. These details are not addressed here.

The t_D term in Equation 19.14 is the time needed if ash layer resistance were in control:

$$t_D = \frac{R^2 C_{B0}}{\nu_B C_{Ab} D_{eA}} \int_{\lambda^*}^{1} \lambda^*(1 - \lambda^*) d\lambda^* \qquad (19.16)$$

Finally t_R term is the time if the reaction were in control:

$$t_R = \frac{RC_{B0}}{\nu_B C_{Ab} k_s} \int_{\lambda^*}^{1} d\lambda^* \qquad (19.17)$$

The final results are presented in terms of the conversion of the solid, defined as

$$X_B = 1 - (\lambda^*)^3$$

Students may wish to verify the following results for the three contributions to the total time needed to achieve a given conversion.

Time for a given conversion for three regimes

External mass transfer control

$$t_m = \frac{RC_{B0}}{3\nu k_m C_{Ab}} X$$

Ash diffusion control

$$t_D = \frac{R^2 C_{B0}}{6\nu D_{eA} C_{Ab}} \left(1 - 3(1 - X)^{2/3} + 2(1 - X) \right)$$

Reaction control

$$t_R = \frac{RC_{B0}}{\nu k_s C_{Ab}} \left(1 - (1 - X)^{1/3} \right)$$

The following diagnostic criteria given by Levenspiel (1999) are useful for finding the controlling regime:

- The time for complete conversion is proportional to R^2 if ash diffusion is controlling.
- The time for complete conversion is proportional to R if surface reaction is controlling.
- The time for complete conversion is proportional to R^n where n is between 1.5 to 2.0 if external mass transfer is controlling. This is because k_m is a function of the radius. In this case the conversion should also change if gas velocity is changed.

Hence experiments with particles of different initial solid radius are indicative of the prevailing control regime for the given experimental conditions. Note that the controlling regime may change during the course of the reaction. Thus in the initial stages the process may be reaction controlled since there is no significant ash layer. In the later stages the process may be ash diffusion controlled. These changes are not captured in the above simple criteria and parameter fitting with tools available in MATLAB or other similar softwares may have to be used.

19.2   Volume Reaction Model

Volume reaction models apply when the particle porosity is large. Since the solid now contains a significant amount of pores, the gas can diffuse into the interior of the pellet and react over the interior surface of the pore rather than at a sharp interface. The schematic mechanism is shown in Figure 19.3, where we show a representative pore located between any differential control volume between r and $r + \Delta r$.

The gas diffuses into the pores, is adsorbed on the reactant surface, and reacts with B at the surface. The adsorption and reaction are lumped together and a "pseudo-homogeneous" rate of reaction is defined on the volume of the solid+gas locally at any location r. Note that a product layer can form on the surface of the solid. This layer is assumed to porous and the resistance due to the gas diffusion through this product layer is usually neglected at the first go at modeling. More detailed models account for this effect (see for example Christman and Edgar, 1983). Also the pore size changes due to product formation, which in turn affects the internal diffisivity.

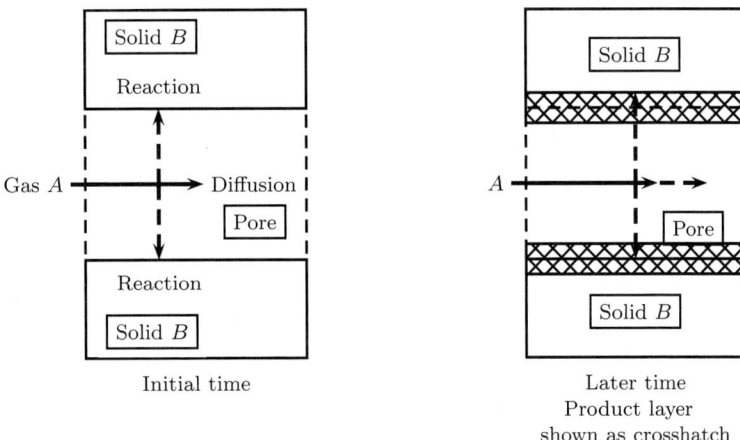

Initial time

Later time
Product layer
shown as crosshatch

Figure 19.3 Illustration of pore level phenomena envisaged in the volume reaction model.

19.2.1 Kinetic Model

The local rate of production of A by reaction is modeled as a power law kinetics:

$$R_A = -kC_A^n C_B^m$$

where n and m are the order of reaction with respect to A and B respectively. Here C_B is the local concentration of the solid reactant. The rate is defined as moles produced per unit total volume (pore + solid) per unit time.

The following remarks on the kinetic model shown above are useful:

- Note the negative sign for R_A since A is consumed rather than produced by reaction.
- The rates are often based on the mass of the solid $R_{A, m}$ in which case the units will be in mol/kg sec rather than mol/m^3 s. If so, we should multiply this by ρ_s to get R_A, the volume-based rate of reaction used previously.
- Rate can also be defined using the surface area of the solid, in which case the initial surface area per unit volume of the solid must be known.
- The order of reaction with respect to the solid reactant is often a fitted parameter rather than a mechanism-based value. Usually it has a value ranging from 0 to 1.

- The reactions often follow a two-step scheme with adsorption of A on the solid first followed by reaction with B at the surface. This is often modeled by a Langmuir kinetics of the following form:

$$R_A = -\frac{kC_A}{1 + K_A C_A} C_B^m$$

19.2.2 Concentration Profile for Gas and Solid

A concentration profile develops for the gas similar to that for a catalytic reaction. This profile can be calculated by a diffusion-reaction model for the gaseous reactant:

$$\epsilon_p \frac{\partial C_A}{\partial t} = \frac{1}{r^2} \frac{d}{dr} \left(D_{eA} r^2 \frac{dC_A}{dr} \right) + R_A \qquad (19.18)$$

where ϵ_p is the internal pore volume of the solid.

The diffusion coefficient is assumed to be a constant for further discussion. This parameter may change in time due to a change in the pore structure and due to reaction of solid B. Variable diffusivity analysis is briefly presented in the Section 19.3.1.

The corresponding rate of reaction of B is then incorporated into a transient mass balance for B (over a differential volume) and provides an equation to calculate the B versus t profile at each radial location:

$$\frac{\partial C_B}{\partial t} = \nu_B R_A \qquad (19.19)$$

The concentration of B is defined based on the total volume of the pellet (pore + solid) and hence the solid holdup term does not appear on the left-hand side. The R_A is assumed to be given by a power law model: $R_A = -kC_A^n C_B^m$. Observed features of the solution for reaction, which is first order in B ($m = 1$) is shown next followed by a zero-order reaction ($m = 0$).

19.2.3 First-Order Reaction in B

In this section a first-order reaction with respect to the solid and gas is considered. Dimensionless formulation and some illustrative results are presented. The following dimensionless concentration variables are used: $c_A = C_A/C_{Ab}$ and $c_B = C_B/C_{Bi}$. Here C_{Bi} is the initial concentration of the solid. Distance is scaled using the radius of the solid as $\xi = r/R$. Time is scaled as

$$\tau = (\nu k C_{Ab})t$$

This provides the following dimensionless equations to be solved for the gas phase and the solid phase:

$$\kappa \frac{\partial c_A}{\partial \tau} = \frac{1}{\xi^2} \frac{d}{d\xi} \left(\xi^2 \frac{dc_A}{d\xi} \right) - \phi^2 c_A c_B \tag{19.20}$$

and

$$\frac{\partial c_B}{\partial \tau} = -c_A c_B \tag{19.21}$$

Two parameters that arise in the dimensionless formulation are

$$\phi = R \sqrt{\frac{kC_{Bi}}{D_{eA}}} \tag{19.22}$$

This is a Thiele modulus parameter based on the initial concentration of B and

$$\kappa = \nu k C_{Ab} \frac{\epsilon_p R^2}{D_{eA}} \tag{19.23}$$

This is a capacity term representing the ratio of the reaction time constant to the time constant for internal diffusion.

The pseudo-steady state hypothesis is often used. The term containing κ in Equation 19.20 is then dropped. This is usually a good assumption for gas–solid systems since C'_{Ab} is usually much smaller than C_{Bi}. Hence the accumulation in the pore phase is often small and this term can be dropped. The assumption may not be valid for liquid–solid reaction. For further discussion, we use the pseudo-steady state hypothesis.

An illustrative plot of the concentration profile of gas A and solid B within the pellet is shown in Figure 19.4.

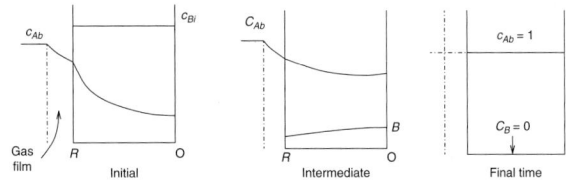

Figure 19.4 Schematic of the concentration profiles for gas A and reactant B for a (1,1) order reaction. The profile of gas A depends on the Thiele modulus and can be nearly flat (low Thiele) or very steep (large Thiele): an intermediate case is shown.

The following discussion is useful to understand the qualitative features of the volume reaction model and to identify the regime of operation. Two cases are normally distinguished: low Thiele and high Thiele. For low values of the Thiele parameter the concentration profile of A is nearly uniform in the pellet. The reaction of the solid takes place uniformly all over the pellet. This is often referred to as a homogeneous reaction model or uniform reaction model.

Uniform Reaction Model

If the Thiele modulus is small, the change in concentration of A in the solid is small. All of the solid is exposed to a nearly uniform concentration. The dimensionless concentration of A, c_A, may be approximated as one here leading to a case where the reaction occurs at a constant rate throughout the pellet:

$$c_B = \exp(-\tau) \tag{19.24}$$

where τ is the dimensionless time.

Intermediate Range of ϕ

For intermediate Thiele modulus values there is a moderate drop in concentration as shown in Figure 19.4. The reaction of B takes place all over the pellet with the interior reacting somewhat slowly due to less availability of A due to diffusion limitation. The concentration drop gets smaller as time progresses for A and the concentration of A increases. The final concentration is unity in dimensionless units. The concentration of B drops with time. The extent of the reaction of B near the surface is larger than at the center at the beginning of the process and the concentration variation of B tends to become more uniform with the progress of time.

Large Thiele Case

For large Thiele modulus values a steep profile develops in the solid. Here a (nearly) gas depleted zone develops near the center and the reaction is confined to a zone near the surface. This zone moves inward with time. Since the reaction is first order with respect B, the concentration of B can never become negative, although it can take infinitesimally small values. Hence there is no mathematical need to consider a separate ash layer formation, unlike the case of a zero-order reaction, which is discussed in the next subsection.

An illustrative result for a (1,1) order reaction simulated numerically using the a modified MATLAB PDEPE solver is shown in Figure 19.5. The time required for a given conversion is smaller if the Thiele modulus is small and the uniform reaction model holds.

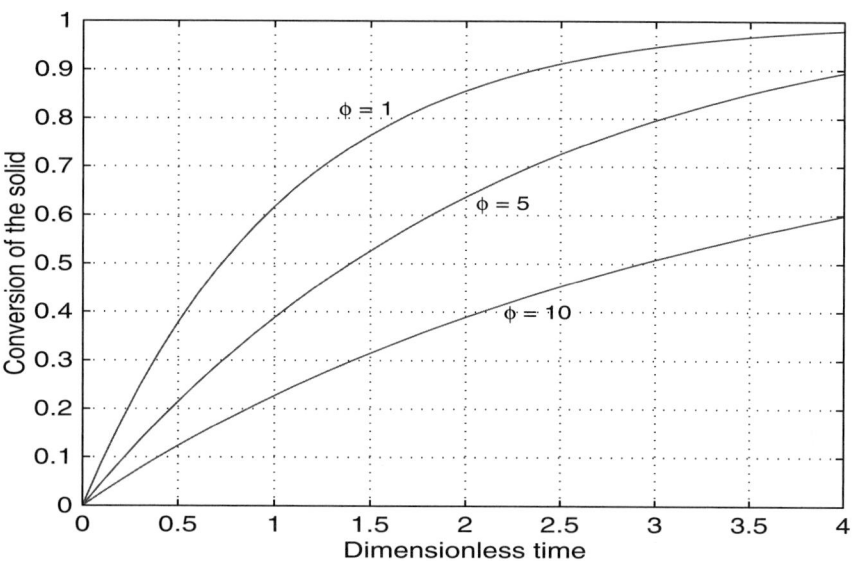

Figure 19.5 Conversion-time plot for a (1,1) order reaction based on the volume reaction model.

An approximate analytical solution can be obtained using the concept of cumulative concentration; this is presented as an exercise problem.

19.2.4 Zero-Order Reaction

Reactions that are zero order can exhibit an ash layer formation and hence they are treated somewhat differently compared to a first-order reaction. The concentration of B can become zero near the surface and hence it is necessary to distinguish two regions. The corrsponding concentration profiles are shown in Figure 19.6.

In the initial stage the concentration of B is above zero everywhere. The model equations for the gas and the solid are

$$\frac{1}{\xi^2}\frac{d}{d\xi}\left(\xi^2\frac{dc_A}{d\xi}\right) = \phi^2 c_A \tag{19.25}$$

and

$$\frac{\partial c_B}{\partial \tau} = -c_A \tag{19.26}$$

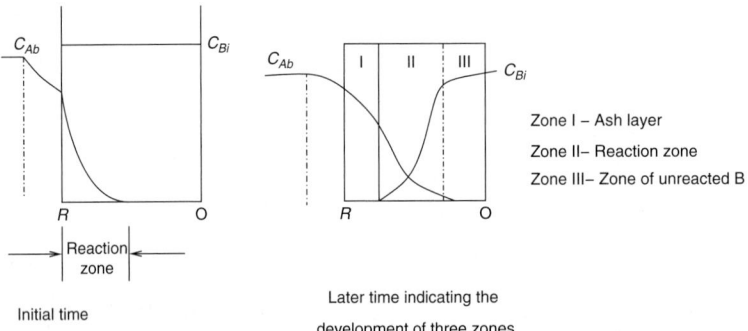

Figure 19.6   Schematic of the concentration profiles for gas A and reactant B for a (1,0) order reaction. An ash layer develops after the initial stage and the reaction zone moves inward with time.

The gas concentration can be solved independently since the reaction is first order. An effectiveness factor for the drop in gas concentration is used in the solid balance. Upon integration this yields the following expression for the conversion of the solid as a function of time:

$$X_B = \tau \left\{ \frac{3}{\phi^2} \left(\phi \coth(\phi) - 1 \right) \right\} \tag{19.27}$$

This solution is valid until the concentration of B remains finite at the surface.

In the later stages of the reaction a B depleted layer develops near the surface of the solid. The concentration of B becomes zero at a dimensionless location λ in the solid and the reaction can now take place only in the region from 0 to λ. The previous model equations now need a minor modification. Equation 19.25 has to be solved separately for the region of no B, which is from λ to 1. The rate term is set to zero in this region. The reaction zone is from λ to 0 and the rate term is retained for this region. Matching of the concentration and the flux of A on either side of λ has to be used as additional conditions. The solution for the concentration profile and the conversion of the solid with this modified model is not very complicated, but the details are not shown here. The following solution for conversion can be derived for this stage of the reaction:

$$X_B = 1 - \lambda^3 + \frac{3\lambda}{\phi^2} [\phi\lambda \coth(\phi\lambda) - 1] \tag{19.28}$$

where λ is given by the following equation:

$$kt = 1 + \frac{\phi^2}{6}(1 + 2\lambda) + (1 - \lambda)[\phi\lambda \coth(\phi\lambda) - 1] \tag{19.29}$$

The reaction zone thickness depends on the Thiele parameter. It is on the order of 0.039 for $\phi = 100$. Thus the sharp interface model can be used as an approximation for such cases. Note that the sharp interface model can be viewed as a limiting case of the volume reaction model for large Thiele modulus values. Details leading to Equation 19.29 are shown in the paper by Ishida and Wen (1968).

19.3 Other Models for Gas–Solid Reactions

In this section, we briefly consider some other related models proposed for gas–solid reactions. These models incorporate additional details such as pore volume change due to a reaction and non-uniform distribution of pores in the solid. The discussion is qualitative and for more details the references cited should be consulted. First we consider the modifications to the volume reaction model to account for structural changes.

19.3.1 Effect of Structural Changes

The product formation can lead to a change in the volume of the pore and the associated changes in the diffusion coefficient. The extent of this change depends on the nature of the product formed and the relative volume the product occupies relative to the solid reactant B. A relative molar volume ratio parameter is defined:

$$Z_V = \frac{\bar{V}_P}{\bar{V}_B} = \frac{\nu_p M_P / [\rho_p (1 - \epsilon_p)]}{\nu_b M_B / \rho_B}$$

This parameter is a measure of the ratio of the volume of product formed per unit volume of the reactant. If Z_V less than one, the pore volume increases, while the pore volume decreases if Z_V is greater than one.

The porosity variation can then be modeled as

$$\epsilon_p = \epsilon_{p0} + C_{B0} \bar{V}_B (1 - Z_V) \left(1 - \frac{C_B}{C_{B0}}\right)$$

Correspondingly the diffusion coefficient is assumed to vary as

$$D_{eA} = D_{eA,i} \left(\frac{\epsilon_p}{\epsilon_{p0}}\right)^n \tag{19.30}$$

where n has a range of values from between 2 and 3 depending on the porosity and tortuosity of the solid. A value of 2 for n holds for the random pore model with a single size distribution of the pore volume. This variation of diffusivity is then incorporated into the gas phase mass balance equation and the diffusivity in Equation 19.18 is no longer treated as a constant.

One of the effects of change in porosity is that an incomplete conversion of the solid can occur if the pore mouth gets closed due to the product occupying a larger volume than the starting reactant. The maximum conversion of the solid is shown to be given by the following equation:

$$X_B(max) = \frac{\epsilon_{p0}}{(Z_V - 1)(1 - \epsilon_{p0})}$$

The initial solid porosity has to greater than $(Z_V - 1)/Z_V$ if complete conversion of the solid has to occur. Data on hydrofluorination of UO_2 and sulfation of limestone shows these effects.

Single Pore Model

How the structural changes affect the rate of reaction was analyzed by Ramachandran and Smith (1977) by considering a single pore as a representative unit cell for the solid. A schematic of the conceptual basis on which the model is built is shown in Figure 19.7.

The model focuses attention on the whole pellet, which is modeled as a single pore of length $R/(3\sqrt{\epsilon_0})$, which is supposed to be representative of changes taking place in the pellet. The structural changes are accounted for

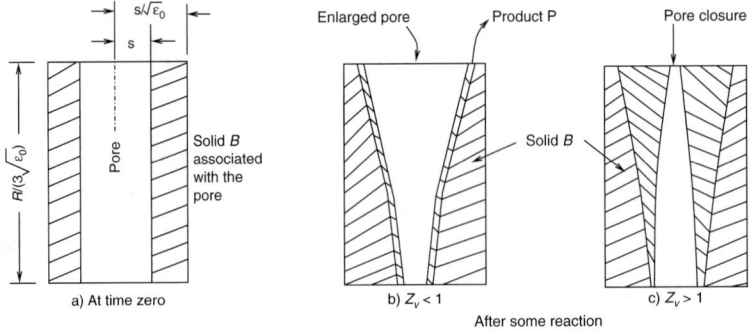

Figure 19.7   Effect of the Z_V parameter on the porosity change in the pellet with time. Pore mouth closure can occur for some range of Z_V values, leading to an incomplete conversion of the solid. Based on Ramachandran and Smith, 1977.

in a simple manner using some simple parameters. These are average pore radius, radius of the associated solid, effective pore length, effective diffusivity through the porous layer, and reaction rate constant. For further details and application of the model please refer to Ramachandran and Smith (1977) and the follow-up citations for that source.

Particle-Pellet Model

In this model the solid pellet is visualized as consisting of a number of small particles or grains. These grains are surrounded by macropores, which provide the pathway for the gas to reach the grains. The reaction occurs at the surface of each grain according to a sharp interface model. A product layer forms on the outer region of the grain and this offers an additional ash layer diffusion. The schematic representation is shown in Figure 19.8. Detailed analysis is presented in Calvelo and Smith (1970), Szekely and Evans (1971), and Sohn and Szekely (1972). Grains can change in size depending on the Z_V parameter. This in turn can lead to a decrease or increase in the macropore volume and thereby change the effective diffusion coefficient in the system. Pore mouth closure and incomplete conversion of the solid can occur for certain conditions if Z_V is larger than one.

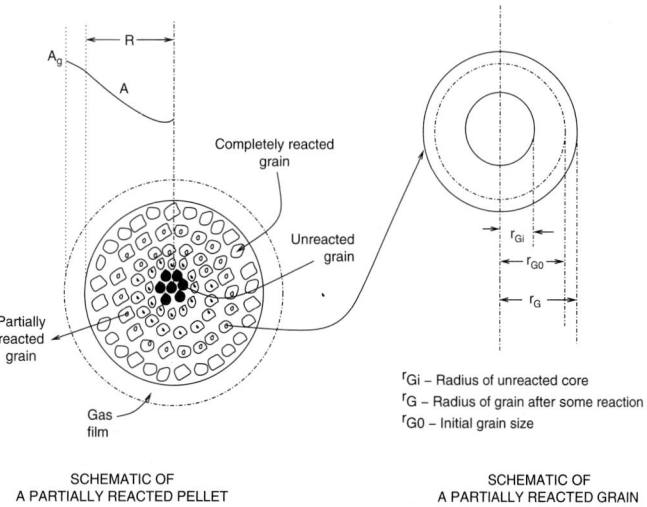

Figure 19.8 Schematic of the concepts used in the particle-pellet model, also known as the grain model. Based on Ramachandran and Doraiswamy, 1977.

Distributed Pore Models:

These models are similar in concept to the single pore model but take into account the effect of pore size distribution. Bhatia and Perlmutter (1981), Reyes and Jensen (1987), and Christman and Edgar (1983) provide illustrative examples and are useful sources for further study of this topic. The additional feature of these models is the simultaneous solution of the population balance equation for the pore size distribution. For a detailed review of various models, Ramachandran and Doraiswamy (1982) is a useful reference. An earlier work by Szekely, Evans, and Sohn (1976) is also another useful source.

19.4   Solid–Solid Reactions

Solid–solid reactions are of importance in a number of industrially important processes. A few examples are given below followed by a very brief discussion on modeling of such reactions.

Carbothermic reduction of ores (e.g., NiO + C) is one of the important examples in metallurgical industry. Cement production (tricalcium aluminate formation) is another large-scale application.

19.4.1   Classical Models

Avrami (1939, 1940, 1941) was one of the earliest investigators to study this and proposed a simple model based on nucleation theory. Transformations are often seen to follow a characteristic s-shaped, or sigmoidal, profile where the transformation rates are low at the beginning and the end of the transformation but rapid in between. The initial slow rate can be attributed to the time required for a significant number of nuclei of the new phase to form and begin growing. During the intermediate period the transformation is rapid as the nuclei grow into particles and consume the old phase while nuclei continue to form in the remaining parent phase. Once the transformation begins to near completion there is little untransformed material for nuclei to form in and the production of new particles begins to slow. Further, the particles already existing begin to touch one another, forming a boundary where growth stops.

Model development leads to the following equation for the conversion as a function of time:

$$X = 1 - \exp(-kt^n)$$

The value of $n = 4$ was assigned in the original Avrami model. The rate constant k was interpreted as

$$k = \pi \bar{N}(\bar{G})^3/3$$

where $\bar{N}$ is the nucleation rate and $\bar{G}$ is the growth rate of the nuclei.

The equation is also known as the Johnson-Mehl-Avrami-Kolmogorov, or JMAK, equation since it appears to have been first derived by Kolmogorov in 1937 but was popularized by Avrami in a series of articles in the *Journal of Chemical Physics*.

19.4.2 Dalvi-Suresh Contact Point Model

Dalvi and Suresh (2011) proposed a contact point model, the schematic of which is shown in Figure 19.9.

In this model the reaction initiates at points where the reactant B is in contact with A. An ash layer develops in the interior of A near the surface. Further reaction proceeds through the diffusion in the ash layer. If there are a large number of contact points the ash layer overlaps and grows over a region near the surface rather than as discrete islands. In such case the behavior of the system is close to the shrinking core model. On the other hand for a low number of contact points the reaction front consists of several moving fronts within the solid A. Thus the contact point model is able to cover the two limiting

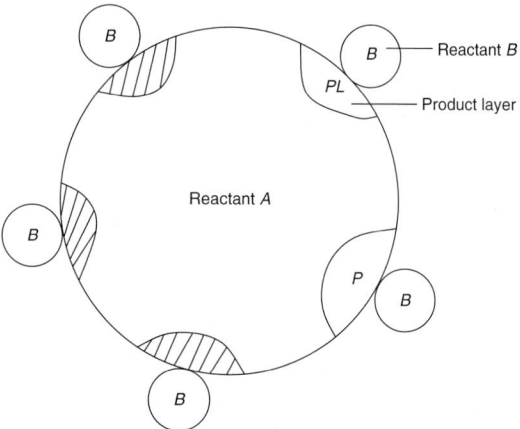

Figure 19.9 Contact points–based model for solid–solid reactions. Based on Suresh and Ghoroi, 2009.

cases of reaction control and ash diffusion control depending on the value of the parameters such as the number of contact points and the rate of growth at these contact points.

Summary

- Reaction of a gas with a solid is important in several industrial situations. Similar to the gas–catalytic reaction, the rate is affected by mass transfer parameters such as film diffusion and internal diffusion; the modeling approach follows a similar pattern to gas–solid catalytic reactions.

- The important difference between catalytic and non-catalytic reactions is that the solid reactant concentration variation needs to be added in the model and the concentration profiles of the gas and solid are changing with time, unlike the case of a catalytic solid.

- If the solid is non-porous the reaction is assumed to take place at a sharp interface between the gas (A) and the solid reactant (B) because the gas A does not penetrate into the interior of the non-porous B. Depending on whether a gaseous product or a porous solid product is formed two variations of this model are generally used.

- If the product is gaseous, then the shrinking solid model is used where the external film mass transfer and reaction at the solid surface are the two steps that need to be considered. A pseudo-steady state profile for the diffusing gas is usually assumed.

- A shrinking core model is also applied when the product of the reaction is a porous solid. A product layer called the ash layer forms on the original solid reactant and adds an extra resistance to mass transfer. Hence three mass transfer steps, film mass transfer, diffusion through the ash layer and reaction at the solid surface, need to be considered in the calculation of the rate.

- The change in B concentration is then calculated using a transient mass balance for B. The time versus conversion of the solid can be calculated on the basis of this model. The time for complete conversion for a given particle radius depends on the controlling regime. Time is proportional to particle radius if the reaction kinetics is controlled and it is proportional to radius squared if the ash diffusion is controlling.

- A solid can react non-uniformly under certain conditions due to surface irregularity and orientation of the crystal planes in the surface. This can lead to anisotropic growth or etching or wormhole formation. The shrinking core model is not applied for such cases and some alternative models proposed in the literature can be used. An example is etching of semi-conductor wafers with potassium hydroxide solutions.

- For porous solids the reaction takes place in the interior surface of the solid. The gas species profile in the solid due to intraparticle diffusion needs to be included in the model. These models are generally called volume reaction models to contrast with the sharp interface models for non-porous solids.

- In volume reaction models the reaction does not take place at a sharp interface but is confined to a finite volume. The extent of this volume depends on parameters similar to the Thiele modulus for gas–solid catalytic reactions. If the Thiele modulus is small the gas concentration is nearly uniform in the pellet and reaction takes place all over the solid. Reactant B is consumed uniformly in the pellet. The model based on this is also referred to as a "homogeneous" reaction model.

- In volume reaction models, if the Thiele parameter is large, the reaction is initially confined to a zone near the surface of the solid. Reactant B is consumed only here and not near the center. Over time as the B near the surface is consumed, the zone moves inward and finally shifts to a region near the center.

- Reactions that are zero order with respect to B exhibit features similar to the catalytic zero-order reaction. If the Thiele parameter is large, reaction takes place in a reaction zone near the surface of the solid. This reaction zone does not move with time until the surface concentration of B drops to zero. Then a B-depleted zone develops near the surface similar to an ash layer and the reaction zone moves inward. Analytic solutions can be developed for the initial time and the later time where the movement of the ash layer has to be tracked as well.

- The molar volume ratio parameter is an important parameter in the context of the volume reaction model. If the parameter is large, the pores get smaller with the extent of reaction of the solid. The pore size reduction has to be accounted for by a corresponding change in the effective diffusion coefficient. A number of models have been proposed for this and all these models differ only in the way the pore structure and the variation in diffusion coefficient are handled.

- For a certain range of the molar ratio parameter, pore plugging can result and incomplete conversion of the solid is observed. An example is the reaction of SO_2 with CaO where pore plugging leads to incomplete utilization of the solid reactant.

- Solid–solid reactions have certain additional complications, but concepts from solid–gas reactions are often applied. The classical model of Avrami (1940) discussed in the text is often used in practice. Many of these are reaction networks, and not single-step reactions as normally assumed. There is no complete theoretical framework available for the analysis of such systems. The importance of contact points between the two solids is studied in a number of newer models. For the case where there is a large set of contact points, the solid can react according to the shrinking core model, which provides some justification to the use of solid–gas models even for solid–solid systems.

Review Questions

19.1   When would you use the shrinking solid model?

19.2   When would you use the volume reaction model?

19.3   How does the time for complete conversion depend on the particle radius for the shrinking core model if the reaction is controlling?

19.4   How does the time for complete conversion depend on the particle radius for the shrinking core model if ash layer diffusion is controlling?

19.5   What is meant by the law of addition of time?

19.6   When can you expect wormhole formation or anisotropic etching?

19.7   If the starting solid is relatively porous what type of model would you use?

19.8   Why is the ratio of the molar volume of the product and the reactant important in modeling gas–solid reactions?

19.9   What are the main differences between a gas–solid reaction and a solid–solid reaction?

19.10   What are the common models useful for solid–solid reactions?

Problems

19.1   **Cynamide formation from carbide.** Calcium carbide reacts with nitrogen to give calcium cynamide by the following reaction:

$$N_2(g) + CaC_2(s) \rightarrow CaCN_2(s) + C(s)$$

Particles of 2 mm reacted in 3 hours while those of 4 mm reacted in 6 hours. Suggest a controlling regime and develop a rate equation.

19.2   **Time scales in the shrinking core model.** Spherical particles of a reacting solid of 4.8 mm are completely converted to products in 100 min. If the ash layer diffusion coefficient is 1×10^{-6} m^2/s and the surface reaction constant is 0.01 m/s, find the contribution of the ash later and reaction to the time needed for the conversion. Estimate the time for the reaction of a particle of 9.6 mm.

19.3 **Reduction of magnetite with hydrogen.** Consider reduction of magnetite ore with hydrogen, which is important in direct routes for iron production. The kinetic data is given as

$$k_s = 1930 \exp(-E/R_g T)$$

with $E = 100,000$ J/mol and k_s in m/s. The effective diffusion coefficient is 3×10^{-6} m^2/s in the ash layer. The reaction is carried out in pure hydrogen at 1 atm pressure and 600 K. The initial particle radius is 5 mm and the density of magnetite is 4600 kg/m^3. Plot the conversion versus time data and indicate the operating controlling regime.

19.4 **Langmuir kinetics.** Topochemical reactions with Langmuir kinetics are often encountered when gas A is strongly adsorbed. Assume that the reaction takes place at a sharp interface and the rate of the surface reaction is then represented as $kC_A(1 + K_A C_A)$. Develop a shrinking core model for this case.

19.5 **Temperature effects on conversion.** What is the effect of temperature on conversion? How does the controlling resistance change? The effect of temperature on MgO conversion with carbon was reported by Sohn and Han (2011). The rate increases first with temperature and then decreases with temperature above 975 K. Provide a suitable explanation to this anomalous effect.

19.6 **Parameter estimation for the shrinking core model.** Find the controlling mechanism for the following time verus conversion data:

Time, Hour	0.18	0.347	0.453	0.567	0.733
Conversion	0.45	0.68	0.80	0.95	0.98

19.7 **Cumulative concentration method for conversion.** The model equations for a (1,1) order reaction can be solved in an approximate analytical manner using the concept of cumulative concentration. This exercise walks you through the steps of this method. The analysis borrows the generalized Thiele modulus concept used on gas–solid catalytic reactions. The equations to be solved are repeated here for convenience. Using the pseudo-steady state is represented in slab geometry as

$$\frac{d^2 c_A}{d\xi^2} - \phi^2 c_A c_B = 0 \qquad (19.31)$$

The solid phase balance is

$$\frac{\partial c_B}{\partial \tau} = -c_A c_B \qquad (19.32)$$

Define a cumulative concentration variable, ψ, as

$$\psi(\tau, \xi) = \int_0^\tau c_A d\tau$$

Show that the concentration of B can now be represented as $c_B = \exp(-\psi)$. Show that the diffusion-reaction model for the gas, Equation 19.31, can now be written as

$$\frac{d^2\psi}{d\xi^2} = \phi^2[1 - \exp(-\psi)] \tag{19.33}$$

and state the boundary conditions to be used. This has a resemblance to a diffusion-reaction problem; hence, show that the following expression for the generalized Thiele modulus can be derived:

$$\Lambda = \frac{\phi(1 - \exp(-\tau)}{\sqrt{2(\tau + \exp(-\tau) - 1}}$$

Correspondingly a "conversion" effectiveness factor can be defined:

$$\eta_s = \frac{\tanh \Lambda}{\Lambda}$$

Show that the conversion of the solid at any time is given as the conversion in the absence of diffusion times the above effectiveness factor.

19.8 **Effect of solid residence time in a fluid bed roaster.** Fluid bed reactors are commonly used in metallurgy. The large-scale units are operated continuously. Since there is spread in residence time, the conversion of the solid depends on the E-curve. Show that the following expression can be derived for the conversion:

$$1 - \bar{X}_B = \int_0^{t_f} (1 - X_B)E(t)dt$$

Here t_f is the time needed for complete conversion for a single particle. If one assumes that the solids are well mixed but segregated, what is the expression for the E-curve? Use this in the previous equation and show that the following equation holds:

$$1 - \bar{X}_B = \int_0^{t_f} [1 - X_B(t)]\frac{\exp(-t/\bar{t})}{\bar{t}}dt$$

Use the model to predict the exit conversion for roasting of pyrrhoites if the mean residence time is 60 min and the time for complete conversion for a single particle is 20 min. Assume that the time for complete conversion is proportional to $R^{1.5}$.

CHAPTER 20

Gas–Liquid Reactions: Film Theory Models

Learning Objectives

After completing this chapter, you will be able to:

- Model systems where a gas is absorbed into a reacting liquid based on the film model.
- Understand the role of a key dimensionless parameter, the Hatta number, in determining the extent of reaction in the film (fast reactions) relative to that in the bulk liquid (slow reactions).
- Distinguish the various regimes of absorption for a fast reaction and examine the effect of operating parameters for these cases.
- Extend and study the modeling of more complex systems by taking an example of simultaneous absorption of two gases.
- Couple the film model with macro- or meso-models for the reactors in order to set up design equations for the absorber or stripper.
- Understand the role of dissolving and reacting fine particles in enhancing the rate of gas–liquid mass transport.

Gas–liquid reactions are common in industrial chemical production and in gas purification processes. Examples in chemical production can be found in liquid phase oxidation reactions such as production of cyclohexanol by oxidation of cyclohexane, p-xylene oxidation to produce terephthalic acid, chlorination of liquid benzene, and so on. Additional important examples can be found in gas purification processes and environmental cleaning, including CO_2 removal from various gas streams such as natural gas, synthesis gas, SO_2 removal from refinery gases and so on. In gas purification, the amount of gas that can be removed by physical absorption is limited due to equilibrium considerations. The rate can be enhanced significantly by absorbing in a reactive liquid. Carbon dioxide removal by absorbing into amine solutions is one such example.

In general gas absorption in a reacting liquid involves mass transfer from gas to liquid followed by reaction in the liquid phase. The goal of this chapter is to use the fundamentals of mass transfer and to calculate the rate of absorption for a wide range of process conditions.

Mass transfer analysis is done on the basis of the film model here. Note that the film model postulates the existence of a stagnant film near the gas–liquid interface. For fast reactions, species A (dissolved gas) may react in this film itself and simultaneous diffusion and reaction considerations are needed for this. If the reactions are slow, however, the mass transfer and reaction occur in series and the model is simpler. Hence we first show the criteria to distinguish the two, the extent of film reaction versus bulk reaction. The analysis is based on a first-order reaction of the dissolved gas.

The variation of the concentration of the second reactant, that is, the liquid phase reactant, is another consideration and depending on some parameter values, it may deplete in the film thereby affecting the rate of reaction in the film. The *no depletion* case is shown first. Then we show how the rate of absorption may be predicted if depletion occurs. The analysis for multiple reactions or simultaneous absorption of two gases is then shown, mainly using numerical tools.

Gas–liquid reactor level models are set up by coupling the local rate of mass transfer predicted by the film model to a meso- or macro-model for the equipment. Examples are shown to illustrate how the two levels are coupled and these will provide an introduction to set up reactor level models. A brief section is also provided for the case of mass transfer in a gas–liquid–dissolving solid system.

The discussions and the theoretical underpinnings also apply with some modifications to a liquid–liquid reaction and a brief section on this is also provided together with the pertinent references.

Gas absorption systems can also be modeled based on the penetration model for mass transfer rather than the film model. This involves the solution of the transient diffusion with reaction model. This is presented in Chapter 21, which can be studied together with the current chapter.

20.1   First-Order Reaction of Dissolved Gas

Consider a gas A that dissolves into a liquid and undergoes a first-order reaction in the liquid phase. We analyze the process based on the film model. The

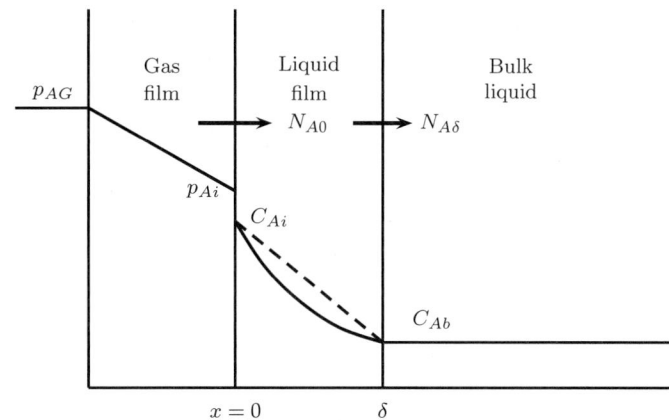

Figure 20.1 Schematic of the film model and the concentration profile of the dissolving and reacting gas in the liquid film. The dashed line represents the case of no reaction in the film.

concentration profiles envisioned in the film model are shown schematically in Figure 20.1.

The steps shown in Figure 20.1 are the following. Gas A diffuses from the bulk gas to the interface through the gas film, dissolves into the liquid at the interface, undergoes simultaneous diffusion and reaction in the liquid film, and crosses over the bulk liquid. Dissolved A can further react in the bulk liquid. There is no concentration profile in the bulk liquid that is assumed to be uniformly mixed. The dashed line in Figure 20.1 represents the concentration profile if no reaction occurred in the film.

The governing equation for species in the film is the diffusion-reaction equation:

$$D_A \frac{d^2 C_A}{dx^2} = k_1 C_A \tag{20.1}$$

The solution of this gives the concentration profile and the flux of A at the interface and into the bulk liquid. The reaction rate is assumed to be first order in A with a rate constant k_1, which has units of s^{-1}.

20.1.1 Boundary Conditions

The boundary conditions used for this problem are the following:

- A simple Dirichlet condition is used at the interface. At $x = 0$, $C_A = C_A^*$ where $C_A^* = p_{Ai}/H_A$ (equilibrium at the interface). This is valid if there is no appreciable resistance on the gas side of the interface. Then p_{Ai} is nearly the same as the (known) bulk gas partial pressure of A, denoted in Figure 20.1 as p_{AG}.

If there is significant gas-side resistance as well, then the model has to be coupled with gas-side transport and a Robin boundary condition has to be used at the interface. Note that the flux balance in the gas film gives the following relation between the bulk gas and the interfacial partial pressures:

$$p_{Ai} = p_{AG} - \frac{N_{A0}}{k_G} \qquad (20.2)$$

Here k_G is the gas-side mass transfer coefficient. This can be used to set p_{Ai} in an iterative manner.

- At $x = \delta$ (the edge of the film thickness) $C_A = C_{Ab}$, which is the dissolved concentration of gas A in the bulk. The bulk concentration of A cannot be independently assigned. The calculation of the dissolved gas concentration in the bulk involves coupling of the film model with a macroscopic model for the bulk liquid. (A Robin boundary condition can be derived rather than the Dirichlet used here.) This is explained in Section 20.2. We use a fixed value here for bulk concentration in order to focus on what is happening in the liquid film. Also, it may be noted that the bulk concentration often approaches zero if the process in the bulk liquid consumes the dissolved gas at a reasonably fast pace. In that case, the bulk liquid balance for dissolved A is not needed.

20.1.2 Dimensionless Version

The dimensionless form is obtained by defining a dimensionless concentration $c_A = C_A/C_A^*$ and a dimensionless distance $\xi = x/\delta$:

$$\frac{d^2 c_A}{d\xi^2} = Ha^2 \, c_A \qquad (20.3)$$

The dimensionless parameter appearing in the model is named the Hatta number squared. This is defined as

$$Ha^2 = \delta^2 \frac{k_1}{D_A}$$

Note that this has the same grouping of quantities as the Thiele modulus squared, introduced in the context of reactions in porous catalysts. Also note that the film thickness is often not known. What is generally known is the liquid side mass transfer coefficient k_L. The film thickness can be estimated from this parameter since $k_L = D_A/\delta$. Hence an alternative representation of Hatta number is

$$Ha = \frac{\sqrt{D_A k_1}}{k_L}$$

It can be shown that the Hatta number squared is a measure of the relative rate of reaction to diffusion or the ratio of the time scale for diffusion in the film to the time scale of reaction. Thus a large Hatta number implies a fast reaction and correspondingly a significant drop in the concentration of A in the film.

The boundary conditions in dimensionless form are $c_A(\xi = 0) = 1$ (no gas-side resistance case) and $c_A(\xi = 1) = c_{Ab}$, the dimensionless bulk concentration of A. The solution for the concentration profile with these boundary conditions is

$$C_A = C_A^* \left[\cosh(Ha\,\xi) - \frac{\sinh(Ha\,\xi)}{\tanh Ha} \right] + C_{Ab} \frac{\sinh(Ha\,\xi)}{\sinh Ha} \tag{20.4}$$

You may wish to verify the mathematical detail leading to this.

20.1.3 Flux Values at the Interface and into the Bulk

Flux at the interface is obtained by using Fick's law at the interface:

$$N_{A0} = -D_A \left(\frac{dC_A}{dx} \right)_{x=0} = -\frac{D_A C_{A^*}}{\delta} \left(\frac{dc_A}{d\xi} \right)_{\xi=0} \tag{20.5}$$

Using the expression for the concentration profile the interfacial flux is given as

$$N_{A0} = \frac{D_A}{\delta} \frac{Ha}{\tanh Ha} \left(C_A^* - \frac{C_{Ab}}{\cosh Ha} \right) \tag{20.6}$$

Note that the flux is not proportional to $C_A^* - C_{Ab}$, unlike the case of physical absorption. The above result was first derived by Hatta (1932).

The quantity of dissolved gas going into bulk is computed from the flux at the edge of the film (δ) (i.e., use Fick's law at this point) and the resulting expression is

$$N_{A\delta} = \frac{D_A}{\delta} \frac{Ha}{\tanh Ha} \left(\frac{C_A^*}{\cosh Ha} - C_{Ab} \right) \tag{20.7}$$

Where Does the Reaction Occur? Film or Bulk

In most cases the concentration in the bulk can be set to zero when $Ha > 1$. Only in exceptional cases where the volume of the film is comparable to the total volume of the liquid would there be a finite dissolved concentration in the bulk liquid when $Ha > 1$. The value of C_{Ab} can in general be estimated by a mass balance for the bulk liquid. Hence it is appropriate to take C_{Ab} as zero for further discussion and examine where the reaction is taking place.

A useful quantity is fractional extent of reaction in film:

$$\frac{N_{A0} - N_{A\delta}}{N_{A0}} = 1 - \frac{1}{\cosh Ha} \tag{20.8}$$

This is the measure of how much of the reaction occurs right when A is diffusing across the film. (This expression is obtained by combining Equations 20.6 and 20.7.)

The following values can be calculated for illustration:

- If $Ha = 0.2$, then the fractional extent of reaction in the film is only 1.97%. Note that C_{Ab} may not be zero in these cases and an estimation of this value is given in the next subsection.
- If $Ha = 1$ then the fractional extent of reaction in the film is about 0.3519. The assumption that $C_{Ab} = 0$ is adequate here in most cases and for larger values of Ha.
- If $Ha = 3$ the fractional extent of reaction in the film is nearly 90%.
- If $Ha = 5$ the fractional extent of reaction in the film is nearly 98.6%.
- If $Ha = 10$ all reaction is in the film itself. Nothing reaches the bulk.

A plot of the concentration profile for various values of Ha is shown in Figure 20.2. The plot is in agreement with the discussion on the fractional extent of reaction in the film as well.

20.1.4 Enhancement Factor

The flux at the interface is also a measure of the quantity of gas absorbed per unit interfacial area per unit time and is expressed as dimensionless flux, which is also known as the enhancement factor, E. This is defined as

$$E = \frac{N_{A0}}{D_A C_A^* / \delta}$$

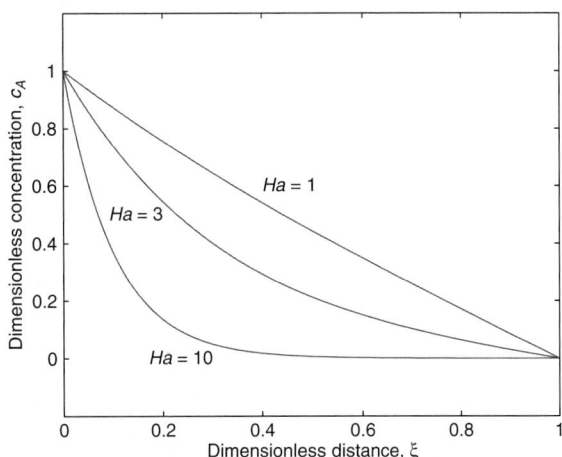

Figure 20.2 Concentration profiles for A in the film where it undergoes a first-order reaction. Note the slopes at ξ of 0 and 1, which is the measure of the local fluxes at these points and are representative of mass crossing the interface and moving into the bulk, respectively.

This is a measure of the gas absorbed from the liquid in a reacting case compared to a non-reacting case. Using the earlier expression for N_{A0}, we have:

$$E = \frac{Ha}{\tanh Ha} \qquad (20.9)$$

Note that this expression assumes C_{Ab} to be zero.

Following simplified expressions for the enhancement factor for a fast reaction is useful and widely used in many applications. The fast reaction case applies if $Ha > 3$. All the reaction takes place in the film for a fast reaction. For $Ha > 3$, the $\tanh Ha$ term can be approximated as one. Hence $E = Ha$.

Using this in the definition of E, the following expression is obtained for the flux of A at the interface:

$$N_{A0} = C_A^* \sqrt{D_A k_1}$$

Note that δ cancels out and the film thickness no longer matters; the rate of absorption does not depend on δ and k_L, the value of the mass transfer coefficient.

Correspondingly, the volumetric rate of absorption R_A^V is given as

$$R_A^V = a_{gl} C_A^* \sqrt{D_A k_1}$$

It depends only the gas–liquid interfacial area. The measured volumetric rate of absorption can be then used to estimate the interfacial area under these conditions, a technique widely used in many studies. (This presumes that the values for D_A and k_1 are known from independent measurements.)

20.2 Bulk Concentration and Bulk Reactions

We now look at how the dissolved concentration of A in the bulk liquid may be estimated and the extent to which the bulk reaction takes place. The bulk composition is determined by a mass balance for the bulk liquid. The control volume can be either a meso- or macroscopic element of the bulk liquid (depending on the extent of mixing in the liquid). An illustrative control volume for the analysis presented here is shown in Figure 20.3.

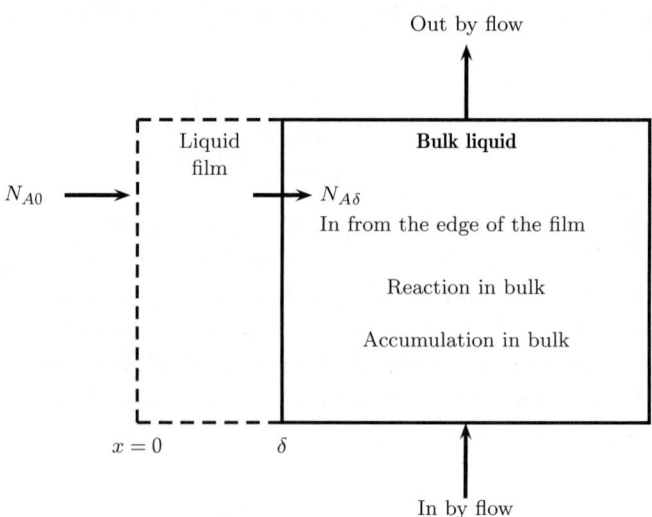

Figure 20.3 Control volume (solid rectangle) and the terms needed for the mass balance in the bulk liquid.

20.2.1 Bulk Concentration

The concentration of dissolved gas in bulk can be calculated starting from a mass balance over this control volume:

In crossing at the edge of the film at δ + in from adjacent bulk control volume − out into adjacent control volume = consumed in bulk + accumulation.

For batch systems where there is no flow in or out of the bulk, it may be appropriate to simplify the balance as

In at δ = Consumed in the bulk.

Here we also neglect accumulation and focus on any instant of time where the input term balances the reaction. This is equivalent to assuming that the bulk liquid is at a pseudo-steady state at a given instant of time.

In at δ, the edge of the film, is given by Equation 20.7. We multiply $N_{A\delta}$ by the gas–liquid interfacial area a_{gl} to get the volumetric transfer rate. Hence

$$R_{A,\delta}^{V} = k_L a_{gl} \frac{Ha}{\tanh Ha} \left(\frac{C_A^*}{\cosh Ha} - C_{Ab} \right) \qquad (20.10)$$

This balances the reaction in the bulk, which is given as

$$R_{A,\delta}^{V} = k_1 \epsilon_L' C_{Ab} \qquad (20.11)$$

Here

$$\epsilon_L' = \epsilon_L - \delta a_{gl}$$

This represents the volume of the bulk liquid per unit reactor volume. Note that this is often approximated as ϵ_L since the thickness of the film is usually very small.

The value of C_{Ab} can now be obtained by solving Equations 20.10 and 20.11 simultaneously. The volumetric absorption rate in turn is obtained by using this value of C_{Ab} in any one of these equations. The additional parameter needed in this model is $\epsilon_L'/\delta a_{gl}$.

The dimensionless bulk concentration is shown in Figure 20.4 as a function of Hatta number for two values of the $\epsilon_L'/\delta a_{gl}$ parameter. The bulk concentration is nearly zero if $Ha > 0.6$.

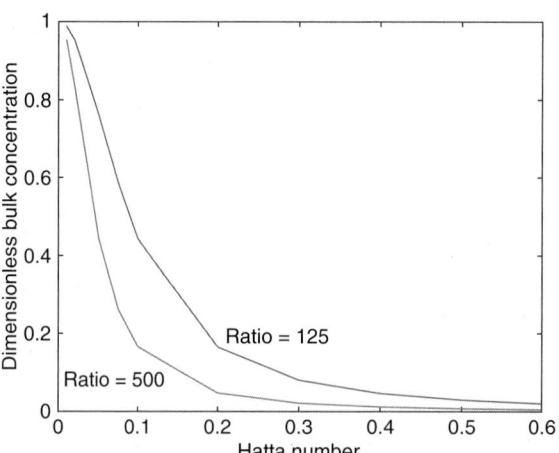

Figure 20.4   The bulk concentration values for a batch absorber as a function of Hatta number for two values of the volume ratio parameter.

20.2.2 Absorption Rate Calculation for $Ha < 0.2$

If $Ha < 0.2$ most of the reaction occurs in the bulk and the film reaction can be neglected. The expression for C_{Ab} can be simplified for this case and a correspondingly simpler expression for the volumetric rate of absorption can be obtained, which is presented in this section.

We note here for $Ha < 0.2$, the $\cosh(Ha)$ term will be nearly one and the $\sinh Ha$ and $\tanh Ha$ terms are nearly equal to Ha. Hence Equation 20.10 simplifies. Combining the resulting expression with Equation 20.11 leads to the following result for the bulk concentration.

$$C_{Ab} = \frac{C_A^*}{1 + [\epsilon_L'/\delta a_{gl}]Ha^2} \tag{20.12}$$

Correspondingly the expression for the rate of absorption simplifies to

$$R_A^V = \frac{(\epsilon_L' k_1)(k_L a_{gl})}{\epsilon_L' k_1 + k_L a_{gl}} C_A^*$$

This can also be expressed in the two resistance form by taking the reciprocal on both sides:

$$\frac{1}{R_A^V} = \left[\frac{1}{\epsilon_L' k_1} + \frac{1}{k_L a_{gl}} \right]^{-1} C_A^*$$

Two limiting cases of this expression will now be examined.

The first is the case of a kinetically limited reaction. The condition for this to be applicable is $k_1\epsilon_L << k_L a_{gl}$, that is, the reaction rate is smaller than the mass transfer rate and the reaction rate is rate limiting. The liquid will remain at the saturation concentration:

$$R_A^V = k_1 C_A^*(\epsilon_L - \delta a_{gl}) \tag{20.13}$$

The reaction rate is simply governed by the kinetics of the reaction. The bulk concentration of A will be now the same as the saturation concentration of the gas.

The second is the mass transfer limited reaction; the condition for this to be applicable is $k_1\epsilon_L' >> k_L a_{gl}$, that is, the reaction rate is much larger than the mass transfer rate. Mass transfer from interface to bulk liquid is rate limiting. The bulk liquid concentration is nearly zero due to the rapid rate of mass transfer. The rate of absorption now becomes

$$R_A^V = k_L a_{gl} C_A^* \tag{20.14}$$

The rate is controlled by the rate at which A arrives from the interface to the bulk liquid. The bulk concentration is nearly zero for this case. The two cases are sketched in Figure 20.5.

A brief summary of the various regimes of absorption are as follows. Given the data on the rate constant and diffusion coefficient, the mass transfer coefficient should be independently measured or estimated. This permits us to

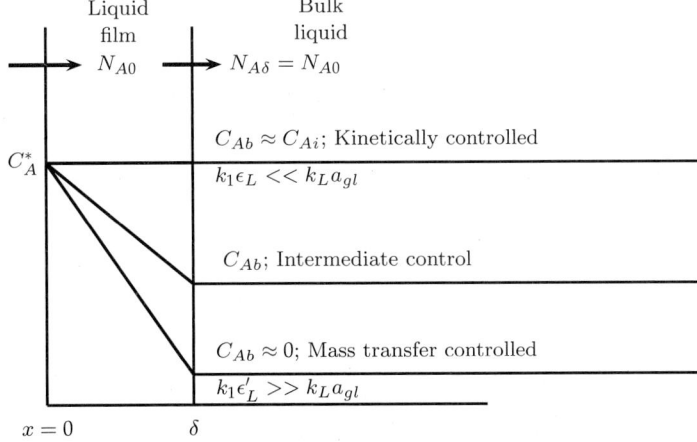

Figure 20.5   Kinetically controlled and mass transfer controlled cases. Concentration in the bulk liquid compared to the interfacial concentration for a case of no reaction in the film.

calculate Ha and see if the reaction is mainly in the film or mainly in the bulk. The following regimes can then be identified depending on the magnitude of Ha:

- $Ha > 3$: Reaction occurs entirely in film. The rate of absorption depends only on the gas–liquid interfacial area for a first-order or pseudo-first-order reaction. The intrinsic mass transfer coefficient does not play a role.
- $0.5 < Ha < 3$: For this range some reaction occurs in film and some in bulk. But the bulk balance leads to a nearly zero concentration of A in the bulk and this assumption can be used to simplify the rate calculation.
- $Ha < 0.5$: No reaction occurs in the film. Two subcases can be noted now depending on the relative values of $k_1\epsilon'_L$ and $k_L a_{gl}$:
    - $k_1\epsilon'_L \gg k_L a_{gl}$: Mass transfer to bulk is rate controlling. The rate of absorption depends on the volumetric mass transfer coefficient. The bulk concentration will be nearly zero here.
    - $k_1\epsilon'_L \ll k_L a_{gl}$: The chemical reaction is rate limiting. The rate of absorption does not depend on the mass transfer coefficient. It depends on the rate constant and the liquid holdup. Bulk concentration is nearly equal to saturation solubility under these conditions.

20.3   Bimolecular Reactions

In this section we consider the case where the reaction is first order in both gases A and B. The reaction scheme considered is a bimolecular reaction, represented as

$$A(g \rightarrow l) + \nu B(l) \rightarrow \text{Products}$$

Examples of such a system include CO_2 absorption in various reacting solvents such as amine solutions and many other gas treating processes. We consider a case where some or most reactions are taking place in the film and examine the effect of transport of B from the bulk liquid to the interface. The diffusion-reaction differential equation is now applied to both species A and B.

The governing equations are as follows:

$$D_A \frac{d^2 C_A}{dx^2} = k_2 C_A C_B \tag{20.15}$$

$$D_B \frac{d^2 C_B}{dx^2} = \nu k_2 C_A C_B \tag{20.16}$$

Here x is the actual distance into the film with $x = 0$ representing the gas–liquid interface.

20.3.1 Dimensionless Representation

We now introduce the relevant dimensionless parameters. The dimensionless concentration that will be used is

$$c_A = C_A/C_A^*$$

C_A^* is the equilibrium solubility of gas A in the liquid corresponding to the partial pressure of A at the gas phase:

$$C_A^* = p_{AG}/H_A$$

Here H_A is the Henry's law constant for species A in units of atm m^3/mol. Note the definition and the unit since there are many ways of defining the Henry constant.

Similarly the dimensionless concentration for B is defined as

$$c_B = C_B/C_{BL}$$

where C_{BL} is the bulk liquid concentration of B.

Finally ξ is used for the dimensional distance in the film and defined as $= x/\delta$.

With these variables, the governing equations are the following:

$$\frac{d^2 c_A}{d\xi^2} = Ha^2 c_A c_B \tag{20.17}$$

and

$$\frac{d^2 c_B}{d\xi^2} = Ha^2 c_A c_B/q \tag{20.18}$$

Two dimensionless quantities appear in the dimensionless formulation. These are

$$Ha^2 = \delta^2 \frac{k_2 C_{BL}}{D_A} \tag{20.19}$$

and

$$q = \frac{D_B C_{BL}}{\nu D_A C_A^*} \tag{20.20}$$

Noting that $k_L = D_A/\delta$, the Hatta number can also be expressed as

$$Ha^2 = \frac{D_A k_2 C_{BL}}{k_L^2} \tag{20.21}$$

The Hatta number (squared) represents the ratio of diffusion time to reaction time and hence a large drop in concentration in the film can be expected for large values of this number. The q parameter is the measure of the concentration ratio of B to A. A relatively large drop in concentration of B in the film can be expected if this parameter is small.

Boundary Conditions

The boundary conditions for species A are as follows. At the interface, the boundary condition will depend on whether the gas film resistance is included or not and will be of the Dirichlet or Robin type. The two cases are as follows:

Case 1: No gas film resistance. Here at $\xi = 0$, $C_A = C_A^*$ and Hence $c_A = 1$. The boundary condition is now of the Dirichlet type.

Case 2: Gas film resistance included. A balance over the gas film provides the boundary condition:

$$k_G(p_{AG} - p_{Ai}) = -D_A \left(\frac{dC_A}{dx}\right)_{x=0} \tag{20.22}$$

Also note that p_{Ai}, the interfacial partial pressure of A, is related to the interfacial concentration of A in the liquid by Henry's law. Thus $C_A(x = 0) = p_{Ai}/H_A$. Using these, the boundary condition at $\xi = 0$ can be expressed in dimensionless form as

$$Bi_G(1 - c_A) = -\left(\frac{dc_A}{d\xi}\right) \tag{20.23}$$

where $Bi_G = k_G H_A/k_L$, a Biot type of number for gas–side mass transfer. The boundary condition is now of the Robin type.

At $\xi = 1$, the edge of the film, $C_A = C_{Ab}$ is some specified value depending on the bulk processes. This value C_{Ab} will depend on the extent of bulk reactions, convective and dispersive flow into the bulk and so on, similar to the model in Section 20.2.1. But even for moderately fast reactions, the bulk concentration of dissolved gas turns out be zero and we take this value to be zero. Hence $c_A = 0$ at $\xi = 1$ will be used as the second boundary condition.

The validity of this assumption can be checked after a solution is obtained by a bulk balance for A and modified if needed.

The boundary condition for species B are specified as follows. At $\xi = 0$ we use $dc_B/d\xi = 0$ since B is non-volatile and the flux is therefore zero. At $\xi = 1$, we have $c_B = 1$. This completes the problem definition.

The results are generally presented in terms of an enhancement factor E, defined as

$$E = -\left(\frac{dc_A}{d\xi}\right)_{\xi=0} = -p_0 \qquad (20.24)$$

This is a measure of the flux at the interface. Here we use p for the concentration gradient $dc_A/d\xi$.

Since the profiles of A and B are now coupled this calls for a numerical solution. We will study the numerical solution using a MATLAB solver in a later section. But the essential details of the concentration profiles that can be expected are shown in Figure 20.6.

Rather than solving numerically we present here various limiting cases that gives us a feel for the problem and the results that could be anticipated from the MATLAB solution. Three regimes can be classified and analytical solutions can be obtained for these cases. These regimes are pseudo-first order, second order, and instantaneous. The regimes depend on the interfacial concentration of B and this is first estimated in the following section.

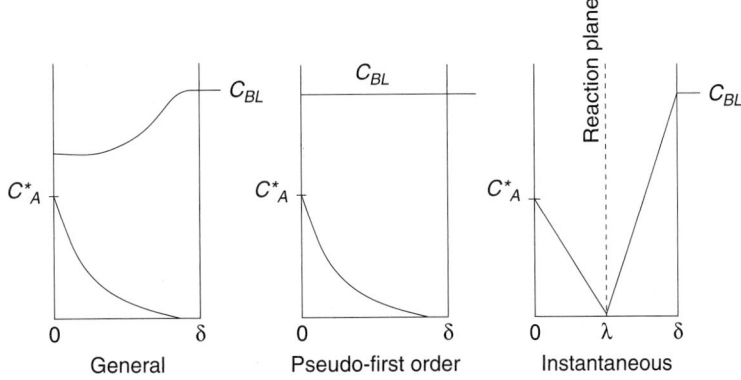

Figure 20.6 Concentration profiles in the film for a gas–liquid reaction, showing three regimes of absorption. Gas film resistance is not included here.

20.3.2 Invariance Property of the System

A relation between the concentration profiles of A and B exists purely based on the stoichometry of the reaction; this is known as the invariance property and is derived in this section.

The rate terms can be eliminated between Equation 20.17 and Equation 20.18 to give

$$\frac{d^2 c_A}{d\xi^2} = q\frac{d^2 c_B}{d\xi^2} \tag{20.25}$$

This can be integrated once to give

$$\frac{dc_A}{d\xi} = q\frac{dc_B}{d\xi} + p_0$$

Here the integration constant has been assigned as the dimensionless concentration gradient of A at the interface. We note that the no flux boundary condition for B at the interface is satisfied by this choice for the constant.

A second integration and use of boundary condition at $\xi = 1$ gives

$$c_A = qc_B + p_0\xi + A_1$$

From the boundary conditions at $\xi = 1$, we can show that $A_1 = -p_0 - q$. Hence the concentration profile of B is related to that for A as

$$c_B = 1 + \frac{c_A}{q} + \frac{p_0}{q}(1 - \xi)$$

This is the invariant property that ties together the A and B concentrations.

The important quantity is the interfacial concentration of B, denoted as c_{Bi}, as it determines which of the previuosly indicated regimes are likely to exist. This is obtained readily by setting $\xi = 0$ in the above expression: this is an invariant of the system. All numerical results should show this result; otherwise the numerical procedure is wrong.

Since the enhancement factor E is equal to $-p_0$, the resulting invariant equation is written in terms of E as

$$\boxed{c_{Bi} = 1 + \frac{1}{q} - \frac{E}{q}} \tag{20.26}$$

The three regimes are now classified as follows:

- Pseudo-first-order reaction: This occurs when c_{Bi} is nearly equal to one. Analysis for this case is similar to that in Section 20.1 and is presented in Section 20.3.3.
- Instantaneous reaction: This occurs when c_{Bi} is nearly equal to zero. Analysis for this case is presented in Section 20.3.4.
- Second-order reaction: The interfacial concentration of B is quite lower than one but does not approach zero. This is also known as the depletion regime. Analysis for this case is presented in Section 20.3.5.

It may be worthwhile to estimate the conditions where c_{Bi} is nearly unity. This will happen if $E << q$. But we do not know E. Let us proceed further assuming that the interfacial concentration of B is nearly unity. In this case Equation 4.18 can be solved analytically. We call this a pseudo-first-order model.

20.3.3 Analysis for Pseudo-First-Order Case

In this case $c_{Bi} = 1$ and c_B in Equation 20.18 can be set to one. The differential equation for A is now linear and analytic solution for C_A obtained in 20.1 for a first-order reaction is applicable. The corresponding enhancement factor derived in Section 20.1.4 is applicable. Hence:

$$E = \frac{Ha}{\tanh Ha} \tag{20.27}$$

Using this value of E in Equation 20.26 we find that if $Ha << q$ then the concentration of B will be nearly one at the interface, leading to a pseudo-first-order approximation. Hence the condition for the applicabiity of the pseudo-first-order reaction is that $Ha << q$.

The following results are of importance when the reaction occurs under pseudo-first-order conditions:

- The volumetric rate of absorption depends only on the gas–liquid interfacial area and not on the composite parameter $k_L a_{gl}$.
- The observed dependency on the concentration of B is half order although the reaction is intrinsically first order in B.

20.3.4 Analysis for Instantaneous Asymptote

The second limit can be analyzed easily as well. There are two ways of doing this. The first is that mathematically we set $c_{Bi} = 0$ in Equation 20.26. We find

$$E = 1 + q \tag{20.28}$$

One can show that if $Ha \gg q$, the instantaneous asymptote is reached, while if $Ha < q$, the pseudo-first-order regime will be observed.

The second approach is based on a physical consideration that postulates that there is a reaction plane λ that separates the A and B regions. In other words we assume that A and B cannot co-exist in the film. The physical situation prevailing is shown in Figure 20.6 earlier in the chapter.

The rate of absorption and reaction is controlled by the rate of transport A from the interface to the reaction plane:

$$N_{A0} = \frac{D_A C_A^*}{\lambda}$$

This is balanced by the rate of transport of B from the bulk to the reaction plane which is

$$N_{A0} = \frac{N_B \lambda}{\nu} = \frac{D_B C_{BL}}{\nu(\delta - \lambda)}$$

Equating these two expressions for N_{A0} one can solve for both λ and N_{A0}, Based on N_{A0} the enhancement factor can be calculated. The result is the same as Equation 20.28.

The approximation $E = q$ is commonly used in Equation 20.28 since q is usually large (roughly 10 or so). In such case the volumetric rate of absorption is given as

$$R_A^V = k_L a_{gl} \left(C_A^* + \frac{D_B C_{BL}}{\nu D_A} \right)$$

Usually C_{BL} is larger than C_A^* and the first term is dropped. Then

$$R_A^V = k_L a_{gl} D_B C_{BL} / \nu D_A$$

This is a rather interesting result that states that the rate is not a function of the gas concentration.

20.3.5 Second-Order Case: An Approximate Solution

For $Ha \approx q$ a second-order reaction case is reached where both equations (profiles for A and B) have to be solved together. This can be done numerically.

However, a good analytical approximation was proposed by Hikita and Asai (1963). They claimed that the pseudo-first-order asymptote can still be used provided the Ha parameter is now based on the interfacial concentration of B rather than the bulk concentration. The rationale for doing this is that species A reacts in a zone near the interface and the concentration seen by A is mostly equal to C_{Bi}. The fact that C_B varies in the film is not of much significance as far as the rate at which it reacts. Thus R_A local $= k_2 C_{Bi} C_A$ is good enough. We thus can still use Equation 20.27 but with a modified Hatta defined as

$$Ha(\text{Modified}) = Ha\sqrt{c_{Bi}}$$

This leads to the following expression for E:

$$E = \frac{Ha\sqrt{c_{Bi}}}{\tanh[Ha\sqrt{c_{Bi}}]} \tag{20.29}$$

where c_{Bi} is given by Equation 20.26. Note that c_{Bi} is a function of E and hence we have to solve for both E and c_{Bi} simultaneously. Simple MATLAB code to do this is given in Listing 20.1. The enhancement factor is calculated given the two dimensionless groups Ha and q. A third group Bi_G is also to be considered if there is appreciable resistance on the gas film. The code also includes the effect of this parameter on c_{Bi}.

Listing 20.1   Enhancement Factor for Second-Order Reaction

```
% Hikita-Asai model solver. Hikita2.m
% Dimensionless parameters
Q = 50.; HA =50.00;  Bi_G = 100.;
% Case 1: No gas side resistance case
Bi_fun = @(X)  (1.0 + 1./Q -HA * X^0.5 *coth(HA * X^0.5) /Q - X );
 bi0= 0.008 % trial value
bi = fsolve (Bi_fun, bi0)
  E1 = HA * bi^0.5 * coth(HA * bi^0.5)
  % Case 2: With gas side resistance
  Bi_fun = @(X)  (1.0 + 1./Q -HA * X^0.5 *coth(HA * X^0.5) /Q ...
   *(1.+ 1./Bi_G)  /(1+HA * X^0.5* coth(HA * X^0.5)/Bi_G) - X );
  bi = fsolve (Bi_fun, bi0)
  E = HA * bi^0.5 * coth(HA * bi^0.5)/ ...
    (1. + HA * bi^0.5 * coth(HA * bi^0.5)/Bi_G)
```

An example of calculation of the interfacial concentration of B and the enhancement factor is shown next in Example 20.1.

Example 20.1   Effect of Ha on the Interfacial Concentration

Using Listing 20.1 show the effect of Ha on c_{Bi} and E for a fixed value of $q = 50$. Also show the effect of gas film resistance for $Ha = 5$.

Solution

If $Ha = 5$ running the code we get an interfacial concentration of 0.9239 and a corresponding enhancement factor E of 4.806, which is close to a pseudo-first-order reaction.

   If $Ha = 50$, we find the interfacial concentration of B is 0.3931 and the corresponding enhancement factor is 31.3471. The regime is second order and appreciable depletion of B in the film occurs.

   If $Ha = 500$ we get a c_{Bi} that is close to zero and hence the instantaneous regime is approached. The enhancement factor is 50.49, which is close to instantaneous value of 51.

   Thus by increase of Ha, the regime of absoprtion shifts from pseudo-first order to second order and finally to instantaneous.

   The above results are in the absence of gas film resistance for illustration of the regime change.

   If gas-side resistance were included with $Bi_G = 10$, $Ha = 50$, and $q = 50$, we find the following results: the interfacial concentration of B is 0.4945 and the corresponding enhancement factor is 26.01. The interfacial concentration is higher but the enhancement factor is lower since there is some concentration drop of A in the gas film.

An illustrative application to carbon dioxide absorption in sodium hydroxide solution is presented next in Example 20.2.

Example 20.2   CO$_2$ Removal with NaOH Solution

Carbon dioxide is absorbed into a solution of NaOH in a packed column absorber. Locally at a given point in the absorber the concentration of NaOH is 1.0 M and the partial pressure of CO$_2$ is 1 atm. Find the rate of absorption. Neglect the effect of gas-side resistance. Parameters needed are shown as part of the solution.

Solution

The physico-chemical parameters needed are listed below together with the values specific to this problem. The parameter values are taken from the work of Danckwerts (1960). Solubility of CO$_2$ at 1 atm pressure = 30 mol/m^3.

Diffusivity of gas A (CO_2) and liquid reactant B (NaOH) are $D_A = 1.8 \times 10^{-9}$ m²s and $D_B = 1.7 D_A$. The rate constant for the reaction is $k_2 = 10$ m³/mole.s

The hydrodynamic parameter needed is the liquid-side mass transfer coefficient. We use $k_L = 1 \times 10^{-4}$ m/s for illustration. Gas-side resistance is neglected here.

The dimensionless parameters Ha and q are first calculated from Equations 20.21 and 20.20 and the values are found to be 42 and 21.25, respectively.

The next step is to solve Equations 20.26 and 20.29 simultaneously for the interfacial concentration of B and the E value. We find $c_{Bi} = 0.186$. Hence this represents a condition where there is a significant depletion of the reactant at the interface.

The enhancement factor is calculated as 18.3. Correspondingly the rate of absorption is $k_L C_A^* E$ and is equal to 0.0539 mol/m² s.

20.3.6 Instantaneous Case: Effect of Gas Film Resistance

The regime of instantaneous absorption is often accompanied by considerable gas film resistance. In some cases, the process can become entirely limited by the gas-side resistance and the reaction plane shifts to the interface. The reaction of the gas then takes place directly at the interface. This section examines the effect of gas film resistance for an instantaneous reaction and the conditions when the reaction plane will shift to the interface.

The concentration profile for an instantaneous reaction including the drop in concentration in the gas film is shown in Figure 20.7. Also shown are three possible cases that are discussed in the following.

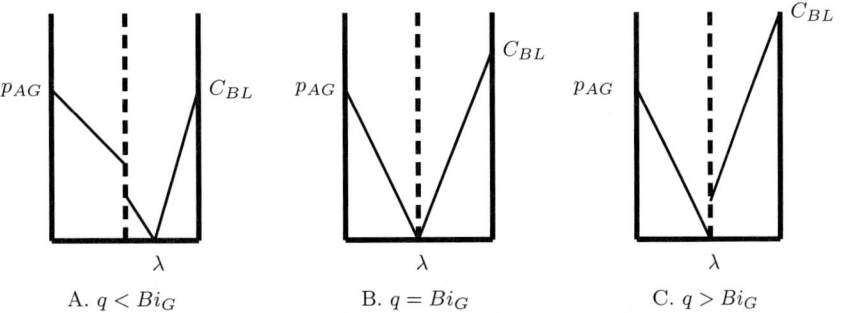

Figure 20.7 Concentration profiles of A and B for an instantaneous reaction with gas film resistance included; note the shift in controlling regime.

The analysis starts with various transport steps and the corresponding rate equation. For the gas film transport of A we use

$$N_{A0} = k_G(p_{AG} - p_{Ai})$$

(20.30)

The transport of A from the interface to a distance λ is given as

$$N_{A0} = \frac{D_A(p_{Ai}/H_A)}{\lambda}$$

(20.31)

Finally from the transport consideration for B from the bulk liquid to the interface we have

$$N_{A0} = \frac{-N_{B,\delta}}{\nu} = \frac{1}{\nu}\frac{D_B C_{BL}}{\delta - \lambda}$$

(20.32)

We now have three equations for three unknowns: N_{A0}, p_{Ai}, and λ. The final result can be expressed in terms of the enhancement factor, now defined as

$$E = \frac{N_{A0}}{k_L(p_{AG}/H_A)}$$

(20.33)

It is easier to solve for E, if the various expressions are in term of $1/E$. From the transport equations of A (gas film plus 0 to λ) we can show that

$$\frac{1}{E} = \frac{1}{Bi_G} + \frac{\lambda}{\delta}$$

Here Bi_G is the Biot number defined as $k_G H_A/k_L$.

From the transport considerations for B we find

$$\frac{1}{E} = \frac{1}{q}[1 - \frac{\lambda}{\delta}]$$

where q is now defined as $D_B C_{BL}/(\nu D_A p_{Ag}/H_A)$.

Equating the two expressions for $1/E$ and solving for λ we obtain

$$\frac{\lambda}{\delta} = \frac{1}{1+q}\left[1 - \frac{q}{Bi_G}\right]$$

(20.34)

It should be noted here that this equation is valid only if $q < Bi_G$ since λ cannot be negative. (The lambda has to be positive or at the most can be equal to zero. If it is zero, the reaction plane would be right at the interface.) In such case the value of E is simply equal to Bi_G; the process becomes controlled entirely by gas film resistance.

Using this value λ in either of the equations we find the enhancement factor (for the case where $q < Bi_G$) can be calculated after some "minor" algebra as

$$E = \frac{1+q}{1+1/Bi_G} \text{ for } q < Bi_G \tag{20.35}$$

and for $q > Bi_G$ as

$$E = Bi_G \text{ if } q > Bi_G \tag{20.36}$$

It is useful to illustrate the calculations to a numerical problem, which we do in Example 20.3.

Example 20.3   H$_2$S Absorption in NaOH Solution

Hydrogen sulfide absorption in amine solutions may be assumed to be instantaneous. At a point in the absorber the total pressure is 20 atm and the gas contains 1% H$_2$S. The ratio D_B/D_A is 0.64 and the amine concentration is 0.25 M. The mass transfer coefficients are $k_L = 0.01$ cm/s and $k_G = 2 \times 10^{-5}$ mol/cm^2 s atm. The Henry's law constant for H_2S is 1950 Pa m^3/mol.

Find the rate of absorption at this point. Also find the amine concentration when the process becomes entirely controlled by gas-side mass transfer.

Solution

We need two dimensionless numbers, q and Bi_G, to use the model for instantaneous reaction with gas-side resistance. It is useful to convert the Henry's law constant to atm to simplify the calculations. The value is $1950/1 \times 10^5 = 0.0195$ atm m^3/mole.

The required parameters are calculated as

$$q = \frac{0.64 \times (0.25 \text{ mol/L} \times 10^3 \text{ L/m}^3)}{20 \text{ atm} \times 0.01/0.0195 \text{ m}^3 \text{ atm /mole}} = 15.6$$

$$Bi_G = \frac{(2 \times 10^{-1} \text{ mol/m}^2 \text{ s atm}) \times (0.0195 \text{ m}^3 \text{ atm /mole})}{1 \times 10^{-4} \text{ m/s}} = 39$$

The enhancement factor is then calculated as $(1+q)/(1+1/Bi_G)$ and is equal to 16.18.

The rate of absorption is calculated as $k_L(p_{AG}/H_A) \times E$ and is found to be $0.0166 \text{ mol/m}^2.\text{s}$.

If the concentration of amine is increased the rate is increased and reaches a case where it becomes gas-side controlled. This concentration is reached when $q = Bi_G$. The concentration is found to be 625 mol/m^3. The rate of absorption at this point is calculated as $k_G p_{AG}$ and is equal to $0.04 \text{ mol/m}^2.\text{s}$. If the amine concentration is increased further, the rate stays at this point since the reaction plane has already shifted to the interface. These are useful guidelines for choosing the optimum operating conditions depending on the mass transfer coefficient prevailing in the contactor, which is discussed briefly next.

20.3.7 Choice of Contactor Based on the Regimes of Absorption

The contactor to be used depends on the regime. Thus in a pseudo-first-order scenario, a contactor that develops a large interfacial area is important. For a slow reaction, a large volume of liquid is needed and intense agitation to create a large interfacial area is not needed. For the instantaneous case the overall mass transfer coefficient is important and contactors that provide this are needed.

20.4   Simultaneous Absorption of Two Gases

In this section, we analyze a case where two gases dissolve simultaneously and react with a common liquid phase reactant. Many complex reaction schemes can be analyzed using a similar approach and hence it is useful to study this example. Simple CHEBFUN-based code is also given for the two gas absorption case (Listing 20.2) and the code can be modified and used for many other applications.

The reaction scheme considered in this section is represented as

$$A + \nu_A C \rightarrow \text{Products}$$

$$B + \nu_B C \rightarrow \text{Products}$$

where A and B are the two gases being absorbed and C is the common liquid phase reactant.

Each reaction is assumed to be second order overall (1,1 order) kinetics with rate constants of k_A and k_B for the first and second reaction.

20.4.1 Model Equations

Governing equations follow from the diffusion-reaction model and are as follows:

$$D_A \frac{d^2 C_A}{dx^2} = k_A C_A C_C \tag{20.37}$$

$$D_B \frac{d^2 C_B}{dx^2} = k_B C_B C_C \tag{20.38}$$

$$D_C \frac{d^2 C_C}{dx^2} = \nu_A k_A C_A C_C + \nu_B k_B C_B C_C \tag{20.39}$$

Boundary conditions at $x = 0$ are

$$C_A = C_A^*; \quad C_B = C_B^*; \quad dC_C/dx = 0$$

The gas-side resistance is not included here but can be added by incorporating a Biot number dependency and changing the gas A and B boundary conditions at $x = 0$ to a Robin condition.

Boundary conditions at $x = \delta$ for a fast reaction that consumes A and B in the bulk are

$$C_A = 0; \quad C_B = 0; \quad C_C = C_{CL}$$

In general a bulk balance is needed to estimate the bulk concentrations of the dissolved gas similar to single gas absorption. These refinements are also not addressed here in order to focus on the ramifications of the film diffusion-reaction part of the model.

20.4.2 Dimensionless Representation

We now introduce the following dimensionless parameters: distance in the film $\xi = x/\delta$; concentrations of the gaseous species: $c_A = C_A/C_A^*$ and $c_B = C_B/C_B^*$; concentration of the liquid phase reactant $c_C = C_C/C_{CL}$.

The dimensionless forms are then obtained and are as follows:

$$\frac{d^2 c_A}{d\xi^2} = M_A c_A c_C \tag{20.40}$$

$$\frac{d^2 c_B}{d\xi^2} = M_B c_B c_C \tag{20.41}$$

and

$$\frac{d^2 c_C}{d\xi^2} = q_A * M_A c_A c_C + q_B * M_B c_B c_C \qquad (20.42)$$

The dimensionless quantities needed are as follows

$$M_A = \text{ Hatta for first reaction} = \delta^2 \frac{k_A C_{CL}}{D_A} \qquad (20.43)$$

$$M_B = \text{ Hatta for second reaction} = \delta^2 \frac{k_B C_{CL}}{D_B} \qquad (20.44)$$

$$q_A = \text{ concentration ratio for A} = \frac{D_C C_{CL}}{\nu_A D_A C_A^*} \qquad (20.45)$$

$$q_B = \text{ concentration ratio for B} = \frac{D_C C_{CL}}{\nu_B D_B C_B^*} \qquad (20.46)$$

The boundary conditions are

$$\text{At } \xi = 0: \quad c_A = 1; \quad c_B = 1; \quad dc_C/d\xi = 0$$

$$\text{At } \xi = 1: \quad c_A = 0; \quad c_B = 0; \quad c_C = 1$$

20.4.3 CHEBFUN Solution

The solution was implemented numerically using CHEBFUN in MATLAB, and the code is presented in Listing 20.2. This code is illustrative of two gases being absorbed but can be modified easily for other cases of mass transfer accompanied by complex reactions. (See paper by Ramachandran and Sharma (1971) for many examples of absorption of gases followed by complex reactions).

Listing 20.2   Code for Simulation of Absorption of Two Gases with Reaction in the Film

```
%% Simultaneous gas absorption with fast second-order reaction
% parameters
MA = 1.E+04; qa =0.1 ; MB = 10.0; qb = 0.2;
% declarations
x = chebfun ('x', [0,1])
a = chebfun ('a', [0,1])
```

```
b = chebfun ('b', [0,1])
c = chebfun ('c', [0,1])
N = chebop (0,1)

N.op = @(x,a,b,c)[ diff(a,2)—MA*a.*c , ...
                diff(b,2)—MB*b.*c , ...
                diff(c,2)—qa*MA*a.*c—qb*MB*b.*c ]
N.lbc = @(a,b,c)[ a—1,b—1,diff(c) ];
N.rbc =   @(a,b,c) [ a,  b,  c—1]
% N.guess = [0*x,0*x];
sol = N\0
diffsol = diff(sol) % derivative of the solution
% trial solution for the next case
N.guess = sol(:,:)

 plot(sol)
 xlabel ( 'Dimensionless distance in the film ' )
 ylabel ( ' dimensionless concentrations ' )
 fluxA  = —diffsol(0,1)
 fluxB  = —diffsol(0,2)
..
```

Illustrative results are shown in Figure 20.8 for a chosen set of parameters.

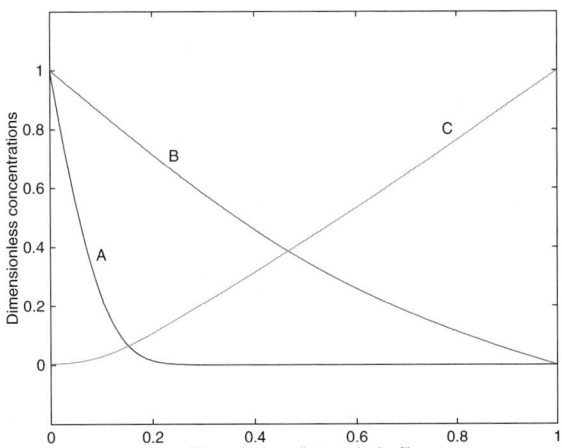

Figure 20.8 Concentration profiles for A, B, and C in the film for simultaneous absorption of two gases; the parameters used are $M_A = 100$, $M_B = 5, q_A = 10, q_B = 5$.

Note that A drops very fast near the interface and the instantaneous reaction condition is nearly approached. B reacts near δ and some of B goes into the bulk and reacts there as well. The bulk conditions are assumed to be such that the concentration of B is zero. The enhancement factor for B is now 1.440. If gas A were not present then the factor would be 2.66. The program enables us to calculate the rate of absorption as a function of various parameters. Normally this will be put as a submodel at film scale in the overall reactor model.

20.5 Coupling with Reactor Models

Some examples are shown here where we couple the film model with the reactor model. These examples provide a clear illustration of how the local differential model for rate of transfer is incorporated into a larger scale reactor or equipment model. The approach is the same as that for physical absorption and models for reacting systems are extensions with the reaction terms added to the macro- or meso-balances. For example the model for a packed bed gas–liquid reactor is developed on similar lines to that in Section 4.3.2.

20.5.1 Semibatch Reactor

A common mode of operation is a semibatch reactor where a gas containing a solute A is bubbled into a pool of agitated liquid where it (A) dissolves and reacts with a species present in the liquid phase. The schematic is the same as in Figure 3.3 where we considered absorption of oxygen in a non-reacting pool of liquid. The model is set up using species conservation laws applied to A in the gas phase, dissolved A in the bulk liquid, and liquid phase reactant B. The mass transfer model provides the expression for the rate terms at any instant of time. A general formulation should include the film and bulk reaction and is shown in this section. Simplifications can be made appropriately when one of the processes is more dominant. The mass balance equations follow. The following equation applies for the concentration of the dissolved gas in the bulk liquid:

$$\epsilon'_L \frac{d}{dt}(C_{Ab}) = N_{A\delta} a_{gl} - \epsilon'_L C_{Ab} C_{BL} \tag{20.47}$$

The liquid phase reactant concentration is given by

$$\epsilon'_L \frac{d}{dt}(C_{BL}) = N_{B\delta} a_{gl} - \nu \epsilon'_L C_{Ab} C_{BL} \tag{20.48}$$

Here $N_{B\delta}$ is the flux of B into the film. This represents the B needed to satisfy the stoichiometry for the reaction of A occurring in the film. This is given as

$$N_{B\delta} = -\nu(N_{A0} - N_{A\delta}) \tag{20.49}$$

This equation states that the B reacted in the film is the difference in the flux of A at the interphase and that to the bulk. The film model gives the values for these quantities, N_{A0} and $N_{A\delta}$. It is solved at each point in time and these values are updated during the time marching scheme.

The representative concentration to be used for the gas phase needs some additional consideration. Two cases need to be distinguished depending on the level of mixing in the gas phase and modeled accordingly.

1. Gas is backmixed: The representative concentration to be used here is the exit gas concentration. This needs to calculated by a gas phase balance, which is represented in the following equation:

$$G_i y_{Ai} - G_e y_{Ae} = N_{A0} a_{gl} V_R \tag{20.50}$$

Expressing N_{A0} in term of the enhancement factor E, which is in turn a function of y_{Ae} and C_{BL}, the gas phase mass balance becomes

$$G_i y_{Ai} - G_e y_{Ae} = k_L a_{gl} E(y_{Ae} P / H_A) V_R \tag{20.51}$$

The inert balance is used to relate G_e to G_i. Since there is no net change in the inert gas flow rate we have

$$G_i (1 - y_{Ai}) = G_e (1 - y_{Ae}) \tag{20.52}$$

Using this in Equation 20.51 we obtain

$$\frac{y_{Ae}}{1 - y_{Ae}} = y_{Ai} - \frac{k_L a_{gl} V_R}{G_i} E y_{Ae} P / H_A \tag{20.53}$$

This gives a quadratic equation for y_{Ae} and the solution gives the representative concentration to find the interface concentration and rate of absorption.

The enhancement factor depends on the value of y_{Ae} and C_{BL} at the current value of time. This is solved in a separate subroutine, for example, using the Hikita-Asai model (for example, using Listing 20.1).

2. Gas is in plug flow: The gas phase balance is a differential balance for this mixing pattern:

$$\frac{dG y_A}{dz} = -N_{A0} a_{gl} A_c \tag{20.54}$$

This is solved simultaneously with the liquid phase balance to find the exit concentration of the gas. The representative concentration is the log mean of the inlet and the exit concentration.

The model is general and is applicable for the intermediate case where reaction can occur in both the film and the bulk. Simplified models can be used for fast reactions as discussed in the following.

Fast Reactions

For fast reactions the dissolved gas concentration in the bulk is not needed. This simplification is allowed because all reactions are in the film and therefore $N_{A\delta} = 0$.

The liquid phase reactant concentration can also be simplified as

$$\epsilon_L \frac{d}{dt}(C_{BL}) = -\nu k_L a_{gl} C_A^* E \tag{20.55}$$

where $C_A^* = y_{Ae} P / H_A$ is the saturation solubility. This is applicable if the gas is well mixed. If the gas is in plug flow the log mean mole fraction of A should be used to find the saturation solubility.

Note that E in Equation 20.50 is a function of the current value of the liquid phase composition and not a constant. Thus

$$E = E(C_{BL}) = f(Ha, q, Bi_G)$$

where Ha and q are based on the values at the current time. This variation should be included in the integration procedure and E should not be treated as a constant.

Levenspiel suggests tabulating the value of E as a function of C_{BL} and then using a graphical integration using a rearranged form of Equation 20.55. Thus a formal solution for batch time t_B to achieve a required conversion is

$$t_B \left(\nu k_L a_{gl} C_A^* / \epsilon_L\right) = \int_{C_{BL,0}}^{C_{BL,f}} \frac{dC_{BL}}{E(C_{BL})}$$

However it is easier to do this computationally using time marching. Using the simple first-order explicit finite difference for the time derivative in Equation 20.55 we have the following scheme for the simulation:

$$C_{BL}(t + \Delta t) = C_{BL}(t) - \nu k_L a_{gl} C_A^* E(C_{BL}) \Delta t / \epsilon_L \tag{20.56}$$

Here an explicit Euler scheme is used, which is sufficiently accurate. The time step has to be chosen such that the concentration changes only by 5% to 10% at each time step.

The computational scheme is therefore as follows for a semibatch liquid with a well-mixed gas phase.

1. Use the condition at t where we know C_{BL}. Calculate Ha.
2. Assume y_{Ae}. This sets the q parameter and is calculated next.
3. Using Ha and q, estimate E.
4. Use E to update y_{Ae} using Equation 20.51 until convergence is obtained. Both of these calculations can be done together using the FSOLVE routine in MATLAB, where both E and y_{Ae} are solved simultaneously.
5. Update C_{BL} to time $t + \Delta t$ by using a finite difference version, Equation 20.56, and continue time marching.

The computational scheme for the case where the gas is in plug flow is similar. The representative concentration to be used in the calculation of the enhancement factor is now the log mean average value. A differential balance is now used for the gas phase instead of Equation 20.51. The liquid phase balance remains the same (Equation 20.55) with the difference that E is now based on the log-mean gas mole fraction. The computational procedure is similar to that for the backmixed case. Step 4 of the computational step is modified and the gas concentration profile is estimated by solving the plug flow equation at the current time.

20.5.2 Packed Column Absorber

Packed column absorbers are usually modeled by assuming plug flow for both phases. Bulk gas concentration is generally zero if $Ha > 1$. Reactions may occur in the film or the bulk liquid but if the bulk gas concentration is zero, the bulk liquid balance for A is not needed.

Model equations where the concentration of A in the bulk liquid is zero are presented in the following.

For species A the material balance is

$$\frac{d(Gy_A)}{dz} = -N_{A0}a_{gl}A_c \tag{20.57}$$

Here A_c is the area of cross-section of the column. It is more convenient to write material balance equation in terms of the enhancement factor:

$$\frac{d(Gy_A)}{dz} = -k_L a_{gl}(y_A P/H_A)E A_c \tag{20.58}$$

The molar flow rate of gas can vary in the column for concentrated gas mixture and an equation for its variation is obtained by an inert balance:

$$\frac{d[G(1 - y_A)]}{dz} = 0 \tag{20.59}$$

Hence

$$G_i \frac{d}{dz}\left(\frac{y_A}{1 - y_A}\right) = -N_{A0} a_{gl} A_c \tag{20.60}$$

For dilute systems the $1 - y_A$ term can be approximated as one and hence

$$G_i \frac{dy_A}{dz} = -N_{A0} a_{gl} A_c \tag{20.61}$$

The liquid phase balance for reactant B for countercurrent flow is

$$Q_L \frac{dC_{BL}}{dz} = \nu k_L a_{gl}(y_A P/H_A) E A_c \tag{20.62}$$

Equations 20.61 and 20.62 are now solved simultaneously as a boundary value problem. At each step the value for E should be calculated based on the local bulk gas and bulk liquid conditions:

$$E = E(y_A, C_{BL})$$

A MATLAB-based simulation scheme can be set up for this but the details are not presented here due to space limitations.

20.6   Absorption in Slurries

Gas absorption with chemical reaction in slurries is encountered in a large class of industrially important processes, such as gas treating, pollution control, oxidation, chlorination, hydrogenation, and so on. Examples of such systems can also be found in the field of product engineering, for instance, in making products such as calcium carbonate, magnetic particles such as geothite, and so on. A particularly important case arises when a gas is absorbed in a slurry containing fine particles (where the particle size is smaller than the diffusion film thickness). The particles then modify the concentration profile in the film thus affecting directly the rate of mass transfer.

Ramachandran and Sharma (1969) provided a starting point in this field which has led to many further works in both modeling and applications. A brief overview is provided here on how fine particles present in the film can enhance the rate of mass transfer. The comprehensive review of the development and more details are summarized by Ramachandran 1993.

The system analyzed here is that of a gas dissolving into a slurry of liquid + solid particles. The slurry consists of soluble particles that dissolve in the liquid and react with the dissolved gas. The reaction scheme can be represented by the following scheme:

$$A(g) \rightarrow A(aq)$$

$$B(s) \rightarrow B(aq)$$

$$A(aq) + \nu B(aq) \rightarrow \text{Products}$$

An example is absorption of CO_2 in a slurry containing $Ca(OH)_2$ particles.

20.6.1 Particle Size Effect

Two cases can be distinguished depending on the relative size of the diffusion film and the particle:

- Large diameter particles (compared to film thickness) such that the particle concentration in the film is small. Here the dissolution takes place in the bulk of the liquid. The dissolved species B diffuses toward the film and reacts with A and the reaction is often complete in the film itself for fast reactions. The enhancement in the gas absorption is only due to the homogeneous reactions taking place in the film and can be calculated by the models described in Section 20.2. The criteria for this regime to hold was derived as

$$\frac{k_{sl}a_{sl}}{4k_L^2}\frac{D_A^2}{D_B} << 1 \tag{20.63}$$

- Particle size is smaller than the film thickness. These are referred to as fine particles or more generally as "microphase." Here the solid dissolution occurs in parallel to the reaction taking place in the film. The rate of gas absorption is enhanced above that calculated from the pure gas–liquid reaction theory.

The conceptual basis for the enhancement of the mass transfer rate due to the presence of fine particles is shown in Figure 20.9.

The concentration profiles for species A and B are shown in the presence (solid lines) and absence of particles (dotted lines) in the film.

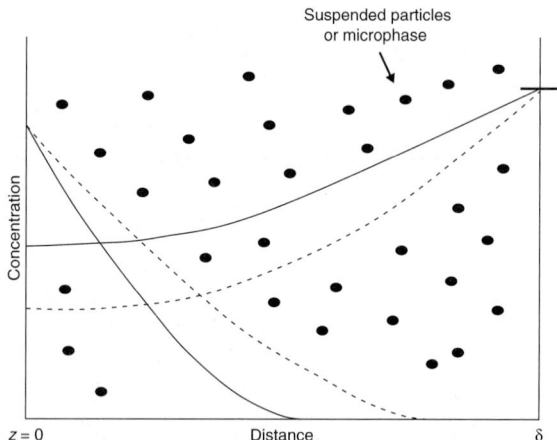

Figure 20.9   A general schematic of the effect of suspended particles on the concentration profiles in the liquid film. The dotted line is for the case of absence of the particles while the solid line is for the particles present in the film.

The simultaneous dissolution of B in the film contributes additional liquid phase reactant and makes the concentration profile of A steeper. An additional enhancement in the the rate of absorption follows as a consequece. Since the rate of absorption is proportional to the concentration gradient at the interface, we find that rate is enhanced in the presence of the fine particles dissolving in the film. The extent of depletion of B in the film is also reduced, which contributes further to the increase in the rate.

For many systems of industrial importance (such as SO_2 in lime slurry), the reaction itself can be treated as instantaneous and hence this case is treated here based on the film model.

20.6.2 Instantaneous Reaction Case

The concentration profiles for species A and B for an instantaneous reaction are presented in Figures 20.10.

Since the reaction is instantaneous, we have a reaction plane at $x = \lambda$ where the reaction occurs. The location of the reaction plane is closer to the interface in the presence of solids as shown by the solid line. The dotted line shows the case where the solid dissolution in the film is negligible. In either

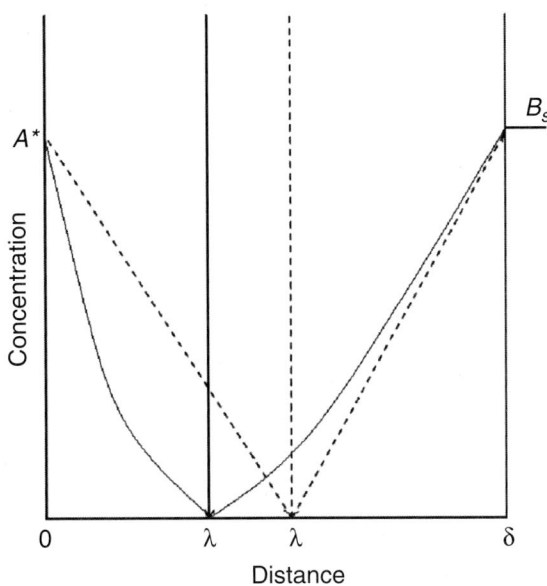

Figure 20.10 Schematic profiles for A and B for an instantaneous reaction case; λ is the location of the reaction plane. The dotted line denotes a case of no particles while the solid line represents the presence of particles.

case for an instantaneous reaction, we have a A-only region from 0 to λ and a B-only region from λ to δ.

For the B-only region, λ to δ, the following equation holds for species B:

$$D_B \frac{d^2 C_B}{dx^2} = -k_{sl} a_{sl}(C_{Bs} - C_B) \tag{20.64}$$

The C_{BS} is the saturation solubility of the species B. The previous equation can be expressed in terms of dimensionless variables as

$$\frac{d^2 c_B}{d\xi^2} = -\beta_{sf}^2 (1 - c_B) \tag{20.65}$$

Here β_{sf} is a dimensionless group that characterizes the solid dissolution in the film and is defined as

$$\beta_{sf} = \delta \sqrt{k_{sl} a_{sl} / D_B} \tag{20.66}$$

It represents the square root of the ratio of the time scale for film diffusion to the solid–liquid mass transfer.

For small particles, a limiting value of $k_{sl} = D_B/R$ can be used since the Sherwood number, Sh_R, for solid–liquid mass transfer is equal to one in the limiting case of a stagnant particle. In addition, $a_{sl} = 3\epsilon_s/R$. Hence β_{sf} can also be expressed as

$$\beta_{sf} = \frac{\delta}{R}\sqrt{3\epsilon_s} \tag{20.67}$$

Hence the range of values for β_{sf} depends on the particle size, R, and particle loading, ϵ_s, and is on the order of 0 to 4 for many practical systems.

For the A-only region, 0 to λ, the following equation holds for the concentration distribution of species A, assuming that the particle dissolution rate is equal to the local rate of reaction:

$$D_A\frac{d^2C_A}{dx^2} = k_{sl}a_{sl}C_{Bs}/\nu \tag{20.68}$$

This can be expressed in terms of dimensionless variables as

$$\frac{d^2c_A}{d\xi^2} = q\beta_{sf}^2 \tag{20.69}$$

where the concentration ratio parameter q is defined as

$$q = D_BC_{Bs}/(\nu D_A C_A^*) \tag{20.70}$$

The quantity of interest is usually the rate of absorption, R_A, which is defined more conveniently in terms of an enhancement factor, E, given as

$$E = \frac{N_{A0}}{k_L C_A^*} = -\left(\frac{dc_A}{d\xi}\right)_{\xi=0} \tag{20.71}$$

The position of the reaction plane is obtained by a flux balance at the reaction plane:

$$-D_A\left(\frac{dC_A}{dx}\right)_{x=\lambda} = \nu D_B\left(\frac{dC_B}{dx}\right)_{x=\lambda} \tag{20.72}$$

or in terms of dimensionless variable as

$$-\left(\frac{dc_A}{d\xi}\right)_{\xi=\lambda^*} = q\left(\frac{dc_B}{d\xi}\right)_{\xi=\lambda^*} \tag{20.73}$$

Solution of Equations 20.69 and 20.65 yields the concentration profiles from which the gradients can be calculated and used in Equations 20.71 and 20.73. The final results are as follows:

For the enhancement factor we obtain

$$E = \frac{1}{\lambda^*} + q\beta_{sf}^2 \frac{\lambda^*}{2} \tag{20.74}$$

and for the (dimensionless) reaction plane we have

$$\frac{1}{\lambda^*} - q\beta_{sf}^2 \frac{\lambda^*}{2} = q\beta_{sf} coth[\beta_{sf}(1 - \lambda^*)] \tag{20.75}$$

In the limiting case of large q, the value of E can be shown to be equal to $q\beta_{sf}$. Using Equation 20.67 for β_{sf} and 20.70 for q, the following expression for the rate of absorption is obtained in the limiting case:

$$N_{A0} = \frac{D_B C_{Bs}}{\nu R} \sqrt{3\epsilon_s} \tag{20.76}$$

N_{A0} becomes inversely proportional to the particle size and directly proportional to the square root of the particle loading, which has also been confirmed in some experimental studies.

20.7   Liquid–Liquid Reactions

Liquid–liquid reactions provide another example where mass transfer effects are important and need to be included in the analysis. In such systems, due to mutual solubility, the reaction can occur in one or the other phase or in both phases. This is unlike a gas–liquid reaction where (for most cases) there is no reaction in the gas phase. The film model is again useful to visualize the processes and to develop a mathematical model.

Examples of liquid–liquid reactions are shown in Table 20.1.

Table 20.1   Some Examples of Liquid–Liquid Reactions

Reaction	Phase I	Phase II
–	organic	aqueous
nitration	benzene	nitric acid
hydrolysis	alkyl acetate	NaOH
alkylation	butane	isobutylene
eserification	fatty acids	methanol
polymerization	trimesolyl chloride	diamine
desulfurization	organic sulfur	peroxyacetic acid

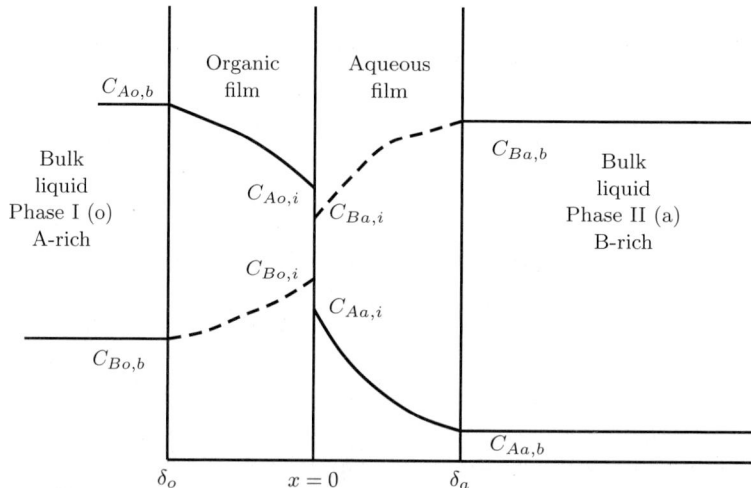

Figure 20.11   Two-film model representation of a liquid–liquid reaction between A and B. The solid line is the profile of A and the dotted line is the profile of B.

The concentration profile according to the film model is shown for two components A and B reacting with each other in Figure 20.11.

The film model is developed by proceeding with a formulation similar to that given in Section 20.2. The differential equations for the two components for both phases are used, leading to four differential equations. The equilibrium conditions (concentration jump) are used as an interfacial boundary condition together with the flux continuity condition at the interface. In some cases analytical solutions are obtained.

The reactor level models are then developed by coupling the local models with the macroscopic or mesoscopic models for the two phases depending on the mixing pattern in each of the phases. Vasudevan and Sharma (1984) have studied the modeling of common types of liquid–liquid reactors.

Summary

- Gas–liquid reactions are another important area with many applications where mass transfer analysis is needed and effectively used to understand and design such systems. The film model is often used as the basis for modeling these systems.

- The basic model where gas A dissolves and undergoes a first-order reaction should be studied carefully. Two regimes can be identified depending on the magnitude of a key dimensionless parameter, the Hatta number. The Hatta number is a measure of relative rate of reaction to diffusion. Large Hatta numbers lead to a reaction being confined to a thin region near the gas–liquid interface. Small Hatta numbers lead to species A crossing the film without appreciable reaction and then reacting in the bulk of the liquid.

- The mass transfer rate is enhanced due to reaction, especially for $Ha > 1$, the film reaction case. An enhancement factor is defined and used to characterize this enhancement. For large Ha and a first-order or pseudo-first-order reaction, this enhancement factor turns out to be equal to Ha. The volumetric rate of reaction is then dependent on only the gas–liquid interfacial area under these circumstances and is not dependent on the intrinsic mass transfer coefficient k_L.

- For small Ha most of the reaction takes place in the bulk liquid. A bulk liquid balance is then used to find the dissolved gas concentration in the bulk liquid. Correspondingly, two regimes can be identified for this case: a mass transfer controlled case where the bulk concentration is nearly zero, and a reaction controlled case where the bulk is almost at the saturation solubility of the gas.

- The identification of the controlling regime is important for making the correct choice of the equipment to be used. If the reaction is mostly in bulk, then equipment with large bulk volume is used. If the reaction is in film, thin film contactors may be sufficient since the bulk volume is not being utilized in this case.

- The liquid phase reactant concentration should also be modeled for bimolecular reactions. For an appreciable extent of the reaction in the film, species B concentration can deplete in the film and in some cases the reactions can become instantaneous, in which case the species B concentration falls to zero in a region near the gas–liquid interface. The model proposed by Hikita-Asai is useful to estimate the level of depletion (the interfacial concentration of B) and to use this to calculate the enhancement factor by an approximate analytical solution.

- Depending on the concentration of B at the interface three regimes can be identified, which are sketched in Figure 20.6. These are B concentration nearly the same as bulk, leading to a pseudo-first-order model; moderate depletion of B, leading to the second-order model; and complete depletion, leading to the instantaneous reaction model. The relative magnitude of Ha and q determines the extent of depletion.

- For a pseudo-first-order case, the rate turns out to be square root dependency on B concentration, even though the reaction is intrinsically first order on B. The rate is also proportional to the square root of the rate constant.

- For an instantaneous reaction the rate of absorption is solely determined by mass transfer considerations, that is, how fast the reactant can reach the reaction plane. The chemical kinetics plays no role under these conditions. The rate is observed to be strongly dependent on the B concentration and not affected significantly by gas concentration.

- The gas-side resistance can be included in the differential model for dissolved gas transport as a Robin condition. An additional parameter, the Biot number for gas-side mass transfer is needed. The boundary condition at the interface is switched to a Robin condition using this parameter.

- Simultaneous absorption of two gases with a common liquid phase reactant is an industrially important problem in gas treating and also an example of absorption with multistep reactions. Hence it is important to understand the methodology of model development for such reactions. For the particular scenario examined, we find the rate of absoprtion of gas B is severely affected by the presence of gas A due to the fast reaction of A with C, thereby making C unavailable to react with B in the regions near the interface.

- Two examples of coupling the film model with the reactor model were illustrated in the text. The mixing pattern of the phases need to be assumed and macroscale or mesoscale model for the reactor scale need to be set up. The local rate of absorption (film model) is then incorporated as a submodel in the reactor model.

- Absorption in a liquid containing dissolving particles is of importance in many applications. Fine particles can dissolve in the film itself, raising the local concentration of B, and thereby enhancing the rate of gas absorption, providing additional enhancement in absorption. For larger particles the dissolution is mainly in the bulk and the gas–liquid models directly apply since the solid dissolution and reaction occur in series rather than simultaneously.

- Models for liquid–liquid reactions follow a similar pattern to gas–liquid reactions. The reactions can now take place in one or both phases depending on the relative solubility of the reactants. The general procedure is to set up the diffusion-reaction model for one or both phases and solve these equations. Reactions can in some cases take place at the interface itself for very fast reactions and when the solubility of one component in the other phase is limited. This is similar to a gas–liquid instantaneous reaction shifting to the interface.

Review Questions

20.1 What is the physical significance of the Hatta number?

20.2 Express Hatta (squared) as a ratio of two time constants.

20.3 How can the extent of reaction taking place in the liquid film be calculated?

20.4 State some approximate condition when the bulk liquid concentration can be approximated as zero.

20.5 Explain on physical grounds why the absorption rate is not dependent on the film thickness for a first-order reaction if $Ha > 3$.

20.6 State conditions when the volumetric rate of reaction does not depend on the value of the intrinsic mass transfer coefficient.

20.7 What is the bulk liquid concentration for a purely kinetic controlled reaction?

20.8 State the three dimensionless parameters needed to model a bimolecular gas–liquid reaction.

20.9 What is meant by the invariance property of a system?

20.10 How can you calculate the interface concentration of the liquid phase reactant B?

20.11 State conditions when the reaction may be assumed to be instantaneous.

20.12 State conditions when the reaction may be assumed to be pseudo-first order.

20.13 Compare the effect of concentration of B on the observed rate of absorption if the reaction is under pseudo-first order conditions versus if the reaction is instantaneous.

20.14 When does an instantaneous reaction become completely controlled by the rate of gas-side mass transfer?

20.15 What is the difference in mass transfer mechanism if a gas is absorbed in a slurry with fine particles versus that with large particles?

Problems

20.1 **Flux expressions for absorption with a first-order reaction.** Verify the details leading to Equations 20.6, 20.7, and 20.8.

20.2 **Gas absorption in an agitated tank with a first-order reaction.** Oxygen is absorbed in a reducing solution where it undergoes a first-order reaction with a rate constant of 6 s^{-1}. The conditions are such that the liquid-side mass transfer coefficient k_L is equal to 8×10^{-3} cm/s. Will there be appreciable reaction in the film? Use $D_A = 1.5 \times 10^{-5}$ cm^2/s for dissolved oxygen. What would be rate of absorption/reaction if the oxygen partial pressure in the gas phase is 0.21 atm? Use the same Henry constant as that for oxygen in water. Neglect the bulk concentration of dissolved oxygen. Also find the flux of O_2 going into the bulk liquid and calculate the percentage of O_2 that reacts in the film itself.

20.3 **Carbon dioxide absorption with first-order reaction.** CO_2 is absorbed into a liquid under conditions such that the liquid-side mass transfer coefficient is 2×10^{-4} m/s. The diffusion coefficient of CO_2 in the liquid is 2×10^{-9} m²/s. The interfacial concentration of CO_2 can be found using Henry's law. The pressure is 1 atm and the temperature is 300 K. Assume that CO_2 reacts with a dissolved solute in the liquid with a rate constant of $1\ s^{-1}$. Also assume that the bulk concentration of CO_2 is zero. Find the Hatta number. Find the flux of CO_2 at the interface. Find the flux of CO_2 going into the bulk liquid. What percentage of CO_2 reacts in the film itself?

20.4 **Gas absorption with a bimolecular second-order reaction.** Carbon dioxide is absorbed into a solution of NaOH in a packed column absorber. Locally at a given point in the absorber the concentration of NaOH is 1.0 M (1000 mol/m³) and the partial pressure of CO_2 is 1 atm. The parameters needed are listed as follows: Henry's law constant = 1500 bar (in $p = Hx$ form); diffusivity of gas A (CO_2) and liquid reactant B (NaOH) are $D_A = 1.8 \times 10^{-9}$ m²/s, $D_B = 1.7 D_A$; rate constant for the reaction $k_2 = 10$ m³/mole.s; liquid-side mass transfer coefficient $k_L = 1 \times 10^{-4}$ m/s. Neglect gas-side resistance. Find the dimensionless parameters Ha and q. Compare these values and indicate what regime applies for gas absorption. Find the rate of absorption per unit interfacial area. If the interfacial area is 200 m^{-1} find the volumetric rate of gas absorption.

20.5 **Change in regime due to partial pressure and concentration changes.** For the previous problem, examine the effects of changing (1) the partial pressure of CO_2 and (2) the concentration of liquid phase reactant. State the range of conditions where the reaction is expected to be under pseudo-first-order conditions. State the conditions under which the reaction can be treated as instantaneous. Note that the following rough rules apply: the reaction can be treated as pseudo-first order if $Ha < q/5$, and the reaction can be treated as instantaneous if $Ha > 5q$.

20.6 **Absorption with instantaneous reaction.** Hydrogen sulfide absorption in amine solutions may be assumed to be instantaneous. At a point in the absorber the total pressure is 20 atm and the gas contains 2% H_2S. The ratio D_B/D_A is 0.64. The mass transfer coefficients are $k_L = 0.03$cm/s and $k_G = 6 \times 10^{-5}$ mol/cm²s.atm The Henry's law constant for H_2S is 1000 Pa m³/mol. Find the rate of absorption for amine concentrations in the range of 0.25 to 1.5 M. Also plot the location of the reaction plane as a function of the amine concentration.

20.7 **Lo-cat process for hydrogen sulfide removal.** In this process hydrogen sulfide bearing gas is absorbed in a chelated iron Fe^{+3} solution where it undergoes a reduction to form elemental sulfur:

$$H_2S + 2Fe^{+3} \rightarrow 2Fe^{++} + 2H^+ + S$$

At a certain point in the reactor the partial pressure of H_2S is 5% and the chelate concentration is 60 mol/m³. Find the rate of absorption given the following data: Henry's law constant = 1950 Pa m³/mol; diffusivity of gas A (H_2S) = 1.44×10^{-9} m²/s; diffusivity of liquid reactant: $D_B = 0.54 \times 10^{-9}$ m²/s, rate constant for the reaction $k_2 = 9 \text{ m}^3/\text{mole.s}$; liquid-side mass transfer coefficient $k_L = 2 \times 10^{-4} \text{ m/s}$; gas-side mass transfer coefficient $k_G = 2 \times 10^{-4} \text{ mol/Pa m}^2\text{s}$; gas liquid transfer area $= 200 \text{ m}^{-1}$.

20.8 Absorption with a reversible reaction. Consider a simple reversible reaction:

$$A \rightleftharpoons P$$

$-R_A \text{ (local)} = k_1(C_A - C_P/K)$ where K is the equilibrium constant for the reaction. Use the following boundary conditions. At the interface $C_A = C_A^*$ and there is no flux for P (P is non-volatile). At the bulk $C_A = C_{Ab}$ and $C_P = KC_{Ab}$, that is, bulk liquid composition is at equilibrium. Set up the models and solve the equations. Derive an expression for the enhancement factor. **Hint:** P concentration can be expressed in terms of A concentration, thereby reducing the model to a single differential equation for A only.

20.9 Ozone absorption with reaction. Wastewater is contacted with an ozone containing gas in a CSTR. The gas phase ozone concentration is 3% and the gas volumetric flow rate is 3 cu.m/sec. Find the extent of consumption of ozone in water and in the exit gas composition. The parameters needed are as follows. The diffusion coefficient is $2 \times 10^{-9} \text{ m}^2/\text{s}$. The rate constant (first-order reaction) 4 s^{-1}. The process is liquid-side mass transfer controlled with a coefficient of $4 \times 10^{-4} \text{ m/s}$. The interfacial area is $200 \text{ m}^2/\text{m}^3$.

20.10 Simultaneous absorption of two gases with one gas reacting instantaneously. An example of this is in simultaneous absorption of H_2S and CO_2 in amine solution. The reaction of H_2S is instantaneous and hence there is a C-depleted zone near the interface of thickness λ. This thickness is also to be determined as part of the solution.

For the region 0 to λ there is no C. Show that the following equations hold:

$$D_A \frac{d^2 C_A}{dx^2} = 0 \tag{20.77}$$

and

$$D_B \frac{d^2 C_B}{dx^2} = 0 \tag{20.78}$$

For the region λ to δ there is no A and the following equations hold:

$$D_B \frac{d^2 C_B}{dx^2} = k_B C_B C_C \tag{20.79}$$

and

$$D_C \frac{d^2 C_C}{dx^2} = \nu_B k_B C_B C_C \tag{20.80}$$

State the boundary conditions needed to solve the problem. State the additional conditions needed to find λ. Set up the model in dimensionless form and find the effect of the presence of A on the rate of absorption of B and vice versa.

20.11 **Simulation of a batch absorber with reaction.** Gas A is bubbled into an agitated liquid containing a liquid phase reactant B. Find the time needed to reduce the concentration of B starting from 5000 to 500 mol/m^3. Use following parameters for transport parameters: $k_L = 2 \times 10^{-4}$ m/s; $a_{gl} = 100$ m^2/m^3 of dispersion volume; $D_A = 2 \times 10^{-9}$ m^2/s; also use $D_B = D_A$; stoichiometric coefficient $\nu = 2$. Values of physico-chemical parameters at the operating temperature of 300 K are as follows: Henry constant: 1000 m^{3}Pa/mol; rate constant $k_2 = 10$ m^3/mol.s, assuming a (1,1) order reaction. The operating pressure is 1 atm.

20.12 **Simulation of a packed column.** A gas stream containing 2.3% chlorine at a molar velocity of 100 mol/m^{2}s is treated with NaOH solution that is flowing countercurrent to the gas at the rate of 250 mol/m^{2}s. The concentration of NaOH in the feed liquid is 2736 mol/m^3. The reaction is instantaneous and the other parameters are $H = 125 \times 10^6$ atm m^3/mol, $k_G a_{gl} = 133$ mol/m^3 hr atm, and $k_L a_{gl} = 45$ hr^{-1}. Find the height of the tower needed to achieve 90% removal of chlorine.

20.13 **Absorption in slurries with fine particles.** Calculate the enhancement due to the presence of the particles for the following case: Particle size = 10 μm; $C_A^* = 4$ mol/m^3; $C_{Bs} = 20$ mol/m^3; $D_A = 2 \times 10^{-9}$ m^2/s; $D_B = 2 \times 10^{-9}$ m^2/s; $\nu = 1$; $k_L = 1 \times 10^{-4}$ m/s. Particle loading is 10%. The reaction is assumed to be instantaneous.

20.14 **Liquid–liquid reaction.** Biodiesel are methyl esters of fatty acids and are produced by reacting triglycerides (TG) with methanol (ME). The reaction scheme is represented as

$$TG + 3ME \rightarrow E + G$$

where G is the side product, glycerol. The reaction is (2,1) order and the rate constant is $1.2 \times 10^{-4} \, C_T^2 C_M$. The solubility of triglycedride in methanol is represented as C_T (methanol phase) $= 25 C_T$ (oil phase). The mass transfer coefficient of TG into methanol is 3×10^{-5} m/s. The reaction was carried out in a microchannel flow reactor with two phase flowing concurrently into the reactor. The reactor diameter used was 0.5 mm and the length was 2.5 m. A mole ratio of 6 to 1 for oil to methanol was used. The flow rate of was 2.92 ml/min TG and 0.57 for ME and the interfacial area for mass transfer was 30,000 m^2/m^3. Simulate the reactor and estimate the exit concentrations of TG and M.

CHAPTER 21

Gas–Liquid Reactions: Penetration Theory Approach

Learning Objectives

After completing this chapter, you will be able to:

- Model gas absorption with reaction on the basis of the penetration theory.
- See the differences between the predictions of the film model and the penetration model.
- Understand the use of laboratory-scale contactors with well-established hydrodynamic patterns to find physico-chemical parameters such as rate constant and diffusivity.

Analysis of gas–liquid reactions based on the film model was studied in Chapter 20. An alternative model for mass transfer from a gas–liquid interface is the penetration theory model, which was introduced in Section 8.8. Here we present an analysis of gas–liquid reactions based on the penetration model. It is essentially an exercise in transient diffusion with chemical reaction. Important results and applications result from these models, in particular, in the experimental methods for parameter estimation, and we illustrate these with some examples.

Penetration theory is based on the concept of surface renewal where the surface elements are exposed to gas for a certain time and then replenished by the eddies or flow from the bulk liquid. Thus the concept of stagnant film from the film model is not used and mass transfer from the gas is modeled as a transient process. The enhancement due to reaction can be modeled based on this concept. We illustrate the mathematics to do so, present illustrative results, and show the implications in design and parameter estimation.

The key parameter in the penetration model is the contact time or the exposure time. This is the time for which a gas–liquid interface remains in contact before the eddies sweep the interface and cause mixing at the interface. Laboratory gas–liquid contacters can be designed so that the contact time is precisely known. This permits the determination of other parameters such as diffusion coefficient as well as the rate constant, which can often be difficult to measure, especially for fast reactions. The methodology is illustrated here where we use the penetration theory model as the basis for determination of these parameters.

21.1   Concepts of Penetration Theory

We first briefly review the concepts in penetration theory for mass transfer and follow this with a study of the effect of first-order reactions on the rate of mass transfer based on this theory.

The film model assumes a stagnant film and steady state diffusion with reaction in the film. In actuality the gas–liquid interface is constantly renewed and the steady state profile envisioned in the film model is only an approximation. Nevertheless it has provided a useful tool for correlation of absorption with reaction data and in the design of equipment.

The penetration model is an attempt to improve the situation where we assume a surface element of liquid remains in contact with the gas for a certain time during which a transient diffusion with reaction takes place. A transient concentration profile starts to develop near the interface. After the elapse of a certain amount of time, called the contact time or exposure time, the surface element is assumed to be replaced by an element coming from the bulk element. This sweeps the concentration profile established near the surface and the process starts all over again. See also the discussion in Section 8.8 and the expression for the mass transfer coefficient based on this model for the no-reaction case (Equation 8.49).

21.1.1   First-Order or Pseudo-First-Order Reaction

From the above discussion showing the transient nature of exposure of gas and liquid it is logical to start with the transient diffusion-reaction equation:

$$\frac{\partial C_A}{\partial t} = D_A \frac{d^2 C_A}{dx^2} - k C_A \qquad (21.1)$$

Film theory uses a steady state model with a finite film thickness. A semi-infinite region is assumed for the penetration model. The boundary and initial conditions are therefore as follows:

- At $x = 0$, $C_A = C_A^*$, a prescribed interface concentration. This will be equal to the saturation solubility of the gas if there is no gas-side resistance. This is applied for $t > 0$.
- As $x \to \infty$, $C_A = C_{Ab}$. Usually the bulk concentration will be zero for reasonably fast irreversible reactions.
- The initial condition is $C_A = C_{Ab}$ for all x for $t > 0$, the bulk concentration.

The solution is calculated for times 0 to t_E, the exposure time. The instantaneous and average rate of absorption can then be calculated from the concentration gradient at the interface, $x = 0$:

$$N_{A0}(t) = -D_A \left(\frac{\partial C_A}{\partial x} \right)_{x=0}$$

The average rate is the integral average of this from 0 to t_E:

$$\bar{N}_{A0} = \frac{1}{t_E} \int_0^{t_E} N_{A0}(t) dt$$

These are the primary quantities of interest.

The solution to this problem was derived by Danckwerts (1950a) and holds a prominent place in the absorption with chemical reaction literature. Also the books by Astarita (1967) and Danckwerts (1970) are useful resources for a detailed study of gas–liquid reactions. The Laplace transform is useful for such problems and we will study the use of this method here.

21.1.2 Laplace Transform Method

We define $\overline{C}_A$ as the Laplace transform, which is a function of the transform variable s:

$$\overline{C}_A = \int_0^\infty C_A(t) \exp(-st) dt$$

This reduces the PDE in Equation 21.1 to an ODE:

$$D_A \frac{d^2 \overline{C}_A}{dx^2} - (s + k)\overline{C}_A = 0 \tag{21.2}$$

where we have used an initial condition of zero, that is, $C_{Ab} = 0$, which is valid for fast reactions.

The boundary condition at the surface in the Laplace domain is C_A^*/s, which holds for $x = 0$. (Note that this is for a step change in the concentration at the surface, the problem being studied here.) Hence the solution to Equation 21.2 is

$$\overline{C}_A(x, s) = \left(\frac{C_A^*}{s}\right) \exp\left(-x \sqrt{\frac{s+k}{D_A}}\right) \tag{21.3}$$

Students should verify the algebra leading to this equation at this stage. All that remains is to find the inverse transform to find the solution in the time domain. Easier said than done! Tables of inverse transforms are available for a large class of problems including this one and can be directly used. Computer algebra based on MAPLE or MATHEMATICA can also be used. We will skip the mathematics and present the final solution in the time domain:

$$\begin{aligned}
\frac{C_A(x, t)}{C_A^*} &= \frac{1}{2} \exp\left(-x\sqrt{k/D_A}\right) \operatorname{erfc}\left(\frac{x}{2\sqrt{D_A t}} - \sqrt{kt}\right) \\
&+ \frac{1}{2} \exp\left(x\sqrt{k/D_A}\right) \operatorname{erfc}\left(\frac{x}{2\sqrt{D_A t}} + \sqrt{kt}\right)
\end{aligned} \tag{21.4}$$

The profiles with and without reaction are illustrated in Figure 21.1 for two values of the rate constant for a fixed value of time equal to 1 sec.

The interesting behavior is that there is not much difference between physical and chemical absorption for small values of the rate constant ($k = 0.1 \text{ s}^{-1}$ in the figure). Hence we will not expect much enhancement in the rate due to reaction. This is similar to the slow reaction case anticipated by the film model.

For a rate constant of $k = 1 \text{ s}^{-1}$ we see that the profiles are different for the reaction (marked with *) and no-reaction cases (solid line). In particular the absolute value of the slope of the profile at the interface is larger, showing that the reaction enhances the rate of mass transfer.

Figure 21.2 shows the effect of a larger value of the rate constant, $k = 10 \text{ s}^{-1}$, for two values of time. The concentration front for the reaction case hardly moves inward toward the liquid bulk and appears to be stagnated in contrast to the no-reaction case. Here the profiles for a reaction are close to

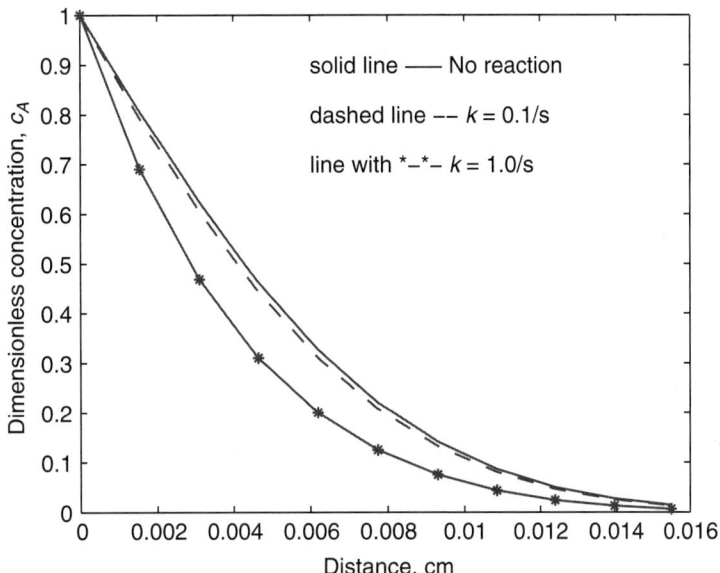

Figure 21.1 Illustrative concentration profiles for transient diffusion with reaction in a semi-infinite region; the slow reaction case shows little difference between the reacting and non-reacting cases. $D_A = 2 \times 10^{-5}$ cm^2/s and $t = 1$ sec for the calculations.

a steady state solution and are represented as

$$\frac{C_A(x,t)}{C_A^*} = \exp\left(-x\sqrt{k/D_A}\right) \tag{21.5}$$

In this case the effect of contact time will not be to be of importance and the concentration profile is established mainly by a balance of diffusion and reaction at each spatial location. The steady state–based film model would have given the same prediction. The value of the mass transfer coefficient is not important here, similar to the pseudo-first-order case in the film model. These differences are important in analyzing experimental data based on the penetration theory as discussed later. Thus by changing the contact time or time of exposure of the system, we can determine both the diffusion coefficient and the rate constant depending on the parameter that has the most dominant effect on the rate of absorption.

The key conclusions are that at low contact times, reaction does not enhance the rate of mass transfer, while at large contact times, the reaction

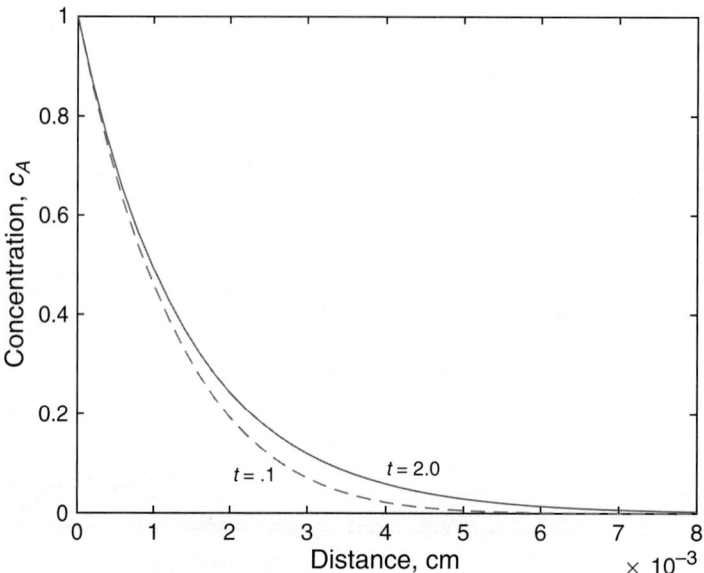

Figure 21.2 Illustrative concentration profiles for transient diffusion with reaction in a semi-infinite region for a fast reaction: $k = 10/\text{s}$; $D_A = 2 \times 10^{-5}$ cm²/s. Solutions are presented for $t = 0.1$ sec and $t = 2$ sec and show that the rate movement of the concentration front is rather small. Diffusion balances reaction here and a nearly steady state profile is reached quickly.

is balanced by diffusion and the contact time does not play a significant role. The characteristic dimensionless parameter is kt_E in the penetration model. The effect of this parameter on the flux is shown next.

21.1.3 Flux and the Average Rate of Mass Transfer

The flux of A at the surface in the presence of reaction is of more practical importance and can be calculated from the following formula derived from Equation 21.4 by applying Fick's law at the interface:

$$N_{A0}(t) = C_A^* \sqrt{D_A k} \left[\text{erf}(\sqrt{kt}) + \frac{\exp(-kt)}{\sqrt{(\pi k t)}} \right] \qquad (21.6)$$

This is the instantaneous flux and it decreases with increase in time.

The more useful result is the average flux over a time interval t_E. This is calculated as an integral of the instantaneous flux with respect to time and then dividing by t_E. The result is

$$\bar{N}_{A0} = C_A^* \sqrt{\frac{D_A}{kt_E^2}} \left[(kt_E + 1/2)\text{erf}(\sqrt{kt_E}) + \exp(-kt_E)\sqrt{\frac{kt_E}{\pi}} \right] \qquad (21.7)$$

This provides the rate of absorption based on the penetration theory for reaction with a first order-reaction.

The enhancement factor is obtained by dividing by the no reaction value given by Equation 8.43, which is repeated here for ease of reference:

$$\bar{N}_{A0} \text{ no reaction} = \left(2\sqrt{\frac{D_A}{\pi t_E}} \right) C_A^*$$

The bracketed quantity on the right-hand side is the value of the mass transfer coefficient according to the penetration theory. Hence the enhancement factor for a first-order (or pseudo-first-order) reaction according to the penetration model is

$$\boxed{E = \frac{\sqrt{\pi}}{2\sqrt{kt_E}} \left[(kt_E + 1/2)\text{erf}(\sqrt{kt_E}) + \exp(-kt_E)\sqrt{\frac{kt_E}{\pi}} \right]} \qquad (21.8)$$

Note that the dimensionless parameter kt_E plays the role of the Hatta number in the penetration model. Hatta is not used here unlike the film model since there is no well-defined film thickness because we deal with a semi-infinite region in the penetration model. The two models will be compared next.

21.1.4 Relation between Film Theory and Penetration Theory

If k_L from penetration theory is used in the definition of the Hatta number the following relation is obtained:

$$Ha^2 = \frac{\pi}{4}(kt_E) \qquad (21.9)$$

This can be used to connect the two models for reacting systems.

The enhancement factor when Ha is small based on the film theory can be approximated as

$$E = 1 + \frac{Ha^2}{3} \text{ for } Ha \approx < 1$$

This is close to that predicted by the penetration theory, which gives the following value (see exercise problem 21.2):

$$E = \frac{1}{2}\left(1 + kt_E\right) \text{ for } kt_E \approx< 1$$

The differences between the film and penetration models become even smaller for larger Ha. The enhancement factor based on the film theory is equal to Ha for $Ha > 3$. For this case penetration theory also predicts the same value if a Hatta number is defined as per Equation 21.9. Overall the two approaches lead to very similar results and the model choice is merely a matter of preference for reacting systems.

21.2   Bimolecular Reaction

Now we consider a case where rate depends on the concentration of both A and B. Species A (dissolved gas) diffuses from the interface to the bulk while species B (the liquid phase reactant) counter-diffuses toward the interface. Hence the model is set up by using the transient diffusion equation for both species A and B.

The rate term in Equation 21.1 is now modified with a second-order term and the differential equation for A is

$$\frac{\partial C_A}{\partial t} = D_A \frac{d^2 C_A}{dx^2} - k_2 C_A C_B \tag{21.10}$$

An additional equation for B is used in conjunction with this:

$$\frac{\partial C_B}{\partial t} = D_B \frac{d^2 C_B}{dx^2} - \nu k_2 C_A C_B \tag{21.11}$$

A numerical solution is needed but gets more involved due to a sharp-moving type of front near the interface. The equations are solved using dimensionless variables. Let us see what dimensionless variables should be used.

21.2.1 Dimensionless Form of the Model

Scaling of the concentration is the same as before in the film theory with c_A and c_B defined as dimensionless concentrations. Since the solution is sought for time up to t_E this can be used as the reference time scale. Hence $\tau = t/t_E$.

The distance is now scaled with $\sqrt{D_A t_E}$. Hence $\xi = x/\sqrt{D_A t_E}$ is defined as the dimensionless distance. The dimensionless representation is therefore

$$\frac{\partial c_A}{\partial \tau} = \frac{d^2 c_A}{d\xi^2} - M c_A c_B \tag{21.12}$$

where M is defined as

$$M = k_2 C_{BL} t_E \tag{21.13}$$

This plays a role similar to the Hatta number.

An additional equation for B is used in conjunction with Equation 21.12:

$$\frac{\partial c_B}{\partial \tau} = \frac{D_B}{D_A} \frac{d^2 c_B}{d\xi^2} - \frac{M c_A c_B}{q} \tag{21.14}$$

Here q is the additional dimensionless group needed and is defined as the concentration ratio parameter:

$$q = \frac{C_{BL}}{\nu C_A^*} \tag{21.15}$$

Note that this does not include the diffusivity ratio D_B/D_A (unlike the film model). The diffusivity ratio appears as a separate parameter. Hence three dimensionless quantities are involved. The film model needs only two since the diffusivity ratio does not appear as an independent parameter.

The quantity of interest is $\bar{N}_{A0}$, the time-averaged value of flux over a period of t_E. The results are expressed in terms of an enhancement factor:

$$E = \frac{\bar{N}_{A0}}{2C_A^*} \sqrt{\frac{\pi t_E}{D_A}}$$

21.2.2 Illustrative Results

An illustrative solution for the concentration profiles of A and B is presented in Figure 21.3, which illustrates how the profiles evolve with time.

Numerical solutions for this figure were generated using the PDEPE solver introduced in Section 8.11. Due to steep profiles near the interface a large number of mesh points in space is needed. Further, the semi-infinite space domain needs to truncated with a finite domain (usually an ξ of 5 is adequate). Also, to find the average flux and the enhancement factor, sufficient evaluation points in time are also needed so that the time integration is also accurate. The enhancement factor for the data used in Figure 20.3 was computed as 7.25 using numerical flux values calculated at equal-spaced times

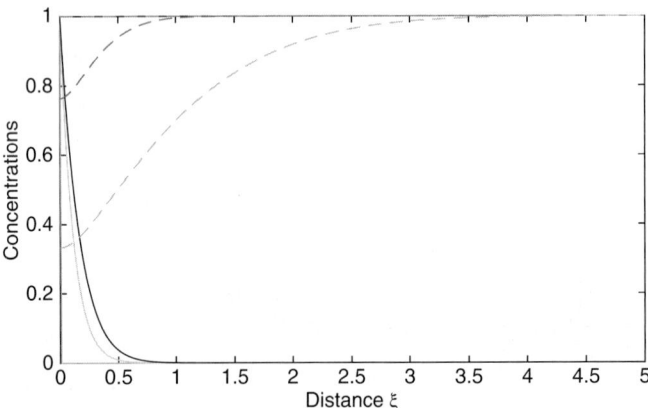

Figure 21.3 Evolution of the concentration profiles for A (solid lines) and B (dotted lines) with time for a second-order (1,1) reaction simulated by the PDEPE solver. $M = 10$ and $q = 10$. Profiles shown for dimensionless times $\tau = 0.1$ and 10.

and then using the trapezoidal rule. This is close to the film model solution value of 7.07.

21.3   Instantaneous Reaction Case

A special case of the bimolecular reaction is reached for large values of M (at a fixed q value) and the instantaneous reaction asymptote is reached. This is similar to the film model analyzed in Section 20.3.2 where the instantaneous reaction asymptote is reached for large values of the Hatta number. Illustrative concentration profiles are shown in Figure 21.4 when the reaction is instantaneous.

The reaction front is assumed to be of thickness λ, similar to the film model. Thus we have a no B zone from 0 to λ and a no A zone from λ onwards, similar to the film theory. The main difference here is that the front moves with time inward. The second difference is that a semi-infinite domain is assumed and the domain of the solution for B is not fixed.

The expression for the front movement can be derived by a mass balance at the interface. The solution was derived by Danckwerts (1951). The reaction front separates species A and B. Only A is present in the region 0 to λ. Transient diffusion with no reaction applies here. It can be shown that the following

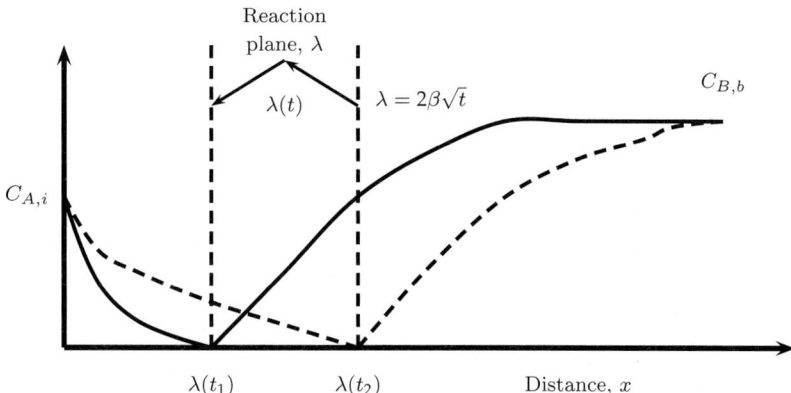

Figure 21.4 Concentration profiles for A and B for an instantaneous reaction based on the penetration model for mass transfer. Solid lines represent the profiles line at time t_1 and the dashed lines at a larger time t_2.

equation satisfies both the differential condition and the boundary condition at $x = 0$:

$$c_A = 1 - A_2 \text{erf}(x/2\sqrt{D_A t}) \tag{21.16}$$

Here A_2 is an integration constant. Now using the boundary condition that $c_A = 0$ at $x = \lambda$ we can evaluate the integration constant:

$$A_2 = \frac{1}{\text{erf}(\lambda/2\sqrt{D_A t})} \tag{21.17}$$

If A_2 has to remain constant and should not vary with time, then λ must be proportional to the square root of time and we can suggest that the following relation should hold for the front movement:

$$\lambda = 2\beta\sqrt{t} \tag{21.18}$$

Here β is a constant to be determined.

A similar analysis for the B-rich zone gives the following equation for the concentration profile of B:

$$c_B = \frac{\text{erfc}(x/2\sqrt{D_A t}) - \text{erf}(\beta/\sqrt{t})}{\text{erfc}(\beta/\sqrt{t})} \tag{21.19}$$

The fluxes at λ from both sides should match and this can be now used to find β. The following implicit expression is obtained after some algebra:

$$\exp\left(\beta^2/D_B\right)\operatorname{erfc}(\beta/\sqrt{D_B}) = \frac{C_{BL}}{\nu C_A^*}\sqrt{\frac{D_B}{D_A}}\exp\left(\beta^2/D_A\right)\operatorname{erf}(\beta/\sqrt{D_A}) \quad (21.20)$$

The flux of A at the interface can now be calculated by applying Fick's law and the result is

$$N_{A0}(t) = \frac{C_{A*}}{\operatorname{erf}(\beta/\sqrt{D_A})}\sqrt{\frac{D_A}{\pi t}} \quad (21.21)$$

The result can be used to derive the following expression for the enhancement factor:

$$E = \frac{1}{\operatorname{erf}(\beta/\sqrt{D_A})} \quad (21.22)$$

Simplified results apply when $E \gg 1$. In such case we have

$$E = \sqrt{\frac{D_A}{D_B}} + \frac{C_{BL}}{\nu C_A^*}\sqrt{\frac{D_B}{D_A}}$$

It can be shown that the enhancement factor predicted by the film and penetration models are identical if $D_A = D_B$. A somewhat approximate but useful generalization follows if the film theory parameter q is corrected as follows: the film theory parameter q defined as $(D_B/D_A)(C_{BL}/\nu C_A^*)$ is modified by taking the square root of diffusivity and is defined as

$$q = \frac{C_{BL}}{\nu C_A^*}\sqrt{\frac{D_B}{D_A}}$$

If the expression $E = 1 + q$ in the film theory model is used with this modified q, then the film model solution is nearly the same as the penetration model. This is known as the square root diffusivity correction of the film model. An application of the model is illustrated in Example 21.1

Example 21.1   H$_2$S Absorption in an Amine Solution

H_2S is absorbed in a 0.1 M solution of the amine, MEA, at 25°C and the reaction can be taken as instantaneous and irreversible. The solubility coefficient of H_2S is 0.1 mol/L atm, the diffusion coefficient is 1.48×10^{-5} cm^2/s and $D_B/D_A = 0.64$. The gas and liquid are in contact for a time of 0.1 seconds. Find up to what point the reaction front has moved, the value of the enhancement factor and the rate of absorption. Compare with the value obtained from the film model.

Solution

Equation 21.20 is solved in MATLAB using the function FZERO. It is easier to solve in terms of $\beta/\sqrt{D_A}$ as a parameter. The value is found as 0.511 and hence $\beta = 0.002$. The reaction front therefore moves with the square root of time with a proportionality constant of 2β. If the liquid is exposed for 0.1 seconds, a B-depleted layer of 0.02 cm will develop near the surface.

The enhancement factor is calculated using Equation 21.22 as 1.8859.

The film theory model will predict with a corrected value of q an enhancement factor of 1.64.

The rate of absorption is calculated as $k_L C_A^* E$.

The k_L in turn is calculated from the penetration model as $2 \left(D/\pi t_E \right)^{0.5}$.

The results for a H_2S partial pressure of 1 atm are as follows:
$k_L = 1.84 \times 10^{-4} \text{cm/s}$ and $N_A = 3.25 \times 10^{-8} \text{mol/cm}^2 \text{ s}$.

Having discussed the penetration model in some detail, we next discuss ideal contactors for measurement of experimental data in gas–liquid reacting systems.

21.4   Ideal Contactors

Ideal contactors refer to gas–liquid absorption equipment where the hydrodynamics is known and the time of exposure can be calculated independently. This permits the evaluation of diffusion and reaction rate constants. We present some of the commonly used ideal contactors and illustrate how they are useful for parameter evaluation.

In general they can be classified into two broad categories (adapted from Mills et al., 1992):

- **Category A:** Both the hydrodynamics and interfacial area are theoretically known. Examples are the laminar jet apparatus and wetted wall and wetted sphere columns discussed in this section.
- **Category B:** Only the interfacial area is known but the mass transfer coefficient is not well characterized. An example is the stirred cell.

A brief description of these categories of contactors is provided in the following section.

21.4.1 Laminar Jet Apparatus

In this device, a jet of liquid enters a gas space through a circular hole and leaves out a slightly larger hole. The time of exposure of the liquid is calculated by assuming the jet is a cylindrical rod with a uniform velocity, v:

$$t_E = \frac{L}{v} = \frac{\pi d^2 L}{4Q}$$

Here L is the length of the jet, d its diameter, and Q the volumetric liquid flow rate. The time of exposure can be varied by changing L or Q or both. A modified design where the gas is mildly agitated without disturbing the flow in the jet is presented in Figure 21.5.

The mild agitation of the gas reduces the gas-side resistance and makes the data interpretation less sensitive to the value of the gas-side mass transfer coefficient. Note that if pure gas is used then the gas-side resistance is non-existent but this may not be always feasible to do so due to other considerations, for example, temperature rise due to heat of the solution and absorption.

21.4.2 Wetted Wall Column

In the wetted wall column, the liquid flows down as a film under the influence of gravity over a surface, usually a rod or a tube. A number of

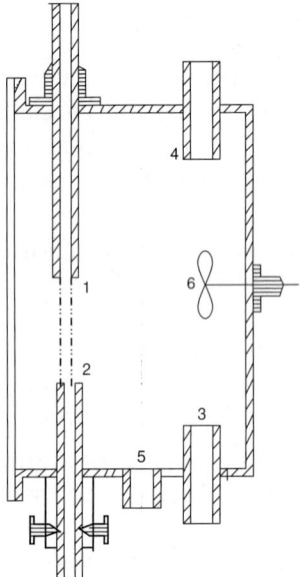

Figure 21.5 Sketch of the laminar jet apparatus. A modified design from Bjerle (1972) is shown where the gas is mixed with mild agitation using a propeller (6). This reduces the gas-side resistance. 1. Transmission nozzle; 2. receiver nozzle; 3. gas inlet; 4. gas outlet; 5. drain.

design variations have been used in the literature. One design of Roberts and Danckwerts (1962) is shown in Figure 21.6.

The mass transfer coefficient can be predicted by the contact time at the interface. This in turn is equal to the length of the tube over which the liquid is flowing and the velocity at the surface. The liquid velocity at the surface is v_{max} for the falling film given in Section 10.5.2. This depends on the film thickness, which in turn depends on the liquid flow rate. Combining all these, the following expression is obtained for the velocity at the surface:

$$v_s = \frac{3}{2}\left(\frac{Q}{\pi d}\right)^{2/3}\left(\frac{\rho g}{3\mu}\right)^{1/3}$$

where d is the diameter of the tube over which the liquid flows. The expression assumes that the film flow is in the laminar regime.

The contact time is therefore equal to L/v_s. The interfacial area is equal to the geometric area of the gas–liquid interface in the absence of waves or ripples. This is equal to πdL.

The range of k_L values is 2×10^{-5}m/s to 2×10^{-4}m/s, comparable to that in a stirred cell. The area to volume parameter is small and on the order of 2×10^{-2} m^{-1} Hence it can be used as a model for simulation of packed columns since the coefficient values in a packed column are in a similar range.

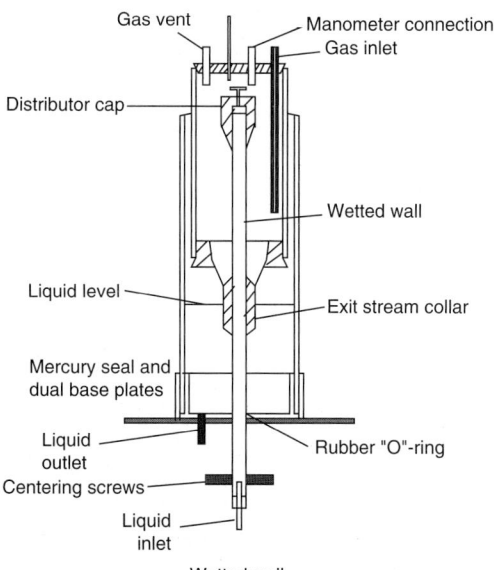

Figure 21.6 Sketch of a wetted wall column.

21.4.3 Wetted Sphere

A contactor similar in concept is the wetted sphere column, shown in Figure 21.7.

Here the liquid flows over a spherical surface rather than on a rod. The advantage over the wetted wall column is that ripples as well as the end effects are reduced. The disadvantage is that data interpretation is complicated due to the converging-diverging flow over the sphere. A detailed solution to the velocity profile was presented by Davidson and Cullen (1957) and this work can be used to find the contact time. Mashelkar and Soylu (1974) used this to measure the diffusion coefficient of solutes for several non-Newtonian and viscoelastic solvents.

21.4.4 Stirred Cells

In this contactor, shown in Figure 21.8 (a design due to Levenspiel and Godfrey, 1974), the gas and liquid are separated by an interface of known area.

In this design there are two flanged sections, the upper section for the gas and the lower for the liquid. Both sections are mildly agitated with the stirrers provided for each section. A plate with holes separates the two sections. The interfacial area can be varied by using plates with varying numbers of holes with varying diameters and is therefore known precisely.

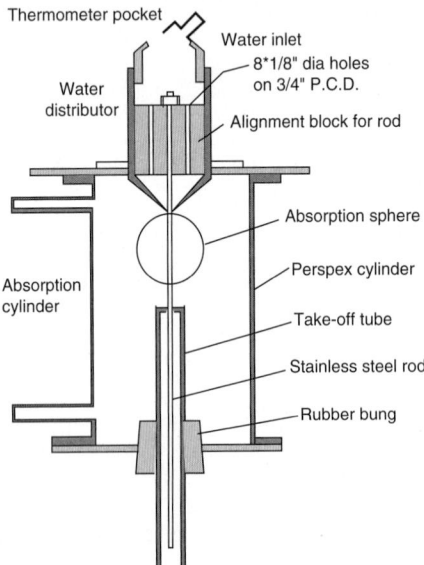

Figure 21.7 Sketch of a wetted sphere apparatus.

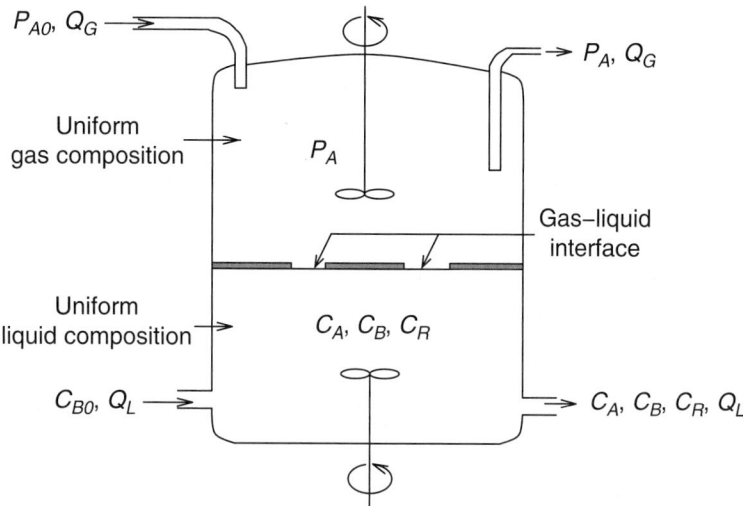

Figure 21.8 Sketch of a stirred cell. The modified design by Levenspiel and Godfrey (1974) is illustrated here.

The mass transfer coefficient is not known and has to be determined separately by independent experiments. The k_L values are seen to be in the range of 2×10^{-5} m/s to 2×10^{-4} m/s, which is comparable to a wetted wall column. The area to volume parameter is small and on the order of 2×10^{-2} m^{-1}.

If the reaction is in pseudo-first-order conditions this can be verified easily since the rate is then proportional only to the interfacial area, which is known. The parameter $\sqrt{D_A k_1}$ is then obtained directly using the experimental rate of absorption. The rate constant in turn can be found if the diffusion coefficient is either known or independently measured.

Summary

- Penetration theory is an attempt to improve the film model by assuming a transient concentration profile near the gas–liquid interface rather than a steady state profile.
- The key parameter is the exposure time or contact time. The mass transfer coefficient in a non-reacting system can be directly predicted if the exposure time is known.

- For reacting systems a transient diffusion with reaction problem is solved to find the concentration profiles near the interface and the corresponding values of the time-dependent flux.
- The time-dependent flux is integrated with respect to time and averaged over the exposure time to get the value of the (averaged) rate of absorption. This is the penetration model for gas–liquid reacting systems.
- An analytical solution (lengthy) can be derived for first-order or pseudo-first-order reactions. The average rate is divided by the rate for the no-reaction case to get Equation 21.8 for the enhancement factor.
- For a bimolecular reaction, species A and B counter-diffuse toward each other. The transient diffusion with reaction is solved for both species. A solver such as PDEPE is useful although it may be noted that the concentration gradients near the interface are rather sharp and hence a large number of mesh points may be needed near the interface. Other "tailor made" methods may be preferable.
- If the bimolecular reaction can be treated as instantaneous, a simpler approach can be used. A reaction front is postulated and this moves inward as a function of time. The location and the movement of the front can be calculated analytically and the enhancement factor can be predicted without a numerical solution.
- Penetration theory predictions are often close to film theory unless the diffusivity ratio of A to B is large. Therefore film theory is often still used after replacing the diffusivity ratio in the q parameter by its square root. The simpler film model then gives results close to penetration theory.
- Experimental values of rate constants are difficult to measure, especially if the diffusional effects interfere. Hence contactors with known values of mass transfer coefficient and/or interfacial area are to be preferred for laboratory studies. These are referred to as "ideal" gas–liquid reactors and some designs of these contactors are provided in the chapter.
- Penetration theory analysis of gas–liquid reactions is an elegant application of the transient diffusion-reaction model and is well worth the study even if you are not going to deal with gas–liquid reactions in your work or prefer to use the film model as an alternative.

Review Questions

21.1   What is meant by exposure time or contact time?

21.2   Does mass transfer coefficient increase or decrease with exposure time?

21.3 What is the dependency of the mass transfer coefficient on the diffusion coefficient when working with the penetration model?

21.4 What form does Equation 21.4 reduce to for $k \to 0$?

21.5 What is the key dimensionless parameter in the penetration model for a first-order reaction? How is the Hatta number related to this parameter?

21.6 What is the difference in enhancement factor for the film and penetration models for $Ha = 3$? Assume a pseudo-first-order reaction.

21.7 What are the key dimensionless parameters in the penetration model for a bimolecular (second-order) reaction with a moderate depletion of B near the interface?

21.8 What are the key dimensionless parameters in the penetration model for an instantaneous reaction?

21.9 How does the location of the reaction plane change with time for an instantaneous reaction?

21.10 State the use of the square root diffusivity correction for the film model.

21.11 What is meant by an ideal gas–liquid contactor?

21.12 What is the range of k_L values that can be expected in a wetted wall column, strirred cell, and jet apparatus?

Problems

21.1 **Verification of the penetration theory solution.** Verify that Equation 21.4 satisfies the required boundary conditions as a well as the initial conditions. MAPLE or MATHEMATICA may turn out to be helpful.

21.2 **Instantaneous flux directly from Laplace solution.** From the Laplace solution given by Equation 21.3 show that the instantaneous flux in the Laplace domain is

$$\bar{N}_{A0}(s) = \left(\frac{C_A^*}{s} \right) \sqrt{D_A s + k}$$

Look up a table of inverse transforms or invert it using MATHEMATICA or similar tools to verify the expression for the instantaneous flux in the time domain given by Equation 21.6.

21.3 **Limiting value of the fluxes for low kt.** Show that the flux expression in Equation 21.7 for the rate gas absorption can be simplified for small values of kt as

$$\bar{N}_{A0} = C_A^* \sqrt{\frac{D_A}{\pi t}} (1 + kt) \text{ for } kt \ll 1$$

Show that the solution error is on the order of 5% for $kt = 0.5$. Find the total quantity of gas absorbed to t_E for this case. Find the average rate of absorption.

If the average rate is divided by $k_L C_A^*$ we can obtain an expression for the enhancement factor. Here k_L can be calculated using the penetration model.

21.4 **Limiting value of the fluxes for large** kt**.** Show that the flux expression in Equation 21.7 for gas absorption can be simplified for large values of kt and the result is

$$\bar{N}_{A0} = C_A^* \sqrt{D_A k} \text{ for } kt \gg 1$$

Show that the solution error is on the order of 3% for $kt = 2.0$. Find the total quantity of gas absorbed from time 0 to t_E for this case. Find the average rate of absorption. Compare with the film model. State why there is no difference between the penetration model and the film model.

21.5 **Use of penetration model in finding the rate constant.** A gas stream with CO_2 at 1 atm partial pressure is exposed to a liquid in which it undergoes a first-order reaction for a time of 0.01 s. The total amount of gas absorbed during this time was measured as 1.5×10^{-4} moles/m^2. Estimate the rate constant for this reaction. Use $C_A^* = 30 \text{ mol/m}^3$ and $D_A = 1.5 \times 10^{-9} \text{ m}^2/\text{s}$.

21.6 **Bimolecular reaction: numerical solution.** Examine the effect of time of exposure on the rate of absorption of CO_2 in ammonia solution for the following set of parameters. Find the enhancement factor for these values of time. The reaction is second order with a rate constant of 30 L/mol.s. The reaction is assumed irreversible and the dissolved concentration of CO_2 in the bulk liquid is zero. The diffusion coefficients are $D_A = D_B = 2 \times 10^{-5} \text{ cm}^2/\text{s}$. The saturation solubility for the given condition is $4 \times 10^{-5} \text{ mol/m}^3$ and the ammonia concentration in the bulk is 1 M.

21.7 **Instantaneous reaction: analytic solution for reaction front.** State the differential equation and the boundary conditions for the A-rich zone and verify that Equation 21.16 holds for the concentration of A in this zone. Similarly verify that Equation 21.19 for B satifies the governing differential equation and the boundary condition. State the matching condition for the fluxes at λ from both sides and use this to derive the expression for β given in the text. Show that the flux of A at the interface is given by

$$N_{A0}(t) = \frac{C_A^*}{\text{erf}(\beta/\sqrt{D_A})} \sqrt{\frac{D_A}{\pi t}}$$

Finally, find the average flux and confirm that the enhancement factor is given by Equation 21.22 in the text.

21.8 **Penetration theory analysis for absorption of two gases.** Simultaneous absorption of two gases with a common reacting liquid phase species was studied by Goettler and Pigford (1971) based on the penetration theory. Set up the model for this, identify the dimensionless parameters, and set up a computational module in MATLAB. Also calculate the enhancement factor for each gas and the ratio of absorption rate of A relative to B for selected values of the dimensionlesss groups.

21.9   Penetration model for two gases, both reacting instantaneously. Extend the penetration model for two species reacting instantaneously. The anticipated concentration profile is shown in Figure 21.9.

Assume the solution on either side of the interface can be expressed in terms of an error function that obviously satisfies the differential equation. Fit the boundary conditions for each side of the interface. Now use the flux balance at the reaction plane and derive an expression for λ as a function of time. How does this model compare with the film model? Compare the results with the work of Goettler and Pigford (1971).

21.10   Danckwerts surface renewal model. Danckwerts modified the Higibe model by suggesting that there is a distribution of surface ages rather than a single age equal to the contact time. He proposed that the chance of a surface element being replaced is independent of the time for which it has been exposed. Based on this hypothesis the fraction of the surface that has been exposed for time t and $t + dt$ can be shown to equal $s \exp(-st)$, where s is the fraction of the surface replaced with fresh liquid in unit time. Verify the following integrals, which are the zeroth and first moment of the surface age distribution function:

$$\mu_0 = \int_0^\infty s \exp(-st) = 1$$

$$\mu_1 = \int_0^\infty ts \exp(-st) = \frac{1}{s}$$

This parameter s is therefore representative of the frequency of surface renewal or the reciprocal of the average age of a surface. Derive the following expression for the instantaneous rate of absorption:

$$\bar{N}_{A0} = \int_0^\infty s \exp(-st) N_{A0}(t) dt$$

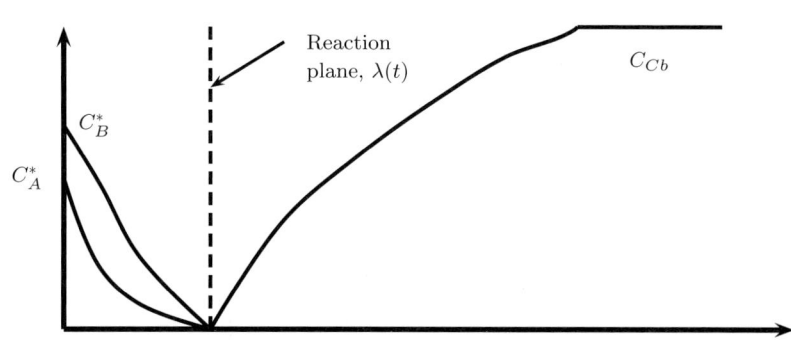

Distance from interface, x

Figure 21.9   Concentration profiles for two gases both reacting instantaneously with a common liquid phase reactant B.

Show that the use of these relations leads to the following expression for the mass transfer coefficient for a non-reacting system:

$$k_L = \sqrt{D_A s} \tag{21.23}$$

Show that for a reacting system the enhancement factor is equal to $\sqrt{1 + Ha^2}$ based on this model for a pseudo-first-order reaction.

21.11 **Laminar jet: physical absorption.** Show that the total quantity of gas absorbed in a physical absorption during contact time t_E is given by the following expression:

$$\mathcal{M}_A = 4C_A^* \sqrt{D_A Q L}$$

21.12 **Laminar jet apparatus: pseudo-first-order reaction.** Derive a similar expression for moles absorbed if the absorption is under a pseudo-first-order reaction and show that the number of moles absorbed is dependent on the diameter of the jet.

21.13 **CO_2 absorption in a laminar jet.** CO_2 is absorbed in a jet of surface area $1.5 \, \text{cm}^2$. For a contact time of $11.3 \, \text{ms}$, the total absorption rate over the exposure time was $15 \times 10^{-7} \, \text{mol/s}$ while for a contact time of $50 \, \text{m}$ it was $10.5 \times 10^{-7} \, \text{mol/s}$. Use the results to estimate the diffusivity and the rate constant for the these data.

21.14 **Absorption in a wetted wall column.** Water flows down a vertical rod of $2.5 \, \text{cm}$ i.d. that is $15 \, \text{cm}$ tall. Calculate the film thickness, the surface velocity, the exposure time, and the rate of absorption of CO_2 at $1 \, \text{atm}$ pressure and into $0.5 \, \text{M}$ NaOH solution with CO_2 at $0.02 \, \text{atm}$ partial pressure.

CHAPTER 22

Reactive Membranes and Facilitated Transport

Learning Objectives

After completing this chapter, you will be able to:

- Calculate the change in the transport rate of the solute due to a carrier present in the membranes.
- Distinguish between the two common regimes in carrier transport: slow diffusion and fast diffusion asymptotes.
- Derive expressions for the flux for these regimes and examine parametric effects.
- Examine some complex examples of reactive membranes for transport of multiple solutes.
- State some examples of applications of the reactive membranes.

This chapter analyzes the transport across a membrane containing a reacting species known as the carrier. The following reaction scheme is illustrative of many common cases:

$$A + B \rightleftharpoons C \qquad C \text{ is a complex of } A \text{ and } B \qquad (22.1)$$

In this scheme, species A, the diffusing solute, reacts with B to form a complex C and is transported in both the free (A) and combined forms (C). An important example is hemoglobin in blood RBC (red blood cells) and myoglobin in tissues, which binds strongly to oxygen, By itself the oxygen solubility in blood is small and life as we know it for human beings cannot be sustained by this small solubility. Here is where the hemoglobin comes into action. Oxygen is carried in free and combined forms and is desorbed by the reverse reaction to provide adequate oxygen concentration for metabolism in the tissues.

In industrial practice there are various ways of immobilizing a carrier, as shown in Section 22.3; the system then acts as a reactive membrane. Higher fluxes can be obtained in these systems than possible if there were no reaction in the membrane. In some cases very selective separation is also possible. The process is also known as facilitated diffusion since the carrier enhances the transport of A.

The system exhibits a number of complexities compared to ordinary diffusion and these are primarily due to the simultaneous reaction taking place. One of the goals in this chapter is therefore to demonstrate the application of the diffusion-reaction model to calculate the enhancement on the transport of A due to the above complex formation.

Maximum enhancement is obtained when the reaction given in Equation 22.1 is fast relative to diffusion. This is known as the instantaneous reaction regime and is more commonly encountered in practice. An asymptotic solution can be obtained for this case similar to that for an instantaneous reaction in gas–liquid systems. The kinetics of reaction plays no part in the model for such cases. Some parametric dependency of the flux on the concentration gradient and the equilibrium constant will be shown in this chapter based on the analytical solution for this case.

Facilitated diffusion with more than one solute diffusing across the system can have many complex features. The two solutes can compete for the same carrier; this is sometimes referred to as counter-transport. The rate of transport of one solute can be curtailed by the other solute and can even occur counter to its own gradient.

In some cases the second solute can bind together to the same carrier along with the first solute; this is referred to as co-transport. The presence of the second solute can then enhance the transport of the first solute. In some cases the first solute can be transported even when its concentration gradient is zero. These effects are indicated.

A general model for reactive membranes in which multiple equilibrium reactions are taking place is then shown and implemented in MATLAB. This can be used to simulate membrane systems with complex chemistry including counter- and co-transport and is a useful tool to examine parametric effects and to fit experimental data for such reactive membranes.

The chapter concludes with some discussion on industrial applications of reactive membranes. The discussion is brief but some useful references are provided as a starting point for further study of this topic.

22.1   Single Solute Diffusion

In this section we analyze the simple case of a reactive membrane where a chemical reaction of the type shown in Equation 22.1 takes place.

Here A is the species being transported across a membrane of thickness L while B and C are the free and the bound carrier. These stay within the membrane. Species A is thus transported in free form as well as bound form as C from one side to the other side of the membrane. The model to compute the flux of A across the membrane is presented in the following section.

22.1.1   Model Equations

The model equations for the carrier mediated transport are formulated by adding a reaction term to the diffusion equation:

$$D_A \frac{d^2 C_A}{dx^2} = k_2(C_A C_B - C_C/K) \tag{22.2}$$

$$D_B \frac{d^2 C_B}{dx^2} = k_2(C_A C_B - C_C/K) \tag{22.3}$$

$$D_C \frac{d^2 C_C}{dx^2} = -k_2(C_A C_B \quad C_C/K) \tag{22.4}$$

Here k_2 is the rate constant for the forward reaction and K is the equilibrium constant.

Boundary conditions are set as follows:

For species A we can specify the concentration at the two sides of the membrane $C_A(x = 0) = C_{A0}$ and $C_A(x = L) = C_{AL}$. Note that these are concentrations in the membrane phase and not in the external fluid near these sides. If the external concentrations are known, then the membrane phase concentrations have to be obtained by multiplying the external concentrations by the partition coefficient. Thus for example, $C_{A0} = K_{pA} C_{A0,f}$, where K_{pA} is the solubility coefficient of A in the membrane and $C_{A0,f}$ is the concentration in the fluid phase adjacent to the membrane.

Species B and C are bound to the carrier and hence they cannot diffuse out the membrane. Hence the no flux conditions are used for both these species at both $x = 0$ and $x = L$. Since no flux is specified at both sides, there is no base concentration to calculate the concentration profiles and the flux of A. Hence the problem is not uniquely specified. We have to augment the problem by an integral condition that states that the total concentration of B in free form and bound form is fixed and equal to some value C_{BT}, the total concentration of B+C in the membrane. This leads to the following integral constraint on the concentrations of B and C:

$$C_{BT} = \frac{1}{L} \int_0^L (C_B(x) + C_C(x))dx \tag{22.5}$$

Once C_{BT} is specified by the above integral constraint, the problem has a unique solution. This equation is simply a mass balance for the total carrier concentration within the membrane.

22.1.2 Dimensionless Representation

The above set of equations can be made dimensionless by using the following variables: $c_A = C_A/C_{A0}$, $c_B = C_B/C_{BT}$, $c_C = C_C/C_{BT}$, and $\xi = x/L$.
The set of dimensionless equations are as follows:

$$\frac{d^2 c_A}{d\xi^2} = M^2(c_A c_B - c_C/K^*) \tag{22.6}$$

$$\frac{d^2 c_B}{d\xi^2} = \frac{M^2}{q}(c_A c_B - c_C/K^*) \tag{22.7}$$

and

$$\frac{d^2 c_C}{d\xi^2} = -\frac{M^2}{q}(c_A c_B - c_C/K^*) \tag{22.8}$$

Note that in the above representation we have used $D_C = D_B$, that is, equal diffusivity for the free and bound carrier, for simplicity. The values would generally be close to each other. Also note the similarity in the model to the gas–liquid reactions discussed in Chapter 20. The integral constraint on B is the only additional modeling feature to be added here.
The dimensionless parameters needed in the model are as follows:

- A Hatta type of modulus parameter representing the ratio of the diffusion time to the reaction time:

$$M^2 = k_2 C_{BT}(L^2/D_A)$$

- A parameter that is the measure of the relative concentrations of B and A and their diffusivity values:

$$q = \frac{D_B C_{BT}}{D_A C_{A0}}$$

- Dimensionless equilibrium constant:

$$K^* = K C_{A0}$$

Thus the performance of the system is characterized by three dimensionless parameters, M, q, and K^*.

Results are often presented in terms of the enhancement factor, which is defined as $-dc_A/d\xi$ calculated at the surface. For no carrier this will be equal to one and hence this factor shows the augmentation of transport due to the presence of the carrier. An illustrative solution to the model is presented in terms of the key dimensionless parameters in Figure 22.1.

This is for large M of 1000 and was generated using BVP4C. Note that the concentration of all three species changes significantly within the membrane. The enhancement factor is simply the negative of the slope of A in dimensionless concentration. It shows to what extent the faciliated diffusion

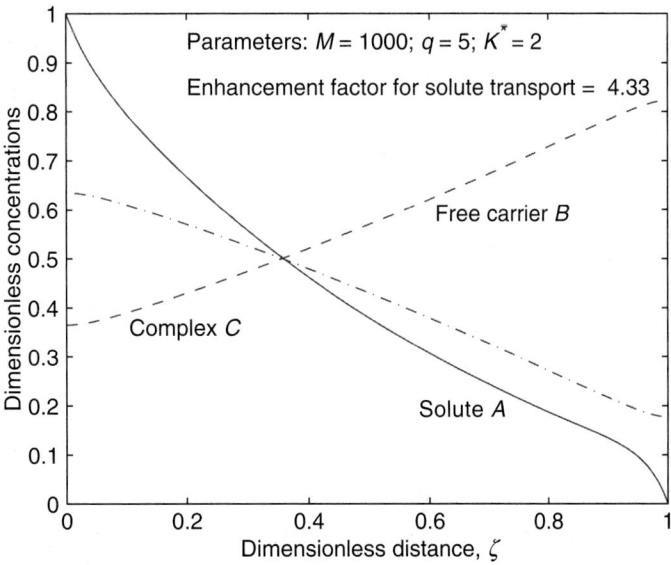

Figure 22.1 Concentration profiles in a reactive membrane for high values of M.

enhances the rate of transport of A. The value for the given set of parameters was found to 4.33.

We now study two asymptotic features of the solution: the large M case where the reaction is treated as instantaneous and the small M case where a pseudo-first-order approximation can be made. Before we proceed it may be in order to look at the invariant of the system.

22.1.3 Invariant of the System

As Equations 22.7 and 22.8 have a common reaction term, there exists an invariance between the concentration of species B and C. The common rate term can be eliminated to give

$$\frac{d^2 c_C}{d\xi^2} = -\frac{d^2 c_B}{d\xi^2} \tag{22.9}$$

Integrating twice and using the total constraint condition given by Equation 22.5 (in dimensionless form) yields:

$$c_C = 1 - c_B \tag{22.10}$$

This is nothing but a restatement of the combined mass balance for the total carrier, that is, $c_C + c_B$ must be one. The carrier must be either in free or combined form since it cannot escape the membrane. The model equations can be reduced from three to two now by eliminating, say, c_C from the rate terms in Equations 22.6 and 22.7.

22.1.4 Instantaneous Reaction Asymptote

An analytical solution for the enhancement can be obtained for reactions with large M. The reaction is assumed to be in equilibrium at all points in the membrane. This is similar to the equilibrium models in a staged process where the mass balance coupled with the equilibrium condition (law of mass action here) is sufficient to model the system. The kinetics of the reaction becomes unimportant in this case.

Combining Equations 22.6 and 22.8 we find

$$\frac{d^2 c_A}{d\xi^2} + q \frac{d^2 c_C}{d\xi^2} = 0$$

Integrating once we find that the total flux of species A is

$$\frac{dc_A}{d\xi} + q\frac{dc_C}{d\xi} = \text{Constant} = p_0 \qquad (22.11)$$

The constant p_0 is the gradient of A at the surface and hence dimensionless flux of A diffusing across the membrane is equal to $-p_0$. A second integration gives an invariant relation:

$$c_A + qc_c = p_0\xi + A_2 \qquad (22.12)$$

Here A_2 is the second integration constant. If Equation 22.12 is applied at $\xi = 0$ we have

$$c_A(0) + qc_c(0) = A_2$$

If Equation 22.12 is applied at $\xi = 1$ we have

$$c_A(1) + qc_c(1) = p_0 + A_2$$

Combining these the expression for the dimensionless flux can be obtained and related to the end point values of the concentration:

$$-p_0 = c_A(0) - c_A(1) + q[c_c(0) - c_c(1)] \qquad (22.13)$$

Further simplification can be made by using the equilibrium assumption $c_C = K^*c_Ac_B$ and the total balance on the substrate $c_B = 1 - c_C$. These conditions lead to

$$c_c = \frac{K^*c_A}{1 + K^*c_A}$$

This can be used at the two ends of the membrane to find $c_C(0)$ and $c_C(1)$. Also we use $c_A(0) = 1$. Substituting all this in the expression for combined flux in Equation 22.13 we find

$$\text{Dimensionless flux } = 1 + \frac{qK^*}{1 + K^*} - c_A(1) - \frac{qK^*c_A(1)}{1 + K^*c_A(1)} \qquad (22.14)$$

In the more common case of $c_A(1) = 0$ (the receiving end is at zero concentration of the solute) this simplifies to

$$\text{Dimensionless flux of A} = 1 + \frac{qK^*}{1 + K^*}$$

The flux in terms of dimensional variables is

$$N_{A0} = N_{AL} = \frac{D_A C_{A0}}{L}\left(1 + \frac{D_B}{D_A}\frac{KC_{BT}}{1 + KC_{A0}}\right) \qquad (22.15)$$

This can be used to illustrate some parametric features of the system. The first term is the ordinary diffusion, also called the passive diffusion, and is the flux in the absence of the carrier. The second term on the right-hand side can be viewed as the enhancement in transport due to the presence of a the reacting carrier for the case of a rapid reaction asymptote. This is the contribution of the facilitated transport.

　　Additional features are discussed in the following paragraphs.

Large Total Carrier Concentration

If the first term on the right-hand side is smaller compared to the second (large carrier concentration with respect to the solute concentration) then Equation 22.15 reduces to

$$N_{A0} = N_{AL} = \frac{D_B C_{BT}}{L} \frac{K C_{A0}}{1 + K C_{A0}} \tag{22.16}$$

The flux depends now on the diffusive transport of the carrier from one side to the other side of the membrane. Also the equation shows that the $1/N_{A0}$ versus $1/C_{A0}$ plot should be linear, which can be used to fit the experimental data.

Large Value of the Equilibrium Constant

Flux attains a constant value at large values of K. For example Equation 22.16 becomes

$$N_{A0} = N_{AL} = \frac{D_B C_{BT}}{L} \tag{22.17}$$

The rate of transport depends on how fast B can move from one side to the other. The extent of binding of A with the carrier makes no difference now. The explanation is that species A binds so strongly to the carrier that no reverse reaction takes place at the other end to release back the solute. This has important implications in choosing the carrier. It should not bind very strongly to the solute. The equilibrium of the solute–carrier complexation must provide a delicate balance. The forward complexation rate must be sufficiently high to obtain a large amount of complex and hence, a high enhancement effect. On the other hand, the reverse rate must also be large enough to readily reverse the complexation step at the receiving end so that the solute can be recovered and the carrier diffused back to pick up more solute from the feed end.

22.1.5 Pseudo-First-Order Reaction Asymptote

When the diffusion is fast compared to the reaction a second asymptote (the pseudo-first-order reaction asymptote) is reached. This is also known as the fast diffusion asymptote. The condition is that M is small, usually less than 10. An illustrative concentration profile for this case is shown in Figure 22.2.

The concentration of the substrate and the complex is observed to be nearly constant in this case and the model can be simplified by using a constant concentration of B and C in the membrane. Hence the differential equation for A reduces to a linear equation and the solution can be derived analytically for this case. The details are not shown here but exercise problem 22.5 addresses some of these calculations. In general the enhancement factors for the fast diffusion case are lower than the fast reaction case. Hence one should seek carriers with fast reactions so that the maximum benefit of binding with the carrier can be obtained.

The results of a parametric study of varying the M parameter are shown in Table 22.1.

The results in this table confirm the discussion that at low M the enhancement factor is low and is nearly equal to one. It also shows that as M increases the enhancement factor approaches the instantaneous reaction asymptotic value.

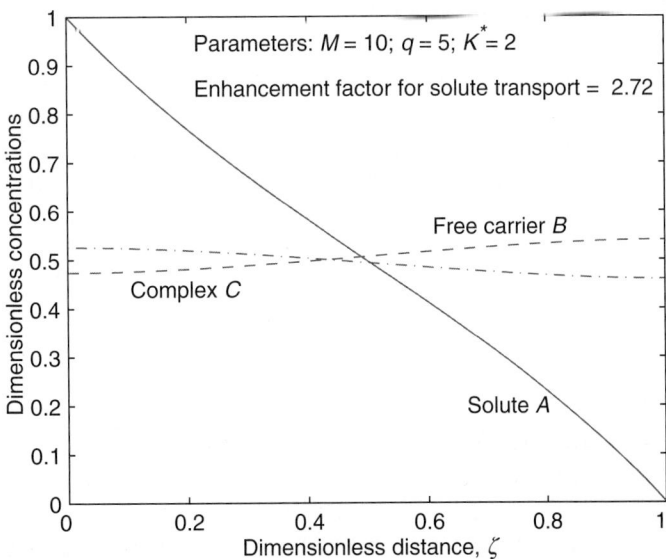

Figure 22.2 Concentration profiles in a reactive membrane for low values of M.

Table 22.1 Effect of M Parameter on the Enhancement Factor for Solute Transport across a Reactive Membrane for $q = 5$ and $K^* = 2$

M	1	10	100	1000	10,000	∞
E	1.04	2.72	3.92	4.28	4.32	4.33

22.2 Co- and Counter-Transport

Membranes with carriers exhibit many other complexities and some discussion is provided in this section. If two solutes compete for the same substrate, a phenomena known as counter-transport is often observed. The reaction scheme for such cases can be schematically represented as

$$\text{Solute } 1 + \text{Substrate S} \rightleftharpoons \text{Complex S+1}$$

$$\text{Solute } 2 + \text{Substrate S} \rightleftharpoons \text{Complex S+2}$$

Solutes 1 and 2 can both diffuse in and out of the membrane while the substrate, complex 1, and complex 2 cannot leave the membrane. Under certain conditions, species 1 (or 2) can be transported against its own concentration gradient. Hence the phenomena is known as counter-transport. Examples are common in biological systems. Thus oxygen and CO can both bind with hemoglobin, leading to counter-transport of oxygen and CO poisoning. On a similar note the cure for methanol poisoning is ethanol, which displaces methanol from the liver enzyme. Additional examples of counter-transport are shown by Cussler (2009) and Noble and Koval (2005).

Co-transport occurs when the following type of overall reaction scheme holds:

$$\text{Solute } 1 + \text{Solute } 2 + \text{Substrate S} \rightleftharpoons \text{Complex S+1+2}$$

In this case solute 1 can be transported by solute 2 or vice versa. Transport of solute 1 can occur even if there is no concentration gradient or even against its own concentration gradient. Thus solute 1 can carry solute 2 along with it and hence the phenomena is known as co-transport.

An *in vitro* example would be a combination of an anion (solute 1), a cation (solute 2), and a membrane with some amine as a carrier (substrate). Additional *in vivo* examples can be found in biological systems as indicated by Cussler (2009).

Co-transport can enhance the transport of one solute due to it being carried by the second solute. An example is presented by Caracciolo et al. (1975) where sodium ions are moved even when the concentration gradient is zero due to co-transport with Li ions. This has important applications in metal recovery and wastewater treatment, especially for the dilute metal solution commonly encountered in hydrometallurgy.

Detailed analysis of both counter- and co-transport can be done by setting up the differential models for each of the species, identifying the dimensionless groups, and computing the results. If the reactions can be assumed to be fast then the equilibrium assumption for each of the reactions can be invoked, leading to analytical solutions. Models are now presented in Sections 22.2.1 and 22.2.2 followed by a general model in Section 22.3 for fast multiple reactions where all the reactions are assumed to be in equilibrium.

22.2.1 Model for Counter-Transport

The model for counter-transport is briefly discussed below. The reaction scheme is represented as

$$A + B \rightleftharpoons C \qquad \text{Complex of A and B} \qquad (22.18)$$

$$P + B \rightleftharpoons E \qquad \text{Complex of P and B} \qquad (22.19)$$

The equations for species A and C are the same as those in section 22.1.1 while that for B is modified by adding an extra term for reaction with P:

$$D_B \frac{d^2 C_B}{dx^2} = k_2(C_A C_B - C_C/K_A) + k_p(C_P C_B - C_S/K_P) \qquad (22.20)$$

Here K_A and K_P are the equilibrium constants for the two reactions. The small ks are the two rate constants.

Two additional equations for P and E are to be added. These are presented as the following pair of equations:

$$D_P \frac{d^2 C_P}{dx^2} = k_p(C_P C_B - C_S/K_P) = -D_E \frac{d^2 E_S}{dx^2} \qquad (22.21)$$

The equations can be solved numerically although it represents a stiff set of differential equations, especially for large values of the Hatta parameters.

The equilibrium approach is presented in further detail in the following.

Fast Reaction: Equilibrium Model

The flux of A is equal to the flux of A in free form and in combined form as species C:

$$\text{Flux of A} = N_{A0} = N_{AL} = \frac{D_A}{L}(C_{A0} - C_{AL}) + \frac{D_C}{L}(C_{C0} - C_{CL})$$

The formal proof of this is similar to that for single instantaneous reaction discussed in Section 22.1.4.

The concentration of C at the two ends is now determined by stoichiometry and the equilibrium consideration. First we need an expression for the concentration of unreacted carrier B. This is an invariant of the system and represents the fact that the carrier B must exist as free, combined as C, or combined as E. The resulting expression can be shown to be

$$C_B = \frac{C_{BT}}{1 + K_A C_A + K_P C_P} \tag{22.22}$$

This can be then used in the equilibrium condition applied at the two edges of the membrane. For example at $x = 0$, $C_C(0) = K_A C_{A0} C_{B0}$. Hence

$$C_C(0) = K_A C_{BT} \frac{C_{A0}}{(1 + K_A C_{A0} + K_P C_{P0})}$$

A similar equation holds for C_{E0} and at the other end $x = L$ as well for both C and E.

The final result for the flux for the common case where the receiving end is at zero concentration, that is, $C_{AL} = 0$, is

$$N_{A0} = N_{AL} = \frac{D}{L} C_{A0}(1 + \mathcal{F})$$

where $\mathcal{F}$ is the enhancement or facilitation factor, which now includes the interaction of the solute P. This is given by

$$\mathcal{F} = \frac{K_A C_{BT}}{(1 + K_A C_{A0} + K_P C_{P0})}$$

A similar equation holds for the flux of P. These models are based on equal diffusion coefficients for all the species. The analysis can be modified for variable diffusivity in a similar manner.

22.2.2 Model for Co-Transport

A model for co-transport can be set up in a similar manner. The scheme analyzed here is

$$A + B + S \rightleftharpoons C$$

where S is the carrier and C is a complex of A, B, and S. The result of the equilibrium model can be derived as

$$\text{Flux of A} = N_{A0} = N_{AL} = \frac{D}{\text{\L}}(C_{A0} - C_{AL})$$

$$+ \frac{DKC_{ST}}{L}\left(\frac{C_{A0}C_{B0}}{(1 + KC_{A0}C_{B0})} - \frac{C_{AL}C_{BL}}{(1 + KC_{AL}C_{BL})}\right) \quad (22.23)$$

Note that the product of the concentrations of A and B now appears in the facilitation term.

22.3 Equilibrium Model: A Computational Scheme

For equilibrium reactions a general approach is presented here that can be used for various complex reaction schemes including the counter- and co-transport scenarios studied in the earlier section.

The diffusion-reaction equation for species j is written in general form as

$$D_j \frac{d^2C_j}{dx^2} = \sum_i \nu_{i,j}\mathcal{R}_i$$

where $\nu_{i,j}$ is the stoichiometric coefficient for species j in the ith reaction. The equation can be formally integrated once from 0 to x to give

$$D_j \frac{dC_j}{dx} = \sum_i \nu_{i,j} \int_0^x \mathcal{R}_i(s)ds - N_{j0} \quad (22.24)$$

where $N_{j0} = -D_j(dC_j/dx)_{x=0}$, the flux of A at the surface. Here s is a dummy variable for integration. This flux term is set to zero for all the species that do not leave the membrane and is retained only for the solutes that cross the membrane.

A second formal integration of Equation 22.24 from 0 to L provides a relation between the concentration of species j at $x = 0$ and $x = L$:

$$C_{j0} - C_{jL} = \frac{1}{D_j} \sum_i \nu_{i,j}Q_i - \frac{N_{Aj}L}{D_j}$$

Q_i results from the second integration and is obainted as the following constant, one for each reaction:

$$Q_i = \int_0^L \left(\int_0^x \mathcal{R}_i(s)ds \right) dx$$

Each reaction provides an integral of this form corresponding to its rate function. These integral constants can be viewed as the tie or mapping between the end point concentrations. They are evaluated by applying the equilibrium relations.

Now the equilibrium constraints are used at the two ends. This generates an additional $2 * nr$ equations, which completes the set of equations. Here nr is the total number of independent reactions (for example 2 for the counter-transport considered in Section 22.2.1). An additional constraint is the total balance equation for the carrier. This completes the set of equations needed to execute the computations. The concentration relations, equilibrium relation, and the total balance constrains are solved together as a set of algebraic equations. The computer program to do this is presented in Listing 22.1.

Listing 22.1 Simulation of Facilitated Transport of General Network of Equilibrium Reactions

```
% counter—transport membrane simulation
function nonlinear
global nu K1 K2 bTotal a0 a1 p0 p1
% stoichiometry
%%%    A   P  B   C  E
nu = [ —1   0 —1   1   0;
        0 —1 —1   0   1];
 K1 = 5; K2=50;
 a0= 1; a1= 0.0;
 p0 = 0.20 ; p1=0.00;
 bTotal = 5
y0 = [ 1 1 1 1 2 1 1 1 1 1 ];
     y = fsolve(@myfun, y0)
        fluxA =  y(9)
        fluxB = y(10)
   %%%%%%%%%%%%%%%%%%%%%%%%%%%%%%%%%%%%
     function F = myfun(y)
        global nu K1 K2 bTotal a0 a1 p0 p1
% unwrap the variables
b0 = y(1); b1= y(2); c0 = y(3);
c1= y(4); e0 = y(5); e1=y(6);
Q1= y(7); Q2 = y(8);   fluxa = y(9); fluxp = y(10);
% set up equations
```

```
% 1 to 5 concentration tie-ups
F(1) = a0-a1 + nu(1,1)*Q1 - fluxa  ; % A
F(2) = p0-p1 + nu(2,2)*Q2 - fluxp  ; % A
F(3) = b0-b1+nu(1,3)* Q1 +  nu(2,3)*Q2 ; % species B
F (4) = c0-c1+nu(1,4)* Q1 % c;
F(5) = e0-e1+nu(2,5)* Q2 ;
F(6) = b0+c0+e0 -bTotal % balance constraint
% equlibrium relations
F(7) = c0-K1*b0*a0;
F(8) = e0-K2*b0*p0;
F(9) = c1-K1*b1*a1;
F(10) = e1-K2*b1*p1;
%%%%%%%%%%%%%%%%%%%%%%%%%%%%%%%%%%%%%%%%%%
```

22.3.1 Illustrative Results

Some illustrative results based on this model are shown below. Examples of some complexities in counter-transport are shown by simulation using Listing 22.1.

For the counter-transport discussed, the unknowns are the concentrations of B, C, and E at the two end points of the membrane (total of six), the two tie parameters Q_1 and Q_2, and the flux of the solutes A and P across the membrane. Thus there are ten unknowns. Ten algebraic equations are available and can be set up and solved with FSOLVE. Upon running the program for the parameters, the following result is obtained for the fluxes of A and B:

$$N_A = 2.5625 \text{ and } N_P = 3.3250. \text{ Scaled units are used here.}$$

You may wish to verify that these are also consistent with the analytical solution presented earlier.

If the equilibrium constant for the second reaction is set to zero, the following results are obtained:

$$N_A = 5.1667 \text{ and } N_P = 0.2. \text{ Scaled units are used here.}$$

Here the passive diffusion value for flux is obtained for P since $K_P = 0$ and the single solute value is obtained for A.

Comparing the two cases shown previously, we find the presence of P, although in a smaller concentration (one fifth of A), has caused a significant drop in flux of A for the first case because the equilibrium constant for P (50) is much larger than that for A (5). The poisoning effect of the strongly complexing solute is illustrated here.

Osmotic Transport

Another interesting effect that can be demonstrated is the osmotic diffusion where a species A diffuses even when the concentration gradient is zero. Here we keep the end point concentration of A the same, equal to one. The program produces the following result for the fluxes:

$$N_A = -2.6042 \text{ and } N_P = 3.3250. \text{ Scaled units are used here.}$$

The flux of A is now negative although its concentration gradient is zero. The solute P complexes strongly with B and the equilibrium for C shifts to the left, releasing A at $x = 0$.

An example of osmotic diffusion is NaCl transport in membranes with crown ethers as a carrier. This was studied by Caracciolo et al. (1975). Here one side of the membrane was maintained at 1 M NaCl and 1 M LiCl. The other side was 1 M NaCl and no LiCl. The LiCl was assumed to be a passive diffuser merely providing extra chloride ions so that electroneutrality was not violated. The sodium ion was found to diffuse against its own concentration gradient until a certain level of concentration difference was achieved. The complex of Na ion, Li ion, and the crown ether was responsible for this osmotic diffusion phenomena. For more details and experimental data on this system, please refer to the preceding reference. The program in Listing 22.1 can be adapted and used to model this and similar systems.

Co-Transport

Many systems show multiple reaction steps. For example the co-transport shown as a single reaction is often a combination of multiple reaction steps. The code shown can be used to incorporate these and develop a more refined model. It is also useful to examine the effect of non-equal diffusion coefficients for the various species. The model in the text used the same diffusivity for A and B.

22.4 Reactive Membranes in Practice

In this section some application examples of reactive membranes in practice are discussed. For such membranes to be useful for practical applications the following properties should be considered:

- There should be no side reactions that will lower the transport of A.
- The carrier must stay in the membrane phase and should not be leached out from it.

- The carrier must remain as a solid solution in the membrane phase and must not form a precipitate or some related phase change within the membrane.
- The reaction between the carrier and the solute should be relatively fast.
- The equilibrium constant must be within an intermediate range. A value that is too low does not provide enough enhancement. Too high a value causes an irreversible binding with the result that the solute does not get released at the lean end of the membrane.

Several configurations of membranes have been developed, each with its own domain of application. Some common ones are briefly discussed in the following.

22.4.1   Emulsion Liquid Membranes (ELM)

In an ELM arrangement two liquid phases are mixed together to form an emulsion. A surfactant is added to stabilize the emulsion. This emulsion is dispersed in a continuous phase, usually an aqueous phase. The solute moves from the aqueous phase to the organic part of the emulsion and then to an inner phase (usually aqueous) that contains the carrier, which binds reversibly with the solute. A schematic of this is presented in Figure 22.3.

The advantages are creation of a large surface to volume ratio for mass transfer. Difficulties are associated with emulsion stability and the process of de-emulsification to recover the receiving phase.

ELM has been applied for recovery of metals from lean leach solution. An example is the extraction of nickel using D2EHPA (di-2ethyl-hexyl

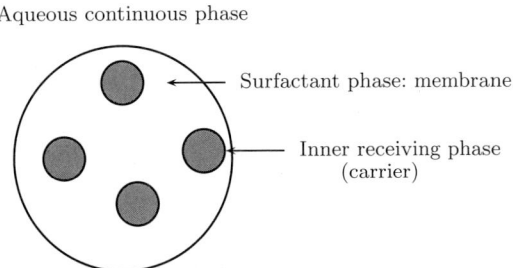

Figure 22.3   Schematic of emulsion liquid membrane (ELM).

phosphoric acid) as a carrier, studied by Kulkarni et al. (1999). The reaction scheme is an exchange of a nickel ion for two hydrogen ions:

$$Ni^{++} + 2RH \rightarrow NiR_2 + 2H^+$$

22.4.2 Immobilized Liquid Membranes (ILM)

In ILM the liquid membrane phase is immobilized in a porous matrix as shown in Figure 22.4.

CO_2 in amine solution is an example of ILM and the work of Marzouki et al. (2005) is an illustrative reference on this topic.

Another use of the ILM approach is the use of high-temperature, inorganic supports to immobilize molten salts (range of temperaure of 575 K or so). The molten salts can reversibly complex specific gases (i.e., NH_3, CO_2) and act as a barrier to other gases (e.g., H_2, He). Carrier loss and solvent loss can be an issue in the practical application of this idea. Solvent loss occurs when the solvent evaporates or is forced from pores by large transmembrane pressures. Carrier loss can occur due to irreversible reactions with the impurities, and when the solvent, which humidifies the feed stream, condenses on the membrane and leaches out carrier.

22.4.3 Fixed-Site Carrier Membranes

In the fixed-site carrier membrane configuration, the carrier is covalently bound to the membrane phase. The advantage is that the carrier is fixed and cannot leak out of the system. The disadvantages are the irreversible

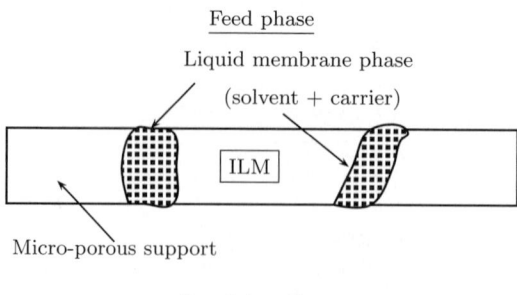

Figure 22.4 Schematic of an immobilized liquid membrane (ILM).

deactivation of the reactive sites with time and low permeability due to limitations of solid state diffusion.

An application of fixed-site membranes has been reported for transport of oxygen through membranes containing various metal-prophyrin compounds (Nishide et al., 1987); a transport analysis of this system has been done by Noble (1991).

Summary

- Membranes containing a reactive compound are common in biological systems and are finding applications in separation processes as well. A solute transported across this membrane combines with the reactive compound (called the carrier) and forms a complex. Hence part of the solute is transported as this complex and the rate of transport is enhanced. The phenomena is also known as facilitated diffusion.

- Models can be set up by using diffusion reaction equations for the solute, carrier, and complex. The methodology is therefore the same as that for gas–liquid reactions. Three differential equations result. No flux conditions are used for the carrier and complex at both ends of the membrane while the prescribed concentration conditions at the two ends are used for the solute.

- For the simple reaction scheme illustrated by Equation 22.1 in the text, three dimensionless parameters are needed. The reaction-diffusion effects are reflected in a Hatta type of parameter, M^2, which is the ratio of diffusion time to reaction time. The concentration ratio together with the diffusivity ratio, q, is the second parameter and the dimensionless equilibrium constant K^* is the third.

- The system behavior approaches a nearly equilibrium condition for large values of diffusion time compared to the reaction time (large value of M). The resulting flux of A can then be calculated as the sum of the flux due to A in free form and to A in combined form which is equal the flux of C diffusion across the membrane. The flux of A depends on the imposed concentration conditions at the end points. The flux of C can be calculated from the equilibrium conditions using the law of mass action at the two ends of the membrane.

- For low values of M, the concentrations of the carrier and the complex are nearly the same. Hence a pseudo-first-order model can be used for the transport of A and analytical solutions can be obtained. For the intermediate case a numerical solution is useful.

- Some variations of the problem when there are two solutes diffusing across a reactive membrane are also important. If the two solutes compete for the carrier, various complex phenomena such as reverse diffusion and so on can occur. The second solute can decrease the transport rate of the first solute, especially if it gets strongly coupled with the carrier. If it combines with the same carrier to form a "triple" complex, a co-transport occurs and the transport is enhanced due to the second solute.

- Common arrangements of reactive membrane for practical applications are the ELM, ILM and fixed-site configurations, each with its own domain of merit. A number of difficult separations can be accomplished by the use of reactive membranes.

- The study of reactive membranes is an elegant example of diffusion-reaction analysis where both analytical and numerical methods can be used as appropriate. If all the reactions are at equilibrium a general analytic model can be determined and the simple MATLAB code shown in Listing 22.1 is useful for complex reaction schemes.

Review Questions

22.1 What is meant by facilitated transport?

22.2 Give some examples of facilitated transport.

22.3 State the key dimensionless parameters for a simple reaction scheme in a reactive membrane.

22.4 How does the change in Hatta parameter affect the concentration profiles in the reactive membrane?

22.5 When can we assume that nearly equilibrium conditions hold everywhere in the membrane?

22.6 How can the flux of the solute be calculated under equilibrium conditions?

22.7 What is the effect of simultaneous diffusion of two solutes in a reactive membrane with a common carrier?

22.8 Can a species diffuse against its concentration gradient in a reactive membrane?

22.9 What is co-transport?

22.10 What are the merits and demerits of ELM?

22.11 What are the merits and demerits of ILM?

22.12 What are the merits and demerits of fixed-site membranes?

Problems

22.1 **Numerical solution of facilitated diffusion.** Solve the problem of facilitated diffusion for a single solute given in Section 22.1 using BVP4C or other numerical tools and compare with the results given in Table 22.1.

22.2 **Glucose uptake by red cells.** Data for glucose uptake in erythrocyte are given by Cussler (2009) and presented in the following:

C_{A0}	1	2	3	4.3	5.0
N_{A0}	0.09	0.14	0.20	0.25	0.28

The data appear to show that the process is a facilitated diffusion. One way to verify this is to plot $1/N_{A0}$ versus $1/C_{A0}$. When would this plot be linear? Test this on the data and provide a model fit to the data.

22.3 **Flux calculation in facilitated diffusion.** $LiCl_2$ diffuses across a membrane containing a reactive species. On one side of the membrane, the Li concentration is 0.1 M and on the other side it is zero. The membrane has a carrier with concentration 6.8×10^{-3} M of a crown ether complex that can bind to Li ions. The equilibrium constant for Li is 290 L/mol and the partition coefficient is 4.5×10^{-4}. Find the flux of Li across a membrane of thickness 32 μm.

22.4 **Effect of reaction stoichiometry.** The model shown in Section 22.1 for the equilibrium reaction case was for the case where the stoichiometry produced one mole of product. The law of mass action depends on the stoichiometry. Consider the following scheme:

$$A + BX \rightleftharpoons BA + X$$

This scheme, for example, applies to copper separation in liquid ion membranes. The law of mass action would give the concentration of BA as KC_AC_{BX}/C_X. This will change the expression for the flux of A in combined form, that is, carried as BA. Use the equation for the equilibrium and derive an expression for the flux of A as a function of the total carrier concentration. How should the data be plotted so that a linear plot is obtained?

22.5 **Analytical solution for pseudo-first-order case.** Solve Equation 22.6 assuming a pseudo-first-order condition where c_B and c_C are replaced by average values. Show that the equation to be solved then is

$$\frac{d^2c_A}{d\xi^2} = M^2[c_A\bar{c}_B - (1 - \bar{c}_B)/K^*]$$

where the invariance relation between B and C is also used. Use the boundary conditions of 1 and 0 for c_A at ξ of 0 and 1, respectively. Now use the A

profile in the B-balance equation (22.7) and show that the average concentration of B is

$$\bar{c}_B = \frac{1}{1 + K_A \bar{c}_A}$$

where $\bar{c}_A$ is the average concentration of A in the system. Derive an expression for the enhancement factor. Test your analytical solution for the parameter values given in Figure 22.2 where the results of a numerical solution are shown.

22.6 **Combined flux equation: validation.** Consider the scheme

$$A + B \rightleftharpoons C$$

Apply the general model shown in Section 22.3 and show that the following relations are applicable for A and C: for A

$$C_{A0} - C_{AL} = -Q_1/D_A - \frac{N_{A0}L}{D_A}$$

and for C

$$C_{C0} - C_{CL} = Q_1/D_C$$

Eliminate Q_1 between the two and verify that the flux of A (N_{A0}) is the sum of the flux of A in free and combined forms. Note that the flux of A is given by the traditional Fick's law applied at $x = 0$; thus N_{A0} is still $-D_A(dC_A/dx)$ evaluated at $x = 0$. But the flux across the system has contributions of the flux of A in free and combined forms.

22.7 **Reverse and osmotic diffusion.** Run the code given in Listing 22.1 for various concentrations of A in the receiving end in the range of 0 to 1, keeping other parameters the same as in the code. Plot the flux of A as a function of the concentration difference. Mark the regions of reverse diffusion in your plot. Also find the concentration difference at which no A diffuses across the membrane.

22.8 **Simulation of co-transport.** Modify the program given in Listing 22.1 for co-transport. Apply this to a case where there is no concentration gradient for A but a finite concentration gradient for B. Show that there is a flux of A, although the concentration of A is the same on the both sides. The flux of B drags A along with it by coupling with A to form a triple complex.

CHAPTER 23

Biomedical Applications

Learning Objectives

After completing this chapter, you will be able to:

- Explain the role of mass transfer in biomedical systems.
- State and use the Hill equation for oxygen-hemoglobin equilibrium.
- Model oxygen transport in lungs and calculate the oxygen profile in an alveolar capillary.
- Understand the Krogh model for oxygen transport in tissues and identify the oxygen starved region.
- Develop pharmacokinetic models to predict drug metabolism.
- Set up and solve a mesoscopic model for a dialysis system and link it to model a patient-hemodialyzer system.

Mass transfer principles find important applications in biomedical engineering. For a biomedical engineer a mechanistic understanding of processes such as oxygen transport in lungs and tissues is important *per se* and also in evaluating medical conditions and adjusting the treatment protocols. For pharmacokinetic scientists who deal with the study of distribution of a drug and its metabolism in the body, the modeling of mass exchange between various compartments/organs is the main tool needed. For the design of many extra-corporal devices, for example, dialysis, membrane oxygenators, artificial liver, and so on, it is necessary to evaluate the role of transport in the equipment and adjust the operating conditions accordingly. Hence we note that modeling of these systems where mass transport plays a key role is an important component of biomedical engineering. The goal of this chapter is to take some of these examples, illustrate how fundamental transport considerations are useful to model, and analyze these processes.

The analysis presented in this chapter is at a simplified level, making some assumptions but nevertheless providing important information and indicating the main factors affecting the process. Upon studying this chapter you will be able to see how the concepts learned in Part I of the book can be applied to prototypical biomedical problems. This broadens your skill in applying mass transfer principles to a wider class of problems than in traditional chemical engineering.

Transport of gaseous species (in dissolved form) is encountered in a number of organs and is shown schematically in Figure 23.1. Two main mass transfer steps are oxygen exchange in lung to blood and oxygen transfer from blood to tissues. Mass transfer–based models to describe these are discussed in Sections 23.1 and 23.2 respectively. Analysis of distribution of drugs together with their metabolism can be modeled by the compartmental models that were introduced in Section 13.8. Additional information and some

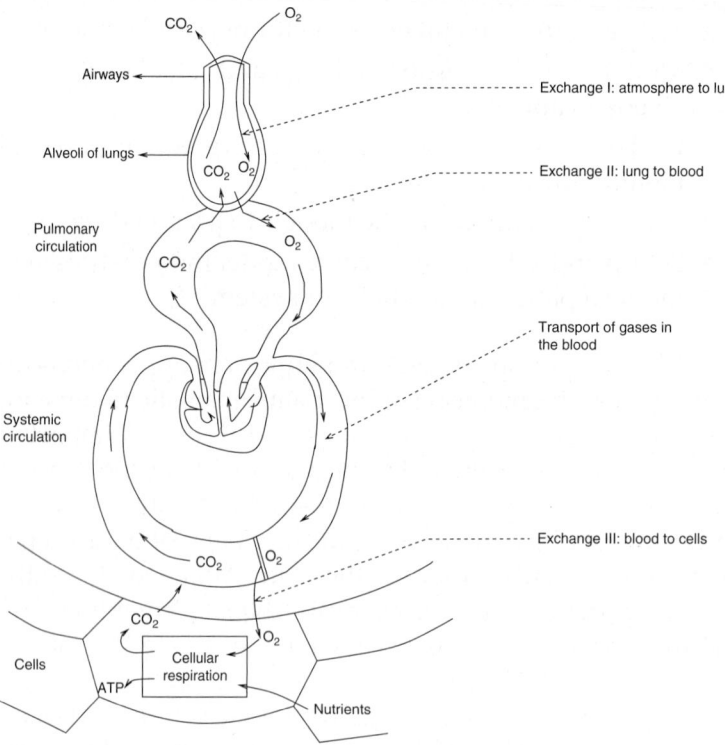

Figure 23.1   Overview of oxygen and carbon dioxide mass transfer in human body. Based on Silverthorn D. U, (2001). *Human Physiology, An Integrated Approach, Second Edition.* Prentice Hall, 2nd ed, Upper Saddle River, N.J.

more detailed modeling trends are then shown in Section 23.3. Modeling of extracorporal devices is illustrated by a mesoscopic-based model for a hemodialyzer in Section 23.4. Here we also show how this model can be linked to a compartment model to simulate the patient blood concentration during a dialysis treatment.

23.1   Oxygen Uptake in Lungs

In this section we consider the modeling of oxygen uptake in lungs (Exchange II in Figure 23.1). First we consider the equilibrium concentration that can be reached in blood for the reaction of oxygen and hemoglobin (abbreviated as Hb). Then we enumerate the transport steps and derive an expression for local transport at any point. This is then included in a mesoscale model for the capillary to calculate the oxygen concentration versus length along the alveolar capillary. One can clearly see the similarity with the mass exchanger modeling studied in earlier chapters.

23.1.1   Oxygen-Hemoglobin Equilibrium

A simple way to calculate equilibrium concentration in blood in contact with air is to apply the law of mass action to the reaction of oxygen and Hb:

$$O_2 + Hb \rightleftharpoons O_2Hb$$

Using the law of mass action, we obtain

$$K = \frac{[O_2Hb]}{[Hb][O_2]}$$

where the square bracket shows the concentration. K is the equilibrium constant for the reaction.

Using Henry's law $[O_2] = p_{O_2}/H_{O_2}$. Let S be the fraction of Hb existing in oxygenated form. Then $[O_2Hb] = C_T S$ and $[Hb] = (1 - S)C_T$ where C_T is the total concentration of Hb in oxygenated plus free form. Using these in the law of mass action, we have

$$\frac{K p_{O_2}}{H_{O_2}} = \frac{S}{(1 - S)}$$

Solving for S, we get

$$S = \frac{K p_{O_2}/H_{O_2}}{1 + K p_{O_2}/H_{O_2}} \qquad (23.1)$$

The above model based on a single reaction scheme does not fit the actual data, which shows an S-shaped curve as shown in Figure 23.2. This is ascribed to oxygen binding to more than one site in a cooperative manner. There are four heme groups in Hb. Binding of oxygen to the first heme group facilitates binding to subsequent sites, resulting in the S-shaped curve shown in Figure 23.2.

The equilibrium curve is therefore fitted as

$$S = \frac{(K p_{O_2}/H_{O_2})^n}{1 + (K p_{O_2}/H_{O_2})^n} \qquad (23.2)$$

This is referred to as Hill's equation. The equation is often written as

$$S = \frac{(p_{O_2}/p_{50})^n}{[1 + (p_{O_2}/p_{50})^n]} \qquad (23.3)$$

Here p_{50} is the partial pressure at which the oxygen saturation is 50%.

The exponent should be four if oxygen binds consecutively to the four sites. The data, however, fits with an exponent of 2.7. The value of p_{50} is 26 mm Hg.

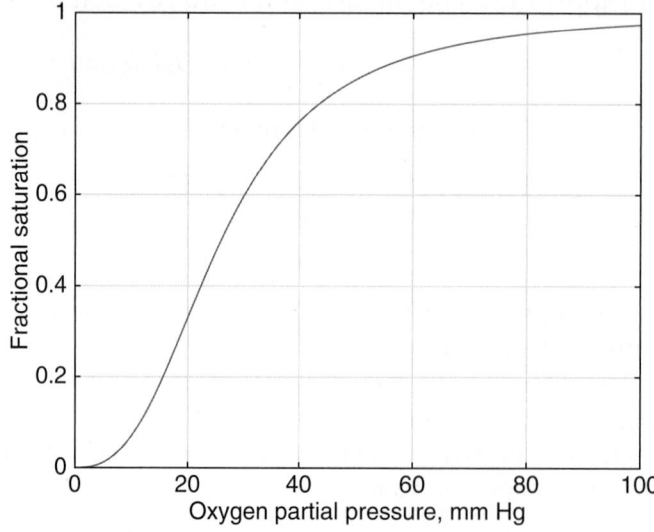

Figure 23.2   Oxygen-hemoglobin equilibrium plot according to Hill's equation.

For oxygen binding to myoglobin (a carrier found in muscles) the single site saturation curve given by Equation 23.1 fits the data well. Hill's model with $n = 1$ and $p_{50} = 5.3$ mm Hg is used for this case.

Effect of Dissolved CO_2

The O_2–Hb equilibrium is affected by pH, which in turn is affected by CO_2 concentration in blood. This has interesting synergic effects that are worth noting. An increase in CO_2 concentration shifts the S-curve to the right. The release of oxygen is facilitated by this shift to the right. This happens in the tissues where CO_2 is the metabolic product and its concentration increases as blood moves through the tissue capillaries. This causes more oxygen to be released for metabolism.

The opposite happens in the lungs where CO_2 is transferred to air. The accompanying decrease in CO_2 causes a shift to the left and increases saturation capacity for oxygen in blood. Thus nature has fine-tuned the equilibrium to accommodate both absorption and desorption!

23.1.2 Transport Steps for Oxygen Uptake

The schematic of transfer of oxygen from the gas in the alveolus to the capillary is shown in Figure 23.3. The transport rate of oxygen in the lungs can be analyzed by considering the following steps:

1. Transport across the alveolar gas phase
2. Transport across the alveolar epithelium and capillary endothelium
3. Diffusion and convection in the blood plasma
4. Diffusion across the red cell membrane
5. Diffusion with reaction in red blood cells (RBCs)

Each step except the last is in series and can be modeled using a transport coefficient. The last step (diffusion and reaction in parallel) can be modeled using the concepts borrowed from gas–solid catalytic reactions using an effectiveness factor. The contribution to the total resistance for each of the four steps occurring in series is shown next. An overall transport coefficient is calculated for steps 1 to 4 and incorporated into the diffusion-reaction model for RBCs (red blood cells) by defining an overall effectiveness factor. The rate of uptake of oxygen can then be calculated.

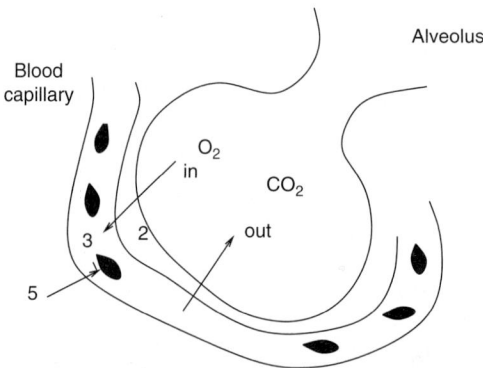

2. Transport across alvelolar epithelium and capillary endothelium
3. Diffusion and convection across the blood plasma
5. Diffusion with reaction in the red blood cells (RBCs)

Figure 23.3   Schematic of mass transport process in the lung. Steps 1 and 4 described in the text are not rate limiting and therefore not shown in this figure.

Magnitudes of the Resistances

The magnitudes of the resistances of the first four steps shown previously are as follows:

1. Resistance due to diffusion in the gas phase is usually small and can be neglected due to the order of magnitude difference in the diffusivity in the gas phase compared in the liquid phase. This step is therefore not included in further analysis.

2. Diffusion across the alveolor membrane and the endothelium can be described by a permeability P_M. This is equal to the diffusion coefficient divided by the thickness of the membrane. A value of 0.387 cm/s for P_M is indicated by Truskey et al. (2004).

3. The transport coefficient in plasma, k_P, should be based on a convection-diffusion model. The Sherwood number is used as the measure of the transfer coefficient. This is defined as $k_p L/D$, where L is the thickness of the plasma layer, which is on the order of 0.35 μm to 1.5 μm. D is the diffusion coefficient of oxygen in plasma, which has a value of 2.4×10^{-5} cm^2/s. Based on the literature data, Truskey et al. (2004) suggest a value of 1.468 for the Sherwood number. Hence the mass transfer coefficient in the plasma is estimated as 0.44 cm/s.

4. Red cell membrane (endothelium) is about 10 nm thick. The solubility of oxygen is also about double that in plasma. Consequently the permeability of this membrane is about 48 cm/s. This value is much larger than the values for step 2 and 3. This resistance due to this transport can therefore be neglected.

From this discussion, we find steps 2 and 3 offer the main resistances for oxygen to reach the red blood cells from the alveolar gas space. These steps occur in series and can be combined into a single overall transfer coefficient, P_T:

$$\frac{1}{P_T} = \frac{1}{P_M} + \frac{1}{k_p} \tag{23.4}$$

Transport in red blood cells with reaction can be characterized using an overall effectiveness factor. If RBC (red blood cell) is treated as an equivalent slab with a characteristic length parameter of L_{RBC} and a first-order uptake reaction with oxygen can be assumed, then the effectiveness factor is given by Equation 18.8 in Chapter 18:

$$\eta = \frac{\tanh \phi}{\phi} \left(\frac{Bi}{Bi + \phi \tanh(\phi)} \right) \tag{23.5}$$

The two parameters required in the preceding equation are as follows. First is the Thiele modulus, defined as

$$\phi = L_{RBC} \sqrt{\frac{k}{D_{RBC}}}$$

This is in the range of 0.6 to 2.2 and hence there is some drop of oxygen concentration in the interior of the red blood cell.

The second is the Biot number, defined as

$$Bi = \frac{P_T L_{RBC}}{D_{RBC}}$$

where P_T is defined in Equation 23.4. The Biot number, in turn, can be shown to be in the range of 5 and hence the "external" mass transfer contributes to some extent as well (about 20% of the overall resistance).

The rate of oxygen uptake is then calculated as

$$\text{Rate per unit volume of RBC} = k_1 \eta (C_A^* - C_A)$$

where C_A^* is the saturation solubility of oxygen in blood, which is equal to p_{O_2}/H_{O_2}, H_{O_2} being the solubility coefficient. C_A is the total oxygen concentration in the RBC.

The rate of transport per unit volume of the capillary is then obtained by multiplying this by the hematocrit value (ratio of volume of RBC to volume of capillary), denoted as f_{HB} here:

$$\text{Rate per unit volume of capillary} = f_{HB}k_1\eta(C_A^* - C_A)$$

The rate provides the local model, which can then be used in an overall balance model for the entire length of the capillary. This is addressed next.

23.1.3 Meso-Model for the Capillary

The local model for oxygen uptake in the lung can be linked to a meso-model for the capillary to find the oxygen profile and the length of the capillary needed to achieve nearly complete saturation.

The total concentration of oxygen (in plasma and bound to RBCs) in the capillary is treated as a variable. Note that this is a cup-mixed radial average concentration. The mass balance for oxygen is

$$\text{In} - \text{out} + \text{transferred across the alveolus} = 0$$

In mathematical terms:

$$(\pi R^2 v C_A)_z - (\pi R^2 v C_A)_{z+\Delta z} + f_{HB}k_1\eta(C_A^* - C_A)\pi R^2 \Delta z = 0 \qquad (23.6)$$

Here v is the velocity of blood flow in the capillary. This leads to the following differential equation for the variation of the oxygen concentration along the length of the capillary:

$$v\frac{dC_A}{dz} = f_{HB}k_1\eta(C_A^* - C_A) \qquad (23.7)$$

The solution to this differential equation with the inlet condition of C_{Ai} at $z = 0$ provides the expected exponential profile for oxygen concentration along the length of the capillary:

$$C_A = C_{Ai}\exp\left(\frac{-f_{HB}k_1\eta}{v}z\right) + C_A^*\left[1 - \exp\left(\frac{-f_{HB}k_1\eta}{v}z\right)\right]$$

The group $v/[f_{HB}k_1\eta]$ has units of length and can be viewed as the characteristic distance for oxygen uptake. Over 99% of the saturation can be expected for a distance equal to five times the characteristic distance. The order of value of the characteristic distance can be shown to be 40 μm. Thus saturation occurs in a distance of 200 μm or so, which is much smaller than the length for mass transfer, which can be approximated as the circumference of an alvelous (754 μm). Hence the lung is an overdesigned piece of equipment with a size more than double that required for saturation.

23.2   Transport in Tissues: Krogh Model

In this section we discuss oxygen transport from blood to tissues (Exchange III in Figure 23.1). A classical model developed by Krogh in 1921 is widely used to calculate the rate of oxygen transport for this case. The conceptual basis of the model is illustrated in Figure 23.4.

In this model, each capillary is assumed to be cylindrical with a radius R_C. A tissue region of radius R_0 is associated with each capillary. This radius R_0 is taken as the half-distance between the center of two capillaries. Note from the figure that the model fails to account for some tissue areas near R_0. But the assumption is needed so that a simple unit cell concept can be used to simulate the problem. In practice the effect of this small region is expected to

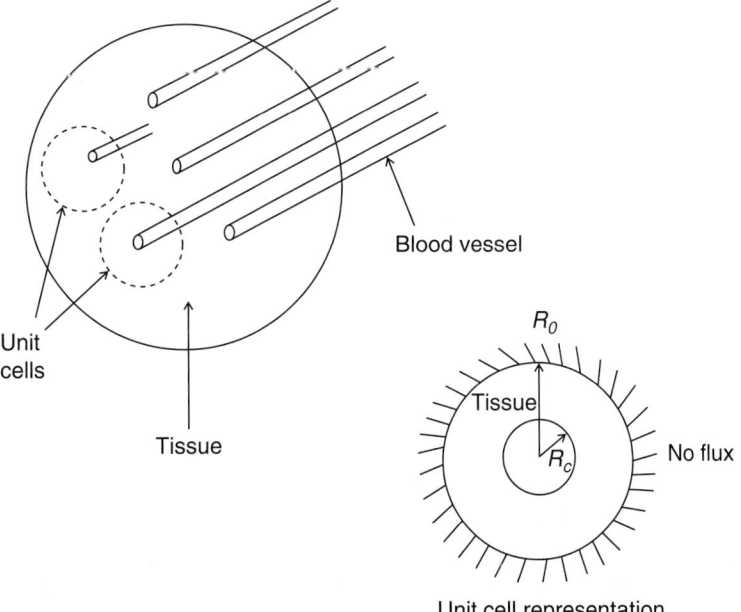

Figure 23.4 Krogh cylinder model for oxygen transport in tissues.

be minor. Also note that the model is not applicable to a complex arrangement of capillaries such as in brain tissue.

The Krogh model assumes that there is a zero-order consumption of oxygen in the tissues leading to

$$\frac{D_e}{r}\left[\frac{d}{dr}\left(r\frac{dC_A}{dr}\right)\right] = k_0 \tag{23.8}$$

The boundary conditions are saturation oxygen concentration in blood of C_{As} at $r = R_C$ and no flux of oxygen conditions at $r = R_0$.

The dimensionless representation is achieved by defining the following reference scales: reference length = R_0; reference concentration = C_{As} the concentration in the blood. The corresponding dimensional variables are as follows: the distance ξ is defined as r/R_0 and the concentration c_A is defined as C_A/C_{AS}. Then the parametric representation of the problem can be shown to be

$$c_A = c_A(\xi; \phi_0^2, \kappa)$$

Here ϕ_0^2 is the Theile modulus for a zero-order reaction, defined as

$$\phi_0^2 = \frac{R_0^2 k_0}{D_e C_{As}} \tag{23.9}$$

κ is the ratio R_C/R_0.

A solution can be obtained analytically for the concentration profile. The critical value of the parameter ϕ_0^2 at which the oxygen concentration drops to zero at $r = R_0$ is obtained as

$$\phi_0^2 \text{ critical } = \frac{4}{(\kappa^2 - 1 - 2\ln\kappa)} \tag{23.10}$$

For ϕ_0 values larger than this, there is a region of zero oxygen concentration referred to as an anoxic region at the end region of the tissue (near R_0). If such a region persists for long time, a part of the tissue will die, leading to a condition known as necrosis. Although necrotic conditions are rare in normal conditions, they can arise in tumors due to a higher rate of metabolism (higher k_0 values for tumor cells).

Some limitations to the Krugh model are now briefly indicated:

- The unit-cell model assumes close-packed tissue and excludes about 21% of the tissue. It therefore underestimates oxygen consumption. Other unit-cell models are considered in later studies, such as hexagonal arrangement of capillaries, but any such arrangement other than the cylinder requires a numerical solution.

- The model does not assume the discrete nature of the red cells and oxygen is assumed to be released uniformly in the plasma.
- The role of myoglobin, which binds with oxygen in the tissue, is not explicitly accounted for.

Details of modeling these issues are not considered here and interested students may wish to see the references cited by Truskey et al. (2004) in this area.

23.2.1 Oxygen Variation in the Capillary

The original model of Krogh did not consider oxygen concentration variation along the length of the capillary. This can be included using a mesoscopic model. The balance is analogous to that in the lung:

$$\text{In} - \text{out} - \text{transferred to the tissues} = 0$$

The differential equation for this is then

$$\pi R_C^2 \frac{dC_A}{dz} = -\pi (R_0^2 - R_C^2) k_0 \eta \tag{23.11}$$

where C_A is the oxygen concentration in the capillary and η is the effectiveness factor for the tissue. η is equal to 1 as long as the Thiele parameter is less than the critical value at all points along the length of the capillary.

A rather interesting consequence of zero-order behavior is observed here. The concentration C_A decreases linearly along a length until a critical length, L^*, is reached. This critical length corresponds to a position along the capillary where the outer corner of the tissue becomes anoxic. that, the length at which ϕ_0 becomes equal to ϕ_0 critical given by Equation 23.10. The effectiveness factor in Equation 23.11 is taken as one until this point.

For further increase in length, an anoxic core (called the lethal corner) develops near the outer radius of the tissues (see Figure 23.5). The behavior in this region is more like a half-order reaction with respect to oxygen concentration. The effectiveness factor has to be calculated as the region of the total tissue volume minus the anoxic region volume for $L > L^*$ and is used in the integration of Equation 23.11 for lengths greater than L^*. These complexities are not addressed here but you should be in a position to model these following the zero-order reaction discussions in Chapter 18.

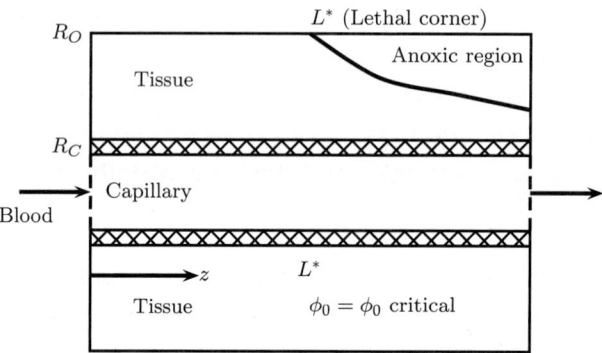

Figure 23.5 Meso-model for oxygen transport to tissue showing some complexities associated with a zero-order uptake.

If one ignores this complexity and assumes a zero-order metabolism throughout the tissue, then the following expression is obtained for the length at which the oxygen concentration becomes zero:

$$L = \frac{C_{Ai} R_C^2 v}{k_0 (R_0^2 - R_C^2)}$$ (23.12)

This provides an approximate estimate of the length at which the oxygen concentration drops to zero.

We now switch gears to study models for drug metabolism in the body.

23.3   Compartmental Models for Pharmacokinetics

Pharmacokinetics is a study of the rate of change of concentration of compounds administered externally with time in various regions of the body, for example orally, intravenously, and so on. Compartmental models provide the basic mathematical platform to interpret experimental data of drug metabolism, toxicity levels, and other useful indicators of drug efficiency. We have already discussed an example in Section 12.2, which should be revisited at this point. The goal of this section is to continue the discussion and provide an example of an improved compartmental model.

23.3.1 Basic Framework

In the basic framework of compartmental modeling, the body is modeled as a network of compartments connected in a prescribed manner. Compartmental models can be developed at various levels of complexity. The simplest is the one-compartment model, which provides an overall lumped approximation to the system. Often with limited data only one compartmental model can be fitted, although the predictions may not be very accurate.

The second level of approximation is the two-compartment model shown in Figure 1.15 and simulated by a worked problem in Example 12.2. It can be shown that the one-compartment model can be obtained from this model by adding the two equations in the two-compartment model and then assuming that the exchange coefficient is rather large. Drugs that are lipid soluble often exhibit such behavior. These are often referred to as flow limited or perfusion limited since the drug concentration is determined by the flow at which it is injected and the metabolic rate for drugs that are lipid insoluble the exchange is often limited by the permeability of the capillary membrane, which is referred to as diffusion-limited behavior. It is useful to know these two limiting cases.

The third level of modeling is the physiologically based compartments model, which is discussed at the end of this section.

The compartments are assumed to be well mixed. The assumption that the blood compartment is well mixed may be puzzling to some readers. One has to compare the blood circulation time to the half-life time of a drug in the body. The cardiac output rate is about 5 L/min and the blood volume is 4 L and hence the circulation time is about 48 sec. A typical time for a drug to metabolize is on the order of hours and assuming that the blood compartment is well mixed is therefore a reasonable assumption.

Rate Model

The metabolism is often modeled by a first-order process, although the following Michaelis-Menten kinetic model is more realistic:

$$-R_A = \frac{kE_O C_A}{K_M + C_A}$$

Here k is the rate constant, K_M is the Michaelis constant, and E_0 is the total concentration of the enzyme. E_0 is total enzyme concentration and C_A is the drug concentration.

The Michaelis-Menten kinetics reduces to a first-order equation for low concentration of the drug, that is, when $C_A << K_M$. In that case

$$-R_A = \frac{kE_O}{K_M}C_A = k_1 C_A$$

This explains the use of the first-order kinetic model in many studies. The first-order kinetics also permits the use of powerful matrix-based solution methods for the system of differential equations. Hence a larger number of compartments can be used in the model without adding significantly to the computational efforts.

23.3.2 Physiologically Based Compartments

In the simpler version of the compartmental models, the compartments do not correspond to a specific organ or part of the body. These are simply black boxes to demark the concentration variations in different parts of the body. The main utility of such models is to model and examine the concentration variation data on the plasma as a function of time. An improvement was suggested by Bischoff and Dendrik, (1980) where each compartment represented a specific organ. These models are called physiologically based pharmocokinetic (PBPK) models and hold a prominent place in this field. Hence a brief discussion is provided here in order to familiarize you with this approach.

Number of compartments to be used depends on the organs where the drug accumulation and metabolism is significant. In other words, organs where the drug concentration is expected to be low are omitted in the model in order to keep the model manageable. Further improvement is that each organ is divided into three subcompartments as illustrated in Figure 23.6.

The subcompartments used are the vascular region, which is the plasma perfusing the organ, the interstitial space, and the cellular space. Mass exchange is assumed between the vacular and the interstitial space as well as between the interstitial and the cellular space. A reaction is assumed to take place in all the three subcompartments. Thus three differential equations are set up for each organ included in the model. Further details are not presented here. An illustrative study on methotrexate, a cancer drug, has been presented by Bischoff et al. (1971) and this topic is also discussed by Truskey et al. (2004).

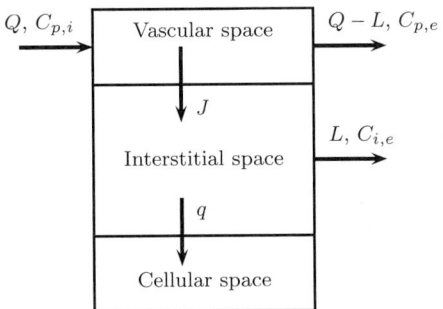

Figure 23.6 Three subcompartments used for a typical organ in the PBPK model. Q = plasma flow, L = rate of lymph flow uptake. J and q are exchange parameters.

23.4 Model for a Hemodialyzer

A hemodialyzer is an extracorporeal unit used for purifying the blood of a person whose kidneys are not working normally. The principle of hemodialysis is the same as other methods of dialysis or membrane separation. It involves diffusion of solutes across a semipermeable membrane. Hemodialysis utilizes countercurrent flow, where the dialysate is flowing in the opposite direction to blood flow in the extracorporeal circuit and hence it is essentially a countercurrent mass exchanger. The models shown in Section 4.3 can therefore be used to calculate the extent of solute removal that can be achieved in this unit. This can be linked to a compartmental model to ascertain the level of purification being achieved and to adjust the dialyzer operating conditions. An example of modeling the dialyzer and a model linking this to a patient model are illustrated in this section, adapted from Ramachandran and Mashelkar (1980).

The schematic of the dialyzer is shown in Figure 23.7 where the blood stream is flowing countercurrent to a dialysate stream. A hollow fiber arrangement is shown in Figure 23.7.

A mesoscopic model is illustrated here. Since the flow in the blood side is usually under laminar considerations, dispersion effects are added in the model. The dialysate side is modeled using plug flow. Model equations based on these flow pattern assumptions are shown next.

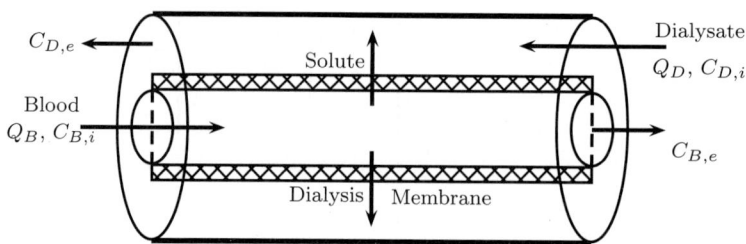

Figure 23.7   Schematic of a hemodialyzer modeled as a counter-current mass exchanger.

23.4.1 Model Formulation

The mass balance on the blood side follows the axial dispersion model and the following equation holds for the concentration of the solute in the blood stream. (Note that all concentrations are implied to be the cup-mixed average concentration and no special tag is shown in the notation.)

$$D_E \frac{d^2 C_B}{dx^2} - v_B \frac{dC_B}{dx} = \frac{s+1}{R} \dot{M}_{bd} \tag{23.13}$$

where $\dot{M}_{bd}$ is the transfer rate of the solute from the blood side to the dialysate side per unit perimeter and v_B is the blood flow velocity. s is a shape parameter equal to zero for channel flow with R being half the width of the channel, while s is equal to one for a hollow fiber circular channel with R being the radius of the fiber.

The mass balance on the dialysate side is based on the plug flow and leads to

$$v_B \frac{Q_D}{Q_B} \frac{dC_D}{dx} = -\frac{s+1}{R} \dot{M}_{bd} \tag{23.14}$$

where Q_D and Q_B are the volumetric flow rate of the dialysate and blood, respectively. C_D is the concentration of the solute in the dialysate side.

The transfer rate is modeled using resistances in series. The transfer involves three steps in series with each step offering its own resistance:

1. Transfer from the bulk of the blood stream to the inner surface of the membrane. This can be modeled using a mass transfer coefficient k_L with $1/k_L$ being the resistance.
2. Transfer across the membrane to the dialysate side. This can be modeled using a membrane permeance P_M with $1/P_M$ being the resistance.

3. Transfer from the walls of the membrane on the dialysate side to the bulk dialysate stream. This can be modeled using a mass transfer coefficient k_D with $1/k_D$ being the resistance.

The overall resistance $\sum \mathcal{R}$ is the sum of these three resistances:

$$\sum \mathcal{R} = \frac{1}{k_L} + \frac{1}{P_M} + \frac{1}{k_D}$$

Hence the transfer rate is given by

$$\dot{M}_{bd} = \frac{(C_B - C_D)}{\sum \mathcal{R}}$$

This completes the model formulation. The Danckwerts boundary conditions are used for the blood side since the dispersion model is used here while the inlet concentration of the solute is used at the inlet side of the dialysate. The model can be solved analytically after suitably combining Equations 23.13 and 23.14. Some details of how to proceed are indicated in exercise problem 23.13 and also in Ramachandran and Mashelkar (1980). The theoretical predictions for the parallel plate hemodialyzer were checked with the experimental data of Grimsrud and Babb (1966) and the data of Ramirez et al. (1971) and a close fit was observed. The sensitivity of model parameters were also investigated. In particular the diffusion coefficient of urea in blood is shear rate dependent (Hyman, 1975) due to the non-Newtonian nature of the blood. Hence sensitivity studies of model parameters are important in general. In this particular case, the sensitivity to variations in diffusivity of the solute was found to be less than 20%.

23.4.2 Model for Patient-Dialyzer System

The model was then coupled to simulate an artificial kidney system. A two-compartment model was used for the patient with part of the blood stream flowing through the dialyzer and recycled back to the patient. The schematic of this model is shown in Figure 23.8.

The mesoscopic model for the dialyzer shown earlier provides the clearance rate of the plasma compartment, which is equal to $Q_B(C_{B,i} - C_{B,e})$. The sensitivity to the exchange parameter was also demonstrated. Once calibrated, the model is useful to formulate a treatment protocol to control the concentration of urea in both the blood stream and in the tissue compartment.

One interesting factor was that the rate of removal has to be restricted so that significant differences in concentration in the two compartments do not

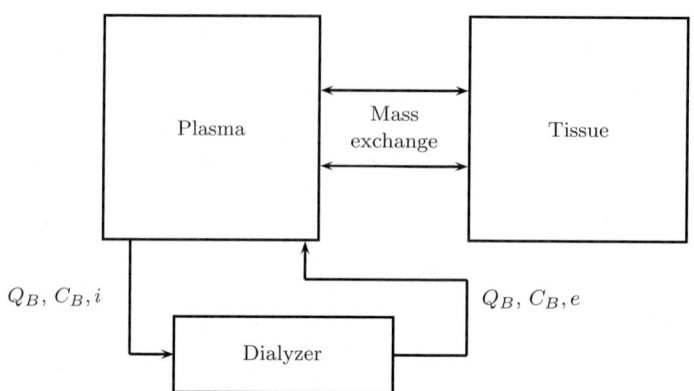

Figure 23.8   Simple two-compartment model for patient linked to a dialysis unit.

develop. This can be achieved by decreasing the dialysate flow rate, increasing the solute concentration in the dialysate, or increasing the blood flow. Detailed results are not presented in the interest of brevity; please refer to Ramachandran and Mashelkar (1980) for the various parametric plots.

Summary

- Models for many systems of biomedical importance can be built using the methodology shown in Part I of the book. Examples provided in this chapter should give you a good feel for development of such models and provide a useful collection of case study examples.

- Oxygen transport in lungs is an important problem and also a good example of mass transfer modeling. The oxygen-hemoglobin equilibrium follows an S-shaped curve that is modeled by Hill's equation. Oxygen binds cooperatively to multiple sites in hemoglobin; this is responsible for the S-shaped nature of the equilibrium curve.

- The transport rate of oxygen in lungs can be calculated using a diffusion-reaction model to account for the reaction with hemoglobin. Transport through the capillary endothelium and convective mass transport in plasma are the two additional steps that are included in the model. These can be treated using an overall external mass transfer coefficient, which leads to a Biot number for external mass transfer. The internal transport and reaction in the red blood cells can be modeled using a Thiele modulus. This permits the calculation of an overall effectiveness factor. Hence the modeling approach is quite parallel to that for gas–solid catalytic reactions.

- The local rate of oxygen transport can then be coupled to a mesoscopic model for the capillary to find the oxygen profiles as a function of the distance along the capillary and the distance needed to achieve a nearly complete saturation.

- The transport rate of oxygen in tissue is another important problem and can be modeled by assuming the tissue is an annular cylinder covering the capillary. A zero-order reaction is normally assumed. We find that there is a critical Thiele modulus above which an oxygen starved region can develop at the outer radius of the tissue.

- The local model for tissue consumption of oxygen can then be coupled with the mesoscopic model for the capillary to find the oxygen profile in the tissue and the length at which the oxygen concentration becomes nearly zero. For regions of low concentration of oxygen in the capillary, a lethal corner (an anoxic region) can develop at the outer radius of the annular cylinder representation of the tissue.

- Distribution and metabolism of drugs in the body is the focus of pharmacokinetic analysis and compartmental models of various levels of detail are commonly used, the simplest being a one-compartment model. A two-compartment model consisting of a plasma and a tissue compartment is quite useful and simple. In further refinement, each compartment can be composed of three subcompartments and hence the models can be made fairly comprehensive and organ specific.

- A dialysis unit is an example of a extracorporeal device and can be modeled as a countercurrent mass exchanger. Analytical solutions can be obtained for the exit concentration of the solute if a mesoscopic approach is taken and the clearance rate calculated from such a model can be incorporated in a two-compartment model for the patient to track the concentrations with time. The two-compartment model provides an estimate of the concentration in the tissue compartment, which is difficult to measure. Hence the model is useful to control both the tissue and the plasma concentration during the progress of the treatment.

Review Questions

23.1 What is Hill's equation?

23.2 How does the carbon dioxide concentration in Hb affect oxygen dissociation?

23.3 What causes carbon monoxide poisoning?

23.4 What are the limiting transport steps in oxygen absorption in the lungs?

23.5 Explain how the effectiveness factor concept is useful to calculate the local rate of absorption of oxygen in lungs.

23.6 Define the characteristic distance for oxygen uptake and its meaning.

23.7 What is the physical picture upon which the Krogh cylinder model for oxygen transport in tissue is based?

23.8 When can an anoxic region develop in the tissue?

23.9 What is the lethal corner and what is the cause of it?

23.10 Justify the assumption that the blood (plasma) compartment is well mixed although it is in circulation inside blood vessels.

23.11 For pharmacokinetic modeling, when is a one-compartment model suitable and when it is not?

23.12 What is the PBPK model?

23.13 Show the similarity in the hemodialyzer modeling and modeling of a conventional chemical engineering separation process.

23.14 Discuss briefly how the patient-dialyzer linked model is useful in clinical applications.

Problems

23.1 **Thermodynamic derivaton of Hill's equation.** Hill's equation is based on oxygen binding to n sites simultaneously, which can be represented by the following reaction scheme:

$$L + nA \rightleftharpoons AL_n$$

The notation A is used for oxygen and L is used for Hb (the ligand). Apply the law of mass action to this scheme and show that

$$[AL_n] = K[A]^n[L]$$

where the square bracket represents concentration. The total balance for the ligand requires $n[AL_n] + [L] = [L_T]$ with L_T being the total ligand concentration. Combine the two equations and shown that Hill's equation results for oxygen saturation.

23.2 **Total oxygen concentration in blood.** Hematocrit refers to the volume fraction of RBCs in blood and is on the order of 0.45. Find the oxygen concentration in plasma using Henry's law. Find the concentration of oxygen bound to RBCs using Hill's equation. Find the total concentration of oxygen if the atmospheric pressure is 760 mm Hg at a sea level locations and 670 mm at a higher elevation of 1 mile.

The solubility coefficient of oxygen in plasma is equal to 1.125×10^{-11} mole/cm³Pa. Note that the oxygen mole fraction in the lung is less than 0.21 since the lungs are saturated with water vapor at 37°C. Please correct for this as well.

23.3 **Oxygen saturation in anemia patients.** People with anemia have lower Hb concentration. Determine the level of oxygen saturation if the Hb level is 0.1 g/mol compared to the normal level of 0.14.

23.4 **Overall effectiveness factor for oxygen uptake.** Calculate the effectiveness factor for the oxygen uptake in RBCs based on the following parametric values suggested by Truskey et al. (2004):
   – Permeability of the capillary walls = 0.387 cm/s
   – Mass transfer coefficient in plasma = = 0.44 cm/s
   – Effective length of RBC = 1.35 μm
   – Diffusion coefficient of oxygen in RBC = 6.0×10^{-6} cm²/s
   – Rate constant for first-order reaction ≈ 1000 s^{-1}.

23.5 **Length needed for saturation.** Calculate the characteristic parameter and the length needed for a capillary to be saturated with oxygen for the following conditions: blood flow velocity = 2.5 mm/sec; hematocrit value = 0.45. For other parameters use the values from the previous problem.

23.6 **Application of Krogh model.** Apply the Krogh model to the following data (from Truskey et al., 2004) to find the maximum intercapillary radius for no anoxic region formation. Metabolic rate = 1×10^{-7} mol/cm³s, a zero-order reaction; capillary radius = 1.5 to 4 μm; $D_{O_2} = 2 \times 10^{-5}$ cm²/s in tissue; oxygen concentration in blood at the given point: 4.05×10^{-8} mol/cm³.

23.7 **Oxygen variation with length in a capillary near tissue.** Solve Equation 23.11 for C_A versus z for the data in the previous example. The velocity of blood is 2 mm/s. The inlet concentration of oxygen is 19.88×10^{-3} M. Consider two cases. First, assume no anoxic region develops in the tissue. Find the length at which the oxygen concentration becomes zero. Second, assume an anoxic region forms and find the length at which the lethal corner can develop.

23.8 **Model for NO distribution in blood and tissues.** NO is a vasodilator and is produced in the endothelium of the blood vessel and the tissue epithelelium. NO then diffuses into the capillary as well as into the tissues, and also undergoes a reaction. A schematic of the process is illustrated in Figure 23.9.

   The concentration profile of NO is important to assess the distance over which NO acts in the tissue. The NO concentration should remain above a critical value for it to be effective. The Krogh cylinder model is useful here. Assume the reaction is first order in NO. Show that the following equation is then applicable:

$$\frac{D_e}{r}\left[\frac{d}{dr}\left(r\frac{dC_A}{dr}\right)\right] = k_1 C_A \qquad (23.15)$$

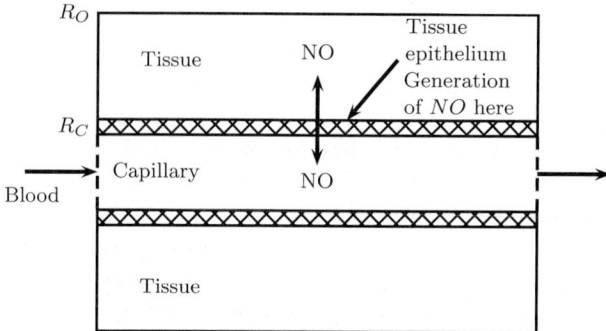

Figure 23.9   Schematic of a model for NO generation
and diffusion+reaction in both tissue and the capillary.

Show that this applies both to the capillary (0 to R_C) and the tissue (R_C to R_0).
However, the equation needs to be solved separately for the two regions since
the reaction rate is not the same in the two regions. The boundary condition at
$r = R_C$ also has to be matched for the two regions to get the complete solution.
The boundary conditions at $r = 0$, the center of the capillary, and $r = R_0$ the
outer edge of the tissue, are the no flux conditions. At $r = R_C$ the concentration is continuous. Note that the thickness of the endothelium is ignored. The
generation in the endothelium $\dot{q}_{NO}$ provides the second matching condition at
$r = R_C$. Show that the following condition holds at $r = R_C$:

$$\dot{q}_{NO} = D_{NO,B}\frac{dC_{A,B}}{dr} - D_{NO,T}\frac{dC_{A,T}}{dr}$$

Simulate and plot the NO profile in the blood and capillary for the following set
of parametric values taken from Truskey et al. (2004):

– Production rate = 5.5×10^{-12} mol/cm²s
– Rate constant for reaction in blood = 1000 s⁻¹
– Rate constant for reaction in tissue = 0.01 s⁻¹
– Diffusion coefficient in blood = 4.5×10^{-5} cm²/s
– Diffusion coefficient in tissue = 3.3×10^{-5} cm²/s
– Radius of the capillary - 25 μm
– Outer radius of the tissues = 100 μm

23.9   **Nutrient limitation in tumor growth.** Tumors pull nutrients needed for their
growth from the surrounding tissues. If the tumor size becomes large, a region
can develop inside the tissue where the nutrient concentration drops below a
critical value. The growth is then limited by lack of nutrients. An illustrative
concentration profile is shown in Figure 23.10.

Solve the model for a zero-order reaction in spherical coordinates. Develop an
expression for the radius of the tumor at which the center concentration drops
below a critical value. Note that the size of the tumor will be limited by this

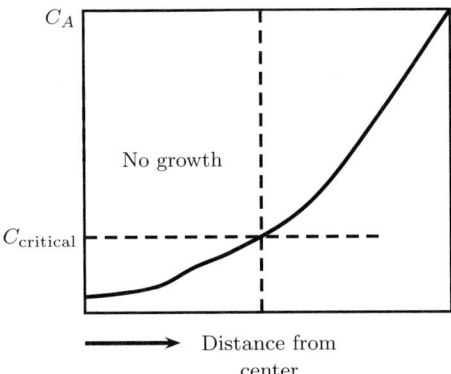

Figure 23.10 Concentration profile for nutrient diffusion with a zero-order reaction in a tumor showing the no growth region within the tumor.

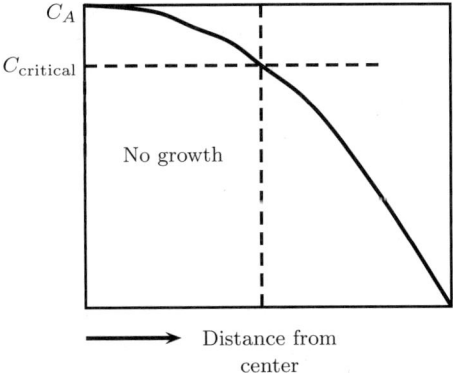

Figure 23.11 Concentration profile for inhibitor-limited growth in a tumor. Tumor growth is confined to the outer shell.

value because nutrient starvation occurs above this value. The model applies to the avascular stage of growth, that is, when the tumor has not developed its own blood vessels.

23.10   **Tumor growth: inhibitor limited model.** Tissues generate inhibitors, which are needed to control growth and other processes. Tumors also do this. Develop a diffusion-reaction model for a **spherical** shaped tumor that generates an inhibitor by a zero-order reaction and also destroys it by a first-order reaction. Solve for the concentration profile. An illustrative profile of the type shown in Figure 23.11 will be obtained.

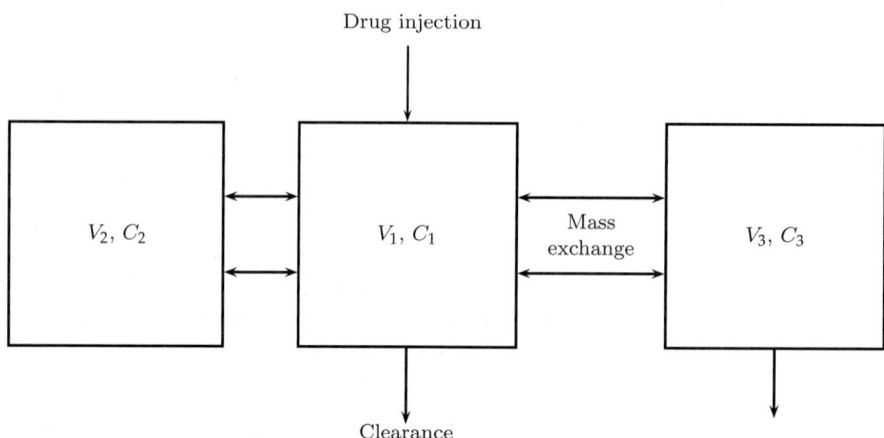

Figure 23.12   A three-compartment model. The drug is injected and cleared from the central compartment (1), which exchanges mass with two peripheral compartments (2 and 3).

The boundary condition at the edge of the tumor can be a Robin or Dirichlet type condition. The Dirichlet condition with zero concentration applies if the permeability of the tumor membrane is rather large. Set up and solve the model for both cases. The tumor growth will stop in regions where the concentration of the inhibitor is above a critical concentration (see Figure 23.11). Solve and plot the profiles for concentration for various radii and mark regions where the tumor ceases to grow, assuming a value for the critical concentration.

23.11   **Solution for a three-compartment model.** The plasma concentration of small molecules is best fitted by a three-compartment model. Truskey et al. (2004) suggests a central compartment where a drug is injected and cleared and two peripheral compartments as shown in Figure 23.12.

Set up the model and solve for the central compartment concentration as a function of time using the matrix method. Show that the concentration in the plasma (central compartment) can be represented as a triexponential solution of the following form:

$$C_p = C_{p0} \left[\alpha_1 \exp(-\lambda_1 t) + \alpha_2 \exp(-\lambda_2 t) + (1 - \alpha_1 - \alpha_2) \exp(-\lambda_3 t) \right]$$

Explain the relation of the constants α s and λ s to the volume parameters, clearance rate, and exchange rates.

23.12   **Analytical solution to the dialyzer model.** Equations 23.13 and 23.14 can be combined and the transfer rate term can be eliminated. The resulting equation can be integrated to get a relation between the dialysate concentration and the blood concentration. Show that the following equation results if this procedure is applied:

$$C_D = C_{D,i} + \frac{Q_B}{Q_D}(C_B - C_{B,e}) - \frac{Q_B}{Q_D}\frac{D_E}{v_B}\frac{dC_B}{dx}$$

Now use this in Equation 23.13 to get a differential equation for C_B. This equation is very similar to the axial dispersion model; the only difference is that the coefficients are modified to include Q_B/Q_D ratio terms. The solution has the same form as the axial dispersion model and provides an analytical solution to the dialyzer model. Set up this equation and solve for the exit concentration of the solute in the blood stream.

CHAPTER 24

Electrochemical Reaction Engineering

Learning Objectives

After completing this chapter, you will be able to:

- Understand basic definitions used in electrochemical engineering and their relation to equivalent concepts in chemical reaction engineering.
- Evaluate the minimum voltage requirements from thermodynamic considerations of the reactions taking place at electrodes.
- Review two common kinetic models for electrochemical reactions: the Butler-Volmer model and the Tafel model.
- Show how mass transfer affects the rate and understand the notion of the limiting current.
- Set up a model and design a simple electrochemical process, copper electrowinning from a leach solution.
- Show how the current versus voltage relation in a fuel cell is related to mass transfer and the reaction taking place in the system.
- Show the various transport steps in a Li-ion battery and their importance in modeling these systems.

This chapter deals with electrochemical reaction engineering, an often neglected topic in undergraduate chemical engineering and textbooks. This is a classical as well as emerging area and has important applications in energy engineering and chemical industries. Sodium hydroxide was probably the first chemical to be made on a large industrial scale by electrolysis of sodium chloride solution and hence electrochemical engineering can be considered to be older than chemical engineering. The interest in electrochemical routes to make chemicals on an industrial scale continues and electro-organic synthesis is emerging as a growth area, especially in the pharmaceutical industry. In the modern context of energy production and storage, electrochemical engineering has become a major design tool. Hence it is important for chemical engineers to understand the basic concepts of these systems and to learn how one goes about the analysis of them. Specialized books are available but a primer such as given in this chapter is very useful to get started and to develop basic knowledge of modeling of electrochemical reactors.

The chapter is organized in the following manner. First we show the basic definitions needed to understand the system. Then the thermodynamics of electrode reactions and the use of oxidation and reduction potentials is discussed, followed by models for kinetics of electrode processes. We then integrate the kinetics and transport effects illustrated in Chapter 16 (Section 16.5 in particular) and show how they can be coupled with basic conservation laws to develop models for electrochemical systems. Simple applications of the kinetics and mass transfer are then presented for three industrially important systems. Overall the study of this chapter will provide a basic understanding of electrochemical reaction engineering and you will be in a position to understand specialized textbooks and papers in this area.

24.1   Basic Definitions

Electrochemical reaction engineering introduces several new concepts and terminologies not so familiar to chemical engineers and there is a need to define these and demonstrate how they relate to classical reaction engineering terms. This permits the translation of the vocabulary used by electrochemical engineers to chemical engineers.

A typical electrochemical reactor or cell consists of an electrolyte with two electrodes on either side forming an electrochemical cell: an anode and a cathode where an oxidation and a reduction reaction takes place, respectively. The electrons generated or consumed in these reactions flow through an external circuit. The ions move within the electrochemical cell across the electrodes. Hence a current flows through the system. The current flow is taken in the opposite direction to the electron flow. A voltage is applied to maintain the current flow in an electrolyzer. A voltage will be generated if the reaction is spontaneous in which case the system acts as a fuel cell.

24.1.1 Anodic and Cathodic Reactions

Reactions are generally represented as oxidation and reduction. Oxidation is the release of electrons. Electrons are the product and appear on the right-hand side of the reaction scheme. This reaction is also referred to as an anodic reaction. The electrode where an oxidation reaction is taking place is called the anode.

Reduction is the reaction where a electrons are the reactant. The electrode where a reduction reaction is taking place is called the cathode.

A schematic diagram illustrative of the processes taking place at the electrode is shown in Figure 24.1. Here a single-step electron transfer process is shown but often multiple electron transfers will be involved in many practical situations. R is the species undergoing oxidation to produce O.

24.1.2 Half Reactions and Overall Reaction

Two halves make a whole. In electrochemical reactors two half reactions (anodic and cathodic) have to occur so that a current flow can occur in the external circuit across the cell. These reactions should add up to an overall reaction.

One of the half reactions may even be a parasitic reaction leading to no useful products, but is a necessary part of the overall process. In some cases both reactions can lead to useful products; such reactions are known as paired synthesis. Here both the anodic and cathodic processes lead to useful products. Industrial applications of paired synthesis are not so common; this is often called the electrochemist's dream. In inorganic chemical industry, chlor-alkali is example where both the anodic and cathodic products lead to useful products. An illustrative sketch of this process is shown in Figure 24.2. The two half reactions are also shown and the overall reaction is the sum of the two:

$$NaCl + H_2O \rightarrow NaOH + \frac{1}{2}Cl_2 + \frac{1}{2}H_2$$

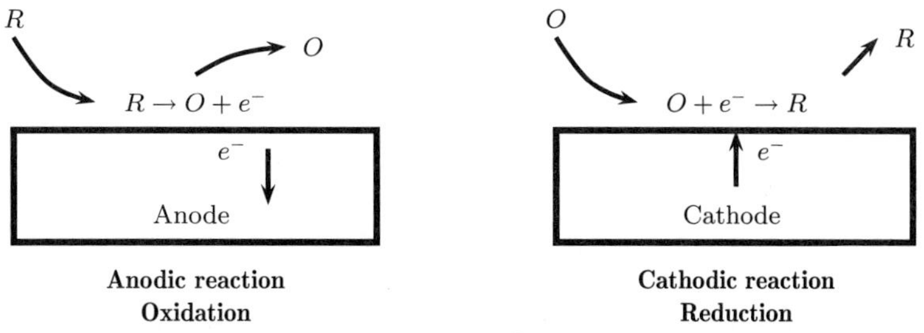

Figure 24.1 Illustration of a simple electrode reaction taking place at an anode and a cathode. A single electron transfer reaction is shown.

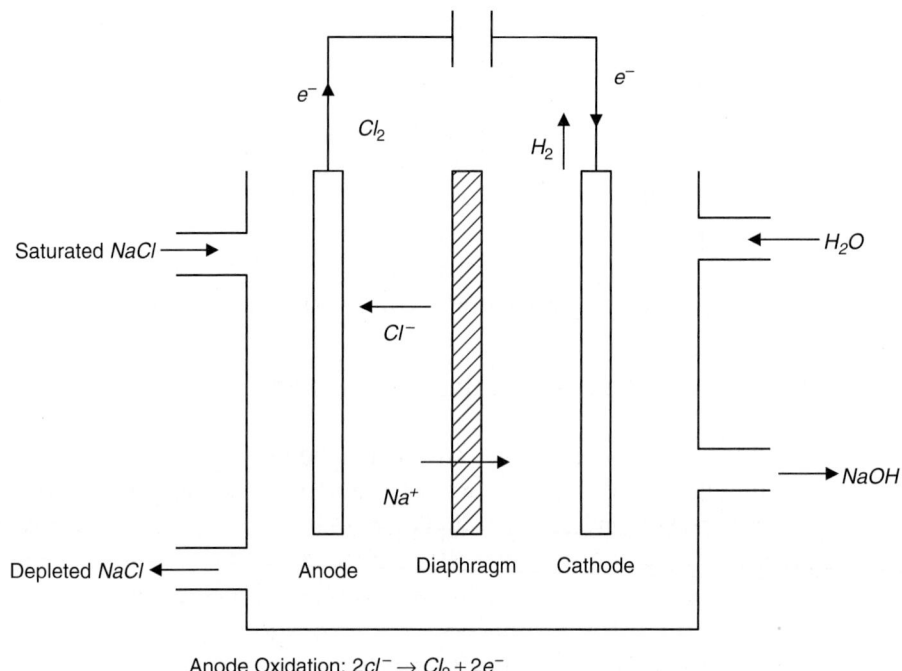

Anode Oxidation: $2cl^- \rightarrow Cl_2 + 2e^-$

Cathode: Reduction: $2H_2O + 2e^- \rightarrow H_2 + 2\,OH^-$

Figure 24.2   An illustrative electrochemical reaction: NaCl electrolysis. Note that a diaphragm or membrane is used to separate the anodic and cathodic regions.

This is the sum of the two half reactions shown in Figure 24.2. The overall reaction has a positive change in free energy and is therefore not spontaneous. Hence a voltage has to be applied in order to drive the reaction from the left to right. How much voltage is needed depends on the rate of reaction (or equivalently the current in the system) and is addressed by a voltage balance shown later.

A second example of half reactions and the overall reaction is the fuel cell shown in Figure 24.3. Here hydrogen enters the anode side and air or oxygen flows on the cathode side and the overall reaction is

$$H_2 + \frac{1}{2}O_2 \rightarrow H_2O$$

This is split into two half reactions, which are shown in Figure 24.3. The electrons flow from the anode to the cathode in the external circuit. Positive ions (H^+) are generated at the anode due to the oxidation of the H_2 gas and are

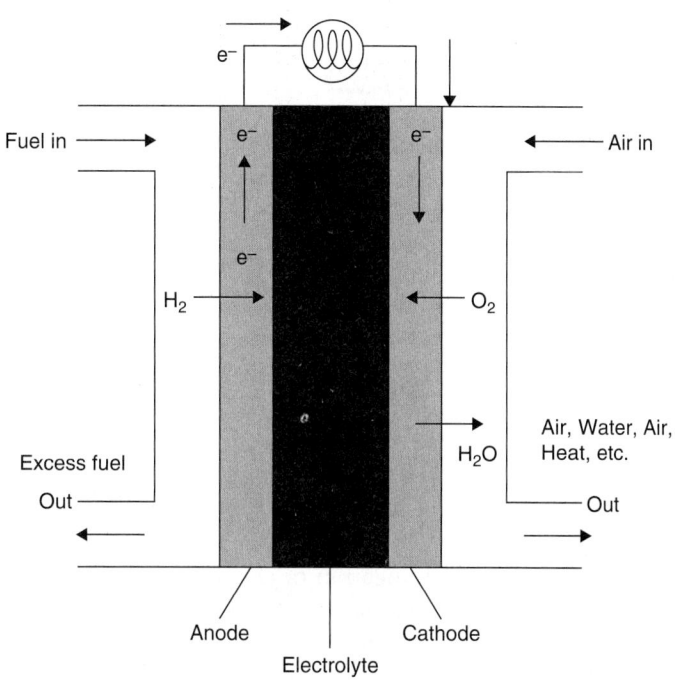

Anode oxidation: $H_2 \rightarrow 2H^+ \ 2e^-$
Cathode: Reduction: : $\frac{1}{2} O_2 + 2H^+ + 2e^- \rightarrow H_2O$

Figure 24.3   Half reactions and current and ion flow in a fuel cell.

transported from from anode to cathode across a proton exchange membrane (PEM). This membrane separates the anode and the cathode and allows passage of only hydrogen ions to the cathode. The hydrogen ion then participates in the reaction with oxygen at the cathode to form water.

The free energy change associated with the overall reaction in a hydrogen fuel cell is negative and hence the reaction is spontaneous. It therefore generates a voltage. The value of the voltage depends on how much current is being drawn from the system and is addressed later.

24.1.3 Classification of Electrode Reactions

Depending on the nature of the product formed in the electrode reaction, the process can be broadly classified into four types, which are shown schematically in Figure 24.4. These are shown for a process taking place at

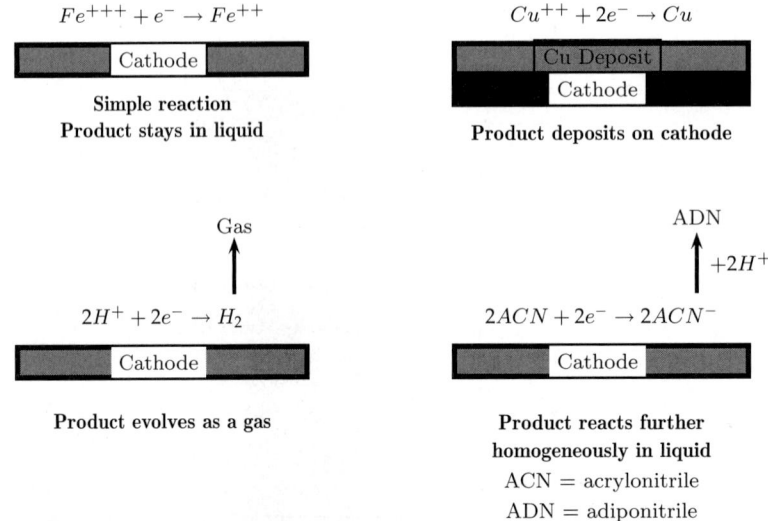

Figure 24.4 Classification of electrode reaction depending of the nature of the product formed by electrochemical reaction.

a cathode, that is, where a reduction is taking place. A similar type of classification applies to an anodic process as well. The reduced product in a cathodic reaction can belong to four categories, as shown in Figure 24.4.

- The simplest case is where both the reactant and product stay in the liquid phase. A common example is the ferrous from ferri reduction. This is often used as a test reaction to measure mass transfer coefficients from the bulk solution to the electrode.
- The second case is where the product is solid that deposits on the electrode; this has important applications in metal refining industries.
- The third example is where the product is a gas. The gas evolution has important consequences as it leads to a two-phase mixture in the system, leading to reduced electrical conductivity and other associated design issues. It can also provide the benefit of increased mass transfer rate at the electrode due to convective mixing caused by the gas bubbles.
- Finally the last case is where the electrode supplies enough energy to provide a radical, which reacts further with a liquid reactant (near the electrode) to form the useful product. Such examples are found in electro-organic synthesis.

Each case needs some adjustments and modifications in the engineering of the process and the associated reactor design. Further details are not addressed here.

24.1.4 Primary Variables

Two primary variables are needed in order to analyze an electrochemical process: the current and the potential or the applied voltage, as already discussed earlier. These are reiterated in the following.

Current: This is a measure of the rate of flow of electrons in the external circuit and the flow of ions within the electrolyte. The units are Coloumb/sec, abbreviated as ampere, A. The current density is the flow per unit area of the electrode; the units are A/m^2. It is important to note that the current is a measure of the overall reaction rate at the electrode surface. The following relation holds for a single reaction:

$$i = nFN_{As} \tag{24.1}$$

where N_{As} is the rate of surface reaction (electrode reaction) per unit area, i is the current density, and n is the number of electrons transferred in the electrode reaction. F, the Faraday constant, is the conversion factor from number of electrons to moles. It represents the amount of electric charge carried by one mole of electrons; the value is 9.65×10^4 C/mol of electrons.

Voltage: This is a measure of the potential energy associated with a unit charge. If a positive charge of 1 C has been moved by one unit of length of 1 m it acquires a potential energy of 1 V. Hence 1 V is equal to 1 J/C. An electrochemical reaction with an almost zero current needs (or produces if it is a fuel cell) a minimum (maximum) voltage, which is dictated by thermodynamic considerations. The calculation of this minimum voltage is addressed in the following section.

Calculation of the current versus voltage relation require consideration of transport effects and kinetic models for the electrode reactions. This is addressed in Section 24.5.

24.2 Thermodynamic Considerations: Nernst Equation

Consider an electrode in contact with a solution with the following reaction taking place on the electrode surface:

$$R \longleftrightarrow O^{n+} + ne^-$$

The chemical potential difference drives the reaction.

Note that at least one of the species R or O has to be a charged species or both of them can be charged species to balance the negative charge due to electrons. The point to note is that the net charge on the left-hand side must balance the net charge on the right-hand side. An example, where O is a charged species and R is not, is

$$Ag \longleftrightarrow Ag^+ + e^-$$

An example where both are charged is

$$Fe^{++} \longleftrightarrow Fe^{+++} + e^-$$

which is the ferri–ferrous reaction often used as a test model reaction in electrochemistry.

The equilibrium potential (denoted as E_{eq}) is defined as the potential difference between the electrode ϕ_m and the solution ϕ_s under conditions of no reaction or equivalently no net current. It should be noted that the ϕ's do not have an absolute value and are determined in accordance with a reference electrode whose potential, is assigned a zero value. It is similar to the gravitational potential, which uses a reference elevation whose potential is set to zero. E_{eq}, being a difference of two potentials, has an absolute value. An equation for this will now be derived.

The forward reaction and the backward reaction take place at the same rate under the condition of equilibrium. Hence the sum of the chemical potential of all species on the left-hand side must balance the sum of the chemical potential of the species on the right-hand side. An equation for E_{eq} can be derived based on this concept, which is the famous Nernst equation in electrochemistry.

The net driving force in the forward direction, that is, toward the oxidation reaction or the anodic reaction, is taken as the difference in the chemical potential between the reactants and the products:

$$\text{Driving force} = \mu_R - (\mu_O + n\mu_e)$$

where μ_e is the chemical potential of the electrons at the metal surface. The chemical potential in turn is related to the concentration of the species. Thus for species R the chemical potential is represented as

$$\mu_R = \mu_R^0 + R_g T \ln C_R \quad \text{(Molar unit)}$$

where R_g is the gas constant, μ_R^0 is the chemical potential in the standard state, and C_R is the concentration expressed as mol/L. This presumes that the standard state is taken as 1 M solution. We assume that R is not charged but the final result will be same if R is also charged (due to the sotchiometric balance of charges on both sides of the equation).

Similarity for species O the chemical potential is

$$\mu_O = \mu_O^0 + R_g T \ln C_O + nF\phi_s$$

where ϕ_s is the electric potential of the solution adjacent to the electrode. Here we have included the electrochemical potential for species O since this is a charged positive species.

The chemical potential of the electrons depends on the potential of the metal or the electrode that is in contact with the solution and can be varied by adjusting the potential of the metal or the electrode ϕ_m. Then the chemical potential of the electrons is equal to

$$\mu_e = \mu_e^0 - F\phi_m$$

Note that both ϕ_s and ϕ_m are relative as mentioned earlier as well. We assume here that these are measured with reference to a standard electrode that has a stable and reproducible potential. Of course the final result is not affected since it is the difference between the metal and the solution potential that will matter.

Net driving force (d.f.) is therefore equal to

$$\text{net d.f.} = (\mu_R^0 + R_g T \ln C_R) - (\mu_O^0 + R_g T \ln C_O + nF\phi_s)$$
$$- n(\mu_e^0 - F\phi_m) \tag{24.2}$$

For no reaction to take place the driving force has to be zero. This represents the condition for equilibrium. The corresponding potential of the metal with respect to the adjacent electrolyte solution is referred to as E_{eq}:

$$E_{eq} = \phi_m - \phi_o$$

Setting the net d.f. as zero in Equation 24.2, we get

$$nF E_{eq} = (\mu_O^0 + n\mu_e^0 - \mu_R^R) + R_g T (\ln(C_O/C_R) \tag{24.3}$$

We define E^0 as

$$E^0 = \frac{(\mu_O^0 + n\mu_e^0 - \mu_R^0)}{nF}$$

This is the free energy change (expressed in electrical potential units) when all the species are at their standard state.

Equation 24.3 is written in terms of E^0 as

$$\boxed{E_{eq} = E^0 + \frac{R_g T}{nF} \ln(C_O/C_R)} \tag{24.4}$$

This is the Nernst equation. If all species are in their standard state then $E_{eq} = E^0$.

The above represents the equilibrium state of the system. No current flows in such cases. Any reaction needs a departure from equilibrium conditions; this is accomplished by applying a potential at the electrode different from the equilibrium value.

Note that there is no unique value to E^0. Hence this is referred to some comparative values. The hydrogen (H_2) to hydrogen ion H^+ reaction is given a value of zero. Then a value can be assigned to other electron transfer reactions. These values, known as oxidation potential, are tabulated in many books. Some typical values are given in Table 24.1.

Note that some books tabulate the reduction potential, that is for the reverse reactions shown in Table 24.1. These values will then be the negative of the values in the table. Hence caution has to be exercised in using these values.

Note that in the table E^0 has the same sign as ΔG (the free energy change under standard conditions). Hence reactions with negative values are favorable to the right (anodic) while those with positive values are cathodic and tend to occur to the left. For example, sodium prefers to stay as Na^+ since the standard potential is minus 2.714. Similarly chlorine prefers to be Cl^- since the potential is plus 1.3595. The ionic bond keeps them together and thus we find only NaCl in nature.

Table 24.1 Oxidation Potential for Some Standard Half Reactions. Reactions at the Top are Thermodynamically Favorable while for those Near the Bottom the Reaction is Favorable in the Reverse Direction

Reaction	Standard Potential
$Na \rightarrow Na^+ + e^-$	−2.714
$Al \rightarrow Al^{+3} + 3e^-$	−1.66
$1/2 H_2 + OH^- \rightarrow H_2O + e^-$	−0.828
$Zn \rightarrow Zn^{+2} + 2e^-$	−0.763
$H_2 \rightarrow 2H^+ + 2e^-$	−0.
$Ag \rightarrow Ag^+ + e^-$	0.7991
$Cu \rightarrow Cu^{+2} + 2e^-$	0.337
$2H_2O \rightarrow O_2 + 4H^+ + 4e^-$	1.229
$2Cl^- \rightarrow Cl_2 + 2e^-$	1.3595

24.2.1 Equilibrium Cell Potential

In an electrochemical reactor two half reactions, one at an anode and one at a cathode takes place. The net potential difference at equilibrium is found from the difference:

$$E_{R,eq}^0 = E_{Eq}^0 \ (anode) + E_{Eq}^0 \ (cathode) \qquad (24.5)$$

If $E_{R,eq}$ is negative, then the reaction is spontaneous. The system can be used as a fuel cell to generate energy. In this case, the magnitude of $E_{R,eq}$ will be the maximum voltage that can be generated in a (single stack) fuel cell for the given reaction. This will apply to a zero current situation.

Example 24.1 shows the application of thermodynamic calculations in electrochemistry.

Example 24.1   Minimum Voltage Calculations

Using the oxidation potential in Table 24.1 find the minimum potential to decompose NaCl in a diaphragm cell. Assume the concentration is 1 M, that is, use standard state values. Repeat for a hydrogen + oxygen reaction in a fuel cell.

Solution

The overall reaction is

$$NaCl + H_2O \rightarrow NaOH + \frac{1}{2}Cl_2 + \frac{1}{2}H_2$$

We need to write this as two half reactions, which are shown in Figure 24.2. From the table we find that the anodic reaction of Cl^- going to Cl_2 has a potential of 1.3595 V. The cathodic reaction of water going to hydrogen and OH^- has a potential of 0.828 V. Note that the value shown in Table 24.1 is if the reaction is anodic but the reaction is cathodic and hence in the reverse direction. We therefore changed the sign on the potential.

The minimum voltage is the sum of anodic and the cathodic potentials and hence

$$E^0 = 1.3595 + 0.828 = 2.1865V$$

This is the minimum potential needed to drive the reaction. Additional voltage needs to be applied to drive the reaction; this will depend on the kinetics of the reaction (as well as the mass transfer steps), which is discussed later. If the NaCl concentration is not 1 M, then a concentration correction is applied using the Nernst equation, Equation 24.4.

For the hydrogen fuel cell the overall reaction is

$$H_2 + \frac{1}{2}O_2 \rightarrow H_2O$$

This needs to be written as two half reactions, one corresponding to the anode reaction or the oxidation and the second corresponding to the cathode reaction, which is reduction.

At the anode hydrogen gas is oxidized to H^+ions; this has a potential of zero, if hydrogen is at the standard condition of 1 atm pressure.

At the cathode the reduction of oxygen is the reaction taking place:

$$\frac{1}{2}O_2 + 2H^+ + 2e^- \rightarrow H_2O \tag{24.6}$$

The potential for this is E cathode = -1.229 V (reverse of the value in Table 24.1 since it is now from right to left). This potential applies if the oxygen partial pressure is 1 atm. Hence $E_0 = 0 - 1.229V$ under standard conditions. As this is negative, the reaction is spontaneous and generates energy.

If air is used, the oxygen partial pressure is 0.21 atm. Now using the Nernst equation (Equation 24.4) $(R_gT/nF)\ln(0.21)$ is added, giving $E_0 = -1.119$ V.

The effect of temperature on the cell potential is indicated here. The equilibrium potential calculated at the standard condition of $T_{ref} = 298$ K is corrected for other temperatures as follows using a linear relation:

$$E_{eq}(T) = E_{eq}(T_{ref}) + (T - T_{ref})\frac{\partial E^0}{\partial T}$$

From thermodynamics, we have

$$\frac{\partial E^0}{\partial T} = \frac{\Delta S^0}{nF}$$

Here ΔS^0 is the standard entropy change for the overall reaction.

24.3   Kinetic Model for Electrochemical Reactions

Now depending on the actual applied potential of the electrode (metal) the reaction can be driven either to the right or to the left. Let us define the difference between the actual applied potential and the equilibrium potential as η, the surface overpotential:

$$\eta = E - E_{eq}$$

If η is greater than zero, the driving force term is positive and hence oxidation is favored. If η is less than zero, the driving force term is negative and hence the

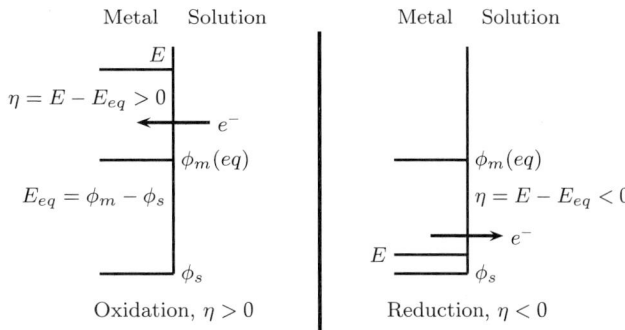

Figure 24.5 Effect of electrode potential on the direction of electron transfer. The left side represents the metal or the electrode while the right side represents the solution adjacent to it.

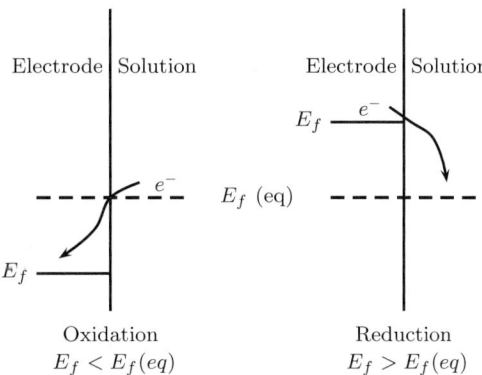

Figure 24.6 Fermi level interpretation of the effect of applied electrode potential on the direction of electron transfer.

reverse reaction (reduction) is favored. A kinetic model should reflect these effects of the applied potential. The model should also be thermodynamically consistent.

The effect of electrode potential is shown schematically in Figure 24.5. The forward reaction (oxidation) is an electron removal process from the solution; the rate increases as the electrode potential becomes more positive, since the electrons are attracted to the positively charged electrode. The reverse (reduction) happens when the electrode potential is reduced; the electrons prefer to leave the electrode as shown in Figure 24.5.

The process can also be interpreted using the Fermi level of the electrons in the electrode; this is illustrated in Figure 24.6. The Fermi level is a measure of the chemical potential of the electrons. At equilibrium conditions the Fermi levels of the electrons in the electrode and in the HOMO (higher occupied molecular orbital) of the solution are the same as shown by the dashed line in Figure 24.6. An increase in electrode potential causes the Fermi level of electrons in the electrode to decrease, causing a jump of electrons from the solution to the electrode. Oxidation takes place. The opposite happens when the electrode potential is reduced.

24.3.1 Butler-Volmer Equation

The rate for oxidation reaction, r_f or r_a (where a stands for anodic), is assumed to be exponentially proportional to βE where β is a factor between 0 and 1 called the symmetry factor:

$$r_f = r_a = k_f C_R \exp(nf\beta E)$$

where f is used as an abbreviation for $F/R_g T$ and has a value of 44.2 V at 298 K. This is similar to $E/R_g T$ in the Arrhenius equation. n is the number of electrons transferred.

The rate of reverse reaction should decrease with an increase in E. Hence

$$r_b = r_c = k_b C_O \exp(-nf(1-\beta)E)$$

The net rate is therefore

$$\text{Rate} = k_f C_R \exp(nf\beta E) - k_b C_O \exp(-nf(1-\beta)E)$$

At equilibrium we have

$$0 = k_f C_R \exp(nf\beta E_{eq}) - k_b C_O \exp(-nf(1-\beta)E_{eq})$$

Hence

$$k_b = k_f \frac{C_R}{C_O} \cdot \frac{\exp(nf\beta E_{eq})}{exp(-nf(1-\beta)E_{eq})} = k_f \frac{C_R}{C_O} \exp(nfE_{eq}) \qquad (24.7)$$

Also, the ratio C_R/C_O is given by Nernst equation as we are looking at equilibrium conditions. Hence rearranging the Nernst equation we have

$$\frac{C_R}{C_O} = \exp[-nf(E_{eq} - E^0)]$$

Substituting this into Equation 24.7, we find that the backward constant is related to the forward constant as

$$\boxed{k_b = k_f \exp(nfE^0)} \qquad (24.8)$$

The rate constants k_f and k_b are not independent but are related to each other. This is because the rate should be zero at equilibrium. For ordinary chemical reactions $k_f/k_b = K_{eq}$ and the above relation is the electrochemical equivalent to this classical relation.

Hence the rate is

$$\text{Rate} = k_f C_R \exp(nf\beta E) - k_f \exp(nfE^0)C_O \exp[-nf(1-\beta)E]$$

It is customary to define a standard rate constant k_0 defined as follows:

$$k_0 = k_f \exp(-nf\beta E^0)$$

or

$$k_f = k_0 \exp(nf\beta E^0)$$

Note: k_0 has the same units as the heterogeneous rate constant for a first-order catalytic reaction, m/s.

Hence the rate can be expressed as

$$r = k_0 \left[C_R \exp[nf\beta(E - E^0)] - C_O \exp[-nf(1-\beta)(E - E^0)] \right]$$

Multiplying both sides by nF we get the following equation for the current density i, which is equal to nFr:

$$\boxed{i = nFk_0\{C_R \exp[nf\beta(E - E^0)] - C_O \exp[-nf(1-\beta)(E - E^0)]\}} \quad (24.9)$$

This is the Butler-Volmer kinetic model for an electrode reaction in the classical form. This model is also thermodynamically consistent. At equilibrium the rate or correspondingly the current density i can be shown to be zero.

Overpotential Form

Let R^* and O^* be two reference concentrations. These could, for example, be the bulk concentrations or the inlet concentration to the electrolyzer. Let E_{eq} be the corresponding equilibrium potential, which can be calculated for the Nernst equation. Then let

$$i_0 = nFk_0C_R^* \exp[nf\beta(E_{eq} - E^0)]$$

which is also equal to

$$i_0 = nFk_0C_O^* \exp[-nf(1-\beta)(E_{eq} - E^0)]$$

since the rate in the forward direction is the same as the backward direction under equilibrium conditions.

The parameter i_0 is known as the exchange current. Equilibrium is viewed as a dynamic process here with equal current flowing back and forth between the solution and the electrode; the exchange current is a measure of this process. Clearly the exchange current may be thought of as the intrinsic electro-catalytic activity of the electrode. It is a measure of the charge transfer process between the electrode and a solution at a given reference concentration. Therefore it can be linked to, for example, the work function of the metal. Note that i_o is often reported using the standard conditions as the reference concentration, but this is not always stated clearly.

The equation for the current i given by Equation 24.9 can then be represented as

$$i = i_0 \left\{ \frac{C_R}{C_R^*} \exp[nf\beta(E - E_{eq})] - \frac{C_O}{C_O^*} \exp[-nf(1 - \beta)(E - E_{eq})] \right\}$$

Let η be the overpotential defined as $E - E_{eq}$. Then

$$i = i_0 \left\{ \frac{C_R}{C_R^*} \exp[nf\beta\eta] - \frac{C_O}{C_O^*} \exp[-nf(1 - \beta)\eta] \right\}$$

which is the overpotential form of the Butler-Volmer equations.

If C_R and C_O (concentrations in the solution adjacent to the electrode) are themselves chosen as the reference concentrations then this reduces to

$$i = i_0 \left[\exp(nf\beta\eta) - \exp(-nf(1 - \beta)\eta) \right]$$

This form of the Butler-Volmer equation is applicable, for example, when there is no mass transfer resistance, that is, the bulk and the surface concentrations are the same.

If $\beta = 1/2$ the Butler-Volmer equation can be written in hyperbolic form:

$$i = 2i_0 \sinh(nf\eta/2) \tag{24.10}$$

Usually the magnitude of η is high so that the reaction is driven to only one side. The reverse reaction can then be dropped, leading to yet another form of the equation:

$$i = i_0 \exp(nf\beta\eta) \tag{24.11}$$

Some typical values for i_0 are shown in Table 24.2:

Table 24.2 Typical Range of Values for the Exchange Current

Process	Conditions	i_0 A/m^2
H$_2$ evolution on Pt	0.5 M H2SO4	10
H$_2$ evolution on Ag	7 M HCl	0.012
Cu deposition	1M CuSO4	0.2

Temperature Effect

The values of i_0 measured at temperature T_{ref} are corrected to any other temperature T by Arrhenius's law:

$$i_0(T) = i_0(T_{ref}) \exp\left(-\frac{E_a}{R_g}\left[\frac{1}{T} - \frac{1}{T_{ref}}\right]\right) \tag{24.12}$$

Here E_a is the activation energy for the electrode reaction.

24.3.2 Tafel Equation

The Tafel equation is a simplified form of Equation 24.11 and is obtained by taking the logarithm on both sides. This is usually written as

$$\eta = a + b\log(i)$$

where a and b are the kinetic parameters of the Tafel model.

24.4 Mass Transfer Effects

If there is a mass transfer resistance near the electrode then the surface concentration and the bulk concentrations will be different. The mass transfer effects are usually accounted for by the film model. Consider the reduction reaction scheme O going to R. An example would be copper ions going to a cathode and reacting there to deposit Cu. The rate at which species O is transported from the bulk (which we assume to be at a concentration of $C_{O,b}$) to the electrode is

$$N_A = k_L f_d(C_{O,b} - C_{O,s})$$

Here $C_{O,s}$ is the surface concentration. f_d is the enhancement in mass transfer due to electric field. k_L is the mass transfer coefficient in the absence of electric field.

The preceding equation for the rate of mass transfer can be expressed in terms of current (multiply by nF):

$$i_m = nFk_Lf_dC_{O,b}\left(1 - \frac{C_{O,s}}{C_{O,b}}\right) \tag{24.13}$$

Let us define i_L, termed the limiting current, as

$$i_L = nFk_Lf_dC_{O,b}$$

Then the rate of mass transfer (expressed as a current) is

$$i_m = i_L\left(1 - \frac{C_{O,s}}{C_{O,b}}\right) \tag{24.14}$$

The electrochemical reaction occurs in proportionality to the surface concentration and exponentially to the electrode overpotential. The current can be written using the Butler-Volmer equation:

$$i_R = i_0\exp(nf\beta|\eta|)\frac{C_{O,s}}{C_{O,b}} \tag{24.15}$$

The reverse current $(1 - \eta)$ term is assumed to be small and neglected here. Also, the absolute value for η is used since we are looking at a reduction here. Note that η is negative for a reduction reaction.

At steady state the two currents are equal:

$$i_m = i_R = i$$

Hence Equations 24.14 and 24.15 can be set equal to each other and $C_{O,s}/C_{O,b}$ can be eliminated. Then either of Equation 24.14 or 24.15 can be used to find i. The final expression can be found easily as

$$i = \left[\frac{1}{i_L} + \frac{1}{i_0\exp(nf\beta|\eta|)}\right]^{-1}C_{O,b} \tag{24.16}$$

which is a version of the law of addition of resistances.

Two limiting cases can be identified:

- Mass transfer control: $i_L << i_0\exp(nf\beta|\eta|)$
  The current is equal to i_L.
- Electrode reaction control: $i_L >> i_0\exp(zf\beta|\eta|)$
  The current is now given by Equation 24.15 with $C_{O,s}$ set the same as $C_{O,b}$.

The analysis is similar to the heterogeneous reaction discussed in Section 6.4.1.

At low overpotential the process is likely to be electrode controlled. As the overpotential increases the rate of electrode reaction increases and mass transfer resistance starts to become important. Beyond this point further increase in overpotential will not cause any further increase in rate.

24.4.1 Concentration Overpotential

Note that concentration drop due to mass transfer creates a potential gradient between the electrode and the bulk solution. This is known as the concentration overpotential or the concentration polarization. The expression for this was presented in Section 16.5, which should be revisited now. The expression was presented for a simple binary electrolyte MX with M reacting at the cathode. The anion X did not participate in any reaction at the cathode and hence its combined flux was set to zero. The migration part of the combined flux of X set up an electric potential in order to cancel out the diffusional part. The expression is repeated here:

$$\phi(b) - \phi(s) = \frac{R_g T}{F} \ln \left(\frac{C_X(b)}{C_X(s)} \right)$$

Role of Supporting Electrolyte

The expression is valid only if there is a single electrolyte, MX here. If there is a supporting electrolyte NX the expression for the potential difference has to be calculated using the Nernst-Planck equation for each of the ions, M^+, N^+, and X^-, and the electroneutrality conditions. Exercise problem 14 in Chapter 16 addressed this. In general the concentration overpotential gets smaller in the presence of a supporting electolyte.

24.5 Voltage Balance

The voltage balance shows how the applied voltage is distributed among the various transport steps and reaction steps. It is an important component in the analysis of electrochemical reactions. It is similar to energy balance in classical chemical reactions.

    The minimum voltage needed to carry out an electrochemical reaction is given from thermodynamic considerations and is equal to $E_{R,eq}$, given by Equation 24.5. The actual voltage to be supplied depends on the current carried in the external circuit.

    To generate an anodic current we need an overpotential η_a, which can be calculated from the Butler-Volmer equation. This is the additional voltage needed to produce a current. The current at the cathode is the same as that in the anode. Additional overpotential is needed for this current, which is the magnitude of the cathodic overpotential. In addition there is a solution voltage drop due to the migration of ions in the bulk liquid from one electrode to other. Additional voltage drop occurs if there is membrane or diaphragm separating the anodic and cathodic regions. Hence the total voltage that needs to be provided to maintain a certain value of the current is

$$V = E_{R,eq} + \eta_a + \phi_a(C) + (-\eta_c) + \phi_c(C) + IR(\text{solution})$$
$$+ IR(\text{membrane})$$

(24.17)

Here $\phi_a(C)$ is any additional potential drop caused by mass transfer effects in the film near the anode. $\phi_c(C)$ is a similar term for the potential drop in the cathodic diffusion layer. An illustrative diagram is shown in Figure 24.7.

As an example, the potential needed for the various steps in NaCl electrolysis is presented in Table 24.3 at a current density of 2000 A/m^2. We find the cathode reaction consumes more energy than the anodic reaction. The membrane also adds to a significant voltage drop. The total voltage to be applied is on the order of 3.65 V. A more detailed calculation of voltage balance is undertaken in the next section by taking a specific case of copper electrowinning.

If the reaction is spontaneous, then $E_R(eq)$ will be negative. The system can be used as a fuel cell in that case. Let V_0 be the magnitude of this, which is a positive quantity. Then V_0 is the maximum voltage the fuel cell can generate and this will be at zero current. Any current needs an overpotential and can cause a voltage drop. Hence the voltage balance in a fuel cell is written as

$$V = V_0 - \eta_a - |\eta_c| - IR(\text{electrolyte})$$

(24.18)

An example of this calculation is shown in Section 24.7.

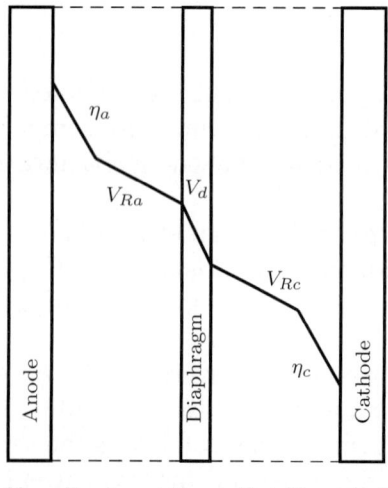

$$V_T = E_{eq} + \eta_a + V_{Ra} + V_d + V_{RC} + |\eta_c|$$

Figure 24.7   Voltage distribution in an electrochemical reactor. The concentration overpotential is not shown and should be added if there are significant mass transfer effects.

Table 24.3   Illustration of Various Voltage Drops for NaCl Electrolysis

Equilibrium	2.17
Anodic overpotential	0.03
Diaphragm drop	0.60
Solution drop	0.35
Cathodic overpotential	0.30
Metal hardware	0.20
TOTAL	**3.65 V**

24.6   Copper Electrowinning

The various concepts are put together and applied in this section. A specific application to copper eletrowinning is considered here but the methodology is general and is applicable to a wide variety of electrosynthesis processes.

The overall reaction in Cu processing from leach solutions of Cu salts can be represented as

$$Cu^{++}(aq) + H_2O(aq) \quad \rightarrow \quad Cu(m) + 2H^+(aq) + 1/2O_2(g)$$

This is split into two half reactions. The anodic reaction is

$$H_2O(aq) \rightarrow 2H^+(aq) + 1/2O_2(g) + 2e^- \quad E_0 = 1.229V$$

This is the oxygen evolution reaction while the cathodic reaction is the desired reaction of copper deposition:

$$Cu^{++}(s) + 2e^- \rightarrow Cu(s) \quad E_0 = -0.337V$$

The thermodynamic cell potential is calculated as the sum of these potentials.

24.6.1   Operating Current Density

The diffusion limiting current (DLC) calculation is an useful start; this is calculated from mass transfer considerations. It sets the maximum production rate that can be achieved. The corresponding voltage can also be calculated and operating above this value will not produce any further increase in the rate of reaction. From the discussion in Chapter 16 we note that

$$i_L = 2nF\frac{D_M C_{Mb}}{\delta_f}$$

The factor 2 (the migration factor) is replaced by 1 in the presence of a supporting electrolyte (usually H_2SO_4). n is the number of electrons transferred, which has a value of two here. D_M/δ_f is replaced by the mass transfer coefficient, k_m, and the limiting current can be calculated as

$$i_L = nFk_m Cu^{++}(bulk) \tag{24.19}$$

A value for the mass transfer coefficient is needed for further calculations and a number of correlations are available for this purpose. A limiting Sherwood number of 7.54 can be used as a simple approximation for flow between two parallel plates. For a more detailed correlation see Section 9.5.3, which is applicable if the flow is laminar. Hence $k_m = 7.54D/2h$. Using $D = 2 \times 10^{-9}$ m^2/s and a gap width of 5 mm we find $k_m = 1.5 \times 10^{-6}$ m/s. Bulk copper concentration is taken as 600 mol/m^3 in this example. Using these values in the equation for the limiting current we find $i_L = 174$ A/m^2.

The actual value of DLC is expected to be higher since fully developed mass transfer may not have developed; hence the above value is an underestimate. A number of similar correlations have been proposed and these may be used as appropriate. The effect of gas evolution is to increase the rate of mass transfer as well; this has not been reflected in our calculations. The value calculated is, however, close to that reported by Beukes and Badenhorst (2009).

In operating reactors a current lower than the DLC is used so as to achieve a good adherent product. We take a value of 0.8 times the DLC for further calculations and calculate the voltage needed to achieve this current. Hence the current i is as 140 A/m^2.

24.6.2 Voltage Balance

The thermodynamic potential shown earlier is for standard conditions (1 M solution). This is corrected for concentration dependency using the Nernst equation and the value is obtained as 0.892 V since the concentration is 0.6 M in this case study.

Anodic overpotential is calculated using the Tafel equation for a lead anode. The Tafel constants are taken from Beukes and Badenhorst (2009):

$$\eta_a = 0.303 + 0.12 \log_{10}(i)$$

This is calculated as 0.5604 V. The cathodic overpotential is calculated using the Butler-Volmer equation:

$$i = i_0(Cu_s/Cu_b) \exp\left(-n\frac{F}{R_gT}\eta_c\right)$$

where we ignore the anodic part of the reaction (the reverse reaction term) and take the β factor as one. A first-order dependency is used here. i_0 is the exchange current and is a function of the copper concentration:

$$i_0 = k[Cu]_s^{0.5}$$

In SI units the value of the exchange current, i_0, is 245 A/m² using $k = 1\ Am^{-2}(m^3/mol)^{-.5}$ as reported by Lapique and Storck (1985).

The surface concentration can be eliminated using the limiting current:

$$\frac{Cu_s}{Cu_b} = \frac{i_L - i}{i_L}$$

Using this in the Butler-Volmer equation and taking the logarithm on both sides, the following equation can be obtained for the cathodic overpotential:

$$\eta_c = \frac{R_gT}{n\beta F}\left[\ln(i_0) - \ln(i) - ln\left(\frac{i_L}{i_L - i}\right)\right]$$

This is a combined equation and includes the overpotential due to the cathodic reaction and due to the concentration drop in the Nernst diffusion layer (film) near the cathode. The value is calculated as 0.335 V.

The ohmic potential is calculated as

$$V_{ohm} = \frac{h}{\kappa}i$$

where κ is the specific conductivity and is calculated as

$$\kappa = F\sum_i z_i^2\mu_iC_i$$

For the given system $\kappa = 78.47\ S/m$. Hence the ohmic drop for the given current is only 0.05 V. The supporting electrolyte minimizes the voltage drop across the solution. The total potential is then found as 2.18 V and is in the range of values used commercially.

24.6.3 Meso-Model for the Electrolyzer

All the factors of importance have been put together and the local rate of reaction (in terms of i for electrochemical engineers) can be related to the required voltage for a given copper concentration.

This can be put into a mesoscopic model to track the variation of copper concentration as a function of length in a parallel plate electrolyzer.

The model is set up as

$$(Wh)v_x\frac{dC_A}{dx} = -\frac{i}{nF}W$$

where i/nF is the local rate of reaction. Here W is the cell width and x is the coordinate along the cell length. One slight problem is that i is not known at each position and what is known usually is the voltage applied across the system. Hence i should be iteratively calculated at each location. Detailed modeling is shown by Lapique and Storck (1985).

24.7 Hydrogen Fuel Cell

We illustrate the current–voltage relation for a hydrogen fuel cell briefly in this section. The voltage generated at any operating current i can be found using the following relation, which is the voltage balance equation:

$$V = E_{eq} - \eta_a - |\eta_c| - V_M \tag{24.20}$$

where E_{eq} is the thermodynamic potential; η_a is the anodic potential, which includes the concentration overpotential; η_c is a similar term for the cathode; and V_M is the voltage drop in the membrane. Equations to calculate each of these terms are now presented. The various voltage drops are schematically shown in Figure 24.8.

The open circuit potential is calculated from thermodynamic considerations and the following equation is used:

$$E_{eq} = 1.23 - 0.9 \times 10^{-3}(T - 298) + \frac{R_g T}{4F}\ln(p_{H_2}^2 p_{O_2})$$

The units for partial pressure are in atm here and depend on the gas mole fractions and the total pressure. The polymeric membrane cannot become dehydrated in practical operation and hence the gases fed to the fuel cell are humidified. Hence the water vapor pressure at the corresponding temperature is subtracted to get the oxygen partial pressure for the cathode side.

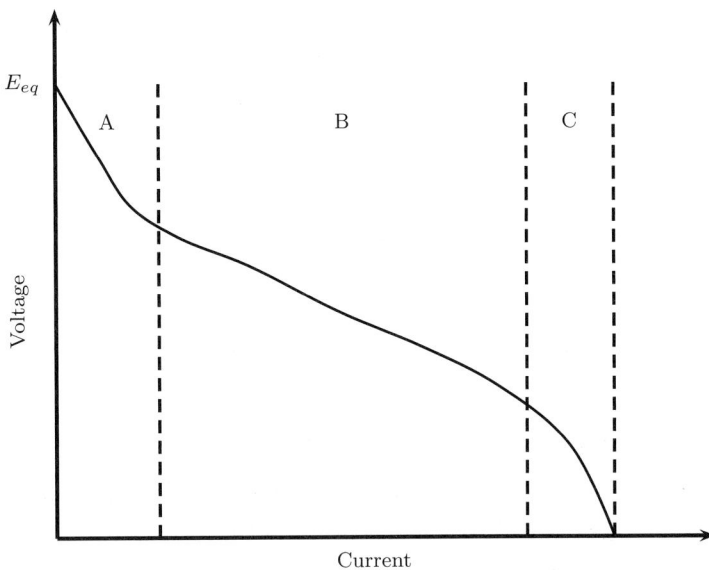

A: Activation polarization

B: Ohmic polarization

C: Concentration polarization

Figure 24.8 Current versus voltage relation in a fuel cell. The goal of modeling is to predict this relationship and use the model for design and operation.

The Butler-Volmer equation in hyperbolic form is commonly used to find η_a and η_c. The anodic overpotential can be calculated from the following equation, which uses the above hyperbolic form and also corrects for mass transfer effects. Details of the derivation are presented by Thampan et al. (2001).

$$\eta_a = \frac{R_g T}{\alpha_a F} \sinh^{-1} \left[\frac{1}{2} \left(\frac{i/i_{a0}}{1 - i/i_{aL}} \right) \right]$$

where α_a is effective transfer coefficient of the anode reaction (usually taken as $1/2$), i_{a0} is the exchange current at the anode, and i_{aL} is the diffusion limiting current. The anodic overpotential is usually small since the exchange current for hydrogen reduction is on the order of $1 \text{ mA}/\text{cm}^2$.

A similar equation holds for the cathodic potential:

$$\eta_c = \frac{R_g T}{n \beta_a F} \sinh^{-1} \left[\frac{1}{2} \left(\frac{i/i_{c0}}{1 - i/i_{cL}} \right) \right]$$

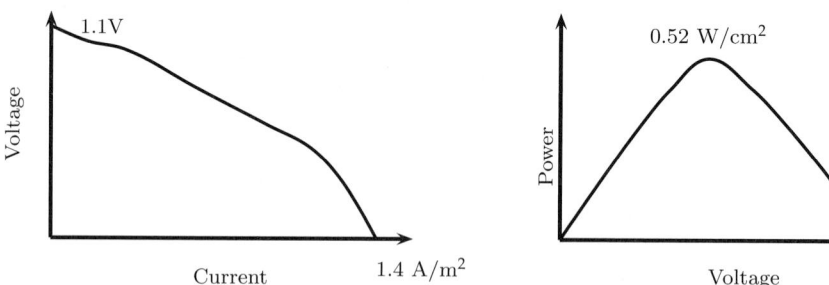

Figure 24.9   An illustrative plot of current density and power density versus voltage in a fuel cell. Temperature = 80 °C, pressure = 2.2 atm. Thampan. T., Malhotra, S. Zhang J and Datta. R; PEM Fuel cell as a membrane reactor, *Catalysis Today*, Volume 67, 2001.

Illustrative current–voltage calculations were determined by Thampan et al. (2001) and are shown in Figure 24.9.

Voltage decreases as current is increased as expected from the various polarization phenomena in the system. At large currents the voltage drops to almost zero value since the mass transfer resistance clicks in (more of a flowrate limitation in fuel cells). The power density increases at first with an increase in voltage, reaches a maximum, and then starts to fall off with further increase in voltage. The optimum power density is noteworthy and important in determining the optimum operating voltage.

24.8   Li-Ion Battery Modeling

In this section, we discuss briefly the working of a Li-ion battery and the transport steps involved in the system. The discussion in this section is qualitative and the goal is to reiterate the importance of mass transfer processes. In a Li-ion battery, energy is stored by converting electrical energy into chemical energy (specifically by storing lithium in a host material). This is the charging cycle of the battery. During the discharge Li moves from a high energy configuration to a low energy configuration and produces a current that does useful work. Unlike the fuel cell, the battery needs charging and a discharging cycles. Hence transient models are usually implied and steady state models are not applicable.

The process of insertion of lithium into a host is known as intercalation. Graphite is a good host material and Li can be inserted due to the small size of the lithium ions. The insertion material during discharging is usually inorganic metal oxides denoted generally as MO_2, where M is a metal atom and is often Co, Mn, or Fe. The two insertion materials are separated by an electrolyte, which facilitates the transport of ions across the two electrolytes. The reactions in the charging and discharging cycles are described briefly next.

24.8.1 Charging

During charging Li ions are released from the anode metal oxide matrix (where they were stored during the discharge cycle). The ions diffuse across the electrolyte and reach the graphite cathode, where they are inserted into the host by the following reaction:

$$C + xLi^+ + xe^- \rightarrow Li_xC$$

The transport processes and the reactions during a charging cycle are shown in Figure 24.10. The electrodes are porous and diffusing Li ions undergo a reaction at the carbon surface, picking up electrons from the external power supply. The Li atoms then diffuse into the carbon matrix and occupy host sites within the matrix.

It should be noted that the reaction taking place at the carbon is reduction and that in the Li-loaded metal oxide is oxidation. The electrodes are marked opposite and this can look confusing. The marking is because by convention the electrodes are marked correctly for the discharge cycle, but the name is retained for the charging cycle even though the reactions taking place switch direction during the charging cycle.

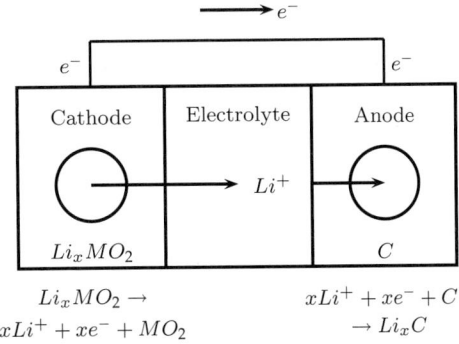

$$Li_xMO_2 \rightarrow$$
$$xLi^+ + xe^- + MO_2$$

$$xLi^+ + xe^- + C$$
$$\rightarrow Li_xC$$

Figure 24.10   Transport and reaction processes taking place during the charging of a Li-ion battery. The cathode matrix gets depleted of Li and the carbon anode gets enriched in Li.

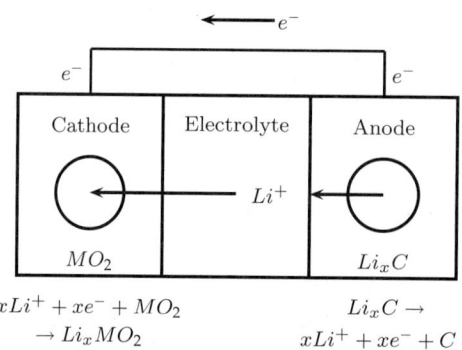

$$xLi^+ + xe^- + MO_2$$
$$\to Li_xMO_2$$

$$Li_xC \to$$
$$xLi^+ + xe^- + C$$

Figure 24.11 Transport processes taking place during the discharging of a Li-ion battery.

24.8.2 Discharging

The reverse reaction takes place during discharging. Lithium ions are released now from Li_xC and move to the cathode where they react as

$$xLi^+ + xe^- + MO_2 \to Li_xMO_2$$

The transport and reaction steps are shown in Figure 24.11.

    The transport mechanism were modeled by Jiang and Peng (2016) using time constants for the key transport steps: solution phase diffusion, solid phase diffusion, and charge transport were identified as the limiting phenomena. A transient diffusion reaction model including the migration due to electric field was formulated and solved to simulate the charging and discharging process. Ramadesigan et al. (2012) also provide detailed information on modeling where they use a multiscale modeling approach. The references cited in these two sources are numerous and provide additional resources on modeling of Li-ion batteries. Further details are not provided here due to space limitations.

Summary

- Electrochemical processes have a wide range of applications in the energy sector as well as in manufacture of chemicals. In the energy sector, there is a need to store solar energy since this can only be produced during daytime. Battery design becomes therefore an important part of the overall solar energy production industry. Similarly the automobile industry needs large capacity batteries. Fuel cells are central to hydrogen-based energy generation and provide another important application area of electrochemical processes.

- An example in the field of chemical production is the production of chlorine and sodium hydroxide by electrolysis of sodium chloride; this precedes the chemical industry. These are large volume chemicals. Other applications can be found in organic synthesis and pharmaceutical production. Adiponitrile production is one of the large-scale applications of electrochemistry for production of organic chemicals.

- Analysis of electrochemical systems starts by writing the overall reactions as two half reactions: the anodic reaction where an oxidation is taking place and a cathodic reaction where a reduction takes place. Each reaction is assigned an oxidation potential, which is based on the free energy change of the reaction under standard conditions. The half reactions with negative oxidation potentials are favorable and take place at the anode while the second half reactions become the cathodic reaction since they are favorable in the reverse direction. The overall potential is the sum of the anodic (oxidation) and cathodic (reduction) potentials and represents the minimum voltage needed to carry out the reaction.

- If the minimum potential is negative, the reaction is spontaneous and the two half reactions can be used to generate energy. This is the basis of, for example, the hydrogen fuel cell. The absolute value of the minimum potential is called the open circuit voltage and represents the maximum voltage that can be generated in the fuel cell. The maximum voltage applies when zero current is withdrawn from the cell.

- The minimum potential applies when the reactants and the products are in standard conditions, usually 1 atm pressure for gases and 1 M solution for liquids. If they are not at standard conditions the Nernst equation is used to correct for the concentration effects.

- The difference between the applied electrode potential and the equilibrium potential is defined as the overpotential. An anodic reaction takes place at the electrode if the overpotential is positive and *vice versa*. The rate of reaction is an exponential function of the overpotential and a kinetic model can be derived based on this concept, which is the Butler-Volmer equation.

- The exchange current form of the Butler-Volmer relation is commonly used. The exchange current is actually a kinetic parameter and can be thought of as a measure of the intrinsic activity of the electrode to charge transfer. The values are reported for standard reference concentrations and temperature. These can be corrected to other concentrations using the order of the reaction and to other temperatures using the Arrhenius equation.

- The Tafel equation is an inverted form of a simplified Butler-Volmer equation and is often used as an approximate representation of the kinetics of the electrode reactions. The overpotential needed to create a given current can be calculated using this equation if the Tafel constants are known for the given reaction.

- Mass transfer rate from the bulk to the electrode surface can be a limiting step and is usually modeled based on film theory, where the migrational contribution is also included (as a correction factor) in addition to the diffusional transport. For fast electrode reactions, the process can be entirely limited by the mass transfer rate and the resulting current is called the limiting current. The concentration drop in the film also causes a difference in potential across the film; this is called the concentration overpotential.

- The voltage balance shows how the applied voltage is utilized in various processes in the system: the anode reaction, anode concentration overpotential, solution or membrane drop, cathodic concentration overpotential, cathode reaction, and voltage drop in the external circuit. The applied voltage for an electrochemical process for a given current is calculated from the voltage balance. If the system is a fuel cell, the voltage balance shows how much voltage is actually produced compared to the thermodynamic maximum. Examples provided in the text show the details of these calculations.

- A Li-ion battery operates by storage of Li in a "carbon hotel." The process is known as intercalation. The Li moves from the carbon anode during the discharge cycle where it gets stored within a metal oxide matrix. Once the Li is depleted, the battery is put into a storing cycle where an external power source moves the Li back from the metal oxide to the carbon where it gets stored. Thus the Li-ion batteries are operated in a cycling manner unlike a fuel cell. Modeling requires the application of the Butler-Volmer kinetics, migration effects, and transient solid state diffusion.

Review Questions

24.1   Faraday's constant is the charge on an electron (C/electron) times the number of electrons per mole. The latter is equal to the Avogadro number. What is the charge on an electron? Using this calculate the value of Faraday's constant.

24.2 What is meant by a half reaction?

24.3 What reaction (oxidation or reduction) takes place at the anode?

24.4 A current density of 100 A/m² is measured in a simple single step electrochemical reaction. What is the rate at which the reaction is taking place in mol/m²s?

24.5 What is meant by standard oxidation potential?

24.6 What is meant by standard reduction potential?

24.7 What data is needed to find the overall equilibrium cell potential for an electrochemical reaction?

24.8 State the Nernst equation for the equilibrium potential.

24.9 When could a pair of half reactions be combined and used as a fuel cell?

24.10 Define the Butler-Volmer equation.

24.11 What is exchange current?

24.12 Define the Tafel equation.

24.13 By what factor is mass transfer enhanced due to electric field? Assume no supporting electrolyte.

24.14 What is the meaning of limiting current?

24.15 What is concentration overpotential?

24.16 What factors are included in the voltage balance?

24.17 When is the power maximum in a fuel cell, at low or high current?

24.18 What is intercalation?

24.19 Describe what happens when a Li-ion battery is charged.

24.20 Describe what happens when a Li-ion battery is being used.

Problems

24.1 **Classification of electrode processes.** Discuss the engineering issues associated with the four types of electrochemical reactions shown in Figure 24.4. What adjustments in the contacting pattern or design may be appropriate?

24.2 **Cu deposition: stoichiometry.** Consider the reaction

$$Cu^{++} + 2e^- \rightarrow Cu$$

If the deposition rate is 3 mole/sec find the current flowing through the system. Find the minimum potential needed for the if (impure) copper dissolution is the reaction at the anode. The cathode will be coated with pure copper.

24.3 **Open circuit voltage in a hydrogen fuel cell.** Find the maximum voltage in a hydrogn fuel cell as a function of oxygen partial pressure and plot the data. Also show the effect of temperature.

24.4   **Butler-Volmer equation: thermodynamic consistency.** Rearrange Equation 24.9 for zero current and show that it is equivalent to the Nernst relation. Hence verify that the equation is thermodynamically consistent.

24.5   **Metal recovery from waste streams.** Metal salts are frequently found in wastewater and industral waste streams. These can be electrochemically processed to recover the metals. Thus we not only treat the waste streams but also generate value from waste. The potential for metal recovery and some statistical data are shown in Allen and Rosselot (1997). From their data we see that the metals in many waste streams are significantly underutilized. Thus there is considerable scope for process development and optimum design in this area.

Consider as an example the recovery of Ag from waste, which can be represented by the following overall reaction:

$$2Ag^+ + H_2O \rightarrow 2Ag + 2H^+ + 1/2O_2$$

Write this as cathodic and anodic half reactions and find the equilibrium potential for a silver ion concentration of 0.01 M. The actual potential depends on the current density. Assume that at 100 A/m² the overpotential is 1 V. Find the energy required and compare with the silver cost.

24.6   **Cobalt recovery.** Cobalt is to be recovered from a solution of cobalt sulfate at a concentration of 0.005 M. Find the reaction potential and also find the potential for the competing reaction of hydrogen evolution at the cathode if the pH of the solution is 1. The standard potential for oxidation of cobalt is 0.277 V.

Note: The reaction potential for hydrogen evolution is 1.17 V at low pH, which is comparable to that for Co deposition. Hence both reactions are likely to occur at this pH. From an application point of view, a relatively high pH is needed for the Co deposition to become the favored reaction. Thus if the wastewater is acidic then Co recovery is difficult using an electrochemical process.

24.7   **Parallel plate reactor.** A parallel plate reactor is used for deposition of a metal from a solution of concentration 50 mol/m³. The reactor is undivided and has an interelectrode gap of 5 mm and width of 1 m. The electrolyte is supplied with a velocity of 0.1 m/s. Assume a mean current density of 360 A/m². The electrode kinetics is represented by a Butler-Volmer equation as

$$i = nFkC_M \exp(\beta E)$$

where $k = 0.2$ and $\beta = 0.5$. Find the limiting current. Find the voltage applied and the residence time needed to convert 95% of the metal ion. There is no supporting electrolyte.

24.8   **Batch electrochemical mode.** A schematic of an electrochemical batch reactor is shown in Figure 24.12 and the goal is to set up a simple model to calculate the batch time. Two modes of operation are to be examined as well.

The first is the potentiostatic mode where the potential difference between the electrode is kept constant. The current will vary with time here. The second is

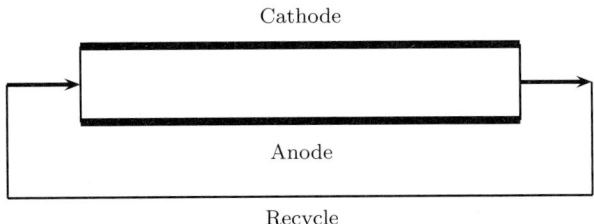

Figure 24.12 Schematic diagram of a parallel plate flow electrolyzer operated as a batch reactor.

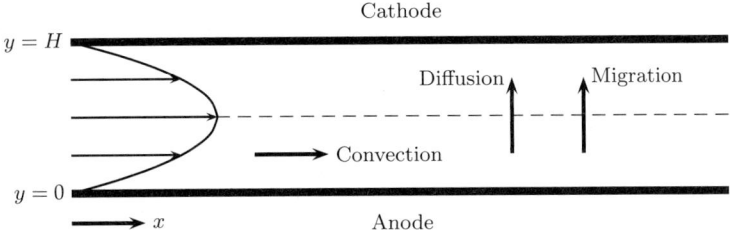

Figure 24.13 Schematic diagram of a parallel plate laminar flow electrolyzer (operated in a continuous mode) showing the three key mass transport steps.

the galvanostatic mode where the current between the electrode is kept constant. The applied voltage will have to vary with time here. Simulate the concentration versus time for an assumed set of parameters, and also show how the current or voltage would vary with time depending on the mode of operation.

24.9 **Convection-diffusion model for a parallel plate electrolyzer.** A parallel plate electrolyzer can be modeled by a convection-diffusion model. This is particularly suitable for laminar flow and preferred rather than the simple mesoscopic model. The steps in involved are shown in Figure 24.13.

The model is similar to the convection-diffusion model shown in Section 10.1. The migration in the cross-flow direction is an additional term that should be included. The electroneutrality condition provides an additional equation to calculate the electric potential. The boundary condition at the electrode is that flux is equal to i/nF. Set up the model equations and develop a simulation model for this mode of operation.

24.10 **Voltage balance in adiponitrile production.** Adiponitrile (ADN) is the first major organic chemical to be made by an electrolytic process. The reaction scheme is

$$2CH_2 = CH - CN + H_2O \rightarrow NC(CH_2)_4CN + \frac{1}{2}O_2$$

Find the free energy change of the reaction and report it as E_{eq}. Write this as two half reactions. The reaction scheme is complicated, involving both homogeneous and heterogeneous reactions, and a detailed model is given in Suwanvaipattana et al. (2006). In this exercise assume a simple Tafel kinetics for both reactions. Use the following values for the Tafel constants. For water oxidation $a = 0.303$, $b = 0.12$ and for acrylonitrile (ACN) conversion $a = 0.6$, $b = 2.2$. Perform a voltage balance and plot the results as a current versus voltage curve.

PART III

Mass Transfer–Based Separations

CHAPTER 25

Humidification and Drying

Learning Objectives

After completing this chapter, you will be able to:

- Understand the process of humidification and the simultaneous heat and mass transfer involved in that process.
- Explain the basis of operation of a cooling tower.
- Develop and solve simple models for sizing a cooling tower.
- Understand different mechanisms for drying of a solid, the so-called constant rate versus falling rate periods.
- Calculate the moisture content versus time during a batch drying process.
- Examine the role of pore diffusion versus capillary flow in drying of solids.

Processes where simultaneous heat and mass transfer occur is common in chemical process industries. The simplest example is evaporation from a pool of water to relatively dry air. The definitions of dry (air temperature) and wet (water temperature) bulb temperatures are important in this context; this is the first topic examined here. The evaporation causes some cooling of the water to supply the latent heat of evaporation. This causes the water temperature to be lower than the air temperature. This cooling phenomena is commonly used to cool hot water from process streams by contacting it with air. This is the basic principle of operation of a cooling tower. We will study how the principles of heat and mass transfer can be used to design this equipment.

Drying of a wet solid is another example of simultaneous heat and mass transfer. This is a commonly practiced unit operation and one of the most energy intensive processes. Hence an understanding of the transport processes in drying is a prerequisite to energy optimization and the design of next-generation drying equipment. The chapter also provides an overview of different mechanisms of moisture transport from the solid being dried to the gas phase.

25.1   Wet and Dry Bulb Temperature

Wet bulb and dry bulb thermometers provide a means to measure the moisture content of air. In the simplest form, a wet bulb thermometer is an ordinary thermometer covered with a wetted wick. The water evaporation causes the wet bulb to cool and the temperature recorded by this thermometer is different from that measured by one with a dry bulb. In this section we model a system based on simultaneous heat and mass transfer considerations and show how the temperature difference between the dry and wet bulbs is related to the moisture content of the air. The system modeled is shown schematically in Figure 25.1.

The mass transfer rate (liquid to gas) is calculated using a mass transfer coefficient, k_m°, assuming a low flux model. The mole fraction of vapor corresponds to the saturation condition at the surface, denoted by y_s, and the mole fraction in the gas is y_b. Hence the mass transfer or evaporation rate is

$$N_A = k_m^\circ C(y_s - y_b)$$

Here y_s is the equilibrium mole fraction at the interface, which depends on T_s. Note $y_s = P_{vap}(T_s)/P$, where P_{vap} is the vapor pressure and P is the total pressure.

Heat transfer from gas at a temperature of T_b to the liquid–gas interface at T_s is modeled using a heat transfer coefficient h:

$$q = h(T_b - T_s)$$

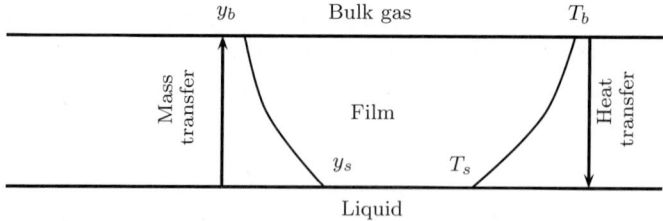

Figure 25.1 Schematic of heat and mass transfer from an evaporating surface. Note: Profiles would be modeled as a linear function for the low flux mass transfer case.

where q is heat flux toward the interface. Assume an adiabatic process such that the latent heat of evaporation ΔH_v comes from the heat transfer in the gas film. Then

$$q = N_A \Delta H_v$$

Equating the two expressions, we get an expression for surface temperature:

$$k_m^\circ C(y_s - y_b)\Delta H_v = h(T_b - T_s) \tag{25.1}$$

25.1.1 The Lewis Relation

The heat transfer coefficient and mass transfer coefficients can be related by the Lewis relation, which is based on the analogy between the two processes. We show how the humidity of the air can be related to the wet and dry bulb temperature.

We have from the film model $k_m = D/\delta_f$ while $h = k/\delta_t$ where δ_f and δ_t are the film thicknesses for mass and heat transfer, respectively. From a scaling analysis of boundary layer equations for heat and mass transfer theory (discussed briefly in Chapter 11) we can use the following relation:

$$\delta_m \propto Sc^{-1/3}$$

A similar relation holds for heat transfer with Pr, the Prandtl number, replacing the Schmidt number, Sc. The Prandtl number is defined as ν/α, where α is the thermal diffusivity $k/\rho c_p$. The thermal boundary layer scales as

$$\delta_t \propto Pr^{-1/3}$$

Hence

$$\frac{h}{k_m} = \frac{k}{D}\left(\frac{Sc}{Pr}\right)^{-1/3}$$

Also, $k/D = \rho c_p (Sc/Pr)$. Hence

$$\frac{h}{k_m} = \rho c_p \left(\frac{Sc}{Pr}\right)^{2/3} \tag{25.2}$$

The ratio Sc/Pr is referred to as the Lewis number, Le:

$$Le = \frac{Sc}{Pr} = \frac{\alpha}{D} = \frac{k}{\rho c_p D} = \frac{\text{thermal diffusivity}}{\text{mass diffusivity}}$$

Therefore we get the following relation between the two transfer coefficients:

$$\boxed{\frac{h}{k_m} = \rho c_p Le^{2/3}}$$

(25.3)

This relation between the heat and mass transfer coefficients is called the Lewis relation. Using this in Equation 25.1 and rearranging we have

$$(y_s - y_b) = (T_b - T_s)\frac{\rho c_p}{C\Delta H_v}Le^{2/3}$$

(25.4)

Since ρ/C is equal to the molecular weight of the gas and $M_W c_p$ is equal to C_p, the molar specific heat, this equation can be written as

$$(y_s - y_b) = (T_b - T_s)\frac{C_p}{\Delta H_v}Le^{2/3}$$

(25.5)

This permits us to calculate y_b and thereby the corresponding value of the humidity in the vapor phase based on the wet bulb temperature of the evaporating liquid. If the wet bulb and dry bulb temperatures are known, the calculation of the moisture content is straightforward. If the unknown is the wet bulb temperature, an iterative calculation is needed since y_s is based on the wet bulb temperature T_s, which is not yet known.

The system can be plotted in terms of an operating line and equilibrium line, as typically done in mass transfer operations. An illustrative plot is shown in Figure 25.2. The Lewis number is an important parameter in determining the slope of the operating line.

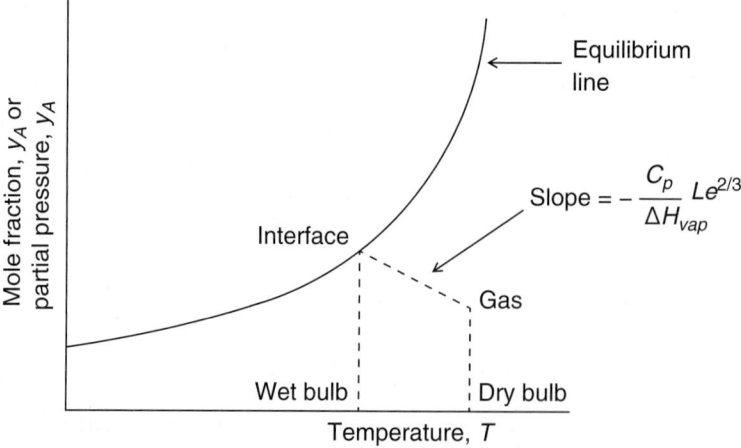

Figure 25.2   Representation of a humidification process in terms of an equilibrium curve and operating curve.

Problems of this type are also of importance in a number of related situations, including fog formation in environmental applications. A simple numerical example is presented in Example 25.1 to get an appreciation for the numbers.

Example 25.1   Humidity Calculation from Wet Bulb Temperature

Find the gas composition and relative humidity if the (dry) gas temperature is 31.8 °C and the wet bulb temperature is 26.8 °C. The total pressure is 1.01 bar. The following values of physical properties evaluated at 300 K are used: $\rho = 1.1769$ kg/m^3; $c_p = 1006$ J/kg.K; $\nu = 1.57 \times 10^{-5}$ m^2/s. $Pr = 0.713$; Le = 0.847.

Solution

The vapor pressure of water at 26.8 °C is computed using the Antoine equation and equal to 26.35 mm Hg. Hence y_s has a value of 26.34/760 or 0.0347. Using Equation 25.5, y_b is determined to be 0.0313.

The results are normally reported in terms of relative humidity, which is defined as follows:

$$\text{Relative humidity} = \frac{\text{mole fraction in air}}{\text{mole fraction if the air were saturated}}$$

In the above example, if the air were saturated at the temperature of 31.8 °C it would exert a vapor pressure of 35.17 mm Hg. Hence the mole fraction would be 35.17/760 or 0.0463 while the actual value is 0.0313. The relative humidity is therefore 0.0313/0.0463 or 67.6%.

In the next section we present an important application of evaporative cooling, which is the principle behind the cooling towers used widely in process industries.

25.2   Humidification: Cooling Towers

Humidification refers to transfer of a liquid phase component (usually water) into a gas stream. Water temperature drops upon evaporation to dry air when the process is operated in an adiabatic manner. This principle can be used to cool large quantities of hot water exiting from heat exchangers in order to recycle the water back to these process units. This cooling process is done is a cooling tower, which finds extensive application in power plants, nuclear reactors, refineries and petrochemical plants. In this section we study the application of heat and mass transfer principles to develop a basic design equation for the cooling tower.

25.2.1 Classification

Depending on how the air is forced through the system, cooling towers can be broadly classified into two types:

- Natural draft: This utilizes the buoyancy of warm, moist air compared to the outside dry cooler air. The buoyancy difference causes a density difference, which produces an upward flow of air through the tower. These are suitable for large power plants.
- Forced draft: The air flow is induced by a blower or fan. These are more suitable for small capacity systems, such as cooling in air conditioning units.

The natural draft cooling tower is usually of hyperboloid design; an illustrative diagram is shown in Figure 25.3. The hyperbolid design is effective from a mechanical strength point of view and also for minimum material cost. It also aids in increasing the upflow velocity of air, thereby improving the gas flow rate.

A second classification is based on the flow pattern of water and air. In the counterflow type, water and air flow countercurrent to each other. The

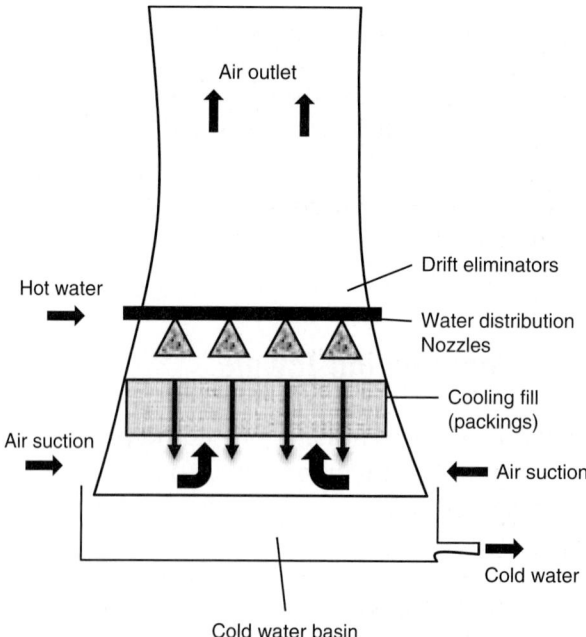

Figure 25.3 Counterflow cooling tower with natural draft.

flow patterns in a natural draft cooling tower by design are countercurrent. In the cross-flow type, liquid flows downward but the gas flow is in horizontal direction, that is, perpendicular to liquid flow.

25.2.2 General Design Considerations

The concept of wet and dry bulb temperature was introduced in Section 25.1 and is important in the design of cooling towers. The dry bulb temperature is the actual temperature of the gas at a given water vapor composition (or given humidity) while the wet bulb temperature is the temperature if the gas were saturated with the vapor. If the gas (dry bulb) temperature is the same as the wet bulb temperature, the gas is saturated and no mass transfer from liquid can take place. Hence the local water temperature has to be above the wet bulb temperature of the air in order to have any cooling. If the gas is saturated then there is no cooling.

Usually cooling towers are designed so that the discharge temperature of water is about 3 to 5 °C above the wet bulb temperature corresponding to the inlet gas conditions. This provides an adequate driving force for mass and heat transfer. This temperature difference (outlet water temperature minus the wet bulb temperature of the inlet air) is called the approach. The change in water temperature from inlet to outlet is called the range.

The mass flow ratio $\dot{m}_L/\dot{m}_G$ is an important parameter that affects the performance of the cooling tower. The mass transfer coefficient is normally correlated in terms of the flow ratio parameter. Typical mass velocities of water in cross-flow arrangements are in the range of 1.8 to 2.7 kg/m^{2}s and the superficial air velocity falls in the range of 1.5-4 m/sec (Mills, 1993). The gas and liquid flow rates can be independently varied in this arrangement. However, in a natural circulation type of arrangement they can not be varied independently. The buoyancy producing the air flow is determined by the exit condition of the air leaving the tower. Hence an iterative solution is needed to calculate the air flow rate for this type of arrangement for a given liquid rate and the specified level of cooling.

25.3   Model for Counterflow

Figure 25.3 showed an equipment schematic of a natural draft counterflow cooling tower. Water flows as thin films down a suitable packing while the air flows upwards. The driving force for air flow is caused by the density difference between the moist and the warm air. The governing equations are based

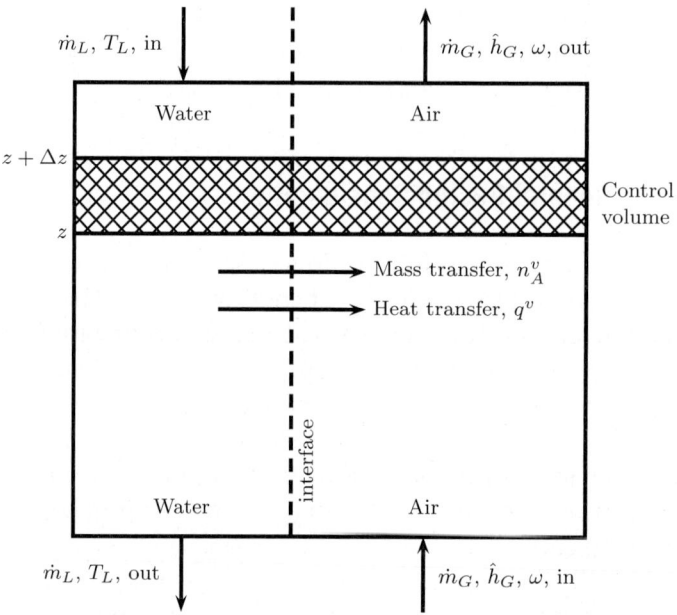

Figure 25.4 Schematic of the control volume for the gas and water phase to model a counterflow cooling tower.

on heat and mass balances on a mesoscopic control volume and use the plug flow assumption. The control volume is shown in Figure 25.4.

25.3.1 Mass Balance Equations

Mass units are commonly used and the driving force for mass transfer is based on the mass fraction difference.

The mass balance for water flow rate is

$$\frac{d\dot{m}_L}{dz} = An_A^v \tag{25.6}$$

The mass balance for gas is

$$\frac{d\dot{m}_G}{dz} = An_A^v \tag{25.7}$$

In these equations, A is the cross-sectional area and n_A^v is the mass transfer rate of water based on unit volume. The latter is given by the mass transport relation and is equal to

$$n_A^v = k_\omega a_{gl}(\omega_s - \omega) \tag{25.8}$$

Here k_ω is the intrinsic mass transfer coefficient based on the mass fraction driving force and a_{gl} is the transfer area per unit tower volume. Note that a low flux model is implied and no drift flux correction is used; these approximations are valid in view of the low water vapor pressures at the operating conditions. ω is the actual mass fraction of water vapor in the gas phase. ω_s is the saturation mass fraction of water corresponding to the interface temperature, and can be calculated as

$$\omega_s = \frac{18y_s}{18y_s + 29(1 - y_s)} \tag{25.9}$$

where y_s is the mole fraction of the saturated vapor given as p_{vap}/p_{total}. p_{vap}, the vapor pressure, in turn can be calculated using the Antoine equation corresponding to the liquid temperature.

Water vapor mass balance in the gas phase air completes the set of mass balance equations:

$$\frac{d}{dz}(\omega \dot{m}_G) = A n_A^v \tag{25.10}$$

The next set of equations are the enthalpy balances for the gas and liquid phases.

25.3.2 Enthalpy Balance Equations

The enthalpy balance for water is written as

$$\frac{d(\dot{m}_L \hat{h}_L)}{dz} = A Q^v \tag{25.11}$$

Q^v is the heat transferred from water to air per unit volume. $\hat{h}_L$ is defined as the enthalpy of the liquid per unit mass.

The heat balance for air is

$$\frac{d(\dot{m}_G \hat{h}_G)}{dz} = A Q^v \tag{25.12}$$

Here $\hat{h}_G$ is the specific enthalpy of the gas phase, which is a mixture of air and water vapor.

Transport relation is now needed for Q^v, which are presented later after defining equations for the enthalpies.

Auxiliary Relations for Enthalpy

Additional auxiliary conditions for enthalpy are needed to close the model, which we will discuss next.

A reference temperature and a reference enthalpy have to be chosen in order to calculate the gas phase and liquid phase enthalpies. It is convenient to choose the reference temperature T_{ref} as zero $0\ °C$. (All temperature from now on in the model will be in degrees Celsius). Enthalpy of water at $0\ °C$ is taken as the reference enthalpy.

The enthalpy of liquid water at any other temperature then is:

$$\hat{h}_L = c_{pL}T_L \qquad (25.13)$$

The enthalpy term in the gas phase in turn is given by the following equation:

$$\hat{h}_G = \bar{c}_{pG}T_G + \omega\hat{h}_{lg} \qquad (25.14)$$

where $\hat{h}_{lg}$ is the latent heat of vaporization for water at $0\ °C$. $\bar{c}_{pG}$ is the specific heat of the mixture, given as

$$\bar{c}_{pG} = (1 - \omega)c_{p,a} + \omega c_{p,w}$$

$c_{p,w}$ is the specific heat of the water vapor and $c_{p,a}$ is the specific heat of air.

Hence the enthalpy of the gas phase can be related to the temperature of the gas as follows:

$$\hat{h}_G = (1 - \omega)c_{p,a}T_G + \omega(c_{p,w}T_G + \hat{h}_{lg}) \qquad (25.15)$$

Rate of Heat Transfer

The rate of heat transfer from the gas–liquid interface to the bulk gas consists of two terms: the convective heat transfer from the interface to the bulk gas due to temperature difference and the enthalpy carried by the flux of water vapor to the gas phase.

Convective heat transfer is given as $ha_{gl}(T_i - T_g)$. Here h is the heat transfer coefficient from the interface to the bulk gas.

The enthalpy due to mass transfer is equal to $n_A^v(\hat{h}_{wi})$. Here $\hat{h}_{wi}$ is the enthalpy of the water vapor at the interface conditions.

The temperature difference in the liquid is assumed to be small and the assumption that $T_i = T_L$ is made here. Hence $\hat{h}_{wi}$ can be approximated as $\hat{h}_{ws}$, the enthalpy of the water vapor at liquid temperature.

Hence the heat transfer rate Q^v is given as

$$Q^v = n_A^v \hat{h}_{ws} + ha_{gl}(T_L - T_g) \tag{25.16}$$

This completes the model formulation. The complete set of equations contains five primary variables: mass flow rates of water and gas, water vapor mole fraction in the gas, and gas and liquid enthalpies. Three mass balances and two enthalpy balances provide the five equations. In addition there are three auxiliary variables: T_L, T_G, and ω_s. Two enthalpy relations (Equations 25.13 and 25.14) and the vapor–liquid equilibrium relation (Equation 25.9) provide three relations and hence the model set is correctly specified. A complete solution is involved and can be done numerically, but is not normally used for design due to uncertainty in the values of the heat and mass transfer coefficients. A simpler and more practical approach is to combine the model equations into a single equation after making some simplifying assumptions. This is the focus of the next section.

25.3.3 Merkel Equation

The set of equations can be reduced to a single equation for enthalpy as shown in the following analysis. We have many invariants and this permits the number of equations to be reduced. In fact one equation is sufficient to calculate the performance of the cooling tower. This equation, derived in 1925, is called the Merkel equation. The derivation leading to this equation is presented in the following section.

We start with the gas phase enthalpy balance given by Equation 25.6 and expand the derivative on the left-hand side:

$$\dot{m}_G \frac{d\hat{h}_G}{dz} + \hat{h}_G \frac{d\dot{m}_G}{dz} = AQ^v \tag{25.17}$$

Further, from Equation 25.7 we have

$$\frac{d\dot{m}_G}{dz} = An_A^v$$

Using this in the second term on the left-hand side, the previous equation is changed to

$$\dot{m}_G \frac{d\hat{h}_G}{dz} + An_A^v \hat{h}_G = AQ^v \tag{25.18}$$

Using Equation 25.16 for Q_v and moving the n_A^v term to the right-hand side gives

$$\dot{m}_G \frac{d\hat{h}_G}{dz} = A\,n_A^v(h_{ws} - h_G) + A\,h a_{gl}(T_L - T_g) \tag{25.19}$$

The heat transfer coefficient h is taken as $k_\omega \bar{c}_{p,G}$, which is an analogy useful for a Lewis number of one. This is the first assumption in the Merkel approach. Hence the previous equation simplifies to

$$\dot{m}_G \frac{d\hat{h}_G}{dz} = A k_\omega a_{gl} \left[(\omega_s - \omega)(\hat{h}_{ws} - \hat{h}_G) + \bar{c}_{p,G}(T_L - T_g) \right] \tag{25.20}$$

An analysis of the magnitude of the various terms in the above equation indicates that it can be simplified to

$$\dot{m}_G \frac{d\hat{h}_G}{dz} = A k_\omega a_{gl}(\hat{h}_s - \hat{h}_G) \tag{25.21}$$

Here $\hat{h}_s$ = enthalpy of saturated gas at water temperature (the equation for this is given later as 25.25) and $\hat{h}_G$ = enthalpy of gas at gas temperature. This is given by Equation 25.14.

A similar manipulation for the liquid gives

$$\dot{m}_L \frac{d\hat{h}_L}{dz} = A k_\omega a_{gl}(\hat{h}_s - \hat{h}_G) \tag{25.22}$$

The assumption that the evaporation rate is small and therefore the water flow rate remains constant is used in arriving at this result; this is the second assumption in the Merkel method. Using $\hat{h}_L = c_{pL} dT_L$ the previous equation can be represented as

$$\dot{m}_L c_{pL} \frac{dT_L}{dz} = A k_\omega a_{gl}(\hat{h}_s - \hat{h}_G) \tag{25.23}$$

The integrated form of this is called the Merkel equation; it provides the liquid temperature distribution and the exit water temperature in the tower.

The enthalpy of the gas at the gas temperature is needed in the integration of Equation 25.23. This is not easy to use since it requires the simultaneous calculation of the mole fraction of water vapor in the gas. However, the enthalpy of the gas need not be calculated from this equation, but can be related to the liquid temperature as shown below. This is yet another computational simplification.

An overall heat balance between any cross-section at z and the bottom of the tower gives

$$\hat{h}_G = \hat{h}_{G,in} + \frac{\dot{m}_L}{\dot{m}_G}(\hat{h}_L - \hat{h}_{L,out})$$

or equivalently in terms of the liquid temperature

$$\hat{h}_G = \hat{h}_{G,in} + \frac{\dot{m}_L}{\dot{m}_G}c_{pL}(T_L - T_{L,out}) \qquad (25.24)$$

Hence the $\hat{h}_G$ can be calculated at any given T_L and used in Equation 25.24.

The h_s is based on saturation conditions at water temperature and can be calculated from the following equation:

$$\hat{h}_s = (1 - \omega_s)c_{p,a}T_L + \omega_s c_{p,w}T_L + \hat{h}_{lg}\omega_s \qquad (25.25)$$

The vapor pressure data is needed to find ω_s at a given value of T_L, which can be calculated from Equation 25.9 with the help of the Antoine equation for vapor pressure. Thus all the quantities needed in the Merkel equation are known.

The Merkel equation is often arranged into an HTU (height of a transfer unit) and NTU (number of transfer units) form similar to that for gas absorption. Thus the height of the tower is calculated as

$$H = \text{HTU} \times \text{NTU}$$

where HTU is defined as

$$\text{HTU} = \frac{\dot{m}_L c_{pL}}{A k_\omega a_{gl}}$$

and NTU is defined as

$$NTU = \int_{T_{L,out}}^{T_{L,in}} \frac{dT_L}{\hat{h}_s - \hat{h}_G}$$

Example 25.2 provides an illustration of the calculation of the NTU parameter.

Example 25.2   NTU Calculation for a Cooling Tower

A cooling tower is to be designed to cool water from 40 °C to 26 °C. The inlet air is at 10 °C and saturated. A liquid to air mass flow ratio of one is used. Calculate the number of transfer units needed.

Solution

The data needed for the various enthalpy calculations are as follows: $c_{p,w} = 1864$ J/kg K, $c_{p,a} = 1005$ J/kg K, and $\hat{h}_{lg} = 2.5 \times 10^6$ J/kg.

The saturation vapor pressure for any given temperature can be calculated using the Antoine equation:

$$\log_{10} p_{sat}(\text{mm Hg}) = A - \frac{B}{T + C}$$

and the constants for water are as follows: $A = 8.07131$, $B = 1730.63$, $C = 233.246$, and T is in °C.

The mole fraction y_s at saturation is $p_{sat}/760$, assuming a total pressure of 760 mm Hg. The mass fraction at saturation ω_s is then calculated using Equation 25.9.

This information is used to calculate the saturation enthalpy and enthalpy of the gas at several values of the liquid temperature between 26 °C and 40 °C. The saturation enthalpy is calculated using Equation 25.25 in the text.

The gas enthalpy is calculated using Equation 25.24. For this we need to calculate the value of $\hat{h}_{G,in}$. The inlet air vapor mass fraction is known since it is assumed to be saturated. Hence Equation 25.15 can be used to find the inlet air enthalpy. The value is 29 kJ/kg; you should verify the calculation.

The results for $\hat{h}_s$ and $\hat{h}_g$ at some selected liquid temperatures are as follows:

T_L,C	$\hat{h}_s$, kJ/kg	$\hat{h}_G$ kJ/kg
26	79	29
33	112	58
40	158	87

The NTU can be found by the area under the curve of $1/(\hat{h}_s - \hat{h}_G)$ versus T_L or by using the Simpson rule after generating a table similar to the above in the temperature interval from the exit to the inlet. The value obtained using the Simpson rule is 1.06. It is easy to set up an Excel spreadsheet or a MATLAB m-file to do these calculations.

25.4   Cross-Flow Cooling Towers

In a cross-flow arrangement, the air and water flow in cross direction. The schematic is shown in Figure 25.5. A cross-flow arrangement is more suitable for small capacity plants.

The balance equations (using Merkel assumptions similar to that in counterflow) lead to the following equations for predicting the performance of this equipment. For gas enthalpy, which varies in the x-direction, we get

$$\frac{\partial \hat{h}_G}{\partial x} = \frac{k_\omega a_{gl}}{G}(\hat{h}_s - \hat{h}_G) \tag{25.26}$$

G is the superficial mass velocity of the gas, equal to $\dot{m}_G/HW$.

For liquid enthalpy, which varies in the y-direction (marked vertically downward) we have

$$\frac{\partial \hat{h}_L}{\partial y} = -\frac{k_y \omega a_{gl}}{L}(\hat{h}_s - \hat{h}_G) \tag{25.27}$$

L is the superficial liquid velocity, $\dot{m}_L/(WX)$.

The liquid enthalpy equation is commonly expressed in terms of liquid temperature as

$$c_{pL}\frac{\partial T_L}{\partial y} = -\frac{k_\omega a_{gl}}{L}(\hat{h}_s - \hat{h}_G) \tag{25.28}$$

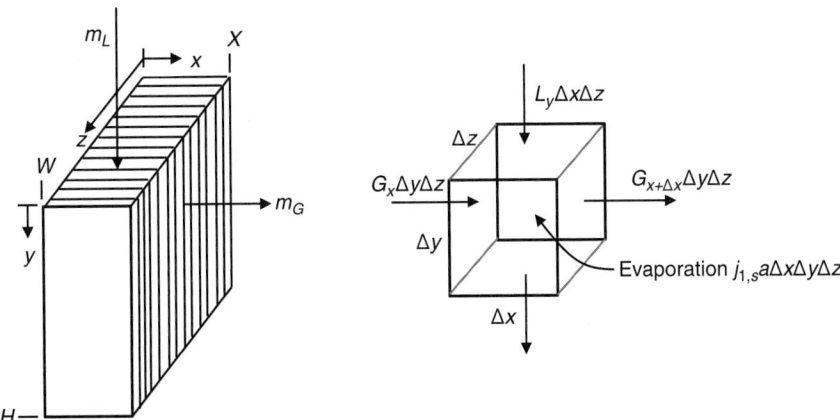

Figure 25.5   Schematic of a cross-flow cooling tower. Adapted from Mills, *Heat and Mass Transfer*. Erwin Publications, page 1105.

The required boundary conditions are

- At $x = 0$, the gas entrance, $\hat{h}_G = \hat{h}_{G,in}$
- At $y = 0$, the liquid, $\hat{h}_L = \hat{h}_{L,in}$ or $T_L = T_{L,i}$.

The balance equations are a pair of first-order partial differential equations now. These require a numerical solution. For example, a suitable finite difference approximation can be set up for the derivatives and the set of algebraic equations can be set up at each mesh point. The equations can be solved in a sequential manner starting at the top row.

Suppose the system is discretized into meshes in the x- and y-directions. See Figure 25.6, which shows an N X M discretization.

The mesh size in the x-direction is W/N and the mesh size in the y-direction is H/M. Each mesh defines a cell, which are marked and numbered in the figure. The entrance conditions for both the gas and the liquid are known at the top corner cell (1,1) on the left of the figure. The conditions on the liquid side are known for all the cells in the top row. Hence starting from cell (1,1) we can successively solve the equation for this entire row for the gas enthalpy values and the liquid exit temperature values from this row. The calculated exit liquid conditions for this row are used as the inlet conditions for the next row. For the gas, the conditions at the entrance of the inlet cell (2,1 now) are known. Hence we can sweep across row 2 to complete the solution for this row. This provides all the data required for row 3, and so on.

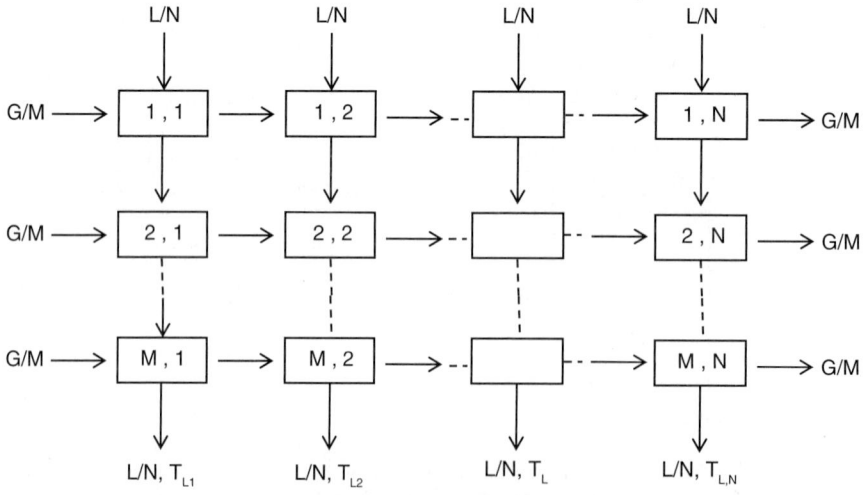

Figure 25.6 A cell-by-cell approach to simulate the cross-flow cooling tower.

The liquid exit from the last row is finally calculated for all the cells in that row. An average value is then calculated and this will represent the liquid exit temperature.

The gas and liquid flows entering each cell are assumed to be constant. This applies since the extent of liquid evaporated in a cooling tower is usually small. Note that this was also the assumption used in the Merkel model in the previous section. The gas mass velocity in each cell is to be taken as G/M, where G is the total gas velocity. Similarly the liquid velocity in each cell is L/N. The computational implementation is not shown here and is left as an exercise problem.

25.5 Drying

Drying, as the name implies, refers to the process of removal of moisture (or other liquid) content of a solid by heating or by contacting with a hot gas. It is often the final step of a series of operations and the dried solid is usually ready for final packaging. Examples include crystalline salts to produce a free flowing material, food to prevent spoilage, detergent, pharmaceutical capsules, wood, soap, paper, and many other products.

25.5.1 Types of Dryers

One classification is based on whether the solid is in direct contact with hot gas or is heated indirectly from a hot metal surface. The first type are also called adiabatic or direct dryers while the second type are known as indirect or nonadiabatic dryers. Drying can also be done by radiation heat transfer or microwave energy.

Dryers may also be classified into two types: batch dryers where the solid is stagnant and continuous dryers where the solid is moving in the dryer.

Batch dryers with stagnant solids are generally tray dryers where the solid to be dried is kept in a tray with a suitable arrangement for the flow of the hot drying gas. The gas flow pattern can be either parallel flow or crossflow. A batch dryer with a moving solid is the fluidized bed dryer, where the solids are separated from each other and kept suspended in motion by the gas. The rotary dryer is another arrangement where the solids are in motion.

A continuous dryer is the flash dryer, where the solids are transported by the gas. Fluid bed dryers can also be operated in a continuous manner. Spray dryers are another type and suitable when the solid to be dried is presented as a dilute solution in a liquid. The various types of dryers are shown in Figure 25.7.

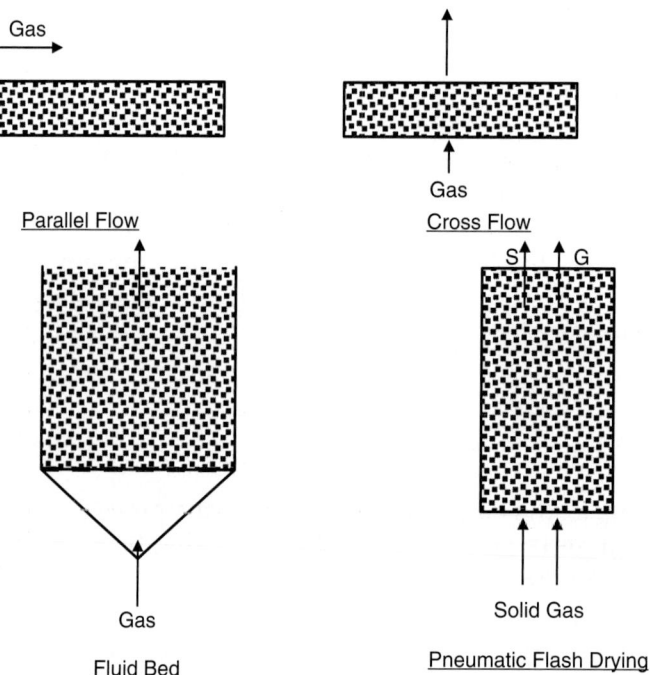

Figure 25.7 Common types of adiabatic dryers (direct heat dryers) and the contacting pattern of gas and solid.

Dryers are seldom designed by the user and usually bought from specialized equipment dealers. Nevertheless the analysis of drying represents an elegant example of simultaneous heat and mass transfer and also illuminates the different models for moisture transport in porous media (which has applications in other fields as well, for example, groundwater transport, flow in porous beds. and so on). Hence it is useful to study the basic mechanism of drying and examine how it relates to mass transfer theory. This is the focus of the rest of this chapter.

25.5.2 Types of Solids

Solids being dried are generally of two types:

- Category 1: Granular or crystalline solids where the moisture is in the open pores between particles. For these type of materials, the structure of the solid is not changed by the moisture removal.

- Category 2: Amorphous, fibrous, or gel type of solids: These are materials such as vegetables, wood, soap, and so on. Moisture is held mostly in the interior of the solid. Solid properties change with drying and considerable shrinkage can occur.

Solids may also be classified as hygroscopic or non-hygroscopic. The former contains both unbound (moisture held on the surface) and bound moisture (held chemically with the solid). Non-hygroscopic materials have only unbound moisture. The classification is important in interpretation of the drying rate and the phenomena of drying of these materials.

Moisture content denoted as X is commonly expressed as mass of moisture per gm of bone dry solid. This is same as mass fraction on a dry basis and is convenient since the mass of dry solid remains constant through the drying process. Mass percent is also used in many papers.

25.5.3 Constant and Falling Rates

Consider a porous solid that is being dried in presence of hot air. The process is again one of simultaneous heat and mass transfer. Heat is transferred to the solid and moisture is removed from the solid. An illustrative plot of moisture content of the solid as a function of time is shown in Figure 25.8. The region A to B is the preheating period where the solid is heated to the wet bulb

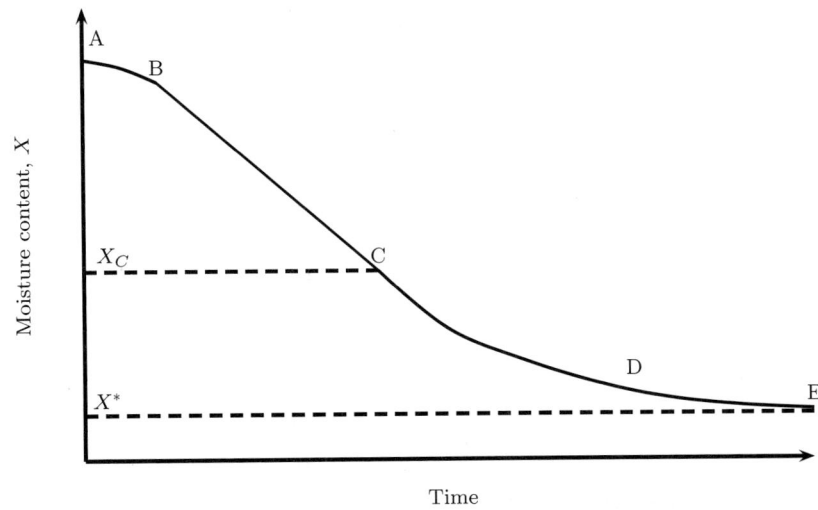

Figure 25.8 Illustrative plot of moisture content versus time showing various regions for drying. AB = initial heating period, BC = constant rate period, CD = falling rate period, DE = second falling rate period.

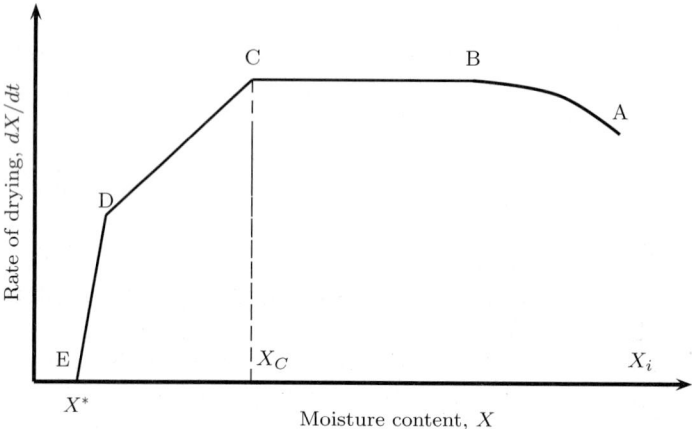

Figure 25.9 Illustrative plot of rate of drying as a function of the moisture content. C − critical moisture content; E = equilibrium moisture content.

temperature of the gas. No appreciable loss of moisture takes place during this time.

After this the solid temperature remains constant for a while at the wet bulb temperature and the moisture content decreases as a linear function of time (BC). This is called the constant rate period. The constant rate period lasts until a critical moisture content (marked as C in the figure) is reached. At this point dry patches appear on the solid and after this period, the rate of drying starts to fall. This is referred to as the falling rate period. The rate decreases even further after point D is reached. This is called the second falling rate period while region CD is called the first falling rate period. The final moisture content after a long period of time is referred to as the equilibrium moisture content. Some solids may exhibit only one falling rate period.

Results of this nature were generated by Professor Sherwood as early as 1929. If the X versus t curve is differentiated with time we get the rate of drying; this is more illustrative of the four periods described earlier (see Figure 25.9).

The mechanism and models to calculate the rate for these periods will now be discussed.

25.6 Constant Rate Period

The transport of heat and mass in the gas film surrounding the solid determines the rate of drying. In the constant rate period, the internal diffusion or capillary flow is fast enough to maintain a constant surface concentration. The surface of the solid is assumed to be wetted and covered by a liquid film and external transport in the boundary layer near the surface is assumed to be the rate controlling step (see Figure 25.10).

The rate of drying can be predicted by either mass transfer considerations or heat transfer considerations. The following equations apply respectively:

$$\dot{m} = k_\omega A(\omega_s - \omega_g)\mathcal{F} \qquad (25.29)$$

or

$$\dot{m} = \frac{hA(T_g - T_s)}{\hat{h}_{lg}} \qquad (25.30)$$

Either of the equations can be used to predict the drying rate. In one case we need to know the mass transfer coefficient from the surface to the bulk gas while in the other the heat transfer coefficient from the bulk gas to the solid surface is required. The equation from heat transfer considerations is more commonly used since the heat transfer data are usually more accurate than that for the mass transfer coefficient. The heat transfer coefficient (for parallel to the solid) is in turn usually calculated using the Mujumdar (1995) correlation:

$$Nu = 0.037 Re^{0.8} Pr^{1/3}$$

where Nu and Re are defined using $L_{ref} = L$, the length of gas flow over the surface. This correlation is similar to the Dittus-Boelter correlation with the

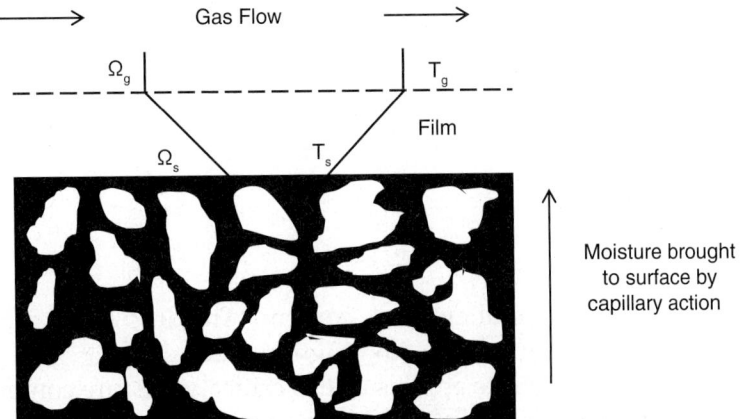

Figure 25.10 Mechanism and rate of transport in the constant rate period. Moisture is available at the surface due to rapid internal transport and the process is limited by the rate of external mass or heat transfer.

leading coefficient somewhat higher. Seader et al. (2011) presented a simplified version for flow parallel to the solid:

$$h = 0.0204G^{0.8}$$

Here G is kg/m²hr and h is the traditional SI unit of W/m²K.

Another arrangement is packed beds with through circulation of the gas. The following correlation is suggested by Mujumdar (1995) and also by Seader et al. (2011):

$$h = 0.151G^{0.59}/d_p^{0.41} \text{ for } Re \geq 350$$

and

$$h = 0.214G^{0.49}/d_p^{0.51} \text{ for } Re < 350$$

The Reynolds number is defined in the usual manner for packed beds as $d_p G/\mu$. The units are kg/m²hr for G and meter for d_p.

Correlations are available for various other dryer arrangements (e.g., fluidized beds, spouted beds, etc.) and a compact tabular summary is available in Table 18.6 Seader et al. (2011) and can be used for first-level design calculations.

It is common to use the rate of drying parameter dX/dt. For the constant rate period, this is a constant equal to R_c, which is related to $\dot{\mathcal{M}}_A$ as follows:

$$R_c = \dot{m}A/m_s$$

where m_s is the mass of bone dry solid. The change in moisture constant of the solid can then be calculated using the rate of drying:

$$\frac{dX}{dt} = -R_c \tag{25.31}$$

This gives

$$X = X_0 - R_c t \text{ for } t < t_c$$

where X_0 is the initial moisture content. (The drop in the initial moisture during the preheating period, AB, is ignored here.)

The constant rate applies until a critical moisture content, X_c, is reached. This critical value is difficult to predict and depends on many parameters including the thickness of the solid, initial porosity, the extent of shrinkage, and pore size changes during drying. There is discussion on the prediction of this parameter is later in this chapter.

The falling rate starts at $X = X_c$ and hence the time at which the falling rate starts is

$$t_c = \frac{X_0 - X_c}{R_C} \tag{25.32}$$

25.7 Falling Rate Period

Falling rate is caused by lack of moisture at the surface due to the formation of a patch of dry solid at the gas–solid surface. The thickness of the dry patch increases with time and moves inward to the solid. The rate of drying is now determined by the rate of heat and mass transport in this region. The governing equations are transient mass and heat transport in the dry region together with external transport in the boundary layer. An illustration of the situation and mechanism in the falling rate period is shown in Figure 25.11.

With further drying, a completely moisture-free region develops near the surface of the solid; this leads to the second falling rate period shown in Figure 25.12.

The method of calculation for the falling rate period depends on whether the solid is porous or non-porous. The moisture diffusion in the solid is now modeled by a diffusion type of equation for porous category 2 solids. If it is non-porous (categroy 1 granular solid) a capillary transport model is used. Some details on these two types of models are provided in later sections.

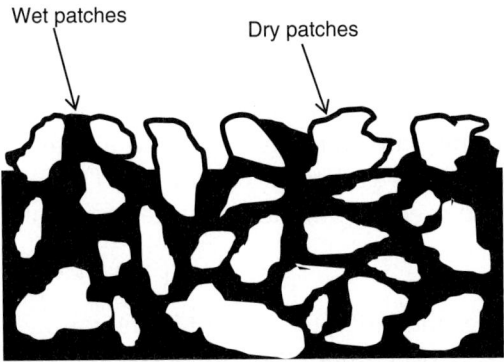

Figure 25.11 Phenomenological representation of the first falling rate period. The internal transport rate is low and not able to keep up with the evaporation rate; this results in dry patches on the surface.

Figure 25.12   Phenomenological representation of the second falling rate period. A dry region and a wet region develop in the solid; the thickness of the dry region increases with time.

However, often simple rate models are used, similar to those used in the power-law kinetics of reaction. These types of models are sometimes referred to as empirical models since they do not attempt to consider any physics of the drying into the model. They are more of a fitting type of model but nevertheless quite useful in practical applications.

25.7.1   Empirical Models

A commonly used model for the rate of drying in the falling rate period states that the rate of drying is proportional to the moisture content. Hence the change in moisture content is given as

$$\frac{dX}{dt} = -k_d X \qquad (25.33)$$

k_d is defined as the rate constant for the falling rate period.

The falling rate starts at $t = t_c$ and at this time the moisture content is equal to the critical moisture content X_C. This provides the initial conditions for the previous equation; the integrated form of the rate equation is

$$X = X_c \exp[-k_d(t - t_c)] \qquad (25.34)$$

This describes the moisture content in the falling rate period. This is coupled to the initial period where the rate of drying is constant. Combining the two periods and rearranging the equations, the time to reach a specified final content of X_f is therefore

$$t_f = \frac{X_0 - X_c}{R_C} + \frac{1}{k_d} \ln\left(\frac{X_c}{X_f}\right)$$

If the rate constants k_d and R_C are available the equation can be used for estimation of the time needed for drying. Alternatively, the rate constant can be estimated from the measured data and used to model dryer performance for other conditions. Note that R_C can be related to heat transfer coefficient. However, k_d is an empirical parameter in this model and is not correlated to any transport models.

Other rate models are also used. For example a combination of a first-order and second-order model is used in many studies (see, for example, Seader et al., 2011):

$$-\frac{dX}{dt} = k_{d1}X + k_{d2}X^2 \qquad (25.35)$$

This model needs two rate constants, k_{d1} and k_{d2}, and is useful for cases where there are two distinct falling rate periods. In the first falling period a linear model with a first-order rate constant is used while the second falling rate period where a combination of first- and second-order rate constants is used. The X versus t behavior of this model essentially involves integration of the dX/dt equation. The integration is done in two parts now corresponding to the two falling rate periods.

25.7.2 Diffusion Type of Models

In the diffusion model, the rate of drying is modeled as a pore diffusion limited process in the solid. In particular, such models are suitable for relatively non-porous solids. For such solids, most of the moisture is in the microvoids inside the solid matrix and a concentration profile is calculated by a transient diffusion model. This type of model was introduced in the drying field by Professor Sherwood himself in 1929 and has found wide applications in the drying of food and pharmaceutical products.

The model equations use the transient diffusion model shown in Chapter 8:

$$\frac{\partial X}{\partial t} = D\frac{\partial^2 X}{\partial z^2}$$

No flux condition is used at $z = 0$, the impermeable bottom of the cake or the midplane if drying is taking place from both sides.

The boundary condition at the surface, $x = L$, depends on whether there is a constant rate period or not. If there is no constant rate period, a Dirichlet condition is used, as discussed next.

Dirichlet Problem

Many solids of category 2 do not have significant constant rate period. The surface moisture attains the equilbrium value of X^* very quickly for such solids. Hence the Dirichlet condition, $X = X^*$, for the equilibrium moisture content is used as the boundary condition at $x = L$ if there is no constant rate period. This is reasonable for many category 2 type of solids where most of the moisture is held internally. Alternatively $t = 0$ can be taken as the start of the falling rate period. The internal distribution of moisture during the constant rate period is often ignored and a constant initial condition of $X = X_0$, the initial moisture content, is often used. With these conditions, the solution for transient diffusion in Section 8.2 can be readily adapted. The moisture distribution as a function of z and t is similar to Equation 8.10.

Average Moisture Content

The average moisture content at any particular time is of interest. This is given by adaptation of the equation shown in Chapter 8:

$$\frac{\langle X \rangle - X^*}{X_0 - X^*} = A_1 B_1 \exp(-\lambda^2 \tau) \tag{25.36}$$

A one-term approximation is shown, which is reasonable except for very short times, say $\tau > 0.1$. Using the values of A_1, B_1, and λ_1 from Chapter 8, the drying rate is represented as

$$\frac{\langle X \rangle - X^*}{X_0 - X^*} = 0.8106 \exp(-2.4674 Dt/L^2) \tag{25.37}$$

The data can be fitted to the above equation to find the diffusion coefficient. Then the value of D can be used to find the drying rate for other conditions. The model assumes a constant diffusion coefficient throughout the period of drying, which may not be valid in many situations as indicated by Sherwood. This change can be caused by shrinkage, case hardening, and many other factors. If the diffusion coefficient varies with time a numerical solution is called for.

Initial Period: Neumann Problem

A constant rate period is often observed for a category 1 solid and even for category 2 in many cases. The explanation for the latter case is that there is a surface film of liquid for which the evaporation is controlled by gas phase mass transfer. The boundary condition at the surface is now modified to a constant flux (Neumann type) condition:

$$-D \left(\frac{dX}{dz} \right)_{z=L} = R_C L \tag{25.38}$$

where R_C is the rate of drying in the constant rate period. An analytical solution to the Neumann problem can be derived and follows the method used for heat transfer with constant flux at the surface:

$$X(\xi, \tau) = X_0 - \frac{R_C L}{D} \left\{ \frac{\xi^2}{2} - \frac{1}{6} + \tau - \frac{2}{\pi^2} \sum_{m=1}^{\infty} \frac{(-1)^m}{m^2} \exp(-\lambda^2 \tau) \cos(\lambda \xi) \right\}$$

where the dimensionless parameters are standard as in Chapter 8 for the transient diffusion problem. These are defined as $\xi = z/L$ and $\tau = Dt/L^2$. The eigenvalue for the Neumann problem is $\lambda_m = m\pi$. The series part of the solution is important only at the start and may be ignored for $\tau > 0.1$.

The constant rate period ends when the surface moisture reaches the equilibrium value. Hence the model provides a method to find the critical moisture content. Empirical models on the other hand need experimental data for this parameter. Thus the time at which the initial rate ends is found by

$$\tau = \frac{1}{3} + \frac{X_0 - X^*}{R_C L/D}$$

The corresponding average moisture content can be calculated from the Neumann solution and is representative of X_c, the critical moisture content.

The model solution is further continued to find the time to dry the solid all the way to the equilibrium value. The surface boundary condition is now switched to the Dirichlet condition with a value of X^* at $z = L$. A parabolic profile for the moisture would have developed inside the slab during the initial rate period. This is used as the initial condition for the solution in the falling rate period.

A solver like PDEPE is useful for simulation and the two periods can be solved in a sequential manner. One can start off with the Neumann boundary condition to simulate the initial rate period. Once the surface moisture value reaches X^*, one switches the boundary condition to Dirichlet with this value and continues the simulation for the falling rate period. The time at which this happens is the end of the constant rate period. The internal moisture content at this time (usually a parabolic profile) is already available as part of the simulation and is automatically used as the initial condition for the falling rate period.

Often the diffusion coefficient is not a constant and changes as a function of the moisture content. Such cases are best handled numerically, for example, using the PDEPE solver.

25.7.3 Capillary Flow Models

The capillary flow model is useful when moisture is present as free moisture in the interstices of the particles rather than in the interior pores of the solid. The capillary model assumes a bed of non-porous spheres surrounding a pore space. Hence it is suitable for granular solids (category 1). Here the liquid is assumed to move by capillary action similar to a wick in a candle. The local variation liquid holdup is assumed to drive the flow of the liquid to the surface. Thus the flow is defined as

$$j_L = -K_L \rho_L \frac{d\epsilon_L}{dz}$$

where K_L is an effective mass conductivity in the porous media. Similarly the vapor flux is modeled as

$$j_V = -D_V (\epsilon_G - \epsilon_L) \frac{d\rho_G}{dz}$$

where ρ_G is the vapor density in the gas phase. These "transport models" are then coupled to the conservation model, leading to two equations: one for the liquid holdup profile and the second for the vapor concentration profile in the solid. Details of the model are presented in Chen and Pei (1989). These authors also distinguished two falling rate periods separately and developed models for each case.

25.7.4 Choosing a Model

Whether the diffusion or the capillary model should be used depends on the nature of the solid. Perry and Green (2007) give the following guidelines.

The diffusion model is useful for single-phase solid systems such as gelatin, soap, and glue. It is useful for other solids such as wood and can also be used in the later stages of drying of materials such as textiles, paper, and so on.

The capillary model is useful for coarse grain solids such as sand, paint, and pigments. In view of the complex transport mechanism associated with capillary flow, it is more common to use diffusion models with a fitted value of the effective diffusivity even for these cases.

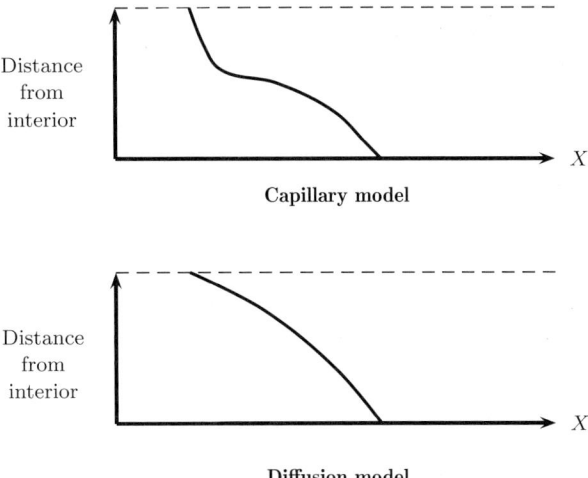

Figure 25.13 Profiles of internal moisture content for two types of models.

The internal moisture profiles are different for these two types of models. Capillary flow is associated with an internal moisture profile showing a double curvature and a point of inflection. The diffusion model is associated with a smooth curve. The differences in the two models are shown in Figure 25.13. The measurement of the internal moisture profile at various stages of drying may provide some information on which type of model is more suitable.

Summary

- Humidification and drying are two important processes where the analysis of simultaneous heat and mass transfer considerations are needed. Humidification refers to transfer of moisture from a liquid to an inert gas such as air. Drying refers to transfer of moisture from a wet solid to air.
- A liquid exposed to a gas containing an unsaturated vapor will reach a lower temperature than the gas due to evaporative cooling. This phenomena can be used to cool large volumes of liquid and finds application in cooling towers. Large-scale industrial applications are found in power plants. Countercurrent flow and cross-flow are the two common types of operation.
- Models for cooling towers are developed using the mass and energy balances together with transport equations for heat and mass. Design

models are based on a simplified version of these equations and the Merkel equation is commonly used for design.

- The difference between the enthalpy of a saturated gas at liquid temperature and the actual enthalpy of the gas at the gas temperature is taken as the driving force in the Merkel equation. The heat balance in the liquid can then be modeled using this driving force multiplied by a mass transfer coefficient. The integrated form of the enthalpy difference can be expressed in terms of a NTU parameter. The transport coefficient part of the model is usually recast into a HTU parameter and the data for this parameter is available for a number of internal designs of the cooling tower. The height of the packed part of the tower is then the product of the HTU and NTU, a similar concept to that used for absorption columns.

- The model for cross-flow design follows an approach similar to counterflow and usually the Merkel equation is used. However, due to the change in flow directions, the resulting equations are a set of partial differential equations for the temperature of the liquid in the downward direction and the enthalpy of the gas in the flow direction. This set of equations can be solved by a finite difference mesh.

- Drying, as the name implies, refers to the process of removal of moisture (or other liquid) content of a solid by heating or by contacting with a hot gas. It is often the final step of a series of operations and the dried solid is usually ready for final packaging. Direct contact and indirect contact are the two common modes of drying in practice.

- Drying of solids is characterized by parameters related to the type of solid being dried; these parameters determine the various period of drying. A constant and a falling rate period are usually observed.

- In the constant drying period there is enough moisture even near the surface of the solid. The surface of the solid remains at the wet bulb temperature corresponding to the conditions in the gas phase. The rate of drying can be determined from the mass transfer rate from the surface of the solid to the gas or from the heat transfer rate from the gas to the solid.

- The rate of drying in the falling period is determined by the transport of moisture from the interior of the solid to the surface. Two models are used to characterize the internal transport: the simple diffusion type of model and the capillary flow model. What model to use depends on the type of solid, whether it is an amorphous gel type of solid with the moisture trapped in fine interior pores or of a granular type where the moisture is held in open pores between the particles.

- Simple power-law models are often used for the falling rate period in lieu of a detailed transport phenomena–based model. These empirical models are fitted to data and then used to examine the parametric effects for the time needed for drying.

Review Questions

25.1 What is meant by the wet and dry bulb temperature?

25.2 State when the wet and dry bulb temperature can be the same.

25.3 State the Lewis relation.

25.4 What is meant by range and approach temperature in a cooling tower?

25.5 Up to what temperature can water be cooled in a cooling tower given the inlet conditions of the gas phase?

25.6 What is the air flow mechanism in a natural circulation cooling tower?

25.7 Can the water to air flow ratio be fixed *a priori* in a natural circulation tower?

25.8 How is the model for countercurrent flow different from cross-flow in a mathematical sense?

25.9 What is meant by critical moisture content of a solid?

25.10 What is meant by equilibrium moisture content of a solid?

25.11 Why does the temperature of the solid remains constant in the initial period of drying?

25.12 What causes the falling rate period in drying?

25.13 Distinguish between the first falling rate period and the second falling rate period.

25.14 What are the two main theories applied to model the falling rate period?

Problems

25.1 **Preliminary design calculation for a cooling tower.** 2000 kg/minute of water is to be cooled from 60 °C to 25 °C in a cooling tower by countercurrent contact with air. The air flow rate is 20 mol/sec with a dry bulb temperature of 30 °C and a dew point of 10 °C. Determine the size of the cooling tower. Use an HTU value of 2m.

25.2 **NTU calculation using the Merkel equation.** A cooling tower is used to cool 50 °C water to yield an approach temperature of 5 °C when the entering air wet bulb temperature is 25 °C. The liquid to gas mass flow ratio was taken as 1.25. Find the Merkel integral and the NTU value.

25.3   Tower NTU and parametric effects. A cooling tower is designed for an approach temperature of 5 °C and a cooling range of 10 °C when the wet bulb temperature is 20 °C. Air flow is estimated as 10,000 m³/min and water flow is 18 m³/min. The tower has 2 m packed section and is 1.8 by 1.8 m in cross-section. Determine the inlet air enthalpy. Find the NTU and HTU. If a new design to handle a wet bulb of 3 °C for the air and an approach of 3 °C is needed, what modifications would you suggest?

25.4   Approach and range temperature calculations. A cooling tower operates with inlet and exit water temperature of 40 °C and 28 °C. Entering air has a temperature of 25 °C and a relative humidity of 35%. The tower has 1.2 m plastic fill and the mass velocities are $G = 2000$ kg/m²sec and $L = 2200$ kg/m²s for gas and liquid, respectively. Find the number of transfer units, the height of the transfer unit, and the temperature approach. If the cooling tower load remains the same and the air temperature drops to 20 °C with a humidity of 70%, calculate the range and the exit temperature of the water.

25.5   Simulation of a cross-flow tower. Simulate a cross-flow cooling tower using the cell-by-cell approach for the following conditions: height $H = 5$ m, depth $L = 2$ m, width $W = 20$ m, liquid flow rate = 100 kg/sec at a temperature of 38 °C, air flow rate is 100 kg/s at 28 °C and a relative humidity of 60%. Using the simulation model, examine the effect of air temperature and humidity on the outlet water temperature.

25.6   Heat transfer coefficient and time for constant rate period. A filter cake of $CaCO_3$ is being dried in a tray dryer with air at 80 °C and a relative humidity of 10% with air flowing parallel to the cake. The air velocity is 4 m/s. The surface area is 1.5 m². Initial moisture content is 30% on a dry basis and the critical moisture content is 10%. Find the heat transfer coefficient and time for the constant rate period.

25.7   Time for drying. A porous solid is in the form of a cake that is 52 mm thick with an area 610 square mm. It is exposed to air with a wet bulb temperature of 26 °C and a dry bulb temperature of 71.1 °C. Air flows parallel to the cake with a velocity of 2.44 m/s. The dry density of the cake is 1922 kg/m³. The critical moisture content is 9%, and the initial moisture content is 20%. Find the time needed to dry the solid to 1%.

25.8   Parameter fitting and estimation of time of drying. A porous solid is dried in a batch drier under constant drying conditions. The critical moisture content of the solid was determined to be 20% and the equilibrium moisture content was 4%. The moisture content was reduced from 35% to 20% in two hours and from 35% to 10% in seven hours. Find the time needed to reduce the moisture further to 5%.

25.9   Mass transfer coefficient correlation. A catalyst with a moisture content of 20% is to be dried and loaded into a packed column with air blown through it. The air enters at 100 °C with a relative humidity of 20%. Assume that the rate of drying in the initial period is controlled by heat and mass transfer from the surface.

Find the j_D factor and the mass transfer coefficient. Use this to calculate the rate of drying. Repeat for a fluid bed dryer where the bed is fluidized with a void fraction of 0.5.

25.10 **Estimation of the diffusion coefficient in the diffusion model.** The following data were obtained in a material being dried:

Time, hour	0.9	2.89	4.02	5.82	8.98
Moisture content, X	1.53	0.267	0.245	0.218	0.183

The initial moisture content was 39.7% on a dry basis and the equilibrium moisture content was 5%. The material was 1.9 cm. Find the diffusion coefficient of moisture in the system by fitting the diffusion model. Assume that all the drying takes place in the falling rate period. Use the data to find the drying time to reduce the moisture to 6%.

25.11 **Critical moisture content.** In the constant rate period, the diffusion model with the Neumann condition holds. Consider the drying of a material with an initial content of 0.3, an equilibrium content of 0.05, and a density of bone dry solid of 1600 kg/m^3. The rate of drying in the constant rate period was measured for a half-thickness of 1.0 cm as 0.2 g/cm^{2}hour. Find the critical moisture content and the time for drying in the constant rate period. Assume a diffusion coefficient of 0.5 cm^2/hour.

25.12 **Drying simulation with PDEPE solver.** Consider drying of a material with the following properties: initial content of 0.3, equilibrium content of 0.05, density of bone dry solid = 1600 kg/m^3, rate of drying in the constant rate period = 0.3g/cm^{2}hour, half thickness of the slab = 1 cm. Set up the PDEPE solver with the Neumann condition to simulate the moisture profile in the initial constant rate period. Execute the simulation up to the time for which the surface moisture reaches a value equal to the equilibrium content. The diffusion coefficient was estimated as 0.3 cm^2/hour.

Stop the simulation at this time as you need to switch to the Dirichlet condition (the falling rate period). The initial profile is available from the simulation using the Neumann model. Set up the Dirichlet model to simulate the falling rate period. Continue the simulation until the average moisture content reaches a value of 0.06.

25.13 **Comparison of the rate of drying for the diffusion model and the capillary model.** Show that for the diffusion model, the rate of drying varies inversely as the square of the solid thickness. The transport of moisture in a capillary model can be modeled on the basis of a laminar flow in a capillary from the interior to the surface. Show that the rate of drying based on this concept is linearly proportional to the average moisture content. Further show that the rate of drying is now inversely proportional to the solid thickness.

CHAPTER 26

Condensation

Learning Objectives

After completing this chapter, you will be able to:

- Understand the Nusselt model for prediction of the heat transfer coefficient for film condensation.
- Calculate the condensation rate of a pure vapor.
- Model the local rate for condensation of a mixture of vapor and an inert gas.
- Identify the conditions for fog formation in general and in condensing equipment in particular.
- Model the heat and mass transfer processes in the condensation of a binary mixture of two condensing species.
- Analyze the condensation of a ternary mixture consisting of two condensing species with an non-condensing gas.
- Write equations for calculation of the performance of condensing equipment based on a stagewise description of the condenser.

Condensation refers to a process where a species (or multiple species) from a vapor phase is transported to a cold surface or cold liquid and condenses and forms a liquid phase (usually called the condensate) at the surface. The direction of mass transfer is from the vapor to the liquid. This is the opposite of drying, where the mass transfer is from the liquid to the vapor phase. A temperature difference between the vapor and the condensing phase is needed for the process to occur. Hence a simultaneous mass and heat transfer from the vapor to the liquid is involved. Thus condensation represents another example of a system where an analysis of simultaneous heat and mass transfer is required.

Various types of equipment can be used for condensation. Commonly a shell and tube type of arrangement is used similar to that for heat exchangers. The vapor is usually on the shell side with the coolant flowing on the tube side. Either vertical or horizontal arrangements can be used. In both cases, it is necessary to find the local rate of condensation at any point in the system. It is the goal of this chapter to examine the local mass and heat transfer effects and calculate the rate of condensation for a given vapor composition and for a given condition of the condensing film or the surface. The local model can then be incorporated as a submodel in the design of the equipment. Thus the basic learning objective is to understand the effects of simultaneous heat and mass transfer in a condensing system and use these to calculate the rate of condensation for a given condition. A simple example of coupling the local model to the equipment-level model is given, but the complete design of condensation equipment is not considered here.

Four cases are analyzed here: condensation of a pure vapor, condensation of a vapor and inert gas (non-condensible) mixture, condensation of a binary gas mixture, and a ternary system with two condensing species and an inert gas. The last case is modeled best by Stefan-Maxwell relations.

A related phenomena is fog formation, which is important per se in environmental applications. It has also been observed in industrial condensing equipment and we briefly discuss the role of mass transfer in fog formation based on the film model and examine conditions under which fog formation can be expected during the condensation of a vapor with an inert non-condensing component.

26.1   Condensation of Pure Vapor

This section is more on the heat transfer aspects of condensation and does not include mass transfer considerations since a pure vapor is involved. The condensing temperature depends only on the pressure and is treated as a constant.

Condensation can be of two types: dropwise or filmwise. Dropwise condensation refers to a process where the condensed vapor forms drops on the surface rather than a continuous film. Since a large portion of the surface is exposed to the gas, the condensation rates are rather high. Heat transfer coefficients as high as $250,000$ W/m^2 can be achieved. However, there are two difficulties in using this effectively: it is difficult to maintain for a sustained period of time and the resistance on the coolant side is often the controlling resistance, which means the heat transfer rate is then determined by how fast the heat can be removed on the coolant side. The high value of the heat transfer coefficient from vapor to drop no longer matters!

In filmwise condensation the surface over which condensation occurs is blanketed by a layer of liquid film. This liquid film provides a resistance to heat transfer and therefore the heat transfer coefficients are much lower compared to dropwise condensation. The film thickness also increases as a function of the height or the distance along the condensing surface. Hence the heat transfer coefficient decreases along the height. In practical applications the filmwise condensation is usually encountered due to difficulties in maintaining the dropwise condensation for a prolonged period of operation. Further discussion in this section pertains to filmwise condensation and dropwise condensation is not discussed.

Condensate forms a film on the condensing surface and the liquid flows down as a thin film. It is useful to characterize the various flow patterns of film flow at this stage to give students a clearer understanding of the range of applications of the various correlations and/or models that are commonly used in practice. The flow regimes are laminar, wavy, and turbulent. The transition from laminar to wavy is around a Reynolds number (defined for film flow as $Re = 4 \langle v \rangle \delta / \nu_L$) of 30 while the flow is turbulent beyond a Reynolds number of 1500. **Note:** The Reynolds number for film is usually written as $4\Gamma/\mu_L$, where Γ is the mass flow rate per unit perimeter of the condensing surface (since $\Gamma - \delta \langle v \rangle \rho_L$ per unit perimeter).

26.1.1   Laminar Regime: Nusselt Model

The problem of laminar condensation was analyzed by Nusselt (1916). The schematic basis of the model is shown in Figure 26.1 where a condensate film of thickness δ is assumed to form on the condensing wall.

The film thickness increases as a function of x, the vertical distance along the wall. The local heat transfer coefficient, h_x, depends on the local film thickness and can be calculated as

$$h_x = \frac{k_l}{\delta(x)} \tag{26.1}$$

where k_l is the thermal conductivity of the liquid.

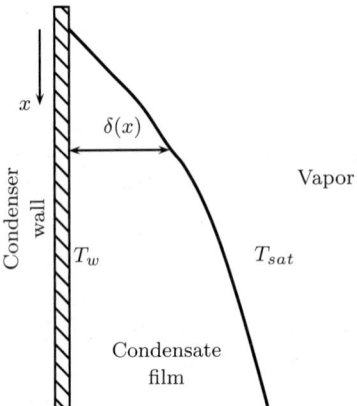

Figure 26.1 Schematic for the model used in Nusselt analysis for filmwise condensation in a laminar falling film.

The film thickness can be calculated using the mass and momentum balance if the operating conditions are in laminar mode:

$$\delta(x) = \left[\frac{4x\mu_L k_l(T_{sat} - T_w)}{g\rho_L(\rho_L - \rho_v)\hat{h}_{lg}} \right]^{1/4} \tag{26.2}$$

The model is called the Nusselt model for condensation. Details leading to this equation are left as an exercise problem. Note that $\rho_L - \rho_v$ is usually approximated as ρ_L. This approximation is used from here forward.

Hence the local heat transfer coefficient is given as

$$h_x = \left(\frac{k_l^3 g\rho_L^2 \hat{h}_{lg}}{4\mu_L(T_{sat} - T_w)x} \right)^{1/4} \tag{26.3}$$

The average rate over a length of a tube L is given as

$$h_L = \frac{1}{L} \int_0^L h_x dx \tag{26.4}$$

Integrating, the following relation can be derived:

$$h_L = 0.943 \left(\frac{k_l^3 g\rho_L^2 g\hat{h}_{lg}}{L(T_{sat} - T_w)\mu_L} \right)^{1/4} \tag{26.5}$$

The equation is valid up to a Reynolds number of 30. (**Note:** The factor $(4/3)/4^{1/4}$ arises from the integration and has a value of 0.943, which is the leading constant in the above equation for h_L.)

The condensate flow rate (per unit perimeter) at the end of the tube is of interest and can be calculated from an overall heat balance:

$$\Gamma = h_L(T_{sat} - T_w)L/\hat{h}_{gl}$$

The Reynolds number, $4\Gamma/\mu_L$, corresponding to this has to be calculated in order to check if the flow is laminar and has to be less than 30. The flow becomes wavy if the Reynolds number is larger than 30.

Corrections for Subcooling and Superheat

The Nusselt model described above does not account for the subcooling of the liquid. A correction is usually applied. The latent heat of vaporization term, $\hat{h}_{gl}$, in Equation 26.2 is usually modified to include the liquid subcooling and is replaced by $\hat{h}_{gl}^*$, which is defined as

$$\hat{h}_{gl}^* = \hat{h}_{gl} + 0.68c_{pL}(T_{sat} - T_w)$$

If the vapor enters as a superheated fluid at a temperature of T_v instead of at T_{sat}, a further correction to the latent heat is applied:

$$\hat{h}_{gl}^* = \hat{h}_{gl} + 0.68c_{pL}(T_{sat} - T_w) + c_{pv}(T_v - T_{sat})$$

26.1.2 Wavy and Turbulent Regime

The wavy regime occurs for $Re > 30$. The effect of waves on the surface is to increase the rate of condensation. However, a detailed model-based analysis is complicated and analytical solutions are not possible unlike the case of laminar flow. The increase in heat transfer is on the order of 20% but can be as high as 50%. The laminar equation can still be used for a conservative design as it will underpredict the heat transfer coefficient. A correction factor of $0.8Re^{0.11}$ was recommended by McAdams (1954). The vapor velocity is assumed to small in the Nusselt analysis and its effect is neglected.

Turbulent Regime

For the turbulent regime, empirical correlations are commonly used. A simple correlation of McAdams (1954) is often used:

$$h = 0.0077 \left[\frac{\rho_L g(\rho_L - \rho_g)k_l^3}{\mu_L^2} \right]^{1/3} Re^{0.4} \qquad (26.6)$$

The local and average heat transfer coefficients are the same in turbulent films. Wavy and turbulent regimes can be modeled in detail by using computational fluid dynamic tools.

26.2   Condensation of a Vapor with a Non-Condensible Gas

In this section we consider the condensation of a species present in the gas along with an inert non-condensing gas. The species has to diffuse from the bulk gas to the interface and hence mass transfer considerations are to be included in addition to heat transfer models. The film model is widely used for local mass and heat transfer analysis. The species mole fraction and temperature profiles are shown in Figure 26.2. The mole fraction varies only in the vapor film since the condensate is pure liquid. The temperature varies in both the vapor film and liquid and also on the coolant side as shown in Figure 26.2.

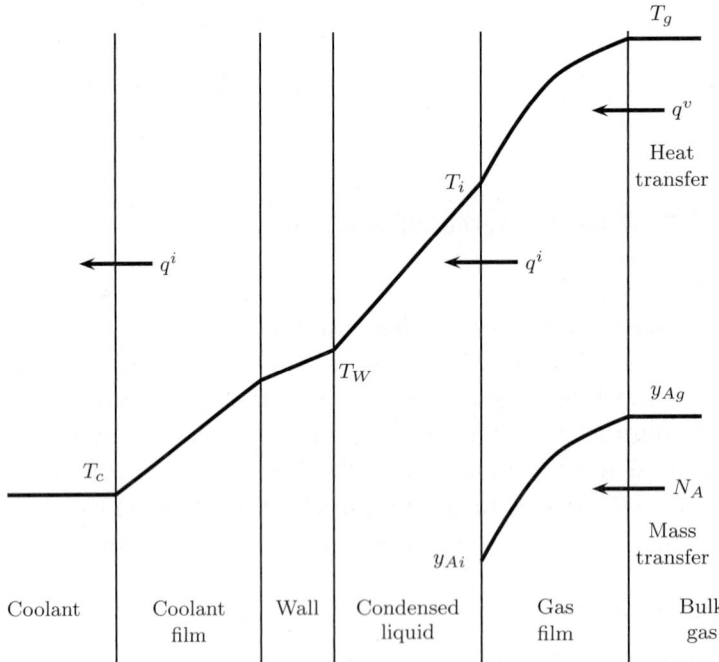

Figure 26.2   Film model for condensation showing the mole fraction profile in the vapor film and the temperature profiles in the liquid and the coolant. Adapted from Thermopedia, A–Z Index.

26.2.1 Mass Transfer Rate

The mass transfer in the vapor film is modeled as a UMD (unimolecular diffusion) process and the rate of mass transport is given as

$$N_A = C k_m^\circ \ln\left(\frac{1 - y_{Ai}}{1 - y_{Ag}}\right) \qquad (26.7)$$

y_{Ag} is the local mole fraction in the bulk gas. y_{Ai} is the mole fraction at the interface and will depend on the interface temperature, which has to be calculated based on heat transfer consideration. The equation for the calculation of the interface temperature will be developed shortly in this section.

26.2.2 Heat Transfer Rate and Ackermann Correction Factor

The heat transfer rate is enhanced by mass transfer due to diffusion-induced convection. An augmentation factor ϕ is used to calculate the enhancement. Hence the heat transfer rate in the vapor film is

$$q^v = h_g^\circ(T_g - T_i)\mathcal{F}_h \qquad (26.8)$$

$\mathcal{F}_h$ is the augmentation factor, which applies for the total heat transfer and h_g^0 is the heat transfer coefficient if there were no augmentation due to convection. An expression for augmentation can be derived by including convection caused by mass transfer in the heat conduction in the film. The resulting expression is

$$\mathcal{F}_h = \frac{\phi \exp(\phi)}{\exp(\phi) - 1} \qquad (26.9)$$

where ϕ is the blowing factor caused by mass transfer and is defined as

$$\phi = \frac{N_A C_{pg,A}}{h_g^\circ}$$

The heat transfer rate from the bulk gas to the interface is often written in the literature in the following alternate form:

$$q^v = h_g^\circ(T_g - T_i)\mathcal{E}_h + N_A C_{pg,A}(T_g - T_i) \qquad (26.10)$$

The first term on the right-hand side accounts for the conduction part of the flux (now enhanced due to mass transfer) while the second term accounts for the sensible heat change across the film. The factor $\mathcal{E}_h$ is the augmentation factor for the (film) conductive part of the flux only and not for the total heat transfer. This factor is called the Ackermann correction factor. Comparing the two expressions, Equations 26.8 and 26.10, we find that

$$\mathcal{E}_h = \frac{\phi}{\exp(\phi) - 1}$$

The use of this factor and the use of Equation 26.10 is more common in some of the older works on condensation heat transfer. The preceding discussion shows that the two formulations are entirely equivalent. Either factor can be used.

The limiting cases of the Ackermann correction factor should also be noted:

- For low ϕ, $\mathcal{E}_h$ equals one and the heat transfer coefficient is not enhanced by mass transport.
- For large ϕ, $\mathcal{E}_h$ equals zero and the heat transfer rate depends only on the sensible heat change from T_g to T_i.

26.2.3 Interface Temperature Calculations

The condensation equipment design requires the estimation of the local mass flux, N_A, and the local rate of heat transport in the vapor film, q^v. The value for the interface temperature is required; this is determined by a heat balance at the interface.

The total heat released at the interface is equal to the rate of heat transport in the gas film plus the energy released at the interface due to the latent heat of condensation. The latter depends on the rate of mass transfer, the rate at which condensation is taking place. The total heat released is therefore equal to

$$q^i = q^v + N_A \Delta H_c \tag{26.11}$$

where ΔH_c is the molar latent heat of condensation. A positive value for $\hat{h}_{gl}$ is implied here.

The heat transferred at the interface is then equal to the heat transfer in the liquid to the walls of the condenser:

$$q^i = h_l(T_i - T_{wall}) \tag{26.12}$$

Hence the interfacial temperature is determined by equating the two equations for q^i:

$$h_g^\circ(T_g - T_i)\mathcal{F}_h + N_A \Delta H_c = h_L(T_i - T_{wall})$$

where h_L is the heat transfer coefficient in the liquid film.

Again the wall temperature may not be known and only the temperature of the cooling liquid may be known. In this case Equation 26.12 is replaced by

$$q^i = U_i(T_i - T_c) \qquad (26.13)$$

where U_i is the overall heat transfer coefficient from the interface to the cooling liquid and T_c is the (local) temperature of the coolant at the point under consideration.

Sensible Heat Correction

The analysis does not include the sensible heat change of the condensed liquid from T_i to T_{wall}. An additional term has to be added to account for this. The term can be neglected as a first approximation. More precisely, the calculation of the sensible heat loss requires the calculation of the average temperature of the liquid film. A simple way to account for this is to augment the value of the latent heat of vaporization by adding a correction term as shown below:

$$\Delta H_c = \Delta H_c + (3/8)C_{pL}(T_i - T_{wall})$$

The augmented value, ΔH_c, is then used in place of ΔH_c to account for the sensible heat change in the liquid. The factor 3/8 can be viewed as the contribution from the average temperature of the liquid. The condensation rate calculation is illustrated in Example 26.1

Example 26.1   Rate of Condensation from a Water + Air Mixture

Water is condensing on a surface at 310 K. The gas mixture has 65% water vapor and is at a temperature of 370 K. The total pressure is 1 atm. Calculate the rate of condensation.

Use the following data for the physical properties: specific heat of water vapor $C_{pg,L} = 37.06$ J/mol K; specific heat of liquid water = 75.13 J/mol K; heat of vaporization = 43,000 J/mol.

The Antoine equation constants values for water are $A = 18.3036$, $B = 3816.44$, and $C = 46.13$ with temperature in °C.

Use the following values for the transport parameters: gas-side heat transfer coefficient = 12 W/m²K; gas-side mass transfer coefficient = 0.1 m/s; liquid-side heat transfer coefficient = 400 W/m²K.

Note that the liquid side mass transfer coefficient is not needed since the liquid is a pure component.

Solution

The interface temperature is dictated by the relative rates of heat and mass transfer. This is unknown. Once the interface temperature is found the problem is essentially solved. Hence the solution procedure is simply solving for the interface temperature in an iterative manner. The steps are outlined here.

We start with an assumed interface temperature, T_i. Let $T_i = 314$ K. The vapor pressure at the interface is calculated using the Antoine equation. Note that the units are in mm Hg and need to be converted to pascals. $P_{vap} = 58.325$ mm Hg. This sets the mole fraction of water at the gas side of the interface: $y_i = p_{vap}/P = 58.325/760 = 0.0768$.

The mass transfer rate across the gas film can now be calculated using Equation 26.7. The total concentration needed here is calculated using the ideal gas law at an average film temperature: $T_f = (T_i + T_g)/2 = 342$ K.

$$C = \frac{P}{R_g T_f} = 35.61 \text{ mol/m}^3$$

The total concentration based on the bulk gas temperature would be 32.93. Hence no significant errors would be expected in using any of these values for the temperature. Now using Equation 26.7, we have $N_A = 0.2979 \text{ mol/s.m}^2$.

The heat transfer to the interface can now be calculated using Equations 26.9 and 26.8. The following values are found; students should verify these: Peclet factor, $\phi = 0.92$, $\mathcal{F}_h = 1.5296$, and finally $q = h(T_g - T_i)\mathcal{F}_h = 1050 \text{ W/m}^2$.

The rate of total heat release at the interface can now be calculated. This is the sum of the rate of heat transfer to the interface plus the heat released by the latent heat of condensation. The latter is proportional to the mass transfer rate: $q(total) = 1698 \text{ W/m}^2$.

The heat released at the interface is equal to the heat transferred across the liquid film to the cold walls: $q(total) = h_L(T_i - T_{wall})$.

The interface temperature can therefore be calculated as

$$T_i(calculated) = T_{wall} + \frac{q(total)}{h_L}$$

The calculated value is 314.2 K, which differs from the assumed value by only 0.2 °C. The calculations can be repeated if needed if the initial guess of the interface temperature is different from the calculated value.

Note: The effect of subcooling of the liquid was neglected here. It can be shown that it adds only an additional 2% to the total heat flux at the interface and hence does not introduce significant errors in the interface temperature.

26.2.4 Condenser Model

The local film transport model can be incorporated into the models for the vapor phase, molar flow rate in the liquid, and coolant temperature. It is easier to set up the mass and energy balance in finite difference models for each differential section of the condenser and solve these sequentially. The model equations are similar to those for humidification shown in Chapter 25. An additional equation for coolant temperature is included. The model equations are presented in Section 26.5 in a more general setting. It is fairly straightforward to set up a MATLAB program to use these equations and march in the z-direction. A local model to calculate N_A, q^v (which in turn needs T_i) and q^i is set up separately and used in the marching scheme.

26.3   Fog Formation

What leads to fog formation? The vapor can cool below the dew point temperature in the film, leading to supersaturation. The vapor then condenses as tiny mist or fog rather than condensing into existing liquid film. This leads to fog formation. Thus if a hot vapor cools faster than the rate of mass transfer and maintains its humidity, it can start condensing in the vapor phase itself, leading to a mist or fog. The key parameter is the Lewis number, Le, the ratio of thermal diffusivity to mass diffusivity. If $Le > 1$, vapor cools faster than the rate of mass transfer and fog formation can occur.

The temperature and concentration during the process of cooling of a vapor–inert gas mixture is shown in Figure 26.3.

Three possibilities can arise: the p_A versus T values in the vapor film can touch the equilibrium line tangentially, which occurs for $Le = 1$; the line can be beneath the equilibrium curve for $Le < 1$ and no fog formation can be expected; and the curve can intersect and cross the equilibrium line before the interface point is reached. In this case there is a region where the vapor is supersaturated; this can lead to fog formation in the vapor film near the interface. This happens for $Le > 1$. For example, in condensing organic vapors from air, the Schmidt number is in general larger than the Prandtl number. This leads to $Le > 1$; the cooling proceeds faster than condensation, leading to the danger of fog formation.

For $Le < 1$ the local temperature in the film is sufficiently high to keep y below the corresponding saturation value and fog formation does not occur. For example for an air–water system, Le is 0.87 and fog formation is not likely in condensing such a mixture.

It is also useful to look at this from a film theory point of view. Recall from Chapter 25 that the Lewis number (Le) is a dimensionless number defined as the ratio of thermal diffusivity to mass diffusivity. Hence the thermal film

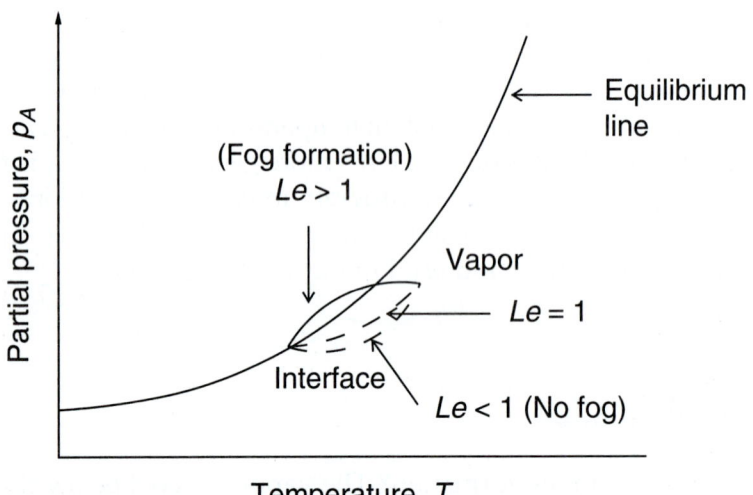

Figure 26.3 Concentration and temperature profiles during cooling of a vapor, indicating the conditions for fog formation.

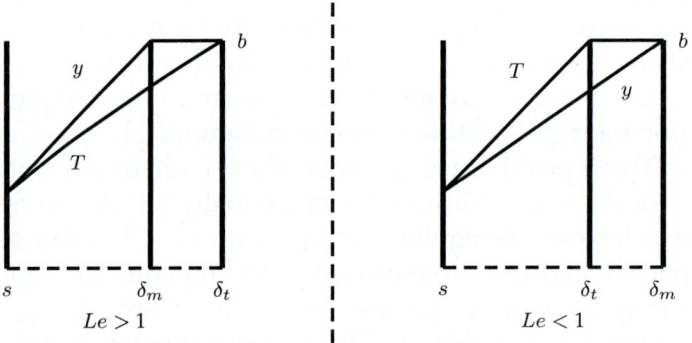

Figure 26.4 Mole fraction and temperature profile in the film for $Le > 1$ and $Le < 1$. Note the film thicknesses for heat and mass transfer are not equal when $Le \neq 1$. b denotes the bulk gas and s denotes the liquid–vapor interface. Note that the temperature profile is shown assuming no enhancement in heat transfer due to mass transfer.

thickness is larger than the mass transfer film thickness if the Lewis number is greater than one. The expected concentration profile is shown in Figure 26.4.

For $Le > 1$, the temperature profile in the film would be lower than that for the case of $Le = 1$. The vapor mole fraction at some point in the film can therefore exceed the saturation mole fraction. This can cause a local supersaturation and lead to local condensation in the gas phase as fog. Note that an excursion into the supersaturation zone does not necessarily imply fog formation. A critical supersaturation ratio has to be reached and the presence of nucleating aerosols may be needed. Thus fog formation is less likely in clean gases compared to a dusty mixture. The fine dust particles act as nucleation sites.

The most comprehensive work on fog formation in condensers is Amelin (1967) in which process calculations, nucleation, and the droplet growth of pure substances are described. Steinmeyer (1972) elucidated the industrial significance of fog formation in condensers and gave rules for minimizing fog problems. Brouwers (1992) presented detailed models.

26.4   Condensation of Binary Gas Mixture

The problem of condensation from a binary mixture of species A and B is considered next. Both species are assumed to be condensable. An example would be the condensation of a vapor leaving the top of a distillation column.

The mass transport part in the vapor film is now solved as a two-component diffusion problem. Thus both N_A and N_B need to be calculated. The liquid phase composition at the vapor–liquid interface is needed for this calculation (in addition to an estimate of the interfacial temperature). This is often fixed by the relative rates of condensation. Thus

$$x_{Ai} = \frac{N_A}{N_A + N_B} = \frac{N_A}{N_t} \tag{26.14}$$

The mole fraction of B is equal to $1 - x_{Ai}$ for a binary system.

This assumption is often referred to the unmixed film model assumption (Colburn and Drew, 1937; Taylor and Krishna, 1993) and implies a large resistance to mass transfer on the liquid side or low condensation rate. More details leading to the assumption of the unmixed model are provided in Section 26.5. The model is useful for the horizontal type of condenser where the condensate is formed from the vapor phase and there are no additional flow terms that cause concentration changes in the bulk liquid. This is also valid at the top of a vertical condenser where the downward liquid flow rate is small. Since the interfacial concentration is directly related to the condensation rates, the values of liquid-side mass transfer coefficient are not needed when using this assumption.

An alternative assumption is called the completely mixed film. Here x_{Ai} is taken as the bulk liquid value x_{AL} and bulk liquid concentrations are specified or calculated from the overall mass balance in the bulk liquid. This would imply no resistance on the liquid side for mass transfer. Again the values of liquid-side mass transfer coefficients are not needed. We proceed with the unmixed film assumption here and derive an equation for the calculation of the condensation rates from equilibrium considerations at the interface.

26.4.1 Condensation Rates: Unmixed Model

The calculations start off by assuming an interface temperature. This is needed to assign the equilibrium composition at the interface and complete the mass flux calculations. The assumed interface temperature is then iteratively corrected in order to satisfy the heat balance requirement.

First the K values are calculated from the vapor pressure data based on the (assumed) interface temperature. The vapor composition at the interface is related to the interfacial liquid concentration by a suitable equilibrium relation. Raoult's model is used here:

$$y_{Ai} = K_A x_{Ai} \tag{26.15}$$

This is valid for ideal solutions while for non-ideal systems an activity coefficient correction is used:

$$y_{Ai} = \gamma_A K_A x_{Ai}$$

where γ_A are the activity coefficients, which are a function of liquid composition. A similar equation holds for B. The activity coefficients are usually calculated by the Margules equation with the needed parameters obtained from Gmehling and Onken (1984) or other data sources. For more details on γ, standard texts on thermodynamics (e.g., Smith et al., 2005; Sandler, 2006) should be consulted. We use the ideal solution for our discussion.

Applying Equation 26.15 to species B we have

$$y_{Bi} = K_B x_{Bi} \tag{26.16}$$

The sum of the mole fraction is equal to one. Hence

$$K_A x_{Ai} + K_B x_{Bi} = 1$$

Since x_{Bi} is equal to $1 - x_{Ai}$ this expression reduces to

$$x_{Ai} = \frac{(1 - K_B)}{(K_A - K_B)} \tag{26.17}$$

All the compositions are known now.

The mass transfer model discussed in Section 6.3 holds where the effect of diffusion-induced convection is also included. This section should be reviewed at this point. The relevant equation is Equation 6.34 for the flux of A; there is a similar equation for the flux of B. If these are added we get the following equation for the total flux:

$$\exp(N_t^*) = \left(\frac{x_{Ai} - y_{Ai}}{x_{Ai} - y_{Ag}} \right) \tag{26.18}$$

where y_{Ai} is equal to the equilibrium value $K_A x_{Ai}$. Here N_t^* is the dimensionless total flux, defined by $N_t = k_m C N_t^*$.

The individual fluxes can now be calculated as $x_{Ai} N_t$ and $x_{Bi} N_t$ on the basis of the unmixed model in Equation 26.14.

Example 26.2 illustrates the calculation of the condensation fluxes through a simple example where we assume that the interface temperature is known.

Example 26.2   Rate of Condensation of a Methanol + Water Mixture

A mixture of methanol and water at 90 °C containing 40% by mole of methanol is in contact with a cold wall. Find the rate of condensation and the composition of the condensate formed at this point. Use the unmixed model for the liquid composition. Assume an interface temperature of 85 °C. Use the following data (methanol = 1; water = 2): specific heat of components in the gas phase: $C_{pg1} = 45$ J/mol.K, $C_{pg2} = 34$ J/mol.K; heat of condensation: $\Delta H_{gl,1} = 36000$ J/mol, $\Delta H_{gl,2} = 43000$ J/mol. We have the following Antoine constants with pressure in pascals and temperature in K: for methanol, $A_1 = 23.402$, $B_1 = 3593.4$, and $C_1 = -34.92$; for water, $A_2 = 23.196$, $B_2 = 3816.4$, and $C_2 = -46.13$. The gas-side heat transfer coefficient is 60 W/m^{2}K. The gas-side mass transfer coefficient is 0.08 m/s.

Solution

The liquid-side composition at the interface based on the assumption of the unmixed model is proportional to the condensation rate. Hence

$$x_1 = N_1/N_t$$

A similar relation applies for x_2.

Using the Antoine equation, the vapor pressures are computed as follows: vapor pressure of methanol = 2.153 bars; vapor pressure of water = 0.575 bars.

The corresponding K values are $K_1 = 2.1528$ and $K_2 = 0.5745$.

The liquid composition at the interface can now be calculated using Equation 26.17. The values are $x_{1i} = 0.27786$ and $x_{2i} = 1 - x_{1i} = 0.7221$.

The composition in the vapor at the interface is calculated using Raoult's law and an ideal solution: for methanol, $y_1 = K_1 x_1 = 0.5904$ and for water $y_2 = K_2 x_2$ or $1 - y_1 = 0.4096$.

All the quantities needed to find the total flux are known. Thus using Equation 26.18 we have $N_t^* = 0.9394$.

The individual fluxes in dimensionless units are $N_A^* = 0.2610$ and $N_B^* = 0.6784$.

The actual fluxes are calculated by multiplying by $k_m C$ the scaling factor. Here C is the total concentration, calculated as 34.75 mol/m^3. This is based on the average temperature of the gas film.

The species fluxes are now calculated: for methanol we have 0.7057 mol/m^2 s, and for water we have 1.8344 mol/m^2 s.

Note that the methanol flux is lower as expected. Methanol accumulates at the vapor side of the interface. The mole fraction is close to 0.6, compared to that in the vapor of 0.4. Water accumulates on the liquid side of the interface. The mole fraction at the interface is 0.4 compared to 0.6 in the bulk.

26.4.2 Calculation of the Interface Temperature

The equation for the calculation of the interfacial temperature is similar to that for a single condensing case. The heat of condensation for both A and B is now included in the interfacial heat balance; the equation is

$$h_g^\circ(T_s - T_i)\mathcal{F} + N_A \Delta H_{c,A} + N_B \Delta H_{c,B} - h_L(T_i - T_{\text{wall}}) = 0 \qquad (26.19)$$

The correction factor for heat transfer augmentation ($\mathcal{F}$) is based on ϕ, which in turn now depends on both N_A and N_B now:

$$\phi = \frac{[N_A C_{pgA} + N_B C_{pgB}]}{h_g^\circ} \qquad (26.20)$$

Equation 26.19 assumes that the wall temperature is known. If only the coolant temperature (T_c) is known rather than the condensing wall temperature, the last term in that equation is replaced by $U(T_i - T_c)$, where U is the overall (condensate side + condensing wall) heat transfer coefficient. The relation $1/U$ is equal to $1/h_L + 1/h_c$ is used to get U, where h_c is the heat transfer coefficient on the coolant side.

The maximum and minimum values of the interfacial temperature are the dew point and the bubble point of the binary gas mixture at the bulk gas

Unmixed Model

The first assumption is when L_{s-1} is small and the term on the left-hand side can be neglected. This leads to

$$x_{A,s} = \frac{N_A}{\sum_{j=1}^{ns} N_j} = \frac{N_A}{N_t} \tag{26.25}$$

This is referred to as the completely unmixed model by Taylor et al. (1986) and other sources. Thus the unmixed model applies when there is no flow contribution to the mole balance in the liquid or when the flow contribution is small.

Mixed Model

A second limiting case occurs if the flux entering the bulk liquid is small compared to the flow rate of liquid entering the stage. Here we write Equation 26.21 in terms of component flow rate as

$$l_{A,s-1} + N_A P \Delta z = L_s x_{A,s} \tag{26.26}$$

where l is the component flow rate, defined as

$$l_{A,s-1} = L_{s-1} x_{A,s-1}$$

If the flow rate is large compared to mass transfer rate the second term on the left-hand side of Equation 26.26 is dropped and we have

$$x_{A,s} \approx \frac{l_{A,s-1}}{L_s} \tag{26.27}$$

The composition in the liquid is now fixed by the flow rate considerations and not by the rate of condensation. This is the basis for the mixed liquid model used in the condensation literature. The term "mixed" should not be interpreted as a completely backmixed model and implies that the flow from the previous stage gets mixed with the condensate and provides the composition for the next stage.

   Equation 26.23 or 26.24 is more general and includes both limiting cases. The equations are supplemented by the vapor balance and the energy balance.

Vapor Phase Mass Balance

The vapor balance is as follows and is similar to the liquid balance. For species balances A, B, C, and so on, the following balance equation can be derived:

$$V_{s-1} y_{A,s-1} - V_s y_{A,s} = N_A P \Delta z \tag{26.28}$$

Here y_A is the mole fraction in the bulk vapor. V is the total vapor molar flow with s denoting the stage number.

Energy Balances

The energy balance for the vapor phase is

$$V_{s-1}\overline{C}_p T_{v,s-1} - V_s \overline{C}_p T_s = -q^v P \Delta z \tag{26.29}$$

The following equation holds for the energy balance for the liquid:

$$L_{s-1} C_{pL} T_{L,s-1} - L_{s-1} C_{pL} T_S = (q^i - q_w) P \Delta z \tag{26.30}$$

Here q_i is the heat transferred to the liquid at the interface, which is given as

$$q_i = q^v + \sum_j N_j \Delta H_{c,j}$$

q_w is the heat flux across the tube wall to the coolant.

Coolant energy balance leads to the following relation:

$$L_c C_{pc}(T_{c,s-1} - T_c, s) = \pm q_w \tag{26.31}$$

The plus sign is used if the coolant is in cocurrent flow with the vapor; the minus sign is used otherwise.

q_w is given as

$$q_w = h_c(T_w - T_c)$$

Here T_w is the wall temperature, which is given as

$$h_L(T_L - T_w) = h_c(T_w - T_c)$$

It is fairly straightforward to set up a MATLAB program to use these equations and march in the z-direction. A local model to calculate N_A, q^v (which in turn needs T_i), and q^i is set up separately and used in the marching scheme. It may be also noted that the stagewise model in this section is equivalent to an implicit Euler method applied to a differential model for the condenser. Taylor et al. (1986) and Furno et al. (1986) also provide more details on the numerical aspects of the model and provide illustrative results.

26.6 Ternary Systems

Binary condensation in the presence of an inert gas is of importance in many practical applications. This is a ternary system for mass transfer modeling. Sardesai and Webb (1982) provide experimental data for a number of systems.

The modeling of such a system can be completed using the Stefan-Maxwell equations as shown by Krishna and Panchal (1977). In this section we summarize the model equations and also show an interesting example where a chemical reaction occurs at the interface.

26.6.1 Stefan-Maxwell Model

The mass transfer of the species in the gas film is modeled by the Stefan-Maxwell model studied in Chapter 15. Note that this is an inverted model and the mole fractions are the explicit variables. The following equation applies for species i with $i = 1$ to 3 for a ternary system. Extension to systems with more than three components is straightforward:

$$\nabla y_i = -\sum_{j=1}^{ns} \frac{y_j \mathbf{N}_i - y_i \mathbf{N}_j}{C D_{ij}} \tag{26.32}$$

The boundary conditions are known in the bulk gas. The boundary condition at the interface has to be set up. The mole fraction in the liquid is needed for this purpose. The unmixed model is commonly used for species 1 and 2, which are assumed to be condensable species. (Species 3 is the inert non-condensable component of the mixture.)

The mole fraction in the liquid side of the interface is therefore equal to $N_i/(N_1 + N_2)$. The corresponding mole fraction in the vapor near the interface is given as $K_i x_i$. Hence $y_{i,s} = K_i N_i/(N_1 + N_2)$ where the subscript s is used for the interface.

The determinacy conditions have to be used as an additional equation. Species 3 is non-condensing and hence N_3 is set to zero. The zero divergence of N_1 and N_2 provides the additional equations. The computational procedure shown in Chapter 15 can be used to solve for both the mole fraction profiles in the vapor film as well as the condensation rate. In particular direct solution using BVP4C is quite useful. Details are left as an exercise problem.

26.6.2 Condensation with Reaction

In some cases a chemical reaction of the condensed species can occur in the liquid film. A model for this was developed by Ramaswamy and Ramachandran (2013) for the reaction scheme:

$$B + C \rightleftharpoons A$$

where all the species are present in the vapor. C (species 3) is treated as a non-condensing species while A and B are condensing. A three-component Stefan-Maxwell model was used for the vapor film.

Species C is assumed to react instantaneously at the vapor–liquid interface. Hence N_3 is not set to zero but is set to $k_s C y_{s,3}$. The rate of reaction of C at the interface produces a flux condition at the interface. An industrially important example is the condensation of a gas mixture consisting of acetic acid (B), ketene (C), and acetic anhydride (A). In this application, the reaction of ketene during the condensation process has to be reduced since the ketene has to be recovered as the vapor phase product. The models provided some basis for the design of condensers to mitigate the ketene loss due to reaction.

Summary

- For problems with simultaneous heat and mass transfer, the convection due to mass transfer can cause a flow and thereby provide an additional mode for heat transfer. The heat transfer rate is then enhanced (or retarded if mass transfer is in opposite direction to heat transfer) due to mass transfer. An augmentation factor can be derived and used to predict these effects. Such models are useful in the design of condensation equipment and in many other applications.

- For a vapor consisting of a pure component, the temperature is equal to the dew point, which is constant. There are no mass transfer limitations and the process is governed by the rate of heat transfer in the condensing liquid film.

- Condensation can be classified as dropwise or filmwise. Heat transfer coefficients are an order of magnitude higher in dropwise condensation. However, industrial applications usually operate in the filmwise region since dropwise condensation is difficult to maintain.

- The liquid flow pattern in filmwise condensation can be classified into a laminar, wavy, or turbulent regime depending on the range of the Reynolds number. For laminar flow a theoretical analysis of Nusselt is useful to predict the heat transfer coefficient. For wavy and turbulent flow, empirical correlations or CFD models are useful.

- Practical problems involve condensation of a vapor in the presence of a non-condensible vapor or simultaneous condensation of two components, for example, in the condenser in a distillation column. Heat transfer considerations alone are not sufficient to design these systems and a study of simultaneous heat and mass transfer is needed.

- The local model for the heat and mass transfer rate can be derived on the basis of the film model for a vapor plus an inert gas mixture. The mass transfer rate is modeled as a UMD model. This includes diffusion-induced convection. The heat transfer rate is also enhanced due to diffusion-induced convection. An Ackermann correction factor is used to calculate this enhancement.

- The film thicknesses for heat and mass transfer are generally assumed to be equal to each other. However, this is valid only if the Lewis number is close to one. For organic vapors the Lewis number can be greater than one. This can cause a faster rate of cooling compared to the rate of mass transfer, causing supersaturation in the film. This can lead to fog formation in the condensing equipment, which is undesirable. Models using different film thickness values can provide conditions leading to fog formation and can also suggest preventive strategies.

- Condensation of two components follows a similar methodology to that for a single component. But since the liquid composition depends on the relative rate of condensation, some assumptions are needed in order to fix the local liquid composition. Two assumptions are that the liquid is well mixed and the liquid is segregated.

- Models for condensing equipment can be set up using mass and heat balances over a differential control volume. If the control volume is discretized by an implicit finite difference, a stagewise model is obtained, which is easier to solve. The model equations are summarized in Section 26.5.

- The condensation of a ternary mixture can be modeled along similar lines as the binary case. The mass transfer part of the model is formulated now as a multicomponent diffusion problem and the Stefan-Maxwell model described in Chapter 15 can be used to compute the mass flux of all the condensing components. Again the assumption of an unmixed liquid or completely mixed liquid is needed to set the interface vapor mole fractions. The unmixed assumption is simpler to use. Calculation of the interface temperature is done by a heat balance similar to the binary case. The Ackermann correction is used to calculate enhancement in heat transfer due to mass transfer.

Review Questions

26.1 What is meant by filmwise condensation?

26.2 What is dropwise condensation?

26.3   Which type of condensation is common in industry?

26.4   What is the basis for the Nusselt theory of condensation?

26.5   How does the local heat transfer coefficient in a liquid film vary with distance along a vertical condensing surface?

26.6   What is meant by modified latent heat of condensation? Where is it used?

26.7   Explain why heat transfer rate can be enhanced by simultaneous mass transfer.

26.8   What is the augmentation factor for heat transfer?

26.9   What is the Ackermann correction factor?

26.10 State the equations to calculate the rate of mass transfer to condense a vapor with some non-condensable gas.

26.11 State the equations to calculate the rate of heat transfer in the vapor film.

26.12 How is the interface temperature calculated?

26.13 State the conditions that can lead to fog formation.

26.14 If $Le > 1$, which film (heat or mass) is thicker?

26.15 What would be the design implications if fog forms in the vapor space of the condenser?

26.16 What additional assumptions are needed to model condensation of a mixture of two condensable gases?

26.17 How does the assumption of an unmixed liquid fix the liquid composition for condensation of two components?

Problems

26.1 **Derivation of the Nusselt model for condensation.** Consider a vapor condensing on a wall and forming a liquid film. Show that the change in the mass flow rate per unit perimeter, Γ, is related to the film thickness as per the following equation:

$$\frac{d\Gamma}{dx} = (\rho g \delta^2 / \mu)\frac{d\delta}{dx}$$

From a heat balance show that the change in mass flow rate is given as

$$\hat{h}_{lg}\frac{d\Gamma}{dx} = \frac{k_l}{\delta}(T_{sat} - T_w)$$

Equate the two expressions for the change in Γ and derive an expression for $d\delta/dx$. Integrate this expression and derive the expression given by Equation 26.2 for δ as a function of x. Find an expression for the local heat transfer coefficient and verify that it is proportional to $x^{-1/4}$. Find an expression for the average heat transfer coefficient for a wall of height L by integration of the local value and hence Equation 26.5.

26.2 **Flow regimes in a condenser.** Saturated steam at 356 K condenses on a 5 cm vertical tube whose surface is maintained at 340 K. Find the height at which the flow becomes wavy. Calculate and plot the flim thickness and the heat transfer coefficient up to this height and also find the average heat transfer coefficient. At what height would the film become turbulent?

26.3 **Condensation of steam: design for tube length.** Saturated steam at 55 °C is to be condensed at a rate of 10 kg/hr on the outside of a 3 cm diameter vertical tube by maintaining the surface at 45 °C. What is the operating pressure? Find the tube length required for this purpose.

26.4 **Condensation of pure ethanol.** Ethanol is condensing at 1 atm pressure on the outside of a tube that is 3 m long with a 2.5 cm diameter. The tube is maintained at 25 °C by circulating cooling water inside the tube. Calculate the quantity of ethanol condensed.

26.5 **Concentration and temperature profiles in the vapor film.** Warm air at 90% humidity at 5 °C is diffusing into cold air at 5 °C and 70% humidity. Sketch the compositions of the $y - T$ diagram for $Le = 0.5$, $Le = 1.0$, and $Le = 2$. Indicate if fog formation is likely.

26.6 **Condensation of a vapor with a non-condensable gas.** Benzene vapor with 25% benzene and the rest air is condensing on a tube at a temperature of 300 K. Find the rate of condensation. Use $h_g = 10$ W/m^2 K for the heat transfer coefficient and $k_m = 0.08$ m/s for the mass transfer coefficient.

26.7 **Condensation of two components.** Methanol and water are condensing on a tube of diameter of 25 mm with a methanol content of 50%. The vapor inlet temperature is 360 K. The coolant temperature at the top is 308 K with a coolant flow rate of 0.06 kg/s. Estimate the interface temperature and the condensation rate of methanol and water at the top of the condenser. Assume the unmixed case. The calculations should be put in an m-file or subroutine for use in the next part of the problem.

26.8 **Simulation of a condenser for methanol-water mixture.** Simulate the condenser for the data in previous example. Use the stagewise model shown in Section 26.5.

26.9 **Condensation of two components with a third inert gas.** Krishna and Panchal (1977) simulated the condensation of a methanol–water system in the presence of air. The composition at the inlet was 0.7, 0.2, and 0.1 for methanol, water, and air, respectively. Assume as a interface temperature of 310 K. Set up the model for the mole fraction of the species as a function of distance along the diffusion film using the Stefan-Maxwell model. Find the flux of the components. Use a film thickness of 2 mm. Use the unmixed model for the ratio of fluxes.

CHAPTER 27

Gas Transport in Membranes

Learning Objectives

After completing this chapter, you will be able to:

- Understand some industrial applications of membrane separation of gas mixtures.
- Model the local transport rate of gaseous species across membranes using Fick's law.
- Understand some nonlinear rate models for membrane transport.
- Relate pore structural parameters to the diffusivity in a membrane.
- Couple a local transport model to an overall model for a gas permeator device.
- Develop models for performance on the permeator for three flow patterns for gas phase mixing.

Membrane-based separations are becoming important in many contexts. Examples can be found in gas separations, dialysis, reverse osmosis, and so on. An additional example is the unit operation of pervaporation, which is becoming important in many applications, including the production of ethanol for biofuel applications.

For pedagogical convenience we study separately the transport of gases and liquids across the membranes in two separate chapters. This chapter introduces the analysis of transport effects in various gas transport membrane processes and the models for such systems.

Transport of a gas mixture across a membrane is done with a view of separating a gas mixture. The Fick's law type of models are generally used to characterize the rate of transport. But many terminologies and jargon, specific to this field and often confusing, are used in the literature. We define these common terms, such as permeance, permeability, selectivity, barrer, and so on.

Some complexities (nonlinear effects) encountered in membrane transport are briefly presented.

Membrane transport rate provides the local model. This is then coupled to the process model for the separation unit. The process model needs some assumption about the mixing pattern in the gas phase. These features are explained and various flow patterns are schematically shown. Finally, models for membrane permeator units are presented as examples of the coupling of local and equipment-level models. These models are useful to evaluate the performance of gas separation equipment or to do a preliminary design of the process.

The final section provides an example where a membrane separation is coupled to a reactor, an example of a reactor–separator combo. The equilibrium limitations can be overcome to some extent by the simultaneous removal of the product and some applications of membrane reactors are indicated.

27.1   Gas Separation Membranes

In the membrane process for separation of gas mixtures, the membranes act as a barrier though which some species move faster than others. The feed mixture, usually at a high pressure, is fed to one side of the membrane and is separated into a retentate (part of the feed that does not pass through the membrane) and a permeate (part of the gas that has passed through the membrane). The permeate side is usually at a lower pressure compared to the feed side. A general schematic of a membrane separation process is presented in Figure 27.1. In some cases a sweep gas is fed on the permeate side in order to promote the flow of the permeate. However, this may dilute the permeate stream and cannot be used in some applications.

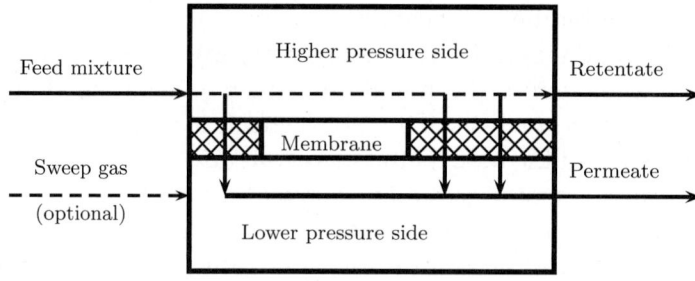

Figure 27.1   Schematic flowsheet of a membrane separation process.

Gas separation using membranes is employed in a number of industrial applications for separation of gas mixtures. A few examples are indicated here:

- Recovery of hydrogen from product streams of ammonia plants, refinery process steams, or mixtures with inert gases such as nitrogen or methane
- Enrichment of air to produce oxygen for medical or metallurgical purposes or use in oxidation processes in chemical industries
- Removal of CO_2 and/or H_2S from natural gas

27.1.1 Membrane Classification

There are a number of ways to classify membranes. The usual classification for gas separation is based mainly on the pore size, and membranes can be porous versus non-porous.

Polymers in a glassy state are examples of non-porous membranes. A solution–diffusion model is used as the mechanism of transport in such membranes. The transport rates of different gases are then determined by the differences in their solubility and diffusivity, as discussed later in this section. Since diffusivity in a non-porous matrix is low, the transport rate is low, but selectivity can be high due to solubility differences between the gases in the membrane. Hence the selection of membranes is often a compromise between the rate of transport and the selectivity to a specified gas. Examples of common polymeric membranes include cellulose triacetate, polyisoprene (natural rubber), polycarbonates, polysulfones, and teflon.

Porous membranes are more commonly used for gas separation. The pore diameter must be smaller than the mean free path of gas molecules. Thus the pore size has to be on the order of 10 nm or so. Gas flux through a porous membrane is much higher (three to five orders of magnitude) than for the non-porous case, but the separation selectivity is lower since both gases become closely permeable.

Membranes can also be asymmetric. This represents a compromise between a non-porous and a porous membrane. Here a dense but thin non-porous layer is bonded to a microporous layer. The dense layer provides the selectivity while the porous layer provides increased structural strength.

Although polymeric membranes are common, in special cases materials other than polymers can be utilized. Microporous α-alumina is one example. A second example is the palladium membranes that permit transport solely of hydrogen. Hydrogen is strongly adsorbed over Pd in a dissociative manner and hence the selectivity of such membranes toward hydrogen is very high. The operating temperatures of such inorganic membranes are higher than polymeric membranes.

27.1.2 Transport Rate: Permeability

Consider a membrane of thickness L with a concentration of the diffusing solute of C_{A0} at $x = 0$ and C_{AL} at $x = L$ as shown in Figure 27.2.

We can use the Fick's law type of model for transport across the membrane itself:

$$N_A = D_A \left(\frac{C_{A0} - C_{AL}}{L} \right) \tag{27.1}$$

where N_A is the flux of the diffusing species across the membrane. Note that the low flux mass transfer model used here and N_A is assumed to be the same as the Fick's law flux. D_A is the diffusion coefficient of A in the membrane, which is assigned a constant value.

Here C_{A0} and C_{AL} are the concentration on the membrane side of the interface. This concentration is usually not directly known. What is known is the concentration in the fluid phase on either side of the membrane. The concentration in the fluid is related to the concentration in the membrane by

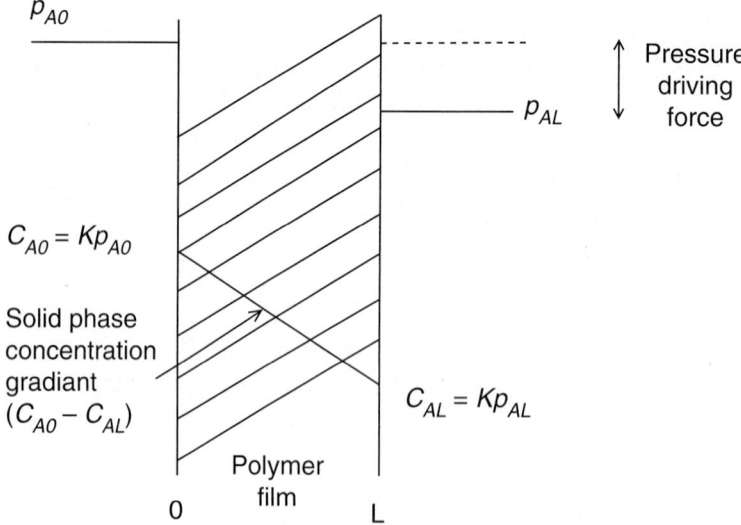

Figure 27.2 Concentration profiles of A across a membrane; note the concentration jump across the interface.

an equilibrium relationship, Thus if $C_{A0,G}$ is the concentration in the gas at $x = 0$ then the concentration in the membrane C_{A0} is given as

$$C_{A0} = K_A C_{A0,G} \tag{27.2}$$

where K_A is an equilibrium solubility constant, also called the partition coefficient. A similar equation holds at $x = L$. Hence

$$C_{AL} = K_A C_{AL,G} \tag{27.3}$$

Using these equilibrium relations in Equation 27.1, we obtain the flux in terms of the concentration driving force based on the fluid concentrations:

$$N_A = \frac{D_A K_A}{L} (C_{AG,0} - C_{AL,G}) \tag{27.4}$$

Hence the overall transport rate depends on the product of D and K (as well as the thickness L) and not merely on the value of the diffusion coefficient. This composite parameter DK is defined as the permeability of the membrane. This has units of m^2/s, similar to that of diffusivity, but it includes the partition coefficient as well.

The driving force can also be based on partial pressure units and a permeability $P_{M,A}$ can be defined by the following equation:

$$N_A = \frac{P_{M,A}}{L} \Delta p_A \tag{27.5}$$

where Δp_A is the partial pressure difference across the membrane. The permeability defined in this manner has units of $mole - m/m^2$ s Pa this case. The two are related by

$$P_M = \frac{DK}{R_g T}$$

Barrer

The unit of permeability as defined above is $mole - m/m^2$ s Pa. However, in membrane-related publications, the permeability is often reported in units of barrer, defined as follows:

$$one\ barrer = 10^{-10}\ cm^3\ STP.\ cm/cm^2\ s\ Hg - cm$$

The conversion factor to traditional SI units is as follows:

$$one\ barrer = 3.348 \times 10^{-16} mole - m/m^2\ s\ Pa$$

The barrer unit is named after early work by Barrer (1951, 1984) in this area. One reason for the persistence of this unit is that the permeability for

Table 27.1   Permeability in Barrer of Common Gases in Typical Membranes at 25 °C

	H_2	O_2	N_2	CO_2
Polyethylene	9.8	2.93	0.97	12.66
Polymethacrylate	–	1.18	0.23	5.05
Polyvinlchloride	1.73	0.046	0.12	0.16
Butyl rubber	7.24	1.30	0.32	5.19

many membranes is on the order of 1 to 10 barrer and hence the values can be easily remembered. Typical values are shown in Table 27.1 from more extensive data provided by Seader et al. (2011).

27.1.3 Transport Rate: Permeance

Usually the thickness of the membrane L is also not precisely known. The partition coefficient K also may not be usually measured and thus may not be known as well. Hence the effects of these terms are combined into an overall term called the permeance $\mathcal{P}$ of the membrane:

$$\mathcal{P} = DK/L \tag{27.6}$$

The transport rate across the membrane can thus be described (in terms of the fluid phase concentration difference as the driving force) as

$$N_A = \mathcal{P}^{(C)}(C_{A0,G} - C_{AL,G}) \tag{27.7}$$

Note that the permeance has the same units as the mass transfer coefficient (m/s) and is hence a measure of transport efficiency of a membrane. The driving force is also defined in this equation in terms of the gas phase concentration difference. The superscript (C) on permeance is used to indicate this.

The permeance of a membrane $\mathcal{P}$ is a composite parameter that depends on the diffusion coefficient of the membrane itself, the solubility or the partition coefficient of the solute in the membrane, and finally the thickness of the membrane itself. Thus the values of $\mathcal{P}$ may vary from one type of membrane to another and cannot be correlated in a theoretical sense unless all three effects can be quantified separately. However in practice the permeance is often used to describe the transport rate and in membrane module analysis as a matter of convenience. Often the permeance values are directly reported in the literature, since they can be directly calculated from the measured rate data.

The partial pressure difference is often used in Equation 27.7 instead of the concentration difference. The flux is given as

$$N_A = \mathcal{P}_A(p_{A0} - p_{AL}) \tag{27.8}$$

where $\mathcal{P}_A$ (with no superscript) is now based on the partial pressure difference. The permeance has units of $\text{mol/m}^2.\text{s.Pa}$ (or $\text{mol/m}^2.\text{s.atm}$ if the pressure is expressed in atm). The units are the same as k_G, the gas-side mass transfer coefficient used in absorption studies. We will use this definition and the partial pressure difference as the driving force for local flux calculation in permeator scale modeling in Section 27.3.

The effect of temperature is to increase permeance. An increase in temperature causes an increase in diffusivity while the effect of solubility may act in either direction. Overall, a modest increase in permeance with temperature is observed.

27.1.4 Selectivity

The relative rate of transport of two gases is the measure of the selectivity of the membrane to one gas relative to the second. Hence it is useful to define a selectivity parameter, S_{12} as follows:

$$S_{12} = \frac{\mathcal{P}_1}{\mathcal{P}_2}$$

or

$$S_{12} = \frac{K_1}{K_2} \cdot \frac{D_1}{D_2}$$

Thus selectivity depends on both the solubility ratio and the diffusivity ratio.

27.1.5 Sievert's Law: Dissociative Diffusion

Gases such as hydrogen can dissociate into H atoms in inorganic membranes with noble metals such as Pd and in such cases a nonlinear transport model with a square root dependency is used:

$$N_A = \mathcal{P}\left(C_{AG0}^{1/2} - C_{AGL}^{1/2}\right) \tag{27.9}$$

This equation is called Sievert's law and describes hydrogen transport in Pd-based membranes. The square root dependency arises due to the law of mass action: the concentration of H atoms (the diffusing species) is proportional to the square root of H_2 concentration.

27.1.6 Nonlinear Effects in Membrane Transport

The discussion in Section 27.1.2 was based on the simple Fick's law concept based on constant values of solubility and diffusivity. However, complexities in the transport law often arise due to local differences in both solubility and diffusivity, both of which can be functions of concentration. Stern (1994) made the following classifications:

- Constant diffusion and constant Henry's law (K): This is the simplest case and the model in subsection 27.1.2 is applicable. Diffusion of permanent gases in elastomers and many harder polymers are often described by such a model.
- Diffusivity as a function of position: Membranes prepared from composite materials of two or more layers show different values of D for each layer.
- Concentration-dependent diffusivity D but constant K: This phenomena is exhibited by gases with critical temperatures near the ambient to $200°$ C, for example, C_4 hydrocarbons in rubbery membranes.
- Variable D and variable K: Gases with high critical temperatures, organic vapors, and so on, are examples of systems showing this pattern.
- Time-dependent effect, that is, $D(C_A, t)$: Here a time- and history-dependent diffusion phenomena is observed. These are seen in polymers with longer relaxation times, for example, organic vapor in ethyl cellulose.

Further discussion of these complex effects is not considered here. The main point is for you to be aware of these phenomena in practical applications and use specific modifications in the model as needed. The models for all the preceding cases can be set up using the methodology described in Chapter 6 for variable diffusivity problems. Numerical computation is usually needed to integrate the local model across the membrane thickness.

Dual Mode Transport

Another model that shows a nonlinear dependency is the dual mode transport model first proposed by Barrer et al. (1958) and then by Koros and Paul (1980). The model is useful for rubbery polymers. In this model, the solution is determined by using Henry's law (linear) in the polymer chains and by the Langmuir isotherm (nonlinear) in holes or sites between chains of glassy polymers. The overall permeance is a weighted average of the linear and nonlinear part of the solution–diffusion model for the two sites. For further details consult the mentioned references.

27.2   Gas Translation Model

The gas translation model, also called the activated Knudsen model, is often used to calculate the permeance and the selectivity parameter. A brief overview of this model is given here. The model was first developed by Xiao and Wei (1992) and later modified by Shelelkhin et al. (1995) and Nagasawa et al. (2014). The model is applicable to microporous membranes where diffusion is the main controlling parameter. The partition coefficient is not much different for many gases in such membranes (unlike polymeric membranes) and hence the selectivity depends mainly on the diffusivity in this case. (The partition coefficient K will be taken as one in the following discussion.)

The gas translation model can be viewed as an extension of the Knudsen diffusion concept. In small micropores the pore diameter is smaller than the molecular mean free path and hence diffusion is through the Knudsen mechanism. Note that the Knudsen diffusion coefficient in a cylindrical straight pore (Section 7.5.1) is given as

$$D_{K,A} = \frac{1}{3} d_p \sqrt{\frac{8 R_g T}{\pi M_A}}$$

where d_p is the diameter of the pore. The term within the square root is the mean free path of the diffusing gas.

The corresponding permeability is given by using the porosity and a tortousity factor:

$$P_A = \left(\frac{\epsilon}{\tau L} \right) D_{KA}$$

The gas translation model is a modification of the above relationship, which was based on the Knudsen diffusion concepts. In this model, the permeate diffusion is assumed to be restricted from the potential field of the membrane wall and a correction factor, p_i, is applied to the this formula. The permeabiity equation is then multiplied by the following factor:

$$P_A = \frac{1}{3} \left(\frac{\epsilon}{\tau L} \right) \left[d_p \left(\sqrt{\frac{8 R_g T}{\pi M_A}} \right) \right] p_i \qquad (27.10)$$

The factor p_i represents the probability factor, which represents the fraction of molecules that have the energy to overcome the activation barrier; this is modeled by an Arrhenius type of equation:

$$p_i = p_0 \exp \left(\frac{-E_i}{R_g T} \right)$$

p_0 is the pre-exponential factor and E_i is the activation energy barrier for transport through the pores.

The concepts behind the modified gas translation model are shown in Figure 27.3. Since the center of the permeating molecules cannot approach the pore wall, the effective cross-sectional area is modeled as a circle with a diameter equal to the effective diffusion length $d_p - d_A$, as shown in this figure.

d_A is the effective molecular diameter of species A. The ratio of this area to the physical area is used as the pre-exponential factor: hence the

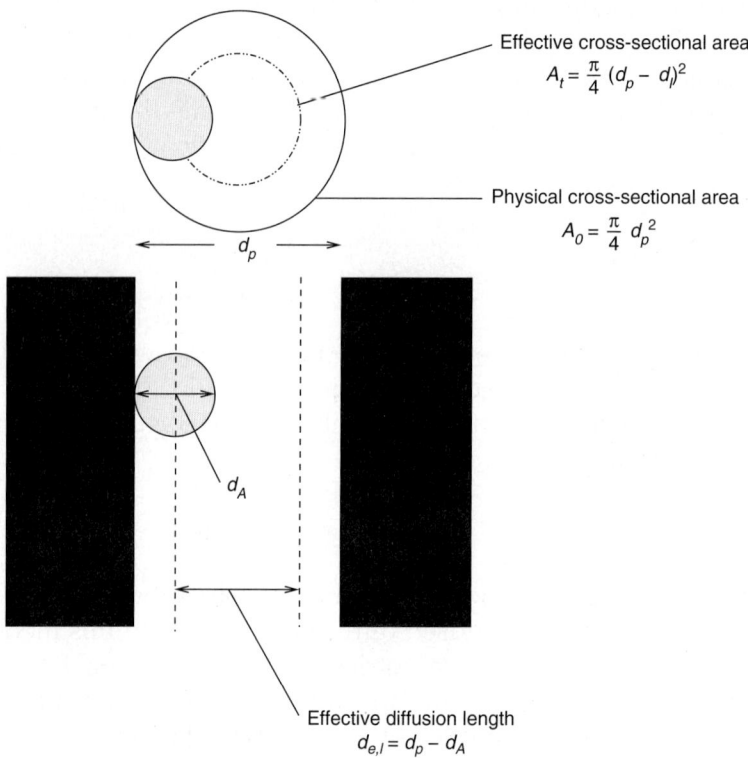

Figure 27.3   Schematic of membrane permeation in a pore and effect of size of the permeate molecule. From Nagasawa et al. 2014.

pre-exponential factor, p_0, is modeled as

$$p_0 = \frac{(d_p - d_A)^2}{d_p^2}$$

Introducing this into Equation 27.10, the permeability can be written as

$$P_A = \frac{1}{3}\left(\frac{\epsilon}{\tau L}\right)\frac{(d_p - d_A)^2}{d_p^2}(d_p - d_A)\sqrt{\frac{8R_gT}{\pi M_A}}\exp\left(\frac{-E_A}{R_gT}\right) \tag{27.11}$$

This provides a useful expression to correlate data on the various gases in terms of structural parameters and an activation energy parameter.

The model also provides a way to assess the selectivity parameter for two gases based on the gas properties. Taking the ratio of the permeability for two gases, the selectivity parameter is given by

$$S_{12} = \sqrt{\frac{M_2}{M_1}} \cdot \frac{(d_p - d_1)^3}{(d_p - d_2)^3}$$

d_1 and d_2 are the molecular diameters of species 1 and 2 respectively. This equation indicates a dependency on the molecular size as well in addition to the molecular weight. The Knudsen model would have predicted only the effect of molecular weight. The second difference is in the temperature effect. The classical Knudsen predicts a square root relation while the gas translation model predicts an additional exponential effect. Thus the temperature dependency is

$$P_A \propto \sqrt{T}\exp\left(\frac{-E_A}{R_gT}\right)$$

27.3   Gas Permeator Models

Here we focus on modeling at the equipment level. Membrane separation systems are often modeled based on idealized flow patterns. Four common patterns are shown in Figure 27.4.

Basic aspects and useful relations common to all four models are first presented. A binary feed consisting of A and B is considered.

The following notation is used here: x is the mole fraction of A in the retentate and y is the mole fraction of A in the permeate (i.e., species subscript A is dropped.) These are also local values at any position in the bulk gas.

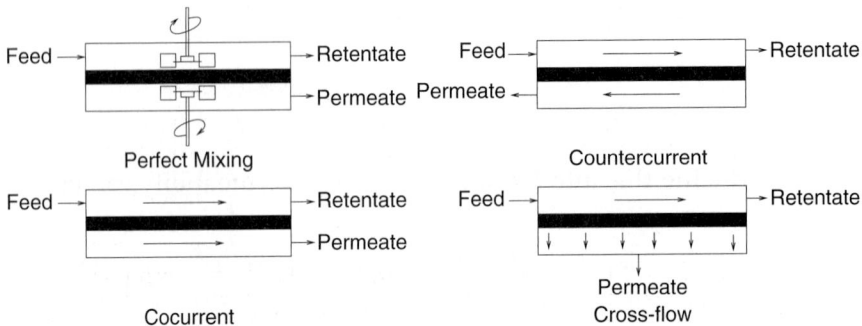

Figure 27.4   Four idealized flow patterns in membrane separation unit. Based on Seader, J. D. Henley, E. J. and Roper D. K. (2011). *Separation Process Principles with Applications using Process Simulators*, John Wiley & Sons, Hoboken, NJ.

27.3.1 Flux Relations

The flux across the membrane is represented as

$$J_A = \mathcal{P}_A(xP_1 - yP_2) \tag{27.12}$$

where P_1 is the pressure on the membrane side and P_2 is the pressure on the pemeate side. $\mathcal{P}_A$ is the permeance of species A.

Similarly for B the flux is given as

$$J_B = \mathcal{P}_B\left[(1-x)P_1 - (1-y)P_2\right] \tag{27.13}$$

The pressure ratio parameter and selectivity parameter are introduced to simplify the above equations.

The selectivity parameter is defined as

$$S = \frac{\mathcal{P}_A}{\mathcal{P}_B}$$

The pressure ratio parameter is defined as

$$P^* = \frac{P_2}{P_1}$$

With these definitions, the fluxes are given as

$$J_A = \mathcal{P}_B P_1 S(x - yP^*) \tag{27.14}$$

$$J_B = \mathcal{P}_B P_1\left[(1-x) - (1-y)P^*\right] \tag{27.15}$$

27.3.2 Local Concentration

It is useful to define a local mole fraction y' in the permeate side. This definition is useful to set up mass balances on the permeate side:

$$y' = \frac{J_A}{J_A + J_B}$$

Using the fluxes the following relation can be shown for the local permeate composition, y':

$$y' = \frac{S(x - P^*y)}{S(x - P^*y) + (1 - x) - (1 - y)P^*} \qquad (27.16)$$

This equation provides the local composition of A crossing the membrane and it can then be used to model the performance of a membrane separator. The material balance equations are to be coupled with these flux relations. The material balances are different for different flow patterns as shown in Figure 27.4. Hence the effect of flow patterns needs to be accounted for; these are studied next.

Before we show these details of permeator performance calculations, it is worth noting two limiting cases of Equation 27.16. The first is when P^* equal to zero. This will give the maximum local mole fraction in permeate:

$$y'_{max} = \frac{Sx}{1 + (S - 1)x}$$

The second limiting case is when $P^* = 1$. The value of y' is now zero since there is no driving force for diffusion when the pressures are equal on both sides. Separation is now possible only by adding a third component on the permeate side, which is called the sweep gas. This is used in some applications to improve separation. However, the sweep gas dilutes the permeate gas and hence may not be suitable in some applications.

There is also a minimum value of P^* that is independent of the value of S. This arises due to the requirement that the partial pressure of A in the permeate does not exceed that in the feed:

$$P_1 x_f \geq P_2 y_{max}$$

Since y_{max} can not exceed one, the operating pressure ratio has to be less than the following minimum value:

$$P^* \leq x_f$$

For example if the feed is 40% A, the pressure ratio has to be less than 0.4 in order to get a nearly pure A (in the permeate stream) under ideal operating conditions. This is similar to the minimum reflux concept in distillation operation. Here we have a minimum retentate side pressure criteria.

Gas permeator models are considered next. We examine the simplest case (backmixed-backmixed) first.

27.3.3 Backmixed-Backmixed Model

Here we provide a simple model to evaluate the performance of a gas permeator. The model assumes the gas is well mixed on both the high and low pressure side. The basis of the model is shown schematically in Figure 27.5.

The following notation is used in this section: F_f is the molar flow rate of the feed with x_f the mole fraction of A in the feed. x is the exit mole fraction in the retentate; y is the exit mole fraction on the permeate side. Due to the assumption of complete mixing on the permeate side, the mole fractions x and y are also representative of the mole fractions in the permeater itself.

It is customary to use a cut parameter, denoted as Θ. This is defined as the fraction of the feed that becomes the permeate:

$$\Theta = \frac{F_p}{F_f}$$

Here F_p is the permeate molar flow rate.

The mass balance for A gives

$$F_f x_f = (F_f - F_p)x + F_p y \qquad (27.17)$$

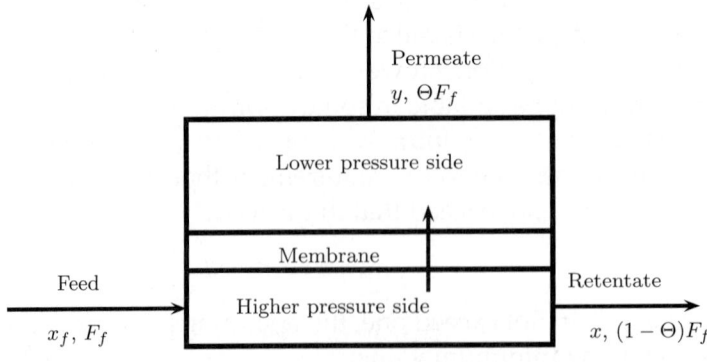

Figure 27.5 Simple backmixed-backmixed model for a membrane separator. Θ is the cut parameter.

This can be expressed using the cut parameter and rearranged to give

$$x = \frac{x_f - \Theta y}{1 - \Theta} \qquad (27.18)$$

This has the status of an operating line since it is based on mass balance and relates x to y. A second relation between these two is given by transport considerations, which are shown next.

The mole fraction in the retentate side y is now equal to the local mole fraction y' since there is no other flow in the retentate (no sweep gas). Hence Equation 27.16 relates y and x. This is reproduced below with the left-hand side changed to y now:

$$y = \frac{S(x - P^*y)}{S(x - P^*y) + (1 - x) - (1 - y)P^*} \qquad (27.19)$$

Equations 27.18 and 27.19 can be solved simultaneously to find the exit mole fractions for a given feed mole fraction with the cut as a parameter. Additional parameters to be specified are the selectivity ratio S and the pressure ratio P^*. Since the cut varies from 0 to 1, a design type of plot can be constructed for the permeate mole fraction as a function of the cut. An illustrative plot of membrane performance is shown in Figure 27.6.

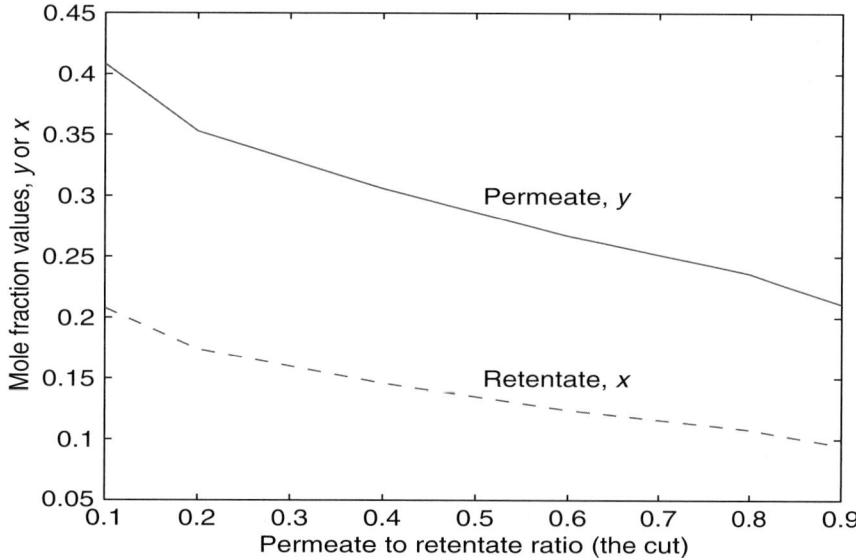

Figure 27.6 Membrane performance as a function of the cut parameter. $x_f = 0.21$, $S = 1000$, $P^* = 100$. Backmixed-backmixed model.

The model is useful for laboratory studies to interpret data and to evaluate permeability values as these are often operated such that the back-mxied assumption is reasonably valid.

The transfer area needed can then be calculated from the following equation:

$$A = \frac{F_f \Theta y}{\mathcal{P}_A P_1 (x - P^* y)}$$

This will provide a conservative estimate since the driving force will be larger in other flow patterns. The flow pattern examined next is the counterflow of the two streams.

27.3.4 Countercurrent Flow

A common mode of operation is the hollow fiber permeater. A schematic of this is shown in Figure 27.7.

Here the membranes are made in tubes or fibers and tied together to form a bundle. The fiber bundle is put into a shell where the feed stream is contacted. The species diffuse into the fiber and the permeate is collected at the open end as shown in the figure. The flow is usually countercurrent since it provides a better degree of separation compared to cocurrent flow. A model for performance analysis is presented in this section.

A differential model is called for since the mole fractions vary along the tube length. However, if the differential model is discretized (e.g., by an implicit Euler scheme) an equivalent stagewise representation is obtained and

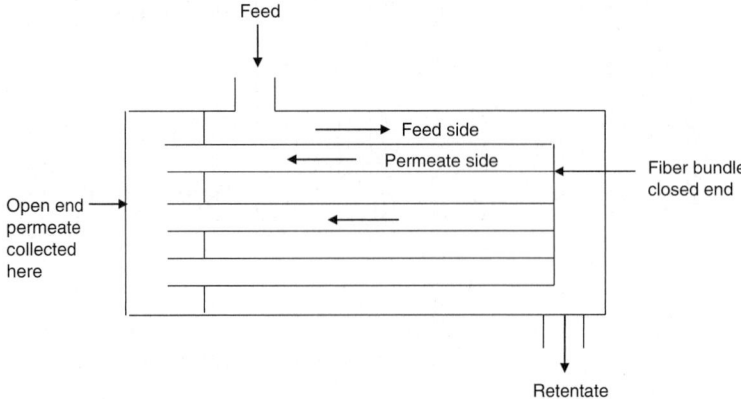

Figure 27.7   Schematic of a hollow fiber membrane permeater.

can be used to simulate the process. A schematic of this is shown in Figure 27.8. The model equations are the species balance and the total flow balance for the permeate and the feed side; these are presented next. The permeater is divided into N stages with stage 1 starting at the closed end of the system. Each stage is assumed to provide an area ΔA_m for mass transfer.

The following notation is used: L is the molar flow rate on the feed side and V is the flow rate on the permeate side. The subscript will be used to denote the stream leaving a stage. The stream entering a stage will have the subscript of the stage from where it came. This notation is similar to that used in distillation. The balance equations using this notation are presented next.

The species balance on the feed side can be derived as

$$L_{k+1}x_{k+1} - L_k x_k = (J_A + J_B)y'_k \Delta A_m$$

The total balance on the feed side is

$$L_{k+1} - L_k = (J_A + J_B)\Delta A_m$$

The following equation holds for the species balance on the permeate side:

$$V_{k-1}y_{k-1} - V_k y_k = -(J_A + J_B)y'_k \Delta A_m$$

Finally, the total balance is needed for the permeate side:

$$V_{k-1} - V_k = -(J_A + J_B)\Delta A_m$$

y'_k and x_k are related by Equation 27.16.

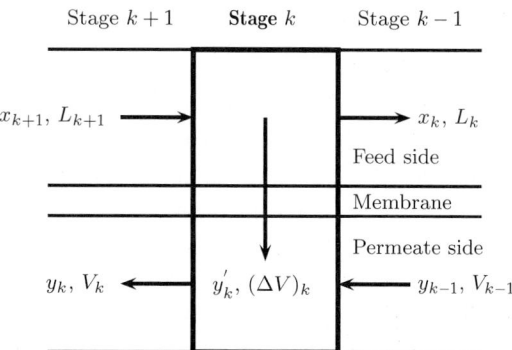

Figure 27.8 Model notation for stagewise computation of a membrane separator with countercurrent flow. Note ΔV is the permeation mole flow and is equal to $J_A + J_B$.

A simple solution procedure is as follows. The computations are usually started at the closed end since the permeate conditions can be specified at this end. The retentate flow and mole fraction are assumed at this end. This is stage one. Hence L_1 corresponds to the specified (or guessed) retentate flow and x_1 is the mole fraction at the exit. For the permeate, V_0 and y_0 are zero in the absence of the sweep gas. All the quantities for starting the calculation for stage one are therefore known and the composition and flow leaving stage one can now be calculated. These are then used for the next stage and we can move across the system to end up in stage N.

What value should be assigned to the number of stages N? This depends on whether it is a simulation or design problem. If it is a simulation problem the total area for mass transfer A_m is known. We can discretize this to any number of stages. The larger N is, the more accurate the results (with more computations) since the differential model is now closely approached. The parameter ΔA_m has to be selected as A_m/N and will depend on the value of N chosen. The calculations are done for N stages and provide the values for the inlet feed conditions and also the exit permeate flow and mole fraction for the chosen exit retentate conditions. Using many different exit conditions, a design chart (a parametric plot) can be prepared.

For a design problem a value of ΔA_m is chosen as an incremental area for simulation. The stagewise calculations are continued until the inlet feed mole fraction is reached. This provides the number of stages (N) needed to achieve the separation. The required membrane area is then calculated as $N\Delta A_m$.

Effect of Pressure Variation

The model shown above assumed that the pressures are constant on both the feed and permeate side. Hence the pressure ratio does not vary across the permeater. A rigorous analysis should include the pressure variations in the feed and permeate sides due to frictional losses. The feed is usually in the shell side in the hollow fiber type of arrangement. The pressure drop on the shell side is usually small and no significant error is caused by assuming that the pressure in the shell is a constant. For the permeate side the pressure gradient is zero at the closed end and increases gradually as we approach the discharge end. The pressure variation can be included in the stagewise model by computing the pressure drop in each stage based on the local velocity at that stage. Note the pressure drop depends on the permeate flow rate at each stage and cannot be assigned a priori. Hence this represents a problem where the momentum transfer is coupled to mass transfer.

The analysis of the cocurrent flow pattern shown in Figure 27.4 involves merely switching the flow arrangement in Figure 27.8 and is not presented here.

27.3.5 Cross-Flow Pattern

Another contacting pattern is the cross-flow. The fibers are often spiral wound with a center pipe as the permeate collector. The feed and the permeate flow in a cross-flow manner as shown in Figure 27.9.

A stagewise model is useful and is shown in Figure 27.10.

The permeate from each stage is assumed to flow into the center pipe. The model is merely a simplification of the countercurrent flow. The V terms in the permeate balance are set as zero. Stage k contributes a permeate molar flow of $(\Delta V)_k$, which is equal to $J_A + J_B$ from that stage. The composition of this stream is the same as the local composition y'. Hence $y_k = y'_k$ is used in the feed species balance equations. The total permeate flow is the sum of the flow from the N stages and the composition is the weighted average:

$$y_{\text{exit}} = \frac{\sum_k y'_k \Delta V_k}{\sum_k \Delta V_k}$$

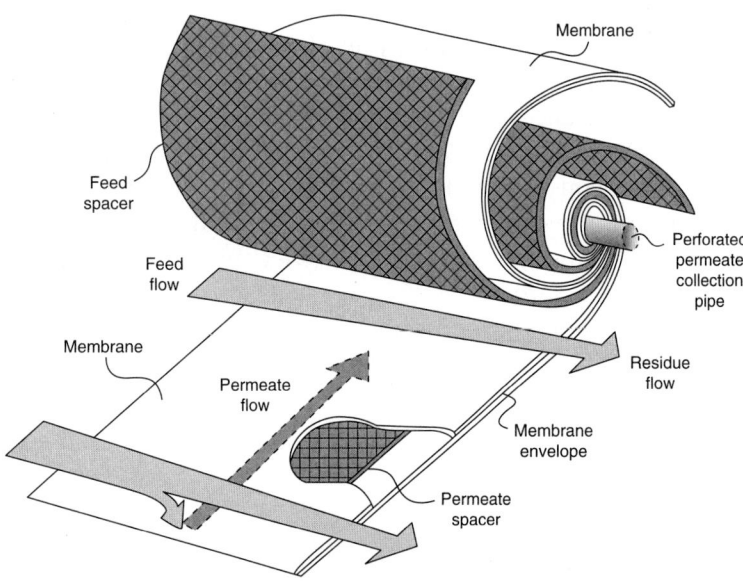

Figure 27.9 Schematic of a spiral-bound membrane permeater. Courtesy Koch Membrane Systems, Inc.

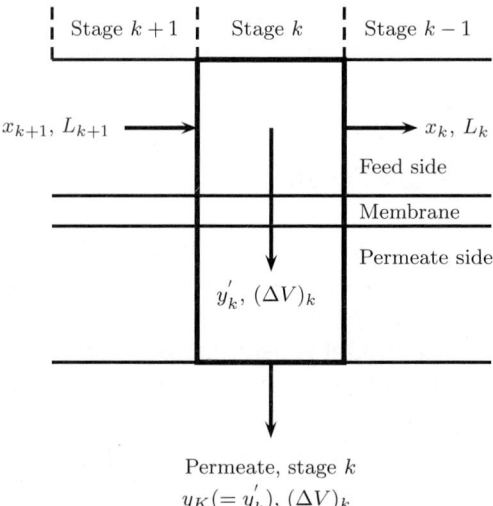

Figure 27.10 Model notation for stagewise computation of a membrane separator with crosscurrent flow.

27.4 Reactor Coupled with a Membrane Separator

This section presents an example of a reactor-separator combo. The walls of the reactor are selectively permeable to a product or reactant. The product removal, for example, overcomes the limitation of an otherwise equilibrium-limited reaction. The arrangement therefore provides a more compact design plus greater conversion. Removal of a product increases the residence time for a given volume of reactor and drives equilibrium-limited reactions toward completion. Another advantage is that the operating range of temperature and pressure is larger.

A schematic of a membrane reactor with product removal is presented in Figure 27.11. One widely studied example is the steam reforming of methane to produce synthesis gas:

$$CH_4 + H_2O \rightleftharpoons 3H_2 + CO$$

The reaction is endothermic and equilibrium limited. The water gas shift reaction also occurs in parallel:

$$CO + H_2O \rightleftharpoons H_2 + CO_2$$

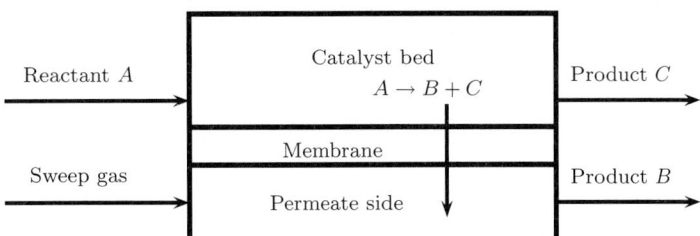

Figure 27.11 Membrane reactor with membrane acting as an product extractant.

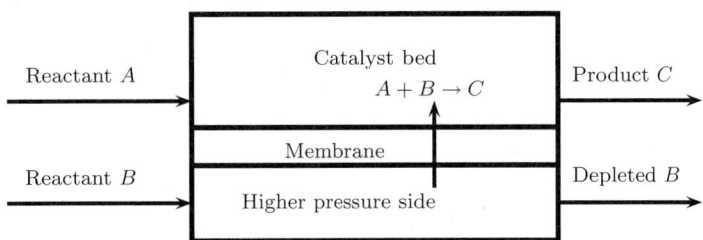

Figure 27.12 Membrane reactor with membrane acting as dosing feed.

This reaction is mildly exothermic. The simultaneous removal of hydrogen shifts the equilibrium to the right. Also it is possible to operate the reactor at somewhat lower temperature.

A second example is the cracking of cyclohexane to benzene:

$$C_6H_{12} \rightleftharpoons C_6H_6 + 3H_2$$

Again equilibrium limitations can be overcome by the use of a membrane reactor. Hydrogen-selective membranes that operate at high temperatures are used in these processes.

Another potential application is in partial oxidation. Here one of the reactants (oxygen) is introduced into the reactor by transport across a membrane. The membrane provides a means of supplying oxygen to the system at a controlled rate. The concentration of oxygen can be adjusted (usually to a low value) so that the secondary oxidation reactions are suppressed. A schematic of a membrane reactor with reactant dosing is shown in Figure 27.12.

Summary

- Membrane-based processes and separations are becoming important in a number of industrial applications and an engineer should be able to analyze and evaluate these processes. Membranes are of course of importance in biological transport in cells and other organs as well.

- For separation of gaseous mixtures, polymer-based membranes are most commonly used. These are essentially non-porous and hence the rate of transport is governed by dissolution accompanied by diffusion. A second class of membranes is nano-porous materials with pore size comparable to the mean free path. In view of the small size of these pores, the Knudsen diffusion is often the operating mechanism in these systems.

- Membranes for gaseous separation in polymeric membranes can be modeled on the basis of the solution-diffusion model. The transport rate is based on the permeability parameter, which is the product of the diffusivity and the solubility divided by the membrane thickness. The relative rate of transport between two gases can be changed by changing the solubility rather than the diffusivity. Thus the solubility parameter plays a key role in the selectivity of the membrane for polymeric membranes.

- Hydrogen diffuses in many membranes after dissociation to H atoms. This is important in many Pd-coated high temperature ceramic membranes. If the equilibrium for dissociation is included, the permeation rate is proportional to the square root of the hydrogen partial pressure rather than being linearly proportional. The resulting expression for permeation is known as the Sievert law.

- For hydrocarbons in polymeric membranes various nonlinear effects are encountered. This can include nonlinear equilibrium relations, nonlinear concentration dependent diffusivity, and time variation in permeability caused by swelling or viscoelastic effects.

- The selectivity for gas separation in microporous membranes is primarily affected by the relative diffusion coefficient ratio. Knudsen diffusion is the main mechanism for diffusion. An improved model is the gas translation model where the Knudsen model is corrected for the diameter of the diffusing species and an activation barrier for pore diffusion.

- There are several ways of arranging the membrane and contacting it with the feed and withdrawal of the permeate. This leads to various flow patterns. Permeator models can be developed by combining species balance and accounting for the effect of the flow pattern.

- The simplest flow pattern to model is the case where both the feed side and the permeate side are well mixed. There is no spatial variation of y and x in this case. The mole fraction in the exit permeate and retentate can be calculated by solving Equations 27.18 and 27.19 in the text simultaneously. The input parameters needed to find the $y-x$ split are the pressure ratio, the selectivity value, and the cut (fraction of the feed flow removed as permeate). Once the transfer rate is known the transfer area can be calculated from the permeability model. This model is useful for laboratory measurements of membrane permeance and selectivity since the interpretation of data is simple and not influenced by the flow pattern. This is reminiscent of differential reactors used for kinetic measurements.

- The membranes are commonly assembled in a hollow fiber arrangement and this is similar to a shell and tube heat exchanger. The flow pattern is usually countercurrent flow in this system. The model is developed using a stagewise model that is similar to the mixing cell model used for reactors. The computations are usually started from the sealed permeate end by assuming an exit retentate flow rate and mole fraction. The calculations are done all the way back to the feed end until the feed composition is reached (design problem) or the prescribed number of stages is reached (simulation problem). This provides the feed flow rate and the cut parameter. The performance (degree of separation) can then be tabulated as a design chart in terms of the cut parameter for various pressure ratios and selectivity values.

- Pressure drop variations (mainly on the permeate side) can be included in the stagewise model by calculating the frictional pressure drop for each stage. This depends on the permeate velocity in each stage.

- Cross-flow is another flow pattern. This is encountered for instance in spiral wound membranes and in radial flow type of arrangements. A stagewise model is again useful from a computational point of view. The local mole fraction is representative of the permeate mole fraction in each stage and the exit mole fraction is calculated as the weighted average of the flow times the mole fraction from each stage.

- Reactors can be combined with membrane separators to go beyond equilibrium conversion with selective removal of product. This is mainly used when hydrogen is the product. Membranes can also act as a controlled feed for a reactant; this is useful to control the selectivity of partial oxidation reactions.

Review Questions

27.1 What are the two streams from a membrane separator called?

27.2 What is meant by a sweep? When is it used?

27.3 What is a barrer?

27.4 What is the effect of temperature on the permeability?

27.5 State Sievert's law. Where is it used?

27.6 State factors that could lead to nonlinear models for membrane transport.

27.7 What are the assumptions in the dual mode transport model in polymeric membranes?

27.8 What additional effects are included in the gas translation model compared to the simple Knudsen diffusion model?

27.9 State four common flow patterns used in membrane separation modeling.

27.10 Define the cut parameter.

27.11 What is the value used for the permeate composition in the cross-flow arrangement?

27.12 State some merits of reactive membranes.

27.13 What is the dosing membrane reactor?

27.14 What is the separating membrane reactor?

Problems

27.1 **Permeability units.** The permeability of a membrane is reported as 100 barrer. Convert this to permeability based on gas-side concentration driving force and gas-side partial pressure driving force expressed in Pa as well as in atm. If the partition coefficient of the diffusing species is 10, determine the diffusivity in the membrane itself.

27.2 **Relative rate based on permeance values.** A nitrogen (20%) and methane (80%) mixture is fed to a membrane at a rate of 1000 kmol/hour. The feed pressure is 550 kPa and the permeate is 100 kPa. The permeance values are 50,000 for nitrogen and 10,000 barrer/cm for nitrogen and methane, respectively. Find the flux of each component across the membrane.

27.3 **Gas translation model.** Estimate the selectivity of hydrogen ($d_A = 0.25$ nm) to nitrogen (0.4 nm) in a membrane with a pore diameter of 0.6 nm. Use the activation energy parameters reported in the paper by Nagasawa et al. (2014) of 4.90 kJ/mole for hydrogen and 7.53 for nitrogen. Compare the values with the selectivity predicted by using the classical Knudsen model. The temperature is 200 °C.

27.4 **Permeabilty from differential data.** Consider a case where air is separated at a feed rate of 20 L/min with $P_1 = 3$ atm and $P_2 = 2.6$ atm gauge. A 3 L/min STP of permeate with 40% O_2 was obtained. The retentate has 17% O_2, which is close to the inlet value. Hence an average value for mole fraction can be used for oxygen. Calculate the permeance of oxygen and the selectivity of the membrane for oxygen. A trial and error calculation is needed since the local mole fraction at the permeate side at the inlet is not known. This needs a selectivity value according to Equation 27.16.

27.5 **Single-stage separation.** Oxygen is separated from air in a membrane that has a selectivity of 8 for oxygen. Determine the maximum separation that can be achieved in a single stage unit. If 60% of oxygen is recovered, find the approximate permeate composition.

27.6 **Hydrogen separation.** A gas containing 70% hydrogen and the rest methane is separated into a nearly pure hydrogen stream in a counterflow membrane permeator that has a selectivty of 100 for hydrogen. The pressure ratio employed is 0.2. Find the fraction of hydrogen recovered in the permeate if the permeate has a purity of 96% for hydrogen and the cut. Model the separator as a countercurrent flow with ten stages.

27.7 **Helium separation from natural gas.** Helium has a mole fraction of 0.82% in a natrual gas and a high permeance with a selectivity factor of 200. Find the helium recovery for different cut values. Assume both the permeate and the retentate are close to backmixed flow in the membrane.

27.8 **CO_2 separation from natural gas.** Separation of CO_2 from natural gas in a hybrid fixed-site carrier type of membrane has been studied by He et al. (2014). Their data indicate a permeance value of 0.2 m^3 STP/m^2 hour bar for CO_2 and a selectivity in the range of 30. The feed contains 10% CO_2 and the retentate should have less than 3% CO_2. A countercurrent unit is used and the pressure ratio is 0.3. Suggest a design by specifying a feed cut. Find the permeate composition.

27.9 **Recovery of VOC from air.** VOC can be separated by using highly selective membrane and Baker et al. (1987) examines this potential. A membrane with permeability of 20,000 barrer for VOC (acetone) and 4 barrer for air was used with a membrane thickness of 2 μm. The feed was 0.5% acetone at atmospheric pressure while a pressure of 3 cm Hg was used in the permeate side. The flow rate was 0.2 m^3/min in a spiral wound type of arrangement. Find the membrane area needed if the retentate has 0.05% mole of acetone and the permeate has 5% mole of acetone.

CHAPTER 28

Liquid Separation Membranes

Learning Objectives

After completing this chapter, you will be able to:

- Classify liquid membrane separation processes based on the pore size.
- Understand the concepts of osmosis and reverse osmosis.
- Model the transport of solvent and solute species across a semi-permeable membrane.
- Examine the concentration polarization effect on the transport rate.
- Develop a simplified model for design of a reverse osmosis system.
- Understand forward osmosis separation and its application.
- Calculate the extent of separation in a pervaporation process.

Separation of liquid mixtures using membranes finds application in many areas. An important application is in wastewater treatment and providing drinking water using desalination. Membranes selective to water transport that exclude the solute are used to purify the water, a process known as reverse osmosis. Other applications include separation of low molecular solutes such as enzymes from fermentation broths, and separation of biobased products such vitamins. The process finds application in the food industry as well in concentration of fruit juices and similar applications.

Transport and separation of components in a liquid mixture by use of membranes is discussed first in this chapter. We classify separation systems based on the pore size of membranes. For micro-sized pores, concepts borrowed from filtration or flow through a capillary can be used. For nano-sized pores, diffusion-based concepts are used. But the concept of osmotic pressure is needed to model the process since the driving force in liquid systems has to be based on the activity rather than the concentration difference. The thermodynamic derivation of osmotic pressure is presented and the effective driving force for the solvent transport is indicated by correction for the osmotic pressure. Application to reverse osmosis is shown. The external mass transfer consideration leads to another important concept in modeling, the concentration polarization effect, which is illustrated next.

Two additional industrially relevant applications are discussed briefly next. The first is the forward osmosis process, where water can be recovered from a wastestream by transport across a selective membrane into a "draw" solution. The second is the pervaporation process, where a solute is transported across a selective membrane but is recovered in the permeate as a vapor.

28.1   Classification Based on Pore Size

Separation of a liquid mixture is similar to filtration (separation of a liquid–solid mixture). Hence similar terminology is often used and the process is classified into three categories depending on the pore size a of the membrane: (i) microfiltration, (ii) ultrafiltration, and (iii) nanofiltration. Microfilters have pore size a in the range of 0.05 to 10 μm and are used for recovery of particles in the range of 0.1 to 20 μm. These filters find applications in bio-separation of animal cells, yeasts, bacteria, and so on, from fermentation broths. These essentially remove insoluble solids.

Membranes for ultrafiltration processes have pore sizes of the order of 100 nm and are used to separate low molecular solutes such as enzymes from high molecular weight solutes such as viruses.

Nanofiltration uses membranes with pore sizes of the order of 10 nm and are used to separate dissolved solutes. An example is purification of salt water to produce potable water; this process is also known as reverse osmosis. Separation of products like glucose, vitamins, and so on, from fermentation broths are other examples of nanofiltration. A molecular weight cutoff in the range of 200 to 1000 daltons is some times used to distinguish between nanofilration and reverse osmosis. Both of these processes can be analyzed in a similar manner, discussed later.

The classification is shown in Figure 28.1. This figure also shows the dependency of the transport mechanism on the pore size relative to the solute size.

For large pores the transport process involves more bulk flow and convection, the microfiltration case. The rate of flow is assumed to be proportional to the pressure gradient, which is referred to as Darcy's law. This states

$$Q_t = P_w \Delta P / t \tag{28.1}$$

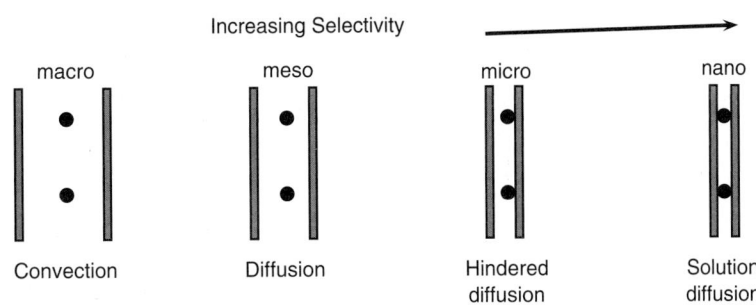

Figure 28.1   Effect of pore size on the transport mechanism of solute in membranes.

Here Q_t is the volumetric flux (velcoity), P_w is a parameter called the Darcy permeability, and ΔP is the applied pressure difference across a membrane and t is the thickness of the membrane. The Darcy permeability is modeled as $d_p^2/32\mu$ based on laminar flow theory, where d_p is the pore diameter for a single straight pore. A correction of ϵ/τ is applied in general to account for the fact that the filter media does not have uniform straight pores. For simultaneous transport of two solutes there is hardly much difference in the rates, leading to no selectivity, as both solutes are carried by the flow.

For smaller pores transport is controlled by diffusion. If now the solute size is comparable to pore size the solute transport is restricted, leading to hindered diffusion. Selectivity is now achieved due to exclusion of larger solutes by a diffusion barrier. This is the situation prevailing in ultrafiltration; the membrane consists of meso- and micro-pores here.

Finally, if the pores are in the nanoscale range, the species dissolves in the membrane and then diffuses across the membrane. A high selectivity can be obtained due to exclusion casued by solubility differences. Often only one type of solute may be transported and the other species may be excluded; such membranes are normally referred to as semi-permeable. The osmotic pressure differences are created by the exclusion of one of the components in the liquid mixture, which is discussed in the following section. The driving force for transport has to be modified to include this factor.

Modeling of microfiltration and ultrafiltration can be done using concepts from flow in porous media, while nanofiltration and reverse osmosis requires the additional concept of osmotic pressure and its influence on the rate of transport. The transport mechanisms depend on the relative magnitude of solute size to pore size. Certain membranes can exclude solutes completely; such membranes are called semi-permeable. The pressure difference minus the osmotic pressure is then used as the driving force for solvent transport, as discussed in the next section.

28.2   Transport in Semi-Permeable Membranes

A semi-permeable membrane is a membrane that permits transport of only one species, usually a solvent such as water. A basic concept in understanding transport in membranes is the osmotic pressure for semi-permeable membranes. Consider a semi-permeable membrane that allows the solvent (water) to diffuse but not the solute (salt). Assume that these (pure water and a salt solution) are separated by a membrane into two compartments. At equilibrium there is a pressure difference between the solution and solvent, as shown in Figure 28.2. This pressure difference is called osmotic pressure and can be calculated based on either thermodynamic or kinetic considerations based on the concept of dynamic equilibrium.

28.2.1   Osmotic Pressure

An expression relating osmotic pressure to the solute concentration is derived in the following paragraphs. This is the minimum pressure that needs to be applied on the salt side in order to prevent an influx of water (solvent) to the salt side of the semi-permeable membrane.

At equilibrium the chemical potential of the solvent is the same on both sides of the membrane. On one side we have pure solvent and let μ_w° represent its chemical potential (the subscript w is used since the solvent is very often water). On the other (salt solution) side, the chemical potential of water is related to its concentration and can be represented as

$$\mu_w = \mu_w^\circ + R_g T \ln a_w + \left(\frac{\partial \mu_w}{\partial P} \right)_{T,C} \Delta P \qquad (28.2)$$

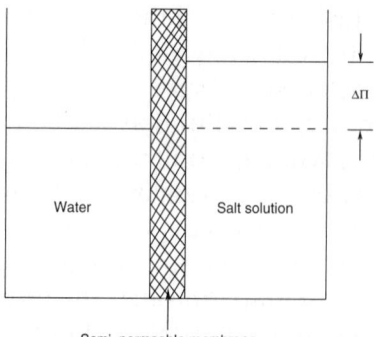

Figure 28.2   Illustration of osmotic pressure difference across a semi-permeable membrane.

Here the second term on the right-hand side is correction due to concentration while the third term is the effect of pressure on the chemical potential. The pressure difference is pressure on the solution side minus that on the pure water side. From thermodynamic, we have

$$\left(\frac{\partial \mu_w}{\partial P}\right)_{T,C} = \tilde{V}_w$$

where $\tilde{V}_w$ is the partial molar volume of water. At equilibrium the chemical potential of water on the solution side μ_w is the same as the chemical potential of pure water:

$$\mu_w = \mu_w^{\circ} \tag{28.3}$$

Hence from Equation 28.2 we can find an expression for the difference in pressure across the membrane:

$$\Delta P = -\frac{R_g T}{\tilde{V}_w} \ln a_w = \Delta \Pi \tag{28.4}$$

This pressure difference is often represented by $\Delta \Pi$, the usual notation for osmotic pressure. Also for dilute solutions we have

$$\ln a_w = \ln x_w = \ln(1 - x_s) \approx -x_s$$

where x_s, the mole fraction of the solute. Hence Equation 28.4 can be written as

$$\boxed{\Delta \Pi = \frac{R_g T}{\tilde{V}_w} x_s = C_s R_g T} \tag{28.5}$$

where $C_S = x_s/\tilde{V}_w$ is the concentration of the solute. This equation, known as the vant Hoff equation, has a resemblance to the ideal gas law. Note that the osmotic pressure is a colligative property, which is defined as any property that depends only on the concentration of the solute and not on the nature of the solute. Correction factors are applied to the previous equation for non-ideal solutions. Also, for solutes such as NaCl, a correction factor of two is used since NaCl exists as Na^+ and Cl^- ions in solutions.

28.2.2 Reverse Osmosis

Now, depending on the applied pressure difference compared to the osmotic pressure difference, three situations can arise, as presented in Figure 28.3.

   If $\Delta P < \Delta \Pi$, there is osmotic flow with the solvent crossing (from the purer side) into the solute side. If the two are equal, we have a situation of

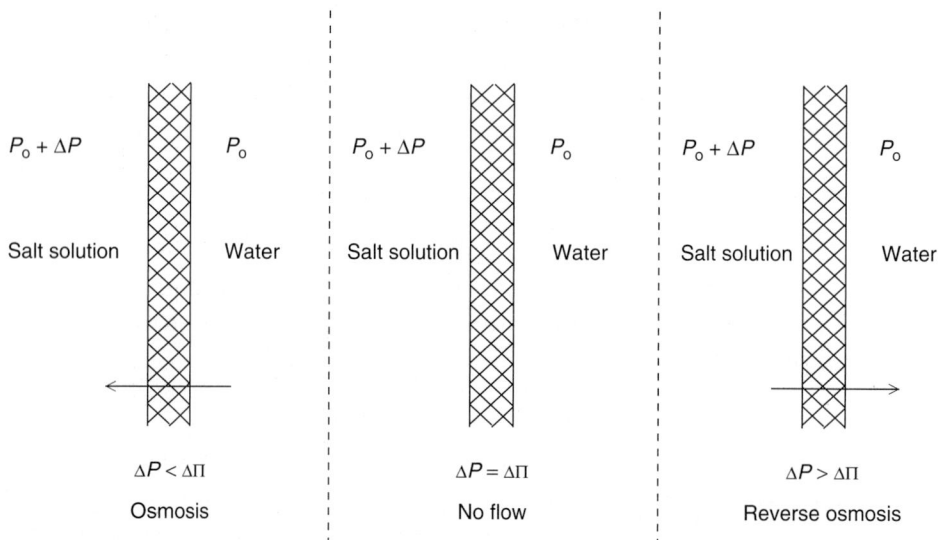

Figure 28.3 Illustration of influence of pressure difference across a semi-permeable membrane: osmosis, no flow, and reverse osmosis.

equilibrium and there is no net flow across the membrane. The third case is when $\Delta P > \Delta\Pi$, which is the condition for reverse osmosis. Here the solvent flows across the membrane (from the solute side) leaving behind an enriched solute solution and pure water as the permeate. This is the condition, for example, for desalination of sea water. The difference $\Delta P - \Delta\Pi$ may be considered the driving force for this case and the flux of the solvent, J_w, is represented as

$$J_w = K_w(\Delta P - \Delta\Pi) \tag{28.6}$$

This is a widely used equation for transport in osmosis and reverse osmosis. Here K_w is a permeability constant for the membrane in units of mole/m^{2}Pa.s.

Partial Solute Rejection

The previous equation for the water flux is applicable under conditions of complete solute rejection. Normally there is some solute diffusion as well, that, complete solute rejection is not achieved. In such cases the flux expression is modified to

$$J_w = K_w(\Delta P - \sigma\Delta\Pi) \tag{28.7}$$

where σ is a factor between 0 and 1. This equation is referred to as the Starling equation in the semi-permeable membrane literature. The case of complete solute rejection corresponds to $\sigma = 1$ and the full impact of the osmotic pressure is left on the rate of transport. The other extreme is $\sigma = 0$, which is the case where the solute is also fully permeable. The parameter σ is known as the Staverman (1952) constant or simply the osmotic reflection coefficient.

For the case where the Staverman constant is not equal to one, an additional equation for the rate of solute transport (across the membrane) is needed. This is modeled by the following equation for the local combined flux:

$$J_s = -D_s C \frac{dx_s}{dz} + x_s J_w (1 - \sigma) \tag{28.8}$$

The first term is the Fick's law diffusion term for the solute while the second term is the solute carried with the solvent flow. x_s is the salt mole fraction. A similar model for partial solute transport is the Kedem-Katchlski model discussed in Section 28.2.4.

We now consider a case with no solute transport ($\sigma = 1$) and show that the solute concentration is larger at the membrane surface compared to bulk liquid.

28.2.3 Concentration Polarization Effects

The convective flow of the solvent causes a corresponding solute flow on the liquid side of the membrane. However, the solute is not transported across the membrane (assuming complete rejection). This causes an increased solute concentration near the membrane surface compared to the bulk liquid. This phenomena is known as concentration polarization, and is illustrated in Figure 28.4.

The effect of concentration polarization is to increase the osmotic pressure at the surface of the membrane and thereby reduce the rate of solvent transport across the membrane. The effect can be modeled as follows by setting the net flux of the solute equal to zero. The solute transport (in the liquid film near the membrane surface) is the combination of diffusive and convective transport given by Equation 28.8. For the case where there is no solute transport, J_s is set as zero. Hence the solute mole fraction in the film is given by the following equation:

$$-DC \frac{dx_s}{dz} = x_s J_W \tag{28.9}$$

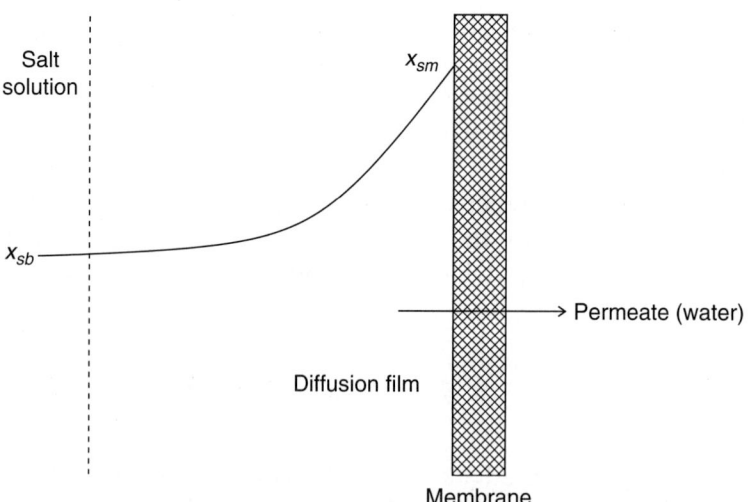

Figure 28.4   Illustration of concentration polarization effect. z is measured from right to left with $z = 0$ at the membrane surface.

The term on the left-hand side is the diffusive transport from the membrane to the bulk liquid. This is counterbalanced by the convective transport caused by the solvent flow, which is represented as the term on the right-hand side.

    The boundary conditions are x_{sm} at $z = 0$ and x_{sb} at $z = \delta$, where δ is the film thickness near the membrane surface. Integrating Equation 28.9 across the film we get

$$-\int_{x_{sm}}^{x_{sb}} \frac{dx_s}{x_s} = \frac{J_w}{CD}\int_0^{\delta} dz$$

Integrating and taking the exponential on both sides gives an expression for the concentration polarization:

$$x_{sm} = x_{sb}\exp(J_W\delta/CD) \qquad\qquad (28.10)$$

The osmotic pressure for solvent transport should therefore be based on x_{sm} rather than x_{sb}. If $\Delta\pi$ is the osmotic pressure based on bulk, then the effective osmotic pressure is

$$\Delta\pi \text{ (Effective)} = \Delta\pi\exp(J_W\delta/CD) \qquad\qquad (28.11)$$

The exponential term is the correction factor for polarization. The transport equation for the solvent given by Equation 28.6 should be modified by multiplying the osmotic pressure by the correction factor. The modified equation (for the case of complete solute rejection) now reads

$$J_w = K_w \left[\Delta P - \Delta \Pi \exp(J_W \delta / CD) \right] \qquad (28.12)$$

This is a transcendental equation for J_w and can be solved by iteration.

Simplified Form

For small values of $J_W \delta / CD$ Equation 28.12 can be simplified by expanding the exponential term and keeping only the linear term. The resulting equation can be explicitly solved for J_w. The final answer shown here has a rather interesting physical meaning:

$$J_w = \frac{[(\Delta P - \Delta \Pi)]}{\left[\frac{1}{K_w} + \frac{\Delta \Pi \delta}{CD} \right]} \qquad (28.13)$$

The denominator term on the right-hand side can be viewed as a sum of two resistances. The reciprocal of this quantity can be viewed as the measured effective permeability:

$$K_w \text{ (Effective)} = \left[\frac{1}{K_w} + \frac{\Delta \Pi \delta}{CD} \right]^{-1} \qquad (28.14)$$

This effective permeability will be a function of flow rate and other parameters (due to the film thickness term) and hence will not be a true measure of the membrane permeability. This is again another effect where the mass transfer effect masks the true parameter value, similar to diffusion in a porous catalyst where the measured kinetic constant $k\eta$ is not necessarily the true kinetic constant k.

Returning back to partial solute transport (which applies if σ is less than one) we present next another model for membrane transport.

28.2.4   Kedem-Katchalski Model

The Kedem-Katchalski model is applicable when solute rejection is not complete. It attempts to provide a simplified representation based on the concept of irreversible thermodynamics and the Onsager reciprocal relations to simplify the number of model parameters. The model also finds important applications in transport in biological membranes.

The model provides the equations for solvent and solute flux across the membranes. A detailed derivation is presented in the original papers by Kedem and Katchalski (1958, 1961) and is not discussed here. A review paper by Jarzynska and Pietruszka (2008) gives the derivation in detail as well and provides details on the interpretation of the model parameters. The final equations of Kedem-Katchalsky model are summarized in the following.

The volumetric flux across the membrane is given as

$$Q_t = P_w[\Delta P - \sigma \Delta \Pi] \tag{28.15}$$

The parameter σ is the solute rejection parameter introduced earlier. $\sigma = 0$ means complete accessibility for the solute transport while $\sigma = 1$ means complete rejection. Usually volumetric flux is commonly used; thus Q_t is in m^3/m^2 s and P_w is a permeance (often referred to as permeability in the osmosis literature) and has units of m/s.Pa.

The solute transport rate is modeled as

$$Q_s = (1 - \sigma)C_{s,l.m}Q_t + \kappa \Delta \Pi \tag{28.16}$$

where $C_{s,l.m}$ is the log mean average of the solute concentration on either side of the membrane.

In this version of the Kedem-Katchalski model the three parameters are P_w, σ, and κ. The significance of these parameters is as follows: P_w is a measure of the Darcy permeability; σ is a measure of osmotic permeability (also called the reflection constant); and κ is a measure of diffusion of the solute in relation to convection.

Estimate of Transport Parameters

The parameters of the Kedem-Katchalsky model were related to fundamental transport parameters in Nakao and Kimura (1982) and Deen (1987). P_w is the Darcy permeability and is related to flow in a cylindrical pore of radius r_p:

$$P_w = \frac{d_p^2}{32\mu} \frac{\epsilon_p}{t}$$

where t is the thickness of the membrane. This applies to the case where there are uniform straight pores. Such membranes are called track-etched membranes. Otherwise a tortuousity factor is added in the denominator.

The parameter σ is related to the ratio of the solute to the pore diameter, denoted as q:

$$\sigma = 1 - S_F \left\{ 1 + \left(\frac{16}{9} \right) q^2 \right\} \tag{28.17}$$

Here S_F is defined as

$$S_F = (1 - q)^2 - (1 - q)^4$$

Finally κ is related to the diffusion coefficient of the solute in the pores:

$$\kappa = D_{bulk} S_D \frac{\epsilon_p}{\mathsf{t}} \qquad (28.18)$$

Here S_D is the correction factor for hindered diffusion:

$$S_D = (1 - q)^2$$

28.2.5 Equipment-Level Model

The local model for solvent and solute transport can be incorporated into a mesoscopic model to find the performance of a reverse osmosis unit or to design this equipment. The model is simple if the solute rejection is complete. Usually a plug flow is assumed for the feed side but the discretized form of this leads to a stagewise model similar to the gas permeator design model in Chapter 27. Either type of approach can be used, but stagewise calculation leads to algebraic equations and is simpler. The calculation procedure follows.

The pressure in each stage has to be calculated from a fluid dynamic equation. The pressure on the permeate side is usually fixed and the pressure drop is small on this side. The pressure difference is therefore calculated. The osmotic pressure value has to be adjusted at each stage corresponding to the salt concentration at that stage. $\Delta P - \Delta \Pi$, the driving force at each stage, is calculated. Equation 28.13 can then be used to find the water flux across the membrane.

The material balance for water and salt on the feed side for this stage is now used to calculate the exit conditions of the stage. Calculations are repeated for the next stage until the end of the membrane unit is reached or the prescribed exit solute concentration is reached. Hence the analysis is similar to that for the gas permeator shown in Section 27.3.5. Further details are left as an exercise.

28.3   Forward Osmosis

Forward osmosis is an osmotic process that uses a semi-permeable membrane to effect separation of water from dissolved solutes. The driving force for this separation is an osmotic pressure gradient between a solution of high concentration, (often referred to as the draw solution) and a solution of lower concentration, referred to as the feed. The osmotic pressure gradient is used to induce a net flow of water through the membrane from the feed into the draw, thus effectively concentrating the feed.

The draw solution can consist of a single or multiple simple salts or can be a substance specifically tailored for forward osmosis applications. The feed solution can be a dilute product stream, a waste stream, or seawater. A schematic of a forward osmosis process is presented in Figure 28.5.

In this sketch a "dirty" stream of brackish water is used as a feed. The draw solution pulls water out of it. For example if the draw solution is concentrated fruit juice, it gets diluted and can be used for drinking purposes. Note that the water is drawn from a dirty source but is in potable form when permeated into the draw solution. This has applications in the food and beverage industry.

In some cases, the draw solution can be regenerated to remove salt (freezing or crystallization) and provide clean water. The draw solution can also be fed to a reverse osmosis system. The advantage (over direct reverse osmosis of the feed stream) is that a relatively clean stream is now fed to the reverse osmosis so that membrane fouling is reduced. Thus there are many applications of the forward osmosis processes. More details are available in Cath et al. (2006) together with illustrative applications.

28.4 Pervaporation

Pervaporation refers to removal of the permeate as vapor and represents an intermediate case between purely gas transport and purely liquid transport in a membrane. The component diffusing through the membrane changes its state from liquid to vapor while it is being transported across the membrane.

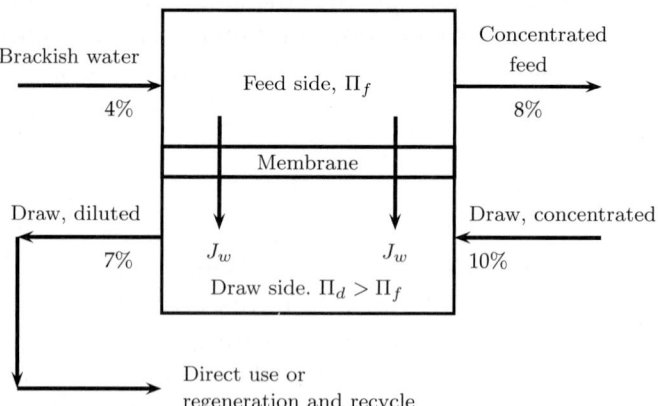

Figure 28.5 Schematic flowsheet for forward osmosis process. Solute % shown are for illustration only.

This requires vacuum equipment on the permeate side and can be costly. However in many cases the overall energy requirement is small compared to distillation and hence it can be an economical alternative, especially for separation of azeotropic mixtures. Composite membranes are used with the dense layer in contact with the liquid and the porous support layer exposed to vapor.

28.4.1 Illustrative Applications

There are a number of applications of pervaporation in practice; two are cited here:

- Separation of ethanol–water mixture: This is a hybrid process designed to produce 99.5% ethanol. A direct distillation cannot be used since an azeotrope is formed. Hence the overhead from the distillation, which has nearly 95% by weight of ethanol, is sent to pervaporation unit. The permeate is rich in water with 25% weight alcohol and is recycled back to the distillation column. The retentate is rich in alcohol and is the required product with 99.5% ethanol. The pervaporation membrane is a hydrophilic material and is selective to water transport. A schematic of this process is shown in Figure 28.6.
- Separation of a volatile organic compound (VOC) from an aqueous solution. Here a hollow fiber module with silicone rubber is used. This

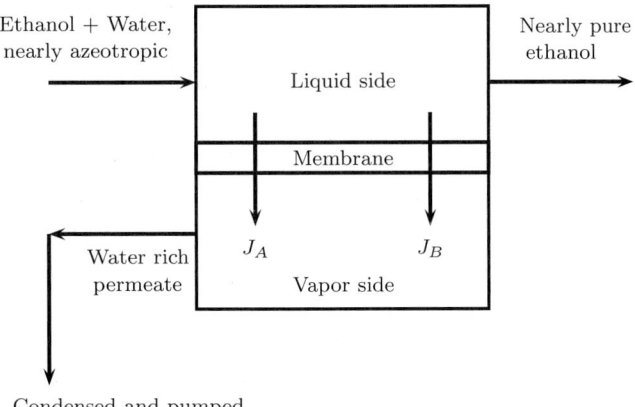

Figure 28.6   Schematic of pervaporation process for purification of ethanol.

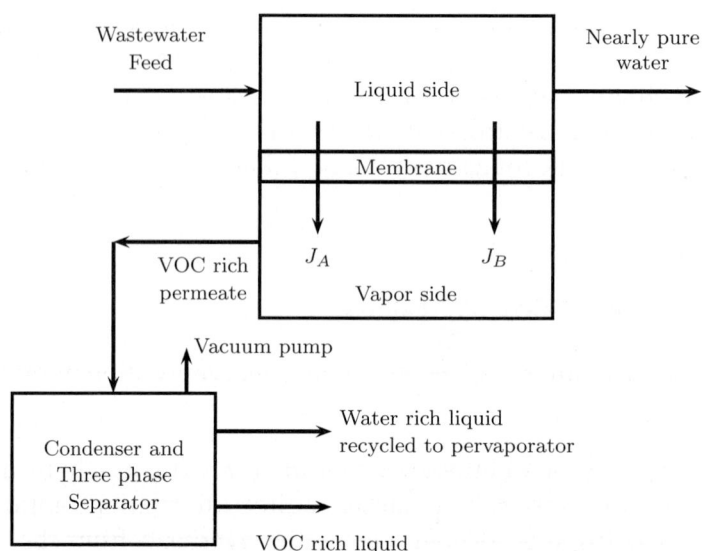

Figure 28.7   Flowsheet for removal of VOC from wastewater using the pervaporation process.

permits the selective separation of the VOC. The retentate is almost pure water with less than 5 ppb of VOC. The permeate after condensation separates into two phases: (i) a nearly VOC-rich liquid and (ii) a water-rich liquid that is recycled back to the membrane unit. A schematic flow-sheet for this is showm in Figure 28.7.

The choice of membrane is critical depending on the application. The membrane can be either hydrophilic or hydrophobic. For water separation, the membrane used is hydrophilic, for example, PVA; water is the main permeate here. Hydrophobic membranes (e.g., silicone rubber, teflon, etc.) are preferred when the organic is the main permeating species, for example, VOC removal.

28.4.2 Model for Permeate Flux

Permeate flux is often modeled by the equation developed by Wijmans and Baker (1993). The fugacity difference is used as the driving force. This concept leads to the following equation for transport across a pervaporation membrane:

$$J_A = \mathcal{P}_A(\gamma_A x_A P_{A,vap} - y_A P_p) \tag{28.19}$$

$\mathcal{P}_A$ is the permeance of A. P_p is the pressure on the permeate side.

A similar equation holds for J_B:

$$J_B = \mathcal{P}_B[(\gamma_B(1 - x_A)P_{B,vap} - (1 - y_A)P_p] \qquad (28.20)$$

It is often difficult to predict the permeance parameters $\mathcal{P}_A$ and $\mathcal{P}_B$ from first principles. Hence this equation is often used to fit these parameters and to compare various PV membrane a rather than as the fundamental law.

The permeance can be a strong function of the feed composition. The temperature variation in the membrane is another complicating factor. A phase change occurs in the membrane and the heat of vaporization is provided by the cooling of the liquid as it permeates the dense layer. This decrease in temperature lowers the permeability and limits the extent of separation that can be achieved in a single unit. The second factor is the concentration dependency of the diffusion coefficient, primarily due to swelling. This increases the diffusion coefficient of both species and thereby alters the selectivity of one species over the other.

28.4.3 Local Permeate Composition

The local permeate composition for a given x_A is related to the ratio of the flux of A, J_A, to the total flux, $J_A + J_B$:

$$y_A' = \frac{J_A}{J_A + J_B} \qquad (28.21)$$

The flux expressions are given in Section 28.4.2 and can be combined and a quadratic similar to that for the gas permeation case can be derived. A modified pressure ratio parameter and a modified selectivity parameter are defined as follows:

The modified pressure ratio parameters for A and B are defined as

$$P_A^* = \frac{P_p}{\gamma_A P_{A,vap}} \qquad (28.22)$$

and

$$P_B^* = \frac{P_p}{\gamma_B P_{B,vap}} \qquad (28.23)$$

The modified selectivity parameter is

$$S = \frac{\mathcal{P}_A \gamma_A P_{A,vap}}{\mathcal{P}_B \gamma_B P_{B,vap}} \qquad (28.24)$$

With these definitions, the expressions for the fluxes can be rearranged as

$$J_A = S\mathcal{P}_B \gamma_B P_{B,vap}(x - yP_A^*) \qquad (28.25)$$

$$J_B = \mathcal{P}_B \gamma_B P_{B,vap}\left[(1-x) - (1-y)P_B^*\right] \tag{28.26}$$

If these fluxes are substituted into the equation for the local mole fraction of y_A' in the permeate side (Equation 28.21) the following nonlinear equation (a quadratic) is obtained:

$$\boxed{y' = \frac{S(x - yP_A^*)}{S(x - yP_A^*) + (1-x) - (1-y)P_B^*}} \tag{28.27}$$

The local mole fraction is also equal to the local permeate mole fraction if there are no additional flows into or out of a local permeate control volume, for example, if the permeate is withdrawn locally and removed. In such cases, y' is the composition of the vapor corresponding to x.

As an example, for 90% by weight of ethanol the local mole fraction of alcohol is 10% for a permeate pressure of 30 mm Hg. The permeate is richer in water. The permeate compositions for various liquid mole fractions of alcohol can be calculated in a similar manner and are shown in Figure 28.8. The membrane used was polyvinyl alcohol and it is observed that the permeate

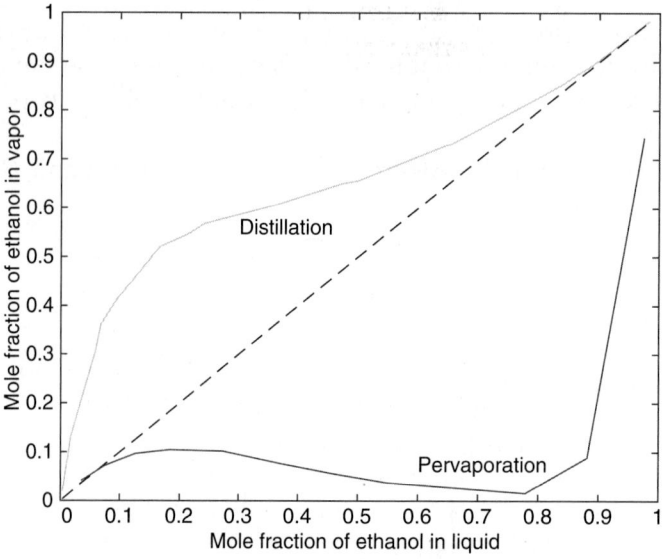

Figure 28.8 Permeate composition for pervaporation with a PVA membrane for an ethanol–water system at 60 °C.

is richer in water than the liquid in contrast to distillation. Note also that the membrane is most selective in the range of alcohol mole per cent between 20 and 85 mole s.

A detailed dynamic model for pervaporation has been proposed by Bausa and Marquardt (2000).

Summary

- Semi-permeable membranes used in many liquid phase separations permit the selective transport of only the solvent. Such membranes find application, for example, in water desalination. At low pressure differences, the water flows from the solvent to the salt side; this is osmosis. At a critical pressure difference equal to the osmotic pressure, the system is at equilibrium and no transport of water occurs in either direction. If the applied pressure is larger than the osmotic pressure then we have conditions for desalination (reverse osmosis; water flows out from the salt solution).

- The rate of transport in a semi-permeable membrane is proportional to the difference between the applied pressure gradient and the osmotic pressure of the solution. The transport equation is given by the Starling equation (Equation 28.7) with a σ parameter known as the Staverman constant, which has a value of one for a completely semi-permeable case. $\sigma = 0$ applies when both the solute and solvent are equally permeable. Thus the Starling equation can be used for both a semi-permeable and completely permeable membrane with the appropriate choice of the σ parameter.

- The solvent flow causes convection effects towards the membrane surface and there is a transport of solute (salt) by convection towards the membrane surface. Assuming there is no net transport of salt across the semi-permeable membrane, the concentration of the salt at the membrane surface has to assume a higher value compared to the bulk liquid value. This is required for the resulting diffusion flux to cancel out the convection flux. The phenomena is known as concentration polarization.

- The effect of concentration polarization is to increase the osmotic pressure locally at the surface (due to higher salt concentration at the surface) thereby reducing the driving force for salt transport. The equation for solvent transport is now given by Equation 28.13. Note that the measured permeability under strong concentration polarization conditions is not the true membrane permeability.

- The Kedem-Katchalsky model is an improvement over the Starling model and provides equations for both solvent and solute transport. The model is based on the concepts from irreversible thermodynamics and involves three parameters: Darcy permeability, osmotic permeability, and a diffusion parameter of the solute in relation to convection.
- Forward osmosis is a process where water is drawn into a draw solution. The draw solution should have a higher solute concentration than the feed solution and the process of water transport is driven by osmotic pressure.
- Pervaporation is a process where a vacuum is applied on the permeate side of a membrane and a vapor is drawn as the product. The retentate side is a liquid. Part of the membrane may be dry and part may be wet depending on the pressure profile in the membrane and the corresponding vapor–liquid equilibrium relationship.
- The simplest model for pervaporation is to use the fugacity difference across the membrane as the driving force. However, the permeance is a strongly concentration-dependent parameter and depends on other factors such as swelling and temperature gradient across the membrane. Preliminary calculations can be done using the fugacity difference model to find the permeate composition.

Review Questions

28.1   What is ultrafiltration?

28.2   What is nanofiltration?

28.3   State Darcy's law.

28.4   What is osmotic pressure?

28.5   How does the osmotic pressure depend on the solute concentration?

28.6   We have a 1 M NaCl and a 1 M KCl solution. Which one has larger osmotic pressure?

28.7   State the Starling equation for solvent transport.

28.8   What is the Staverman constant?

28.9   What is reverse osmosis?

28.10 State the Kedem-Katchalsky model.

28.11 What is meant by concentration polarization?

28.12 What is the effect of concentration polarization on the rate of transport across a reverse osmosis membrane?

28.13 What is forward osmosis?

28.14 What is pervaporation?

28.15 When is a hydrophilic membrane used in pervaporation?

28.16 When is a hydrophobic membrane used in pervaporation?

28.17 What is the commonly used driving force for transport in a pervaporation membrane?

28.18 Given x, the local mole fraction on the feed side, how would you calculate y, the mole fraction on the permeate side?

Problems

28.1 **Minimum energy for separation.** From thermodynamic relations for the chemical potential, calculate the minimum energy required to produce 1 mole of pure water from sea water by reverse osmosis at 300 K. Seawater has an NaCl salt concentration of 35.34 g/1000 g water and has a density of 1025 kg/m^3.

28.2 **Osmotic pressure and freezing point depression.** The freezing point of a salt solution is less than that of water. Derive a relationship between osmotic pressure exerted by a solute and the freezing point depression caused by the solute in the solution. State the principle of thermodynamics that is common between the two phenomena.

28.3 **Osmotic pressure and boiling point elevation.** What is the boiling point of 1 M NaCl solution compared to pure water? The total pressure is 1 atm. State the principle of thermodynamics responsible for this elevation. Explain the relation of osmotic pressure to boiling point elevation

28.4 **Water flux for various pressure gradients.** A semi-permeable membrane has a Darcy permeability of 2.67×10^{-8} mol/cm^2 s atm. The applied pressure difference is 10 atm on side A and 1 atm on side B. Find the water flux and direction of water flow for the following 3 cases where side B is maintained at different NaCl concentrations: side B is at 0, 0.1 M and 0.2 M NaCl concentration.

28.5 **Effect of concentration polarization.** At a certain point in a reverse osmosis module, the salt concentration is 1.8% by weight of NaCl and the applied pressure is 6.5 atm. On the permeate side the pressure is 2 atm. The permeance of the membrane is 1.1×10^{-5} g/cm^2 s atm and there is complete solute rejection. Find the flux of water across the system. Neglect concentration polarization effects. If the mass transfer coefficient in the film near the membrane surface is 0.0025 cm/s, find the salt concentration at the membrane surface and find the reduction in water flux due to concentration polarization effects.

28.6 **Correction for concentration polarization.** A 1 M salt solution is maintained at a pressure of 10 atm above its osmotic pressure and a permeability of 1.5×10^{-5} g/cm^2 s atm is reported based on the measured flux. Find the true permeability if the mass transfer coefficient is 0.0025 cm/s in the film near the membrane surface.

28.7 **Design of a desalination unit.** A salt solution with 3000 ppm of salt at a flow rate of 350 m^3/day is to be desalinated. The feed-side pressure is 80 atm and the permeate pressure is 2 atm. The permeance is 1.1×10^{-5} g/cm^2 s atm and the external mass transfer coefficient is 0.005 cm/s. Design a reverse osmosis system for a 50% recovery of potable water.

28.8 **Transport rate in a inorganic membrane.** The transport of various solutes in inorganic nanofilration membranes was studied by Tsuru et a. (2000) and the following data was given for σ and κ:

Species	σ	κ
Methanol	0.35	4.58
Ethanol	0.55	4.23
Butanol	0.85	1.58

Estimate the fraction of solute rejected for each of these cases.

28.9 **Estimation of parameters in the Kedem-Katchalski model.** The Stokes radius of methanol is estimated as 0.137 nm. Find the σ parameter for a membrane with a pore radius of 1 nm. Find the κ parameter if the diffusion coefficient is 1.79×10^{-9} m^2/s. Assume a porosity of 0.5. The Darcy permeability coefficient for water is 3 m/s Pa.

28.10 **Area calculation for concentration of fruit juice.** Fruit juice is to be concentrated from an initial concentration of 10% to 35%. The operating pressure is 40 atm, which is equivalent to the osmotic pressure of a 42% solution. The permeability is 2×10^{-6} m^3/m^2 min atm. Find the surface area of the membrane needed to achieve the required concentration.

28.11 **Flux and area calculation in forward osmosis.** For the concentrations shown in Figure 28.5, find the membrane area to be used if the water permeability of the membrane is 5 L/m^2 hour bar.

28.12 **Pervaporation permeabilty from measured data.** A pervaporation membrane was exposed to 90% ethanol and the rest water a 60° C and the flux of alcohol was measured as 0.29 kg/m^2 hour. The permeate composition was 7.1% alcohol. The liquid side was at atmospheric pressure while the gas side was at 15 mm Hg. Find the permeability and the selectivity of the membrane for water. Use the Margules equation for the calculation of the activity coefficients (W = water, A = alcohol):

$$\gamma_W = (0.7947 + 1.615 x_W) x_A^2$$

$$\gamma_A = (1.6022 - 1.615 x_W) x_W^2$$

28.13 **Separation calculation for pervaporation.** A membrane used for alcohol–water separation has a permeance of 152 mol/m^2 hour atm for water and a selectivity of 175. Calculate the local permeate composition for 90% alcohol in the feed and a pressure on the vapor side of 30 mm Hg.

28.14 **Permeate composition.** Calculate the local permeate mole fraction for 90% by mass of alcohol over a membrane operated at 30 mm Hg downstream pressure. Use the Margules equation for activity coefficients. Use the following data for the membrane: permeance for water = 152 mol/m^2 hour atm; permeance for alcohol = 0.87 mol/m^2 hour atm; vapor pressure of water = 148 mm Hg; vapor pressure of alcohol = 340 mm Hg.

28.15 **Effect of feed composition and downstream pressure.** Calculate the permeate composition of the previous problem for alcohol mass percents of 95%, 99%, and 99.9%. The following results are shown in McCabe et al. (2010). Verify these results.

% alcohol by mass	x_W	y_w
95	0.1186	0.915
99	0.0252	0.256
99.9	0.00255	0.026

Repeat the calculations of the above problem for a downstream pressure of 15 mm Hg. The results show how the permeate gets relatively richer in alcohol as the alcohol mole fraction in the feed increases and therefore getting to almost pure alcohol gets progressively difficult.

CHAPTER 29

Adsorption and Chromatography

Learning Objectives

After completing this chapter, you will be able to:

- Appreciate typical applications of adsorption and chromatography.
- Understand common types of models used to describe solid–fluid equilibria (adsorption isotherms).
- Understand the sequence of mass transfer steps involved in adsorption and chromatography.
- Set up and solve models for predicting the performance of batch and continuous packed bed adsorption columns.
- Understand how axial dispersion and internal diffusion effects can affect the performance of chromatographic columns.

Adsorption is a mass transfer process by which a chemical component from a fluid phase is transferred and attached to a solid phase. This can be an effective way of removing trace contaminants from either a gas or liquid stream. The species being adsorbed is called the adsorbate while the solid on which it is adsorbed is called the adsorbent. The adsorption process depends on the selective partitioning of the adsorbate in the solid phase. Thus the equilibrium distribution of the adsorbent between the fluid and solid phase is an important thermodynamic property needed in the design of adsorption equipment. We therefore discuss first the various types of adsorption isotherms. Since the species is transferred from the fluid to the solid phase, a number of mass transfer steps are involved in the process. These steps are discussed next. This is followed by models for batch and continuous adsorption processes. For batch adsorbers the fluid phase is modeled using a macroscopic model while for continuous adsorption in a packed column, the mesoscopic model is useful. These models are coupled suitably with the differential model for the adsorbent particle as illustrated in this chapter. Models are needed to calculate the time of operation and design and scale-up of an adsorption column using laboratory data.

Chromatography is a related process where two or more components are separated due to the difference in adsorption property of these species over a solid. One model is to assume that equilibrium exists at all points in the column. This generates the first-level model for the system, the so-called ideal separation model. However, mass transfer effects are important and one of the goals is to examine their effect on the extent of separation and show the deviation from ideal behavior.

29.1 Applications and Adsorbent Properties

Adsorption is useful to separate a solute in small concentrations from a liquid or gas phase and finds application mainly in pollution control. A typical application is in the gas purification process, for example, removal of organics from vent gases. Adsorption from concentrated solutions is also used to accomplish bulk separation of a preferentially adsorbing solute. Typical largescale adsorption-based separation processes are the removal of CO_2 from natural gas, nitrogen–oxygen separation, and separation of normal paraffins from isoparaffins.

The quantity adsorbed is proportional to the surface area of the adsorbent and efficient adsorption needs materials with a large area. This can be achieved only if they are highly porous solids. Activated carbon is one such common adsorbent and has a surface area of $1000 \text{ m}^2/\text{g}$ or more. The process of activation is used to create this high area and consists of two steps: charging and heating of a carbon source such as coconut shells in the range of 500 to 700 °C in a gas phase with a low oxygen concentration, and treating the charred material to an oxidizing gas such as CO or steam to volatilize the tarry material and thereby create a large surface area. A network of micro- and macropores is created by this process. Macropores provide less area but provide pathways to transport to micropores. The micropores provide the large area where most of the adsorption occurs. Hence both macro- and micro-pore volumes are important. The relative proportion of these can be adjusted by choosing suitable starting materials. For example, coconut shells, which are dense, produce activated carbon with more micropores while bituminous coal as a starting material produces an adsorbent more macropores.

Other common materials are silica gel and the zeolites. Silica gel is a porous form of silicon dioxide with typical pore sizes in the 4–7 nm range and a surface area of around 700 m^2/g. It has strong adsorption properties for water and is commonly used as a desiccant. It is also commonly used as a packing in chromatography columns. Zeolites are a class of compounds belonging to the alumina-silicate family of chemicals. They are also known as molecular sieves due to their ability to selectively sort molecules based primarily on a size exclusion process. This is due to a very regular pore structure of molecular dimensions. The maximum size of the molecular or ionic species that can enter the pores of a zeolite is controlled by the dimensions of the channels. For example, zeolite $5°A$ has a nominal pore diameter of 0.5 nm and can separate n-paraffins from branched ones. Zeolites in general have wide applications both as catalysts (e.g., catalytic cracking in the petroleum industry) and as adsorbents (e.g., oxygen separation from nitrogen).

29.2 Isotherms

The equilibrium data for fluid–solid systems often have to be measured in contrast to the vapor–liquid equilibrium for which case models for activity coefficients such as UNIFAC can be used to predict the values computationally. The adsorption data is called an isotherm and can be represented by the models described here.

29.2.1 Langmuir Model

The Langmuir model is the most commonly used model. In this model, adsorption is visualized to occur on active sites of the solid. Let θ be the fraction of sites occupied. Then the rate of adsorption is proportional to the fluid concentration and to the fraction of empty sites $(1 - \theta)$. The rate therefore is equal to $k_a C_A (1 - \theta)$ where k_a is the rate constant for the adsorption process. Simultaneously the desorption rate is proportional to the fraction of occupied sites (θ) and is therefore given as $k_d \theta$, with k_d being the rate constant for the desorption process. The schematic of the basis for the Langmuir model is shown in Figure 29.1.

At equilibrium the two processes balance each other. Therefore

$$k_a C_A (1 - \theta) = k_d \theta$$

Let the ratio of the two rate constants be designated as K_A, an equilibrium constant. Thus let K_A be defined as k_a/k_d. Using this in the above equation

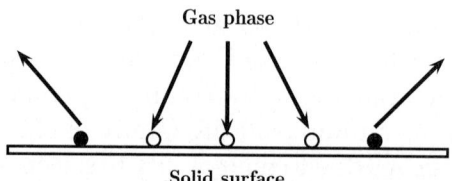

Figure 29.1 Equilibrium viewed as a dynamic balance between adsorption and desorption; dark circles represent the occupied sites and the open circles are available for adsorption. The desorption rate is proportional to the fraction of dark circles.

and solving for θ we get

$$\theta = \frac{K_A C_A}{1 + K_A C_A}$$

The adsorbed concentration of A, q_A, is defined as $q_{max}\theta$, where q_{max} is the concentration at complete coverage corresponding to θ equal to 1. (A monolayer coverage of the sites is implied in the Langmuir model.) Hence

$$\boxed{q_A = q_{max}\theta = \frac{q_{max}K_A C_A}{1 + K_A C_A}} \tag{29.1}$$

which is referred to as the Langmuir adsorption isotherm. This is often rearranged and expressed as

$$q_A = \frac{q_{max} C_A}{\mathcal{K} + C_A}$$

where $\mathcal{K}$ is the reciprocal of K_A. A plot of $1/q_A$ versus $1/C_A$ would be linear if the Langmuir isotherm holds and such a plot is useful to estimate the two constants needed to characterize the model.

Linear Isotherm

At low fluid phase concentrations we have $K_A C_A \ll 1$ and the Langmuir isotherm shows a linear relation:

$$q_A = q_{max}K_A C_A = K_{A,linear}C_A \tag{29.2}$$

where $K_{A,linear} = q_{max}K_A$. The linear model is useful when the solute concentration is low, for example, applications such as pollutant removal. The nonlinear Langmuir model (Equation 29.1) is needed for higher concentrations, for example, for adsorption-based separation processes.

Generally the adsorption capacity K_A decreases with temperature. This is consistent with Le Chatelier's principle since adsorption is an exothermic process.

29.2.2 Competitive Adsorption Isotherm

If two more components are present, they compete for the same vacant sites and hence a modification of the Langmuir model is needed. In such cases the following equation is used for species A in presence of B:

$$q_A = \frac{q_{A,max} K_A C_A}{1 + K_A C_A + K_B C_B} \tag{29.3}$$

A similar equation holds for species B:

$$q_B = \frac{q_{B,max} K_B C_B}{1 + K_A C_A + K_B C_B} \tag{29.4}$$

where $q_{i,max}$ is the maximum amount of adsorption for species i at complete coverage of the all active sites. The model is often called the extended Langmuir model. It ignores the interaction between adsorbed A and B and simply includes the reduction of vacant sites for adsorption of, say A, due to the simultaneous adsorption of B.

29.2.3 Freundlich Isotherms

An equation attributed to Freundlich (1907) is used to describe adsorption equilibria in many cases:

$$q_A = K_{A,f} C_A^{1/n} \tag{29.5}$$

Here $K_{A,f}$ and n are the two parameters needed in this model. The range of n is usually from 1 to 5. $n = 1$ is the linear case. The Freundlich isotherm can be derived by assuming a heterogeneous surface with a non-uniform distribution of the heat of adsorption over the surface.

29.2.4 BET Isotherm

Another model is the BET (Brunauer-Emmet-Teller) isotherm. This is an extension of the Langmuir isotherm. Here the adsorption is not restricted to a monolayer and a multilayer formation is assumed to apply. Brunauer et al. (1938) showed five isotherm patterns as depicted in Figure 29.2.

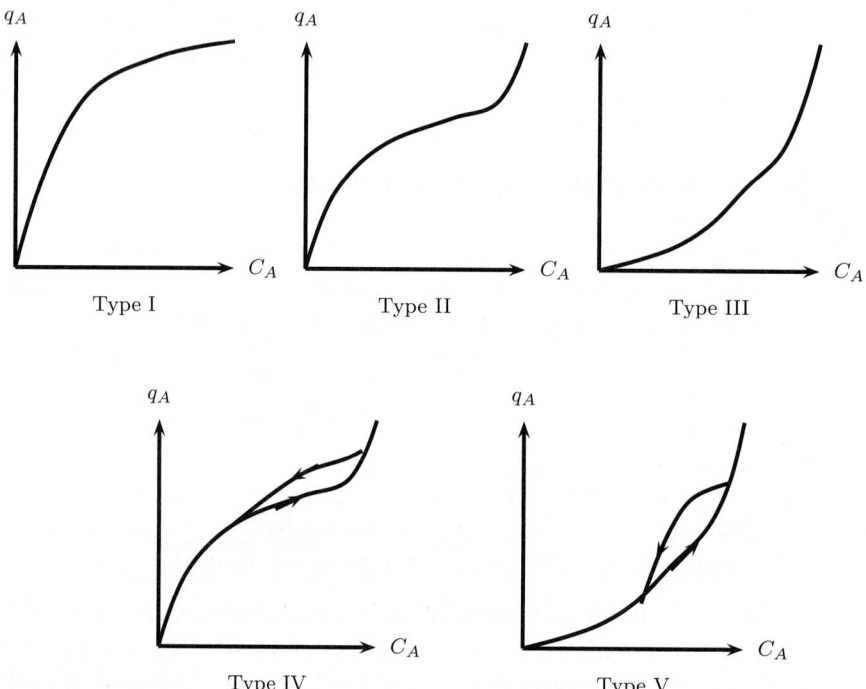

Figure 29.2 Five common types of adsorption isotherms according to Brunauer et al. (1938). Based on Hines and Maddox (1985).

Type I is the standard Langmuir isotherm and shows a monotonic approach to q_{max}; it applies for the case where the adsorption is restricted to a monolayer. Type II occurs when the monolayer adsorption occurs at low pressures and a multilayer condensation type of phenomena occurs at higher pressures. Type III is a case where the condensation type of phenomena is the dominant mechanism. Adsorbents showing this pattern are not useful in practice because the adsorbed concentration is low except at very high concentrations or pressures.

Types IV and V are systems showing a hysteresis phenomena, which is due to capillary condensation. These are capillary condensation versions of Type II and III, respectively.

29.3 Model for Batch Slurry Adsorber

Data from a batch adsorber provide a means of measuring equilibria and can also be used for small-scale processes for gas/liquid purification. Time versus liquid concentration profiles are calculated. In this section we show how mass balance equations can be combined with transport laws and an equilibrium model to predict the transients in this process.

29.3.1 Model Equations

Mass balance on the liquid phase in words is

Loss of solute in the liquid phase = Gain of mass in the solid phase

This can be written mathematically as

$$-V_L \frac{dC_{AL}}{dt} = V_s \frac{d\bar{q}}{dt} \tag{29.6}$$

where $\bar{q}$ is the average adsorbed concentration in the solid. C_A is the concentration in the bulk liquid. The average concentration needs a transient diffusion model for the particle. This particle-level model is common to both the batch slurry adsorber discussed in this section and the fixed bed adsorber discussed in Section 29.4.

29.3.2 Particle-Level Model

The concentration in the particle varies as a function of position due to pore diffusion limitations. The local equilibrium is assumed for the adsorption process and hence the adsorbed concentration at any point in the solid, $q_A(r)$, is equal to $KC_{Ap}(r)$. Here C_{Ap} is the concentration of A in the fluid phase in the pores at any location r.

Note: The notation K is used here for the linear adsorption relation as an abbreviation for the $K_{A,linear}$ in Equation 29.2. This should not be confused with the K_A in the Langmuir model. Also q_A will be abbreviated as q and C_{Ap} is abbreviated as C_p.

The average concentration in the pores (a spherical particle is assumed here) is given as

$$\bar{q} = \frac{3}{R^3} \int_0^R r^2 q(r) dr \tag{29.7}$$

Using the local equilibrium relation, this can be written in terms of the pore level concentration C_p as:

$$\bar{q} = \frac{3K}{R^3} \int_0^R r^2 C_p(r) dr \tag{29.8}$$

The variation of C_p as a function of r can be modeled by the transient diffusion model:

$$\epsilon_p \frac{\partial C_p}{\partial t} + (1 - \epsilon_p)\frac{\partial q}{\partial t} = D_e \left(\frac{\partial^2 C_p}{\partial r^2} + \frac{2}{r}\frac{\partial C_p}{\partial r} \right) \tag{29.9}$$

The two terms on the left-hand side represent the mass capacity in the pores and the mass capacity on the solid. Usually K is large and the second term, the mass capacity over the solid, is much larger and hence in further discussion the first term is dropped. Also note that the surface diffusion (diffusion of adsorbed species along the surface) is ignored here and only the pore diffusion is included.

Since $q = KC_p$ (assuming a linear isotherm here) the PDE can also be written as

$$K(1 - \epsilon_p)\frac{\partial C_p}{\partial t} = D_e \left(\frac{\partial^2 C_p}{\partial r^2} + \frac{2}{r}\frac{\partial C_p}{\partial r} \right) \tag{29.10}$$

Equation 29.10 requires two boundary conditions and one initial condition. The center condition at $r = 0$ is the symmetry condition while the boundary condition at $r = R$ depends on whether external mass transfer (liquid to solid) is important or not. Thus we can use the Dirichlet at the surface if the liquid to solid mass transfer is relatively fast compared to internal diffusion. This provides the following condition for the case of no external resistance:

$$C_p = C_{AL} \text{ at } r = R \tag{29.11}$$

If the external mass transport is included, a balance of flux at the solid surface leads to the following Robin condition:

$$D_e \frac{dC_p}{dr} = k_m(C_{AL} - C_{ps}) \tag{29.12}$$

The initial condition is taken as zero concentration, that is, a fresh adsorbent particle.

The solution of Equation 29.10 can be used in Equation 29.8 to get $\bar{q}$ and $\partial \bar{q}/\partial t$, which in turn is needed in Equation 29.6. This is the exact procedure. It is useful to solve the particle model numerically since the analytical solution is cumbersome in spite of the fact that the PDE is linear. The solution can be done using orthogonal collocation, which appears to be the most efficient for this system. This method is discussed in Section 29.3.5. An approximate analytical solution is based on the linear driving force model, which is presented in the following section.

29.3.3 Linear Driving Force Model

The linear driving force model provides an approximation (LDF approximation). This was first introduced into the adsorption field by Glueckauf (1955). The concentration profile in the particle is assumed to a parabolic function in the derivation of this approximation. The concentration profiles and the LDF approximation to the internal profiles are shown in Figure 29.3.

The quadratic profile assumption can be used to derive an approximate lumped transfer coefficient for internal diffusion. Equivalently its reciprocal is the internal diffusional resistance. The external mass transfer leads to an additional drop in concentration from bulk to the surface. The overall resistance is then the sum of the external solid–liquid resistance and the internal diffusion resistance. The overall driving force is represented as $C_{AL} - \bar{q}/K$. The rate of transfer is represented as a transfer coefficient, k_0, times the driving force. Thus the uptake per particle volume is modeled as

$$\frac{\partial \bar{q}}{\partial t} = K k_0 (C_{AL} - \bar{q}_A/K) \tag{29.13}$$

The exchange resistance $1/k_0$ is modeled as the reciprocal of the sum of two resistances.

An approximate formula for this exchange coefficient is

$$\frac{1}{k_0} = \frac{R}{3k_s} + \frac{R^2}{15D_e} \tag{29.14}$$

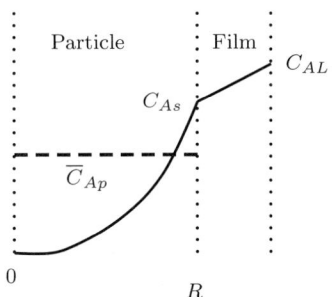

Figure 29.3 Concentration profile in a solid adsorbent and the external film at a given instant of time. The dotted line is the average concentration used in the LDF model.

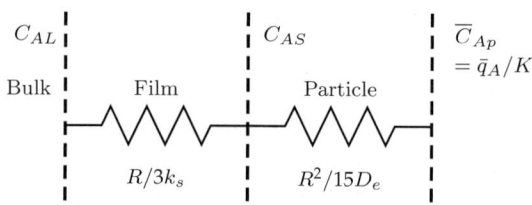

Figure 29.4   Resistances in series as viewed in the LDF model.

The first term on the right-hand side is the resistance to solid–liquid mass transfer and the second term is resistance to internal diffusion.

The equivalent representation in terms of the resistance is presented in Figure 29.4. This figure captures the essence of the linear driving force model.

29.3.4 Calculation of the Slurry Transients

We now link the LDF model to the slurry mass balance model given by Equation 29.6. The average concentration in the adsorbent ($\bar{q}$) can be related to C_{AL} by an overall balance from the start to any instantaneous time:

(Solute in liquid + solute in solid) at time t = Solute in liquid at time zero

Using this material balance, the following equation can be derived:

$$\bar{q}_A = \frac{V_L}{V_S}[C_{Ai} - C_A(t)] \tag{29.15}$$

Here C_{Ai} is the initial concentration of the solute in the bulk liquid.

Using the above relation for $\bar{q}_A$ in Equations 29.6 and 29.13, we can show that

$$-V_L\frac{dC_{AL}}{dt} = Kk_0V_S\left(C_{AL} - \frac{V_L}{KV_S}[C_{Ai} - C_{AL}]\right) \tag{29.16}$$

An adsorption factor α is defined as

$$\alpha = \frac{V_L}{V_S K}$$

Using this factor Equation 29.16 is written as

$$\frac{dC_{AL}}{dt} = -\frac{k_0}{\alpha}[(1+\alpha)C_{AL} - C_{Ai0}] \tag{29.17}$$

Integration of Equation 29.16 provides the concentration profile in the liquid as a function of time. The resulting expression is

$$C_{AL}(t) = \frac{C_{Ai}}{1+\alpha}\{\alpha + \exp[-k_0 t(1+\alpha)/\alpha]\} \tag{29.18}$$

The final concentration in the liquid is obtained in the limit t tending to infinity:

$$C_{A\infty} = C_{Ai}\frac{\alpha}{1+\alpha} \tag{29.19}$$

This can also be verified by an overall mass balance from time 0 to ∞.

The final concentration depends only on α, which can be obtained by changing the volume of the liquid to that of the adsorbent, V_L/V_S. The measured value of the final composition is used to find the parameter α using Equation 29.19, which in turn provides the equilibrium constant, K. The assumption of linear equilibrium can also be tested by this set of experiments since K should be independent of the final concentration for a linear isotherm.

Note also that for each final concentration obtained by changing the V_L/V_S ratio, the corresponding adsorbed concentration is given by Equation 29.15:

$$\bar{q}_{A\infty} = \frac{V_L}{V_S}[C_{Ai} - C_{A\infty}] \tag{29.20}$$

This equation is independent of the type of isotherm since it is purely based on mass balance. Hence q_A versus C_A data can be obtained in a batch slurry experiment by merely changing the solid loading progressively and allowing each set to reach the steady state. The data can then be used to fit a suitable isotherm.

29.3.5 Simulation Using the Collocation Method

In order to simulate the batch slurry transients numerically one needs to discretize the spatial operator in Equation 29.10. One very useful tool is the orthogonal collocation method. We introduce this briefly here and show how the model can be reduced to a set of ordinary differential equations in time.

Equation 29.10 is first made dimensionless in distance by defining $\xi = r/R$, leading to

$$\frac{R^2 K'}{D_e}\frac{\partial C_p}{\partial t} = \left(\frac{\partial^2 C_p}{\partial \xi^2} + \frac{2}{\xi}\frac{\partial C_p}{\partial \xi}\right) = \nabla^2 C_p \tag{29.21}$$

where $K' = K(1 - \epsilon_p)$ is used as an abbreviated notation. Note that the time is not dimensionless here. The right-hand side is the Laplacian defined in dimensionless coordinates and can be approximated using the B matrix given in Section 18.6. This is applied at all the interior points leading to a set of equations for dC_{pi}/dt:

$$\frac{dC_{pi}}{dt} = \frac{D_e}{R^2 K'} \left(\sum_{j=1}^{N-1} B_{ij} C_{pj} + B_{iN} C_{pN} \right) \tag{29.22}$$

Here N is the total number of collocation points and C_{pi} is the concentration of A in the pores at collocation point i.

The point N is the surface of the catalyst ($r = R$) and the concentration at this point, C_{pN}, is assigned by using the the boundary condition given by Equation 29.12. Using the A matrix for the derivative, the collocation version of this boundary condition is

$$\sum_{j=1}^{N} A_{Nj} C_{pj} = Bi(C_{AL} - C_{pN}) \tag{29.23}$$

Here the parameter Bi is defined as $= k_m R / D_e$. Rearranging this equation, the surface concentration is given as

$$C_{pN} = \frac{Bi C_{AL} - \sum_{j=1}^{N-1} A_{Nj} C_{pj}}{A_{NN} + Bi} \tag{29.24}$$

Note that if Bi is large, the Dirichlet condition holds and C_{pN} can be simply approximated as C_{AL}. This is a case where the film mass transfer is rapid compared to internal diffusion.

An additional equation for dC_{AL}/dt can be obtained from Equation 29.6. This needs an average adsorbed concentration, $\bar{q}$. The relation $\bar{q} = K \overline{C}_p$ is used. The integral to calculate the average concentration in the pore, $\overline{C}_p$, is approximated using the w values, the quadrature weights, and therefore

$$\overline{C}_p = 3 \sum_{i}^{N} w_i C_{pi} \tag{29.25}$$

Using this in Equation 29.6 and rearranging we get

$$\frac{dC_{AL}}{dt} = -3/\alpha \sum_{i}^{N} w_i \frac{dC_{pi}}{dt} \tag{29.26}$$

where α is defined as $V_L / V_S K'$.

The initial value equations (Equations 29.22 and 29.26) can be readily integrated, for example, using ODE45. The time evolution of the concentration in the interior points in the pore and in the bulk liquid can then be calculated simultaneously. Further details are left as an exercise.

29.3.6 Additional Complexities

Other effects related to the model caused by additional complexities are discussed qualitatively here. Some useful references for further study are also provided.

Variable Diffusivity

This type of model arises when there is a significant concentration dependence on the intracrystalline diffusivity. An example is diffusion in molecular sieve zeolites. Numerical solutions for constant surface concentration have been presented by Garg and Ruthven (1973). A similar nonlinear dependence also arise for nonlinear adsorption isotherms due to the nonlinear dependency of q on C_i, for example, the Freundlich isotherm. Collocation methods are still suitable with the nonlinear dependent terms added to the B matrix. This presents an interesting case study for students. One consequence is that the adsorption and desorption curves are no longer mirror images of each other. For large concentration values, the adsorption front moves faster than desorption and vice versa.

Bidispered Particles

Many commonly used adsorbents contain both macropores and micropores; these have a grain or particle-pellet structure similar to that discussed in Section 19.3. In this case the macropores provide the transport pathways while the major adsorption takes place in the micropores. Detailed modeling and a solution were presented by Gray and Do (1991) and Liu and Bhatia (2001) and for more details these sources are useful.

29.4   Fixed Bed Adsorption

Fixed bed adsorption columns are commonly used in industrial practice. In this mode of operation the feed containing a solute to be removed is continuously introduced to the system. The column operates in a transient mode and the effluent concentration of the solute varies with time. This concentration is monitored and the operation is stopped when the effluent concentration increases above a threshold value. The adsorbent bed is then regenerated and the operation repeated. Usually two beds are used in tandem, one in adsorption mode and the second in regeneration mode so that uninterrupted continuous operation can be achieved. This section provides an analysis of the transient concentration profiles in fixed bed adsorption and indicates some design considerations.

The model is similar to the tracer response curve in a packed bed studied in Chapter 14. Modeling can be done on many levels of progressive complexity and the approach is discussed in the following.

Models where the concentration profiles in the solid are not solved explicitly are called pseudo-homogeneous models. If the profiles in the solid are also included the model is a called heterogeneous model. The pseudo-homogeneous type of model can again be developed at various levels of detail.

The first category of model is based on local equilibrium and provides an approximate estimate of the column size needed and also the effect of the adsorption isotherm on the outlet concentration. Here no mass transfer resistances are considered nor is the effect of axial dispersion.

A second level of model keeps the local equilibrium assumption but introduces the axial dispersion effect.

The third level includes the mass transfer resistances in addition to dispersion. But the particle-scale model is not solved in detail and instead lumped into an overall mass transfer coefficient based on the LDF concept.

Models where the solid concentration profile is also included and the gas and solid are treated as separate phases are called heterogeneous models. The gas phase is modeled using dispersion and the solid phase is modeled using a transient diffusion-adsorption model; the two are coupled and solved simultaneously.

29.4.1 Equilibrium Model

Plug flow of gas is assumed and local equilibrium between gas and solid is assumed at each point in the adsorber. Transport effects are not included. The gas balance is done by writing the mass balance:

$$\text{in} - \text{out} = \text{accumulation}$$

In = $u_G A (C_A)_x$ where u_G is the superficial velocity of the gas. Similarly, out = $u_G A (C_A)_{x+\Delta x}$.

The accumulation is the time derivative of moles of A in the system. Moles of A are present in the gas phase in the void space plus that adsorbed on the solids. The adsorbed concentration is in equilibrium with the gas concentration. Hence the adsorbed concentration is equal to $K_A C_A$ for linear adsorption. Combining the two, the following expression is obtained for the moles of A in the differential volume:

$$\text{Moles of A} = [\epsilon_B + K_A(1 - \epsilon_B)]\, C_A A \Delta x$$

Here ϵ_B is the bed voidage, that is, the fraction of the bed that is not occupied by the solid.

Putting all these together into the gas phase balance, we can derive the following differential equation for the concentration variation of A in the system:

$$[\epsilon_B + K_A(1 - \epsilon_B)]\frac{\partial C_A}{\partial t} = -u_G \frac{\partial C_A}{\partial x} \tag{29.27}$$

A wave speed w_s can be defined as

$$w_s = \frac{u_G}{\epsilon_B + K_A(1 - \epsilon_B)} \tag{29.28}$$

Using this Equation 29.27 can be expressed in a simpler form as

$$\frac{\partial C_A}{\partial t} = -w_s \frac{\partial C_A}{\partial x} \tag{29.29}$$

This is the simplest PDE of the hyperbolic type. The equation requires an inlet condition $C_{Ai}(0, t)$ (which can be any a specified function of time) and an initial condition (taken as zero here). The solution is simply a time delay:

$$C_A(x, t) = C_{Ai}(0, t)H(x - w_s t) \tag{29.30}$$

Here H is the Heaviside step function.

The inlet value simply moves along the bed with the wave velocity. The inlet response therefore appears at position x after an elapse of time of L/w_s for a step input.

The time constant for a plug flow, also called the retention time (which is also the first moment of the system for a pulse input) is given by

$$\bar{t} = \frac{L}{u_G}\left(\epsilon_B + (1 - \epsilon_B)K\right) \tag{29.31}$$

The second moment can be shown to be equal to zero and therefore there is no spread or dispersion. A pulse introduced at the inlet will appear at the exit intact after the elapse of this retention time. Similarly a square wave input

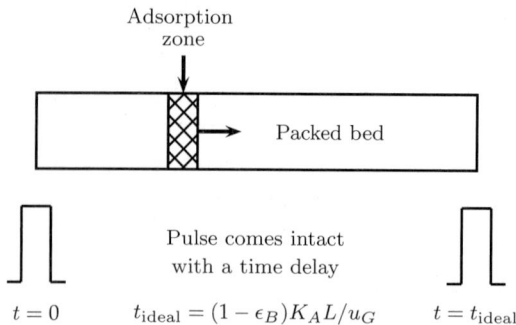

Figure 29.5   Exit response to a pulse injection in an adsorption column modeled as plug flow with an equilibrium model.

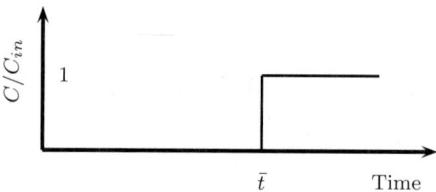

Figure 29.6   Exit concentration for a step input: the ideal breakthrough curve.

at the inlet will appear at the exit as a square wave after this time. A step input will appear as a step input at the exit after the elapse of the retention time.

The exit response is shown schematically in Figure 29.5. (Note that the ϵ_B term in Equation 29.31 is normally dropped as done in this figure.)

The exit response to a step input is shown in Figure 29.6. For a continuous flow of solute with the inlet gas, no solute appears in the exit for time less than the retention time. After this time, the exit concentration is equal to the inlet concentration. Hence for a given length of the bed the maximum time of operation is equal to the retention time.

29.4.2 Axial Dispersion Effects

In reality the response is not a sharp jump as depicted in the previous figure but is instead broadened due to axial dispersion effects. Another factor that can cause spreading is the internal diffusion and in some cases the solid–liquid film mass transfer. Each of these contribute to the variance of the pulse

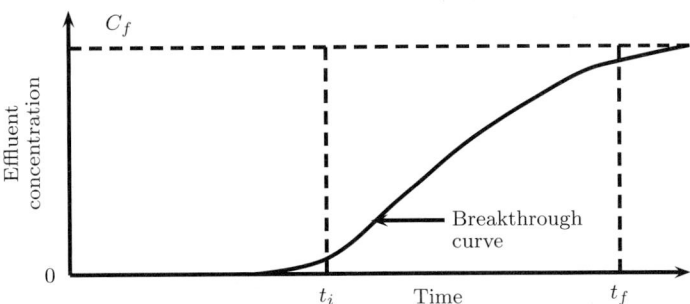

Figure 29.7 Breakthrough curve with exit concentration as a function of time. Data from Hines and Maddox (1994).

response or to the spread of a step response. The characteristic exit then shows a spread around the retention time. The resulting curve is called the breakthrough curve and is sketched in Figure 29.7. Note that a part of the solute appears earlier than the retention time and some appears after the retention time. If a certain cutoff purity is required at the exit, the operating time has to be reduced. This results in a part of the bed not being utilized. If the exit concentration has to be less than, say, C_b then the operating time should be t_b the corresponding time on the breakthrough curve.

The steepness of the breakthrough curve shows how much of the bed is underutilized. If it is very steep a large part of the bed is utilized, approaching an ideal case. Thus prediction of the shape of the breakthrough curve is quite important. It is also useful to know what effects the bed length, bed diameter, and particle diameter have on the shape of the breakthrough curve. The mass transfer models are useful to do these calculations.

The spread due to dispersion is modeled by adding an additional contribution to axial dispersion, but other assumptions in the ideal model are retained. Thus film diffusion and internal diffusion are not accounted for and a local equilibrium is assumed at any location in the adsorber. This, for example, could apply for small particles where the mass transfer coefficients are large and the internal diffusion resistance is also small.

The gas balance now reads

$$[\epsilon_B + K_A(1 - \epsilon_B)]\frac{\partial C_A}{\partial t} = D_E \frac{\partial^2 C_A}{\partial x^2} - u_G \frac{dC_A}{\partial x} \tag{29.32}$$

where the first term on the right-hand side reflects the spread due to axial dispersion. The model now needs one initial condition and two boundary conditions. This is similar to the dispersion model studied in Section 14.3. The time constant is simply adjusted to account for the accumulation in the solid. The response curve would look similar to that plotted in Figure 14.2.

The dispersion coefficient is an important parameter in quantifying the pulse spread. Correlations to calculate these were reviewed in Section 14.6 and are useful to simulate adsorption and chromatographic columns.

29.4.3 Heterogeneous Model

In this model the gas and solid are modeled separately with the solid–liquid mass transfer connecting the two models. The model equations are stated below. See also the discussion in Section 14.5 where the variance of the tracer curve is shown for a packed bed. It may be useful to review this section at this point.

The gas balance equation is now modeled with the accumulation in the particle added as an additional term. The $\bar{q}_A = KC_A$ assumption is no longer used:

$$\epsilon_B \frac{\partial C_A}{\partial t} + (1 - \epsilon_B) \frac{\partial \bar{q}_A}{\partial t} = D_E \frac{\partial^2 C_A}{\partial x^2} - u_G \frac{dC_A}{dx} \tag{29.33}$$

where $\bar{q}_A$ is the average concentration in the solid:

$$\bar{q}_A = \frac{3}{R^3} \int_{0, \bar{q}_A}^{R} r^2 q(r) dr \tag{29.34}$$

The variation of q_A as a function of r can be modeled by the transient diffusion model. This part of the model is given in Section 29.3.2.

The solution of this model needs a coupled computational scheme. At each location x in the bed, the particle model has to be solved to provide the $\bar{q}_A$ information. This provides the required coupling for the gas phase balance. A number of numerical methods can be used. Collocation methods are again suitable but need some additional refinements. Since there are two spatial variables, one in the bed x and the second internal to the particle r, two separate collocation discretizations are needed. The method is sometimes called double collocation. Sampath et al. (1975) provide details of this method. Although this paper is for a reacting solid problem it has the same mathematical structure as the transient adsorption simulation model.

29.4.4 Klinkenberg Equation

A useful approximate analytical solution to calculate the breakthrough curve was developed by Klinkenberg (1962). Here axial dispersion is not included but mass transfer effects are. But the particle model is not solved and the LDF approximation is used to account for mass transfer and internal diffusion. The concentration profile is given by the following equation as a function of dimensionless distance and time:

$$\frac{C_A}{C_{Ai}} = \frac{1}{2}\left[1 + \text{erf}\left(\sqrt{\tau} - \sqrt{\xi} + \frac{1}{8\sqrt{\tau}} + \frac{1}{8\sqrt{\xi}}\right)\right] \tag{29.35}$$

Here ξ is the dimensionless axial distance, defined as

$$\xi = \frac{k_0 K}{v}\frac{1 - \epsilon_B}{\epsilon_B}x \tag{29.36}$$

and τ is dimensionless time corrected for the motion of the fluid, defined as

$$\tau = k_0\left(1 - \frac{x}{v}\right) \tag{29.37}$$

Here k_0 is given by the LDF model (Equation 29.14).

29.4.5 Scale-Up Aspects

Often data from small-scale columns needed to measure adsorption capacity; the data are used to predict the performance of large-scale columns. Crittenden and Thomas (1998) suggested the following scale-down rules:

$$\frac{t_s}{t_l} = \left(\frac{d_{ps}}{d_{pl}}\right)^{2-\alpha}$$

where t_s is the time for estimated breakthrough in a small-scale column and t_L is the value for the large-scale column. d_p is the particle diameter used.

The exponent of the particle diameter α has a value of 0 for a constant diffusion coefficient and is equal to 1 for diffusivity proportional to concentration. The latter value is representative if surface diffusion is the dominant pore transport mechanism. Spread by internal diffusion is the main controlling mechanism implied in this scale-up rule. Since the time constant for the constant diffusivity is proportional to the square of the diameter, the exponent is two for the constant diffusivity case.

The matching of the Reynolds number is used to determine the operating velocity in the small columns. Hence

$$\frac{v_s}{v_l} = \frac{d_l}{d_s}$$

Particle to column diameter should be chosen as 20 or more to avoid channeling near the walls.

29.5   Chromatography

Chromatography is a generic term to denote separation of multicomponent mixtures by using the difference in adsorptivity. The principle behind chromatography is similar to adsorption. Both involve contacting a multicomponent fluid mixture with a solid. The only difference is that the solute is injected as a pulse and usually has many components. An inert carrier gas is used as a base component and this provides the motion of the pulse in the bed. The process is also known as elution chromatography since the pulse is eluted by the carrier stream. Depending on the relative magnitude of the adsorption constants, two components appear at the exit at different times and chromatographic separation is achieved. Chromatography is mainly used as an analytic tool. Large-scale chromatography can be used for separation of liquid mixtures; the separation of xylene isomers is an example.

We first look at the process assuming plug flow and then correct the model for the effect of axial dispersion. The pulse of A will appear intact after the elapse of retention time if there are no factors that cause the broadening of the pulse. The retention time for A is given as

$$\bar{t}_A = \frac{L}{u_G} \left(\epsilon_B + (1 - \epsilon_B) K_A \right)$$

Note that the mean retention time depends on the adsorption constant, as observed from the above equation. If the two solutes have different mean retention times, they appear in the exit stream at different times, causing a separation. This is the principle of chromatography. If the column is modeled as an ideal system, the two components will appear as two sharp peaks at the corresponding mean retention time. In reality, there are dispersion effects, which cause a spread; this is referred to as pulse broadening. An illustrative response curve is shown in Figure 29.8.

If the injected sample is small, the pulse becomes Gaussian and the concentration distribution can be modeled as a Gaussian distribution around the mean:

$$\frac{C_A}{C_{Ai}} = \frac{1}{\sqrt{2\pi\sigma^2}} \exp\left(-\frac{(t - \bar{t})^2}{2\sigma^2} \right) \tag{29.38}$$

The variance depends on the extent of axial dispersion, solid–liquid mass transfer, and the internal diffusion. Each term contributes in an additive manner to the variance. Equation 14.45 can be used to estimate the contribution of the above three factors to the variance. The spread around the mean retention time is about 4σ, as shown in Figure 29.8 for a Gaussian distribution.

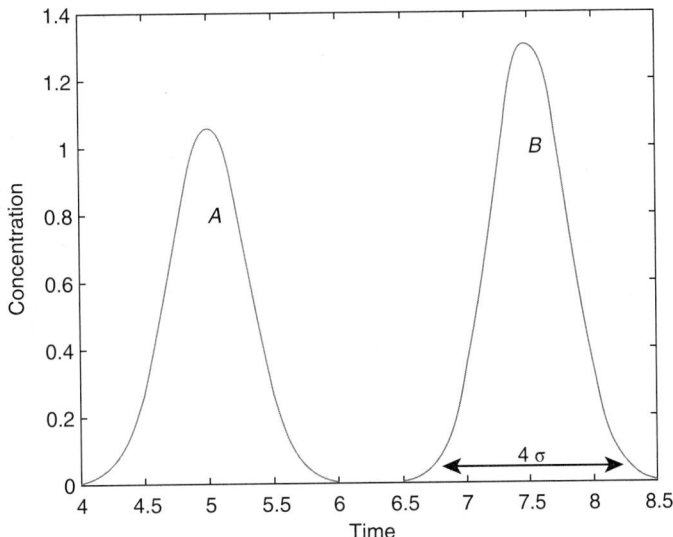

Figure 29.8 Typical response for a binary mixture; the peak spread width Δt corresponds to four times the standard deviation. $\bar{t}_A = 5$ and $\bar{t}_B = 7$. Time is arbitrary units for illustration purposes only.

The area under the curve is a measure of the amount of that component. Overlapping peaks are often noticed. This depends on the σ parameter, which causes the spread. If $\bar{t}_A$ and $\bar{t}_B$ are somewhat close and simultaneously if σ is also high, then overlap will result. Any factor that decreases σ will help in reducing the overlap. For example, increasing the column length can be helpful. Change in temperature, temperature programing, change in carrier gas composition, or change in column material are other tools which can be used to increase the resolution between adjacent peaks.

Summary

- Adsorption is a process where a solute is transferred from a fluid phase to the surface of a solid and retained there. It is generally used for purification of waste streams. Some large-scale processes are also in practice, for example, separation of air into nitrogen and oxygen, and separation of isomers such as xylene.
- Adsorption isotherms show the equilibrium relation between the gas concentration and the adsorbed concentration. For low concentrations

a linear relation is used while for high concentrations, the Langmuir or Freundlich isotherm is commonly used.

- The BET isotherm accounts for multiple layers over the solid and is generally of five types: monolayer adsorption, monolayer adsorption followed by multilayer condensation, multilayer condensation, and capillary condensation with hysteresis of two types (convex upward and convex downward).

- Batch slurry adsorption is used in small-scale systems and also as an experimental tool to determine adsorption rate and equilibrium. The process requires two transport steps: film mass transfer and internal diffusion in the pores. Particle-level transient diffusion models can be set up and coupled to a macroscopic balance for the liquid in order to track the concentration changes in the liquid with time. The coupled model can be solved by the orthogonal collocation method.

- The LDF model provides a simpler approach for the particle-scale model. The diffusion equation for a concentration profile within the particle is not used and lumped into an internal transport coefficient. An overall mass transfer coefficient can be defined on this basis and combined with the solute mass balance in the bulk liquid. Analytical solutions result for the linear adsorption model.

- In an ideal fixed bed operation, the fluid is assumed to be in equilibrium with the local fluid concentration. No other effects are included. The results show that a sharp concentration front develops in the bed. This front moves with a velocity known as the wave velocity, which is given by Equation 29.28. The corresponding time constant is called the retention time. The solute breakthrough is achieved after the retention time and the bed can be operated up to this time.

- When mass transfer effects are included, the ideal response is not seen and instead an S-shaped curve is observed for the exit concentration. This curve is called the breakthrough curve. The time of operation is reduced due to the S-shaped nature and a portion of the bed will remain unused.

- The Klinkenberg model is a useful model to predict the breakthrough curve. The model provides useful information to scale up laboratory data to large-sized commercial units. The model does not include axial dispersion and also assumes a linear adsorption isotherm. Also it uses the LDF approximation for the particle scale transport.

- Chromatography is a generic term to denote separation of multicomponent mixtures by using the difference in adsorptivity. The principle behind chromatography is similar to adsorption. Both involve contacting a multicomponent fluid mixture with a solid. The only difference is that the solute is injected as a pulse and usually has many components.

Review Questions

29.1 What is the mechanistic basis for the Langmuir model?

29.2 How should the experimental data of q versus C_A be plotted in order to test the Langmuir model?

29.3 State the Freundlich isotherm and also the units of the equilibrium constant used in the model.

29.4 How should the experimental data of q versus C_A be plotted in order to test the Freundlich model?

29.5 If a plot of $1/q$ versus $1/C_A$ is linear, what do the slope and intercept represent?

29.6 If a log-log plot of q versus C_A is linear, what do the slope and intercept represent?

29.7 What can cause hysteresis behavior in an adsorption isotherm, that is, are the q versus C_A plots different for adsorption compared to that for desorption?

29.8 What is the LDF approximation?

29.9 What is the assumption used to derive the LDF approximation?

29.10 What are the differences between the pseudo-homogeneous model and heterogeneous model for packed columns?

29.11 What is the ideal pulse velocity in an adsorption or chromatographic column?

29.12 What is the breakthrough curve?

29.13 What principle is used in chromatographic separation?

29.14 What factors cause pulse broadening in a chromatographic column?

Problems

29.1 **Fitting an isotherm to data.** The following data were obtained for adsorption of pure methane on activated carbon by Ritter and Yang (1991):

q, STP cm³/gm	40	165	350	545	760	910	970
p, atm		45.5	91.5	113	121	125	126	126

Fit a suitable isotherm model for the this data. Deterimine which isotherm (Langmuir or Freundlich) provides a better fit to the data.

29.2   **Batch adsorber calculations.** An aqueous solution of phenol with a concentration of 0.01 mol/L is treated with activated carbon in a batch adsorber. An adsorbent loading of 10 g/L is used. The equilibrium constant K has a value of 100. The external mass transfer coefficient is 5×10^{-5} m/s. The internal diffusion coefficient is 2×10^{-9} m^2/s and the particle size used is 1.5 mm diameter. Find the α parameter and use this to find the final concentration of phenol in the liquid. Find the k_0 parameter and use this to track the concentration versus time profile of phenol in the liquid.

29.3   **Laplace transform method.** Obtain the solution to Equation 29.29 in the Laplace domain. Find the first and second moments and verify the expression for the time constant is given in Equation 29.31 and also that the variance is zero. Look up books that provide a table of inverse transforms and verify the time domain solution given in the text.

29.4   **Ideal adsorption model.** An aqueous solution of 3 g/cm^3 each of glucose and fructose is fed into an adsorption column packed with an ion exchange resin. The adsorption isotherm is modeled as a linear equation $q = KC_A$ with $K = 0.26$ for glucose and $K = 0.40$ for fructose. The units are g per 100 cm^3. The superficial velocity is 0.031 cm/s and the bed void fraction is 0.4. If the feed is introduced as a sharp pulse in a column of length 200 cm, determine the time when the two solutes appear at the exit. If a square wave feed pulse lasting 500 seconds is introduced, sketch the exit concentration of both the solutes. Assume the pulse travels as a wavefront with the wave velocity given by Equation 29.28.

29.5   **Wave speed for nonlinear equilibrium.** A wave speed w_s was shown to be given by Equation 29.28 for a linear adsorption isotherm. Show that for a nonlinear isotherm the wave speed can be calculated as

$$w_s = \frac{u_G}{\epsilon_B + (1 - \epsilon_B)(dq/dC)}$$

29.6   **Klinkenberg model for breakthrough.** Air at 25 °C and 1 atm containing 0.9% benzene is fed to an adsorption tower that is 30 cm in diameter and 2 m long. The air flow rate is 5 kg/min. The bed is packed with silica gel with an effective diameter of 2.6 mm. The bed void fraction is 0.5. The isotherm is linear with K equal to 5.2. The mass transfer coefficient k_0 was found from separate experiments to have a value of 0.2 min^{-1}. Calculate the retention time. Use the Klinkenberg model and plot the breakthrough curve. Find the time at which 5% benzene appears in the exit. Find the fraction of the bed that is not utilized. Repeat the calculation if the bed were 8 m long. Determine by what factor the pressure drop would increase. Note that the cost of pumping has to be balanced with the need for more frequent regeneration.

29.7   **Packed bed: size specifications.** A packed column is used to separate a solute from an air stream of 1 L/min flow. A superficial velocity of 4 cm/s is suggested and the column will operate for 4 hours before regeneration. The equilibrium constant is 500. The exit stream should contain no more than 5% of the incoming solute. The particle size is 3 mm and the bed void fraction 0.4. The diffusion coefficient of the solute in air is $0.16 \text{ cm}^2/\text{s}$. Calculate the solid–liquid mass transfer coefficient and the k_0 parameter. Find the column height needed. Find the pressure drop in the column. Examine the sensitivity of your results by changing the superficial velocity and the suggested operation time.

29.8 **Scale-up of a packed adsorber.** A small-scale unit is operating with a particle radius of 0.1 mm at a velocity of 25 m/hour and the breakthrough was observed after 24 sec. Based on this what should be operating velocity if a pilot scale adsorber will operate at 5 m/hour? What would be breakthrough time in this column?

29.9 **Chromatographic separation.** Consider separation of two components with the following properties: $K_A = 1$; $K_B = 2$; $D_A = 2 \times 10^{-4} \text{ m}^2/\text{s}$; $D_B = 8 \times 10^{-4} \text{ m}^2/\text{s}$. The eluting fluid has a flow rate of $1 \text{ cm}^3/\text{min}$ and the column of diameter of 5 mm. The length of the column is 2 m. The particle size is 0.5 mm.

Find σ and test if there is an overlapping of the peaks. Assume that the spread is due to internal diffusion only.

Sketch and plot the exit concentrations of the two solutes as a function of time.

CHAPTER 30

Electrodialysis and Electrophoresis

Learning Objectives

After completing this chapter, you will be able to:

- Understand the principles behind separation by electrodialysis and some of its applications.
- Classify the types of membranes used in the process and understand the operation of a bipolar membrane.
- Design and evaluate a electrodialysis system based on a simple model.
- Understand how two charged species can be separated due to the differences in their relative mobility by the electrophoresis method.
- Review the flow and field arrangements in some common electrophoresis units.
- Develop and solve simple models for electrophoresis and apply these for preliminary design of a device.

Electrodialysis refers to mass transport through stacks of ion-selective membranes under the action of an electric field. It is used to separate an electrolyte feed solution into a dilute and a concentrated solution. A common application is to purify brackish water in connection with desalination, but the technique has found applications in many other fields as well. The use of selective ion-exchange membranes is the separating factor here. In this chapter, we start with the description of the process, application areas, and technological aspects together with some variations. Then a simple model to design or simulate the equipment is shown where we apply mass conservation law, Faraday's law, and Ohm's law to determine the voltage needed for the separation and the power requirements. Transport across the membranes is simply lumped into a resistance in this model. A more detailed model uses the transport across charged membranes using the Nernst-Planck equation and a brief discussion and pertinent references are provided. The concept of limiting current is an important design consideration as well and is also discussed briefly.

Electrophoresis refers to movement of charged particles in solution under the action of an electric field. This field is becoming important in what is now called the subject of proteomics, which requires separation of proteins, DNA molecules, and so on. The difference in mobility is used as the separating agent here. Proteins and colloidal materials carry a surface charge and move under the influence of an electric field with the velocity proportional to the mobility and the direction of motion dependent on the type of charge. If now a flow is superimposed, different species will follow different trajectories due to differences in their surface charge or mobility. Hence they separate into "bands" similar to chromatography. This is the basic idea used in the process and there are many variations and designs used in practice. Three such designs are introduced. A simple model is discussed to predict the bandwidth and the extent of separation.

30.1   Technological Aspects

Electrodialysis (ED) refers to transport across membranes using membranes that are selective only to either the cations or anions. Thus cation exchange membranes (CEMs) exclude anions and permit only the transport of positive ions; the negative ions are rejected. Conversely anion exchange membranes (AEMs) permit transport of only negative ions. By stacking these in an alternating manner it is possible to remove salt from water and produce a solution with a lower salt concentration; a more concentrated salt will be formed in the alternating layer between the membranes. An electrolytic reaction is allowed to take place at the anode and the cathode placed at the end of the stacks of membranes. This provides the current and is necessary to promote the transport of ions across the membrane. Thus the potential applied across the equipment provides the energy for separation. The electrodes are chemically neutral metals with the anode being typically stainless steel and the cathode a platinum- or titanium-coated surface.

An illustrative diagram of the unit is presented in Figure 30.1. The unit shown has two membrane pairs of anion-selective and cation-selective membranes. The working principle of the separation method can now be discussed using this figure as a reference. Note that in practice a large number of membrane pairs arranged into an alternating manner are used and assembled into a "plate and frame" type of design.

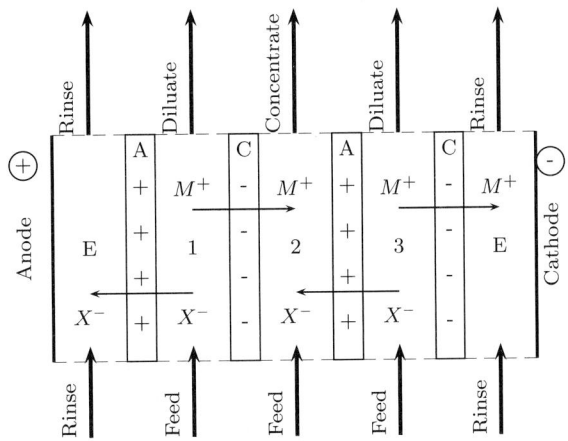

Figure 30.1 Schematic diagram of electrodialysis process: A = anion exchange membrane; C = cation exchange membrane. Note that only two stack pairs are shown but larger scale applications use a large number of stack pairs in a similar alternating arrangement.

In this arrangement, the feed solution gets diluted in alternative compartments marked 1 and 3 in the figure. The presence of an anion-selective membrane on the left side of compartment 1 and a cation-selective one on the right side removes the anion X^- and the cation M^+ in compartment 1. The salt concentration in feed solution gets diluted as it flows through compartment 1. A similar process occurs in compartment 3.

In compartment 2 the feed gets concentrated and the reason for this is because the M^+ coming into compartment 2 from 1 is not able to cross to 3 due to the presence of the anion-exchange membrane as a barrier. Similarly X^- coming from 3 meets a cation exchange barrier and is not able to cross over to compartment 1. Hence a concentrated salt solution results in this compartment.

In addition to the cell pairs, there are two electrodes at the two ends of the of the unit. A electrolyte rinse solution flows in these electrode compartments (marked E in the figure) in order to increase the solution conductivity. (The rinse solution is usually recycled.) Chemical reactions O_2 evolution and H_2 evolution occur at the anode and the cathode, respectively. The anode reaction is

$$2OH^- \rightarrow \frac{1}{2}O_2 + H_2O + 2e^-$$

The following is the cathode reaction:

$$2H^+ + 2e^- \rightarrow H_2$$

These reactions are needed for the current flow along the external circuit.

Chlorine formation at the anode is a possible side reaction if the feed contains Cl^- ions. This is an anodic reaction represented as

$$2Cl^- \rightarrow 2e^- + Cl_2$$

The extent of these reactions are minimal and only a small fraction of the water gets electrolyzed. The overall material balance is that the brackish water gets separated into relatively deionized water and a concentrated solution. This nature of separation is similar to that in reverse osmosis although the separation principles are different. One is based on pressure-driven transport while the other is based on electricity-driven transport.

30.1.1 When to Use Electrodialysis

The following guidelines are useful to decide when to use electrodialysis in lieu of other separation methods:

- Very dilute salt solution: If the range of concentration of the salt is below 1000 mg/L, then use ion exchange.
- Dilute solution: Electrodialysis fits well and is economical if the range of concentration is within a range of 1000 to 10,000 mg/L.
- Not so dilute solution: If the range of concentration of the salt is above 10,000 mg/L, then reverse osmosis is more economical.

Large capacity systems, where about $220,000$ m^3/day of brackish water are treated for potable use, have been designed. Other applications include the food industry (salt removal from cheese whey, soy sauce, etc.), organic chemicals (e.g., organic acids from organic salt wastes), and related applications. Valero et al. (2011) provide more application examples and a review of electrodialysis technology.

30.1.2 Membranes

A brief description of the membranes used in electrodialysis (ED) is in order. ED systems as shown earlier consist of a cation-transfer membrane and anion-transfer membrane arranged in alternating order. The cation transfer allows only positive ions to pass through. These are generally made from crosslinked polystyrene, which has been sulfonated to produce SO_3H groups attached to the polymer. In water this ionizes to produce immobile SO_3^- groups within the membrane. This blocks the transport of negative ions through the membrane.

The anion-exchange membrane allows negative ions to pass through. The membrane matrix usually has fixed positive charge, for example, quaternary ammonium salts of the generic formula RNH_3^+, which repels the positive ions. Chloromethylation followed by amination is done on a polymer matix to create these functional groups.

Both membranes should have the following desirable properties:

- Low electrical resistance
- Insoluble in aqueous media
- Mechanical rigidity
- Resistance to change in pH and to osmotic swelling

The membranes can be homogeneous or heterogeneous. In the heterogeneous structure, the ion-selective membranes are embedded in a neutral membrane matrix. Membrane that are selective to univalent ions are also available, for example, NO_3^- transport is permitted with exclusion or limited transport to SO_4^{--}, and so on. Such selectivity is accomplished by specially treating anion-exchange membranes to provide additional functional groups. Thus a wide range of membranes is available and selection depends on the application areas and relative cost of the equipment. A detailed review on the developments in the preparation, properties, and application of ion-exchange membranes is available in Xu (2005).

30.1.3 Electrodialysis Reversal Process

The electrodialysis reversal (EDR) process is a variation on the ED process that uses electrode polarity reversal to automatically clean membrane surfaces. EDR works the same way as ED, except that the polarity of the DC power is reversed two to four times per hour. When the polarity is reversed, the source water dilute and concentrate compartments are also reversed and so are the chemical reactions at the electrodes. This polarity reversal helps prevent the formation of scale on the membranes. The setup is very similar to an ED system except for the presence of reversal valves. More technological aspects of EDR are discussed by Katz (1977).

30.1.4 Electrodialysis with Bipolar Membranes

Bipolar membranes (BPMs) are composite membranes consisting of layers of both cation-exchange and anion-exchange membranes. These, for instance, permit water separation to H^+ and OH^- ions with no gas evolution. The action of a bipolar membrane is shown in Figure 30.2. The BPM consists of a cation-selective layer, a transition layer, and an anion-selective layer sandwiched together. Th water splitting occurs in the transition layer and H^+ and OH^- formed in this layer move in opposite directions to the bulk fluid toward the respective electrodes.

The mechanism for water splitting is not clear and could be due to the catalytic effect of the transition layer. Fu et al. (2003) discuss some proposed models for water splitting and they found that PEG (polyethylene glycol) has a strong effect on water splitting. The other solvent studied in BPMs is methanol, which dissociates into CH_3O^- and H^+ ions.

A common application is splitting the salt into an acid and base. The arrangement is shown in Figure 30.3. The CEM and AEM permit the separation of the cation and anion of the salt (MX) while the bipolar membrane (BPM) causes water to split into H^+ and OH^- ions. The H^+ ions permeate through the cation-exchange side of the BPM and form acid HX with the X^- ions provided by permeation through the AEM. Similarly on the other side the OH^- ions (shown as R^- in the figure for generality) from the anion exchange side of the BPM form the base MR (MOH) with M^+ being supplied via the CEM. The final result is that a salt solution, MX, in the feed is split into an acid (HX) and a base (MOH). Note that this is one type of cell arrangement and a number of alternative designs are available as reviewed in Tongwen and Weihua (2002).

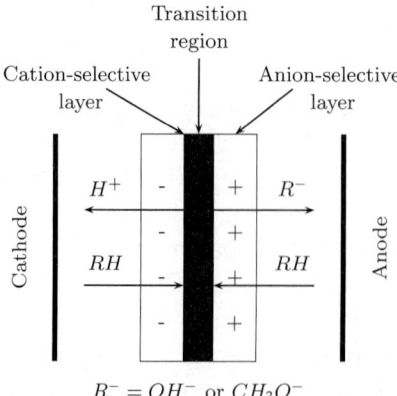

Figure 30.2 Illustration of the action of a bipolar membrane: water splitting occurs at the transition layer and the cations and anions move out on either side.

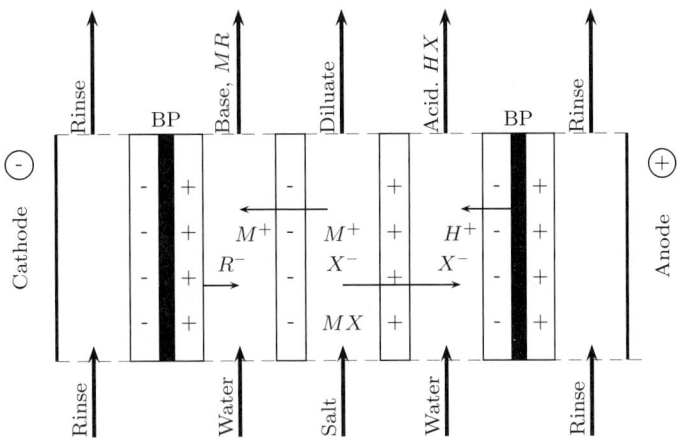

Figure 30.3   Illustration of the use of a bipolar membrane for separation of an organic salt into an acid and a base.

Detailed review of the application of elecrodialysis with a bipolar membrane is presented by Tongwen and Weihua (2002). These membranes are finding applications in wastewater treatment and recovery of value from waste.

30.2   Preliminary Design of an Electrodialyzer

Chemical engineers may be involved in procurement and setup of ED units and hence it is useful to know how equipment can be evaluated. Simple back of the envelope calculations are often useful and are illustrated this section. Manufacturers of ED equipment are always looking for improvement in energy efficiency in which case it is important to know the mass transfer effects in more detail. Nernst-Planck equation–based mass transport models are useful in this regard and are discussed in briefly at the end of this section.

Preliminary design is done using mass balance and Faraday's and Ohm's laws. Detailed concentration profiles in the membrane and in the boundary layer near the membrane surface are not included. In other words, mass transport across the membrane and in the external film near the membrane are not considered. Finally a correction factor, the current efficiency, is applied to allow for these effects. These design aspects are available in Lee et al. (2002) and Korngold (1982) and the discussion follows the ideas developed in these sources.

30.2.1 Current and Voltage

The main parameters to be calculated for prescribed process conditions are the membrane area and power requirement. First the current needs to be calculated using Faraday's law for a given extent of salt separation:

$$I = nFQ\Delta C_A \tag{30.1}$$

where F is the Faraday constant, Q is the flow rate across each cell pair, and ΔC_A is the concentration change from inlet to outlet. The concentration is often expressed in g-equivalent (g-eq per m^3 rather than g mole/m^3). This takes care of the valency of the cation or anion and n (number of electrons transferred) is then set as one. For a univalent electrolyte such as NaCl, g-eq is the same as g-mol.

The membrane area can then be calculated if an operating current density i is chosen:

$$A_m = \frac{I}{i\eta} \tag{30.2}$$

where η is a current efficiency factor to account for the various losses in the system.

The voltage drop across each stack pair in accordance with Ohm's law is

$$\frac{V}{N} = i\left(\sum R\right) \tag{30.3}$$

where $\sum R$ is the total resistance for current flow in the cell pair. V is the applied voltage across the system and N is the number of cell pairs in the complete unit.

The total resistance is composed of four individual resistances shown in Figure 30.4; these are added to get the total resistance.

These resistances in series are the following:

1. Resistance of the diluate solution, R_1. This is calculated as h/κ_d where h is the cell width and κ_d is the specific electrical conductivity of the diluate solution, S/m.
2. Resistance of the AEM membrane, R_A, in Ωm^2.
3. Resistance of the concentrate solution, R_2. This is calculated as h/κ_c where h is the cell width and κ_c is the specific electrical conductivity of the concentrate solution, S/m.
4. Resistance of the CEM membrane, R_C in Ωm^2.

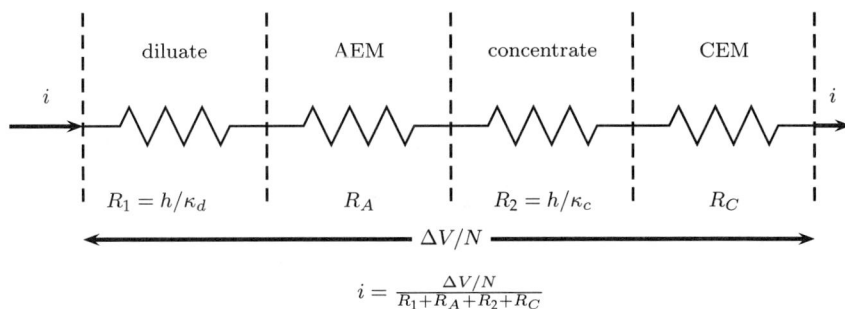

Figure 30.4 Ohm's law diagram for a cell pair showing the current–voltage relation; h is the thickness of one cell. AEM = anion-exchange membrane; CEM = cation-exchange membrane.

The total resistance $\sum R$ is equal to $R_1 + R_A + R_2 + R_C$. Note that these are based on a unit area for transport. Hence the unit is Ωm^2. Also note that S represents Siemens, an unit for conductance that is the same as Ω^{-1} and is denoted as mho in earlier work.

The resistance of the commercial ion-exchange membranes are in the range of 3 to 6 $\times 10^{-4}$ Ωm^2 and depends on the thickness of the membrane that are in the range of 0.3 to 0.6 mm.

The specific conductivity of the salt solution is concentration dependent and is expressed as

$$\kappa = \Lambda C$$

where Λ is the equivalent conductance (S m^2/mol) and C is the concentration in either the diluate or the concentrate compartment. The concentration changes along the reactor length and an average value between the inlet and outlet is used as an approximation. (Note that a mesoscopic model is required for more detail.)

The equivalent conductance of a salt solution is concentration dependent as well. However, a value of 10.86 S m^2/kmol for equivalent conductance appears to be a reasonable average value, as suggested by Lee et al. (2002), and can be used for a preliminary design.

The calculations are demonstrated by Example 30.1.

Example 30.1   Voltage Balance and Power for Salt Separation

$24,000$ m^3/day of salt solution is to be purified from 1500 mg/L to 300 mg/L. A conversion of 50% is assumed, that is, 50% of the feed gets diluted and the rest ends up as the concentrate. One hundred fifty membrane pairs are used, with each membrane having a thickness of 0.65 mm. The operating current density is fixed at 30 A/m^2 and a current efficiency of 90% is assumed. Find the membrane area needed, the voltage to be applied, and the power consumption. Additional data needed: membrane resistance = 7.0×10^{-4} Ωm^2; equivalent conductivity of salt solution = 10.5 S m^2/kmol.

Solution

The flow rate of feed solution is 24,000 m^3/day, which is equal to 0.2778 m^3/s. The flow rate of NaCl across each cell pair is therefore equal to 0.2778/300 = 9.26×10^{-4} m^3/sec. The inlet concentration is 1500 mg/L. The molecular weight of NaCl is 58.5. Hence the inlet concentration is calculated as 25.64 mol/m^3. Similarly the exit concentration is 5.13 mol/m^3. This gives a ΔC_A of 20.5.

Now using a mass balance the exit concentration of the concentrate is found as 46 mol/m^3. The current in the system can now be calculated using Equation 30.1:

$$I = (96500 \text{ C/mol}) \times (9.26 \times 10^{-4} \text{ m}^3/\text{s}) \times 20.5 \text{ mol/m}^3 = 1.83 \times 10^3 \text{ A}$$

The membrane area is then calculated using Equation 30.2:

$$\text{area} = 1.83 \times 10^3 \text{ A}/30 \text{ (A/m}^2)/0.9 = 67 \text{ m}^2$$

To find the power consumption the voltage to be applied across the system has to be calculated. First we find the individual resistances and the total resistance. The specific conductance of the diluate based on the average concentration is 0.1615 S/m. Hence the resistance of the diluate is 0.6×10^{-3} m/0.1615 S/m = .004 Ωm^2. The exit concentration of the concentrate was calculated by a mass balance and is 46 mol/m^3. Hence the specific conductance of the concentrate based on the average concentration is 0.3769 S/m. The resistance of the concentrate is thus 0.6×10^{-3} m/0.3769 S/m = .0017Ω m^2. The total resistance is calculated by summing the individual resistances:

$$\sum \mathcal{R} = 0.004 + 7.0 \times 10^{-4} + 0.0017 + 7.0 \times 10^{-4} = 0.0071 \ \Omega \text{ m}^2$$

Hence the voltage needed is 30 A $\times$ 0.0071 Ωm^2 = 0.2144 V per cell pair. The total voltage across the system is 300 $\times$ 0.2144 = 64.37 V. The power is equal to the voltage time current and is 64.37 V $\times$ 30 A = 117 kW.

30.2.2 Limiting Current

The operating current density is the key value used in the previous example to get the design estimates. The maximum value is limited by mass transfer considerations and hence the limiting current is an important factor in the design of ED systems. The limiting current refers to a situation where the concentration of the diffusing species becomes zero at the membrane surface.

The operating current density cannot be larger than the limiting current. A value of 0.85 times the limiting current is used as an approximate design criteria.

The equation to calculate the limiting current density is

$$i_L = nFk_L C_A$$

where k_L is the mass transfer coefficient near the external film near the membrane. C_A should be the exit concentration since this is where the limiting conditions will happen. It is however difficult to estimate the mass transfer coefficient since it depends on the flow velocity, spacer configuration, and membrane surface roughness. Hence experimentally determined values are used in practical design as suggested by Lee et al. (2002). Typical values are in the range of 30–50 A/m². Additional information is available in Tanaka (2005).

30.2.3 Detailed Models

Mass transfer theory–based models are presented in Kraaijeveld et al. (1995), who also modeled a circulating batch system. (These are commonly used in laboratory testing.) The fluxes across the membranes were calculated using the Stefan-Maxwell model and coupled to the transient equation for the batch reactor. A ternary system, Na^+, H^+, and Cl^-, were used in their study, for which the Stefan-Maxwell model is more appropriate. For a binary system (e.g., NaCl) the transport model in a charged membrane shown in Section 16.4 can be modified and used to find the fluxes across the membrane. The previously cited study shows an example of coupling a macroscopic model (for bulk fluid concentrations in the batch system) with the differential model (membrane transport).

Factors leading to inefficiencies such as diffusion of co-ions through the membranes and water transport across the membranes can also be investigated using the framework of the detailed model. Hence such models provide more phenomenological understanding, which in turn helps in improving the performance of the unit.

30.3 Principle of Electrophoresis

Electrophoresis refers to movement of charged particles in solution under the action of an electric field. This separation method is widely used in bioseparations, for example, separation of proteins, DNA molecules, and so on. Different components in a mixture have different mobilities and hence different migration velocities. Electrophoretic separation exploits this difference to accomplish a separation. In some sense it is similar to chromatography where the differences in the adsorption equilibrium constant are used to achieve separation. The application of this technique has now expanded to separation of a large class of products in addition to proteins, including dyes, colloids, and so on.

It is useful to indicate the direction of the motion of the various types of molecules. This depends on the type of the surface charge, which depends on a number of factors discussed next.

30.3.1 Solutes with Fixed Type of Charge

Colloids are suspensions of fine particles in a liquid media with a particle size generally less than 10 μm and can be classified into hydrophobic or hydrophilic types. Typical examples of hydrophobic colloids are suspensions of metals, metal oxides, colloidal sulfur, and so on, and these move in a fixed direction in an electric field depending on the type of charge carried by these particles. Solutes of metallic particles are positively charged while those of metal oxides, and so on, are negatively charged.

30.3.2 Solutes with Charge Dependent on pH

For proteins, the direction of the motion is not unique and is very sensitive to the hydrogen ion concentration. A typical amino acid that forms the backbone of a protein can be represented as NH_2RCOOH, but is an amphoteric electrolyte and has a bipolar structure represented as $^+NH_3RCOO^-$; it is also called as Zwitterion. In strong acids the following reaction takes place, which makes it positively charged:

$$^+NH_3RCOO^- + H^+ \rightarrow^+ NH_3RCOOH$$

In alkaline solutions the proteins acquire a negative charge due to the following reaction:

$$^+NH_3RCOO^- + OH^+ \rightarrow NH_2RCOO^- + H_2O$$

Hence the nature of the charge of a protein therefore depends on pH. If the pH is increased progressively, the charge changes from positive to zero and then negative. The pH at which the charge is zero is called the iso-electric pH value and the protein will not move in an electric field at that pH. This gives another handle on separation since the relative motion of one protein to another can be controlled by simply changing the pH of the solution. This method is called isoelectric focusing.

30.4   Electrophoretic Separation Devices

We now discuss the common electophoretic separation devices and simple models to track the trajectory of a species injected to the system. Continuous free-flow electrophoretic systems are widely used and there are many variations of the design. A basic electrophoretic device is a flow channel in which an elutant fluid is used to carry the protein (or species to be separated). A voltage is also applied in the device by placing electrodes suitably in the system. The species migrates toward one or the other electrode depending on the type of charge carried by it and is also being carried by the flow depending on the local velocity. Since different species have different mobilities or different types of charge, they get separated at various locations downstream and are collected at various ports downstream in the flow channel. A number of design arrangements have been proposed and we discuss some of these in some detail.

30.4.1   Philpot Design

In this design, we have the elutant flowing in a narrow channel with a voltage gradient applied across the gap as shown in Figure 30.5.

The feed solution containing the proteins to be separated is introduced as a point source at a location near the inlet. This is convected by flow in the x-direction and moves to the electrode at $y = H$ by electrophoretic migration. (Here we assume $y = H$ is of the opposite polarity to the charge on the protein.) The computation of the trajectory of a protein and the point at which it will separate is described next by a simple model. The model is for plug flow of the liquid for simplicity but it is easy to extend this to a parabolic profile, which is characteristic of laminar flow in a channel.

A coordinate system in the direction of the flow vector (x-axis) and perpendicular to the flow (y-axis or the direction of the field) is set up as shown in the figure. The species is convected by flow in the x-direction and hence has an x-velocity of v_x. Here we assume no slip and the velocity of the particle is

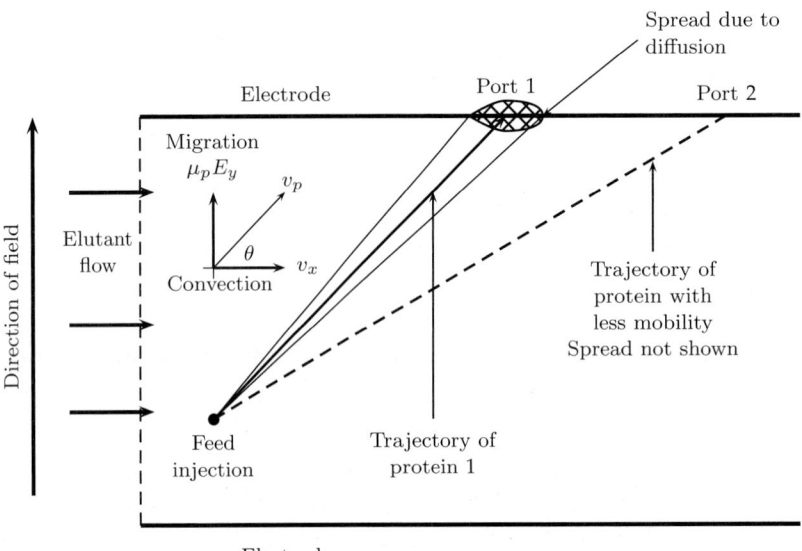

Figure 30.5   Electrophoretic transport of two proteins with different mobilities injected into a flow stream. Philpot design is shown where the applied field is perpendicular to the flow. Protein 1 accumulates at port 1 while the protein with lower mobility accumulates at port 2.

taken to be equal to that of the fluid. The species moves with a y-velocity of $\mu_p E_y$ in the y-direction. Hence the velocity of the particle is

$$\boldsymbol{v} = \boldsymbol{e_x} v_x + \boldsymbol{e_y} \mu_p E_y$$

The angle between the direction of convection and the x-axis can therefore be calculated as

$$\tan \theta = \frac{\mu_p E_y}{v_x}$$

The speed in this direction (the θ-direction), denoted as v_p, is

$$v_p = \sqrt{[v_x^2 + (\mu_p E_y)^2]}$$

This provides the trajectory of the protein due to the action of the flow and migration and is shown in Figure 30.5. Trajectories are shown in this figure for two proteins with different mobility values.

Different proteins have different mobilities and arrive at different locations as shown in the figure. The fractions are collected at suitable locations along the wall (ports 1 and 2 in the figure).

Superimposed on the trajectory there is a spread due to diffusion. This can be described by a Gaussion distribution:

$$C = m_p \sqrt{\frac{v_p}{4\pi D_p x_c}} \exp \left(-\frac{v_p y_c^2}{4 D_p x_c} \right)$$

where m_p is the source strength equal to the mass or moles of protein being injected at the source point per unit width of the slab. x_c is the distance along the trajectory. y_c is the distance perpendicular to the trajectory; this is the direction where the spread occurs due to diffusion.

The standard deviation for diffusion is

$$\sigma = \sqrt{D_p x_c / v_p} \tag{30.4}$$

The separability of two protein species by electrophoresis can be assessed by this simple model. In general the standard deviation of 4 will maintain a concentration difference of 0.1%. Otherwise there will be an overlap and the separation will not be sharp.

This discussion provides a basic model for the process and is useful to provide some guidelines in design and selection of suitable equipment. The factors to consider are as follows:

- The length of the device must be reasonable.
- Multiple species must be separable without overlap.
- Joule heating due to the imposed field causes a temperature rise, which can cause protein denaturation. Hence the design should limit the maximum temperature rise in the system and heat transfer models also become important.
- Other effects such as Taylor dispersion and double-layer distortion affecting the mobility of the particle become important in many cases.

30.4.2 Hannig Design

In the Hannnig design, the direction of the field and the direction of the flow are parallel. A schematic diagram of this arrangement is shown in Figure 30.6.

The velocity of a protein in the flow direction is now the algebraic sum of the convection and electromigration velocity. The time it takes to arrive at a distance L (the length of the device) is L/v_p; this represents the retention time. Two proteins with different retention times can therefore be separated. The analysis is therefore similar to the chromatography studied in Chapter 29.

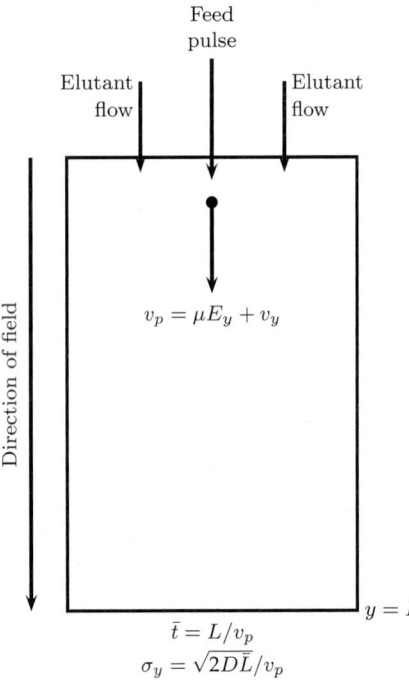

Figure 30.6   Hannig design: the applied field is parallel to the flow. Proteins appear at the exit at different mean retentions, which depend on their mobility.

An ideal pulse will be observed in the exit stream as a dirac delta function after the elapse of the retention time. In reality band broadening will be observed due to both diffusion and dispersion. The diffusion in the flow direction (y) will cause a spread in the response curve in the y-direction while the axial dispersion will cause a peak broadening in the x-direction as well. The difference in the retention time for separation should be about 4σ apart so that the pulse overlap does not interfere with the separation.

Additional band spreading can occur due to electro-osmosis, and hydrodynamic effects caused by thermal gradients. The thermal gradients are due to Joule heating caused by the applied potential. These can cause protein denturation and control of the temperature peak is important.

30.4.3 Rotating Annular Bed

Another design is the continuous flow rotating annular bed electrophoresis column, which is shown in Figure 30.7, The feed and elutant enter at a fixed location in the annular region between two coaxial cylinders. The electric field is applied in the axial direction. The annular bed is slowly rotated about its axis. Each component leaves the column at a different angular position located

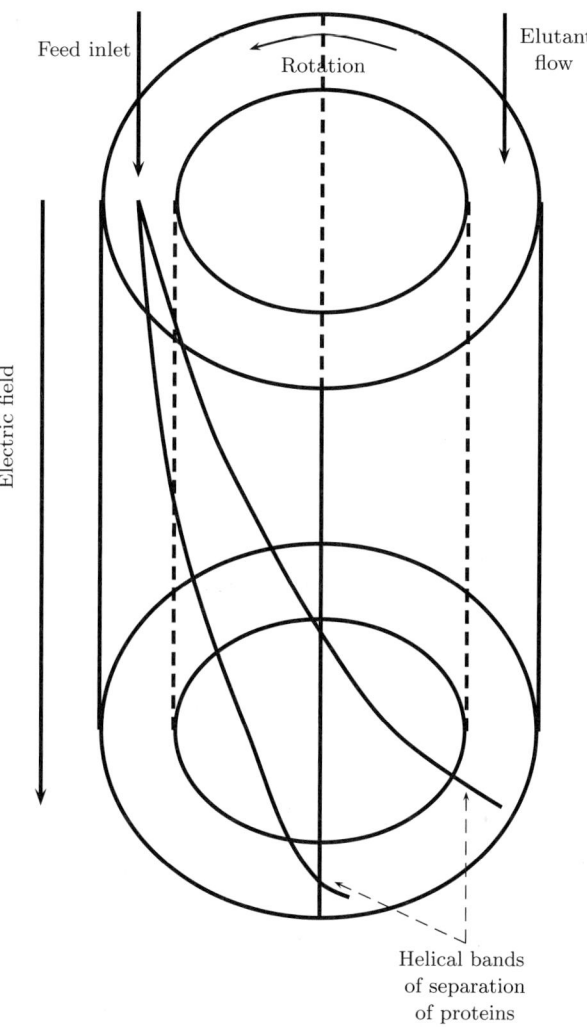

Figure 30.7   Continuous flow rotating bed electrophoresis unit.

at the bottom of the unit. Multiple product ports are provided along the circumference at the bottom of the column. As a result of rotation, separated fractions appear as helical bands. Since separation occurs in the angular direction in addition to the axial direction, the bed thickness can be kept small to reduce the bed temperature rise. This reduces the effect of Joule heating. Both the inner and outer part of the system are kept cold by circulating a coolant to further reduce the temperature rise in the system. Detailed modeling and the band spreading calculations are presented in Yoshisato et al. (1986) and also addressed briefly in exercise problem 30.6 in this chapter.

Other designs include serpentine channels, studied for example by Wang et al. (2004), and the use of microchannel devices.

Summary

- Electrodialysis refers to separation using transport in a charged membrane that permits selective transport of only positive or negative ions.

- Pairs of cation-selective and anion-selective membranes are arranged in series in a typical electrodialysis setup. The feed solution gets depleted in one of these setup cells while it gets concentrated in the adjacent cells.

- An electric potential applied across the cells promotes the current flow. Near the electrodes a electrolyte rinse solution is introduced to improve the solution conductivity and the current flow. Some gas evolution reaction may take place near the electrodes; the extent of gas evolution is small compared to the quantity of feed treated.

- A preliminary model for electrodialysis is based on a mass balance, Faraday's law, and Ohm's law. Resistance to the current flow is taken as the sum of four resistances in series: the resistance of the two membranes and the resistance of the diluate and the concentrated solution. The current density cannot exceed the limiting current and is usually chosen as 80% of the limiting current value. The area of membrane required for a given separation and the power needed can be assessed based on this model.

- Detailed models for electrodialysis include mass transfer effects in the film and in the charged membrane. These models provide detailed concentration profiles in the membrane and permit current–voltage calculations using a detailed Nernst-Planck model rather than the simple Ohm's law. A transport model in a charged membrane together with the Donnan equilibrium at the membrane–solution interface is used as the basic framework in setting up these models.

- Electrophoresis refers to the migration of a charged species relative to a liquid caused by an applied electric field. It finds wide applications in bio-separation including separation of proteins, DNA, and so on.

- A simple migration-convection model is useful to track the particle trajectory in electrophoretic devices. The migration is caused by the electric field while the convection is caused by the elutant flow. The spread due to diffusion is then superimposed as a Gaussian distribution. Hence the separation distance and the spread can be evaluated and the model can be used to assess the separability given the mobility parameter.

- A number of flow arrangements can be used to improve the separation efficiency in electrophoretic devices. These include the Philpot design, the Hannig design, the rotating annular bed, and so on. The relative merits of each setup can be analyzed using simple convection-migration-diffusion models.

Review Questions

30.1 What is electrodialysis?

30.2 State some applications of electrodialysis.

30.3 What are the two exit streams from an electrodialysis system?

30.4 What is electrodialysis reversal?

30.5 What are some of the applications of electrodialysis with bipolar membranes?

30.6 What is the general chemical composition of a cation-exchange membrane?

30.7 What is the general chemical composition of an anion-exchange membrane?

30.8 Define the limiting current and show its importance in electrodialysis.

30.9 What is the principle behind electrophoresis?

30.10 What is an amphoteric electrolyte?

30.11 What is meant by an isoelectric point?

30.12 What data is needed to find the isoelectric point of an amino acid?

30.13 Differentiate between the field arrangements in the Philpot and Hannig designs of an electrophoresis unit.

30.14 What is the effect of Joule heating in electophoresis on separation efficiency?

30.15 What additional benefit is obtained in the rotating bed over the Hannig design?

Problems

30.1 **Current and power calculation in electrodialysis.** A 150 L/hour of salt solution with 3000 ppm of NaCl is to be purified to 400 ppm by electrodialysis. A cell stack of 100 cell pairs are used. Find the current density needed. If the voltage drop per cell pair is 0.3 V find the power needed and the energy needed per liter of water processed.

30.2 **Membrane area needed for separation.** An electrodialysis unit has a feed concentration of 58 g eq./m^3 and the diluate concentration is reduced by 90%. The recovery factor is 0.75. The production capacity of the plant is 350 m^3/day. An operating current density of 40 A/m^2 is suggested with 200 cell pairs. Find the area of membrane needed. The current efficiency is 0.9.

30.3 **Limiting current.** If a limiting current of 50 A/m^2 is measured for a salt concentration of 0.1 M NaCl, estimate the mass transfer coefficient. The data on limiting current at a fixed salt concentration was fitted as $i_L = Av^\alpha$ by Lee et al. (2002) and the exponent α was found to be in the range of 0.41 to 0.52. Determine the exponent if a convection-diffusion model with laminar flow is used.

30.4 **Separation distance for proteins.** It is required to separate two proteins with mobilities of $\mu_1 = 8 \times 10^{-5}$ m.C/N.s and $\mu_2 = 6 \times 10^{-5}$ m. C/N.s. The diffusion coefficient is 6×10^{-10} m^2/s. The flow velocity is 0.2 mm/s. The applied field is 2000 V/m. The electrodes are placed 1 cm apart. The Philpot design is used with the field perpendicular to the flow. Find the trajectory of the "plumes" of the two proteins and the distance at which the two proteins can be separated. First assume plug flow and use the equations suggested in the text. Second assume laminar flow and calculate the trajectory by solving the following pair of equations. The following equation is applicable for the distance along the elutant flow (same for both proteins):

$$\frac{dx}{dt} = v(y) = \frac{G}{2\mu_L}y(d - y)$$

Here the right-hand side represents the velocity for laminar flow in a channel, G the applied pressure gradient, and μ_L the viscosity of the liquid. For distance across the flow (different for different proteins) the following equation is applicable:

$$\frac{dy}{dt} = \mu_p E_y$$

30.5 **Retention time and spread in the Hannig design.** Two proteins with mobilities of $\mu_1 = 8 \times 10^{-5}$ m.C/N.s and $\mu_2 = 6 \times 10^{-5}$ m.C/N.s are to be separated in a Hannig design arrangement. The diffusion coefficient is 6×10^{-11} m^2/s. The elutant velocity is 1 cm/s. Find the retention time and the spread in the bands if the length of the channel is 100 cm.

30.6 **Rotating annular bed.** Show that the distance traveled in the angular direction in the rotating annular bed is $R_f \omega \bar{t}_A$, where $\bar{t}_A$ is the retention time for protein A, R_f is the radial position at which the feed is introduced, and ω is the angular velocity. Show that the separation distance between two components A and B is

$$\Delta S_{AB} = R_f(\omega \bar{t}_A - \omega \bar{t}_B)$$

Determine the expression for the retention time for A and B. Find the separation distance for a design with an inner radius = 25 cm, outer radius = 35 cm, length = 50 cm, applied voltage = 60 V, $\mu_A = 8 \times 10^{-5}$ m.C/N.s and $\mu_B = 4 \times 10^{-5}$ m C/N.s, and rotation speed = 15 rpm.

REFERENCES

Abramowitz, M. and Stegun, J. A. (1964). *Handbook of Mathematical Functions.* Nat. Bureau of Standards, Washington.

Ackerman, G. (1937). "Simultaneous Heat and Mass Transfer with Large Temperature and Partial Pressure Differences." *Ver. deutscher Ig. Forth.*, **8**, 1–16 (in German).

Acrivos, A. and Taylor, T. E. (1962). "Heat and Mass Transfer from Single Sphere in Stokes Flow," *Phys. Fluids*, **5**, 387–394.

Akita, K and Yoshida, F. (1974). "Gas Holdup and Volumetric Mass Transfer Coefficients in Bubble Columns." *Ind. Eng. Chem. Proc. Des. Dev.*, **13**, 84.

Amelin, A. G. (1967). "Theory of Fog Condensation," *Israel Program for Scientific Translations*, Jerusalem. Second edition.

Aris, R. and Amundson, N. R. (1957). "Some Remarks on Longitudinal Mixing or Diffusion in Fixed Beds." *AIChE Journal,* **3,** 280–282.

Aris, R. and Hatfield, B. (1969). "Combined Effect of External and Internal Diffusion for Nonisothermal Case." *Chem. Eng. Sci.*, **24**, 1213–1220.

Astarita, G. (1967) *Mass Transfer with Chemical Reaction.* Elsevier, Amsterdam.

Avdeev, A. A. (2016). "Bubble Growth, Condensation (Dissolution) in Turbulent Flows." In: Bubble Systems: Mathematical Engineering, Springer, Cham.

Avrami, M. (1939). "Kinetics of Phase Change. I. General Theory." *Journal of Chemical Physics*, **7**(12), 1103–1112.

Avrami, M. (1940). "Kinetics of Phase Change. II. Transformation-Time Relations for Random Distribution of Nuclei." *Journal of Chemical Physics*, **8**(2), 212–224.

Avrami, M. (1941). "Kinetics of Phase Change. III. Granulation, Phase Change, and Microstructure." *Journal of Chemical Physics*, **9**(2), 177–184.

Baker, R. W., Yoshioka, N., Mohr, J. M., and Khan, A. J. (1987). "Separation of Organic Vapors from Air." *J. Membrane Sci.*, **31**, 259–271.

Barrer, R. M. (1951). *Diffusion in and through Solids.* Cambridge University Press, Cambridge, England.

Barrer, R. M. (1984). "Diffusivities In Glassy Polymers for The Dual Mode Sorption Model." *J. Membrane Sci.*, **18**, 25.

Battles, Z. and Trefethen, L. N. (2004). "An Extension of Matlab to Continuous Functions and Operators." *SIAM Journal on Scientific Computing*, **25**, 1743–1770.

Bausa, J. and Marquardt, W., (2000). "Shortcut Design Methods for Hybrid Membrane/Distillation Processes for the Separation of Nonideal Multicomponent Mixtures." *Ind. Eng. Chem. Res.,* **39** (6), 1658–1672.

Bejan, A. (2004). *Convective Heat Transfer*. John Wiley & Sons, Hoboken.

Berbente, C. P. and Ruckenstein, E. (1968). "Hydrodynamics of Wave Flow." *AIChE J.,* **14**, 772–782.

Berger, D. and Pei, D. C. T. (1973). "Drying of Hygroscopic Capillary Porious Solids: A Theoretical Approach." *Int. J. Heat and Mass Transfer*, **16**, 293–302.

Beukes, N. T. and Badenhorst, J. (2009). "Copper Electrowinning: Theoretical and Practical Design." *Hydrometallurgy Conference*, African Institute of Mining and Metallurgy, 213–240.

Bhagwat, S. S. and Sharma, M. M. (1988). "Intensification of Solid-Liquid Reactions with Microemulsions." *Chem. Eng. Sci.,* **43**, 195–205.

Bhatia, S. K. and Perlmutter, D. D. (1981). "The Effect of Pore Structure on Fluid-Solid reactions: Application to the SO_2-Lime Reaction." *AIChE Journal*, **27**, 226–234.

Bhattacharya, A. and Ramachandran, P. A. (1986). "'A General Analysis of Mass Transfer Accompanied by Multiple Equilibrium Reactions." *J. Indian Inst. Sci.,* **66**, 583–594.

Bhavaraju, S. M., Russell, T. W. F., and Blanch, H. W. (1978). "Design of Gas-Sparged Devices for Viscous Liquid Systems," *AIChE Journal*, **24**, 454–466.

Bin, A. K. (1983) "Mass Transfer into a Turbulent Liquid Film." *Int. J. Heat and Mass Transfer*, **26**, 981–991.

Bischoff, K. B. (1965). "Effectiveness Factor for General Reaction Rate Forms." *A. I. Ch.E Journal*, **11**, 351.

Bischoff, K. B. and Dedrick, R. L. (1980). "Species Similarities in Pharmacokinetics." *Federation Proc. in Pharamacokinetics*, **39**, 54–59.

Bischoff, K. B., Dedrick, R. L., and Zaharko, D. S. and Longstreth, J. A. (1971). "Methotrexate Pharmacokinetics." J. Phram. Sci., 60, 1128–1133.

Bjerle, I., Bengtsson, S., and Farnkvist, K. (1969). "Absorption of SO_2 in $CaCO_3$-Slurry in a Laminar Jet Absorber." *Chem. Eng sci.,* **27**, 1832–1856.

Blasius, H. (1908). "Grenzschicten in Flussigkeiten mit Kleiner Reibung." *ZAMP Soc*, **56**, 1. (English translation in NACA TM 1256.)

Botemuma, R. H. B., Checchetti, A., Chidichimo, G., and Drioli, E. (1977). "Permeation through a Heterogeneous Membrane: The Effect of the Dispersed Phase." *J. Membrane Sci.,* **128**, 141.

Brian, P. L. T., Hales, H. B., and Sherwood, T. K. (1969). "Transport of Heat and Mass between Liquid and Spherical Particles in an Agitated Tank." *AIChE. Journal,* **14**, 727.

Brouwers, H. J. H. (1992). "Film Models for Transport Phenomena with Fog Formation: The Fog Film model." *Int. J. Heat and, Mass Transfer*, **35**(No. I), 13–28.

Brumfield, L. K., Houze, R. H. and Theofanous, T. G. (1975). "Turbulent Mass Transfer at Free Gas-Liquid Interfaces with Applications to Film Flow." *Int. J. Heat Mass Transfer,* **18**, 1077–1091.

Brunauer, S., Emett, P. H., and Teller, E. (1938). "Adsorption of Gases in Multimolecular Layers." *J. Am. Chem. Soc.*, **60**, 309–319.

Calderbank, P. H., Johnson, D. S. L., and Loudon, J. (1970). "Mechanics of Mass Transfer of Single Bubbles in Free Rise through Some Newtonian and non-Newtonian Liquids." *Chem. Eng. Sci.*, **25**, 235–241.

Calderbank, P. H. and Moo-Young (1961). "Mass Transfer in Bubble Swarms." *Chem. Eng. Sci.*, **16**, 39–44.

Calvelo, A. and Smith, J. M. (1970). "Particle-Pellet Model for Gas-Solid Reactions." *Proceedings of Chemeca* 1970, Butterworths, Sydney, paper 3.1, 15.

Camper, D., Becker, C., Koval, C., and Noble, R. (2005). "Low Pressure Hydrocarbon Solubility in Room Temperature Ionic Liquids." *Ind. Eng. Chem. Res.*, **44**(6), 1928–1933.

Caracciolo, F., Cussler, E. L., and Evans, D. F. (1975). "Membranes with Common Ion Pumping." *AIChE Journal*, **21**, 160–167.

Carniglia, S. C. (1986). "Construction of Tortuosity Factor from Porosimetry." *J. Catalysis*, **101**, 401–418.

Cath, T., Childress, A., and Elimelech, M. (2006). "Forward Osmosis: Principles, Applications, and Recent Developments." *Journal of Membrane Science*, **281**, 70–87.

Cebeci, T. and Smith, A. M. O. (1974). *Analysis of Turbulent Boundary Layers*. Academic Press, New York.

Chakraborthy, S. and Balakotaiah, V. (2002). "Low-Dimensional Models for Describing Mixing Effects in Laminar Flow Tubular Reactors." *Chem. Eng. Sci.*, **57**, 2545–2564.

Chen, S. and Pei, C. T. (1989). "A Mathematical Model for Drying Process." *Int. J. Heat and Mass Transfer,* **32**, 297–310.

Christman, P. G. and Edgar, T. F. (1983). "Distributed Pore-Size Model for Sulfation of Limestone." *AIChE J.,* **29**, 399.

Chuanhui, H. and Tongwen, X. (2006). "Asymmetric Bipolar Membranes in Acid-Base Electrodialysis." *Environ. Sci. Technol.*, **40**(17), 5527–5531.

Churchill, S. W. and Yu, B. (2006). "Effect of Transport on Reactions in Homogeneous Tubular Flow." *Ind. Eng. Chem. Res.*, **45**, 8583–8593.

Clark, M. M. (2009). *Transport Modeling for Environmental Engineers and Scientists, 2nd Edition.* John Wiley & Sons, New York.

Cleland, F. A. and Wilhelm, R. H. (1956). "Diffusion and Reaction in Viscous Flow Reactor." *AIChE Journal*, **2**, 489–494.

Clift, R., Grace, J. R., and Weber, M. E. (1978). *Bubbles, Drops and Particles.* Dover Publications, Mineola, New York.

Colburn, A. P. and Drew, T. N. (1937). "Condensation of Mixed Vapors." *Trans. AIChE.*, **33**, 197–215.

Combest, D. P., Ramachandran, P. A., and Dudukovic, M. P. (2011). "On the Gradient Diffusion Hypothesis and Passive Scalar Transport in Turbulent Flows." *Ind. Eng. Chem. Research*, **50**, 8817–8823.

Conner, J. M. and Elghobashi, S. E. (1987). "Numerical Solution of Laminar Flow Past a Sphere with Surface Mass Transfer." *Numerical Heat Transfer*, **12**, 57–68.

Crank, J. (1975). *Mathematics of Diffusion*. Clarendon Press, Oxford.

Crittenden, B., and Thomas, W. J. (1998). *Adsorption Technology and Design.* Butterworh-Heinemann, Oxford.

Cussler, E. L. (2009). *Diffusion: Mass Transport in Fluid Systems.* Cambridge University Press, Cambridge, UK.

Cussler, E. L. and Moggridge, G. D. (2001). *Chemical Product Design.* Cambridge University Press, Cambridge.

Dalvi, V. V. and Suresh, A. K. (2011). "The Separate Sulfation Rate of MgO Contained in Calcined Calcium Magnesium Acetate (CMA)." *AIChE Journal,* **57,** 1329–1338.

Danckwerts, P. V. (1951). "Significance of Liquid Film Coefficients in Gas Absorption." *Ind. Eng. Chem.,* **43**(1951), 1960–1967.

Danckwerts, P. V. (1970). *Gas-Liquid Reactions.* McGraw-Hill Co., New York.

Dani, A., Cockx, A., and Guiraud, P. (2006) "Direct Numerical Simulation of Mass Transfer from Spherical Bubbles: The Effect of Interface Contamination at Low Reynolds Numbers." *International Journal of Chemical Reactor Engineering* online, **4,** 1542–6580.

Davidson, J. F. and Cullen, E. J. (1957). "The determination of diffusion coefficients of sparingly soluble gases in liquids." *Trans. Instn. Chem. Engrs,* **35,** 51–60.

Davis, H. T. (1977). "Effective Medium Theory of Diffusion in Composite Media," *J. Am. Ceramic Soc.,* **60,** 499.

Deen, W. M. (1987). "Hindered Diffusion of Large Molecules in Liquid Filled Pores." *AIChE Journal,* **33**(9) 1409–1425.

Deen, W. M. (2012). *Analysis of Transport Phenomena. Second Edition,* Oxford University Press.

De Groot, S. R. and Mazur, P. (1968). *Non-Equilibrium Thermodynamics.* Dover Publications, New York.

Delgado, J. M. P. Q. (2006). "A Critical Review of Dispersion in Packed Beds." *Heat Mass Transfer,* **42,** 279–310.

Doraiswamy, L. K. and Sharma, M. M. (1984). *Heterogeneous Reactions: Analysis, Examples and Reactor Design.* Wiley Interscience, New York.

Evangelista, J. J., Katz, S., and Shinnar, R. (1969). "Scale-up Criteria for Stirred Tank Reactors." *AIChE Journal,* **15,** 843–853

Fair, R.B. and Tsai, J. C. C. (1977). "Quantitative Model for the Diffusion of Phosphorus in Silicon and the Emitter Dip Effect." *J. Electrochem. Soc.* **124** (7), 1107–1118.

Fan, L. T. and Hwang, W. S. (1965). "Dispersion of Ostwald-de-Waele fluid in Laminar Flow Through a Cylindrical Tube." *Proc. Roy. Soc. Lond.,* **A283,** 576–582.

Fournier, R. L. (2011). *Basic Transport Phenomena in Biomedical Engineering.* CRC Press, Boca Raton, FL.

Fredd, C. N. and Fogler, H. S. (1998). "Influence of Transport and Reaction on Wormhole Formation in Porous Media." *AIChE Journal,* **48,** 1933–1949.

Freundlich, H. (1907) "ÜBer die Adsorption in Lösungen." *Zeitschrift fur Physikalische Chemie - Stöchiometrie und Verwandschaftslehre,* **57,** 385–470.

Froessling, N. (1938). The Evaporation of Falling Drops *Beitr. Geophysik,* **52,** 170.

Froment, G. F., Bischoff, K. B., and de Wilde J. (2011). *Chemical Reactor Analysis and Design.* John Wiley & Sons Inc., 3rd edition, New York.

Fu, R. Q., Xu, T. W., Wang, G., Yang, Z. X., and Pan J. (2003). "PEG-Catalytic Water Splitting in the Interface of a Bipolar Membrane." *J. Colloid Interface Sci.*, **263** 386.

Fuller, E. N., Schetteler, P. D., and Giddings, J. C. (1966). "New Method for Prediction of Binary Gas-Phase Diffusion Coefficients." *Ind. Eng. Chem.*, **58**(5), 18–27.

Furno, J. S., Taylor, R., and Krishna, R., (1986). 'Condensation of Vapor Mixtures. 2. Comparison with Experiment." *Ind. Eng. Chem. Proc. Des. Dev.*, **25**, 98–101.

Garg, D. R. and Ruthven, D. M. (1973). "On Fixed Bed Sorption Behavior of Gases with Nonlinear Equilibria." *AIChE Journal.*, **19**, 852–853.

Geiger G. H. and Poirer, D. R. (1998). *Transport Phenomena in Material Processing.* John Wiley & Sons, Hoboken.

Ghandhi, S. K. (1983). *VLSI Fabrication Principles: Silicon and Gallium Arsenide.* Wiley-Interscience, New York.

Ghiaasiaan, S. M. (2011). *Convective Heat and Mass Transfer.* Cambridge University Press, Cambridge.

Glasgow, L. A. (2010). *Transport Phenomena: An Introduction to Advanced Topics.* John Wiley & Sons, Honoken, NJ.

Glassman, I., Hansel, J. G., and Eklund, T. (1969). "Hydrodynamic Effects in Flame Spreading, Ignitability and Steady Burning of Liquid Fuels." *Combustion and Flame,* **13**, 99–101.

Glueckauf, E. (1955). "Theory of Chromatography, Part 10: Formula for Diffusion into Spheres and Their Application to Chromatography." *Trans. Faraday. Soc.*, **51**, 1540–1543.

Gmehling, J. and Onken, U. (1984). "Vapor Liquid Equilibrium Data Collection." *DCHEMA Chem. Data. Series*, **1–8**, 197.

Goettler, R. and Pigford, R. L. (1971). "Computational Studies of Simultaneous Chemical Absorption of Two Gases." *AIChE Journal*, **17**, 1213–1220.

Gray, P. G. and Do, D. D. (1991). "Dynamics of Carbon Dioxide Sorption on Activated-Carbon Particles." *AIChE Journal*, **37**(7), 1027–1034.

Gray, W. G. (1983). "Local Volume Averaging of Multiphase Systems Using a Non-Constant Averaging Volume." *International Journal of Multiphase Flow*, **9**, 755–761.

Grimsrud, L. and Babb, A. L. (1966). "Velocity and Concentration Profiles for Laminar Flow of a Newtonian Fluid in a Dialyser." *Chem. Eng. Progr. Symposium Ser. 62*, **66**, 20–31.

Grossman, G. and Heath, M. T. (1984). "Simultaneous Heat and Mass Transfer in Absorption of Gases in Turbulent Liquid Films." *Int. J. Heat and Mass Transfer*, **27**, 2365–2376.

Gupta, A. S. and Thodos, G. (1963). "Direct Analogy between Mass and Heat Transfer to Beds of Spheres." *AIChE Journal*, **9**(6) 751–754.

Haase, R. (1968). *Thermodynamics of Irreversible Processes.* Dover Publishing Company, New York.

Hadamard, J. S. (1911). "Mouvement Permanent lent d'une Sphere Liquide et Visqueuse dans un Liquide Visquex." *Comp. Rend. Acad. Sci.*, **152**, 1735–1738.

Hanna, O. T., Sandall, O. C., and Mazet, P. R. (1981). "Mixing Length Model for Mass Transfer." *AIChE Journal*, **27**, 693–697.

Harriott, P. (2003). *Chemical Reactor Design.* Marcel Dekker, New York.

Harrison, R. G. (2014). "Bioseparations Basics." *Chem. Eng. Progress*, October, 36–42.

Harrison, R. G., Todd, P. W., Rudge, S. R., and Petrides, D. (2002). *Bioseparations Science and Engineering* (Topics in Chemical Engineering), 1st Edition, Oxford University Press, New York.

Hatta, S. (1932). "On the Absorption Velocity of Gases by Liquids." *Technol. Repts. Tohoku Imp. University* **10,** 119–135.

Haynes, H. W. (1984). "Multicomponent Diffusion and Reaction in a Porous Catalyst." *ACS Symposium Series,* Edited by M. P. Dudukovic and P. L. Mills, Vol. 217, p. 407.

He, X., Kim, T., and Hagg, M. (2014). "Hybrid Fixed-Site-Carrier Membranes for CO_2 Removal from High Pressure Natural gas: Membrane Optimization and Process Condition Investigation." *J. Membrane Science,* **470,** 266–274.

Higbie, R. (1935). "Rate of Absorption of a Pure Gas into Still Liquids during Short Periods of Exposure." *Trans AIChE,* 31, 368–389.

Hikita, H. and Asai, S. (1963). "Gas Absorption with (*m,n*)-th Order Irreversible Reaction." *Int. Chem. Eng.,* **4,** 312–340.

Hikita, H., Ishimi, K., and Ohba, A. Y. (1979). "Simultaneous Heat and Mass Transfer in Turbulent Gas Streams in Wetted-wall Columns" *Can J. Chem Eng.,* **57,** 578–581.

Hines, A. L. and Maddox, R. N. (1985). *Mass Transfer: Fundamentals and Applications,* Prentice-Hall, Englewood Cliffs, NJ.

Huang, C. and Xu, T. (2006). "Electrodialysis with Bipolar Membranes for Sustainable Development." *Environ. Sci. Technol.,* **40**(17), 5233–5243.

Hyman, W. A. (1975). "Augmented Diffusion in Flowing Blood." *Trans. ASME,* **58**.

Ishida, M. and Wen, C. Y. (1968). "Comparison of Kinetic and Diffusional Models for Solid-Gas Reactions." *AIChE Journal,* **14,** 311–317.

Jarzynska, M. and Pietruszka, M. (2008). "Derivation of the Practical Kedem-Katchalski Equation." *Old and New Concepts in Physics,* **5**(3), 459–474.

Jiang, F. and Peng, P. (2016) "Elucidating the Performance Limitations of Lithium-Ion Batteries due to Species and Charge Transport through Five Characteristic Parmeters." *Sci. Rep.,* **6,** 326–339.

Kalia, N. and Balakotaiah, V. (2007). "Modeling and Analysis of Wormhole Formation in Reactive Dissolution of Carbonate Rocks." *Chem. Eng. Sci.,* **62**(4), 919–928.

Kalia, N. and Balakotaian, V. (2009). "Effect of Medium Heterogeneities on Reactive Dissolution of Carbonates." *Chem. Eng. Sci.,* **64**(2), 376–390.

Katchalsky, A. and Curren, P. F. (1965). *Non-Equilibrium Thermodynamics in Biophysics,* Harvard University Press, Cambridge, MA.

Katz, W. E. (1979). Electrodialyisis Reversal (EDR) Process. *Desalination,* **28**(1), 31–40.

Kedem, O. and Katchalsky, A. (1958). "Thermodynamic Analysis of the Permeability of Biological Membranes to non-Electrolytes." *Biochim. Biophys. Acta.,* **27,** 229–246.

Kedem, O. and Katchalsky, A. (1961). "A Physical Interpretation of the Phenomenological Coefficients of Membrane Permeability." *J. Gen. Physiol.* **45,** 143–179.

Kendoush, A. A. (1994). "Theory of Convective Heat and Mass Transfer in Spherical Cap Bubbles." *AIChE Journal,* **40,** 1440–1447.

Kilic, S. G. (2008). *Dynamic Fugacity Modeling in Environmental Systems.* Doctoral Thesis, Georgia Institute of Technology.

King, C. J. (1966). "Turbulent Phase Mass Transfer at a Free Gas-Liquid Interface." *Ind. Eng. Chem. Fundamentals,* **5,** 1–8.

Klinkenberg, A. (1954). "Heat Transfer in Cross-Flow Heat Exchangers and Packed Beds." *Ind. Eng. Chem,* **46**, 2285–2289.

Kolmogorov, A. N. (1949). "On the Disintegration of Drops in a Turbulent Flow." *Dokl. Akad. Nauk, SSSR,* **66**, 825–828.

Korngold, F. "Electrodialysis Unit: Optimization and Calculation of Energy Requirement." *Desalination,* **40**, 171–179.

Koros, W. J. and Paul, D. R. (1980). "Sorption and Transport of CO_2 above and below the Glass Transition of Poly(ethylene terephthalate)." *Poly. Engr. and Sci.,* **20**, 14.

Kraaijeveld, G., Sumbervova, N. V., Kuindersma, S., and Wesselingh, H. (1995). "Modeling Electrodialyis Using the Maxwell-Stefan Description." *Chem. Eng. and Biochemical Eng. Journal,* **57**(2), 163–176.

Kreyszig, E. (2011). *Advanced Engineering Mathematics.* Wiley, New York.

Krishna, R. and Panchal, C. B. (1977). "Simplified Mass Tranfer Analysis for Multicomponent Condensation." *Chem. Eng. Sci.,* **32**, 741.

Krogh, A. (1919). "The Number and Distribution of Capillaries in Muscles with Calculation of the Oxygen Pressure Necessary for Supplying the Tissue." *J. Physiol,* **52**, 409–415.

Kulkarni, P. S., Tiwari, K. K., and Mahajani, V. V. (1999). "Studies in Extraction of Nickel by Liquid Emulsion Membrane Process." *Indian J Chem.,* **6**, 329–335.

Kuo, S. (1996). *Transport Phenomena and Materials Processing.* Wiley-Interscience, New York.

Laddha, G. S. and Degaleesan, T. E. (1976). *Transport Phenomena in Liquid Extraction.* Tata McGraw-Hill Publishing Company, New Delhi, India.

Lapicque, F. and Storck, A. (1985). "Modelling of a Continuous Parallel Plate Plug Flow Electrochemical Reactor: Electrowinning of Copper." *Journal of Applied Electrochemistry,* **15**(6), 925–935.

Larson, R. I., and Verazunis, S. (1973). "Mass Transfer in Turbulent Flow." *Int. Journal of Heat and Mass Transfer,* **16,** 121–128.

Lee, H., Sarfert, F. Strathmann, H., and Moon, S. (2002). "Designing of an Electrodialysis Desalination Plant." *Desalination,* **142**(3), 267–286.

Lennard-Jones, J. E. (1924). "On the Determination of Molecular Fields." *Proc. R Soc. Lond A.,* **106**(738), 463–477.

Levenspiel, O. (1999). *Chemical Reaction Engineering.* John Wiley & Sons, New York.

Levenspiel, O. (2013). *Chemical Reactor Omnibook.* Oregon State University Press, Corvallis.

Levenspiel, O. and Godfrey, J. H. (1974). "A Gradientless Contactor for Experimental Study of Interphase Mass Transfer with/without Reaction." *Chem. Eng. Sci.,* **29**, 1723–1730.

Levenspiel, O. and Smith, W. K. (1957). "Notes on the Diffusion-type Model for the Longitudinal Mixing of Fluids in Flow." *Chem. Eng. Sci.,* **6**, 227–233.

Levich, V. G. (1962). *Physicochemical Hydrodynamics.* Prentice-Hall, Englewood Cliffs, NJ.

Lewis, W. K. and Whitman, W. G. (1924). "Principles of Gas Absorption." *Ind. Eng. Chem.* **16**, 1215–1220.

Lightfoot, E. N. (1974). *Transport Phenomena in Living Systems*. John Wiley & Sons Publishers, New York.

Linek, F. and Dudukovic, M. P. (1982). "Representation of Breakthrough Curves for Fixed-bed Adsorbers and Reactors using Moments of the Impulse Response." *Chem. Eng. Jl.*, **23**, 31–36.

Linton, W. H. J. and Sherwood, T. K. (1950). "Mass Transfer from Solid Shapes to Water in Streamline and Turbulent flow." *Chem. Eng. Progress*, **46**, 258–264.

Liu, F. and Bhatia, S. K. (2001). "Application of Petrov-Galerkin methods to Transient Boundary Value Problems in Chemical Engineering: Adsorption with Steep Gradients in Bidisperse Solids." *Chem. Eng. Science*, **56**, 3727–3735.

Mackay, D. (2001). *Multimedia Environmental Models: The Fugacity Approach*, Second Edition. Lewis Publishers, Boca Raton, FL.

Mackay, D. (2001). *Multimedia Environmental Models*. Lewis Publishers.

Mackay, D. and Paterson, S. (1991). "Evaluating the Multimedia Fate of Organic Chemicals: A Level III Fugacity Model." *Environmental Science & Technology*. **25**(3), 427–436.

Mackay, D., Paterson, S., Di Guardo, A., and Cowan, E.C. (1996a). "Evaluating the Environmental Fate of a Variety of Types of Chemicals Using the EQC Model." *Environ. Toxicol. Chem.*, **15**, 1627–1637.

Mackay, D., Paterson, S., Kicsi, G., Cowan, E.C., Di Guardo, A., and Kane, D.M. (1996b). "Assessment of Chemical Fate in the Environment Using Evaluative, Regional and Local-Scale Models: Illustrative Application to Chlorobenzene and Linear Alkylbenzene Sulfonates." *Environ. Toxicol. Chem.*, **15**, 1638–1648.

Mackay, D., Paterson, S., Kicsi, G., Di Guardo, A., and Cowan, C. E. (1996c). "Assessing the Fate of New and Existing Chemicals: A Five Stage Process." *Environ. Toxicol. Chem.*, **15**, 1618–1626.

Mao, Z. and Chen, J. (2004). "Numerical Simulation of Marangoni Effect on Mass Transfer to Single Slowly Moving Drop in Liquid-Liquid Systems." *Chem. Eng. Sci.*, **59**, 1815–1828.

Marzouqi, M. H., Abdulkarim, M. A., Marzouk, S. A., El-Naas, M. H., and Hasanain, H. M. (2005). "Facilitated Transport of CO_2 Through Liquid Membrane." *Ind. Eng. Chem. Res.*, **44**, 9273–9278.

Mashelkar, R. A. and Chavan, V. V. (1973). "Solid Dissolution in Falling Films of non-Newtonian Liquids." *J. Chem. Eng. Japan.*, **6**, 160–167.

Masheklar, R. A. and Soylu, M. A. (1974). "Diffusion in Flowing Films of Dilute Polymeric Solutions." *Chem. Eng. Sci.*, **29**, 1089.

McAdams, W. H. (1954). *Heat Transmission, Third Edition*. McGraw-Hill Book Company, New York.

McCabe, W. L., Smith, J. C., and Harriott, P. (2010). *Unit Operations of Chemical Engineering. Seventh Edition*. McGraw-Hill Publications, New York.

Meadows, D. L. and Peppas, N. A. (1984). "Solute Diffusion in Swollen Membranes. III. Non-Equilibrium Thermodynamic Aspects of Solute Diffusion in Polymer Network Membranes." *Chem. Eng. Commun.*, **31**, 101–119.

Mehta, B. N and Aris, R. (1971). "Internal Diffusion for Fractional Order Reactions." *Chem. Eng. Sci. J.*, **26**, 1699–1711.

Merkel, F. (1925). Verdunstungskuhlung, VDI. *Forschungsarbeiten,* Berlin, **275**.

Merrill, L. S. and Hamrin, C. E. (1970). "Conversion and Temperature Profiles for Complex Reactions in Laminar and Plug Flow." *AIChE Journal,* **16**, 194.

Middleman, S. (1998). *Introduction to Mass and Heat Transfer.* John Wiley & Sons, New York.

Middleman, S. and Hochberg, A. K. (1993). *Process Engineering Analysis in Semiconductor Device Fabrication.* McGraw-Hill Publishing Co., New York.

Mills, A. F. (1993). *Heat and Mass Transfer.* Irwin Publishing Company, Toronto.

Mills, P. L., Ramachandran, P. A., and Chaudhari, R. V. (1992). "Multiphase Reaction Engineering for Fine Chemicals and Pharmaceuticals." *Reviews in Chemical Engineering,* **8**, 1–172.

Mujumdar, A. S. (Ed.) (1995). *Handbook of Industrial Drying.* Marcel Dekker, New York.

Nagasawa. H., Niimi, T., Kanezashi, M. Yoshioka, T., and Tsuru, T. (2014). "Modified Gas-Translation Model for Prediction of Gas Permeation through Microporous Organosilica Membranes." *AIChE Journal,* **60**, 4199–4210.

Nagata, S. (1975). *Mixing Principles and Applications.* Kodansha Ltd. Tokyo/Wiley, New York.

Neogi, P. (1966). *Diffusion in Polymers.* CRC Press, Boca Raton, FL.

Nernst, W. (1904). "Theorie der Reaktionsgeschwindigkeit in Heterogenen Systemen." Published online: https://doi.org/10.1515/zpch-1904-4704.

Neumann, E. B. (1987). *Chemical Reactor Design.* John Wiley & Sons, New York.

Newman, J. and Thomas-Alyea. K. E. (2004). *Electrochemical Reactions.* John Wiley & Sons, Hoboken.

Nikuradse, J. (1932). "Laws of Turbulent Flow in Smooth Pipes." *VDI-Forschungsheft,* **356**, (English translation: NASA TT F-10: 359, 1966).

Noble, R. D. (1991). "Facilitated Transport Mechanism in Fixed Site Carrier Membranes." *J. Membrane Science,* **60**, 297–306.

Noble, R. D. and Koval, C. A. (2006). "Review of Facilitated Transport Membranes" in *Materials Science of Membranes for Gas and Vapor Separation* (Y. Yampolskii, I. Pinnau, and B. Freeman Eds.). John Wiley & Sons, Ltd, Chichester, UK.

Noble, R. D. and Stern, S. A. (Eds.) (1995). *Membrane Separations Technology: Principles and Applications.* Elsevier Science Publishing Company, Amsterdam.

Novosad, Z., and Ulbrecht, J. (1966). "Conversion in Chemical Reactions for Isothermal Laminar Flow of Non-Newtonian Liquids in a Tubular Reactor of Circular Cross Section." *Chem. Engg. Sci.,* **21**, 405–411.

Nusselt, W. (1916). "Die Oberflachenkondensation des Wasserdampfes." *Zeitchr. ver. Deutsch Ing.,* **60**, 541–569.

Olander, D. R. and Reddy, L. B. (1964). "Effect of Concentration Driving Force on Liquid-Liquid Mass Transfer." *Chem. Eng. Sci.,* **19**, 63.

Oldrich, W. E. and Wild, J. D. (1960). "Diffusion from the Free Surface into a Liquid Film in Laminar Flow over Defined Shapes." *Chem. Eng. Sci.,* **24**, 25–32.

Onda. K, Takeuchi, H., and Koyama, Y. (1967). "Effect of Packing Material on the Wetted Surface Area." *Kagaku Kogaku,* **31**, 126.

Onda. K, Takeuchi, H., and Okumoto, Y. (1968). "Mass Transfer Coefficients between Gas and Liquid Phases in Packed Columns." *J. Chem. Eng. Japan,* **1**, 56–62.

Onsager, L. (1945) "Theories and Problems of Liquid Diffusion." *Ann. N.Y. Acad. Sci.,* **46**, 241–265.

Osborne, F. T. (1975). "Purely Convective Models for Tubular Reactors with Non-Newtonian Flow." *Chem. Eng. Sci.,* **30**, 159–166.

Pai, S . I. (1953). "On Turbulent Flow Between Two Parallel Plates." *J. Appl. Mech.,* **20**, 109–114.

Patankar, S. V. (1980). *Numerical Heat Transfer and Fluid Flow* Hemisphere, Washington, DC.

Peppas, N. A. and Meadows, D. L (1983). "Macromolecular Structure and Solute Diffusion in Membranes: An Overview of Recent Theories." *J. Membr. Sci.,* **16**, 361–377.

Perry, R. L. and Green, D. W. (2007). *Perry's Chemical Engineering Handbook,* McGraw-Hill Professional Publications, New York.

Petropoulos, J. H. (1985). "A Comparative Study of Approaches Applied to the Permeability of Binary Composite Polymeric Materials." *J. Polym. Sci., Polym. Phys. Ed.,* **23**, 1309.

Plawsky, J. (2010). *Transport Phenomena Fundamentals.* CRC Press, Boca Raton, FL.

Poling, R. C., Prausnitz, J. M., and O'Connnell. J. P. (2001). *Properties of Gases and Liquids, Fifth Edition.* McGraw-Hill Publishers, New York.

Prandtl, L. (1925). "Bericht uber Untersuchungen zur ausgebildeten Turbulenz." *ZAMM,* **5**, 136.

Pratt, H. R. C. (1975). "Simplified Analytical Design Method for Differential Extractors with Backmixing. I. Linear Equilibrium Relationship." *Ind. Eng.Chem. Process Des. Dev,* **14**, 74.

Proudson, I., and Pearson, J. R. A. (1958). "Convection Cells induced by surface tension." *J. Fluid Mechanics,* **4**(5), 489–500.

Ramachandran, P. A. (1993). "Gas absorption in Slurries Containing Fine Particles: Review of Models and Recent Advances." *Ind. Eng. Chem. Res.,* **46**, 3137–3152.

Ramachandran, P. A. (2013). *Advanced Transport Phenomena: Modeling: Analysis, Modeling and Computations.* Cambridge University Press, U.K.

Ramachandran, P. A. and Chaudhari, R. V. (1983). *Three Phase Catalytic Reactors.* Gordon and Breach Science Publications, New York.

Ramachandran, P. A. and Doraiswamy, L. (1982). "Modeling of Non-Catalytic Gas-Solid Reactions." *AIChE Journal,* **28**(6), 881–900.

Ramachandran, P. A. and Mashelkar, R. A. (1980). "Lumped Parameter Model for Haemodialyser with Application to Simulation of Patient-Artificial Kidney System." *Med & Biol. Eng. Computing,* **18**, 179–188.

Ramachandran, P. A. and Mashelkar, R. A. (1976). "Homogeneous Reaction in Turbulent Pipe Flow." *Chem. Eng. Journal,* **11**, 73–76.

Ramachandran, P. A. and Sharma M. M. (1969). "Absorption with Fast Reaction in a Slurry Containing Soluble Fine Particles." *Chem. Eng. Sci.,* **24**, 1681–1686.

Ramachandran, P. A. and Sharma, M. M. (1971). "Simultaneous Absorption of Two Gases." *Trans. Instn. Chem Engrs. U.K,* **49**, 253–280.

Ramachandran, P. A. and Smith, J. M. (1977). "Single-Pore Model For Gas-Solid Non-catalytic Reactions." *AIChE Journal,* **23**, 353–361.

Ramachandran, P. A. and Smith, J. M. (1979). "Dynamic Behavior of Trickle Bed Reactors." *Chem Eng. Sci.*, **34**, 75–91.

Ramadesigan, V., Northrop, P. W. C., De. S., Santhanagoplan, S., Braatz, R. D., and Subramanian, V. R. (2012). "Modeling and Simulation of Lithium-Ion Batteries from a Systems Engineering Perspective." *J. Electrochem. Soc.*, **159**, R31–45.

Ramaswamy, R. C. and Ramachandran, P. A. (2008). "Multiple Steady State Calculations for Diffusion Problems Using Arc Length Continuation." *J of Research in Eng. and Tech.*, Kasetsart University Publication, **5**(3), 255–275.

Ramaswamy, R. C. and Ramachandran, P. A. (2013). "Modeling of Reactive Condensation Systems." Paper presented at NASCRE conference.

Ramirez, W. F., Mickley, M. C., and Lewis, D. N. (1971). "Mathematical Modeling of a KIIL Haemodialyser." *Chem. Eng. Prog. Symp. Series*, **114**(67), 116–122.

Ramkrishna., D. and Amundson, N. R. (1985). *Linear Operator Methods in Chemical Engineering*, Prentice-Hall, Englewood Cliffs, NJ.

Ranz, W. E. and Marshall, W. R. (1952). "Evaporation from Drops." *Chem. Eng. Prog.*, **48**, 141–146.

Rashidi, M. and Banerjee, S. (1988). "Turbulence Structure in Free Surface Channel Flows." *The Physics of Fluids*, **31**(9), 2491–2503.

Reddy, K. S. and Suresh, A. K. (2012). "Reactions in Solid Particles: A Reappraisal of Model." *AIChE Journal*, **58**, 3161–3166.

Reid, R. C., Prausnitz, J. M., and Poling, B. E. (1987). *Properties of Gases and Liquids* McGraw-Hill Publishing Co., New York.

Renkin, E. M. (1954). "Filtration, Diffusion and Molecular Motion in Porous Cellulosic Membranes." *J. Gen. Physiol*, **38**, 225–243.

Reyes, S. and Jensen, K. F. (1987). "Percolation Concepts in Modelling of Gas-Solid Reactions and Photonic Materials." *Chem. Eng. Sci.*, **42**, 565.

Ritter, T. and Yang, T. (1987). "Equilibrium Adsorption of Multicomponent Gas Mixtures at Elevated Pressures." *Ind. Eng. Chem. Res*, **26**, 1679–1686 .

Roberts, D. and Danckwerts, P. V. (1962). "Kinetics of CO2 Absorption in Alkaline Solutions. – I. Transient Absorption Rates and Catalysis by Arsenite." *Chem. Eng. Sci.*, **17**, 961–969.

Ruckenstein, E. and Berbente, C. (1964). "Convective Instability in a Two-Layer System with an Interfacial Reaction." *Chem. Eng. Sci.*, **19**, 329.

Rybczynski, W. (1911). "Uber die Fortschreitende Bewegung einer Flussigen Kugel in einem Zahen Medium." *Bull Acad. Sci, Carcovie*, A 40–46.

Sampath, B. S., Ramachandran, P. A., and Hughes, R. (1975). "Modeling of Noncatalytic Gas-Solid Reactions: Transient Simulation of a Packed Bed Reactor." *Chem. Eng. Sci.*, **30**(1), 135–43.

Sandler, S. L. (2006). *Chemical and Biochemical and Engineering Thermodynamics, Fourth Edition*, John Wiley & Sons, Hoboken.

Sardesai, R. G., and Webb, D. R., (1982) "Condensation of Binary Vapours of Immiscible Liquids," *Chem. Eng. Sci.*, **37**(4), 529–537·

Schiltting, H. and Gersten, K. (2000). *Boundary Layer Theory*, Springer Publishing Co.

Schneider, P. and Smith, J. M. (1968). "Adsorption Rate Constants form Chromatography." *AIChE Journal*, **14**, 762–768.

Scriven, L. E. (1960). "Dynamics of a Fluid Interface." *Chem. Eng. Sci.*, **12**, 98–108.

Seader, J. D., Henley, E. J., and Roper, D. K. (2011). *Separation Process Principles with Applications Using Process Simulators*. John Wiley & Sons, Hoboken, NJ.

Sharma, K. R. (2010). *Transport Phenomena in Biomedical Engineering: Artificial Organ Design and Development and Tissue Design*, McGraw-Hill Professional, New York, NY.

Shelelkhin, A. B., Dixon, A. G., and Ma. Y. H. (1995). "Theory of Gas Diffusion and Permeation in Inorganic Molecular Sieve Membranes." *AIChE Journal*, **41**, 58–67.

Sherwood, T. K. (1929). "Drying of Porous Solids." *Ind. Eng. Chem.*, **21**, 12–16.

Sherwood, T. K. and Wei, J. C. (1955). "Ion Diffusion in Mass Transfer between Phases." *AIChE Journal*, **1**(4), 522–527.

Sherwood, T. K. and Wei, J. (1957). "Interfacial phenomena in liquid extraction." *Ind. Eng. Chem.*, **49**, 1030.

Silverthorn, D. U. (2001). *Human Physiology, An Integrated Approach, Second Edition*. Prentice Hall, Upper Saddle River, NJ.

Simonsson, D. and Lindman, L. (1979). "On the Application of Shrinking Core Model for Liquid-Solid Reactions." *Chem. Eng. Sci.*, **34**, 31–35.

Smith, J. M. (1981). *Chemical Engineering Kinetics*, McGraw-Hill, New York.

Smith, J. M., Van Ness, H. C. and Abbott, M. M. (2005). *An Introduction to Chemical Engineering Thermodynamics*. McGraw-Hill, New York.

Sohn, H. Y., and Han, D. H. (2011). "Ca-Mg Acetate as Dry SO_2 Sorbent." *AIChE Journal*, **57**, 1329–1338.

Sohn, H. Y. and Szekely, J. (1972). "Structural Model for Gas-Solid Reactions with a Moving Boundary-III A General Dimensionless Representation of the Irreversible Reaction between a Porous Solid and a Reactant Gas." *Chem.. Eng. Sci.*, **27**, 763–768.

Staverman, A. J. (1952). Nonequilibrium Thermodynamics of Membrane Transport. *Trans. Faraday Soc.*, **48**, 176–185.

Steinberger, R. L. and Treybal, R. E. (1960). "Mass Transfer from a Solid Soluble Sphere to a Flowing Liquid Stream." *AIChE Journal*, **6**(2), 227–232.

Steinmeyer, D. E. (1972). "Fog Formation in Partial Condensers." *Chem. Eng. Prog.*, **68**(7), 64–68.

Sten-Knudsen. O. (2002). *Biological Membranes: Theory of Transport*. Cambridge University Press, Cambridge.

Stern, S. A. (1994). "Polymers for Gas Separations: The Next Decade." *J. Membrane Sci.*, **94**, 1–65.

Sternling, C. V. and Scriven, L. E. (1959). "Interfacial Turbulence: Hydrodynamic Instability and Marangoni Effect." *AIChE Journal*, **5**, 514–523.

Stewart, W. E. and Prober, R. (1964). "Matrix Calculation of Multicomponent Mass Transfer in Isothermal Systems." *Ind. Eng. Chem. Fundamentals*, **3**, 224–235.

Suresh, A. K. and Ghoroi, C. (2009). Solid-Solid Reacitons in Series: Modeling and Experimental Studies." *AIChE Journal*, **55**, 2399.

Suwanvaipattana, P., Limtrakul, S., Vatanatham. T., and Ramachandran, P. A. "Modeling of Electro-Organic Synthesis to Facilitate Cleaner Manufacturing: Adiponitrile Production" *J. of Cleaner Production*, **142**, 1296–1308.

Szekely, J., Evans, J. W., and Sohn, H. Y. (1976). *Gas-Solid Reactions* Academic Press.

Szekely, J. and Themelis, N. J. (1991). *Rate Phenomena in Process Metallurgy*. Wiley-Interscience, New York.

Takagaki, N., Kurose, R., Kimura, A., and Komori, S. (2016). "Effect of Schmidt Number on Mass Transfer Across a Sheared Gas-Liquid Interface in a Wind-Driven Turbulence." *Scientific Reports*, **6**, Article number: 37059.

Tanaka. Y. (2005). "Limiting Current Density of an Ion-Exchange Membrane and in an Electrodialyzer." *J. Membrane Sci.*, **266**, 6–17.

Tavlarides, L. L. and Stamatoudis, M. (1981). "Analysis of Interface Reactions and Mass Transfer in Liquid-Liquid Dispersions." *Advances. in Chem. Eng.*, **11**, 199–272.

Taylor, G. I. (1953). "Dispersion of a Solute Matter in a Solvent Flowing Slowly through a Tube." *Proc. Roy. Soc.*, **A219**, 186–203.

Taylor, G. I. (1954). "Conditions under Which Dispersion of a Solute in a Stream of Solvent Can Be Used to Measure Molecular Diffusion," *Proc. Roy. Soc. A.*, **225**, 473–477.

Taylor, G. I. (1954). "Dispersion of Matter in Turbulent Flow through a Pipe." *Proc. Roy Soc.* **223**, 446–467.

Taylor, R. and Krishna, R. (1993). *Multicomponent Mass Transfer*. Wiley InterScience, New York, NY.

Taylor, R., Krishnamurthy, R., Furno, J. S., and Krishna, R. (1986). "Condensation of Vapor Mixtures. I. Nonequilibrium Models and Design Procedure." *Ind. Eng. Chem. Proc. Des. Dev.*, **25**, 83–97.

Thampan, T., Malhotra, S. Zhang, J. X., and Datta, R. (2001). "PEM Fuel Cell as a Membrane Reactor." *Catalysis Today*, **67**, 15–32.

Tien, C. L. and Wasan, D. T. (1963). "Law of the Wall in Turbulent Channel Flow." *Physics of Fluids*, **6**, 144–146.

Tongwen, X. and Weihua, Y. (2002). "Citric Acid Production by Electrodialysis with Bipolar Membrane." *Chem. Eng and Processing*, **41**, 519–524.

Toor, H. L. (1957). "Diffusion in Three-Component Gas Mixtures." *AIChE Journal*, **2**, 198–207.

Trefethen, L. N. (2007). "Computing Numerically with Functions Instead of Numbers." *Mathematics in Computer Science*, **1**, 9–19.

Truskey, G. A., Yuan, F., and Katz, D. F. (2004). *Transport Phenomena Biological Systems*. Prentice Hall, Upper Saddle River, NJ.

Tsuru, T., Kondo, H., Yoshioka, T., and Asaeda, M. (2004). "Permeation of Non-aqueous Solution through Organic/Inorganic Hybrid Nanoporous Membranes." *AIChE Journal*, **50**, 1080–1087.

Valero, F. Barceló. A., and Arbós, R. (2011). *Electrodialysis Technology - Theory and Applications*, In: *Desalination, Trends and Technologies,* Michael Schorr (Ed.), InTech, pp. 3–20.

van Driest, E. R. (1956). On Turbulent Flow near a Wall *J. Aero. Sci.*, **23**, 1007–1001.

van't Reit, K. (1979). "Review of Measuring Methods and Results in Nonviscous Gas-Liquid Mass Transfer in Stirred Vessels." *Ind. Eng. Chem. Proc. Des. Dev.*, **18**, 357.

Vasudevan, T. V. and Sharma, M. M. "Some Aspects of Process Design of Liquid-Liquid Reactor." *Ind. Eng. Chem. Proc. Des. Dev.*, **23**, 400.

Vigness, A. (1966). "Diffusion in Binary Solutions. Variation of Diffusion Coefficient with Composition." *Ind. Eng. Chem. Fundam.* **5**, 189–198.

Villadsen, J. V. and Michelsen, M. L. (1978). *Solution of Differential Equation Models by Polynomial Approximation.* Prentice-Hall, Englewood Cliffs, NJ.

Vrentas, J. S. and Vrentas, C. M. (2012). *Diffusion and Mass Transfer.* CRC Press, Boca Raton, FL.

Wakao, N. and Smith, J. M. (1962). "Diffusion in Porous Solids." *Chem. Eng. Sci.,* **17**, 825.

Wang, Yi., Lin, Q., and Mukherjee, T. (2004). "System-Oriented Dispersion Models of General-Shaped Electrophoresis Microchannels." *Lab on Chip,* **4**, 453–463.

Warsi, Z. U. A. (1999). *Fluid Dynamics: Theoretical and Computational Approaches.* CRC Press, Boca Raton, FL.

Webb, D. R. and Sardesai, R. G. (1981). "Verification of Multicomponent Mass Transfer Models for Condensation Inside a Vertical Tube." *Int. J. Multiphase Flow,* **7,** 507–520.

Welty, J. R., Wicks, C. E., Wilson, R. E., and Rorrer, G. L. (2008). *Fundamentals of Momentum, Heat and Mass Transfer.* John Wiley & Sons, Inc, New York.

Wen, C. Y. and Fan, L. T. (1975). *Models for Flow System and Chemical Reactors,* Marcel Dekker, New York.

Wesselingh, J. A. and Krishna, R. (2000). *Mass Transport in Multicomponent Mixtures.* VSDD Publishers, Amsterdam.

Westermeier, R. (2005). *Electrophoresis in Practice.* Wiley-Blackwell, New York.

White, R. E. and Subramanian, V. (2010). *Computational Methods in Chemical Engineering with MAPLE.* Springer-Verlag, Berlin.

Wijmans, G. and Baker, R. W. (1993). "A Simple Predictive Treatment of the Permeation Process in Pervaporation," *Journal of Membrane Science* **79**, 101.

Wilke, C. R. (1950). "Diffusional Properties of Multicomponent Gases." *Chem. Eng. Progress,* **46**, 95–104.

Wilke, C. R. and Chang P. (1955). "Correlation of Diffusion Coefficients in Dilute Solutions." *AIChE Journal,* **1**, 264–270.

Won, Y. S. and Mills, A. F. (1982). "Correlation of the Effects of Viscosity and Surface Tension on Gas Absorption in Freely Falling Turbulent Films." *Int. J. Heat and Mass Transfer,* **25**, 223–229.

Wyman, C. E. and Carter, L. F. (1976). "Numerical Model for Tubular Polymerization Reactors." *AIChE Symp. Series,* **72**, 1.

Xiao, J. and Wei, J. (1992). "Diffusion Mechanisms of Hydrocarbons in Zeolites." *Chem. Eng. Sci.,* **47**, 1123–1141.

Xu, T. (2005) "Ion Exchange Membranes: State of Their Development and Perspective." *Journal of Membrane Science,* **263**, 1–29.

Yamashita, F. and Inoue, H. (1975). "Gas Holdup in Bubble Columns." *J. Chem. Eng. Japan,* **8**, 334.

York, R. L., Bratlie, K. M., Hile, L. R., and Jang, L. K. (2011). "Dead Zones in Porous Catalysts: Concentration Profiles and Efficiency Factors." *Catalysis Today.* **160**, 204–212.

Yoshisato, R. A., Korndorf, L. M., Carmichael, G. R., and Datta, R. (1986). "Performance Analysis of a Continuous Rotating Annular Electrophoresis Column." *Separation Science and Technology,* **21**(8), 727–753.

INDEX

Absorption
  description, 216
  film theory models
    bulk concentration and bulk reactions, 670–672
    first-order and pseudo-first-order reactions, 707
    first-order reaction of dissolved gas, 668
    two gases, 684–688
  multistage cascades, 467–468
  in slurries
    instantaneous reaction case, 694–697
    overview, 692–693
    particle size effect, 693–694
Accumulation
  evaporation in open containers, 197–198
  flux form equations, 154
  macroscopic models
    batch reactors, 91
    dissolved oxygen concentration in stirred tanks, 104
    macroscopic balance, 90
    reactor–separator combination, 97–98
    single-phase systems, 173–174
    sublimation of spherical particles, 101

silicon oxidation, 208
tracer response in two-phase systems, 505
Acetic acid in water diffusion rate, 257
Acetone in water, diffusion coefficient for, 245
Ackermann correction factor, 851–852
Activated carbon in adsorption, 920
Activation energy
  interstitial diffusion, 246–247
  porous catalyst temperature effects, 612
Active diffusion, 258
Active zones in backmixed assumption tests, 111–112
Activity coefficients
  condensation unmixed model, 858
  liquid-liquid interface, 19
  models for non-ideal liquids, 244–245
Activity correction factor for non-ideal liquids, 243–244
Adiabatic dryers, 827
Adiponitrile (ADN) production, 807–808
Adsorption
  applications and adsorbent properties, 920–921
  batch slurry adsorbers
    collocation method, 929–931

complexities, 931
    linear driving force model, 927–928
    model equations, 925
    overview, 924–925
    particle-level model, 925–926
    transients, 928–929
  fixed bed adsorption
    axial dispersion effects, 934–936
    equilibrium model, 932–934
    heterogeneous model, 936
    Klinkenberg equation, 936–937
    overview, 931–932
    scale-up aspects, 937
  isotherms, 921–924
  overview, 919–920
  problems, 941–943
  review questions, 941
  summary, 939–941
Adsorption constant in tracer response in two-phase systems, 505
Adsorption isotherm in fluid–solid interface, 19
Advection effects for environmental transport, 461
Aeration systems, 148
Age distribution functions in tracer studies, 441–442
Agitated tanks, 325–327

Air
  diffusion volume, 232
  heat balance, 819
  Lennard-Jones constant, 229
Air and water mixture,
    condensation rate in,
    853–854
Air phase in compartmental
    models fugacity of
    pollutants, 460
Ambipolar diffusion, 552
Amine solution
  CO_2 in, 744
  H_2S absorption in, 716–717
Amino acids, 956
Ammonia (NH_3)
  Henry's law constant, 16
  molar volume in liquid, 241
Ammonia (NH_3)
  porous catalysts in production
    of, 586
  selective catalytic reduction of
    NO, 209–210
Amorphous solids, drying, 829
Anemia, 769
Anion exchange membranes
    (AEMs), 946–950
Anions in mass transport in
    electrolytic systems
  across charged membranes,
    554
  of charged species, 545
  across uncharged membrane,
    551
Annular ring catalyst, 596–597
Anodic current and reactions
  definitions, 776–777
  electrochemical reaction
    voltage balance, 793–795,
    796
  half, 777–779
  hydrogen fuel cells, 798
Anoxic core in oxygen transport
    in tissues, 759
Antoine equation for vapor
    pressure, 18, 48
Apparent rate constant for
    porous catalysts
  diffusion-reaction model,
    597–599
  linking with reactor models,
    624

Applied potential in
    electrochemical reactions,
    786–787
Arnold cells for evaporation,
    197, 532–533
Arnold diffusion in film model,
    204
Arrhenius law and equation
  batch reactors, 92
  boron in silicon, 247
  Butler-Volmer equation,
    788–791
  gas translation model, 880
  laminar flow reactor heat
    transfer effects, 578
  porous catalyst temperature
    effects, 612
Arsenic diffusion in silicon,
    247–248, 296
Ash layer effects in shrinking
    core model, 641–644
Asymmetric membranes in gas
    separation, 873
Augmentation factor for
    condensation vapor with
    non-condensible gas,
    851–852
Automobile catalytic converters,
    21
Average absorption rate in
    first-order and
    pseudo-first-order
    reactions, 707
Average concentration
  batch slurry adsorbers, 925,
    927–928, 930
  capillary flow models, 756
  cross-sectional, 490, 576
  differential models, 5
  dispersion model, 478
  fixed bed adsorption, 936
  hemodialyzers, 764
  internal flows, 311
  laminar flow reactors, 571
  long cylinders, 279
  mass exchangers, 134
  multiphase systems, 177
  partial micromixing, 453–454
  pipe flows, 423
  reactor-separation
    combination, 98, 100
  single-phase systems, 174
  slab solutions, 274–275

solid dissolutions from walls,
    39, 126
stirred reactors, 437
transient diffusion processes,
    280–282
tubular flow reactors, 41,
    129–131
2-D problems, 287
wall transfer term, 89–90
Average cross-sectional rate in
    tubular flow reactors, 131
Average flux
  bimolecular reactions, 713
  penetration theory, 711
  transient diffusion processes,
    290
Average moisture content in
    drying falling rate period,
    836–837
Average molecular weight
  calculation, 9–10
  constant-density systems, 167
  from mole fractions, 10
Average pore radius in catalysts,
    587–588
Average rate
  condensation, 848
  diffusion-reaction model, 592,
    597–598
  film theory models, 710–711
  laminar boundary layers, 398
  laminar flow, 356
  macroscopic models
    continuous stirred tank
      reactors, 108
    interface transfer term, 90
Average reaction rate
  collocation method, 615
  porous catalysts, 604
  second-order reactions, 454
  single-phase systems, 173
Average velocity
  diffusion coefficients in gases,
    227
  film flow, 353
  flux form, 155
  kinetic model, 227
  laminar flow in pipes, 337
  mixtures, 156–157, 164
  plug flow, 477
  solid dissolutions from walls,
    39
  system, 152

turbulent flow, 413
wall shear stress, 416
Avrami model for solid–solid
reactions, 654–655
Axial diffusion time in laminar
flow reactors, 570
Axial dispersion
biomedical applications, 764
in channel flow, 493
chromatography, 938
in dispersion model, 479, 482
fixed bed adsorption, 934–936
Hannig design, 960
mesoscopic models, 42–43, 575
packed beds, 494
tracer response, 506
tubular flow reactors, 132–133
turbulent cases, 580–581
in turbulent flow, 493–494
Axial distance
fixed bed adsorption, 937
laminar flow, 337–338, 568
Sherwood number, 344,
352–353
Axial flux values in mesoscopic
models, 39
Axial length in laminar flow
reactors, 569
Axial velocity profiles
mesoscopic models, 38
segregated flow model, 492

Backmixed assumption
stirred reactors, 436–438
test for, 110–112
Backmixed reactor transient
balance, 440
Backmixed–backmixed model
fluid–fluid systems, 462–464
gas permeator models,
884–886
liquid–liquid extraction, 115
Backmixing model for
multistage cascades, 466
Balance equations
batch slurry adsorbers, 925
condenser model, 861, 863
cooling towers, 818–819,
825–826
differential models, 31, 92
fixed bed adsorption, 936
flow over inclined plates, 389,
392

flux form, 154
gas permeator models, 883,
887, 889
gas-solid reactions, 652, 654
hydrogen fuel cells, 798
mass basis, 166
membranes, 740
mesoscopic models
mass exchangers, 136, 138,
140
tracer response, 504–505
mole basis, 168
momentum equation, 371,
377–378
multicomponent systems, 236,
519, 524
packed column absorbers, 691
porous catalysts, 611–612, 622
semibatch reactors, 688
single-phase systems, 174
species, 113–114
stirred reactors, 436–437
tracer studies, 438–439, 442
two-phase systems, 504
Balance of forces in diffusion
coefficients in liquids, 238
Barrer units in gas separation
membranes, 875–876
Barrier diffusion, 236
Batch dryers, 827
Batch reactors
differential equations, 91–93
electrochemical, 806–807
ODE45 with CHEBFUN, 93–96
overview, 90
with product removal, 99–100
Batch slurry adsorbers
collocation method, 929–931
complexities, 931
linear driving force model,
927–928
model equations, 925
overview, 924–925
particle-level model, 925–926
transients, 928–929
Batteries, li ion
modeling, 800–802
transfer and reaction steps, 27
BCFUN function, 300
Bed voidage in fixed bed
adsorption, 933
Benzene
cracking cyclohexane to, 891

system velocity in presence of
diffusion, 160–161
Benzoic acid–coated wall in
contact with flowing
NaOH solution, 381–382
Bernoulli equation for flow over
inclined and curved
surfaces, 388
Bessel functions
steady state mass transfer with
reaction, 72
transient diffusion processes,
278–279
BET (Brunauer-Emmet-Teller)
isotherms, 923–924
Bhavaraju correlation in gas
bubbles, 324
Bidispersed particles in batch
slurry adsorbers, 931
Bimodal distribution
catalysts, 588
fluids in porous solids,
253–254
Bimolecular reactions
gas–liquid reactions in film
theory
dimensionless
representation, 673–675
invariance property, 676–677
overview, 672–673
gas–liquid reactions
penetration theory
dimensionless form, 712–713
illustrative results, 713–714
instantaneous reaction case,
714–717
overview, 712
surface reactions, 209–210
Binary diffusion
in gases, 230–231
multicomponent transport,
518
Binary gas mixtures,
condensation of
interface temperature, 860–861
overview, 857–858
unmixed model, 858–859
Binary liquid mixtures,
evaporation of, 531–533
Binary mixture with UMD,
Stefan-Maxwell model for,
237

Binary pair diffusivity in
multicomponent
diffusivity matrix, 536–539
Biodiesel production, 704
Biological systems
counter-transport, 736
Biomedical applications and
systems
compartmental models, 43–44
examples, 25
hemodialyzers, 763–766
membrane aeration systems,
148
overview, 749–751
oxygen uptake in lungs
meso-model for capillaries,
756–757
overview, 751
oxygen-hemoglobin
equilibrium, 751–753
transport steps, 753–756
pharmacokinetics, 760–763
problems, 768–773
review questions, 767–768
summary, 766–767
transport in tissues, 757–760
Bioseparations, 24
Biot number
film theory models
bimolecular reactions, 674
simultaneous absorption of
two gases, 685
fugacity of pollutants, 460
laminar flow in pipes, 345, 347
laminar flow reactors, 571
oxygen uptake in lungs, 755
porous catalysts
diffusion-reaction model,
591, 594
temperature effects, 612–613
Robin condition, 276–278
transient diffusion processes
cylinders and spheres,
279–282
problems in 2-D, 287
Bipolar membranes (BPMs) in
electrodialysis, 950–951
Bischoff approximation, 605
Blasius approach
dispersion coefficient values,
494
laminar boundary layers

flat plate with low flux mass
transfer, 371–376
flow over inclined and
curved surfaces, 392
high flux analysis, 386–388
Blasius correlation for axial
dispersion model, 581
Blood cells
glucose uptake, 747
hemodialyzers, 763–766
oxygen transport in tissues,
757–760
oxygen uptake in lungs,
753–757
Blowing factor
laminar boundary layer high
flux analysis, 387
steady state diffusion across
slabs, 54
transport rate in presence of
convection, 67
vapor with non-condensible
gas, 851
Bodenstein number in
dispersion model, 479
Boltzmann constant
diffusion coefficients in gases,
227
interstitial diffusion, 247
Bolus input
pharmacokinetic analysis, 471
transient balance, 439–441
Boron in silicon interstitial
diffusion, 247
Boundary conditions
batch slurry adsorbers, 926,
930
concentration profiles in film
model, 203–204
condensation in ternary
systems, 865
convective mass transfer
entry region analysis,
348–349
film flow, 355
turbulent flow, 417
wall boundaries, 362
cross-flow cooling towers, 826
D-D problem in slab geometry,
283–285
description, 171–172
diffusion
with convection, 65–66

with first-order reactions in
long cylinders, 72
drying falling rate period,
835–837
in electrolytic systems, mass
balance for reacting
systems, 550
equations of mass transfer,
171–172
evaporation
binary liquid mixtures, 531
ternary liquid mixtures, 530
flat plate with low flux mass
transfer, 367
gas–liquid reactions in film
theory
bimolecular reactions,
674–676, 715
first-order reaction of
dissolved gas, 663–664
simultaneous absorption of
two gases, 685–686
gas–liquid reactions in
penetration theory, 707
hemodialyzers, 765
heterogeneous reactions, 526
laminar boundary layers
flat plate with low flux mass
transfer, 372, 374
flow over inclined and
curved surfaces, 393
high flux analysis, 386–387
integral balance approach,
377, 381
laminar flow in pipes,
339–340, 344–348
laminar flow reactors, 571
mesoscopic dispersion
model, 576
model equations, 571–572
liquid separation membranes,
904
mass transfer from spheres to
stagnant gas, 192
mass transport in electrolytic
systems, 550
mesoscopic models
dispersion model, 480–483
ideal flow patterns, 500–502
Taylor model, 490
tracer response method, 484,
486–487

oxygen profiles in pools of liquid, 59–60
oxygen transport in tissues, 758
porous catalysts
  diffusion-reaction model, 589–590, 593, 597, 600
  finite difference methods, 621
  orthogonal collocation method, 616
  temperature effects, 611–613
reacting systems, 520–525
single solute diffusion, 729
slab solutions
  Dirichlet case, 269–270
  Robin condition, 276–278
  temporal evolution, 63
steady state mass transfer with reaction, 71–72
transient diffusion
  integral method, 290–293
  problems in 2-D, 286
  with reaction, 301–302
  semi-infinite slab analysis, 289
  in slabs, 62
  variable diffusivity, 296
Boundary layers
  channels and pipes, 421–422
  fluid–solid mass transfer benefits, 201
  film model, 198–201
  laminar. *See* Laminar boundary layers
Breakthrough curve for fixed bed adsorption, 935
Bubble point in binary gas mixtures condensation, 860–861
Bubbles and drops
  convective mass transport, 321–325
  laminar boundary layers
    overview, 396
    rigid bubbles, 396–397
    spherical cap bubbles, 397–399
  penetration theory of mass transfer, 294–295
Buffer zones in turbulent flow
  constant wall flux, 420
  velocity profiles, 414–416

Bulk concentration
  gas–liquid reactions in film theory models, 668–672
  porous catalysts temperature effects, 611
Bulk density in catalysts, 587
Butler-Volmer equation
  hydrogen fuel cells, 799
  kinetic model, 788–791
  mass transfer effects, 792
  mass transport in electrolytic systems, 550
  voltage balance, 793, 797
BVP4C method
  Blasius approach, 387
  condensation in ternary systems, 865–866
  evaporation in liquid mixtures, 531–534
  Falkner-Skan equation, 393
  flat plate with low flux mass transfer, 374
  heterogeneous reactions, 527
  porous catalyst temperature effects, 614
  reacting systems, 520–522
  ternary systems, 865
  transient diffusion processes, 296

Calcium carbide, nitrogen reaction with, 658
Calcium cynamide, nitrogen reaction for, 658
Calderbank correlation
  gas bubbles, 324
  mechanically agitated tanks, 326
Calderbank-Moo-Young correlation, 323
Capacity term in tracer response in two-phase systems, 505
Capillaries
  drying flow models, 838–839
  meso-model for, 756–757
  transport in tissues, 757–760
Capsules, drug release rate from, 282–283
Carbon
  adsorption, 920
  diffusion volume, 231
Carbon dioxide (CO_2)
  in amine solution, 744

diffusion volume, 232
Henry's law constant, 16, 47
Lennard-Jones constant, 229
molar volume in liquid, 241
oxygen-hemoglobin equilibrium, 753
removal with NaOH solution, 680–681
sequestration
  mass transfer in human body, 750
  underground mines, 28
Carbon monoxide (CO)
  catalytic oxidation of, 541
  counter-transport, 736
  diffusion volume, 232
Carbothermic reduction of ores, 654
Cartesian coordinates
  diffusion with convection, 64–67
  divergence operator, 155
  Laplacian of concentration, 190
  steady state diffusion
    across slabs, 52–55
    reactions in slabs, 56–62
    transient diffusion in slabs, 62
Catalysts. *See* Porous catalysts
Catalytic converters, 20–21
Catalytic oxidation of CO, 541
Cathodic current and reactions
  definitions, 776–777
  electrochemical reaction voltage balance, 794, 797
  half, 777–779
  hydrogen fuel cells, 799
  reduced product, 780
Cation exchange membranes (CEMs) in electrodialysis, 946–950, 953
Cations in electrolytic systems
  transport across charged membranes, 554
  transport across uncharged membrane, 551
  transport of charged species, 545
Cauchy-Euler type for laminar boundary layers, 380
Cement production, 654
Center concentration
  laminar flow, 341, 359

Center concentration (*continued*)
limiting cases, 573
porous catalysts
diffusion-reaction model,
600, 603
finite difference methods,
619
spherical catalysts, 595
temperature gradients, 614
spheres, 280, 282, 303
Stanton number, 424
Central difference equations for
porous catalysts, 618–619
Channels
axial dispersion, 493
convective mass transfer,
350–353
description, 577
laminar flow, 315
overview, 316
turbulent flow
constant wall flux, 418–421
correlations, 315
momentum transfer
analogy, 422–423
overview, 417–418
Stanton number for
boundary layers, 421–422
Stanton number for pipe
flows, 423–425
Chapman-Enskog model
diffusion coefficients in gases,
228–231
effective diffusivity, 253
Charge equality in electrolytic
systems, 558
Charge neutrality in electrolytic
systems, 547–548
Charged membranes in
electrolytic systems,
553–556
Charged species diffusion, 258
Charging li-ion batteries, 801
Chebechev polynomials
batch reactors, 94
channel flows with mass
transfer, 352
CHEBFUN solutions
batch reactors, 93–96
channel flows with mass
transfer, 351–353
dispersion model, 482–483
Falkner-Skan equation, 393

laminar flow in pipes, 345–346
macrofluid models, 452
reacting systems, 521–522
simultaneous absorption of
two gases, 684, 686–688
transient diffusion processes,
297–298
Chemical potential in Nernst
equation, 781–782
Chilton-Colburn analogy for
heat and mass transfer, 370
Chlorine formation, 948
Chloromethylation, 949
Chromatography, 938–939
Classification
cooling towers, 816–817
electrode reactions, 779–780
gas separation membranes,
873–874
liquid separation membranes,
898–899
Clift correlation for gas bubbles,
324–325
Closed-closed boundary
conditions in dispersion
model, 480–481
Closure
axial dispersion model, 43
convective mass transfer in
turbulent flow, 407,
411–413
Co- and counter-transport in
reactive membranes
illustrative results, 741–742
overview, 736–739
Cobalt recovery, 806
Coefficient matrix for
evaporation of binary
liquid mixtures, 532
Collision integral for diffusion
coefficients in gases,
230–231
Collocation methods
adsorption and
chromatography
batch slurry adsorbers,
929–931
fixed bed adsorption, 936
porous catalysts, 615–617
Colloids in electrophoresis
principle, 956
Column design in gas
absorption, 141–142

Column extractors in multistage
cascades, 466
Columns, bubble, 322
Combined flux equation for
steady state diffusion, 187
Compartmental models
development, 43–44
drug distribution, 750
environmental transport
fugacity of pollutants, 460
Level I, 460–461
Level II, 461
Level III, 461–462
Level IV, 462
overview, 459
matrix representation,
456–457
overview, 436, 454–456
pharmacokinetics, 457–459,
760–763
Competitive adsorption
isotherms, 923
Completely mixed film
assumption in
condensation, 858
Composite dielectric medium,
256
COMSOL-based solutions
laminar boundary layers, 390
laminar flow reactors,
579–580
Concentration and
concentration profiles
adsorption and
chromatography
batch slurry adsorbers, 925,
927–930
fixed bed adsorption, 933,
936–937
Butler-Volmer equation, 789
capillary flow models, 756
convective mass transfer
channel flow, 316
entry region analysis, 349
external and internal flows,
310–311
film flow, 354, 356–358
turbulent flow, 406–407,
419–421, 424
wall reaction, 347
cross-sectional, 490, 576
differential models, 5

diffusion
  with convection, 66
  across cylindrical shells, 69
  in gases, 226
electrochemical reaction
    voltage balance, 797
electrodialyzer design,
    952–953
film model, 199–200, 203–204
fixed bed adsorption, 936
fog formation, 856
gas permeator models,
    883–884
gas separation membranes,
    874–875, 878
gas–liquid reactions in film
    theory models
  absorption in slurries,
    693–694, 696
  bimolecular reactions, 673,
    675–676, 681, 714–715
  bulk concentration and
    reactions, 668–672
  coupling with reactor
    models, 688–690
  first-order reaction of
    dissolved gas, 665–667
  Laplace transform method,
    709–710
  liquid–liquid reactions, 698
hemodialyzers, 764
internal flows, 311
jump at interface
  fluid–solid interface, 19
  gas–liquid interface, 15–17
  liquid–liquid interface,
    18–19
  nonlinear equilibrium
    models, 19–20
  overview, 15
  vapor–liquid interface,
    17–18
laminar boundary layers
  film model, 384
  flat plate with low flux mass
    transfer, 367–369, 375–376
  flow over inclined and
    curved surfaces, 393
  high flux analysis, 386–388
  integral balance approach,
    379–380
  wall stress, 373
laminar flow in pipes, 339–344

laminar flow reactors, 569, 571,
    576
liquid separation membrane
    osmotic pressure, 900–901
long cylinders, 279
macroscopic models
  continuous stirred tank
    reactors, 107–108
  fluid–fluid systems, 463–465
  macrofluid models, 450–451
  moment analysis, 445
  stirred reactors, 437
  sublimation of spherical
    particles, 102
  tanks in series models, 449
  transient balance, 439,
    443–444
  variance-based models,
    453–454
mass exchangers, 134
mass transfer equations, 160
mass transfer from spheres to
    stagnant gas, 192
mass units, 9–10
mesoscopic models
  dispersion model, 478, 480
  first-order reaction, 481
  ideal flow patterns, 501–503
  plug-backmixed model,
    496–498
  plug flow idealization, 477
  solid dissolutions from
    walls, 38–39, 126
  Taylor model, 488–491
  tracer response method, 487
  tubular flow reactors,
    131–132
mole units, 8–9
multiphase systems, 177
oxygen-hemoglobin
    equilibrium, 751
oxygen profiles in pools of
    liquid, 61
oxygen transport in tissues,
    758–760
oxygen uptake in lungs, 756
partial micromixing, 453–454
partial pressure units, 10
pipe flows, 423
polarization effects in
    semi-permeable
    membranes, 903–905
porous catalysts

diffusion-reaction model,
    588–592, 600–602
finite difference methods,
    619–621
linking with reactor models,
    622
multiple species, 606–607
orthogonal collocation
    method, 616
spherical catalysts, 595, 598
reacting solids, 646, 650
reacting systems, 521
reactive membranes and
    facilitated transport,
    739–740
reactor-separation
    combination, 98, 100
shrinking core model solid
    product, 641
single-phase systems, 174
single solute diffusion,
    729–735
slabs
  illustrative results, 272,
    274–275
  semi-infinite, 288–290
  temporal evolution, 63–64
steady state diffusion
  low flux model, 190
  across slabs, 55
  across spherical shells, 74
transfer rate in diffusion film
    near electrodes, 558
transient diffusion processes
  cylinders and spheres,
    280–281
  integral method, 293
  problems in 2-D, 287
  semi-infinite slab analysis,
    288–291
  spherical coordinates, 76–77
tubular flow reactors, 41
tumors, 770–771
2-D problems, 287
variable diffusivity, 296
wall transfer term, 89–90
Concentration-dependent
    diffusion, 247
Concentration form
  constant-density systems,
    167–168
  mass basis, 166–167
  mole basis, 168–169

Concentration form (*continued*)
  overall continuity, 168
  overview, 166
Concentration overpotential in electrochemical reaction, 793
Concentration ratio parameter for bimolecular reaction, 713
Concurrent flow in ideal flow patterns, 500–501
Condensation
  binary gas mixtures
    interface temperature, 860–861
    overview, 857–858
    unmixed model, 858–859
  condenser model
    liquid and vapor phase balances, 862–864
    overview, 861
    vapor with non-condensible gas, 854–855
  fog formation, 855–857
  overview, 845–846
  problems, 868–869
  pure vapor, 846–850
  review questions, 867–868
  summary, 866–867
  ternary systems, 864–866
  vapor with non-condensible gas
    condenser model, 854–855
    heat transfer rate and Ackermann correction factor, 851–852
    interface temperature, 852–853
    mass transfer rate, 851
    overview, 850
Condensation rate
  methanol and water mixture, 859–860
  water and air mixture, 853–854
Conductivity
  electrolytic systems electric field, 548
  salt solution in electrodialyzer design, 953
Conservation laws and equations
  diffusion with convection, 64–65

flux form equations, 154
macroscopic models
  balance, 87
  continuous stirred tank reactors, 106
  dissolved oxygen concentration in stirred tanks, 104–105
  stirred reactors, 436–437
  sublimation of spherical particles, 101
mass basis, 166
mesoscopic models, 38, 125–128
model development, 28–30
mole basis, 168
multiphase systems, 175
reacting systems, 520, 525
steady state diffusion
  radial, 68
  with slabs, 53–54, 57
  transient diffusion in slabs, 62
  tubular flow reactors, 41–42
Constant and falling drying rates, 829–830
Constant-density systems
  common simplifications, 170
  concentration form, 167–168
Constant rate drying period, 830–833
Constant surface concentration in semi-infinite slab analysis, 288–291
Constant volume in batch reactors, 92–93
Constant wall concentration in laminar flow in pipes, 339–341, 347
Constant wall flux, 418–421
Constitutive model for multicomponent transport, 518–520
Constraint condition in single solute diffusion, 732
Contact point model in solid–solid reactions, 655–656
Contacting efficiency in fluid–fluid systems, 465
Contactors in bimolecular reactions, 684, 717–721
Contaminants in bubbles and drops, 397

Continuity equation for turbulent flow, 407–408
Continuous dryers, 827
Continuous stirred tank reactors (CSTRs)
  first-order reaction, 107
  overview, 106–107
  second-order reaction, 108–110
Continuum assumption, 7
Contracted axial distance parameter for laminar flow in pipes, 338
Control surfaces in macroscopic models, 87, 174
Control volumes
  differential models, 31
  gas–liquid reactions in film theory models, 668
  macroscopic models
    balance, 87–89
    liquid–liquid extraction, 113
    mass transfer equations, 157–158
  mesoscopic models
    description, 37
    mass exchangers, 136
    plug-backmixed model, 496
    tubular flow reactors, 129–130
  model development, 28–29
  multiphase systems, 176
  porous catalysts, 622
  steady state diffusion
    radial, 68
    across spherical shells, 73
Convection
  diffusion-induced, 170, 187–188, 193–198
  diffusion with, 64–67
  limiting cases, 572–574
  Taylor model, 488–489
  transport rate in presence of, 66
Convection-diffusion equation
  convective mass transfer in turbulent flow
    channels and pipes, 417
    gas–liquid interface, 427
    time-averaged equation, 408
  film flow, 357–358
  laminar boundary layers
    film model, 383

flat plate with low flux mass
transfer, 367–368, 374
laminar flow in pipes, 336–337
parallel plate electrolyzer, 807
Convection-diffusion-reaction
(CDR) model for laminar
flow reactors, 568–569
Convective flux
description, 12
diffusion coefficients in gases,
226
Convective heat transfer in
cooling towers, 820
Convective mass transfer for
internal laminar flow
channel flows with mass
transfer, 350–353
entry region analysis, 348–350
film flow
gas absorption from
interface, 357–358
overview, 353–354
solid dissolution at walls,
354–356
laminar flow in pipes
concentration, wall mass
flux, and Sherwood
number, 341–344
constant wall concentration,
339–341
dimensionless form, 337–339
overview, 336–337
overview, 335–336
PDEPE for, 358–359
problems, 361–364
review questions, 361
summary, 359–360
wall reaction, 344–348
Convective mass transfer in
turbulent flow
channels and pipes
constant wall flux, 418–421
momentum transfer
analogy, 422–423
overview, 417–418
Stanton number for
boundary layers, 421–422
Stanton number for pipe
flows, 423–425
closure models, 411–413
gas–liquid interface, 427–430
overview, 403–404
problems, 433–434

review questions, 432
summary, 430–432
time-averaged equation,
408–411
time averaging, 406–408
turbulent flow properties,
404–406
van Driest model, 425–427
velocity and turbulent
diffusivity profiles,
413–417
Convective mass transport
bed absorbers, 327–329
external and internal flows,
310–311
flows
in film, 318–320
in flat plates, 316–318
in pipes and channels,
315–316
gas bubbles, 321–325
key dimensionless groups,
313–315
mechanically agitated tanks,
325–327
overview, 309–310
problems, 331–333
relation to differential model,
311–313
review questions, 330–331
solid spheres, 320–321
summary, 329–330
Cooling towers, 815
classification, 816–817
counterflow model, 817–823
design considerations, 817
NTU calculation, 823–824
Copper (Cu)
electrochemical reaction
voltage balance, 797
electrowinning, 795–798
from leach solutions, 795
Correction factor
diffusion flux, 67
drift flux in diffusion-induced
convection, 195–196
evaporation in open
containers, 198
film model, 202, 204–205
gas–liquid mass transfer in
packed bed absorbers,
327
liquid-filled pores, 251

non-ideal liquids, 243–244
tubular flow reactors, 132
Correlations
condensation, 849
constant rate drying period,
831–832
convective mass transport
gas bubbles, 323–325
gas–liquid mass transfer in
packed bed absorbers,
328–329
low flux value, 312
mechanically agitated tanks,
325–327
dispersion coefficient values,
493–495
mass transfer coefficient
sublimation of spherical
particles, 102–104
solid dissolutions from walls
in pipe flow, 128
Counter-diffusion
equimass, 162–163
equimolar, 161–162
high flux model surface
reactions, 210–212
Countercurrent cascade in
multistage cascades, 467
Countercurrent flow
gas permeator models,
886–889
ideal flow patterns, 502
Counterflow model for cooling
towers
enthalpy balance equations,
819–821
mass balance equations,
818–819, 821
Merkel equation, 821–823
overview, 817–818
Coupled computational scheme
for fixed bed adsorption,
936
Coupling with reactor models,
688–692
Cross-correlation in turbulent
flow, 407
Cross-flow
cooling towers, 825–827
gas permeator models,
889–890
Cross-section averaged models
overview, 37

Cross-section averaged models (*continued*)
    solid dissolutions from walls, 38–41
Cross-sectional area in gas translation model, 880
Cross-sectional average
    dispersion model, 478, 482
    plug flow idealization, 476
    Taylor model, 490–491
    tubular flow reactors, 130–131
Crosscurrent cascade in multistage cascades, 467–468
Crosslinked polystyrene, 949
Crystalline solids, drying, 828–829
Cubic profile in integral balance approach, 378
Cumulative concentration for reacting solids, 659–660
Cup-mixed concentration
    description, 39
    example, 40
    hemodialyzers, 764
    laminar flow in pipes, 342–343
    laminar flow reactors, 571
    mesoscopic models
        mass exchangers, 135
        solid dissolutions from walls, 126–127
        tubular flow reactors, 130–131
    oxygen uptake in lungs, 756
    plug flow idealization, 476
    pure convection model limiting cases, 573
    Taylor model, 490
Current
    diffusion film near electrodes, 559
    electrochemical processes
        Butler-Volmer equation, 789–790
        description, 781
        mass transfer effects, 792
        voltage balance, 793–794
    electrodialysis, 946
    electrodialyzer design, 952–955
    hydrogen fuel cells, 799–800

Curved surfaces, flow over. *See* Flow over inclined and curved surfaces
Cyclohexane to benzene cracking, 891
Cylinders
    diffusion in first-order reactions, 72
    porous catalysts in diffusion-reaction model, 593–594
    transient diffusion in, 73, 278–279
Cylindrical coordinates
    divergence operator, 154–155
    Laplacian of concentration, 190
    overview, 67–68
    radial diffusion in, 68–70, 191
    steady state mass transfer with reaction, 70–72
    transient diffusion in, 73
Cylindrical pore filled with methane
    diffusion, 249–250
    fluids in porous solids, 249
Cylindrical shells, diffusion across, 69–70

D-D problem in slab geometry, 283–285
Dalvi-Suresh contact point model for solid–solid reactions, 655–656
Damkohler number
    laminar boundary layers, 376
    laminar flow reactors
        dispersion model, 576
        model equations, 569
        plug flow model, 574–575
    macroscopic models
        continuous stirred tank reactors, 107–108
        macrofluid models, 452–453
        stirred reactors, 437
        tanks in series models, 449
    mesoscopic models
        dispersion model, 479
        plug flow idealization, 477
        tubular flow reactors, 133
    porous catalysts, 623

Damping of turbulence in gas–liquid interface, 428–429
Danckwerts boundary conditions
    hemodialyzers, 765
    laminar flow reactors, 576
    mesoscopic models
        dispersion model, 480, 482–483
        ideal flow patterns, 500–501
        tracer response method, 484
Danckwerts surface renewal model, 725
Darcy permeability, 906
Darcy's law, 898–899
Darken relation in non-ideal liquids, 244
Dead zones
    backmixed assumption tests, 111–112
    porous catalysts, 630
Debye layer in electrolytic systems, 548
Density
    constant-density systems, 167–168
    mixture, 9–10
Desorption, 216
Determinacy condition
    equimolar counter-diffusion, 533
    evaporation in ternary liquid mixtures, 529
    steady state diffusion, 188–189
Determinacy correction factor for film model, 204–205
Dew point for binary gas mixtures, 860
Diameters of bubble, 323–324
Differential (1-D) balances
    Cartesian coordinates
        diffusion with convection, 64–67
        overview, 52
        steady state diffusion across slabs, 52–55
        steady state diffusion with reaction in slabs, 56–62
        transient diffusion in slabs, 62
    cylindrical coordinates
        overview, 67–68

steady state mass transfer with reaction, 70–72
steady state radial diffusion, 68–70
transient diffusion in cylinders, 73
overview, 51–52
problems, 79–83
review questions, 78–79
spherical coordinates, 73
diffusion and reaction, 75–76
steady state diffusion across spherical shells, 73–74
transient diffusion in spherical coordinates, 76–77
summary, 77–78
Differential equations and models
development, 29
laminar boundary layers, 379
macroscopic models
batch reactors, 91–93
reactor–separator combination, 97
multicomponent systems, 525
overview, 30–32
relation to convective mass transport, 311–313
two streams, 139
Differential species mass balance equation, 168–169
Differential volume in fixed bed adsorption, 933
Diffusion
batch slurry adsorbers, 926
common simplifications, 170
complex effects, 257–258
with convection, 64–67
in cylinders, 73
cylindrical pore filled with methane, 249–250
across cylindrical shells, 69–70
diffusion coefficients. See Diffusion coefficients
electrophoretic separation devices
Hannig design, 960
Philpot design, 959
equimolar counter-diffusion, 161–163
with first-order reactions
example, 59
in long cylinders, 72

fluids in porous solids, 248–254
gas separation membranes, 877–878
heterogeneous media, 254–256
liquid separation membranes, 899
multicomponent diffusivity matrix, 535–539
non-ideal liquids, 243–245
overview, 223–224
polymeric membranes, 256–257
problems, 261–264
review questions, 260–261
single solute diffusion, 729
solid–solid diffusion, 246–248
in spherical coordinates, 75–77
steady state radial, 68–70
summary, 258–260
system velocity in presence of, 160–161
time-averaged equation in turbulent flow, 408
transient. See Transient diffusion
with zero-order reactions, 58
Diffusion balances in integral balance approach, 382
Diffusion coefficients
electrolyte transport across uncharged membrane, 552
fixed bed adsorption, 937
gas separation membranes, 874
gas-solid reaction models, 651–652
in gases
frictional interpretation, 232–234
model based on kinetic theory, 225–232
multicomponent diffusion, 235–236
overview, 224–225
in liquids
ethanol in water, 242
overview, 237–239
Stokes-Einstein model, 239–240
Wilke-Chang equation, 241–242
values, 14
Diffusion-dominated processes

diffusion-induced convection, 193–198
gas–liquid interface in two-film model, 212–217
overview, 185–186
problems, 219–221
review questions, 219
steady state diffusion, 186–192
summary, 217–218
surface reactions, 206–212
Diffusion film near electrodes, transfer rate in, 556–559
Diffusion flux
description, 12–13
diffusivity values, 13–14
dispersion, 15
laminar boundary layers in film model, 383–384
mass transfer equations, 156–157
properties, 163–165
steady state diffusion with slabs, 54, 57
turbulent diffusion, 14–15
Diffusion-induced convection
common simplifications, 170
condensation in unmixed model, 859
conditions for validity of low flux model, 193
description, 187–188
drift flux correction factor, 195–196
mole fraction profiles in UMD, 196–197
UMD analysis, 193–195
Diffusion limiting current (DLC)
calculation in copper electrowinning, 795–798
Diffusion-reaction model
first-order and pseudo-first-order reactions, 706–707
porous catalysts
first-order reaction, 589–598
nth-order reaction, 603–605
overview, 588–589
zero-order reaction, 599–603
reactive membranes and facilitated transport equilibrium, 739
volume reaction model, 646

Diffusion type of models in drying falling rate period, 835–838

Diffusional effects for porous catalysts, 598–599

Diffusive flow field for fully turbulent flow, 405

Diffusivity
batch slurry adsorbers, 931
bimolecular reaction, 713
convective mass transfer in turbulent flow, 413–417
gas separation membranes, 878
pseudo-binary, 165–166
transport of charged species, 544–545

Dilute solutions in electrodialysis and electrophoresis, 948

Dilute system assumption for fluid–fluid systems, 463

Dimensionless boundary conditions in porous catalysts temperature effects, 612–613

Dimensionless groups
convective mass transport, 313–315
laminar flow reactors, 568–572

Dimensionless representation
bimolecular reaction, 712–713
Dirichlet case in slabs, 268–269
dispersion model, 479–480
gas–liquid reactions in film theory models
bimolecular reactions, 673
first-order reaction of dissolved gas, 664–665
simultaneous absorption of two gases, 685–686
internal laminar flow in, 337–339
porous catalysts temperature effects, 611–612
reactive membranes and facilitated transport
instantaneous reaction asymptote, 732–734
invariant of system, 732
pseudo-first-order reaction asymptote, 735–736

single solute diffusion, 730–732

Dirac delta function in transient diffusion processes, 292–293

Direct dryers, 827

Direction of transfer in two-film model, 216–217

Dirichlet boundary conditions, 171–172

Dirichlet case and problem
batch slurry adsorbers, 926, 930
falling rate period, 836–837
gas–liquid reactions in film theory models
bimolecular reactions, 674
first-order reaction of dissolved gas, 663–664
heterogeneous reactions, 526
laminar flow in pipes, 339–341
laminar flow reactors, 571
porous catalysts
diffusion-reaction model, 589, 591–594, 597, 600
orthogonal collocation method, 616–617
reacting systems, 523
slab solutions
average concentration, 274–275
dimensionless representation, 268–269
illustrative results, 272–274
overview, 267–268
series coefficient, 271–272
series solution, 269–271

Discharging li-ion batteries, 802

Discretization formula for porous catalysts, 615

Dispersed phase in multiphase systems, 176–177

Dispersion
axial. *See* Axial dispersion
diffusion, 15
fixed bed adsorption, 934–936

Dispersion closure in tubular flow reactors, 132–134

Dispersion coefficients
common cases, 493–495
Laplace domain, 485
moments of response curve, 485–487

non-Newtonian fluid flow reactors, 578
overview, 484–485
time domain solution, 487–488

Dispersion model
boundary conditions, 480–481
CHEBFUN, 482–483
first-order reaction, 481
negligible dispersion, 483–484
nonlinear reactions, 482
overview, 478–480

Dispersion number
dispersion model, 479
laminar flow reactors, 575

Dissipative flow field in fully turbulent flow, 405

Dissociation diffusion
description, 257
gas separation membranes, 877–878

Dissolved CO_2 in oxygen-hemoglobin equilibrium, 753

Dissolved oxygen concentration in stirred tanks, 104–106

Distillation
film model, 205
ternary mixtures, 534–535

Distributed pore models in gas–solid reaction models, 654

Dittus-Boelter correlation and equation
channel flow in convective mass transport, 315–316
drying constant rate period, 831

Divergence
flux form equations, 154–155
multiphase systems, 177

Domain statements in ODE45 with CHEBFUN, 94

Donnan equation, 553–556

Dopants
interstitial diffusion, 247–248
semiconductor and solar devices, 24–25
silicon with phosphorus, 293

Double collocation in fixed bed adsorption, 936

Drag coefficients
convective mass transfer in turbulent flow, 417

flat plate flow friction, 318
wall stress, 373
Drift flux correction
  convective mass transport, 312
  diffusion-induced convection,
      195–196
Dropwise condensation, 846
Drug release from capsules, 26,
      282–283
Dry bulb temperature, 812–815
Drying. *See* Humidification and
      drying
Dual mode transport in gas
      separation membranes,
      878–879

E-curves in transient balance,
      441
Eddy diffusivity
  convective mass transfer in
      turbulent flow
    gas–liquid interface, 428
    velocity profiles, 415–416
  description, 14
Eddy viscosity
  convective mass transfer in
      turbulent flow
    closure models, 412–413
    gas–liquid interface, 427–428
    van Driest model, 426
  turbulent flow reactors, 580
Effective diffusivity of fluids in
      porous solids, 251–254
Effectiveness factor
  oxygen transport in tissues,
      759
  oxygen uptake in lungs, 755
  porous catalysts
    diffusion-reaction model,
      591–594, 601, 604–605
    finite difference methods,
      619
    linking with reactor models,
      622
    multiple species, 607
    spherical catalysts, 594–595
    temperature effects, 613
Eigenvalues and eigenfunctions
  channel flows with mass
      transfer, 351–352
  Dirichlet case in slab solutions,
      269–272
  laminar flow in pipes, 340–341

matrix representation, 456
Neumann problem, 837
Robin condition in slabs,
      276–278
tracer response method, 487
transient diffusion processes
  CHEBFUN calculations,
      297–298
  cylinders and spheres,
      278–280
  wall reaction, 344–345
Einstein equation for transport
      of charged species, 545
Einstein relation for electric
      field, 548
Electric field in electrolytic
      systems
  charge neutrality, 547
  across charged membranes,
      552
  general expression, 548–549
  Laplace equation for potential,
      549
  mass balance for reacting
      systems, 550–551
  Nernst-Planck equation, 546
  transfer rate in diffusion film
      near electrodes, 558–559
  transference number, 550
  transport of charged species,
      545–546
Electrochemical batch reactors,
      806–807
Electrochemical processes
      applications, 26–27
Electrochemical reaction
      engineering
  copper electrowinning,
      795–798
  definitions
    anodic and cathodic
      reactions, 776–777
    classification of electrode
      reactions, 779–780
    half reactions and overall
      reaction, 777–779
    overview, 776
    primary variables, 780–781
  hydrogen fuel cells, 798–800
  kinetic model
    Butler-Volmer equation,
      788–791
    overview, 786–788

li-ion batteries, 800–802
  mass transfer effects, 791–793
  overview, 775–776
  problems, 805–808
  review questions, 804–805
  summary, 802–804
  thermodynamic
      considerations, 781–786
  voltage balance, 793–795
Electrochemist's dream, 777
Electrodes
  reaction classification, 779–780
  transfer rate in diffusion film
      near, 556–559
Electrodialysis and
      electrophoresis
  electrodialyzer design
    current and voltage, 952–953
    detailed models, 955
    limiting current, 954–955
    overview, 951
  electrophoresis principle,
      955–957
  electrophoretic separation
      devices
    Hannig design, 959–960
    overview, 957
    Philpot design, 957–959
    rotating annular beds,
      960–962
  overview, 945–946
  problems, 963–964
  review questions, 963
  summary, 962–963
  technological aspects
    bipolar membranes, 950–951
    electrodialysis reversal
      process, 949
    guidelines, 948
    membranes, 948–949
    overview, 946–948
Electrodialysis reversal (EDR)
      process, 949
Electrolyte transport across
      uncharged membrane,
      551–553
Electrolytic systems. *See* Mass
      transport in electrolytic
      systems
Electroneutrality across charged
      membranes, 554–555
Electrons in semiconductor and
      solar devices, 24–25

Empirical models for drying
falling rate period,
834–835

Emulsion liquid membranes
(ELMs), 743–744

Energy balances in condenser
model, 864

Enhancement factor
gas–liquid reactions in film
theory models
absorption in slurries, 697
average rate of mass
transfer, 711
bimolecular reactions, 677,
679, 682–683, 713, 716
coupling with reactor
models, 691–692
first-order reaction of
dissolved gas, 666–668
reactive membranes and
facilitated transport
co- and counter-transport,
738
single solute diffusion,
731–732, 736

Enthalpy balance equations in
cooling towers, 819–821

Entry regions
convective mass transfer,
348–350
laminar flow in pipes, 337

Environmental applications, 28

Environmental transport in
compartmental models
fugacity of pollutants, 460
Level I, 460–461
Level II, 461
Level III, 461–462
Level IV, 462
overview, 459

Eotvos number for spherical cap
bubbles, 397

Equilibrium constants
adsorption, 19
nonlinear models, 19–20
reactive membranes and
facilitated transport
co- and counter-transport,
737
single solute diffusion, 734
single solute diffusion, 729,
731

Equilibrium curves for
evaporation, 814

Equilibrium models
adsorption and
chromatography
batch slurry adsorbers, 929
fixed bed adsorption,
932–934

Equilibrium models
environmental transport,
460–461
macroscopic models
fluid–fluid systems, 464–465
multistage cascades, 466–468
reactor–separator
combination, 97
nonlinear, 19–20
plug-backmixed model, 499
reactive membranes and
facilitated transport
co- and counter-transport,
738–739
computational scheme,
739–742

Equilibrium mole fraction in
evaporation, 812

Equilibrium potential in
electrochemical reactions,
781–786

Equilibrium stage model
description, 36
liquid–liquid extraction,
115–116

Equimass counter-diffusion,
162–163

Equimolar counter-diffusion
(EMD), 161–162
convective mass transport, 312
diffusion-induced convection,
193
film model, 204–205
non-reacting systems, 533–534
steady state diffusion, 188–189

Equipment-level model for
semi-permeable
membranes, 907

erfc function for semi-infinite
slab analysis, 290

Erythrocyte, glucose uptake in,
747

Ethane, Lennard-Jones constant
for, 229

Ethanol

diffusivity in water, 242
distillation of, 534–535
for methanol poisoning, 736
vapor phase dehydrogenation
of, 527
in water
mole fraction, 912
separation of, 909
Wilke-Chang equation, 241

Ethylene oxide production,
586–587

Euler schemes in reactor models,
690

Evaporation
diffusion-induced convection,
187–188, 194
film model, 204
multicomponent systems
binary liquid mixtures,
531–533
ternary liquid mixtures,
528–530
in open containers, 197–198
wet and dry bulb temperature,
812

Exact analysis for flat plate with
low flux mass transfer,
371–376

Excess reactants in porous
catalysts multiple species,
606

Exchange current in
Butler-Volmer equation,
788–791

Exit stream in macrofluid
models, 450

expint function, 451

expm function, 527

Exponential matrix concept, 532

External flows in convective
mass transport, 310–311

External mass transfer in porous
catalysts, 591

External resistance in porous
catalysts, 597

External transport effects in
porous catalysts, 592–593

Extraction factor
liquid–liquid extraction, 116
multistage cascades, 467

Eyring theory, 237–238

F-curves in transient balance, 441–442
Facilitated diffusion, 257–258
Facilitated transport and reactive membranes. *See* Reactive membranes and facilitated transport
Facilitation factor in counter-transport, 738
Falkner-Skan equation for flow over inclined and curved surfaces, 390, 393–394
Falling rate period in drying
  capillary flow models, 838
  diffusion type of models, 835–838
  empirical models, 834–835
  model selection, 838–839
  overview, 833–834
Faraday constant for transport of charged species, 545
Faraday's law in electrodialyzer design, 951–952
Fast reactions
  diffusion film near electrodes, 559
  dissolved gas, 667
  film theory models, 690–691
  heterogeneous reactions, 172, 526
  high flux model surface reactions, 211
  homogeneous reactions, 382
  reactive membranes, 738
  silicon oxidation, 207–208
Fate and contaminant transport, 28
Fermi level in electrochemical reaction engineering, 787–788
Fibrous solids, drying, 829
Fick's law
  boundary conditions, 171–172
  combined flux equation, 187
  convective mass transfer, 356, 358
  convective mass transport, 312
  differential models, 31
  diffusion coefficients
    in gases, 226, 234
    in liquids, 239
  diffusion-induced convection problems, 170

diffusion modeling, 12–14
film model, 202
gas separation membranes, 874
gas–liquid reactions in film theory models
  bimolecular reaction, 716
  first-order reaction of dissolved gas, 665
  flux, 710
laminar flow in pipes, 342
mass basis, 167
mass transfer equations, 158–160
mass transfer from spheres to stagnant gas, 192
mass transport in electrolytic systems
  electrolyte transport across uncharged membrane, 552
  transfer rate in diffusion film near electrodes, 558–560
  transport of charged species, 545–546
mole basis, 169–170
multicomponent diffusivity matrix, 536–539
multicomponent transport, 518–519
non-ideal liquids, 243
oxygen profiles in pools of liquid, 61
porous catalysts, 591
pseudo-binary diffusivity, 165
semi-permeable membranes, 903
slab solutions, 273
steady state diffusion
  radial, 69
  with reaction in slabs, 57
  across slabs, 54
  across spherical shells, 74
Taylor model, 488, 491
transient diffusion, 62, 266
Film concept in mass transfer analysis
boundary layer concept for fluid–solid mass transfer, 198–201
concentration profiles, 203–204
determinacy correction factor, 204–205
model approximation, 201–202

overview, 198
porous catalysts, 590
problems, 219–221
review questions, 219
summary, 217–218
Film flow in convective mass transfer
  gas absorption from interface, 357–358
  overview, 318–320, 353–354
  solid dissolution at walls, 354–356
Film model
  condensation, 850
  flux transfer rate in diffusion film near electrodes, 556–559
  gas–liquid reactions in. *See* Gas-liquid reactions in film theory
  laminar boundary layer high flux analysis, 383–384
  overview, 185–186
Film thickness in condensation, 847–848
Filmwise condensation, 846–847
Filters in liquid separation membranes, 898
Finite difference methods for porous catalysts
  central difference equations, 618–619
  Neumann and Robin conditions, 621
  nonlinear kinetics, 620–621
  overview, 617–618
  zero-order reaction, 619–620
First-order reactions
  axial dispersion model for turbulent cases, 581
  continuous stirred tank reactors, 107
  diffusion with, 59, 76
  dissolved gas
    boundary conditions, 663–664
    dimensionless version, 664–665
    enhancement factor, 666–668
    flux values, 665–666
    overview, 662–663
  gas–liquid reactions in film theory, 72

First-order reactions (*continued*)
gas–liquid reactions in penetration theory, 706–707
low flux model in surface reactions, 206–207
mesoscopic models
dispersion model, 479, 481
plug flow idealization, 478
tubular flow reactors, 131, 133
porous catalyst linking with reactor models, 623–624
porous catalysts
diffusion-reaction model apparent rate constant, 597–599
effectiveness factor, 591–592
external resistance, 597
long cylinders, 593–594
overview, 589–591
shape normalization, 595–596
slab external transport effects, 592–593
slab solution for Dirichlet condition, 591
spheres, 594–595
steady state mass transfer with reaction, 71–72
surface reactions in shrinking core model, 638
volume reaction model, 646–649
Fisher-Tropsh synthesis in three-phase catalytic reactions, 608
Fixed bed adsorption
axial dispersion effects, 934–936
equilibrium model, 932–934
heterogeneous model, 936
Klinkenberg equation, 936–937
overview, 931–932
scale-up aspects, 937
Fixed bed type in three-phase catalytic reactions, 608
Fixed-site carrier membranes, 744–745
Flat plate flow
convective mass transport, 316–318

flow over inclined and curved surfaces, 389
laminar, 201
Flat plate with low flux mass transfer
concentration equation, 367–368
exact or Blasius analysis, 371–376
overview, 366–367
scaling results and analogies, 369–370
velocity equations, 368–369
Flow-averaged concentration in laminar flow in pipes, 342–343
Flow over inclined and curved surfaces
Falkner-Skan equation, 393–394
integral balance method, 390–391
overview, 388
pressure variation term, 388–390
rotating disks, 395–396
similarity variable, 392
stagnation point, 394–395
Flow rate in liquid separation membranes, 898–899
Flow ratio parameter
ideal flow patterns, 501
plug-backmixed model, 498
Flow regimes in condensation, 847
Flow weighted concentration in mesoscopic models, 39, 126
Flowing phases in tracer response in two-phase systems, 506–507
Fluid bed dryers, 827
Fluid bed reactors, 660
Fluid–fluid systems
backmixed–backmixed model, 462–464
equilibrium model, 464–465
mixing cell model, 465
overview, 462
Fluids in porous solids, diffusion with
liquid-filled pores, 250–251
overview, 248

porous catalysts, 251–254
single-pore gas diffusion, 248–250
Fluid–solid interface, adsorption isotherm at, 19
Flux and flux models
Blasius approach, 386
condensation, 852, 858–859, 861
convective mass transfer for laminar flow
entry regions, 349
internal and external flows, 310–311
convective mass transfer in turbulent flow
channels and pipes, 418–421
time-averaged equation, 409
van Driest model, 426
definition, 10–12
differential models, 31
diffusion
with convection, 65–66
properties, 163–165
steady state, 53, 187, 189–191
transient processes, 266
diffusion coefficients in gases, 226
diffusion-induced convection, 193–196
diffusion-reaction model
nth-order reaction, 603–604
electrodialyzer design, 955
equations
mass basis, 155
mole basis, 153–155
evaporation of liquids in ternary mixtures, 529–530
film model, 202, 383
flat plates, 366–367
gas permeator models, 882
Flux and flux models
gas–liquid reactions in film theory models, 665–666, 710–711, 716
heterogeneous reactions, 525–526
laminar flow, 341–344, 349
liquid separation membranes
pervaporation, 910–911
pore size, 899

semi-permeable membranes, 902–903, 906

mass transport in electrolytic systems

across charged membranes, 555

transport of charged species, 546

across uncharged membrane, 551–552

multicomponent transport, 518–519

porous catalysts in diffusion-reaction model, 590, 594

reacting systems, 520–522, 525

reactive membranes and facilitated transport

co- and counter-transport, 738–739

equilibrium model, 742

slab solutions, 273–274, 276, 292–293

stirred reactors, 437–438

surface reactions

first-order, 206–207

nonlinear, 209–210

product counter-diffusion, 210–212

thermal diffusion, 258

tracer studies, 439, 442, 445

transient diffusion processes

cylinders and spheres, 280

integral method, 292

semi-infinite slab analysis, 290

turbulent, 409

Flux of species in reactive membranes, 733–734

Flux vector

convection flux, 12

diffusion flux, 12–15

molar and mass flux, 10–12

solid spheres sublimation, 34–35

Fog formation, 855–857

Force balance model for gas bubbles, 323

Forced draft cooling towers, 816

Forward osmosis in liquid separation membranes, 907–908

Fractional extent of reaction in film, 666

Fractional order kinetics in stirred reactors, 437

Fractional order reactions in porous catalysts, 630

Frames of reference in mass transfer equations, 156–163

Free volume theory for polymeric membranes, 256–257

Freundlich isotherms in adsorption, 923

Frictional interpretation in diffusion coefficients in gases, 232–234

Frictional resistance in diffusion coefficients in liquids, 239–240

Froessling-Marshall correlation in solid spheres, 320

Froude number in packed bed absorbers, 329

FSOLVE

backmixed–backmixed model, 115

evaporation of binary liquid mixtures, 532

reactive membranes equilibrium model, 741

stirred reactors, 438

Fuel cells

half reactions, 778–779

overview, 798–800

Fugacity of pollutants, 460

Fuller correlation for diffusion coefficients in gases, 231–232

Fully turbulent flow in convective mass transfer, 405

FZERO function, 717

Galvanostatic mode in electrochemical batch reactors, 807

Gamma function in convective mass transfer, 349

Gas, diffusion coefficients in. *See* Diffusion coefficients

Gas absorption

column design, 141–142

from interface in convective mass transfer film flow, 357–358

Gas balance equation in fixed bed adsorption, 935–936

Gas bubbles in convective mass transport, 321–325

Gas concentration profiles in volume reaction model, 646, 650

Gas film diffusion in shrinking core model, 642

Gas film resistance in bimolecular reactions, 681–684

Gas holdup in gas bubbles, 325

Gas-liquid interface

convective mass transfer in turbulent flow, 427–430

Henry's law, 15–17

two-film model, 212–217

Gas-liquid mass transfer

convective mass transport film flow, 319–320

packed bed absorbers, 327–329

Gas-liquid reactions in film theory

absorption in slurries

instantaneous reaction case, 694–697

overview, 692–693

particle size effect, 693–694

bimolecular reactions, 714

contactor choice, 684

dimensionless representation, 673–675

instantaneous asymptote, 678

instantaneous case, 681–684

invariance property, 676–677

overview, 672–673

pseudo-first-order case, 677

second-order reactions, 678–680

bulk concentration and bulk reactions, 668–672

coupling with reactor models, 688–692

first-order reaction of dissolved gas

boundary conditions, 663–664

Gas-liquid reactions in film theory (*continued*)
dimensionless version, 664–665
enhancement factor, 666–668
flux values, 665–666
overview, 662–663
liquid–liquid reactions, 697–698
overview, 661–662
problems, 701–704
review questions, 701
simultaneous absorption of two gases, 684–688
summary, 698–700
Gas-liquid reactions in penetration theory
bimolecular reaction
dimensionless form, 712–713
ideal contactors, 717–721
illustrative results, 713–714
instantaneous reaction case, 714–717
overview, 712
concepts
average rate of mass transfer, 710–711
film theory and penetration theory, 711–712
first-order and pseudo-first-order reactions, 706–707
Laplace transform method, 707–710
overview, 706
overview, 705–706
problems, 723–726
review questions, 722–723
summary, 721–722
Gas phase balance in macroscopic models
fluid–fluid systems, 463–464
ideal flow patterns, 500–501
plug-backmixed model, 497
reactor–separator combination, 98
Gas-phase driving force in two-film model, 214–215
Gas-solid reaction models, 651–654
Gas transport in membranes
dissociative diffusion, 877–878
gas permeator models

backmixed-backmixed model, 884–886
countercurrent flow, 886–889
cross-flow pattern, 889–890
flux relations, 882
local concentration, 883–884
overview, 881–882
gas separation membranes
classification, 873–874
nonlinear effects, 878–879
overview, 872–873
permeability transport rate, 874–876
permeance transport rate, 876–877
selectivity, 877
gas translation model, 879–881
overview, 871–872
problems, 894–895
reactor coupled with membrane separators, 890–891
review questions, 893–894
summary, 892–893
Gaussian distribution
chromatography, 938
electrophoretic separation devices, 959
tracer response method, 487
Gel solids, drying, 829
Generation term
concentration, 30
plug flow model, 476
porous catalysts, linking with reactor models, 622
silicon oxidation, 208
stirred reactors, 33–34
tubular flow reactors, 41, 130
Glass transition temperature for polymeric membranes, 256
Glass walls, helium leakage rate across, 55–56
Glucose uptake in erythrocyte, 747
Gradients, temperature, in porous catalysts, 613–614
Graetz number and problem
convective mass transfer entry region analysis, 349–350
turbulent flow, 417
laminar flow in pipes, 339–342
PDEPE solutions, 358–359

Grain model for gas–solid reaction models, 653
Granular solids, drying, 828–829
Graphite and li-ion batteries, 800
Grashof number for convective mass transport in solid spheres, 321
Gray's theorem
and Leibnitz rule, 183–184
multiphase systems, 177–178
Green-Gauss theorem, 173
Grotthuss effect for transport of charged species, 545
Groundwater transport, 28
Gupta-Thodos correlation for packed bed reactors, 333

Hadamard- Rybczynski regime in gas bubbles, 322
Half reactions
copper electrowinning, 795
definitions, 777–779
electrochemical reaction, 784
plug flow idealization, 478
Hannig design for electrophoretic separation devices, 959–960
Hatta number in gas–liquid reactions
in film theory, 711–712
bimolecular reactions, 674, 679–680, 714
bulk concentration and bulk reactions, 669
first-order reaction of dissolved gas, 664–665
simultaneous absorption of two gases, 686
in penetration theory, 713
Hatta parameters
co- and counter-transport, 737
single solute diffusion, 730–731
Heat balance
in cooling towers, Merkel equation for, 821–823
porous catalysts temperature effects, 612
Heat transfer
condensation
binary gas mixtures, 860
vapor with non-condensible gas, 851–852

drying constant rate period, 831

flat plate with low flux mass transfer, 370

laminar flow reactors, 578–579

Heat transfer coefficient
condensation, 846–848
cooling towers, 822
drying falling rate period, 835
evaporation, 812

Heat transfer rate in cooling towers, 820–821

Heat transport equations, temperature effects in, 613–614

Heaviside step function for fixed bed adsorption, 933

Helium
diffusion volume, 232
leakage rate across glass walls, 55–56
tracer response in two-phase systems, 506

Hematocrit, 768

Hemodialyzer models, 763–766

Henry parameter for fluid–fluid systems, 463, 465

Henry's law
bimolecular reactions, 673–674
boundary conditions, 171–172
for CO_2, 47
fugacity of pollutants, 460
gas separation membranes, 878
gas–liquid interface, 15–17
oxygen-hemoglobin equilibrium, 751
reactor–separator combination, 98–99
two-film model, 216–217

Heterogeneous media in diffusion, 254–256

Heterogeneous membranes in electrodialysis, 949

Heterogeneous models in fixed bed adsorption, 932, 936

Heterogeneous reactions
multicomponent systems, 525–527
ternary diffusion, 527

Hexachlorobiphenyl, 473

Hiemenz flow, 394–395

High flux analysis for laminar boundary layers
Blasius approach, 386–388
film model, 383–384
integral balance method, 384–385
overview, 383

High flux model for surface reactions product counter-diffusion, 210–212

Hikita-Asai model for reactor model coupling, 689

Hill's equation
oxygen binding, 768
oxygen-hemoglobin equilibrium, 752–753

Hindered diffusion of fluids in porous solids, 250–251

Homogeneous membranes in electrodialysis, 949

Homogeneous reactions in integral balance approach, 381–382

Homogeneous solution in matrix representation, 456

HTU (height of a transfer units)
mass exchangers, 143–144
Merkel equation, 823

Humidification and drying
cooling towers
counterflow model, 817–823
cross-flow, 825–827
overview, 815–817
drying
constant and falling rates, 829–830
constant rate period, 830–833
dryer types, 827–828
falling rate period, 833–839
overview, 827
solid types, 828–829
overview, 811–812
problems, 841–843
review questions, 841
summary, 839–841
wet and dry bulb temperature, 812–815

Hydrocracking of porous catalysts, 608

Hydrodesulfurization of diesel, 608

Hydrodynamic hindrance factor of fluids in porous solids, 251

Hydrogen
diffusion volume, 231–232
Henry's law constant, 16
Lennard-Jones constant, 229
molar volume in liquid, 241
transport in palladium membranes, 873–874
transport rate in presence of convection, 66–67
trickle bed reactors, 21–22

Hydrogen fuel cells
half reactions, 779
overview, 798–800

Hydrogen ion concentration, 956

Hydrogen sulfide (H_2S) absorption
in amine solution, 716–717
in NaOH solution, 683–684
bearing gas, sulfur from, 702–703

Hydrogenation reactions in porous catalysts
three-phase catalytic reactions, 608
zero-order reaction, 599

Hygroscopic solids, 829

Hyperboloid design for cooling towers, 816

Hypergeometric function for laminar flow in pipes, 340

Hysteresis phenomena in adsorption, 924

I-curves for transient balance, 442

ICFUN function, 300

Ideal contactors for bimolecular reaction, 717–721

Ideal flow patterns
non-idealities, 499–503
overview, 495–496
plug-backmixed model, 496–499

Ideal gas law, partial pressure from, 10

Immobilized liquid membranes (ILMs), 744

In and out flow terms
batch reactors, 91
dispersion model, 478

In and out flow terms (*continued*)
macroscopic balance, 88–89
reactor–separator
combination, 97
solid dissolutions from walls,
38–40, 126
stirred reactors, 33
surface reactions, 208
tubular flow reactors, 41, 130
Inclined flat plates
flow over inclined and curved
surfaces, 389
similarity variable, 392
Inclined surfaces, flow over. *See*
Flow over inclined and
curved surfaces
Indirect dryers, 827
Inert gas flow rate in film theory
models, 689
Inhibitors in tissues, 771
Initial value problems (IVP)
matrix representation,
456–457
Instantaneous reaction
asymptote in single solute
diffusion, 732–734
Instantaneous reaction case in
film theory models
absorption in slurries, 694–697
bimolecular reactions,
677–678, 681–684, 714–717
Instantaneous velocity in
turbulent flow, 406–408
Integral balance approach for
laminar boundary layers
flow over inclined and curved
surfaces, 390–391
high flux analysis, 384–385
homogeneous reactions,
381–382
momentum balance, 376–379
no reaction case, 379–381
overview, 376
species mass balance, 379
Integral method in semi-infinite
slab analysis, 290–292
Intercalation in li-ion batteries,
800
Interconnected cells model for
backmixed assumption
tests, 111
Interface flux values in film
theory models, 665–666

Interface temperature
condensation
binary gas mixtures, 860–861
unmixed model, 858
vapor with non-condensible
gas, 852–853
cooling towers counterflow
model, 819
Interface transfer term in
macroscopic balance, 89
Interfacial across charged
membranes, 553–554
Interfacial area for gas bubbles,
325
Interfacial mass transfer, 6
Interfacial partial pressures in
film theory models, 664
Interfacial turbulence in
turbulent flow, 430
Intermedia transport effects in
environmental transport,
461–462
Internal age distribution in
transient balance, 442
Internal diffusion
three-phase catalytic reactions,
610
tracer response in two-phase
systems, 506
Internal laminar flow. *See*
Convective mass transfer
for internal laminar flow
Interstitial diffusion, 246–248
Interstitial velocity in tracer
response in two-phase
systems, 506
Intrinsic volume average in
multiphase systems, 177
Invariance property for
bimolecular reactions,
676–677
Invariant of system in single
solute diffusion, 732
Ions
across charged membranes,
554
transport of charged species,
545
Isoelectric focusing, 957
Isotherms in adsorption,
921–924

j-factor form in convective mass
transport, 318
Jacobi polynomial in orthogonal
collocation method, 615
Johnson-Mehl-Avrami-
Kolmogorov (JMAK)
equation for solid–solid
reactions, 655
Joule heating in rotating annular
beds, 961
Jump model for diffusion
coefficients in liquids, 237

Kedem-Katchalski model, 903,
905–907
Key dimensionless groups
convective mass transport,
313–315
laminar flow reactors, 568–572
Kidneys, hemodialyzers for,
763–766
Kinetics
diffusion coefficients in gases
based on, 225–232
dispersion model, 482
electrochemical reaction
Butler-Volmer equation,
788–791
overview, 786–788
plug flow idealization,
477–478
porous catalysts
diffusion-reaction model, 589
finite difference methods,
620–621
linking with reactor models,
623
Neumann and Robin
conditions, 621
reacting solids, 645–646
shrinking core model, 640
topochemical reactions, 659
volume reaction model,
645–646
King expression for convective
mass transfer in turbulent
flow, 428
Klinkenberg equation for fixed
bed adsorption, 936–937
Knudsen diffusion
fluids in porous solids,
248–250, 252–253
reacting systems, 525

Knudsen model for gas translation, 879–881
Kremser equation for multistage cascades, 467–468
Krogh cylinder problem for diffusion-reaction model, 600
Krogh model for transport of oxygen in tissues, 82, 757–760
Kummer functions for laminar flow, 340, 343

Lagrangian in transient balance, 441
Laguerre polynomials, 448
Laminar boundary layers
bubbles and drops, 396–399
flat plate with low flux mass transfer
concentration equation, 367–368
exact and Blasius analysis, 371–376
overview, 366–367
scaling results and analogies, 369–370
velocity equations, 368–369
flow over inclined and curved surfaces
Falkner-Skan equation, 393–394
integral balance method, 390–391
overview, 388
pressure variation term, 388–390
rotating disks, 395
similarity variable, 392
spheres, 395–396
stagnation point, 394–395
high flux analysis
Blasius approach, 386–388
film model, 383–384
integral balance method, 384–385
overview, 383
integral balance approach
homogeneous reactions, 381–382
momentum balance, 376–379
no reaction case, 379–381

overview, 376
species mass balance, 379
overview, 365–366
problems, 401–402
review questions, 401
summary, 399–400
Laminar flow
convective mass transport, 312–313, 315–316
flat plates
convective mass transport, 317
mass transfer coefficient, 201
internal. *See* Convective mass transfer for internal laminar flow
liquid separation membranes, 899
mesoscopic models
solid dissolutions from walls, 38
tubular flow reactors, 133
non-Newtonian fluids, 513
velocity profile in, 48
Laminar flow dispersion, Taylor model for, 488–491
Laminar flow reactors
examples
axial dispersion model for turbulent cases, 580–581
channel flow, 577
heat transfer effects, 578–579
non-Newtonian fluids, 577–578
turbulent flow reactor, 579–580
key dimensionless groups, 568–572
limiting cases
overview, 572
plug flow model, 574–575
pure convection model, 572–574
mesoscopic dispersion model, 575–577
model equations, 568–572
overview, 567–568
problems, 582–584
review questions, 582
summary, 581–582
Laminar jet apparatus in bimolecular reaction, 718

Laminar regime in condensation, 847–849
Laminar sublayer velocity profiles, 414, 416
Langmuir-Hinshelwood models for shrinking core model, 640
Langmuir isotherm
adsorption, 921–924
gas separation membranes, 879
nonlinear equilibrium models, 19
Langmuir kinetics
topochemical reactions, 659
volume reaction model, 646
Laplace domain in tracer response method, 484–485
Laplace transform
moments from, 446–449
N tanks in series, 447–448
in penetration theory, 707–710
Laplacian equation
electrolytic systems electric field, 549
steady state diffusion, 58, 189–191
Laplacian operator
batch slurry adsorbers, 930
diffusion and reaction, 75
porous catalysts
diffusion-reaction model, 593
orthogonal collocation method, 615
steady state mass transfer with reaction, 70–71
transient diffusion processes, 266–267
Laplacian terms
constant-density systems, 167
convective mass transfer, 348
laminar flow in pipes, 337
Large Thiele modulus in volume reaction model, 648
Law of mass action in oxygen-hemoglobin equilibrium, 751
Le Chatelier's principle, 923
Leibnitz rule
for differentiation, 183–184
transient diffusion processes, 292

Lennard-Jones constants for diffusion coefficients in gases, 229–230

Lethal corner in oxygen transport, 759

Level I model for environmental transport, 460–461

Level II model for environmental transport, 461

Level III model for environmental transport, 461–462

Level IV model for environmental transport, 462

Leveque solution for convective mass transfer, 349

Lewis number
  fog formation, 855
  wet and dry bulb temperature, 813–815

Liebnitz rule in integral balance approach, 377

Limited bulk reactions, 671

Limiting cases in laminar flow reactors
  overview, 572
  plug flow model, 574–575
  pure convection model, 572–574

Limiting current
  copper electrowinning, 796
  electrochemical reaction mass transfer effects, 791–793
  electrodialyzer design, 954–955
  mass balance for reacting systems, 550

Limiting reactants in porous catalysts multiple species, 606

Linear driving force model for batch slurry adsorbers, 927–928

Linear equilibrium case for NTU equation, 144

Linear isotherms in adsorption, 922–923

Linearized model for multicomponent diffusivity matrix, 539

LINSPACE function, 300

Linton-Sherwood correlation for mass transfer in pipe flow, 128

Liquid enthalpy equation for cross-flow cooling towers, 825

Liquid-filled pores, 250–251

Liquid-liquid extraction
  backmixed–backmixed model, 115
  equilibrium stage model, 115–116
  overview, 112–114
  unit operations, 22–23

Liquid-liquid interface, partition constant for, 18–19

Liquid-liquid reactions in film theory models, 697–698

Liquid mixtures, evaporation rate in, 532

Liquid phase balance
  condenser model, 862–864
  fluid–fluid systems, 463–464
  ideal flow patterns, 500–501
  plug-backmixed model, 498

Liquid-phase driving force in gas–liquid interface, 215

Liquid pools, oxygen profiles in, 59–62

Liquid separation membranes
  classification, 898–899
  forward osmosis, 907–908
  overview, 897–898
  pervaporation, 908–913
  problems, 915–917
  review questions, 914–915
  semi-permeable membranes
    concentration polarization effects, 903–905
    equipment-level model, 907
    Kedem-Katchalski model, 905–907
    osmotic pressure, 900–901
    overview, 900
    reverse osmosis, 901–903
    summary, 913–914

Liquid side coefficients for packed bed absorbers, 328

Liquids
  diffusion coefficients in. *See* Diffusion coefficients

evaporation
  in open containers, 197–198
  in ternary mixtures, 528–530

Lithium-ion (Li ion) batteries
  modeling, 800–802
  transfer and reaction steps, 27

Local concentration in gas permeator models, 883–884

Local permeate composition in pervaporation, 911–913

Local volume average in multiphase systems, 176–178

Log mean driving force (LMDF)
  gas absorption column design, 141–142
  mass exchangers, 140–141
  solid dissolutions from walls, 127–128

Long cylinders, diffusion in
  with first-order reactions in, 72
  porous catalysts, 593–594
  transient diffusion processes, 278–279

Long reactors in dispersion model, 484

Low flux mass transfer
  film model, 204
  transient diffusion processes, 266

Low flux model
  diffusion-induced convection, 193
  drift correction, 195–196
  steady state diffusion, 189–191
  surface reactions
    first-order reaction, 206–207
    nonlinear reactions, 209–210

Lungs, oxygen uptake in
  meso-model for capillaries, 756–757
  overview, 751
  oxygen-hemoglobin equilibrium, 751–753
  transport steps, 753–756

Macrofluid models
  continuous stirred tank reactors, 109–110
  macroscopic and compartmental models, 450–453

Macropores
  applications and adsorbent
    properties, 920
  catalysts, 588
Macroscopic models
  batch reactors, 90–96
  continuous stirred tank
    reactors, 106–110
  development, 29
  dissolved oxygen
    concentration in stirred
    tanks, 104–106
  fluid–fluid systems, 462–465
  liquid–liquid extraction,
    112–117
  macrofluid models, 450–453
  macroscopic balance, 87–88
    accumulation term, 90
    in and out terms from flow,
      88–89
    interface transfer term, 89
    rate term, 90
  multistage cascades, 465–468
  overview, 85–87, 435–436
  problems, 119–122, 471–473
  reactor–separator
    combination, 96–101
  review questions, 118–119, 471
  single-phase systems, 172–175
  stirred reactors, 436–438
  sublimation of spherical
    particles, 101–104
  summary, 117–118, 469–470
  tanks in series models, 449
  tracer data moments, 444–449
  tracer experiments, 110–112
  transient balance
    age distribution functions,
      441–442
    overview, 438
    pulse input, 439–441
    step input, 438–439
    tanks in series model,
      442–444
    variance-based models for
      partial micromixing,
      453–454
Macroscopic scale
  equilibrium stage model, 36
  mixer-settler models, 35–36
  overview, 32–33

stirred tank reactors, 33–34
sublimation of solid spheres,
  34–35
Maleic anhydride production,
  587
MAPLE
  diffusion-reaction model, 593
  Laplace transform method,
    708
  matrix representation, 456–457
  slab solutions, 271
Marangoni effect
  gas bubbles, 322
  gas–liquid interface, 429–430
Margules equation
  condensation unmixed model,
    858
  non-ideal liquids, 244–245
Mass balance
  adsorption and
    chromatography
    batch slurry adsorbers, 929
    model equations, 925
  batch reactors, 92–93
  batch slurry adsorbers, 925,
    928–929
  compartmental models, 455
  concentration form, 166–168
  condensation, 858
  condenser model, 862–864
  conservation principle, 29
  control volume, 37
  cooling towers counterflow
    model, 818–819, 821
  differential models, 31
  diffusion in porous solids, 248,
    250
  dispersion model, 478
  dissolved gas, 666
  electrodialyzer design, 951
  evaporation, 197
  film flow, 354
  fixed bed adsorption, 932
  fluid-fluid systems, 463, 465
  flux form, 154
  gas permeator models,
    883–886
  gas-solid reaction models, 652
  hemodialyzers, 764
  high flux analysis, 383
  instantaneous reaction case,
    714

integral balance method, 376,
  379, 391
laminar boundary layers, 386
liquid-liquid extraction,
  114–115
macroscopic models
  stirred reactors, 437
  transient balance, 439–441
mass exchangers, 135,
  139–141, 144
multicomponent systems,
  235–236, 519
oxygen uptake in lungs, 756
plug flow, 477, 496, 498, 500
porous catalysts
  linking with reactor models,
    622
  temperature effects, 611–612
reacting systems electric field,
  550–551
reactor–separator
  combination, 97–98
semibatch reactors, 688–689
shrinking core model, 638–639,
  642
single solute diffusion, 730,
  732
slab solutions, 278
surface reactions, 208
tanks in series model, 449
tracer studies, 507
velocity profiles, 415
volume reaction model, 646
Mass basis
  concentration form, 166–167
  flux form equations, 153–155
  overall continuity, 168–169
Mass diffusion flux in equations
  of mass transfer, 158
Mass exchangers
  NTU and HTU representation,
    143–144
  overview, 134
  single stream, 134–136
  two streams, 136–141
Mass flow ratio in cooling
  towers design
  considerations, 817
Mass flux
  Blasius approach, 386
  condensation, 852, 858, 861
  convective mass transfer, 349,
    426

Mass flux (*continued*)
definition, 10–12
diffusion coefficients in gases,
226
film model, 383
flat plates, 366–367
laminar flow, 341–344, 349
slab solutions, 273–274, 276,
292–293
stirred reactors, 437–438
thermal diffusion, 258
tracer studies, 439, 442, 445
transient diffusion processes,
292
turbulent, 409
Mass fractions
averaged velocity, 156–159,
161
cooling towers counterflow
model, 818–819
definition, 10
diffusion coefficients in gases,
226
mole fractions conversions, 46
weighted velocity, 161
Mass transfer coefficients
convective mass transfer in
laminar flow
channel flows with mass
transfer, 353
film flow, 356, 358
convective mass transfer in
turbulent flow
channels and pipes, 421
gas–liquid interface, 428–429
van Driest model, 426
convective mass transport
film flow, 319
gas bubbles, 323
internal and external flows,
310–313
cooling towers
counterflow model, 819
design considerations, 817
copper electrowinning, 796
electrodialyzer design, 955
evaporation, 813
film model, 202
fixed bed adsorption, 935
flat plates with laminar flow,
201
gas–liquid interface two-film
model, 212–217

gas–liquid reactions in film
theory models
average rate of mass
transfer, 711
bimolecular reaction, 719,
721
first-order reaction of
dissolved gas, 664–665,
667
hemodialyzers, 765
laminar boundary layers
bubbles and drops, 397
flat plate with low flux mass
transfer, 370
laminar flow in pipes, 342–343
macroscopic models
dissolved oxygen
concentration in stirred
tanks, 105
fluid–fluid systems, 464
interface transfer term, 89
liquid–liquid extraction, 117
reactor–separator
combination, 98
sublimation of spherical
particles, 102–104
mass transfer from spheres to
stagnant gas, 192
mesoscopic models
ideal flow patterns, 501
mass exchangers, 138
plug-backmixed model, 497
solid dissolutions from
walls, 40–41, 126–127
oxygen uptake in lungs,
754–755
porous catalysts, 590
shrinking core model
no solid product, 640
solid product, 643
solid spheres sublimation,
34–35
Mass transfer correlations in
pipe flow, 128
Mass transfer equations
boundary conditions,
171–172
common simplifications,
170–171
concentration form, 166–171
diffusion flux, 163–165
flux form, 153–155
frame of reference, 156–163

laminar boundary layers
flat plate with low flux mass
transfer, 374
flow over inclined and
curved surfaces, 393
mass fraction averaged
velocity, 156–158
multiphase systems local
volume averaging,
175–178
overview, 151–152
problems, 181–184
pseudo-binary diffusivity,
165–166
review questions, 181
single-phase systems, 172–175
summary, 179–181
Mass transfer overview
application examples
biomedical, 25
bioseparations, 24
electrochemical processes,
26–27
environmental, 28
metallurgy and metal
winning, 25–26
overview, 20
product development and
engineering, 26
reacting systems, 20–21
semiconductor and solar
devices, 24–25
unit operations, 21–24
causes, 7
concentration
mass units, 9–10
mole units, 8–9
partial pressure units, 10
concentration jump at
interface, 15–20
conservation principle, 29–30
continuum assumption, 7
description, 3–6
flux vector
convection flux, 12
diffusion flux, 12–15
molar and mass flux, 10–12
interfacial, 6
model development, 28–29
problems, 46–49
review questions, 45–46
summary, 44–45

Mass transfer rate
  condensation vapor with
    non-condensible gas, 851
  electrochemical reactions, 792
  evaporation, 812
  macroscopic models
    liquid–liquid extraction,
      114–115
    reactor–separator
      combination, 99
  mass exchangers, 139
  spherical cap bubbles, 398
Mass transfer resistance in
    shrinking core model,
    642–643
Mass transfer–limiting regime
    for surface reactions, 207
Mass transport in electrolytic
    systems
  charge neutrality, 547–548
  charged species
    mobility and diffusivity,
      544–545
    Nernst-Planck equation,
      545–547
    across charged membranes,
      553–556
  electric field, 548–551
  overview, 543–544
  problems, 562–564
  review questions, 561–562
  summary, 560–561
  transfer rate in diffusion film
    near electrodes, 556–559
  across uncharged membrane,
    551–553
Mass units in concentration, 9–10
Mass velocities for cooling
    towers
  cross-flow, 825
  design considerations, 817
Material balance
  batch slurry adsorbers,
    928–929
  condenser model, 861
  dispersion model, 480, 483
  electrodialysis, 948
  film theory models, 691
  fluid–fluid systems, 465
  gas permeator models, 883
  liquid separation membranes,
    907
  packed column absorbers, 691

MATHEMATICA
  Laplace transform method,
    485, 708
  slab solutions, 271
MATLAB program
  backmixed–backmixed model,
    115
  batch reactors, 93–96, 100
  bimolecular reactions, 679
  compartmental models matrix
    representation, 456–457
  condensation
    condenser model, 864
    vapor with non-condensible
      gas, 854–855
  convective mass transfer
    entry region analysis, 349
    film flow, 355
  diffusion coefficients in gases,
    230–231
  evaporation of binary liquid
    mixtures, 532
  H_2S absorption in amine
    solution, 717
  heterogeneous reactions, 527
  laminar boundary layers, 372,
    374
  laminar flow in pipes, 340
  laminar flow reactors, 579
  macrofluid models, 451–452
  plug flow closure in tubular
    flow reactors, 131
  porous catalysts temperature
    effects, 614
  reacting solids, 648
  reacting systems, 520–522
  slab solutions, 272
  stirred reactors, 437
  temporal evolution
    concentration profiles in
      slabs, 63
  tracer response method, 488
  transient diffusion processes
    eigenvalue computations
      with, 297–298
    semi-infinite slab analysis,
      290
    variable diffusivity, 296
Matrix form
  compartmental models,
    456–457
  evaporation of binary liquid
    mixtures, 531–532

Maxwell-Boltzmann
    distribution, 261–262
Maxwell's equation for
    electrolytic system charge
    neutrality, 547
McAdams correlation, 849
Mean area for diffusion across
    cylindrical shells, 70
Mean free path for diffusion
    coefficients in gases, 227
Mean molecular speed for
    diffusion coefficients in
    gases, 226–227
Mean residence time
  laminar flow reactors, 570,
    572–573
  plug flow, 477
  tanks in series, 448
  tracer studies, 439–441, 484,
    487
Measured vs. true kinetics in
    low flux model surface
    reactions, 207
Mechanically agitated tanks,
    convective mass transport
    in, 325–327
Megascopic model for
    environmental transport,
    459
Membranes
  aeration systems, 148
  electrodialysis and
    electrophoresis, 948–949
  gas transport in. *See* Gas
    transport in membranes
  liquid separation. *See* Liquid
    separation membranes
  mass transport in electrolytic
    systems
    charged, 553–556
    uncharged, 551–553
  for oxygen separation, 79
Mercury porosimetry catalysts,
    588
Merkel equation for cooling
    towers, 821–823
Mesh size in cross-flow cooling
    towers, 826
Meso-model
  capillaries, 756–757, 760
  copper electrowinning, 798

Mesoscopic models
  development, 29
  dispersion coefficient values,
    484–488, 493–495
  dispersion in laminar flow
    reactors, 575–577
  dispersion model
    boundary conditions, 480–481
    CHEBFUN, 482–483
    first-order reaction, 481
    negligible dispersion,
      483–484
    nonlinear reactions, 482
    overview, 478–480
  hemodialyzers, 763–766
  ideal flow patterns
    non-idealities, 499–503
    overview, 495–496
    plug-backmixed model,
      496–499
  mass exchangers
    NTU and HTU
      representation, 143–144
    overview, 134
    single stream, 134–136
    two streams, 136–141
  overview, 37, 123–124, 475–476
  plug flow idealization,
    476–478
  problems, 146–149, 510–515
  review questions, 146, 510
  segregated flow model,
    491–493
  solid dissolutions from walls,
    38–41, 124–128
  summary, 145–146, 507–509
  Taylor model, 488–491
  tracer response, 484–488,
    503–507
  tubular flow reactors, 41–43,
    129–134
Metabolism modeling, 761–762
Metal oxide semiconductor
  (MOS) devices, 25
Metal recovery
  co-transport in, 737
  emulsion liquid membranes,
    743–744
  from waste streams, 806
Metal winning, 25–26
Metallurgy
  application examples, 25–26
  fluid bed reactors, 660

Methane
  adsorption on activated
    carbon, 941
  Lennard-Jones constant, 229
  synthesis gas from, 890
Methane-ethane pairs,
  diffusivity of, 230, 232
Methanol (ME)
  biodiesel production, 704
  and water mixture in rate of
    condensation, 859–860
Methanol poisoning, ethanol for,
  736
Methanol synthesis, porous
  catalysts for
  application, 587
  three-phase catalytic reactions,
    608
Michaelis-Menten kinetic model
  for pharmacokinetics,
    761–762
Microfiltration for liquid
  separation membranes,
    898–899
Micromixed reactors with
  complete backmixing, 109
Micromixing, partial, 453–454
Microphase
  absorption in slurries, 693–694
  description, 453
Micropores
  applications and adsorbent
    properties, 920
  catalysts, 588
Minimum liquid flow rate in
  mass exchangers, 141
Mixed condenser model, 863
Mixer-settler models, 35–36
Mixing assumption in matrix
  representation, 456
Mixing cell model for fluid–fluid
  systems, 465
Mixing concentration. *See*
  Cup-mixed concentration
Mixing model for stirred tank
  reactor, 33–34
Mixture average velocity, 157
Mixture density, 9–10
Mixture velocity, weighting
  factors for, 183
Mobility
  electrical, 550
  electrodialysis, 946

electrophoretic separation
  devices, 957–959
  frictional interpretation, 234
  transport of charged species,
    544–546
Model development
  basic methodology, 28–29
  compartmental models, 43–44
  conservation principle, 29–30
  differential models, 30–32
  macroscopic scale, 32–36
  mesoscopic, 37–43
Model equations
  adsorption in batch slurry
    adsorbers, 925
  gas–liquid reactions in film
    theory, 685
  laminar flow reactors, 568–572
  porous catalysts multiple
    species, 606
  reactive membranes, 729–730
Moisture. *See* Humidification
  and drying
Molar flux, definition, 10–12
Molar rate
  dissolved oxygen
    concentration in stirred
    tanks, 105
  gas in film theory models, 692
  gas permeator model
    countercurrent flow, 887
  macroscopic balance, 88
  mass exchangers, 136
  solid dissolutions from walls,
    125–126
Molar units in macroscopic
  balance, 88
Molar volume
  diffusion coefficients in
    liquids, 241–242
  reacting solids, 651
Mole balance
  condenser model, 862–863
  solid dissolutions from walls,
    127
Mole basis
  concentration form, 168–169
  flux form equations, 153–155
Mole fraction
  in averaged velocity for
    equations of mass transfer,
    158–159

condensation
  binary gas mixtures, 857–858
  condenser model, 864
  ternary systems, 865
  unmixed model, 858
  vapor with non-condensible gas, 850
definition, 9
in fog formation, 856–857
gas permeator models, 883–885
gas–liquid interface two-film model, 215–216
gradients
  equations of mass transfer, 159–160
  multicomponent diffusivity matrix, 537
liquid separation membranes
  osmotic pressure, 901
  pervaporation, 912
  semi-permeable membranes, 903
mass fractions conversions, 46
in UMD, 196–197
vapor in evaporation, 812
Mole ratio
  liquid–liquid extraction, 113–116
  mass exchangers, 138
Mole units in concentration, 8–9
Molecular sieves, 921
Molecular speed in Maxwell-Boltzmann distribution, 261–262
Molecular weight
  definition, 9–10
  diffusion coefficients in gases, 228–229
  liquid separation membranes, 898
  mass transfer equations, 153
  spherical particles, 101
  water, 8
Molten salts, immobilized liquid membranes for, 744
Moments
  N tanks in series, 447–448
  tracer data analysis, 444–449
  tracer response, 485–487, 506

Momentum balance in laminar boundary layers
  high flux analysis, 385
  integral balance approach, 376–379
Momentum boundary in layer film model, 200–201
Momentum diffusivity in convective mass transfer in turbulent flow
  constant wall flux, 418
  van Driest model, 426
Momentum equation for flat plate with low flux mass transfer, 371–372
Momentum integral for high flux analysis, 384–385
Momentum transfer
  convective mass transport channels and pipes, 422–423
  film flow, 318–319
  flat plate with low flux mass transfer, 367, 369–370
  laminar boundary layers high flux analysis, 387
Morton number for spherical cap bubbles, 397
Moving reference frame in Taylor model, 490
Mujumdar correlation for drying, 831–832
Multicomponent systems
  constitutive model for, 518–520
  diffusion coefficients in gases, 235–236
  heterogeneous reactions, 525–527
  multicomponent diffusivity matrix, 535–539
  non-reacting systems, 528–534
  overview, 519
  problems, 541–542
  reacting system, 520–525
  review questions, 540–541
  summary, 539–540
Multiphase systems, local volume averaging in, 175–178
Multiple reactions in batch reactors, 92
Multiple species in porous catalysts, 605–607

Multistage cascades
  equilibrium model, 466–468
  overview, 465–466
Murphree vapor efficiency, 514
Myoglobin, oxygen binding to, 753

N tanks in series, moments for, 447–448
Nanofiltration in liquid separation membranes, 898–899
Natural draft cooling towers, 816
Navier-Stokes equation, 409–410
Near wall region in turbulent flow
  constant wall flux, 419–420
  velocity profiles, 414
Necrosis, 758
Negligible dispersion, 483–484
Nernst equation
  electrochemical reactions, 788
  thermodynamic considerations, 781–786
Nernst-Planck equation
  electrodialyzer design, 951
  mass transport in electrolytic systems
    across charged membranes, 554–555
    electric field, 548
    electrolyte transport across uncharged membrane, 551
    transfer rate in diffusion film near electrodes, 557–558
    transport of charged species, 545–547
Net efflux term in differential models, 31
Net mole efflux in flux form equations, 153–154
Neumann boundary conditions
  description, 171
  laminar flow reactors, 571
  oxygen profiles in pools of liquid, 60
  porous catalysts, 621
  reacting systems, 523
Neumann problem in drying falling rate period, 836–838
Newton-Raphson method for orthogonal collocation, 616

Newtonian fluid
  channel flow reactors, 577
  turbulent flow reactors, 580
Newtonian laminar flow in
    Taylor model, 489
Newton's law of viscosity
  laminar boundary layers wall
    stress, 373
  turbulent flow closure models,
    411
Nickel recovery, emulsion liquid
    membranes for, 743–744
Nitric oxide (NO)
  production, 769
  selective catalytic reduction of,
    209–210
Nitrogen (N)
  diffusion volume, 231
  Henry's law constant, 16
No reaction case for laminar
    boundary layers, 379–381
Non-condensible gas, vapor
    with, 850–855
Non-homogeneous problems in
    transient diffusion
    processes, 283–285
Non-hygroscopic solids, 829
Non-ideal liquids, diffusion in,
    243–245
Non-idealities in flow patterns,
    499–503
Non-isothermal batch reactor
    modeling, 120
Non-Newtonian fluids
  flow reactors, 577–578
  laminar flow, 513
  pipe flow, 493
Non-reacting systems,
    evaporation in
  binary liquid mixtures,
    531–533
  ternary liquid mixtures,
    528–530
Nonadiabatic dryers, 827
Nonlinear effects in gas
    separation membranes,
    878–879
Nonlinear equilibrium models,
    19–20
Nonlinear kinetics in porous
    catalyst finite difference
    methods, 620–621

Nonlinear reactions
  dispersion model, 482
  low flux model surface
    reactions, 209–210
NRTL three-constant model for
    non-ideal liquids, 244
NTU (number of transfer units)
  cooling towers, 823–824
  mass exchangers, 143–144
Nusselt model for condensation,
    847–849
Nutrient transport, 28

ODE45 solver
  batch reactors, 93–96
  evaporation of binary liquid
    mixtures, 531
  flat plate with low flux mass
    transfer, 372
  plug flow closure in tubular
    flow reactors, 131
  porous catalysts, linking with
    reactor models, 622–624
Ohmic potential, 797
Ohm's law
  electrodialyzer design,
    951–953
  electrolytic systems electric
    field, 549
Onda correlations for packed
    bed absorbers, 328–329
One-compartment model for
    pharmacokinetics, 761
1-D (differential) balances. *See*
    Differential (1-D) balances
1-D problems, 266–267, 300–304
One-term approximation
  concentration in laminar flow
    in pipes, 341
  transient diffusion processes
    in cylinders and spheres,
    280–282
Onsager reciprocal relations, 536
Open containers, liquid
    evaporating in, 197–198
Open-open boundary conditions
    in tracer response method,
    486–487
Operating current density in
    copper electrowinning,
    795–796
Organs, drug accumulation in,
    762–763

Orthogonal collocation method
  batch slurry adsorbers, 929
  porous catalysts, 615–617
Orthogonality property of F
    function for laminar flow
    in pipes, 340
Oseen model for solid spheres,
    321
Osmosis in liquid separation
    membranes, 907–908
Osmotic diffusion, 236
Osmotic pressure in
    semi-permeable
    membranes, 900–901
Osmotic transport in reactive
    membrane equilibrium
    model, 742
Overall continuity in mass basis,
    168
Overall mass balance
  batch reactors, 93
  batch slurry adsorbers, 929
  concentration form, 168
  condensation of binary gas
    mixture, 858
  fluid-fluid systems, 465
  liquid–liquid extraction, 115
  mass exchangers, 140–141
  trace methods, 439, 507
Overall rate constant in
    shrinking core model
  no solid product, 637–638
  solid product, 642
Overall reaction, definitions,
    777–779
Overpotential form of
    Butler-Volmer equation,
    789–791
Oxidation
  CO, 541
  electrochemical reactions, 782,
    784–788
  li-ion batteries, 801
  partial, 891
  porous catalysts, 608
  reduction, 776–777
  silicon, 207–209
Oxygen
  binding, in Hill's equation, 768
  concentration in water, 17
  counter-transport, 736
  diffusion volume, 231–232

dissolved concentration in
stirred tanks, 104–106
Henry's law constant, 16
mass transfer in human body,
750
molar volume in liquid, 241
in pools of liquid, 59–62
separation, 79, 256
silicon oxidation, 207–209
transient diffusion of, 275
transport in tissues, 82,
757–760
transport with fixed-site
carrier membranes,
744–745
uptake in lungs
meso-model for capillaries,
756–757
overview, 751
oxygen-hemoglobin
equilibrium, 751–753
transport steps, 753–756
Oxygen-hemoglobin
equilibrium, 751–753
Ozone, Henry's law constant for,
16

p-substitution in
diffusion-reaction model
nth-order reaction, 603
Packed beds
dispersion coefficient, 494–495
gas–liquid mass transfer in,
327–329
Packed column absorbers
coupling with reactor models,
691–692
mass exchangers, 136–137
Palladium gas separation
membranes, 873–874
Parallel plate electrolyzer
modeling, 807
Partial micromixing, 453–454
Partial oxidation, 891
Partial pressure difference in gas
separation membranes,
875, 877
Partial pressure units in
concentration, 10
Partial solute rejection for
semi-permeable
membranes, 902–903

Particle-level model for batch
slurry adsorbers, 925–926
Particle-Pellet model for
gas–solid reaction, 653
Particle size effect for absorption
in slurries, 692–693
Partition coefficient
gas separation membranes,
875
liquid–liquid extraction, 114
liquid–liquid interface, 18–19
mass exchangers, 138
Patient-dialyzer system in
hemodialyzers, 765–766
PDEFUN function, 300
PDEPE solver
bimolecular reaction, 713–714
channel flow reactors, 577
convective mass transfer,
358–359
drying falling rate period,
837–838
laminar flow reactors
boundary conditions, 571
heat transfer effects, 579
non-Newtonian fluid flow
reactors, 578
temporal evolution
concentration profiles in
slabs, 63
tracer response method, 488
transient diffusion processes,
299–304
volume reaction model, 648
PDEVAL function, 301
Peclet number
defined, 314
film model, 383–384
laminar flow in pipes,
338–339
laminar flow reactors,
569–570, 576
mesoscopic models
dispersion coefficient,
494–495
dispersion model, 479
ideal flow patterns, 501–502
pipe flow, 493
tracer response method, 484,
486
non-Newtonian fluid flow
reactors, 493, 578
solid spheres, 320

sublimation in pipes, 129
transport rate in presence of
convection, 66–67
PEG (polyethylene glycol) in
water splitting, 950
Pellets in gas–solid reaction
models, 652–653
Penetration depth
doping silicon with
phosphorus, 294
transient diffusion process
integral method, 292
Penetration theory approach
film flow, 319–320
gas–liquid reactions. *See*
Gas-liquid reactions in
penetration theory
transient diffusion processes,
294–295
Performance equations
mass exchangers, 135–136
plug flow idealization, 477
stirred reactors, 437
Permeability
gas transport in membranes
gas separation membranes,
874–876
gas translation model, 879,
881
heterogeneous media, 254
oxygen uptake in lungs,
754–755
polymeric membranes, 256
semi-permeable membranes,
905
Permeance transport rate for gas
separation membranes,
876–877
Permeate flux model for
pervaporation, 910–911
Permittivity in electric field, 547
Pervaporation
application examples,
909–910
local permeate composition,
911–913
overview, 908–909
permeate flux model, 910–911
pH
electrophoresis principle,
956–957
oxygen-hemoglobin
equilibrium, 753

Pharmacokinetics
  bolus injection, 471
  compartmental models,
    457–459, 760–763
Philpot design, 957–959
Phosphorus
  doping silicon with, 293
  in silicon, interstitial diffusion,
    248
Physiologically based
    pharmocokinetic (PBPK)
    models, 762–763
Pipes and pipe flow
  convective mass transfer in
    turbulent flow
    constant wall flux, 418–421
    momentum transfer
      analogy, 422–423
    overview, 417–418
    Stanton number for
      boundary layers, 421–422
    Stanton number for pipe
      flows, 423–425
  convective mass transport,
    312–316
  internal laminar flow
    concentration, wall mass
      flux, and Sherwood
      number, 341–344
    constant wall concentration,
      339–341
    dimensionless form, 337–339
    overview, 336–337
  mass transfer correlations in
    solid dissolutions from
    walls, 128
  non-Newtonian liquid, 493
  segregated flow model,
    491–493
  sublimation in, 129
  tubular flow reactors, 133
Plasma in three-compartmental
  model, 772
Plate and frame design in
  electrodialysis, 946
Plug-backmixed model,
  496–499
Plug flow
  fixed bed adsorption, 933
  laminar flow reactors
    axial dispersion model for
      turbulent cases, 581
    limiting cases, 574–575

mesoscopic models
  closure in tubular flow
    reactors, 131
  idealization, 476–478
  solid dissolutions from
    walls, 38
  tracer response method, 487
  tubular flow reactors, 42
  tracer studies, 440
Plug flow reactors for porous
  catalysts, 622
Poisson's equation for
  electrolytic system charge
  neutrality, 547
Polarization effects for
  semi-permeable
  membranes, 903–905
Pollutants in environmental
  transport, 460
Polymeric membranes
  diffusion, 256–257
  gas separation, 873–874
  hydrogen fuel cells, 798
Polystyrene, 949
Pools of liquid, oxygen profiles
  in, 59–62
Pores
  batch slurry adsorbers,
    925–926
  fluids in porous solids
    liquid-filled, 250–251
    size, 248–250
  gas separation membranes, 873
  liquid separation membranes,
    898–899
Porosity
  gas–solid reaction models,
    651–653
  volume reaction model, 644
Porous catalysts
  diffusion-reaction model
    first-order reaction, 589–598
    nth-order reaction, 603–605
    overview, 588–589
    zero-order reaction, 599–603
  finite difference methods
    central difference equations,
      618–619
    Neumann and Robin
      conditions, 621
    nonlinear kinetics, 620–621
    overview, 617–618
    zero-order reaction, 619–620

fluids in porous solids,
    251–254
  linking with reactor models,
    622–625
  multiple species, 605–607
  orthogonal collocation
    method, 615–617
  overview, 585–586
  problems, 628–633
  properties and applications,
    586–588
  review questions, 627
  summary, 625–627
  temperature effects
    dimensionless boundary
      conditions, 612–613
    dimensionless
      representation, 611–612
    equations for heat and mass
      transport, 610–611
    temperature gradients,
      613–614
  three-phase catalytic reactions,
    607–610
Potential field in electrolytic
  systems, 549, 558
Potentiostatic mode for
  electrochemical batch
  reactors, 806–807
Power consumption in
  mechanically agitated
  tanks, 326
Power input for gas bubbles, 324
Power-law kinetics
  dispersion model, 482
  plug flow idealization,
    477–478
  porous catalysts, linking with
    reactor models, 623
  volume reaction model,
    645–646
Power law model
  non-Newtonian fluids flow
    reactors, 577–578
  variable diffusivity, 296
Prandtl analogy for laminar
  boundary layers
  flat plate with low flux mass
    transfer, 370, 375
  flow over inclined and curved
    surfaces, 391, 395
Prandtl model for eddy
  viscosity, 412–413, 425

Prandtl number
  channels and pipes, 422–423
  evaporation, 813
  in fog formation, 855
Prandtl universal velocity
  profile, 580
Pre-exponential factor in gas
  translation model,
  880–881
Pressure
  flow over inclined and curved
    surfaces, 388–390
  gas permeator models, 882,
    888–889
  liquid separation membranes,
    901
  local permeate composition,
    911–912
  partial pressure units, 10
  semi-permeable membranes,
    907
  in small pores, 250
Pressure diffusion, 258
Primary variables, definitions,
  781
Probability factor in gas
  translation model, 880
Product counter-diffusion for
  surface reactions, 210–212
Product development and
  engineering application
  examples, 26
Product removal, batch reactor
  with, 99–100
Product solution method for
  transient diffusion
  processes, 285–287
Production rate
  batch reactors, 91–92
  reacting systems, 524–525
  volume reaction model,
    645–646
Proteins
  electrophoresis principle, 956
  electrophoretic separation
    devices, 958–960
Proton exchange membrane
  (PEM) in fuel cells, 779
Pseudo-binary diffusion
  coefficient in evaporation
  of liquids, 529
Pseudo-binary diffusivity,
  165–166

Pseudo-first-order reactions
  bimolecular reactions, 677
  penetration theory approach,
    706–707
  reactive membrane single
    solute diffusion, 735–736
Pseudo-homogeneous models
  for fixed bed adsorption,
  932
Pseudo-steady state hypothesis
  for volume reaction
  model, 647
Pulse injection in tracer response
  method, 487
Pulse input
  fixed bed adsorption, 933
  tracer studies, 439–441
Pulse response
  integral method, 292–293
  tracer studies, 441
Pure convection model for
  laminar flow reactors,
  572–575, 578
Pure vapor in condensation,
  846–850

Quadratic profile assumption
  for batch slurry adsorbers,
  927
Radial averaging in laminar
  flow reactors, 575
Radial concentration gradient
  for tubular flow reactors,
  132
Radial diffusion
  in cylindrical geometry, 191
  laminar flow reactors, 570
  steady state, 68–70
  Taylor model, 488–489
  tubular flow reactors,
    132–134
Radial direction in mesoscopic
  models, 37
Radial position in laminar flow
  reactors, 569
Radius dependency in shrinking
  core model, 640
Random flow field in fully
  turbulent flow, 405
RANS (Reynolds-averaged
  Navier-Stokes) equation,
  410
Ranz-Marshall equation, 640

Raoult's law and model
  condensation unmixed model,
    858
  vapor–liquid interface, 17–18
Rate constants
  drying falling rate period, 835
  Laplace transform method,
    708–709
  porous catalysts, 597–599
  shrinking core model
    no solid product, 637–638
    solid product, 642
Rate equations for bimolecular
  reactions, 682
Rate model for
  pharmacokinetics, 761–762
Rate term
  bimolecular reaction, 712
  macroscopic balance, 90
Reacting solids
  gas–solid reaction models,
    651–654
  overview, 635–636
  problems, 658–660
  review questions, 658
  shrinking core model
    no solid product, 637–640
    overview, 636
    solid product, 641–644
  solid–solid reactions, 654–656
  summary, 656–658
  volume reaction model
    concentration profile for gas
      and solid, 646
    first-order reaction, 646–649
    kinetic model, 645–646
    overview, 644–645
    zero-order reaction, 649–651
Reacting systems
  examples, 20–21
  flat plate with low flux mass
    transfer, 375–376
  multicomponent systems,
    520–525
Reaction contribution in
  time-averaged equation,
  410–411
Reaction kinetics in matrix
  representation, 456
Reaction-limiting regime in
  surface reactions, 207

Reaction planes
  absorption in slurries, 694–697
  bimolecular reactions, 678,
    681–682
  shrinking core model solid
    product, 642
Reaction rate in porous catalysts
  diffusion-reaction model, 589
  linking with reactor models,
    622
Reaction time in laminar flow
    reactors, 570
Reactive membranes and
    facilitated transport
  application examples
    emulsion liquid membranes,
      743–744
    fixed-site carrier
      membranes, 744–745
    immobilized liquid
      membranes, 744
    overview, 742–743
  co- and counter-transport
    illustrative results, 741–742
    overview, 736–739
  equilibrium model, 739–742
  overview, 727–728
  problems, 747–748
  review questions, 746
  single solute diffusion
    dimensionless
      representation, 730–732
    model equations, 729–730
    overview, 729
  summary, 745–746
Reactor coupled with membrane
    separators, 890–891
Reactor models
  gas–liquid reactions in film
    theory models, 688–692
  porous catalysts, 622–625
Reactor performance for tanks in
    series models, 449
Reactor types in three-phase
    catalytic reactions, 608–609
Redlich-Kister equation for
    non-ideal liquids, 245
Reduction in oxidation, 777
Reference temperature for
    cooling towers enthalpy
    balance equations, 820
Regions for constant wall flux,
    419

Relative humidity, 815
Relative velocity
  defined, 157
  diffusion coefficients in
    liquids, 238–239
  gas bubbles, 323–325
  multicomponent diffusion, 235
  multiphase systems, 176–178
Resistances
  batch slurry adsorbers, 928
  diffusion coefficients in
    liquids, 239–240
  electrochemical reactions, 792
  electrodialyzer design,
    952–953
  gas–liquid interface in
    two-film model, 216–217
  hemodialyzers, 765
  heterogeneous media, 254
  mass exchangers, 136, 138
  oxygen uptake in lungs,
    754–756
  porous catalysts
    diffusion-reaction model,
      597
  shrinking core model, 642–643
  surface reactions, 206–207
Response curves
  tracer response method,
    485–487
  transient balance, 440, 443–444
Retention time
  chromatography, 938
  fixed bed adsorption, 933–935
  Hannig design, 959–960
Reverse diffusion, 236
Reverse osmosis
  salt water purification, 898
  semi-permeable membranes,
    901–902
Reverse reaction rate in
    Butler-Volmer equation,
    788
Reynolds analogy
  channels and pipes, 422–423
  flat plate with low flux mass
    transfer, 369
Reynolds number
  channel flows with mass
    transfer, 351
  condensation, 847–849
  convective mass transfer in
    turbulent flow

channels and pipes, 423–424
  constant wall flux, 419
  gas–liquid interface, 427–428
  transition criteria, 404–405
  velocity profiles, 416–417
convective mass transport
  dimensionless
    representation, 313–314
  film flow, 319
  gas bubbles, 322
  gas–liquid mass transfer in
    packed bed absorbers, 328
  mechanically agitated tanks,
    326
  packed bed reactors, 333
  solid spheres, 321
dispersion coefficient, 495
drying constant rate period,
    832
fixed bed adsorption, 937
laminar flow in pipes, 338
mass transfer coefficient for
    sublimation of spherical
    particles, 102–103
mass transfer correlations in
    pipe flow, 128
spherical cap bubbles, 397
sublimation in pipes, 129
Reynolds stresses in
    time-averaged equation,
    409–410
Richardson-Zaki correction for
    gas bubbles, 325
Rigid bubbles, 396–397
Robin boundary conditions
  batch slurry adsorbers, 926
  description, 171–172
  in electrolytic systems, mass
    balance for reacting
    systems, 550
  film theory models
    bimolecular reactions, 674
    first-order reaction of
      dissolved gas, 664
    simultaneous absorption of
      two gases, 685
  heterogeneous reactions, 526
  laminar flow in pipes, 344–348
  laminar flow reactors, 571
  porous catalysts
    diffusion-reaction model,
      590, 593

finite difference methods, 621

orthogonal collocation method, 616

slab solutions, 276–278

Rotary dryers, 827

Rotating annular beds, 960–962

Rotating disks, flow over inclined and curved surfaces, 395

Runge-Kutta method for batch reactors, 93

Salt

decomposition, 785–786

electrolysis, 794–795

osmotic transport in membranes, 742

separation

electrodialyzer design, 952–953

voltage balance, 953–954

solutions

electrodialysis, 948

freezing point, 915

splitting, 950

Salt water purification, 898

Scale-up aspects in fixed bed adsorption, 937

Scaled rate of reaction in porous catalyst temperature effects, 611

Scaling results and analogies in flat plate with low flux mass transfer, 369–370

Schmidt number

convective mass transfer in turbulent flow

channels and pipes, 423–425

closure models, 411–412

constant wall flux, 418

velocity profiles, 416–417

convective mass transport dimensionless representation, 313–314

solid spheres, 320

evaporation, 813

film model, 200

fog formation, 855

laminar boundary layers

flat plate with low flux mass transfer, 370, 374–375

high flux analysis, 387

laminar flow in pipes, 338

laminar flow reactors, 580

mass transfer coefficient in sublimation of spherical particles, 102

mass transfer correlations in pipe flow, 128

sublimation in pipes, 129

Second-order reactions

bimolecular reactions, 677–679

continuous stirred tank reactors, 108–110

low flux model surface reactions, 209

macrofluid models, 451

plug flow idealization, 478

porous catalysts, linking with reactor models, 624

tubular flow reactors, 131

Segregated model

continuous stirred tank reactors, 109

laminar flow, 492–493

Selectivity

gas permeator models, 882

gas separation membranes, 877

liquid separation membranes, 899

pervaporation, 911

Self-diffusion coefficient in gases, 227

Semi-infinite slab analysis in transient diffusion processes

constant surface concentration, 288–291

integral method, 290–292

overview, 287

pulse response, 292–293

Semi-permeable membranes

concentration polarization effects, 903–905

equipment-level model, 907

Kedem-Katchalski model, 905–907

osmotic pressure, 900–901

overview, 900

reverse osmosis, 901–903

Semibatch reactors, 688–691

Semiconductor devices

diffusion-reaction analysis, 24–25

power law model, 296

Sensible heat correction in vapor with non-condensible gas, 853

Separation

ethanol–water mixture in liquid separation membranes, 909

multistage cascades, 465

VOCs, 909–910

Series coefficient slab solutions, Dirichlet case, 271–272

Series reaction simulation for batch reactors, 95

Series solution in slab solutions, Dirichlet case, 269–271

Shape normalization in porous catalysts, 595–596, 605

Shear stress

total, 433

turbulent flow closure models, 412

Shell balance approach for transient diffusion in spherical coordinates, 76

Sherwood number

absorption in slurries, 696

channel flow, 316

convective mass transfer

channel flows with mass transfer, 352–353

entry region analysis, 349

convective mass transport dimensionless representation, 313–314

flow in flat plate, 317

copper electrowinning, 796

laminar boundary layers

bubbles and drops, 397

flat plate with low flux mass transfer, 374–375

high flux analysis, 388

integral balance approach, 381

laminar flow in pipes, 315, 341–344, 346–347

mass transfer coefficient for sublimation of spherical particles, 102–103

mass transfer correlations in pipe flow, 128

oxygen uptake in lungs, 754

turbulent flow, 315

Shrinkage rate of subliming spheres, 103–104

Shrinking core model for reacting solids
no solid product, 637–640
overview, 636
solid product, 641–644

Sievert's Law for gas separation membranes, 877–878

Silica gel, 920

Silicon
doping with phosphorus, 293
interstitial diffusion, 247–248

Silicon oxide (SiO_2) in surface reactions, 207–209

Similarity variable for flow over inclined and curved surfaces, 392

Simultaneous absorption of two gases, 684–688

Single flowing phase in tracer response in two-phase systems, 504–506

Single-phase systems, 172–175

Single-pore gas diffusion, 248–250

Single pore model for reacting solids, 652–653

Single solute diffusion
dimensionless representation, 730–732
instantaneous reaction asymptote, 732–734
invariant of system, 732
model equations, 729–730
overview, 729
pseudo-first-order reaction asymptote, 735–736

Single stage in multistage cascades, 467

Single stream in mass exchangers, 134–136

Slabs
concentration profiles in temporal evolution, 63–64
D-D problem in transient diffusion processes, 283–285
Dirichlet case
average concentration, 274–275
dimensionless representation, 268–269

illustrative results, 272–274
overview, 267–268
series coefficient, 271–272
series solution, 269–271

Slabs
porous catalysts
diffusion-reaction model, 591
external transport effects, 592–593
temperature effects, 610–611
Robin condition, 276–278
steady state diffusion across, 52–55
steady state diffusion with reaction in, 56–62
transient diffusion in, 62, 287–293

Slow reactions
dispersion model, 483
high flux model surface reactions, 211

Slurries
absorption in film theory models, 692–697
three-phase catalytic reactions, 608

Slurry adsorbers. *See* Batch slurry adsorbers

Sodium chloride (NaCl)
decomposition, 785–786
electrolysis, 793–795
osmotic transport in membranes, 742

Sodium hydroxide (NaOH)
H_2S absorption in, 683–684
production, 775

Soil, fugacity of pollutants in, 460

Solar devices, 24–25

Solid concentration profiles in volume reaction model, 646

Solid density of catalysts, 587

Solid dissolutions at walls
film flow, 354–356
mesoscopic models, 38–41, 124–128

Solid reactant mass balance in shrinking core model, 638–639

Solid spheres
convective mass transport, 320–321
sublimation of, 34–35

Solid to liquid in film flow, 318–319

Solids
drying, 828–829
reacting. *See* Reacting solids

Solid–solid diffusion
interstitial diffusion, 246–248
vacancy diffusion, 246

Solid–solid reactions in reacting solids, 654–656

Solutes
in electrophoresis principle, 956–957
transport rate in semi-permeable membranes, 906

Solvents in multicomponent diffusivity matrix, 537

Species balance equations
gas permeator models, 889
reacting systems, 524–525
single-phase systems, 174
tracer studies, 438

Species mass balance
compartmental models, 455
gases, 236
integral balance approach, 379
multicomponent systems, 519
plug-backmixed model, 496
porous catalysts
linking with reactor models, 622
temperature effects, 612
reacting systems, 550
tanks in series models, 449
two-phase systems, 500

Species mole fraction
condensation of vapor with non-condensible gas, 850
mass exchangers, 136
reacting systems, 520

Species velocity in mass transfer equations, 156–157

Speed, molecular, in Maxwell-Boltzmann distribution, 261–262

Spheres
bimolecular reaction, 720
convective mass transport, 320–321

dispersion coefficient, 495
flow over inclined and curved surfaces, 395
mass transfer to stagnant gas, 191–192
porous catalysts
diffusion-reaction model, 594–595
sublimation of, 34–35
transient diffusion, 279–280, 303–304
two-point collocation, 616–617
Spherical cap bubbles in laminar boundary layers, 397–399
Spherical coordinates
diffusion and reaction, 75–76
divergence operator, 155
Laplacian of concentration, 190
steady state diffusion across spherical shells, 73–74
transient diffusion in, 76–77
Spherical particles, sublimation of, 101–104
Spherical shells, steady state diffusion across, 73–74
Stage efficiency
fluid–fluid systems, 465
liquid–liquid extraction, 116–117
Stages in condenser model, 861
Stagnant gas in mass transfer from spheres, 191–192
Stagnation point in flow over inclined and curved surfaces, 394–395
Stanton number
convective mass transfer in turbulent flow
boundary layers, 421–422
pipe flows, 423–425
van Driest model, 426
convective mass transport
dimensionless representation, 314
flat plate with low flux mass transfer, 369
Starling equation for semi-permeable membranes, 902–903
Staverman constant for semi-permeable membranes, 903

Steady-on-average flows in turbulent flow, 405–406
Steady state case
porous catalysts
diffusion-reaction model, 588
radial diffusion, 68–70
Steady state diffusion, 186
combined flux equation, 187
determinacy condition, 188–189
diffusion-induced convection, 187–188
low flux model, 189–191
radial, 68–70
with reaction in slabs, 56–62
across slabs, 52–55
across spherical shells, 73–74
Steady state mass transfer with reaction, 70–72
Stefan cells, 197
Stefan-Maxwell model
binary mixture with UMD, 237
condensation in ternary systems, 865–866
constitutive model for multicomponent transport, 519–520
diffusion coefficients in gases
frictional interpretation, 234
multicomponent diffusion, 235–236
electrodialyzer design, 955
equimolar counter-diffusion, 533–534
evaporation
binary liquid mixtures, 531
ternary liquid mixtures, 528–530
heterogeneous reactions, 526–527
multicomponent diffusivity matrix, 535–538
reacting systems, 520, 525
Step input in transient balance, 438–439
Steric partition constant for fluids in porous solids, 251
Stimulus-response studies for transient balance, 438
Stirred cells in bimolecular reaction, 720–721

Stirred reactors
backmixing assumption, 436–438
dissolved oxygen concentration in, 104–106
mixing model, 33–34
Stochastic nature in turbulent flow, 405–406
Stokes-Einstein model
diffusion coefficients in liquids, 239–240
Wilke-Chang extension, 241
Stokes law
convective mass transport in solid spheres, 321
diffusion coefficients in liquids, 240
Stream function for flat plate with low flux mass transfer, 371–372
Stress
total shear, 433
turbulent flow closure models, 412–413
turbulent flow reactors, 580
walls, 373
Stress discontinuity in turbulent flow gas–liquid interface, 429
Stripping factors in multistage cascades, 467–468
Structural changes in gas–solid reaction models, 651–654
Subcooling in condensation, 849
Sublimation
in pipes, 129
solid spheres
mass transfer coefficient, 34–35
rate of shrinkage, 103–104
spherical particles, 101–104
Substitution diffusion, 247
Sucking factor in laminar boundary layer high flux analysis, 387
Sulfur (S)
diffusion volume, 231
from hydrogen sulfide bearing gas, 702–703
Sulfur compounds, removal by porous catalysts, 630
Sulfur dioxide (SO_2)
Henry's law constant, 16
in water, 20

Sulfuric acid (H_2SO_4)
  manufacture, 586
Superheat in condensation, 849
Supersaturation in fog
  formation, 855–857
Supporting electrolytes
  copper electrowinning, 796
  mass transfer effects, 793
Surface concentration
  batch slurry adsorbers, 930
  electrochemical reaction
    voltage balance, 797
  porous catalysts
    diffusion-reaction model,
      600–601
    finite difference methods,
      621
    semi-infinite slab analysis,
      288–290
Surface contaminants in bubbles
  and drops, 397
Surface flux
  diffusion-reaction model
    nth-order reaction,
      603–604
  transient diffusion processes
    cylinders and spheres, 280
    integral method, 292
    semi-infinite slab analysis,
      290
Surface reactions
  electrochemical processes, 781
  high flux model with product
    counter-diffusion, 210–212
  low flux model
    first-order reaction, 206–207
    nonlinear reactions, 209–210
  overview, 206
  shrinking core model
    no solid product, 638–640
    solid product, 642
Surface tension in turbulent flow
  at gas–liquid interface,
    428–429
Surface term in differential
  models, 31
Surface velocity in film flow, 320
Swarms, bubble, 322
Swelling in polymeric
  membranes, 256
Symmetry factor in
  Butler-Volmer equation,
    788

Synthesis gas, 890–891
System velocity in presence of
  diffusion, 160–161

Tafel equation
  Butler-Volmer equation, 791
  copper electrowinning, 796
  mass balance for reacting
    systems, 550
Tanks in series model
  ideal flow patterns, 502–503
  reactor performance, 449
  transient balance, 442–444
Taylor dispersion model for
  laminar flow, 488–491, 575
Temperature and temperature
  effects
  Butler-Volmer equation, 791
  condensation
    binary gas mixtures, 860–861
    vapor with non-condensible
      gas, 850
  conversion effects, 659
  diffusion coefficients in gases,
    231
  fog formation, 855–856
  gas separation membranes,
    877
  laminar flow reactors, 578–580
  polymeric membranes, 256
  porous catalysts
    dimensionless boundary
      conditions, 612–613
    dimensionless
      representation, 611–612
    equations for heat and mass
      transport, 610–611
    temperature gradients,
      613–614
  slab solutions Robin condition,
    277
  wet and dry bulb, 812–815
Temporal evolution of
  concentration profile in
  slabs, 63–64
Ternary diffusion in
  heterogeneous reactions,
    527
Ternary mixtures
  distillation of, 534–535
  liquid evaporation in, 528–530
  pseudo-binary diffusivity in,
    166

Ternary systems
  condensation, 864–866
  multicomponent diffusivity
    matrix, 536–538
Test for backmixed assumption,
  110–112
Thermal boundary layer in
  evaporation, 813
Thermal diffusion, 258
Thermicity group for porous
  catalysts, 612
Thermodynamic considerations
  in electrochemical
  reactions, 781–786
Thermodynamic equilibrium in
  liquid–liquid extraction,
  114
Thermodynamic potential in
  hydrogen fuel cells, 798
Thickness
  film, 354, 373
  laminar boundary layers
    flow over inclined and
      curved surfaces, 395
    integral balance approach,
      378–380
Thiele modulus
  oxygen profiles in pools of
    liquid, 61
  oxygen transport in tissues,
    758
  oxygen uptake in lungs, 755
  porous catalysts
    diffusion-reaction model,
      590, 592, 600–603, 605
    finite difference methods,
      620
    linking with reactor models,
      623–625
    spherical catalysts, 595–596,
      599
    temperature effects, 611
    transient diffusion processes,
      285
    volume reaction model,
      647–648
Three-compartmental model for
  plasma, 772
Three dimensional flow field for
  fully turbulent flow, 405
Three-phase catalytic reactions,
  607–610

Time averaging in turbulent flow, 406–411, 417
Time-concentration plots for slab solution Robin condition, 277
Time constants in laminar flow reactor model equations, 570–571
Time-dependent effect for gas separation membranes, 878
Time domain fitting in transient balance, 444
Time domain solution in tracer response method, 487–488
Time for conversion in shrinking core model solid product, 643–644
Time in temporal evolution concentration profiles in slabs, 63
Time marching in film theory models, 690
Time scales in laminar flow reactor model equations, 570
Tissues
    inhibitors in, 771
    transport of oxygen in, 82, 757–760
Topochemical reactions in Langmuir kinetics, 659
Tortuosity factor for fluids in porous solids, 253
Total concentration
    definition, 8
    single solute diffusion, 734
Total flow balance in gas permeator model countercurrent flow, 887
Total heat released in condensation, 852
Total mass balance in condenser model, 862–864
Total mass flux in turbulent flow, 418
Total molar concentration, 8
Total molar flow rate in macroscopic models, 88
Total shear stress, 433
Tracer response method
    Laplace domain, 485
    overview, 484–485

response curve moments, 485–487
time domain solution, 487–488
two-phase systems, 503–507
Tracer studies
    age distribution functions, 441–442
    moment analysis, 444–449
    overview, 438
    pulse input, 439–441
    segregated flow model, 492
    step input, 438–439
    tanks in series model, 442–444
    test for backmixed assumption, 110–112
Track-etched membranes, 906
Transfer area
    gas permeator models, 886
    packed bed absorbers, 328–329
Transference number for electric field, 550
Transformation in environmental transport, 459
Transient balance in macroscopic models
    age distribution functions, 441–442
    overview, 438
    pulse input, 439–441
    step input, 438–439
    tanks in series model, 442–444
Transient calculations of batch slurry adsorbers, 928–929
Transient diffusion
    batch slurry adsorbers, 926
    bimolecular reaction, 714–715
    cylinders and spheres, 73, 278–282
    drying falling rate period, 835
    eigenvalue computations, 297–298
    non-homogeneous problems, 283–285
    1-D problems, 266–267
    overview, 265–266
    oxygen, 275
    PDEPE Solver computations, 299–304
    penetration theory of mass transfer, 294–295
    problems, 306–308
    with reaction, 284–285

review questions, 305–306
slab solutions. *See* Slabs
spherical coordinates, 76–77
summary, 304–305
2-D problems, 285–287
variable diffusivity, 295–297
Transient effects in environmental transport, 462
Transient species balance equation, 438
Transition criteria for turbulent flow, 404–405
Translocation in environmental transport, 459, 461
Transpiration velocity in diffusion with convection, 65
Transport of charged species in electrolytic systems
    mobility and diffusivity, 544–545
    Nernst-Planck equation, 545–547
Transport of oxygen in tissues, 82, 757–760
Transport rate
    across charged membranes, 554–556
    gas separation membranes, 874–876
    in presence of convection, 66
Trickle bed reactors, 21–22
Triglycerides (TG) in biodiesel production, 704
True kinetics vs. measured kinetics in surface reactions, 207
TSPAN function, 300
Tubular flow reactors
    dispersion closure, 132–134
    overview, 41–43, 129–131
    plug flow closure, 131
Tumors, nutrient limitation for, 770
Turbulent core in constant wall flux, 420–421
Turbulent diffusion, 14–15
Turbulent flow
    axial dispersion, 493–494
    convective mass transfer in. *See* Convective mass transfer in turbulent flow

Turbulent flow (*continued*)
  description, 579–580
  dispersion model, 483
  flat plates, 317
  pipes and channels, 313,
    315–316
  solid dissolutions from walls,
    38
  velocity profile, 49
Turbulent mass flux in
    time-averaged equation,
    409
Turbulent regime in
    condensation, 849–850
Turbulent Schmidt number,
    411–412
Two-compartment models for
    pharmacokinetics,
    456–457, 761
2-D models
  transient diffusion processes,
    285–287
  turbulent flow reactors,
    579–580
Two-film model for gas–liquid
    interface, 212–217
Two-phase flow
  non-idealities, 499–503
  overview, 495–496
  plug-backmixed model,
    496–499
Two-phase systems, tracer
    response in, 503–507
Two-point collocation in
    orthogonal collocation
    method, 616–617
Two streams in mass
    exchangers, 136–141
Two-way coupled problems for
    flat plate with low flux
    mass transfer, 367

Ultrafiltration in liquid
    separation membranes,
    898–899
Uncharged membranes,
    electrolyte transport
    across, 551–553
Uniform reaction model for
    reacting solids, 648
Unimolecular diffusion (UMD)
  binary mixture with, 237

condensation vapor with
    non-condensible gas, 851
convective mass transport, 312
diffusion-induced convection,
    193–195
film model, 204
mole fraction profiles in,
    196–197
steady state diffusion, 188
UNIQUAC model for non-ideal
    liquids, 244
Unit operations, 21–24
Universal velocity profiles in
    turbulent flow, 414–415
Unmixed model for
    condensation
  binary gas mixtures, 857–859
  condenser model, 863

Vacancy diffusion in solid–solid
    diffusion, 246
Valency of ions in transport of
    charged species, 545
Validity of low flux model in
    diffusion-induced
    convection, 193
van Driest model for convective
    mass transfer in turbulent
    flow, 425–427
van Laar equation for non-ideal
    liquids, 244
vant Hoff equation for osmotic
    pressure, 901
van't Riet correlation for
    mechanically agitated
    tanks, 325
Vapor in fog formation, 855–857
Vapor phase balance in
    condenser model, 862–864
Vapor phase dehydrogenation of
    ethanol, 527
Vapor pressure in Antoine
    equation, 18, 48
Vapor with non-condensible gas
    in condensation
  condenser model, 854–855
  heat transfer rate and
    Ackermann correction
    factor, 851–852
  interface temperature
    calculations, 852–853
  mass transfer rate, 851
  overview, 850

Vapor–liquid interface in
    Raoult's law, 17–18
Variable diffusivity
  batch slurry adsorbers, 931
  transient diffusion with,
    295–297
Variance-based models for
    partial micromixing,
    453–454
Vector-matrix form for
    multicomponent
    diffusivity matrix, 537
Velocity and velocity profiles
  bimolecular reaction, 718–719
  channel flow reactors, 577
  channel flows with mass
    transfer, 351
  constitutive model for
    multicomponent
    transport, 518–519
  convective mass transfer
    entry region analysis, 348
    film flow, 353–354, 357
  convective mass transfer in
    turbulent flow, 405–406,
    413–417
    constant wall flux, 420
    time-averaged equation,
    408–410
  convective mass transport in
    gas bubbles, 324–325
  cooling towers
    cross-flow, 825
    design considerations, 817
  diffusion coefficients in gases,
    227
    frictional interpretation, 233
    multicomponent diffusion,
    235
  film flow surface, 320
  flux form, 155
  kinetic model, 227
  laminar boundary layers
    bubbles and drops, 396–397
    flat plate with low flux mass
    transfer, 368–369, 372, 375
    flow over inclined and
    curved surfaces, 393
    high flux analysis, 385–387
    integral balance approach, 378
  laminar flow
    formula, 48
    pipes, 337

mesoscopic models
plug flow idealization, 477
solid dissolutions from walls, 125
tracer response in two-phase systems, 506
tubular flow reactors, 132–134
mixtures, 156–157, 164
non-Newtonian fluids flow reactors, 577–578
solid dissolutions from walls, 39
system, 152
transport of charged species in electrolytic systems, 544–545
turbulent flow
closure models, 412
model equation, 49
turbulent flow reactors, 579–580
Velocity defects in turbulent flow, 415
Vignes equation for diffusion coefficients in liquids, 242
Viscosity
convective mass transfer in turbulent flow
closure models, 411
constant wall flux, 419–420
velocity profiles, 414
laminar boundary layer wall stress, 373
Volatile organic compound (VOC)
recovery of, 895
separation of, 909–910
Voltage
electrochemical processes calculations, 785–786
description, 781
electrodialyzer design, 952–953
hydrogen fuel cells, 800
Voltage balance
electrochemical reactions
copper electrowinning, 796–797
overview, 793–795
hydrogen fuel cells, 798
salt separation, 953–954
Volume averages in multiphase systems, 177

Volume reaction model for reacting solids
concentration profile for gas and solid, 646
first-order reaction, 646–649
kinetic model, 645–646
overview, 644–645
zero-order reaction, 649–651
Volumes
environmental transport, 461
tanks in series models, 449
transient balance, 442–443
Volumetric flow rate
hemodialyzers, 764
in and out flow terms, 88–89
mass exchangers, 135–136
solid dissolutions from walls, 126
Volumetric flux in liquid separation membranes, 899, 906
Volumetric liquid flow rate in bimolecular reaction, 718
Volumetric rate of absorption
bimolecular reactions, 677
first-order reaction of dissolved gas, 668
Volumetric transfer rate in bulk concentration and reactions, 669
von Karman analogy for channels and pipes, 423
von Karman approximation for film model, 199–200

Wakao-Smith model
catalysts, 588
fluids in porous solids, 253–254
Wall flux
boundary conditions, 362
laminar flow in pipes, 341–344
turbulent flow in channels and pipes, 418–421
Wall shear stress relations, 416–417
Wall term in macroscopic balance, 89
Walls
helium leakage rate across, 55–56

laminar boundary layer stress, 373
Robin problem, 344–348
solid dissolutions from, 38–41, 124–128
Water
acetic acid in, 257
acetone in, 245
enthalpy balance, 819–820
fugacity of pollutants, 460
molar volume in liquid, 241
oxygen concentration in, 17
SO_2 in, 20
splitting, 950
Water and air mixture, condensation rate in, 853–854
Water and ethanol mixture, separation of, 909
Water and methanol mixture, rate of condensation in, 859–860
Water vapor in cooling tower counterflow model, 819
Wave speed in fixed bed adsorption, 933
Wavy regime in condensation, 849–850
Weber number in convective mass transport
bed absorbers, 329
gas bubbles, 323–324
Wedge flow over inclined and curved surfaces, 393–394
Weighting factors in mixture velocity, 183
Weisz modulus, 629
Wet bulb temperature, 812–815, 817
Wetted spheres, 720
Wetted wall columns, 718–719
Wilke-Chang equation for diffusion coefficients in liquids, 241–242
Wilke equation for evaporation of liquids in ternary mixtures, 528–529
Wilson equation for non-ideal liquids, 244
x-momentum balance for laminar boundary layers
flat plate with low flux mass transfer, 368–369, 371

x-momentum balance for
laminar boundary layers
(*continued*)
flow over inclined and curved
surfaces, 392
integral balance approach,
572, 580

XMESH function, 300
Xylene isomers, separation of,
938

Yamashita-Inoue correlation for
gas bubbles, 325

Zeolites, 920–921
Zero-order behavior in oxygen
transport in tissues, 759
Zero-order reactions
diffusion-reaction model, 75
diffusion trials, 60
macrofluid models, 451–453
plug flow idealization, 478
porous catalysts
diffusion-reaction model,
589, 599–603
finite difference methods,
619–620

linking with reactor models,
624–625
steady state mass transfer with
reaction, 71
stirred reactors, 437
systems, reaction model,
649–651
Zero total velocity in electrolytic
systems, 544–545
Zeroth moment in tracer
response in two-phase
systems, 507
Zwitterion, 956